The AutoCAD® Tutor
for
Engineering Graphics
Release 12 & 13

by
Alan J. Kalameja

The AutoCAD® Tutor
for
Engineering Graphics
Release 12 & 13

by
Alan J. Kalameja

Press

I(T)P™ An International Thomson Publishing Company

Albany • Bonn • Boston • Cincinnati • Detroit • London • Madrid
Melbourne • Mexico City • New York • Pacific Grove • Paris • San Francisco
Singapore • Tokyo • Toronto • Washington

NOTICE TO THE READER

Several figures used in the problems at the end of the units are from TECHNICAL DRAWING by Goetsch, Neslon and Chalk, Second Edition, © 1989 by Delmar Publishers

Trademarks
AutoCAD® and the AutoCAD® logo are registered trademarks of Autodesk, Inc.
Windows is a trademark of the Microsoft Corporation.
All other product names are acknowledge as trademarks of their respective owners.

Cover Design:Michael Speke

Delmar Staff
Publisher: Michael McDermott
Acquisitions Editor: Mary Beth Ray, CompuServe 73234,3664
Project Development Editor: Jenna Daniels
Production Coordinator: Andrew Crouth, CompuServe 74507,250
Art and Design Coordinator: Lisa L. Bower

COPYRIGHT © 1995
By Delmar Publishers Inc.
an International Thomson Publishing Company
The ITP logo is a trademark under license.

Printed in the United States of America

For more information, contact:

Delmar Publishers
3 Columbia Circle, Box 15015
Albany, New York 12212-5015

International Thomson Publishing Europe
Berkshire House 168-173
High Holborn
London, WC1V 7AA
England

Thomas Nelson Australia
102 Dodds Street
South Melbourne, 3205
Victoria, Australia

Nelson Canada
1120 Birchmont Road
Scarborough, Ontario
Canada, M1K 5G4

International Thomson Editores
Campos Eliseos 385, Piso 7
Col Polanco
11560 Mexico D F Mexico

International Thomson Publishing GmbH
Konigswinterer Strasse 418
53227 Bonn
Germany

International Thomson Publishing Asia
221 Henderson Road
#05-10 Henderson Building
Singapore 0315

International Thomson Publishing--Japan
Hirakawacho Kyowa Building, 3F
2-2-1 Hirakawacho
Chiyoda-ku, Tokyo 102
Japan

Delmar Publishers' Online Services
To access Delmar on the World Wide Web, point your browser to:
http://www.delmar.com/delmar.html
To access through Gopher: gopher://gopher.delmar.com
(Delmar Online is part of "thomson.com", an Internet site with information on more than 30 publishers of the International Thomson Publishing organization.)
For information on our products and services:
email: info@delmar.com
or call 800-347-7707

4 5 6 7 8 9 10 XXX 00 99 98 97

Library of Congress Cataloging-in-Publication Data

Kalameja, Alan J.
    The AutoCAD tutor for engineering graphics release 12 and release 13/
    Alan J. Kalameja—2nd Ed.
        p. cm.
    Includes index.
    ISBN 0-8273-5914-4
    1. Engineering graphics. 2. AutoCAD (Computer file) I. Title.
T385.K345 1995
604.2'0285'5369—dc20

93-48004
CIP

Contents

(For a more detailed contents description, refer to the individual unit opener page)

Preface

Engineering graphics has been around for a long time as a means of defining a product graphically before it is constructed and used by consumers. Previously, the process for producing the drawing involved aids such as pencils, ink pens, triangles, T-squares, etc., to place the idea on paper before making changes and producing blueline prints for distribution. The ability to produce these drawings on a computer may be new; however, the principles and basics of engineering drawing remain the same.

This text uses engineering drawing basics to produce drawings using AutoCAD and a series of tutorial problems that follow each unit. In most cases, the tutorials may be performed using AutoCAD Release 12; a unit introducing AutoCAD Release 13 is provided to illustrate enhancements and differences in AutoCAD Release 12. Following the tutorials, extra problems are provided to add to your skills. A brief description of each unit follows:

Unit 1 - AutoCAD Basics
This unit is provided to introduce AutoCAD concepts such as basic screen elements, use of function keys, opening a drawing file, setting units and limits, using layers, using object snaps, methods of creating selection sets, options of the Zoom command, using the View command, and saving drawings. A layer tutorial follows along with questions on general AutoCAD terms.

Unit 2 - Object Construction and Manipulation
This unit provides a brief explanation of all AutoCAD drawing and editing commands in addition to such topics as coordinate methods of input and grips. A series of tutorials follow to complement the topics covered in the unit. Additional questions and problems are provided at the end of the unit.

Unit 3 - Geometric Constructions
This unit discusses how AutoCAD commands and command options may be used for geometric constructions. A series of tutorial exercises follow along with additional problems at the end of the chapter.

Unit 4 - Shape Description/Multi-View Projection
Shape description and multi-view projection using AutoCAD is the focus of this unit. The basics of shape description are discussed along with proper use of linetypes, fillets and rounds, chamfers, and runouts. One tutorial follows outlining the steps used for creating a multi-view drawing using AutoCAD. Another tutorial approaches multi-view drawing through the use of .XYZ filters. Additional questions and problems are provided at the end of the unit.

Unit 5 - Dimensioning Techniques
Dimensioning techniques using AutoCAD are the topic of this unit. Basic dimensioning rules are discussed before concentrating on all AutoCAD dimensioning options and dimension variables. Tutorials follow, along with additional questions and problems on dimensioning.

Unit 6 - Analyzing 2-D Drawings

This unit provides information on analyzing drawings for accuracy reasons. The commands Area, ID, List, and Dist will be used as analysis tools along with the DDMODIFY dialog box. A series of tutorial exercises follow to allow the user to test their accuracy of drawing. Additional questions and problems are provided at the end of the unit.

Unit 7 - Region Modeling Techniques

This unit provides instruction on using the region modeler as an alternate means of entity construction and analysis. Objects are formed through the boolean operations of union, subtraction, and intersection. Mass property calculations yield area, perimeter, and centroid information of the region. Tutorial exercises and drawing problems are provided at the end of this unit.

Unit 8 - Section Views

Section views are described in this unit including full, half, assembly, aligned, offset, broken, revolved, removed, and isometric sections. Hatching techniques in AutoCAD are also discussed. Tutorial exercises follow, along with additional questions and problems related to section views.

Unit 9 - Auxiliary Views

Producing auxiliary views using AutoCAD is discussed in this unit. Tutorials and drawing problems follow a discussion on auxiliary view basics.

Unit 10 - Isometric Drawings

This unit discusses constructing isometric drawings with particular emphasis on using the Snap-Style option of AutoCAD. In addition to isometric basics, creating circles and angles in isometric is also discussed. Tutorial exercises follow, along with additional problems.

Unit 11 - 3-D Modeling

This unit begins the study of 3-dimensional modeling with the creation of wireframe and surfaced drawings. A brief comparison on the need for 3-dimensional drawings is discussed along with the UCS, 3Dface, ruled surface, tabulated surface, edge surface, and revolved surface commands. A series of tutorials follow, along with additional problems to be drawn as 3-dimensional wireframe and surfaced drawings.

Unit 12 - Solid Modeling

This unit begins the topic of solid modeling and how it compares to wireframe modeling. The Advanced Modeling Extension, (AME), will be discussed along with basic creation of solid primitives. The boolean operations of subtraction, union, and intersection will be demonstrated for further modeling creation. Using the solid model as the basis for producing orthographic drawings is demonstrated. Methods of displaying solid models will also be discussed. Tutorial exercises follow in addition to other problems at the end of the unit.

Unit 13 - Release 13 Features

This unit provides a brief update to the latest drawing and modeling enhancements provided in AutoCAD Release 13. Topics include new Windows user interface, DOS screen enhancements, multiline text and spell checking, dimensioning by inference, new dimension style dialog boxes, dimension families, geometric dimension and tolerancing techniques, associative crosshatching, enhancements to trim-extend and fillet-chamfer, object selection cycling, Osnap-from, Osnap-apparent intersection, object grouping, xlines and rays, mutilines, 3D modeling enhancements, and rendering enhancements.

A disk is provided with this text to supplement some of the tutorials. Each tutorial will alert you to either begin a new drawing or call up an existing drawing provided on the disk and proceed with the remainder of the tutorial.

Acknowledgments

I wish to thank the staff at Delmar for their assistance with this document, especially Michael McDermott, Mary Beth Ray, Andrew Crouth and Jenna Daniels. Special thanks go out to Barbara Savins, who assisted with the problems at the end of the Shape Description/Multi-View Projection chapter and George Moss for reviewing the entire document in addition to producing the solutions to many of the drawing problems. I also wish to thank Roy Baker, Gary Crafts, Frank Dagostino of Trident Technical College, Patrick Malone of Carolina Design, and Ron Nightengale of Industrial Spraying Systems for testing drawing problems associated with Unit 6 (Analyzing 2-D Drawings), Unit 7 (Region Modeling Techniques) and Unit 12 (Solid Modeling). Thanks also go out to Barbara Bowen and Maureen Barrow of Autodesk for permission to use various pages from the *AutoCAD Release 12 Certification Exam Preparation Manual* in this publication. Finally, I wish to thank my wife Linda and daughters Suzanne and Kathryn for their patience during the development of this project.

Alan Kalameja

About the Author

Alan J. Kalameja is the Department Head of Computer-Aided Design at Trident Technical College located in Charleston, South Carolina. He has been at the College for over 13 years and has been using AutoCAD since 1984. He directs the Authorized AutoCAD Training Center at Trident, which is charged with providing industry training to local area companies and firms. Currently, he is an Education Training Specialist with Autodesk, is a member of the AutoCAD Certification Exam Board, and has authored the *AutoCAD Release 12 Certification Exam Preparation Manual* by Delmar Publishers, 1993. He also occupies a seat on the Executive and Technical Committees charged with the development of a generic CAD exam designed around the National CADD Standard, a project directed by the Foundation for Industrial Modernization (FIM), a subgroup of the National Coalition for Advanced Manufacturing (NACFAM) headquartered in Washington, DC.

Conventions

All tutorials in this publication use the following conventions in the instructions:

Whenever you are told to enter text, the text appears in **boldface** type. This may take the form of entering an AutoCAD command or entering such information as absolute, relative, or polar coordinates. You must follow these and all text inputs by striking the Return or Enter key to execute the input. For example, to draw a line using the Line command from coordinate value (3,1) to coordinate value (8,2), the sequence would look like the following:

Command: **Line**
From point: **3,1**
To point: **8,2**
To point: *(Strike Enter to exit this command)*

Instructions in this tutorial are designed to enter all commands, options, coordinates, etc., from the keyboard. You may enter the same commands by selecting them from the screen menu, digitizing tablet, or pulldown menu area.

Instructions for selecting objects are in italic type. When instructed to select an object, move the pickbox on the object to be selected and press the pick button on the mouse or digitizing puck.

If you enter the wrong command for a particular step, you may cancel the command by holding down the "Ctrl" key and typing the letter "C." The translation for this procedure is commonly called "Control-C" and stands for "Cancel." Use this procedure to cancel any command in operation. Cancels can also be found in the pull-down menu area under Assist, on the digitizing tablet under the selection set modes, or on one button if using a 4-button puck.

UNIT 1

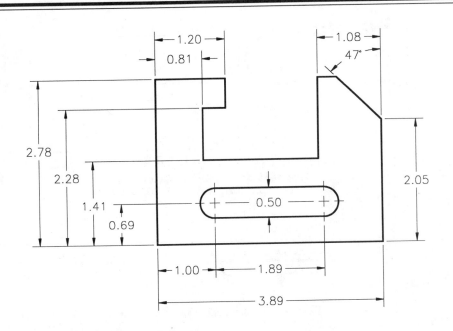

AutoCAD Basics

Contents

Getting Started

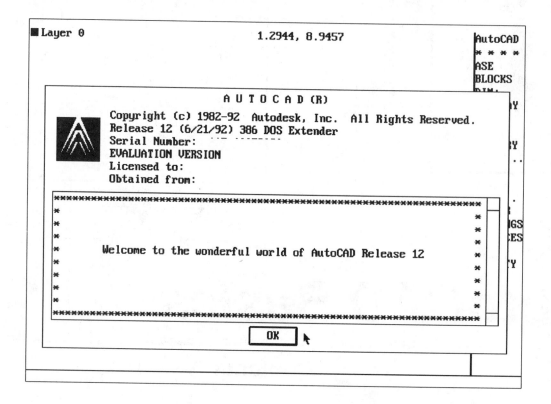

Welcome to the world of Computer-Aided Design and how it is used as a tool to produce all types of engineering graphical drawings whether they be 2-dimensional or 3-dimensional; whether they be architectural, electrical, or mechanical in application. Drafting and design has always been compared to a language such as English or German; however, the design process and the ability to capture a design technically on paper was and still is considered the language of industry. Due to the evolution of the computer, the design has shifted from paper to the video display although paper output through a plotter is still considered in some applications an absolute must. The same tools that were available for the manual production of drawings have changed considerably using the computer and, in some cases, the results may appear unnoticeable therefore casting doubt on the justification of the CAD terminal versus the drawing board. However, if ever there was an instrument of some sort that could have been used to take out the drudgery and tedious nature of the manual drawing board, the CAD terminal has found its place in the modern design office. As drawings were first laid out manually, and if the drawing was not properly centered for appearance purposes, a new calculation was performed, lines were erased (if they were drawn lightly), and new lines were constructed at the new location. With CAD, although it may be argued that basic manual practices such as centering drawings should always be prac-

ticed, most are considered unnessary. In CAD, if a drawing appears off-center, simply exercise a popular CAD command to move the drawing into the proper position. Once dimensions are added and the drawing once again appears off-center, move it again and again to achieve the proper appearance. This is but one of hundreds of time savers used to justify the existence of a computer in the design process; but it just doesn't stop here for 2-dimensional drawings.

The ability to use the computer to model an object in 3 dimensions has always been considered a form of art when performed on the manual drawing board. Not any more. When growing up as children and when asked to draw a house, we all tended to create a picture complete with receding lines and depth because this is what our eyes actually say. Then to our surprise, this ability was taken away as we were told to look at the picture as consisting of a series of primary views and draw them flat. Now with 3-dimensional graphics available on most CAD systems, an individual not only is able to construct in 3 dimensions but is also able to construct a prototype called a solid model to be used for analysis purposes. Use the text in the next series of pages and units to get a better idea of the world of 2D and 3D design and how it has been changed using the tool called the computer.

A Typical AutoCAD Drawing Screen

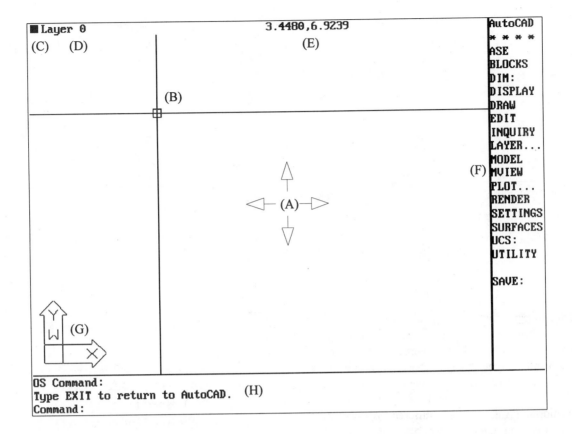

Illustrated above is a sample AutoCAD drawing screen when entered from the DOS prompt. The active screen area at "A" is the part of the display that holds the drawing image. The cursor at "B" is used to create entities or perform display commands such as Zoom. The status line at "C" displays the current layer name at "D" and the current X and Y absolute coordinates at "E" relative to the cursor position. The colored square at the far left shows the current color entities will be drawn in. For selecting commands, a screen menu is provided at "F." In the lower left corner at "G" is the world coordinate system icon, which signifies 0,0 is in the far lower left corner with the X axis traveling horizontally and the Y axis traveling vertically. In the bottom of the display screen at "H" is the command prompt area, which asks for user input depending on the command currently in progress.

Moving the cursor on top of the status area activates the menu strip at "I," illustrated at the right. This menu strip holds numerous pulldown menu commands similar to the example under the "Assist" area at "J." Selecting "Object Snap>" under "Assist" brings up another menu at "K." This menu is referred to as a cascading menu since it is activated by either picking "Object Snap" or by sliding the arrow to the right.

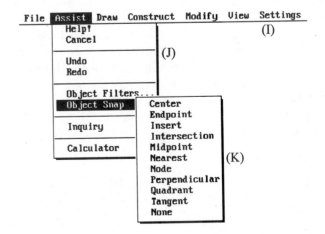

Using Keyboard Function Keys

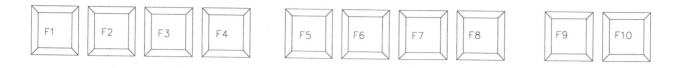

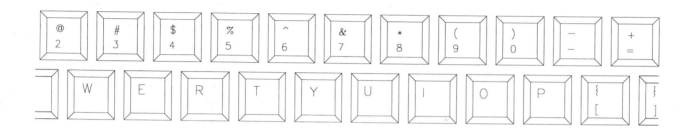

Once in a drawing file, the user has additional aids to control such settings as grid and snap. Illustrated above is a typical enhanced keyboard. Of particular interest are the function keys above the alphanumeric keys. These function keys are labeled F1 through F12. Software companies commonly program certain functions into these keys to assist the user in their application and AutoCAD is no different. All keys act as switches either turning functions on or off. The F1 key when pressed takes the user from the graphics display screen to the text screen. This may be helpful at viewing the previous command sequence in text form. The next key supported by AutoCAD is F6. This key toggles the coordinate display located in the status line on or off. When off, the coordinates update when an area of the screen is picked with a mouse or digitizer puck. When on, the coordinates dynamically move and display the current position of the cursor. The F7 key turns the grid on or off. The actual grid spacing is set by the Grid command and not by this function key. Orthogonal mode is toggled on or off using the F8 key. Use this key to force entities such as lines to be drawn horizontally or vertically. Use F9 to toggle snap mode on or off. The Snap command sets the current snap value. Use F10 to toggle tablet mode on or off. This mode is only activated when the digitizing tablet has been calibrated for the purpose of tracing a drawing into the computer. Function keys F2 through F5 are undefined in AutoCAD; however, they may be easily customized for specific uses.

DOS Version Function Keys

F1 = Toggle Text/Graphics Screen
F2 = Undefined by AutoCAD
F3 = Undefined by AutoCAD
F4 = Undefined by AutoCAD
F5 = Undefined by AutoCAD
F6 = Toggle Coordinates On/Off
F7 = Toggle Grid Mode On/Off
F8 = Toggle Ortho Mode On/Off
F9 = Toggle Snap Mode On/Off
F10 = Toggle Tablet Mode On/Off

Windows Version Function Keys

F1 = Windows Online Help
F2 = Toggle Text/Graphics Screen
F3 = Undefined by AutoCAD
F4 = Undefined by AutoCAD
F5 = Undefined by AutoCAD
F6 = Toggle Coordinates On/Off
F7 = Toggle Grid Mode On/Off
F8 = Toggle Ortho Mode On/Off
F9 = Toggle Snap Mode On/Off
F10 = Toggle Tablet Mode On/Off

Methods of Choosing Commands

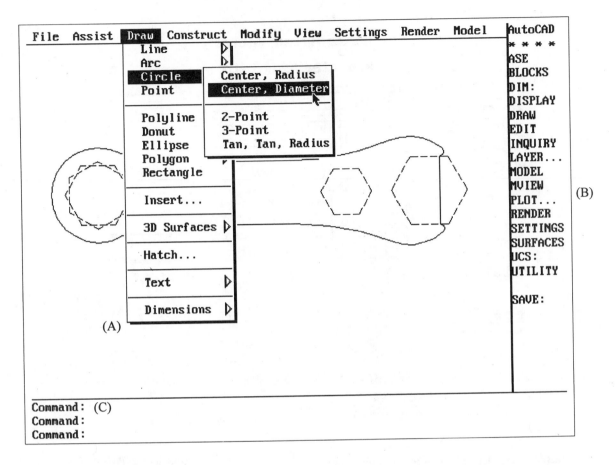

There are four ways to enter or select commands for constructing drawings. Above in the typical screen illustration are three of the four methods of selecting commands. At "A," the pulldown menu is becoming very popular with many users. Pulldown menus are activated by moving the cursor to the top of the status line. When this occurs, a menu bar appears to pick certain areas that hold most AutoCAD commands. At "B" is the screen menu area. This menu has been in existence since the very first version of AutoCAD. Once areas are selected from the root menu, a submenu takes the user to commands; as an example, selecting "Draw" from the screen activates a submenu containing commands specific to "Draw." Commands may also be entered directly from the keyboard. This practice is popular for users familiar with the commands. Illustrated at the right is an example of the fourth method for selecting commands; namely through a custom overlay template on the digitizer.

INQUIRY/DIMENSION

Icon Menus

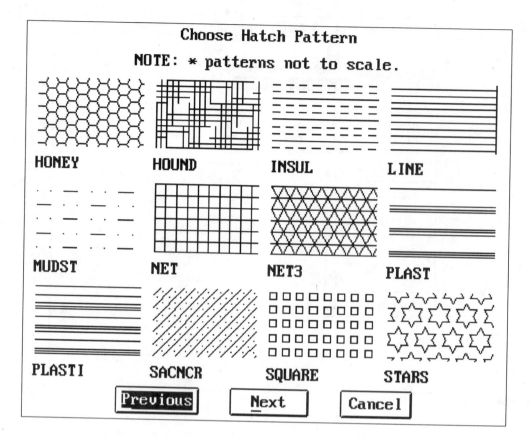

Icon menus display graphical pictures representing commands or command options. The icon menu illustrated above shows numerous hatch patterns. This icon menu is displayed after selecting the "Pattern..." option of the Bhatch command. In the past, an operator had to know by name the hatch pattern to use such as "Plasti", Insul, or "Ansi32". Once an operator entered the hatch pattern from the keyboard, he had no idea in some cases what the hatch pattern would look like until performing the entire Hatch command. Sometimes the hatch pattern proved unsuitable for the particular application. Icon menus solve this dilemma. The graphical picture lets the operator preview what the pattern will look like in appearance before placing in in a drawing.

To select a desired pattern, move the selector arrow to the pattern and pick anywhere inside the pattern. A box will be placed outlining the pattern similar to the illustration at the right. If the wrong pattern is selected, select another pattern and that pattern will be highlighted with-in the box.

Icon menus allow for quick selection of items for use in a drawing because of their graphical nature.

Unselected Pattern

INSUL

Selected Pattern

INSUL

Command Aliasing

Basic Commands.					
A,	*ARC	CONE,	*SOLCONE	CHPRIM,	*SOLCHP
C,	*CIRCLE	CYL,	*SOLCYL	MAT,	*SOLMAT
CP,	*COPY	SPHERE,	*SOLSPHERE	INTERF,	*SOLINTERF
DV,	*DVIEW	TORUS,	*SOLTORUS	MATERIAL,	*SOLMAT
E,	*ERASE			MOV,	*SOLMOVE
L,	*LINE	Complex Solids.		SL,	*SOLLIST
LA,	*LAYER	FIL,	*SOLFILL	MP,	*SOLMASSP
M,	*MOVE	CHAM,	*SOLCHAM	SA,	*SOLAREA
MS,	*MSPACE	EXT,	*SOLEXT	SSV,	*SOLVAR
P,	*PAN	EXTRUDE,	*SOLEXT		
PS,	*PSPACE	REV,	*SOLREV	Documentation commands.	
PL,	*PLINE	SOL,	*SOLIDIFY	FEAT,	*SOLFEAT
R,	*REDRAW			PROF,	*SOLPROF
Z,	*ZOOM	Boolean operations.		SECT,	*SOLSECT
		CUT,	*SOLCUT	SU,	*SOLUCS
Solid Primitives.		UNION,	*SOLUNION		
BOX,	*SOLBOX	SUB,	*SOLSUB	Model representation commands.	
WEDGE,	*SOLWEDGE	DIF,	*SOLSUB	WIRE,	*SOLWIRE
		SEP,	*SOLSEP	MESH,	*SOLMESH

On the previous page, it was stated that one method of executing AutoCAD commands is by entering the commands in at the keyboard. It is always best to be familiar with the location of all major keys before being comfortable with this method. To assist with the entry of AutoCAD commands from the keyboard, certain commands have been shortened and are considered "Aliases." The list above shows all AutoCAD commands that are currently identified as command aliases. These commands work this way; instead of entering the following at the command prompt...

     Command: **Redraw**

...enter the following appropriate substitute for the Redraw command as defined by the list above:

     Command: **R**

The purpose of these aliased commands is to shorten the amount of typing if entering in commands from the keyboard. The list of commands in bold print above is considered basic and is used for all types of entity creation, editing, and displaying. The remaining commands support the solids modeling capability of AutoCAD. These commands will be of use when studying Solid Modeling in Unit 12.

All of the above commands may be changed or new commands added by editing the ACAD.PGP file found in the Support subdirectory.

Dialog Box Elements

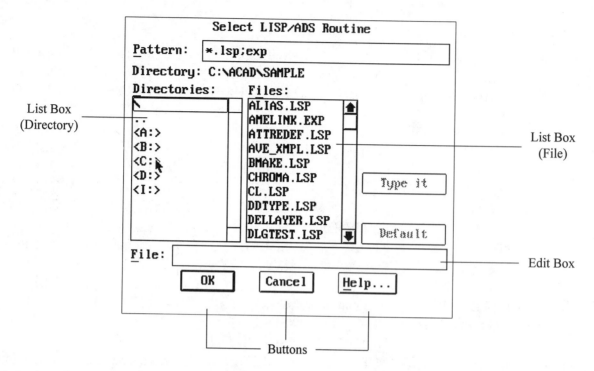

List Box
(Directory)

List Box
(File)

Edit Box

Buttons

Dialog boxes provide a more graphical means of picking command options instead of entering commands or values directly in from the keyboard. Illustrated above is an example of a typical file dialog box used in this case to identify either LISP or ADS routines. Because of the nature of not only this but all dialog boxes dealing with files, the following dialog box components are available to the user:

Buttons

Buttons are primarily used to either accept information entered in at the Edit Box or reject this information and perform another search to obtain the correct information. A Help button allows the user to get assistance in using this box.

Edit Boxes

These areas are set aside for the final name of the file to load or bring up. The file name in this area may be selected from the File List Box area or may be entered in directly from the keyboard.

Radio Buttons

Radio buttons provide for easy selection of various command options. Only one radio button may be selected at a time. In the example at the right, selecting "Normal" removes the radio button from where it was previously marked.

Image Tile

An image tile is a small picture representing a feature of a particular command as in the image tile of the normal hatch style illustrated at the right.

List Box (Directory)

This area is used to view the directory structure of any disk whether it be hard disk or floppy disk. To move from one directory to another, use the current input device and double click on the appropriate directory. The name of the current directory at the top of the dialog box should change to reflect the new directory name.

List Box (File)

Directly to the right of the Directory List Box is the File List Box. As a directory is located, the File List Box searches the directory for the specific pattern identified at the very top of the dialog box and displays its contents in the File List box. In the example above, with a current directory of C:\ACAD\SAMPLE and with a pattern of *.LSP or *.EXP, all files found in the directory are displayed in the File List Box.

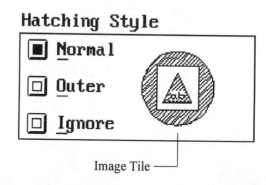

Image Tile

Scroll Bars

Scroll bars are useful in viewing data that does not fit in the entire list box area as in the illustration at the right. Use of the Up or Down arrows steps through each file one by one. As the arrows are selected, the Slider bar updates its position relative to the total number of files contained in the directory. A faster method would be to select the area below the Slider bar titles, the Page-Down area. Marking a point anywhere below the Slider bar displays the files by page instead of individually as with the arrows. As the Slider bar updates itself and moves toward the middle of the scroll bar area, the area above the Slider bar becomes active for the purposes of paging up. For a more dynamic approach, the Slider bar may be selected by pressing and dragging it up or down resulting in a fast method of displaying all files from top to bottom. If a small number of files exist where all files are displayed in the list box, the scroll bar disappears.

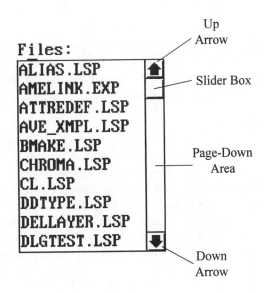

Check Boxes

Check boxes provide a form for toggling functions either On or Off; the presence of a "check" means the function is On; the lack of a "check" means the function is Off. In the illustration at the right, the Solid, Blips, and Highlight functions are currently On while the Ortho and Quick Text functions are Off.

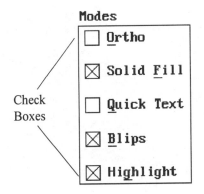

Pop-Up List Box

Illustrated above is an example of a Pop-up List Box used to support the number of decimal places past the zero used by the Units command. The down arrow prompts the user that more information concerning the highlighted value is available as the Pop-up List Box is brought down.

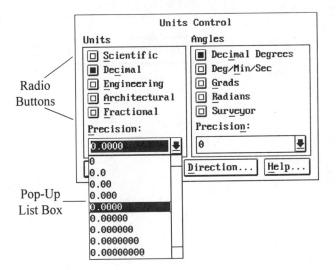

Starting a New Drawing

To begin a new drawing file, select the "New" command from the "File" pulldown menu area illustrated at the right. "New" could also be entered in at the command prompt similar to the following command sequence:

Command: **New**

Either of the above methods used for starting a new drawing will bring up the dialog box illustrated below. The information provided in this dialog box allows the user to follow different methods to create a new drawing file. The file must conform to the DOS standard file size which is limited to 8 characters or less. When this dialog box first appears, a blinking vertical line in the lower edit box prompts the user to enter the name for the new drawing.

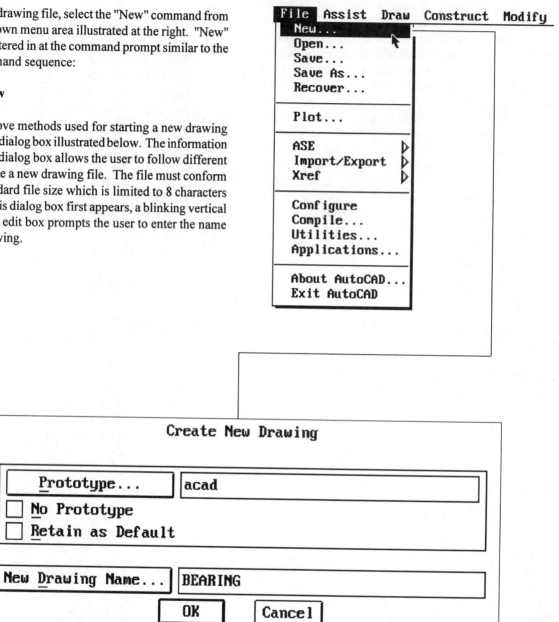

One of the options in the "Create New Drawing" dialog box controls the type of prototype drawing used for the new drawing file. A typical prototype drawing has items such as grid, snap, and limits already set. When beginning a new drawing, and the prototype drawing is referenced for use in the new drawing, it takes on the exact settings from the prototype drawing. Prototype drawings are excellent for creating all layers and dimension styles in addition to the grid and snap settings. A prototype drawing could also hold a title block. Then when the new drawing is created from the prototype drawing, the new drawing has the grid, snap, limits, and layers already set in addition to having a title block present at the start of the drawing. Illustrated at the right is a dialog box that appears if "Prototype..." is selected from the previous dialog box. A listing of all prototype drawings allows the user to control which prototype drawing will be used for the new drawing. At the right, a prototype drawing has been created specifically for metric applications. The prototype file "25-D" could mean to begin a new drawing with this prototype drawing file. The "D" could signify a D-size sheet of paper. The "25" could mean that the scale of the drawing is to be made to the scale of 0.25=12 or 1/4" = 1'-0". Some type of naming techniques should instituted to easily distinguish each type of prototype drawing file.

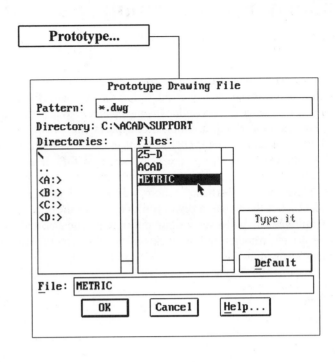

Picking "New Drawing Name..." from the previous dialog box brings up an additional dialogue box illustrated at the right. Use this dialog box to direct the new drawing name to a particular subdirectory.

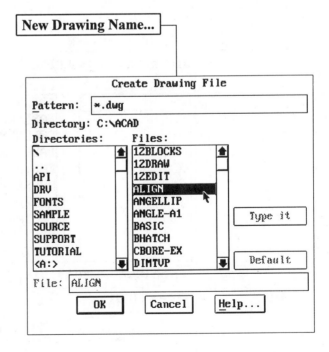

Opening an Existing Drawing

The Open command is used to edit a drawing that already exists or has already been created. Select this command from the "Files" area of the pulldown menu area. When this command is selected, a dialog box appears similar to the illustration below. The "Pattern:" alerts the user that all files with the extension ".dwg" are being searched for. If any drawing files are present in the current subdirectory, they are displayed in the file list box. To choose a different subdirectory, double click the mouse or digitizer pick button on the desired disk drive. This will display all subdirectories associated with the drive. Another double click on the subdirectory will display any drawing files contained in it if they exist.

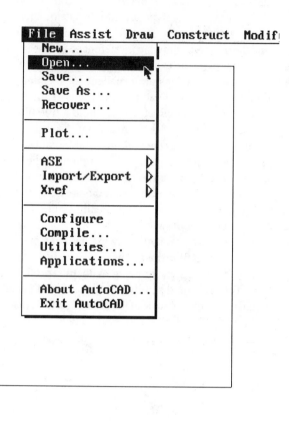

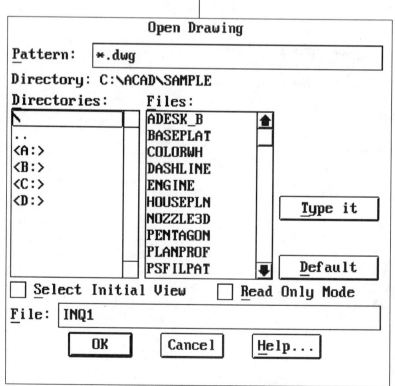

A view is a previously saved portion of the drawing display screen. Illustrated at the right are typical orthographic view names. Instead of opening a drawing file where the entire drawing is displayed, the user has the option of opening up the drawing file and automatically going to one of the views. This is considered good practice; rather than bring up the entire drawing and then zoom into a specific area of the screen, this zoom operation is bypassed by going directly to the named view. If a drawing file has been previously saved containing views, the "Select Initial View" box is checked. When the drawing file loads, the dialog box at the right displays on the screen alerting the user to select the desired view.

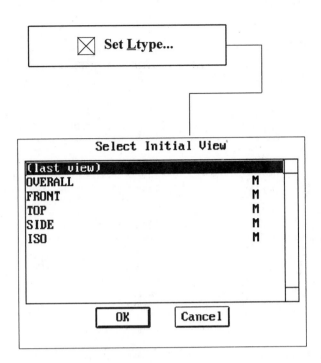

To manually load a drawing file called "Multi.Dwg" and automatically have the file reference a pre-defined view called "Front", enter the following at the DOS prompt:

C:\ACAD> **ACAD Multi,Front**

The above sequence will load the "Multi" drawing file and display the contents of the view "Front" on the screen.

If the "Read Only Mode" is checked, a drawing file is displayed on the screen. Entities may be drawn or edited as with all drawing files. However, none of the changes may be saved because the drawing may only be viewed; this is the purpose of read only mode. If any changes were made and a save is attempted, the alert box illustrated at the right informs the user the current file is write protected and no changes may be permanently made.

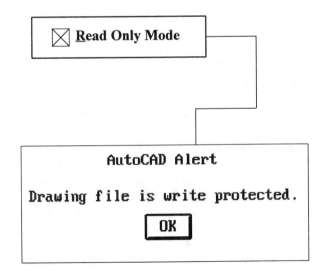

Setting Drawing Units - Decimal

```
Report formats:        (Examples)

  1.  Scientific     1.55E+01
  2.  Decimal        15.50
  3.  Engineering    1'-3.50"
  4.  Architectural  1'-3 1/2"
  5.  Fractional     15 1/2

With the exception of Engineering and Architectural formats,
these formats can be used with any basic unit of measurement.
For example, Decimal mode is perfect for metric units as well
as decimal English units.

Enter choice, 1 to 5 <2>:
Number of digits to right of decimal point (0 to 8) <4>:

Systems of angle measure:       (Examples)

  1.  Decimal degrees         45.0000
  2.  Degrees/minutes/seconds 45d0'0"
  3.  Grads                   50.0000g
  4.  Radians                 0.7854r
  5.  Surveyor's units        N 45d0'0" E

Enter choice, 1 to 5 <1>:
Number of fractional places for display of angles (0 to 8) <0>:


Direction for angle 0:
   East    3 o'clock  =  0
   North  12 o'clock  =  90
   West    9 o'clock  =  180
   South   6 o'clock  =  270
Enter direction for angle 0 <0>:

Do you want angles measured clockwise? <N>
```

The Units command is used for setting systems of units. It may be selected from the Settings area of the Screen menu or entered directly at the command prompt:

Command: **Units**

Use the above illustration to follow the different modes for the Units command. By default, decimal units are set. The number of decimal places past the zero is 4. This is reflected in the coordinate display located in the status line. The following systems of units are available:

Scientific
Decimal
Engineering
Architectural
Fractional

Scientific mode is displayed in exponential formats. Engineering units are displayed in feet and decimal inches. Archi-

tectural mode is displayed in feet and fractional inches. Fractional mode is displayed in fractional inches.

The following methods of measuring angles are supported by the Units command:

Decimal degrees
Degrees/minutes/seconds
Grads
Radians
Surveyor's units

Accuracy of angles may be controlled between zero and 8 places.

By default, the direction for angle zero is at the 3 o'clock position. This direction may be changed. Also by default, angles are measured in the counterclockwise direction. This may be changed to the clockwise direction if desired by the user.

Setting Drawing Units - Architectural

```
1.   Scientific      1.55E+01
2.   Decimal         15.50
3.   Engineering     1'-3.50"
4.   Architectural   1'-3 1/2"
5.   Fractional      15 1/2

With the exception of Engineering and Architectural formats,
these formats can be used with any basic unit of measurement.
For example, Decimal mode is perfect for metric units as well
as decimal English units.

Enter choice, 1 to 5 <4>:

Denominator of smallest fraction to display
(1, 2, 4, 8, 16, 32, or 64) <16>:

Systems of angle measure:       (Examples)

1.   Decimal degrees            45.0000
2.   Degrees/minutes/seconds    45d0'0"
3.   Grads                      50.0000g
4.   Radians                    0.7854r
5.   Surveyor's units           N 45d0'0" E

Enter choice, 1 to 5 <1>:

Number of fractional places for display of angles (0 to 8) <0>:


Direction for angle 0:
   East    3 o'clock  =  0
   North  12 o'clock  =  90
   West    9 o'clock  =  180
   South   6 o'clock  =  270
Enter direction for angle 0 <0>:

Do you want angles measured clockwise? <N>
```

Use the above illustration to use the Units command for changing decimal units to architectural units. The only difference with architectural and decimal units is in the form of accuracy. Decimal accuracy is given in number of places past the decimal point. Architectural accuracy is given in smallest number to display in the denominator of the fraction. Denominator value ranges from 1 to 64. All other modes for architectural units are identical to decimal units.

Using the DDUNITS Dialog Box

The DDUNITS dialog box is available to interactively set the units of a drawing. This dialog box is selected from the pulldown menu under "Settings". Selecting "Units Control..." activates the dialog box illustrated below.

The DDUNITS dialog box is designed to hold all items contained when using the regular Units command where the display screen changes to the text screen. The advantage of using this dialog box over the text screen mode is that there is no order in the dialog box that a user has to use to set the units. If the precision of angles needs to be changed first, select the appropriate area in the dialog box and make the change. If this change is incorrect, then change it back to its original. When using the Units command, changes must be made in the order that the prompts are presented to the user. If a mistake is made, the user must cancel the command, get back to the Units command, and make the change again.

Selecting "Direction..." in the "Units Control" dialog box displays another dialog called "Direction Control". Use this dialog box to control the direction of angle zero in addition to changing the way angles are measured from counterclockwise to clockwise.

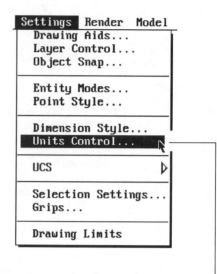

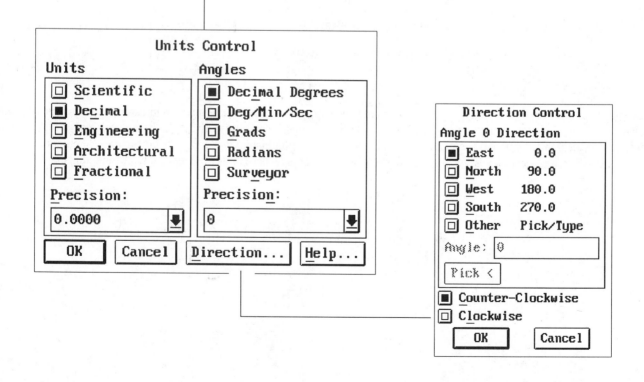

Using the Limits Command

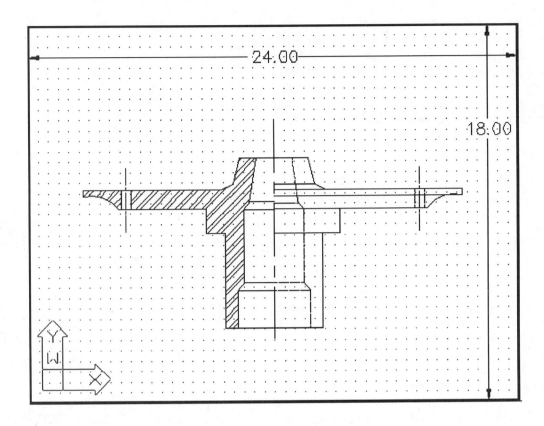

By default, the size of the drawing display screen in a new drawing file measures 12 units in the X direction and 9 units in the Y direction. As this size may be ideal for small objects, larger drawings require a more drawing screen area. Use the Limits command for increasing the size of the drawing area. Select this command from the "Settings" area of the pulldown menu area; the Limits command is also present in the Screen menu at the right of the pulldown menu; this command may also be entered directly at the command prompt. Illustrated above is a section view drawing. This drawing fits in a screen size of 24 units in the X direction and 18 units in the Y direction. Follow the prompt sequence below for changing the limits of a drawing.

Command: **Limits**
ON/OFF/<Lower left corner>: <0.0000,0.0000>:   *(Strike Enter to accept this value)*
Upper right corner <12.0000,9.0000>: **24,18**

Before continuing, issue a Zoom-All to change the size of the display screen to reflect the changes in the limits of the drawing.

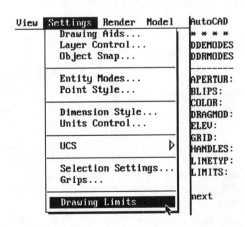

Command: **Zoom**
All/Center/Dynamic/Extents/Left/Previous/Vmax/Window/ <Scale(X/XP)>: **All**

Using the Grid Command

Use grid to get a relative idea as to the size of entities. Grid is also used to define the size of the display screen originally set by the Limits command. The dots that make up the grid will never plot out on paper even if they are visible on the display screen. Grid dots may be turned on or off either by using the Grid command or by pressing the F7 function key. By default, the grid is displayed in 1-unit intervals similar to the illustration at the right. Follow the command prompt sequence below for using the grid command:

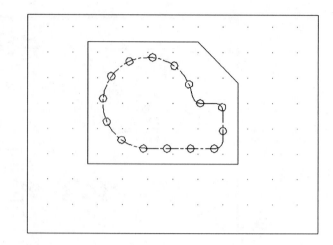

Command: **Grid**
Grid spacing(X) or ON/OFF/Snap/Aspect <0.0000>: *(Enter a value)*

Illustrated at the right is a grid that has been set to a value of 0.50 or one-half its original size. Follow the command prompt sequence below for performing this change.

Command: **Grid**
Grid spacing(X) or ON/OFF/Snap/Aspect <0.0000>: **0.50**

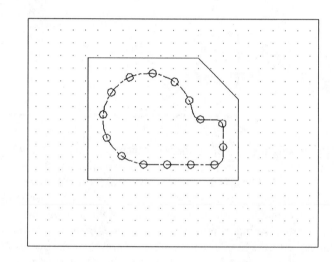

While Grid is a useful aid for construction purposes, it may harm the overall performance of the computer system. If the grid is set to a small value and is visible on the display screen, it takes time for the grid to display. If so small a value is used for Grid, a prompt will be displayed warning that the grid value is too small to display on the screen.

Use the Aspect option of the Grid command to display a grid with different X and Y values. Illustrated at the right, the horizontal grid spacing is 0.50 while the vertical spacing is 0.25. Follow the prompt sequence below for using the Aspect option of the Grid command.

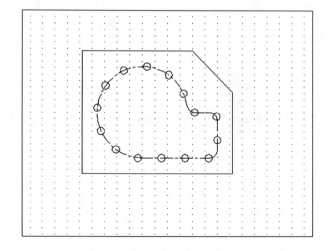

Command: **Grid**
Grid spacing(X) or ON/OFF/Snap/Aspect <0.0000>: **Aspect**
Horizontal spacing(X) <0.50>: *(Strike Enter to accept this value)*
Vertical spacing(X) <0.50>: **0.25**

Using the Snap Command

Illustrated at the right is a sample drawing screen with a grid spacing of 1.0000 units. The cursor is positioned in between grid dots. It is possible to have the cursor lock onto or snap to a grid dot; this is the purpose of the Snap command. By default the current snap spacing is 1.0000 units. Even though a value is set, the snap must be turned on for the cursor to be positioned on a grid dot. This can be accomplished by using the Snap command below or by pressing the F9 function key.

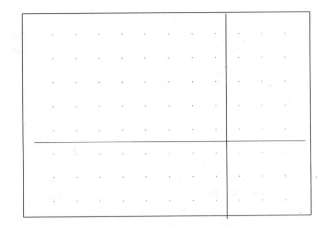

Command: **Snap**
Snap spacing(X) or ON/OFF/Aspect/Rotate/Style <1.0000>:

The current snap value may even affect the grid. If the current grid value is zero, the snap value is used for the grid spacing.

Illustrated at the right are the results of changing the snap value to 0.50 units. With the current grid set to zero, changing the snap value to 0.50 affects the grid and turns Snap mode on. Notice the cursor snapping to the grid dots.

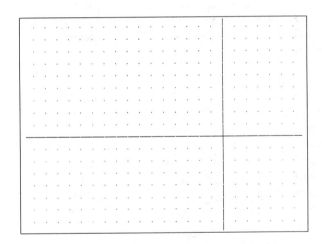

Command: **Snap**
Snap spacing(X) or ON/OFF/Aspect/Rotate/Style <1.0000>:
0.50

Some drawing applications require that the snap be rotated at a specific angular value. Changing the snap in this fashion also affects the cursor. Follow the command prompts below for rotating the snap.

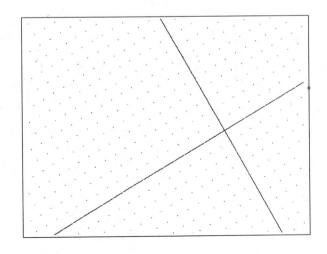

Command: **Snap**
Snap spacing(X) or ON/OFF/Aspect/Rotate/Style <1.0000>:
Rotate
Base point<0.0000,0.0000>: *(Strike Enter to accept this value)*
Rotation angle <0>: **30**

One application of rotating snap is for auxiliary views where an inclined surface needs to be projected in a perpendicular direction. This will be explained and demonstrated in a later unit.

Using the DDRMODES Dialog Box

Selecting "Settings" from the menu strip and then selecting "Drawing Aids..." displays the DDRMODES dialog box illustrated below. This is a helpful dialog box used for making dynamic changes to such commands as Grid and Snap. In addition to these commands, the following command modes may also be changed: Ortho; Solid Fill; Quick Text; Blips; Highlight.

Placing a check in the box provided turns on the specific mode. Checking the box again removes the check turning off the mode. Ortho mode is the ability to force only horizontal or vertical movement. Solid Fill is controlled by the Fill command and affects filled-in entities such as polylines and donuts. Quick Text is controlled by the Qtext command and converts all text on the display screen to rectangles to speed up drawing response time. For plotting purposes, the rectangles are converted back into text. The display of blips is controlled by the Blipmode command. Blips are the small specks or marks made on the screen whenever constructing or editing entities. To clean up blips, a simple Redraw command is used.

Turning blips off prevents them from appearing on the screen. Highlight is controlled by the system variable Highlight. This determines if a selected entity highlights informing the user it has been selected.

Another mode controls whether isometric grid is present or not. Also, three isometric modes may be toggled On or Off.

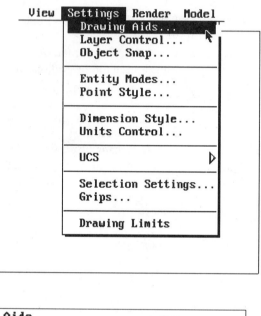

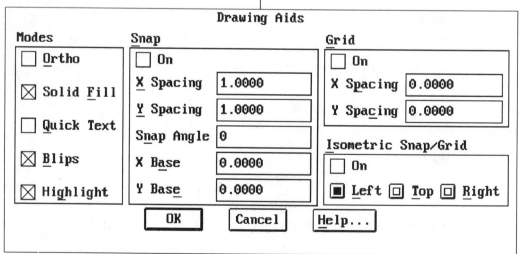

$e^- = -1.6 \times 10^{-19} C$ $k = 9 \times 10^9 \, N \cdot m^2/c^2$

$p^+ = +1.6 \times 10^{-19} C$ $\varepsilon_0 = 8.85 \times 10^{-12}$

$|F| = \dfrac{k|q_1||q_2|}{r^2} \hat{r}$ $\left(kc = \dfrac{1}{4\pi\varepsilon_0}\right)$

$\left(\varepsilon_0 = \dfrac{1}{4\pi kc}\right)$

$\hat{r} \leftarrow$ meters

micro 10^{-6} $\hat{r} = -\cos\theta + \sin\theta$

nano 10^{-9} $\phi = \tan^{-1} \dfrac{y}{x} = \phi^\circ$

$\rho = C/m^3$

$\lambda = C/m$

$\sigma = C/m^2$

Gauss Law:

$|E|(4\pi R^2) = \dfrac{q_{net}}{\varepsilon_0} \leftarrow$ sphere $|E| = \dfrac{q_{net}}{2\pi r L \varepsilon_0}$

$|E| = \dfrac{q_{net}}{4\pi R^2 \varepsilon_0}$ cylinder $\nearrow$ $2\pi r L \varepsilon_0$

planar $|E| = \dfrac{\lambda L}{2\pi r L \varepsilon_0} + \dfrac{\rho(\pi r^2 L - \pi r_1^2 L)}{2\pi r L \varepsilon_0}$

$\nwarrow$ charge·volume

$E = \dfrac{\sigma}{\varepsilon_0}$

Using the Layer Command

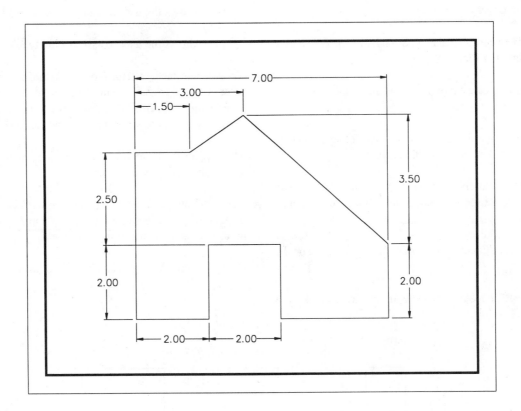

As a means of organizing entities, a series of layers should be devised for every drawing. Layers can be thought of as a series of transparent overlays which combine to form the completed drawing. The illustration above shows a drawing consisting of object lines, dimension lines, and border. Creating these three drawing components is illustrated at the right. Only the drawing border occupies a layer that could be called "Border". The object lines occupy a layer that could be called "Object" and the dimension lines could be drawn on a layer called "Dimension". At times, it may be necessary to turn off the dimension lines for better clarity of the object. Creating all dimensions on a specific layer will allow the user to turn off the dimensions while viewing all other entities on layers still turned on. The following command prompt sequence illustrates the Layer command along with its options:

Command: **Layer**
?/Make/Set/New/ON/OFF/Color/Ltype/Freeze/Thaw/LOck/
Unlock:

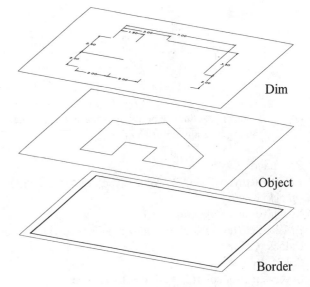

Dim

Object

Border

Options of the Layer Command

Options of the Layer Command:

? - Used to give a complete list or partial listing of all layers in the current drawing file.

Make - Used to create a new layer and automatically set the new layer to the current layer.

Set - Allows the user to change to a new current layer; all new entities drawn will be drawn on this layer.

New - Used to create a new layer or series of new layers. The Set option is then used to change from one layer to a new current layer.

On - Makes all entities created on a certain layer visible on the display screen.

Off - Turns off or makes invisible all entities created on a certain layer.

Color - Allows the user to assign a color to a layer name.

Ltype - Allows the user to assign a linetype to a layer name.

Freeze - Similar to Off; turns off all entities created on a certain layer. Entities frozen will not be calculated when performing a drawing regeneration. Therefore, Freeze is used as a productivity tool to speed up drawing response.

Thaw - Similar to On; turns on all entities created on a certain layer that were previously frozen.

Lock - Allows entities on a certain layer to be visible on the display screen while protecting them from accidentally being modified through an Editing command.

Unlock - Unlocks a previously locked layer.

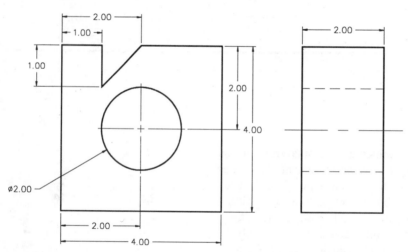

Follow the prompt sequence below for creating a layer called "Object" along with the color "Yellow":

Command: **Layer**
?/Make/Set/New/ON/OFF/Color/Ltype/Freeze/Thaw/LOck/
Unlock: **New**
New layer name(s): **Object**
?/Make/Set/New/ON/OFF/Color/Ltype/Freeze/Thaw/LOck/
Unlock: **Color**
Color: **Yellow**
Layer name for color(2) Yellow <0>: **Object**
?/Make/Set/New/ON/OFF/Color/Ltype/Freeze/Thaw/LOck/
Unlock: *(Strike Enter to exit this command)*

Follow the prompt sequence at the right for creating a layer called "Hidden" along with the color "Red" and the "Hidden" linetype:

Command: **Layer**
?/Make/Set/New/ON/OFF/Color/Ltype/Freeze/Thaw/LOck/
Unlock: **New**
New layer name(s): **Hidden**
?/Make/Set/New/ON/OFF/Color/Ltype/Freeze/Thaw/LOck/
Unlock: **Color**
Color: **Red**
Layer name for color(1) Red <0>: **Hidden**
?/Make/Set/New/ON/OFF/Color/Ltype/Freeze/Thaw/LOck/
Unlock: **Ltype**
Linetype: **Hidden**
Layer name for HIDDEN <0>: **Hidden**
?/Make/Set/New/ON/OFF/Color/Ltype/Freeze/Thaw/LOck/
Unlock: *(Strike Enter to exit this command)*

Using the DDLMODES Dialog Box

Use the DDLMODES dialog box to perform the following
layer operations:

Creating new layers
Making a layer the new current layer
Assigning a color to a layer
Assigning a linetype to a layer
Turning layers on or off
Freezing or thawing layers
Locking or unlocking layers

Activate this dialog box by first selecting "Settings" from the
menu strip; then select "Layer Control..." to display the dialog
box illustrated below. Selecting the layer called "HIDDEN"
illustrated below activates a majority of the dialog buttons
used to control the layer.

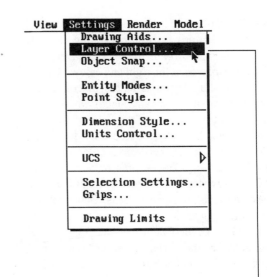

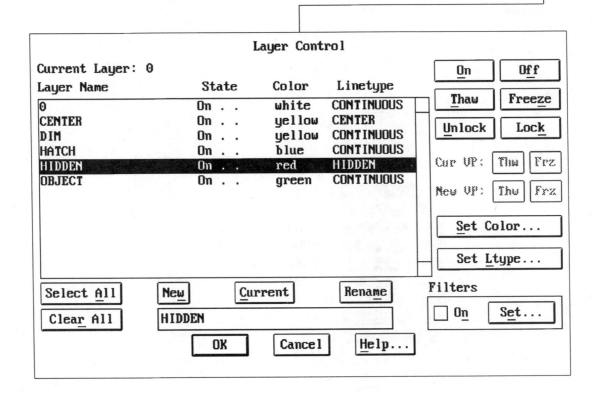

Setting Colors and Linetypes in the DDLMODES Dialog Box

Once a layer is selected from the list box of the DDLMODES dialog box, numerous buttons activate depending on what option the user wishes to perform on the selected layer. Two of these options are displayed at the right and below. One of the advantages of creating layers is to assign colors and linetypes for the purpose of graphically identifying the suggested purpose of an entity. Selecting "Set Color..." from the DDLMODES dialog box illustrates the secondary dialog box illustrated at the right. Select the desired color from the "Standard Colors" area, "Grey Shades" area, or from the "Full Color Palette" area. The standard colors consist of the basic colors available on most color display screens. Depending on the video driver, different shades of standard colors may be selected from the "Full Color Palette". Selecting the "Set Ltype..." button activates the "Select Linetype" dialog box illustrated below. Use this dialog box to dynamically select pre-loaded linetypes to be assigned to various layers. If no linetypes are pre-loaded using the Linetype command, only the continuous linetype appears in the list box area.

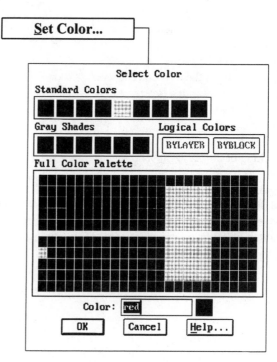

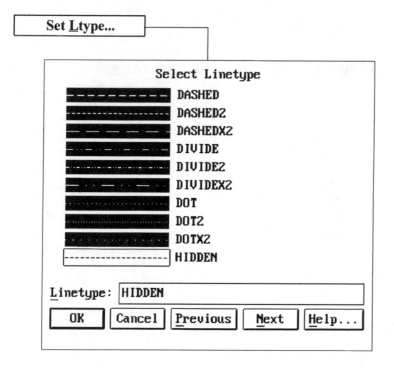

Loading Linetypes Using the Linetype Command

Before linetypes can be selected from the "Select Linetype" dialog box illustrated on the previous page, they must first be loaded using one of two methods. The first method is to assign a linetype manually using the Layer command. This procedure automatically loads that linetype. A second method is to load any number of linetypes using the Linetype command. Linetypes available to the user are displayed at the right. All linetypes may be loaded using the following prompt sequence:

Command: **Linetype**
?/Create/Load/Set: **Load**
Linetype(s) to load: ***** *(For all linetypes)*
(Strike Enter when the "Select Linetype File" dialog box below and to the right appears)
BORDER loaded.
BORDER2 loaded.

....

PHANTOMX2 loaded.
?/Create/Load/Set: *(Strike Enter to exit this command)*

BORDER, __ __. ____ . ____ . ____ . ____ . ____ .
BORDER2, _._._._._._._._._._._._
BORDERX2, ____ ____ . ____ ____ . ____ .
CENTER, ____ ____ _ ____ ____ _ ____ __
CENTER2, ___ _ ___ _ ___ _ ___ _ ___ _ ___
CENTERX2, _____ __ _____ __ _____
DASHDOT, _ . __ . __ . __ . __ . __ . __ .
DASHDOT2, ...
DASHDOTX2, ____ . ____ . ____ . ____ . ____ .
DASHED, __ __ __ __ __ __ __ __ __
DASHED2, _ _ _ _ _ _ _ _ _ _ _ _ _ _ _
DASHEDX2, ____ ____ ____ ____ ____
DIVIDE, ____ .. ____ .. ____ .. ____ ..
DVIDE2, _.._.._.._.._.._.._.._.._.._.._..
DIVIDEX2, _____ .. _____ .. _____ ..
DOT, ...
DOT2, ...
DOTX2, ...
HIDDEN, __ __ __ __ __ __ __ __ __ __
HIDDEN2, _ _ _ _ _ _ _ _ _ _ _ _ _ _ _ _
HIDDENX2, ____ ____ ____ ____ ____
PHANTOM, ____ __ __ ____ __ __ ____ __
PHANTOM2, ____ _ _ ____ _ _ ____ _ _ ____ _
PHANTOMX2, _____ ____ ____

Unfortunately, as all linetypes are loaded, the question must be answered whether all linetypes are used. Unused linetypes are added to the database of a drawing as excess baggage. A better solution would be to use the following prompt sequence, which illustrates the loading of the Hidden and Center linetypes.

Command: **Linetype**
?/Create/Load/Set: **Load**
Linetype(s) to load: **Hidden, Center**
(Strike Enter when the "Select Linetype File" dialog box below appears)
HIDDEN loaded.
CENTER loaded.
?/Create/Load/Set: *(Strike Enter to exit this command)*

```
┌─────────────────────────────────────────┐
│            Select Linetype File          │
│ Pattern:  *.lin                          │
│ Directory: C:\ACAD\SUPPORT               │
│ Directories:        Files:               │
│ \                   ┌ACAD──────┐         │
│ ..                  │          │         │
│ <A:>                │          │         │
│ <B:>                │          │         │
│ <C:>                │          │ ┌────────┐│
│ <D:>                │          │ │Type it ││
│ <I:>                │          │ └────────┘│
│                     │          │          │
│                     │          │ ┌────────┐│
│                     │          │ │Default ││
│ File: ACAD          │          │ └────────┘│
│   ┌────┐  ┌──────┐  ┌──────┐             │
│   │ OK │  │Cancel│  │Help..│             │
│   └────┘  └──────┘  └──────┘             │
└─────────────────────────────────────────┘
```

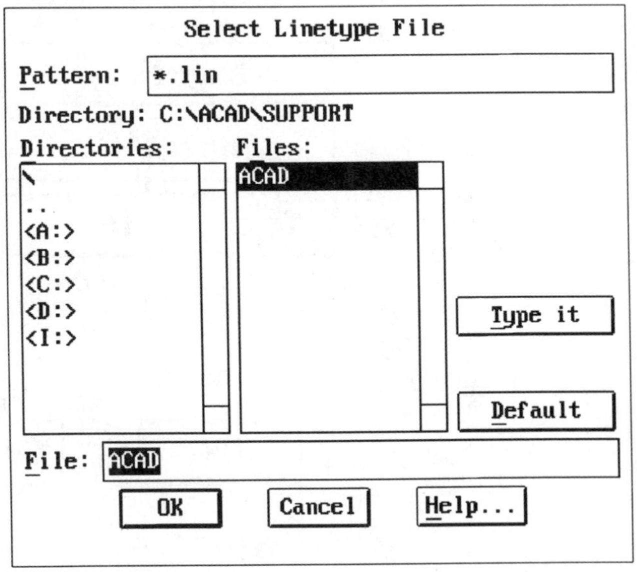

Using the DDEMODES Dialog Box

Use the DDEMODES dialog box to change the following properties of an entity:

> Color
> Layer
> Linetype
> Text Style

Activate this dialog box by first selecting "Settings" from the menu strip; then select "Entity Modes" to display the dialog box illustrated below.

This dialog box is considered a substitute for using the Chprop command.

Selecting "Color..." displays an additional dialog box similar to the one that appears at the top of page 24. Use the "Select Color" dialog box to change the color of a selected entity.

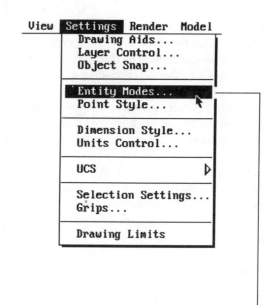

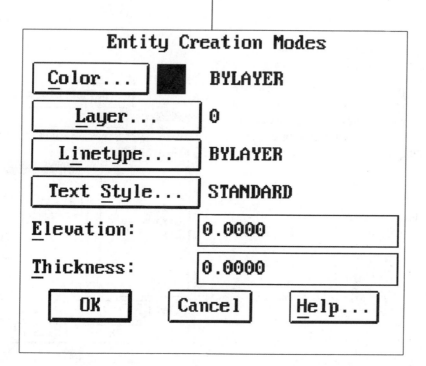

Using DDEMODES to Change the Layer, Linetype, and Text Style of an Entity

Layer...	**0**
Linetype...	**BYLAYER**
Text Style...	**STANDARD**

Selecting the "Layer..." option of "Entity Creation Modes" activates the "Layer Control" dialog box illustrated at the right. This dialog box is identical to the DDLMODES dialog box. Use this dialog box to change the selected entity to a layer chosen from the list at the right.

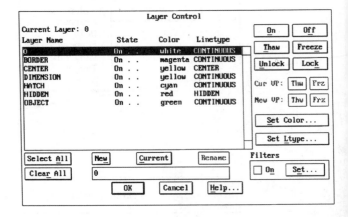

Selecting the "Linetype..." option of "Entity Creation Modes" activates the "Select Linetype" dialog box illustrated at the right. Use this dialog box to change the selected entity to a linetype chosen from the list at the right. Only the linetypes that have been previously loaded using the Linetype command will appear in this dialog box.

Selecting the "Text Style..." option of "Entity Creation Modes" activates the "Select Text Style" dialog box illustrated at the right. Use this dialog box to change to a different text style. Text styles are created and controlled by using the Style command.

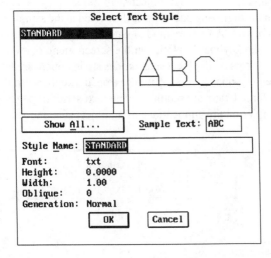

Using the Osnap Command

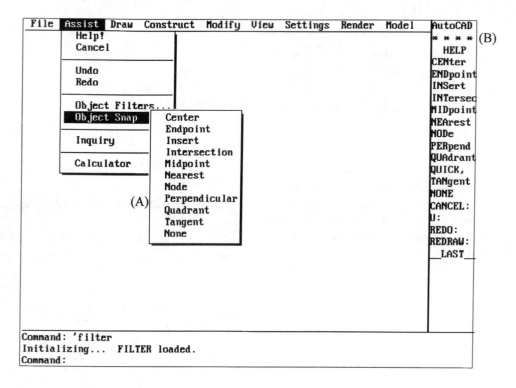

Snap and grid are two modes used to aid the user in creating and editing entities. However all entities cannot be guaranteed that they will fall exactly on a grid dot. Rather than rely entirely on grid or snap, the object snap mode is one of the most important modes for locking onto key entity points. Illustrated at the right are all supported object snap modes. An example of how object snap is used may be in dimensioning applications where exact endpoints and intersections are needed.

Object snap modes may be selected from one of four different areas. Illustrated above is the typical drawing screen area. Selecting "Assist" from the menu strip and then:

Object Snap >

from the pulldown menu displays a cascading menu of all object snap modes at "A". Selecting the four asterisks * * * * from the screen menu at "B" displays all object snap modes. For a graphical format, object snap modes may be selected from the digitizing pad. This is illustrated in the sample tablet area below. A final method of entering object snap modes is from the keyboard. Notice in the screen menu that the first three letters of all object snap modes are in upper case letters. When entering object snap modes from the keyboard, only the first three letters are required. The next series of pages give examples on the applications of all object snap modes.

Center
Endpoint
Insert
Intersection
Midpoint
Nearest
Node
Perpendicular
Quadrant
Tangent
None

| NONE | CEN | ENDP | INSERT | INT | MID | NEAR | NODE | PERP | QUAD | TAN |

Using Options of the Osnap Command

Osnap-Center Option

Use this mode to snap to the center of a circle or arc. To accomplish this, position the object snap aperture box along the edge of the circle or arc similar to the illustration at the right.

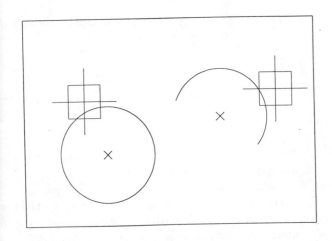

Osnap-Endpoint Option

This is one of the more popular object snap modes that is very helpful in snapping to the endpoints of lines or arcs. One application of osnap-endpoint is during the dimensioning process where exact distances are needed to produce the desired dimension. Position the object snap aperture box along the edge of the entity to snap to the endpoint. In the case of the arc illustrated at the right, the aperture box does not actually have to be positioned at the endpoint; favoring one end automatically snaps to the closest endpoint.

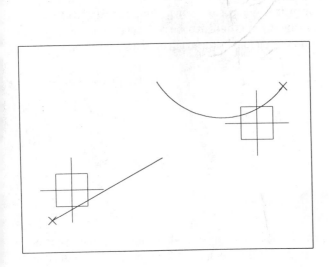

Osnap-Insert Option

This object snap option snaps to the insertion point of an entity. In the case of the text entity illustrated at the right, positioning the aperture box any place on the text snaps to its insertion point, in this case at the lower left corner of the text at "A." The other entity illustrated at the right is called a block. It appears to be constructed of numerous line entities; however, all entities that make up the block are considered to be a single entity. Blocks are then inserted into a drawing. Typical types of blocks are symbols such as doors, windows, bolts, etc., anything that is used numerous times in a drawing. In order for a block to be brought into a drawing, it needs an insertion point, or a point of reference. The osnap insert option, when positioning the aperture box on a block, will snap to the insertion point of a block.

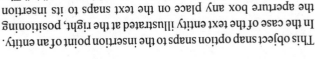

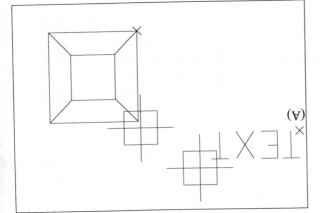

Osnap-Intersection Option

Another very popular object snap mode is intersection. Use this mode to snap to the intersection of two entities. As in the other object snap modes, the exact center of the aperture box does not have to be touching the intersection to snap to; position the aperture box anywhere around the intersection and the snap is made.

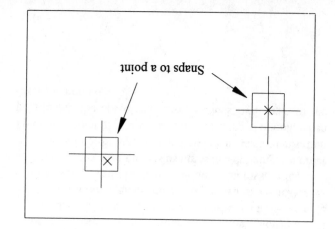

Osnap-Midpoint Option

This object snap mode snaps to the midpoint of entities. Illustrated at the right are line and arc examples. When in the midpoint mode, touch the entity anywhere with some portion of the object snap box; the midpoint will be searched for and found.

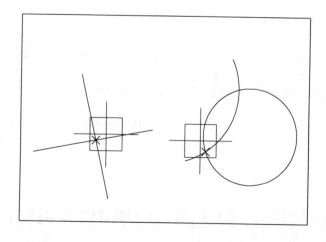

Osnap-Nearest Option

This object snap mode snaps to the nearest point it finds on an entity. Use this mode when you need to grab onto an entity for the purposes of further editing. The nearest point is calculated from the intersection of the crosshairs in the aperture box perpendicular to the entity; or the shortest distance from the crosshairs in the aperture box to the entity.

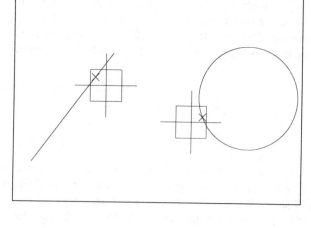

Osnap-Node Option

This object snap mode snaps to a node or point. Touching the point anywhere snaps to its center. The Osnap-Node mode above can also be used to snap to a point.

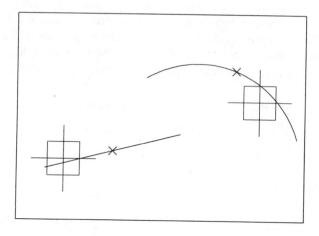

Snaps to a point

Osnap-Perpendicular Option

This is a helpful object snap mode for snapping to a point normal or perpendicular to an entity. Illustrated at the right is a line segment drawn perpendicular from the point at "A" to the inclined line "B." A 90 degree angle is formed with the perpendicular line segment and the inclined line "B." This mode is also able to construct perpendicular to circles also illustrated at the right.

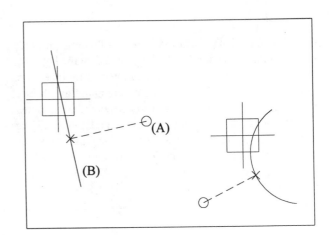

Osnap-Quadrant Option

Circle quadrants are defined as points located at the 0 degree, 90 degree, 180 degree, and 270 degree positions of the circle, as in the example illustrated at the right. Using the osnap-quadrant option will snap to one of these four positions as the edge of a circle or arc is selected. In the example of the circle illustrated at the right, the edge of the circle is selected by the object snap aperture box. The closest quadrant to the aperture box is selected.

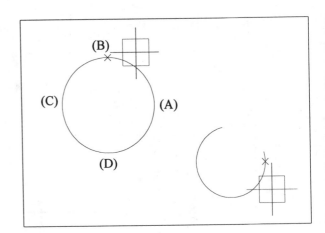

Osnap-Tangent Option

This object snap mode is very helpful in constructing lines tangent to other entities such as the two circles illustrated at the right. In this case, the object snap tangent option is being used in conjuction with the Line command. Follow the command prompt sequence for constructing a line segment tangent to two circles:

Command: **Line**
From point: **Tan**
from *(Select the circle near "A")*
To point: **Tan**
to *(Select the circle near "B")*

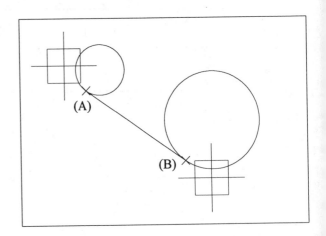

Alternate Methods of Choosing Osnap Options

An additional aid is present to bring up all options of object snap. Holding down the "Shift" key located on the keyboard and pressing the enter button of the mouse or digitizer puck automatically brings up a pulldown menu containing all object snap modes. This menu is displayed wherever the cursor was last positioned. This provides the user with an even quicker way of getting to the object snap modes. The illustration at the bottom of the page shows a typical drawing screen with the object snap modes activated with the "Shift-Mouse Enter" operation.

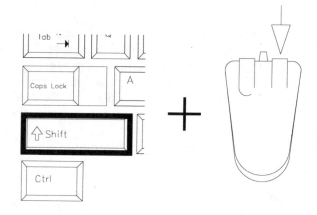

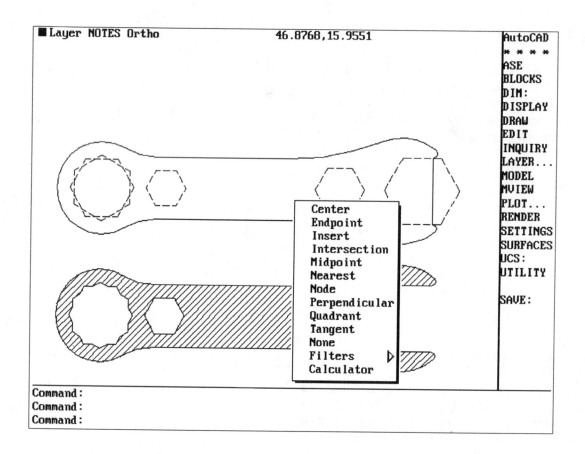

Choosing Running Osnap from the DDOSNAP Dialog Box

Thus far, all osnap modes have continuously been selected from the screen, tablet, or pulldown menus; they could also be entered in at the keyboard. The problem with this method is that if a certain mode was used over a series of commands, the osnap mode had to be selected every time. It is possible to make the object snap mode or modes automatically present during a command sequence; this is called running osnap. Running osnap mode may be selected from the "Settings" menu bar followed by the selection of "Object Snap..." from the pulldown menu. This activates the "Running Object Snap" dialog box illustrated below. One or more object snap modes may be selected by checking their appropriate boxes in the dialog box. These osnap modes remain in effect during the drawing until the Osnap-None option is invoked turning off running osnap mode and returning to the regular method of using Osnap options. While running osnap is invoked, the aperture box appears at the intersection of the graphics cursor. The size of the aperture box is controlled by the Aperture command. Use this command to adjust the size of the box; or use the dialog box below to dynamically change the aperture box using the slider bar under "Aperture size."

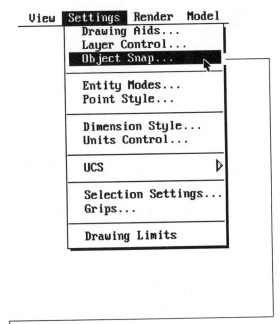

Using the Zoom Command

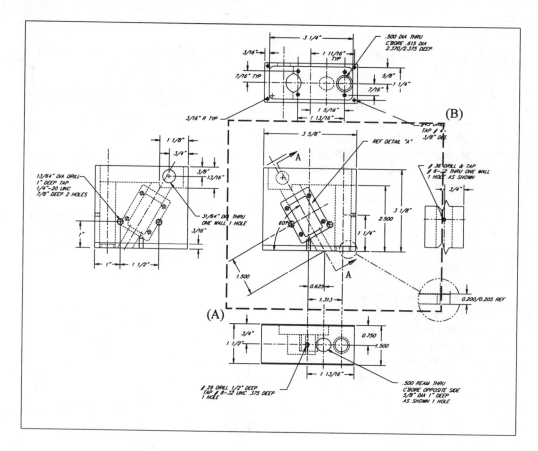

Illustrated above is a somewhat complex drawing of a part consisting of all required orthographic views. To work on details of this and other drawings, the Zoom command is used to magnify or demagnify the display screen. The following are options of the Zoom command:

Command: **Zoom**
All/Center/Dynamic/Extents/Left/Previous/Vmax/Window/
<Scale(X/XP)>: *(Enter one of the listed options)*

Executing the Zoom command and picking a blank part of the screen places the user in automatic Zoom-Window mode. Selecting another point zooms into the specified area. Follow the command sequence below for using this mode of the Zoom command on the object illustrated above:

Command: **Zoom**
All/Center/Dynamic/Extents/Left/Previous/Vmax/Window/
<Scale(X/XP)>: *(Mark a point at "A")*
(Mark a point at "B")

The Zoom-Window option is automatically invoked once a blank part of the screen is selected followed by a second point. The results of the magnified screen are illustrated at the right.

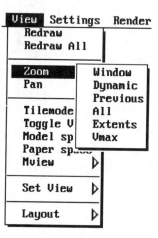

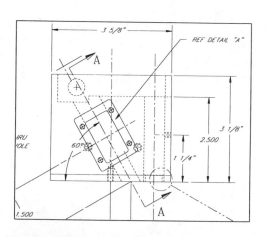

Using Zoom All

In the typical AutoCAD screen, the Zoom command may be picked from the screen, tablet, or pulldown menu by picking "View" in the menu strip followed by "Zoom," which activates the cascading menu consisting of the zoom options.

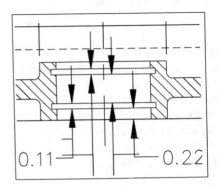

One of the options of the Zoom command is All. Use this option to zoom to the current limits of the drawing as set by the Limits command. In fact, right after the limits of a drawing have been changed, issuing a Zoom-All updates the drawing file to reflect the latest screen size. The Zoom-All option is illustrated below in the following command sequence:

Command: **Zoom**
All/Center/Dynamic/Extents/Left/Previous/Vmax/Window/
<Scale(X/XP)>: **All**
Regenerating drawing.

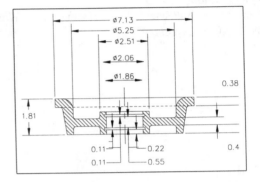

At the right, the top illustration shows a zoomed in portion of a part. Use the Zoom-All option to zoom to the drawing's current limits in the bottom illustration.

Using Zoom Center

The Zoom-Center option allows you to specify a new display based on a selected center point. A window height controls whether the image on the display screen is magnified or demagnified. If a smaller value is specified for the magnification or height, the magnification of the image is increased or the object is zoomed into. If a larger value is specified for the magnification or height, the image gets smaller, or a zoom out is performed.

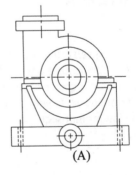
(A)

Command: **Zoom**
All/Center/Dynamic/Extents/Left/Previous/Vmax/Window/
<Scale(X/XP)>: **Center**
Center point: *(Mark a point at the center of circle "A" illustrated above)*
Magnification or Height <7.776>: **2**

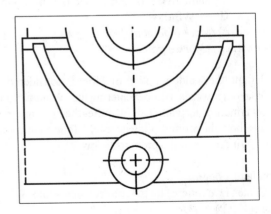

Using Zoom Extents

The image illustrated above of the pump reflects a Zoom-All operation. This displays the entire drawing area based on the drawing limits even if the entities that make up the image appear small. Instead of performing a zoom based on the drawing limits, Zoom-Extents uses only the extents of the image on the display screen to perform the zoom. Illustrated at the right is the largest possible image displayed as a result of using the Zoom command and the Extents option.

Command: **Zoom**
All/Center/Dynamic/Extents/Left/Previous/Vmax/Window/
<Scale(X/XP)>: **Extents**

Drawing Extents

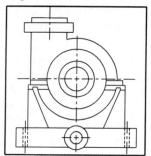

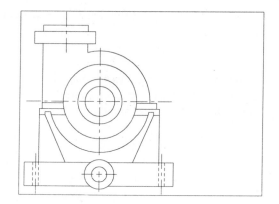

Using Zoom Window

Zoom-Window allows the user to specify the area to be magnified by marking two points representing a rectangle. The center of the rectangle bacomes the center of the new image display; the image inside of the rectangle is either enlarged or reduced. The following prompt sequence illus- trates using Zoom-Window:

Command: **Zoom**
All/Center/Dynamic/Extents/Left/Previous/Vmax/Window/
<Scale(X/XP)>: **Window**
First corner: *(Mark a point at "A")*
Other corner: *(Mark a point at "B")*

By default, the window option of zoom is considered auto- matic; in other words, without entering the "Window" option, a point is marked instead. This identifies the first corner of the window box; the prompt "Other corner:" follows to complete Zoom-Window as in the prompts below:

Command: **Zoom**
All/Center/Dynamic/Extents/Left/Previous/Vmax/Window/
<Scale(X/XP)>: *(Mark a point at "A")*
Other corner: *(Mark a point at "B")*

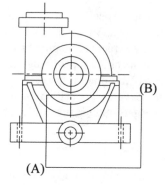

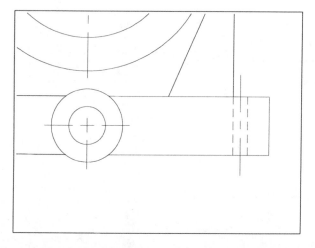

Using Other Options of the Zoom Command

Zoom-Previous

After magnifying a small area of the display screen, use the Previous option of the Zoom command to return to the previous display. The system automatically saves up to ten views when zooming. This means you can begin with an overall display, perform two zooms, and use the Zoom-Previous command twice to return back to the original display. Zoom-Previous is also less likely to create a drawing regeneration.

Command: **Zoom**
All/Center/Dynamic/Extents/Left/Previous/Vmax/Window/
<Scale(X/XP)>: **Previous**

Zoom-Vmax

Use this option of the Zoom command to zoom out as far as possible without forcing a regeneration of the drawing.

Command: **Zoom**
All/Center/Dynamic/Extents/Left/Previous/Vmax/Window/
<Scale(X/XP)>: **Vmax**

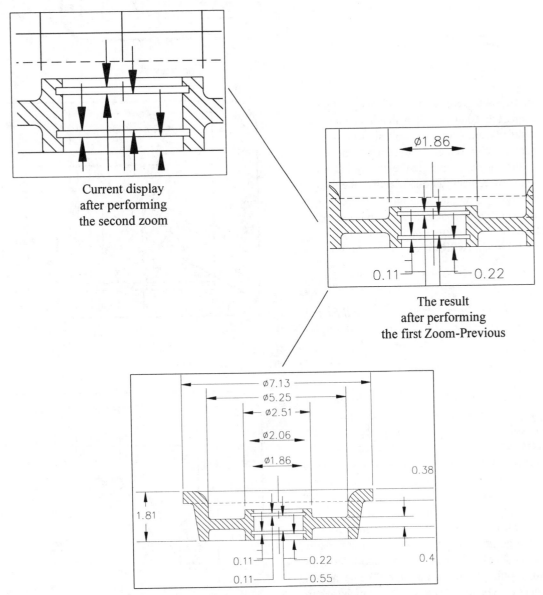

Current display
after performing
the second zoom

The result
after performing
the first Zoom-Previous

The result after performing
the second Zoom-Previous

Using the Pan Command

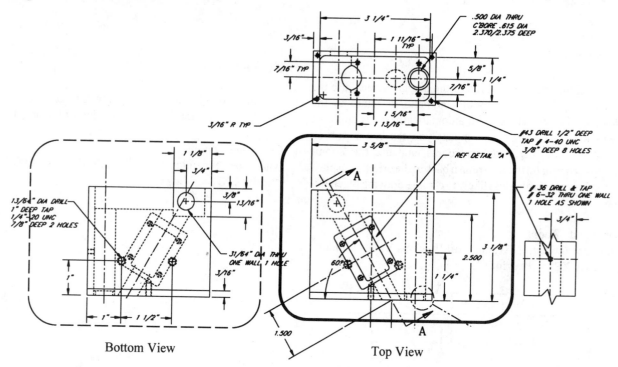

Bottom View   Top View

As an operator performs numerous Zoom-Window and Zoom-Previous operations, it becomes apparent that it would be nice to zoom into a detail of a drawing and simply slide the drawing to a new area or detail without changing the magnification; this is the purpose of the Pan command. In the illustration above, the top view is magnified using Zoom-Window; the result is illustrated at the right. Now, the bottom view needs to be magnified to view certain dimensions. Rather than use Zoom-Previous and then Zoom-Window again to magnify the bottom view, the Pan command will be used; follow the command sequence below:

Command: **Pan**
Displacement: *(Mark a point at "A" illustrated at the right)*
Second point: *(Mark a point at "B" illustrated at the right)*

Top View

Pan requires two points which provide the direction for moving the drawing. The first point is called "displacement" and acts as a point of reference. The second point provides the distance and direction of the pan operation in relation to the original displacement point. In the illustration at the right, the bottom view is now visible after panning the drawing from the top view to the bottom view while keeping the same display screen magnification. Pan can also be used transparently; that is, while in a current command, the Pan command may be selected, which temporarily interrupts the current command, performs the pan, and restores the current command. Pan may be selected from the pulldown menu area under "View" or may be typed in from the keyboard by entering "p," which is short for the Pan command.

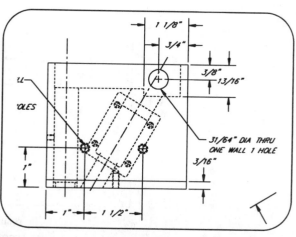

Bottom View

Using the View Command

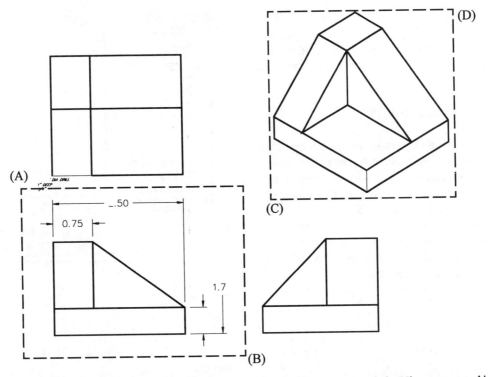

(A)

(B)

(C)

(D)

An alternate method of performing numerous zooms is to create a series of views of key parts of a drawing. Then, instead of using the Zoom command, restore the named view to perform detail work. This named view is saved in the database of the drawing for use in future editing sessions. Using the illustration above as an example, the following command sequence is used to create a series of views:

Command: **View**
?/Delete/Restore/Save/Window: **Window**
View name: **Front**
First corner: *(Pick a point at "A")*
Other corner: *(Pick a point at "B")*

Command: **View**
?/Delete/Restore/Save/Window: **Window**
View name: **Iso**
First corner: *(Pick a point at "C")*
Other corner: *(Pick a point at "D")*

Once views are created, the View command is used along with the Restore option to display the view, as in the prompt sequence below:

Command: **View**
?/Delete/Restore/Save/Window: **Restore**
View name: **Iso** *(Illustrated below)*

Command: **View**
?/Delete/Restore/Save/Window: **Restore**
View name: **Front** *(Illustrated below)*

View Name = ISO

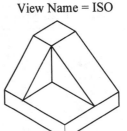

View Name = FRONT

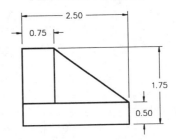

Using the DDVIEW Dialog Box

Selecting "Named view..." from the "Set View" area of the "View" dialog box activates the "View Control" dialog box illustrated below. This same dialog box is activated through the keyboard by entering the following at the command prompt:

Command: **DDVIEW**

The same options of the View command apply to the operation of this dialog box; that is, you may create a new view, restore an existing view, delete an existing view, or be provided with a description of the current view. Simply pick the button that applies to the appropriate View command option. The "View Control" dialog box below provides a more user friendly way to manipulate views. Notice the 5 existing view names below. To display the contents of one of the views below, select the view until it highlights, as in the view "FRONT." Pick the "Restore" button to display the view.

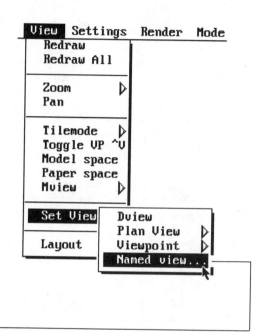

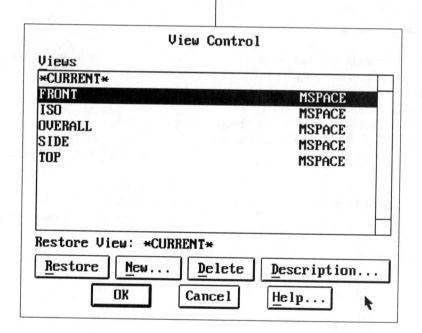

Selecting the "New..." button from the "View Control" dialog box activates the "Define New View" dialog. Use this dialog box to guide you in creating a new view. By definition, a view is created from the current display screen. This is the purpose of the "Current Display" radio button illustrated at the right. Many views are created using the "Define Window" radio button which will create a view of the contents of a window. Selecting the "Window <" button prompts the user for the first corner and other corner required to create a new view by window. When the window has been created, the "Define New View" dialog box redisplays. The absolute coordinate value of the first corner and other corner are displayed in the dialog box. Before selecting the "Save View" button, enter the name of the view you want in the "New Name:" edit box. Finally, select the "Save View" button and the view name and window coordinates are added to the database of the current drawing. This view may now be restored using the "Restore" button located in the "View Control" dialog box illustrated on the previous page.

Selecting "Description..." button from the "View Control" dialog box displays the "View Description" dialog box illustrated at the right. The following information is given about the current view: View Name; Width of the window defining the view; Height of the window defining the view; Twist angle of the view (if any); the center of the view in the world coordinate system; the direction the view is being displayed; if perspective mode is On or Off; If front clipping is On or Off; if back clipping is On or Off; the current lens focal length; if any view offsets are present.

Perspective mode, front clipping, back clipping, the lens length, and offsets are controlled by the Dview command.

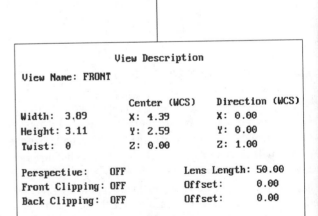

Creating Entity Selection Sets

Selection sets are used to group a number of entities together for the purpose of editing. Applications of selection sets are covered in the next series of pages in addition to being illustrated in the next unit. Once a selection set has been created, the group of entities may all be moved, copied, or mirrored just to name a few of the operations supported through selection sets. An entity manipulation command supports the creation of selection sets if it prompts the user to "Select objects:". Any command displaying this prompt supports the use of selection sets. Options of selection sets, or how a selection set is made, are illustrated at the right. This is a typical screen menu display; directly below is an example of how selection sets may be picked from the digitizing tablet. Below and in the next series a few applications of how selection sets are used for manipulating groups of entities are illustrated.

Add
All
CPolygon
Crossing
Fence
Last
Previous
Remove
Window
WPolygon
Undo

F	CP	WP	W	L	P	C	R	A
FENCE	POLYGON	POLYGON	WINDOW	LAST	PREVIOUS	CROSSING	REMOVE	ADD

Entity Selection by Individual Picks

When prompted at the Command prompt with "Select objects:", a pickbox appears as the cursor is moved back onto the display screen. Any entity enclosed by this small box when picked will be considered selected. To show the difference between a selected and unselected entity, the selected entity highlights on the display screen. In the example illustrated at the right, the small box is placed over the arc segment and picked by the mouse button or digitizing puck. To signify that the entity is selected, the arc turns dotted or highlights.

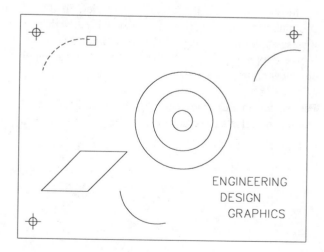

ENGINEERING
DESIGN
GRAPHICS

Entity Selection by Window

The individual pick method outlined above works fine for small amounts of entities; however when numerous entities need to be edited, selecting each individual entity could prove too time consuming. Instead, all entities desired to become part of a selection set are selected by the Window selection mode. This mode requires the user to create a rectangular box by picking 2 points. In the example illustrated at the right, a selection window has been created with point "A" as the first corner and "B" as the other corner. A more interesting observation is made on the highlighted entities; only those entities completely enclosed by the box are selected. The window box selected 4 line segments, two arcs, and two points (too small to display highlighted). The three circles, eventhough they are touched by the window, are not completely surrounded by the window and therefore are not selected.

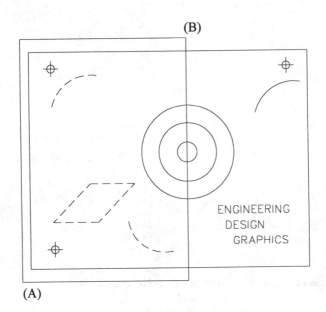

Entity Selection by Crossing Window

In the previous example of window selection sets, the window only selected those entities completely enclosed by it. Illustrated at the right is an example of selecting entities by a crossing window. The crossing window requires two points to define a rectangle as does the window selection option. At the right, the same rectangle was used to select entities; however this time the crossing window was used. The results are illustrated by the highlighted entities. As entities highlight only if completely enclosed by a window, all entities that are touched by the crossing window rectangle are selected. Now, as the crossing rectangle passes throuth the three circles without enclosing them, they are still selected by this entity selection mode.

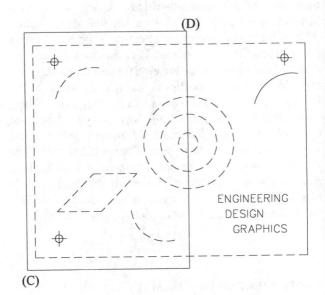

Entity Selection by Fence

Use this mode to create a selection set by drawing a line called a fence. Any entity touching the line will be selected. The fence does not have to end exactly where it was started. In the example illustrated at the right, all lines contacted by the fence are selected as represented by the dashed lines.

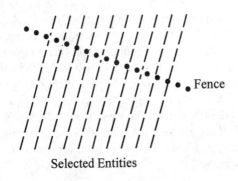

Selected Entities

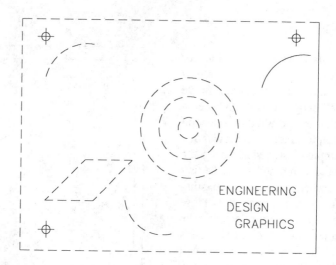

Removing Entities from a Selection Set

All previous examples of creating selection sets have illustrated creating new selection sets; another way to say this is that entities were added to create a selection set of entities. What if too many entities are selected? What if the wrong entity or entities were selected? In previous versions of AutoCAD, the operator had to cancel the selection process and carefully reselect the entities. This is the purpose of the Remove option of selection sets. In the illustration above, notice all of the highlighted entities signifying that they make up a current selection set of entities. However, the large circle was mistakenly selected as part of the selection set. The Remove option allows the operator to remove highlighted entities from a selection set. When a highlighted entity is removed from the selection set, as in the illustration of circle "A" at the right, it regains its original display intensity.

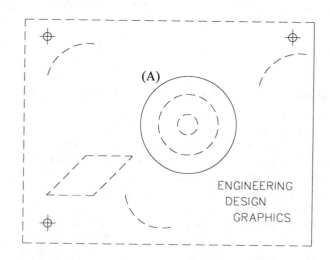

Entity Selection by Crossing Polygon

Whenever using window or crossing window modes of creating selection sets, 2 points specify a rectangular box to select entities. At times, it is difficult to select entities by the rectangular window or crossing box because in more cases than not, extra entities are selected and have to be removed from the selection set. Illustrated at the right is a mechanical part with a "C"-shaped slot. The object is to use the best available selection set mode to select just the slot. Rather than use window or crossing window modes, the crossing polygon mode is used. An operator simply picks points representing a polygon. Any entity that touches or is inside the polygon is added to a selection set of entities. At the right, the crossing polygon is constructed using points "A" through "F." A similar but different selection set mode is the window polygon. Entities are selected using this mode when they lie completely inside the window polygon similar to the regular window mode.

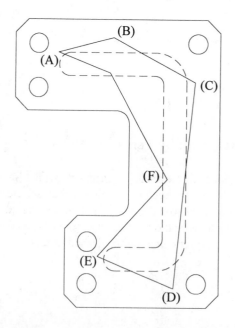

Saving a Drawing File

Drawings may be saved using the Qsave, Save, and End commands. Choose two of the three commands from the "File" pulldown menu area illustrated at the right.

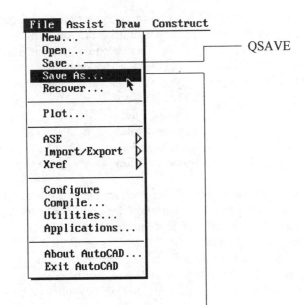

Save As...
Using this command always displays the dialog box illustrated below. Simply pick the "OK" button or strike the Enter key to save the drawing under the current name displayed in the edit box. This command is more popular for saving the current drawing under an entirely different name. Simply enter the new name in place of the the highlighted name in the edit box.

Save...
This command is short for Qsave, which stands for Quick Save. If a drawing file has never been saved and this command is selected, the dialog box below displays. Once a drawing file has been saved, selecting this command saves the drawing file in the quickest fashion by not displaying the "Save Drawing As" dialog box below.

End
Use this commmand to automatically save the drawing file along with exiting AutoCAD.

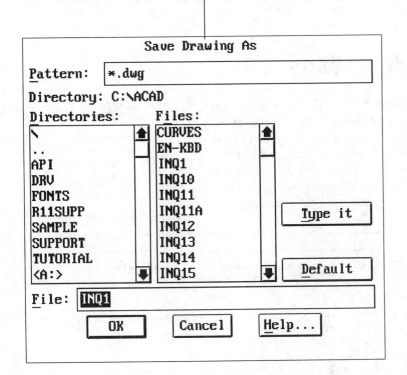

Managing Files Using the DDFILES Dialog Box

Basic Disk Operating functions may be performed without
leaving the AutoCAD environment using the "File Utilities"
dialog box illustrated below. This dialog box appears after
selecting "Utilities . . ." from the "File" pulldown menu area.
The following operating system functions may be performed:

Listing Files
Copying Files
Renaming Files
Deleting Files
Unlocking Files

In all of the above cases, a dialog box appears; the title of the
dialog differs depending on the function used. For example,
when listing files, the user must enter the proper file name and
extension; wild cards such as "?" and "*" may be used. When
copying files a destination dialog box and direction dialog box
appear; this is similar to the operation of renaming files where
a dialog box of the original file first appears followed by a
dialog box of what the new file name will be.

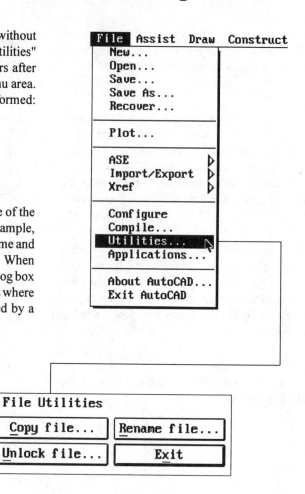

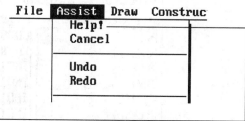

Using the Help Utility

Selecting "Help!" from the "Assist" pulldown menu area
activates the Help dialog box illustrated below. Use this dialog
box to enter any AutoCAD command in at the edit box to
receive help on the command. Browsing the Help dialog box
provides the user with a series of steps to use to get help on a
specific command in addition to explaining the purpose of the
"OK," "Top," "Previous," "Next," and "Index" buttons.

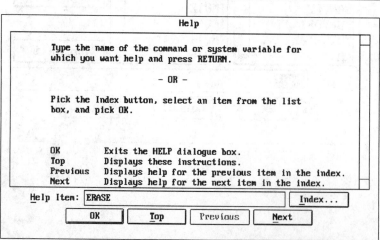

Selecting the "Index" button displays the Help Index dialog box illustrated at the right. The user will recognize commands such as Move, New, Offset, and Open as valid commands to receive help on. The user may not be familiar with MP, Offsetdist, Orthomode, or Osmode. Not only does the Help Index provide information on popular commands, the index also provides information on system variables such as Osmode and Orthomode. Commands listed as MP are aliased commands that have been shortened to one or two letters; in this case, MP is short for Mass Property and is used for performing mass property calculations on a region or solid model. Use the scroll bars to move through the index searching for the command to receive help on.

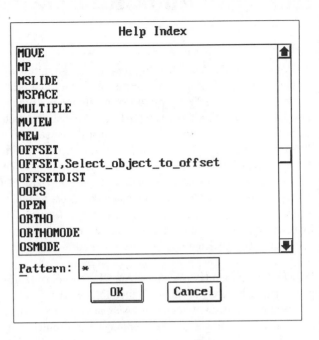

Once a command is discovered in the "Help Index" dialog box, selecting the "OK" button adds the command to the edit box and provides the user with information on the specific command. Illustrated below is an example of the type of help provided for the Erase command. If more help is provided than fits in the designated dialog box area, the scroll bar is activated to browse the screen for the amount of help provided for the specific command.

```
                            Help
The  ERASE  command lets you delete selected entities from the
drawing.

Format: ERASE
        Select objects:  (select)

You can easily erase just the last object you drew by responding to
the "Select objects" prompt with "l".

The OOPS command can be used to retrieve the last thing you erased.



     Help Item:  ERASE                              Index...

         OK           Top        Previous      Next
```

Exiting an AutoCAD Drawing Session

It is considered good practice to properly exit any drawing session. One way of exiting is by selecting the "Exit AutoCAD" option located in the "File" pulldown menu area. Another acceptable way of exiting AutoCAD is by using the End command. This is a save command in that it first saves the drawing file before exiting the program. Another command that exits AutoCAD is Quit. This command should be used with extreme caution because Quit exits the drawing file without saving any changes. This command is useful when just browsing drawing files where no changes were made. Beginning users of AutoCAD are warned not to use this command until sufficient experience is gained by the user.

Whichever command is used to exit AutoCAD, a built-in safeguard gives the user a second chance to save the drawing before exiting especially if changes were made and Save was not performed by mistake. The user may be confronted with three options illustrated in the "Drawing Modification" dialog box below.

If changes were made to the drawing but not saved, the user may want to pick the "Save Changes . . ." button before exiting the drawing.

If changes were made but do not have to be saved, the user may want to pick the "Discard Changes" button. This is commonly practiced in "What if" scenarios.

If changes were made and the exit AutoCAD option was mistakenly picked, choosing the "Cancel Command" button cancels the previous command used to exit AutoCAD and returns the user to the drawing editor and the current drawing.

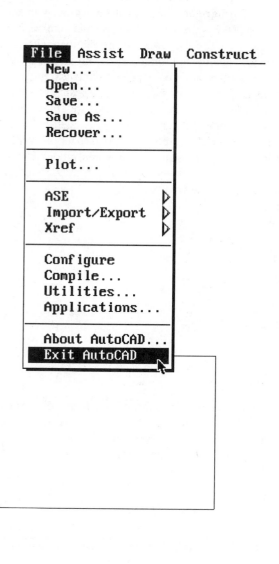

Tutorial Exercise #1
Creating Layers Using the Layer Command

Layer Name - Color - Linetype
Object - White - Continuous
Hidden - Red - Hidden
Center - Yellow - Center
Dim - Yellow - Continuous
Section - Blue - Continuous

Use the command sequences below to create the following layers according to the above specifications:

Command: **Layer**
?/Make/Set/New/ON/OFF/Color/Ltype/Freeze/Thaw/LOck/Unlock: **New**
New layer name(s): **Object,Hidden,Center,Dim,Section**
?/Make/Set/New/ON/OFF/Color/Ltype/Freeze/Thaw/LOck/Unlock: **Color**
Color: **Red**
Layer name(s) for color 1 (RED) <0>: **Hidden**
?/Make/Set/New/ON/OFF/Color/Ltype/Freeze/Thaw/LOck/Unlock: **Ltype**
Linetype (or ?) <CONTINUOUS>: **Hidden**
Layer name(s) for linetype HIDDEN <0>: **Hidden**
?/Make/Set/New/ON/OFF/Color/Ltype/Freeze/Thaw/LOck/Unlock: **Color**
Color: **Yellow**
Layer name(s) for color 2 (YELLOW) <0>: **Center,Dim**
?/Make/Set/New/ON/OFF/Color/Ltype/Freeze/Thaw/LOck/Unlock: **Ltype**
Linetype (or ?) <CONTINUOUS>: **Center**
Layer name(s) for linetype CENTER <0>: **Center**
?/Make/Set/New/ON/OFF/Color/Ltype/Freeze/Thaw/LOck/Unlock: **Color**
Color: **Blue**
Layer name(s) for color 5 (BLUE) <0>: **Section**
?/Make/Set/New/ON/OFF/Color/Ltype/Freeze/Thaw/LOck/Unlock: *(Strike Enter to Exit this command)*

Tutorial Exercise #2
Creating Layers Using the Layer Dialog Box

Create the same layers outlined above in Tutorial Exercise #1 this time using the DDLMODES dialog box to create the layers.

Step #1

Since the Hidden and Center linetypes need to be assigned to certain layers, they must be first preloaded into the database of a drawing. To do this, use the Linetype command and the Load option. When prompted for the linetypes to load, enter the names Hidden and Center. When a file dialog box appears requiring a file search, strike enter to accept the ACAD.Lin file where all linetype definitions are located.

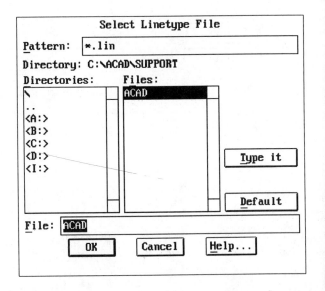

Command: **Linetype**
?/Create/Load/Set: **Load**
Linetype(s) to load: **Hidden, Center**
(Strike Enter when the "Select Linetype File" dialog box at the right appears)
HIDDEN loaded.
CENTER loaded.
?/Create/Load/Set: *(Strike Enter to exit this command)*

Step #2

The previous exercise dealt with creating layers using a manual method where the Layer command is used, prompts are carefully followed, and the result is the desired layer names complete with correct color and linetype assignments. The manual method, however, is open to numerous errors in the assignment of color and linetype. A typical example is to assign the color "Red" to a layer called "Hidden"; unfortunately the prompt states that Layer 0 will be changed to red. This is quite a surprise when the color of Layer 0 changes from "White" to "Red." A more direct method of creating layers is to use the DDLMODES dialog box. This graphically displays the creation of layers; better yet, it is easier to spot mistakes during the layer creation process and even easier to correct these mistakes. Select the DDLMODES dialog box from the "Settings" menu located in the pulldown menu area. Then select "Layer Control..." to display the dialog box illustrated below.

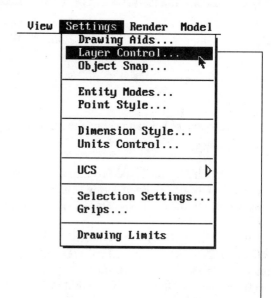

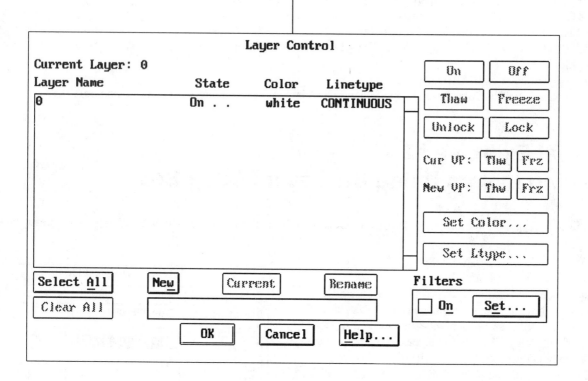

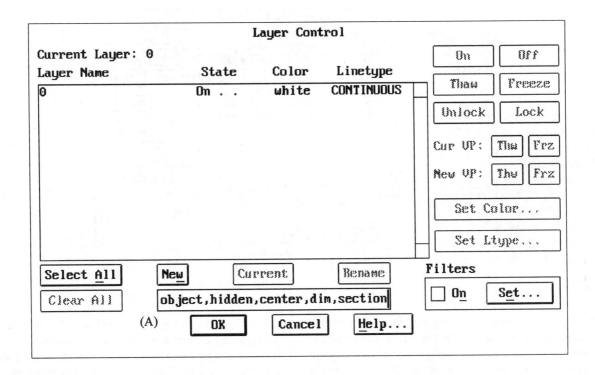

object,hidden,center,dim,section|

Step #3

Once the "Layer Control..." dialog box displays as in the illustration above, notice only one layer is currently displayed in the large edit box, namely Layer 0. This happens to be the default layer; the layer that is automatically assigned to all new drawings. Since it is considered poor practice to construct any entities on Layer 0, new layers will be created not only for object lines but for hidden lines, center lines, dimension components, and section lines as well. To create these layers, pick the edit box identified by "A" in the illustration above. A vertical blinking line appears signifying that this dialog is expecting input from the operator in the form of text for the names of the desired layers. If the blinking vertical bar is already present, begin entering the desired layer names. In the edit box illustrated above, enter the following layer names:

object,hidden,center,dim,section

By separating the layer names with a comma, numerous layers may be created instead of making them individually. If the layers are so numerous that they appear not to fit in the edit box, the box will scroll to the right to accommodate extra layer names.

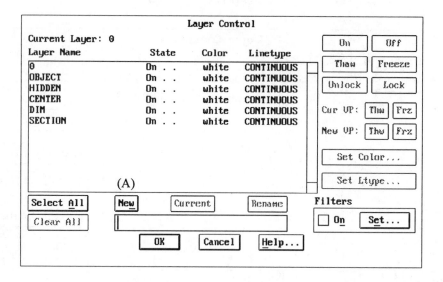

Step #4

As all layers are entered in the edit box, the most common reflex to use to create and display the layers is to strike the Enter key. This may be continuously struck without creating any layers. In fact, take caution at this point not to select the "Cancel" button. All layer names entered in the previous step will be erased, requiring them to be re-entered into the system. The easiest way to create the layers is to select the "New" button from the dialog box located at "A" illustrated above. All layers are created under the names just entered and display in the original order that they were created. If this dialog box is exited when selecting the "OK" button and then returned to, all layers are displayed automatically in alphabetical order.

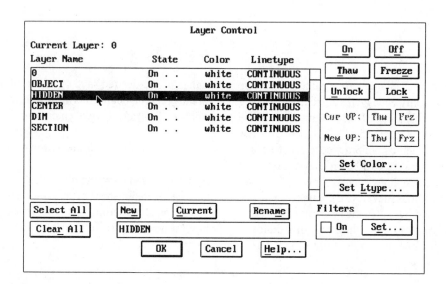

Step #5

As all layers are displayed, the names may be different, but they all have the same color and linetype assignments. At this point, the dialog box comes in real handy to assign color and linetypes to layers in a quick manner. First, highlight the desired layer to add color or linetypes by picking the layer. A long horizontal bar displays signifying that this is the selected layer. Also, notice that a series of buttons on the upper right side of the dialog box become active. Follow the next step to assign the color "Red" to the "Hidden" layer name.

Step #6

Selecting the "Set Color..." button displays the "Select Color" dialog box illustrated at the right. Select the desired color from one of the following areas: Standard Colors; Gray Shades; Full Color Palette. The standard colors represent colors 1 through 9. If shades of grey are desired, select them from the grey shades area. On display terminals with high resolution cards, the full color palette displays different shades of the standard colors, which gives the operator a greater variety of colors to choose from. For the purposes of this tutorial, the color "Red" will be assigned to the "Hidden" layer. Select the box displaying the color red; a box outlines the color and echoes the color in the bottom portion of the dialog box. Pick the "OK" button to complete the assignment; if "Cancel" is selected, the color assignment will be removed requiring this operation to be used again.

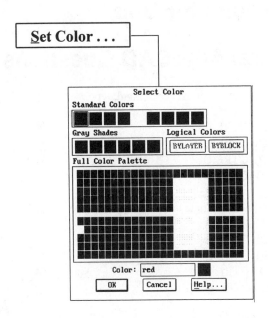

Step #7

Once the color has been assigned to a layer, the next step is to assign a linetype if any to the layer. The "Hidden" layer requires a linetype called "Hidden." Pick the "Set Ltype..." button in the main layer dialog box screen to display the "Select Linetype" dialog box illustrated at the right. Since the linetypes were already pre-loaded back in Step #1, they appear in graphical form in the "Select Linetype" dialog box. Selecting the actual linetype outlines the linetype in a box in addition to echoing it in the edit box in the lower part of the dialog box. Pick the "OK" button to assign the "Hidden" linetype to the "Hidden" layer.

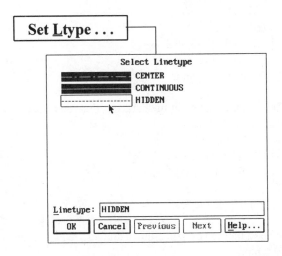

Step #8

The highlighted layer remains active to manipulate such as changing color, turning the layer On or Off, or freezing and thawing the layer. To highlight a second layer, pick the currently highlighted layer. This removes the highlight. Next, use the methods in Steps #5 through #7 to complete the remaining layers until the layers in your dialog box match the layer names, color assignments, and linetype assignments illustrated below.

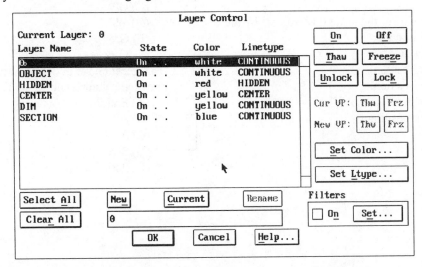

Questions for Unit 1

General AutoCAD Questions

1-1. To cancel an AutoCAD command, type
(A) CTRL A
(B) CTRL B
(C) CTRL C
(D) CTRL D
(E) CTRL E

1-2. To toggle Coordinate mode On or Off, type
(A) CTRL A
(B) CTRL B
(C) CTRL C
(D) CTRL D
(E) CTRL E

1-3. To toggle Grid mode On of Off, type
(A) F7
(B) CTRL F
(C) 'DDEMODES
(D) All of the above.
(E) Only A and C.

1-4. The maximum number of characters that can be used for naming a View is
(A) 8 characters.
(B) 10 characters.
(C) 23 characters.
(D) 31 characters.
(E) 40 characters.

1-5. The best method of cancelling a drawing regeneration is by typing
(A) ENTER
(B) CTRL-C
(C) CTRL-D
(D) CTRL-R
(E) CTRL-ALT-DEL

1-6. The default direction for the angle 0 (Zero), set by the UNITS command is at the
(A) 12 o'clock position.
(B) 3 o'clock position.
(C) 6 o'clock position.
(D) 9 o'clock position.
(E) 11 o'clock position.

1-7. Command aliasing is found in the
(A) ACAD.ALS file
(B) ACAD.PGP file
(C) ACAD.EXE file
(D) ALIAS.COM file
(E) ALIAS.BAT file

1-8. To edit an existing drawing, use the
(A) Files command.
(B) New command.
(C) Open command.
(D) Config command.
(E) Recover command.

1-9. By default, pressing the Enter button of a mouse or digitizing puck while holding down the Shift key activates
(A) POP 0
(B) a pulldown menu containing all Object Snap modes.
(C) the Enter key
(D) Only A and B.
(E) Only B and C.

1-10. The object selection mode that allows the user to select all entities that lie within the boundaries of a polygon is
(A) W
(B) F
(C) CP
(D) WP
(E) C

1-11. The object selection mode that allows the user to select all entities within or touching the boundary of a polygon is
(A) W
(B) F
(C) CP
(D) WP
(E) C

1-12. The selection set option used to select all entities in a drawing including entities on layers turned off is
(A) Entities
(B) All
(C) A
(D) Screen
(E) Both B and C.

Display Questions

1-13. To view a different portion of the drawing without changing its magnification, use the
- (A) Pan command.
- (B) Move command.
- (C) View command.
- (D) Zoom command.
- (E) Copy command.

1-14. The Pan command
- (A) always requires a regeneration.
- (B) decreases the current display magnification.
- (C) increases the current display magnification.
- (D) maintains the current display magnification.
- (E) cannot be used transparently.

1-15. To perform a zoom at 1/10th of the current screen size, issue the Zoom command and type
- (A) 0.01X
- (B) 0.10X
- (C) 0.10
- (D) X0.10
- (E) 0.01

1-16. All of the following special characters can be used in the name of a View except
- (A) $
- (B) *
- (C) - (Hyphen)
- (D) _ (Underbar)
- (E) None of the above

1-17. The following are all options of the Zoom command except
- (A) Previous
- (B) Dynamic
- (C) Right
- (D) Left
- (E) Vmax

1-18. Successive Zoom-Previous commands may be used to restore up to
- (A) 5 previous zoomed views.
- (B) 7 previous zoomed views.
- (C) 8 previous zoomed views.
- (D) 10 previous zoomed views.
- (E) 12 previous zoomed views.

1-19. Using the Zoom command with the Vmax option
- (A) zooms to the drawings extents.
- (B) performs a zoom-all..
- (C) performs a zoom to the maximum size of a named view
- (D) performs a zoom-all using the virtual screen space without causing a regen.
- (E) automaticaly regenerates the drawing.

1-20. The default Zoom option immediately available to the user when entering the Zoom command is
- (A) Center
- (B) Dynamic
- (C) Extents
- (D) Left
- (E) Window

1-21. The fastest way to get back to the last Zoom used is with the
- (A) Previous option of the Zoom command.
- (B) Last option of the Zoom command.
- (C) Window option of the Zoom command.
- (D) Center option of the Zoom command.
- (E) Extents option of the Zoom command.

1-22. When resetting the limits of a drawing, the option of the Zoom command that displays the entire drawing limits is
- (A) Center
- (B) All
- (C) Extents
- (D) Window
- (E) Scale

1-23. The Window option of the View command allows for
- (A) saving a view by means of a window.
- (B) selection of a view by means of a window.
- (C) calling up of a listing within a window.
- (D) changing of the viewpoint.
- (E) control of the number of viewports.

Layer Questions

1-24. To create a new layer and automatically set the layer current, use the Layer command along with the
- (A) Create option.
- (B) Make option.
- (C) Set option.
- (D) New option.
- (E) Current option.

1-25. To list all layers associated with a drawing, use the Layer command along with the
- (A) ? option.
- (B) List option.
- (C) Inquire option.
- (D) Analyze option.
- (E) None of the above.

1-26. All of the following are considered valid options of the Layer command except
- (A) Lock
- (B) Use
- (C) Make
- (D) Freeze
- (E) Set

1-27. The maximum number of characters used to name a layer is
- (A) 8 characters.
- (B) 10 characters.
- (C) 23 characters.
- (D) 31 characters.
- (E) 40 characters.

1-28. The following are valid layer names except
- (A) FRONT
- (B) SIDE_VIEW
- (C) $VIEW
- (D) 1
- (E) All of the above are valid layer names.

1-29. All of the following are valid options of the DDLMODES dialog box except
- (A) Load
- (B) Lock
- (C) On
- (D) Freeze
- (E) Color

1-30. A layer where entities may not be edited or deleted but are still visible on the screen and may be osnapped and dimensioned to is considered
- (A) Frozen
- (B) Locked
- (C) On
- (D) Thawed
- (E) Unlocked

1-31. The option of the Layer command that has the most impact on reducing regen time is
- (A) On
- (B) Off
- (C) Freeze
- (D) New
- (E) Make

1-32. A layer that is turned off will be
- (A) seen but not plotted.
- (B) plotted but not seen.
- (C) the current layer.
- (D) neither seen nor plotted.
- (E) both plotted and seen.

1-33. All of the following operations may be performed when using the Layer command except
- (A) freezing layers.
- (B) changing entities to another layer.
- (C) starting a new layer.
- (D) turning a layer off.
- (E) turning a layer on.

1-34. The option of the Layer command used to display all available layers is
- (A) M
- (B) LT
- (C) ?
- (D) N
- (E) F

Settings Questions

1-35. The following are valid systems of units identified in the Units command <u>except</u>
(A) English
(B) Fractional
(C) Decimal
(D) Scientific
(E) Engineering

1-36. To toggle the Snap mode On or Off
(A) press CTRL B.
(B) press the F9 function key.
(C) type 'DDRMODES and check Snap.
(D) All of the above.
(E) Only A and C.

1-37. To enter a temporary object snap mode, select the mode from the
(A) "****" from the screen menu.
(B) assist area of the pulldown menu.
(C) digitizing tablet Object Snap options area.
(D) keyboard by entering the first three letters of the desired object snap mode.
(E) All of the above.

1-38. The command that forces AutoCAD to create lines parallel to the horizontal or vertical axes is
(A) Horizontal
(B) Ortho
(C) Snap
(D) Parallel
(E) Vert_hor

1-39. The Aperture command adjusts the size of the
(A) Viewing window.
(B) Zoom aperture.
(C) Object snap target box.
(D) Viewing angle.
(E) Object selection pickbox.

1-40. All of the following are valid Osnap modes <u>except</u>
(A) SECtion
(B) NODe
(C) ENDpoint
(D) INSert
(E) QUAdrant

1-41. The command used to change the distance and spacing of a hidden line is
(A) Change
(B) Ltscale
(C) Scale
(D) Space
(E) Dashdist

1-42. When the Snap option is selected in the Grid command
(A) the grid is removed from the screen.
(B) the snap setting becomes the same as the grid value.
(C) the grid spacing becomes the same as the snap setting.
(D) the axis and the grid become the same as the snap setting.
(E) the axis becomes the same as the snap setting.

1-43. By default, an angle due west, or at the 9 o'clock position, would measure
(A) 0 degrees.
(B) 90 degrees.
(C) 180 degrees.
(D) 270 degrees.
(E) 330 degrees.

1-44. The Osnap mode used to snap to the 0, 90, 180, or 270 degree position of a circle or arc is
(A) ENDpoint.
(B) TANgent.
(C) NEArest.
(D) CENter.
(E) QUAdrant.

1-45. The Osnap INSert Mode
(A) snaps to the insertion point of a block.
(B) snaps to the beginning of a polyline.
(C) snaps to the insertion point of text.
(D) Both A and C.
(E) Both B and C.

1-46. To change the horizontal and vertical spacing of grid, use
(A) Grid Rotate
(B) Grid On
(C) Grid OFF
(D) Grid Aspect
(E) Grid Style

1-47. The best command to use immediately used after setting the limits of the drawing is
(A) Units
(B) Grid
(C) Snap
(D) Zoom-All
(E) Regen

Identify the correct object snap mode from the illustrations below. Place the correct answer in the space provided next to each illustration.

1-48.

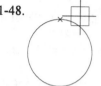

_____

1-49.

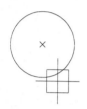

_____

1-50.

Block

_____

1-51.

_____

1-52.

_____

1-53.

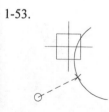

_____

1-54.

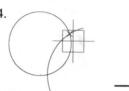

_____

1-55.

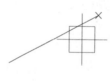

_____

1-56.

_____

1-57.

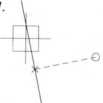

_____

1-58.

_____

1-59.

Point

_____

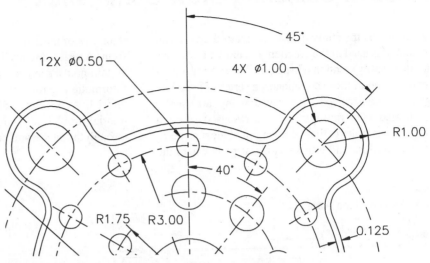

UNIT 2

Object Construction and Manipulation

Contents

Methods of Selecting Draw Commands

This chapter begins the study of the necessary drawing and editing commands used in the creation of a drawing. Blocks and their role in entity creation will also be discussed. Following the explanation of these commands, a series of one view drawings will be constructed. One view drawings are usually less complicated than multiple view drawings or section views. Typical examples of these drawings include such items as automotive gaskets and thin sheet metal parts where the thickness of these items is too thin to represent in a drawing form. A series of tutorial exercises follow giving directions on system preparation, suggested command usage, and plotting information before proceding with the main body of the tutorial. At the end of the chapter are problems related to all tutorial problems. The tutorials are designed to guide you through the successful completion of these problems.

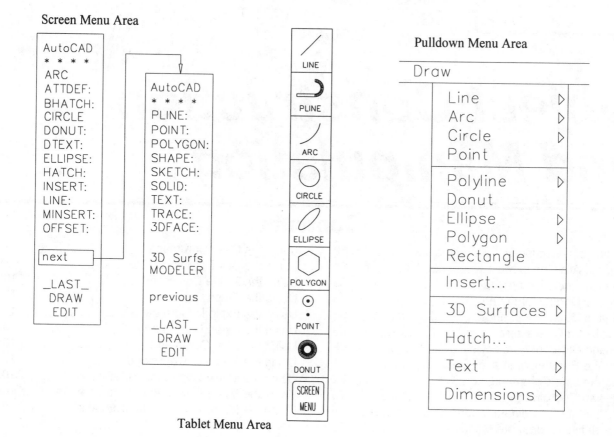

This unit introduces the commands used for entity creation. The following commands will be explained in this section:

Arc
Circle
Donut
Dtext
Ellipse
Line
Point
Polygon
Rectang
Solid

Select these commands from one of the following menu areas which are illustrated above:

Screen Menu
Tablet Menu
Pulldown Menu

Draw commands may also be entered directly from the keyboard by using their entire name such as "Point" for the Point command. The following commands may be entered by using only the first letter of the command:

Enter "A" for the Arc command
Enter "C" for the Circle command
Enter "L" for the Line command

Using the Arc Command

3 Point Arc Mode:

By default, arcs are drawn using the 3 point method. The first and third points identify the endpoints of the arc. This arc may be drawn in any direction. Use the following prompt sequence along with the illustration at the right for constructing a 3 point arc.

Command: **Arc**
Center/<Start point>: *(Mark a point at "A")*
Center/End/<Second point>: *(Mark a point at "B")*
End point: *(Mark a point at "C")*

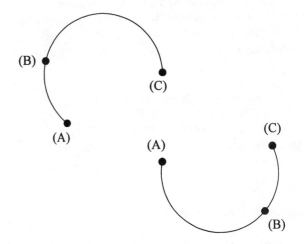

Start, Center, End Mode:

Use this arc mode to construct an arc by defining its start point, center point, and end point. This arc will always be constructed in a counterclockwise direction. Use the following prompt sequence along with the illustration at the right for constructing an arc by start, center, and end points.

Command: **Arc**
Center/<Start point>: *(Mark a point at "A")*
Center/End/<Second point>: **Center**
Center: *(Mark a point at "B")*
Angle/Length of chord/<End point>: *(Mark a point at "C")*

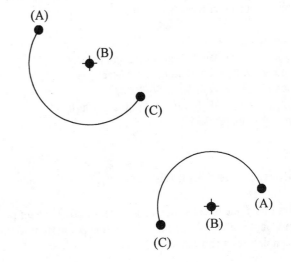

Start, End, Angle Mode:

Use this arc mode to construct an arc by defining its starting point, end point, and included angle. This arc is draw in a counterclockwise direction when a positive angle is entered; it the angle is negative, the arc is drawn clockwise. Use the following prompt sequence along with the illustration at the right for constructing an arc by start point, end point, and included angle.

Command: **Arc**
Center/<Start point>: *(Mark a point at "A")*
Center/End/<Second point>: **Center**
Center: *(Mark a point at "B")*
Angle/Length of chord/<End point>: **Angle**
Included angle: **135**

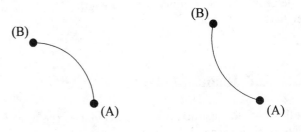

Angle = 90 degrees   Angle = -90 degrees

Start, End, Direction Mode:

Use this method to create an arc in a specified direction. This method is especially helpful for drawing arcs tangent to other entities. Use the following prompt sequence along with the illustration at the right for constructing an arc by direction.

Command: **Arc**
Center/<Start point>: *(Mark a point at "A")*
Center/End/<Second point>: **End**
End point: *(Mark a point at "B")*
Angle/Direction/Radius/<Center point>: **Radius**
Radius: **1.00**

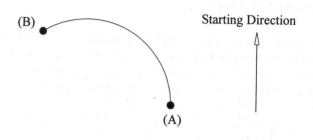

Start, Center, Angle Mode:

Use this mode to construct an arc by start point, center point, and included angle. If the angle is positive, the arc is drawn in the counterclockwise direction; a negative angle constructs the arc in a clockwise direction.

Command: **Arc**
Center/<Start point>: *(Mark a point at "A")*
Center/End/<Second point>: **Center**
Center: *(Mark a point at "B")*
Angle/Length of chord/<End point>: **Angle**
Included angle: **135**

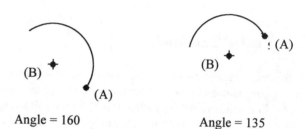

Angle = 160   Angle = 135

Start, Center, Length Mode:

Use this mode to construct an arc by start point, center point, and length of chord. See the illustration at the right to show what defines a chord.

Command: **Arc**
Center/<Start point>: *(Mark a point at "A")*
Center/End/<Second point>: **Center**
Center: *(Mark a point at "B")*
Angle/Length of chord/<End point>: **Length**
Length of chord: **2.50**

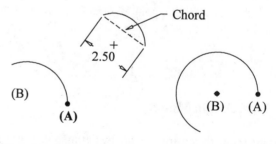

Chord length = 2.50   Chord length = -2.50

Start, End, Radius Mode:

Use this mode to construct an arc by start point, end point, and radius. A positive radius draws a minor arc; a negative radius draws a major arc.

Command: **Arc**
Center/<Start point>: *(Mark a point at "A")*
Center/End/<Second point>: **End**
End point: *(Mark a point at "B")*
Angle/Direction/Radius/<Center point>: **Radius**
Radius: **1.00**

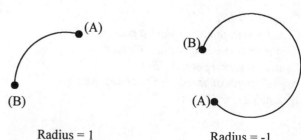

Radius = 1   Radius = -1

Using the Circle Command

Circle by Radius Mode:

Use the Circle command and the Radius mode to construct a circle by a radius value specified by the user. After selecting a center point for the circle, the user is prompted to enter a radius for the desired circle. Study the prompts below and illustration at the right for constructing a circle using the Radius mode.

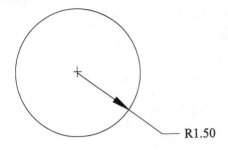

R1.50

Command: **Circle**
3P/2P/TTR/<Center point>: *(Mark the center at "A")*
Diameter/<Radius>: **1.50**

Circle by Diameter Mode:

Use the Circle command and the Diameter mode to construct a circle by a diameter value specified by the user. After selecting a center point for the circle, the user is prompted to enter a diameter for the desired circle. Study the prompts below and illustration at the right for constructing a circle using the Diameter mode.

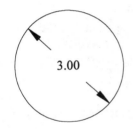

3.00

Command: **Circle**
3P/2P/TTR/<Center point>: *(Mark the center at "A")*
Diameter/<Radius>: **Diameter**
Diameter: **3.00**

3 Point Circle Mode:

Use the Circle command and the 3 Point mode to construct a circle by 3 points identified by the user. No center point is required when entering the 3 Point mode. Simply select three points and the circle is drawn. Study the prompts below and illustration at the right for constructing a circle using the 3 Point mode.

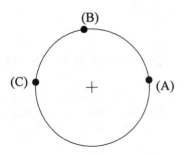

Command: **Circle**
3P/2P/TTR/<Center point>: **3P**
First point: *(Select the point at "A")*
Second point: *(Select the point at "B")*
Third point: *(Select the point at "C")*

2 Point Circle Mode:

Use the Circle command and the 2 Point mode to construct a circle by selecting 2 points. These points will form the diameter of the circle. No center point is required after entering the 2 Point mode. Study the prompts below and illustration at the right for constructing a circle using the 2 Point mode.

Command: **Circle**
3P/2P/TTR/<Center point>: **2P**
First point: *(Select the point at "A")*
Second point: *(Select the point at "B")*

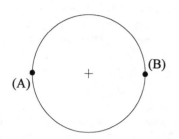

Tangent-Tangent-Radius Mode - Method #1

This mode is very powerful for constructing a circle tangent to two entities. Illustrated at the right is an application of using the Circle TTR mode to construct a circle tangent to two line segments. Study the prompts below for creating this type of circle.

Command: **Circle**
3P/2P/TTR/<Center point>: **TTR**
Enter Tangent spec: *(Select the line at "A")*
Enter second Tangent spec: *(Select the line at "B")*
Radius: **0.75**

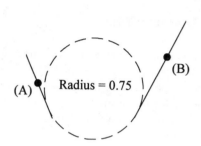

Tangent-Tangent-Radius Mode - Method #2

Illustrated at the right is an application of using the Circle TTR mode to construct a circle tangent to a line segment on another circle. Study the prompts below for creating this type of circle.

Command: **Circle**
3P/2P/TTR/<Center point>: **TTR**
Enter Tangent spec: *(Select the line at "A")*
Enter second Tangent spec: *(Select the circle at "B")*
Radius: **1.00**

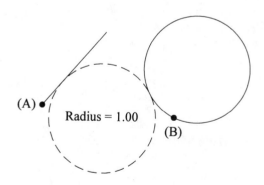

Tangent-Tangent-Radius Mode - Method #3

Illustrated at the right is an application of using the Circle TTR mode to construct a circle tangent to a line segment an another circle. Study the prompts below for creating this type of circle.

Command: **Circle**
3P/2P/TTR/<Center point>: **TTR**
Enter Tangent spec: *(Select the line at "A")*
Enter second Tangent spec: *(Select the circle at "B")*
Radius: **1.00**

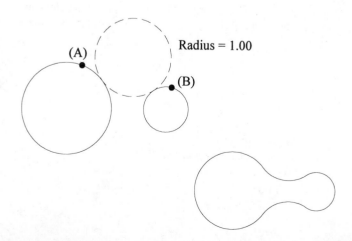

Using the Donut Command

Use the Donut command to construct a filled-in circle. This entity actually resembles a polyline. The illustration at the right is an example of a donut with an inside diameter of 0.50 units and an outside diameter of 1.00 units. When placing donuts in a drawing, the multiple option is automatically invoked. This means you can place as many donuts as you like until another command is selected from one of the three menu areas or a "Cancel" or CTRL-C is issued.

Command: **Donut**
Inside Diameter<0.50>: *(Strike Enter to accept the default)*
Outside Diameter<1.00>: *(Strike Enter to accept the default)*
Center of donut: *(Select a point to place the donut)*
Center of donut: *(Select a point to place the donut or strike Enter to exit this command)*

Setting the inside diameter of a donut to a value of zero (0) and an outside diameter to any other value constructs a donut representing a dot.

Command: **Donut**
Inside Diameter<0.50>: **0**
Outside Diameter<1.00>: **0.25**
Center of donut: *(Select a point to place the donut)*
Center of donut: *(Select a point to place the donut or strike Enter to exit this command)*

Below are two applications of where donuts could be useful; donuts are sometimes used in place of arrows to act as terminators for dimension lines. The other example illustrates how donuts might be used to act as connection points in an electrical schematic.

Donut Displayed with Fill
Mode Turned Off

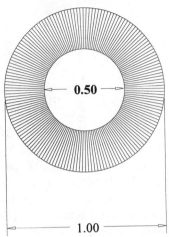

0.50

1.00

Donut Displayed with Fill
Mode Turned Off

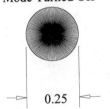

0.25

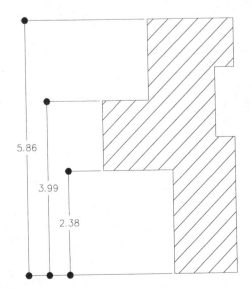

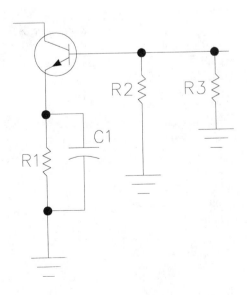

Using the Dtext Command

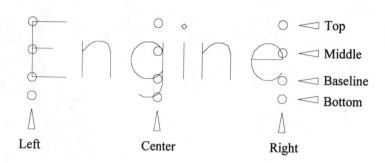

Left   Center   Right

The Dtext command stands for Dynamic Text mode and allows the user to place text in a drawing and have the text viewed as it is typed in. Before getting into a few examples of the Dtext command, study the illustrations above and to the right for the numerous ways of justifying text. By default, text used in the Dtext command is left justified. Enter one of the following initials at the right to justify text in a different location. The following examples illustrate using the Dtext command along with the following justification modes:

Left justified
Center justified
Middle justified
Aligned text
Fit text
Right justified

LEFT	Align Left (Default)
C	Center
M	Middle
R	Right
TL	Top/Left
TC	Top/Center
TR	Top/Right
ML	Middle/Left
MC	Middle/Center
MR	Middle/Right
BL	Bottom/Left
BC	Bottom/Center
BR	Bottom/Right
A	Align
F	Fit

As said above, by default the justification mode used by the Dtext command is left justified. Study the illustration at the right and the prompt sequence below to place the text string "MECHANICAL."

Command: **Dtext**
Justify/Style/<Start point>: *(Pick the point at "A")*
Height <0.20>: **0.50**
Rotation angle <0>: *(Strike Enter to accept this default)*
Text: **MECHANICAL**

MECHANICAL
(A)

The illustration at the right and prompt sequence below is an example of justifying text by a center point.

Command: **Dtext**
Justify/Style/<Start point>: **Center**
Center point: *(Pick a point at "A")*
Height <0.20>: **0.50**
Rotation angle <0>: *(Strike Enter to accept this default)*
Text: **CIVIL ENGINEERING**

CIVIL ENGINEERING
(A)

The illustration at the right and prompt sequence below are an example of justifying text by a middle point.

Command: **Dtext**
Justify/Style/<Start point>: **Middle**
Middle point: *(Pick a point at "A")*
Height <0.20>: **0.50**
Rotation angle <0>: *(Strike Enter to accept this default)*
Text: **CIVIL ENGINEERING**

CIVIL ENGINEERING
(A)

The illustration at the right and prompt sequence below are an example of justifying text by aligning the text between two points. The text height is automatically scaled depending on the length of the points and the number of letters that make up the text.

Command: **Dtext**
Justify/Style/<Start point>: **Align**
First text line point: *(Pick the point at "A")*
Second text line point: *(Pick the point at "B")*
Rotation angle <0>: *(Strike Enter to accept this default)*
Text: **MECHANICAL**

MECHANICAL
(A)   (B)

The illustration at the right and prompt sequence below are an example of justifying text by fitting the text in between two points and specifying the text height.

Command: **Dtext**
Justify/Style/<Start point>: **Fit**
First text line point: *(Pick the point at "A")*
Second text line point: *(Pick the point at "B")*
Height <0.20>: **0.50**
Rotation angle <0>: *(Strike Enter to accept this default)*
Text: **MECHANICAL**

MECHANICAL
(A)   (B)

The illustration at the right and prompt sequence below are an example of justifying text by a point at the right.

Command: **Dtext**
Justify/Style/<Start point>: **Right**
Emd point: *(Pick the point at "A")*
Height <0.20>: **0.50**
Rotation angle <0>: *(Strike Enter to accept this default)*
Text: **MECHANICAL**

MECHANICAL
(A)

Special Text Characters

Special text characters called control codes enable you to apply certain symbols to text entities. All control codes begin with the double percent sign (%%) followed by the special character that invokes the symbol. These special text characters are illustrated below and to the right.

Underscore (%%U)

Use the double percent signs followed by the letter "U" to underscore a particular text item. Illustrated at the right is the word "Mechanical," which is underscored. When prompted for entering the text, type the following:

Text: **%%UMECHANICAL**

MECHANICAL

Diameter Symbol (%%C)

The double percent signs followed by the letter "C" create the diameter symbol illustrated to the right. When prompted for entering text, the diameter symbol is displayed by typing the following:

Text: **%%C0.375**

Ø0.375

Plus/Minus Symbol (%%P)

The double percent signs followed by the letter "P" create the plus/minus symbol illustrated to the right. When prompted for entering text, the plus/minus symbol is displayed by typing the following:

Text: **%%P0.005**

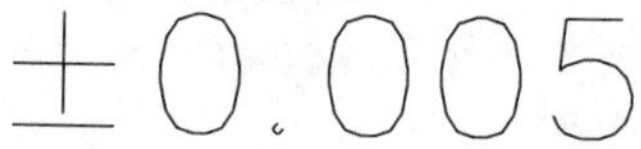

±0.005

Degree Symbol (%%D)

The double percent signs followed by the letter "D" create the degree symbol illustrated to the right. When prompted for entering text, the degree symbol is displayed by typing the following:

Text: **37%%D**

37°

Overscore (%%O)

Similar to the underscore, the double percent signs followed by the letter "O" overscore a text entity. When prompted for entering text, the overscore is displayed by typing the following:

Text: **%%OMECHANICAL**

MECHANICAL

Combining Special Text Character Modes

The Example illustrated at the right shows how the control codes are used to toggle on or off the special text characters. Enter the following at the text prompt:

Text: **%%UTEMPERATURE%%U 29%%D F**

where the first "%%U" toggles underscore mode On and the second "%%U" turns underscore Off.

TEMPERATURE 29° F

Selecting Different Font Styles

Use the Draw pulldown menu area illustrated at the right to create different text styles. Select "Text," then "Set Style" to bring up an icon menu displaying the names and appearance of the text fonts supplied in AutoCAD. All of the fonts displayed on this first icon menu have the extension .SHX, which means the font was compiled from an .SHP source file. Selecting Next at the bottom of the first icon menu displays another icon menu of more supported fonts. Most of the fonts on the second icon menu have the extension .PFB representing fonts in Postscript format. These fonts may be used as is or converted into an .SHX extension using the Compile command. Compiling a Postscript fonts changes the .PFB extension to the .SHX extension. This type of conversion will decrease the size of the original font and make the font appear faster while in a drawing. The Compile command is located under the "Files" area of the pulldown menu area. Illustrated at the bottom and on the next page are the two icon menus displaying all supported fonts.

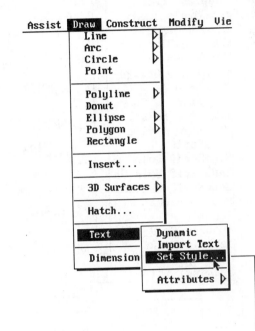

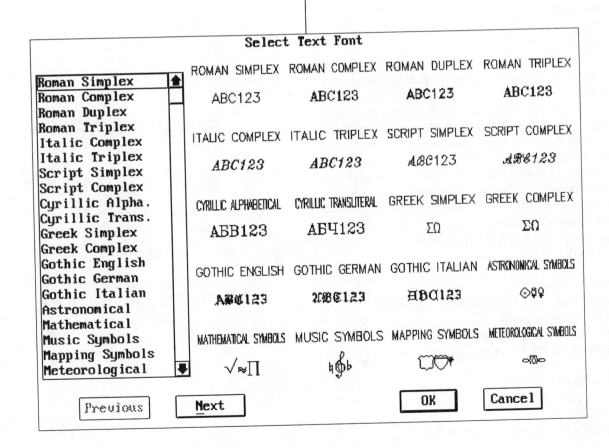

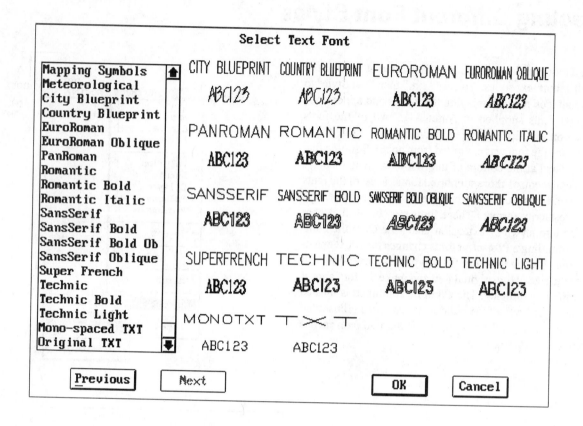

Suppose a new font such as "Roman Simplex" needs to be used inside a drawing. The Style command will be used to organize the appearance of the selected font. Selecting "Roman Simplex" from the icon menu creates a text style name called "Romans." The font file to use is also "Romans." The user has the option of changing the following prompts: Height, Width factor, Obliquing angle, Backwards text, Upside-down text, Vertical text. Entering a value for "Height" places all text under this style at that height. The width factor acts as a separator between letters. Entering a value smaller than 1.00 tightens up the text; entering a value larger than 1.00 spreads out the line of text. The obliquing angle is set if inclined lettering is desired.

```
ROMAN SIMPLEX

   ABC123
```

Command: **Style**
Text style name (or ?) <STANDARD>: **Romans**
Font file <TXT>: **Romans**
Height <0.0000>: *(Strike Enter to accept this default value)*
Width factor <1.0000>: *(Strike Enter to accept this default value)*
Backwards? <Y/N>: **No**
Upside-down? <Y/N>: **No**
Vertical? <Y/N>: **No**
Romans is now the current text style.

Width Factor = 0.50

SIMPLEX

Width Factor = 2.00

SIMPLEX

Using the Ellipse Command

Use the Ellipse command to construct an approximate elliptical shape. Before studying the three examples for ellipse construction, see the illustration at the right to view two important parts of any ellipse, namely its major and minor diameters. Once the ellipse is drawn, it consists of one polyline entity that is made up of short arc segments.

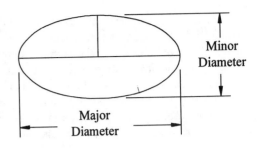

An ellipse may be constructed by marking two points, which specify one of its axes. These first two points also identify the angle with which the ellipse will be drawn. Responding to the prompt "Other axis distance" with another point identifies half of the other axis. The rubber banded line is added to assist you in this ellipse construction method.

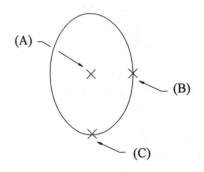

Command: **Ellipse**
<Axis endpoint 1>/Center: *(Mark a point at "A")*
Axis endpoint 2: *(Mark a point at "B")*
<Other axis distance>/Rotation: *(Mark a point at "C")*

An ellipse may also be constructed by first identifying its center. Points may be picked to identify its axes or polar coordinates may be used to accurately define the major and minor diameters of the ellipse. See the illustration at the right and the prompt sequence below to construct this type of ellipse. Refer to page 76 for a more detailed description of the polar coordinate mode.

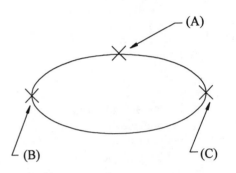

Command: **Ellipse**
<Axis endpoint 1>/Center: **Center**
Center of ellipse: *(Mark a point at "A")*
Axis endpoint: **@1.50<0** *(To point "B")*
<Other axis distance>/Rotation: **@2.50<270** *(To point "C")*

This last method illustrates constructing an ellipse by way of rotation. Identify the first two points for the first axis. Reply to the prompt "Other axis distance/Rotation" with Rotation. The first axis defined is now used as an axis of rotation that rotates the ellipse into a third dimension.

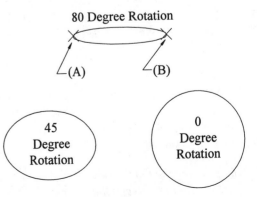

Command: **Ellipse**
<Axis endpoint 1>/Center: *(Mark a point at "A")*
Axis endpoint 2: *(Mark a point at "B")*
<Other axis distance>/Rotation: **Rotation**
Rotation around major axis: **80**

Using the Line Command

Use the Line command to construct a line from one endpoint to the other. As the first point of the line is marked, the rubber band cursor is displayed along with the normal crosshair to help see where the next line segment will be drawn to. The Line command stays active until either the Close option is used or a null response is issued by striking the enter key at the prompt "To point". Study the illustration at the right and prompt sequence below for using the Line command.

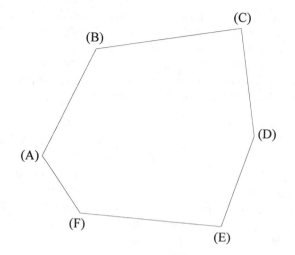

Command: **Line**
From point: *(Mark a point at "A")*
To point: *(Mark a point at "B")*
To point: *(Mark a point at "C")*
To point: *(Mark a point at "D")*
To point: *(Mark a point at "E")*
To point: *(Mark a point at "F")*
To point: **Close**

From time to time, mistakes are made in the Line command by drawing an incorrect segment. Illustrated at the right, segment DE is drawn incorrectly. Instead of exiting the Line command and erasing the line, a built in Undo is used inside of the Line command. This removes the previously drawn line while still remaining in the Line command. Follow the illustration at the right and prompts below for using the Undo option of the Line command.

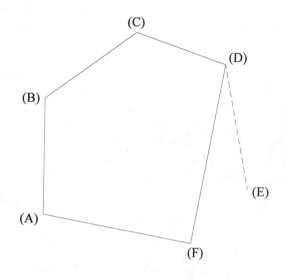

Command: **Line**
From point: *(Mark a point at "A")*
To point: *(Mark a point at "B")*
To point: *(Mark a point at "C")*
To point: *(Mark a point at "D")*
To point: *(Mark a point at "E")*
To point: **Undo** *(To remove the segment from "D" to "E" and still remain in the Line command)*
To point: *(Mark a point at "F")*
To point: **Endp**
of *(Select the endpoint of the line segment at "A")*
To point: *(Strike Enter to exit this command)*

Another option of the Line command is the Continue option. The dashed line segment at the right was the last segment drawn before exiting the Line command. To pick up at the last point of a previously drawn line segment, pick the Continue option from the screen menu.

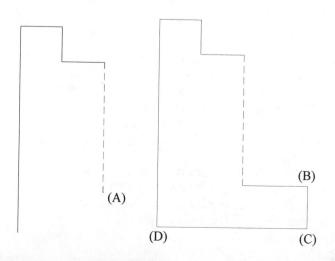

Command: **Line**
From point: *(Select the "Continue" option from the screen menu area)*
To point: *(Mark a point at "B")*
To point: *(Mark a point at "C")*
To point: **Endp**
of *(Select the endpoint of the vertical line segment at "A")*
To point: *(Strike Enter to exit this command)*

Cartesian Coordinates

Before drawing precision geometry such as lines and circles, an understanding of coordinate systems must first be made. The Cartesian, or rectangular, coordinate system is used to place geometry at exact distances through a series of coordinates. A coordinate is made up of an ordered pair of numbers usually identified as X and Y. The coordinates are then plotted on a type of graph or chart. The graph, illustrated at the right, is made up of two perpendicular number lines called coordinate axes. The horizontal axis is called the X-axis. The vertical axis is called the Y-axis.

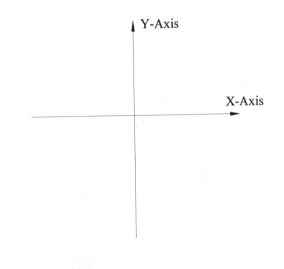

From the illustration at the right, the intersection of the two coordinate axes forms a point called the origin. Coordinates used to describe the origin are 0,0. From the origin, all positive directions move up and to the right. All negative directions move down and to the left.

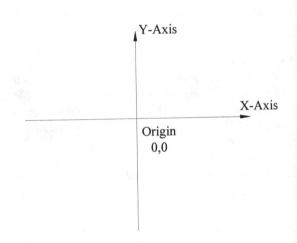

The coordinate axes are divided into four quadrants that are labeled I, II, III, and IV. In Quadrant I, all X and Y values are positive. Quadrant II has a negative X value and positive Y value. Quadrant III has negative values for X and Y. Quadrant IV has positive X values and negative Y values.

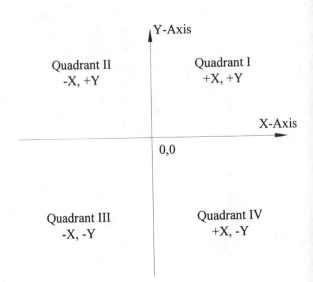

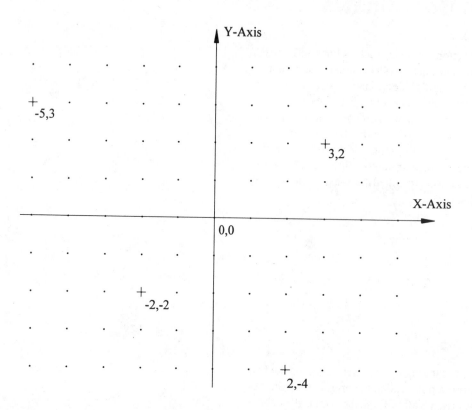

For each ordered pair of (x,y) coordinates, X means to move from the origin to the right if positive and to the left if negative. Y means to move from the origin up if positive and down if negative. The illustration above displays a series of coordinates plotted on the number lines. One coordinate is identified in each quadrant to show the positive and negative values. As an example, coordinate 3,2 located in the quadrant I means to move 3 units to the right of the origin and up 2 units. The coordinate -5,3 located in quadrant II means to move 5 units to the left of the origin and up 3 units. Coordinate -2,-2 located in quadrant III means to move 2 units to the left of the origin and down 2. Lastly, coordinate 2,-4 located in quadrant IV means to move 2 units to the right of the origin and down -4.

When beginning a drawing in AutoCAD, the screen display reflects quadrant I of the Cartesian coordinate system. The origin 0,0 is located in the lower left corner of the drawing screen. The current screen size is measured by the upper right coordinate of the screen which is, by default, 12,9. This value may be changed using the Limits command to accommodate any drawing including architectural and civil engineering.

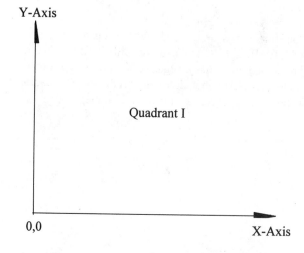

The Absolute Coordinate Mode

When drawing geometry such as lines, a method of entering precise distances must be used especially when accurracy is important. This is the main purpose of using coordinates. The simplest and most elementary form of coordinate values is Absolute coordinates. Absolute coordinates conform to the following format:

x,y

One problem with using absolute coordinates is that all coordinate values refer back to the origin 0,0. This origin on the AutoCAD screen is usually located in the lower left corner of a brand new drawing. The origin will remain in this corner unless it is altered using the Limits command. Study the Line command prompts below along with the illustration above:

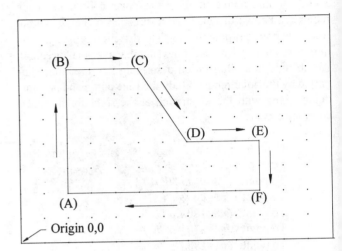

> Command: **Line**
> From point: **2,2** *(at "A")*
> To point: **2,7** *(at "B")*
> To point: **5,7** *(at "C")*
> To point: **7,4** *(at "D")*
> To point: **10,4** *(at "E")*
> To point: **10,2** *(at "F")*
> To point: **C** *(To close the figure)*

As you can see, all points on the object make reference to the origin at 0,0. Even though absolute coordinates are useful in starting lines, there are more efficient ways to continue lines and draw objects.

The Relative Coordinate Mode

In absolute coordinates, the origin at 0,0 must be kept track of at all times in order to enter the correct coordinate. With complicated objects, this is sometimes difficult to accomplish and as a result, the wrong coordinate is entered. It is possible to reset the last coordinate to become a new origin or 0,0 point. The new point would be relative to the previous point and for this reason this point is called a Relative Coordinate. The format is as follows:

@x,y

In the format above, we use the save X and Y values with one exception; The "At" symbol or @ resets the previous point to 0,0 and makes entering coordinates less confusing. Study the Line command prompts below along with the illustration above:

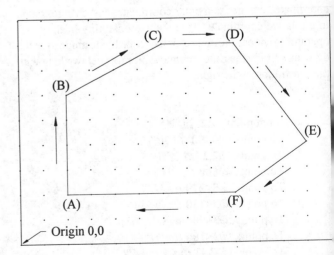

> Command: **Line**
> From point: **2,2** *(at "A")*
> To point: **@0,4** *(to "B")*
> To point: **@4,2** *(to "C")*
> To point: **@3,0** *(to "D")*
> To point: **@3,-4** *(to "E")*
> To point: **@-3,-2** *(to "F")*
> To point: **@-7,0** *(back to "A")*
> To point: *(Strike Enter to exit this command)*

In each example above, the @ symbol resets the previous point to 0,0.

The Polar Coordinate Mode

Another popular method of entering coordinates is by Polar mode. The format is as follows:

@Distance<Direction

As the format above implies, the polar coordinate mode requires a known distance and a direction. The @ symbol resets the previous point to 0,0. The direction is preceded by the < symbol, which reads the next number as a polar direction. At the right is an illustration describing the directions supported by the polar mode. Study the Line command prompts below along with the illustration at the right for the polar coordinate mode:

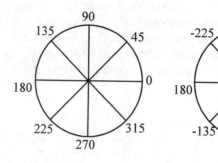

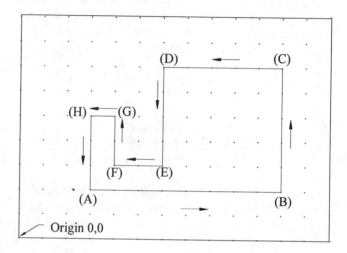

Command: **Line**
From point: **3,2** *(at "A")*
To point: **@8<0** *(to "B")*
To point: **@5<90** *(to "C")*
To point: **@5<180** *(to "D")*
To point: **@4<270** *(to "E")*
To point: **@2<180** *(to "F")*
To point: **@2<90** *(to "G")*
To point: **@1<180** *(to "H")*
To point: **@3<270** *(back to "A")*
To point: *(Strike Enter to exit this command)*

Combining Coordinate Modes

So far, the past three pages concentrated on using each example of coordinate modes (absolute, relative, and polar) to create geometry. At this point, we do not want to give the impression that once you start with a particular coordinate mode you must stay with the mode. Rather, drawings are created using one, two, or three coordinate modes in combination with each other. In the illustration above, the drawing starts with an absolute coordinate, changes to a polar coordinate, and changes again to a relative coordinate. It is the responsibility of the CAD operator to choose the most efficient coordinate mode to fit the drawing. Study the Line command prompts below along with the illustration at the right:

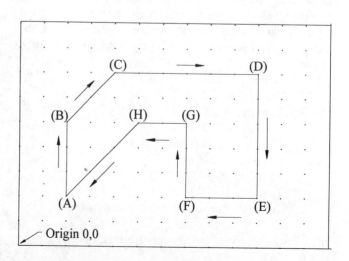

Command: **Line**
From point: **2,2** *(at "A")*
To point: **@3<90** *(to "B")*
To point: **@2,2** *(to "C")*
To point: **@6<0** *(to "D")*
To point: **@5<270** *(to "E")*
To point: **@3<180** *(to "F")*
To point: **@3<90** *(to "G")*
To point: **@2<180** *(to "H")*
To point: **@-3,-3** *(back to "A")*
To point: *(Strike Enter to exit this command)*

Using the Pline Command

Polylines are similar to individual line segments except that a polyline may consist of numerous segments and still be considered as a single entity. Width may also be assigned to a polyline compared to regular line segments, which makes polylines perfect for drawing borders and title blocks. Study both command sequences below for using the Pline command.

Command: **Pline**
From point: *(Select a point at "A")*
Current line-width is 0.0000
Arc/Close/Halfwidth/Length/Undo/Width/<Endpoint of line>:
(Mark a point at "A")
Arc/Close/Halfwidth/Length/Undo/Width/<Endpoint of line>:
Width
Starting width <0.0000>: **0.10**
Ending width <0.1000>: *(Strike Enter to accept default)*
Arc/Close/Halfwidth/Length/Undo/Width/<Endpoint of line>:
(Mark a point at "B")
Arc/Close/Halfwidth/Length/Undo/Width/<Endpoint of line>:
(Mark a point at "C")
Arc/Close/Halfwidth/Length/Undo/Width/<Endpoint of line>:
(Mark a point at "D")
Arc/Close/Halfwidth/Length/Undo/Width/<Endpoint of line>:
(Mark a point at "E")
Arc/Close/Halfwidth/Length/Undo/Width/<Endpoint of line>:
(Strike Enter to exit this command)

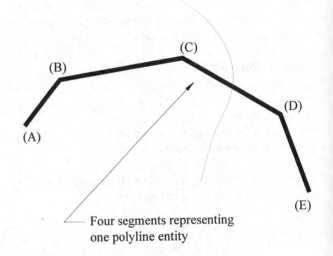

Four segments representing
one polyline entity

Command: **Pline**
From point: *(Select a point at "A")*
Current line-width is 0.0000
Arc/Close/Halfwidth/Length/Undo/Width/<Endpoint of line>:
@1.00<0 *(To "B")*
Arc/Close/Halfwidth/Length/Undo/Width/<Endpoint of line>:
@2.00<90 *(To "C")*
Arc/Close/Halfwidth/Length/Undo/Width/<Endpoint of line>:
@0.50<0 *(To "D")*
Arc/Close/Halfwidth/Length/Undo/Width/<Endpoint of line>:
@0.75<90 *(To "E")*
Arc/Close/Halfwidth/Length/Undo/Width/<Endpoint of line>:
@0.75<180 *(To "F")*
Arc/Close/Halfwidth/Length/Undo/Width/<Endpoint of line>:
@2.00<270 *(To "G")*
Arc/Close/Halfwidth/Length/Undo/Width/<Endpoint of line>:
@0.50<180 *(To "H")*
Arc/Close/Halfwidth/Length/Undo/Width/<Endpoint of line>:
@2.00<90 *(To "I")*
Arc/Close/Halfwidth/Length/Undo/Width/<Endpoint of line>:
@0.75<180 *(To "J")*
Arc/Close/Halfwidth/Length/Undo/Width/<Endpoint of line>:
@0.75<270 *(To "K")*
Arc/Close/Halfwidth/Length/Undo/Width/<Endpoint of line>:
@0.50<0 *(To "L")*
Arc/Close/Halfwidth/Length/Undo/Width/<Endpoint of line>:
Close

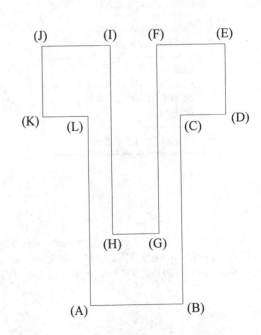

Using the Point Command

Use the Point command to identify the location of a point on a drawing. This point may be used for reference purposes. The Osnap-Node or Nearest options are used to snap to points. By default, a point is displayed as a dot on the screen. This dot may be confused with the existing grid dots already on the screen. To distinguish point entities from grid dots, use the chart at the right to assign a new point type; this is accomplished through the Pdmode system variable. Entering a value of 3 for Pdmode displays the point as an "X." The Pdsize system variable controls the size of the point. Use the prompts below for changing the point mode to a value of 3.

Command: **Pdmode**
New value for variable PDMODE <0>: **3**

Command: **Point**
Point: *(Mark the new position of a point using the cursor or one of the many coordinate systems)*

Use the DDPTYPE dialog box below for dynamically selecting a new point mode and point size.

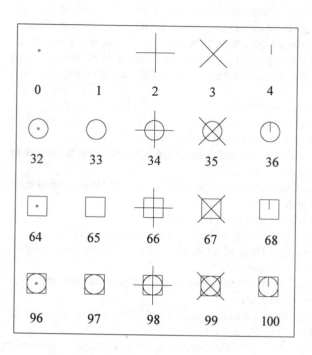

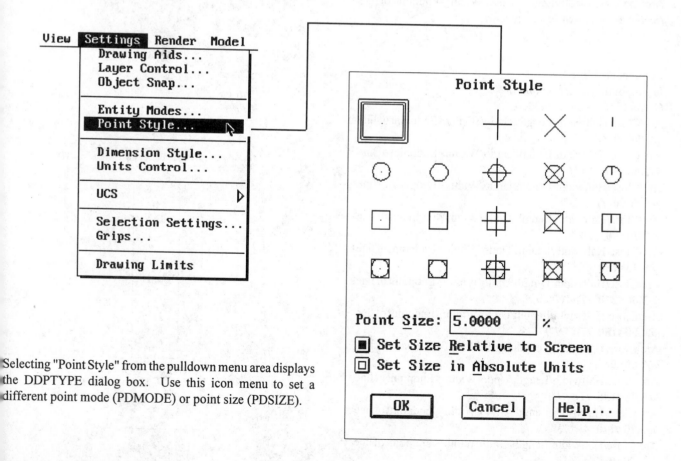

Selecting "Point Style" from the pulldown menu area displays the DDPTYPE dialog box. Use this icon menu to set a different point mode (PDMODE) or point size (PDSIZE).

Using the Polygon Command

The Polygon command is used to construct a regular polygon. Polygons are defined by the radius of circle which classifies the polygon as either being inscribed or circumscribed. Polygons consist of a closed polyline entity with width set to zero. The following prompt sequence is used to construct an inscribed polygon with the illustration at the right as a guide.

Command: **Polygon**
Number of sides: **6**
Edge/<Center of polygon>: *(Select a point at "A")*
Inscribed in circle/Circumscribed about circle (I/C): **Inscribed**
Radius of circle: **1.00**

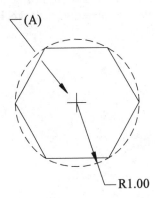

An Inscribed Polygon

The following prompt sequence is used to construct a circumscribed polygon with the illustration at the right as a guide.

Command: **Polygon**
Number of sides: **7**
Edge/<Center of polygon>: *(Select a point at "A")*
Inscribed in circle/Circumscribed about circle (I/C):
Circumscribed
Radius of circle: **1.00**

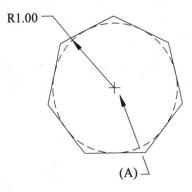

A Circumscribed Polygon

Polygons may be specified by locating the endpoints of one of its edges. The polygon is then drawn in a counterclockwise direction. Study the illustration at the right and the prompt sequence below for constructing a polygon by one of its edges.

Command: **Polygon**
Number of sides: **5**
Edge/<Center of polygon>: **Edge**
First endpoint of edge: *(Select a point at "A")*
Second endpoint of edge: *(Select a point at "B")*

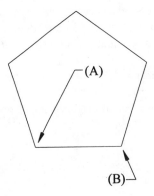

A Polygon by Edge

Using the Rectang Command

Use the Rectang command to construct a rectangle by defining two points. Most Draw commands may be selected from the tablet or screen menu. The Rectang command is found only in the pulldown menu area illustrated at the right. This command may also be entered from the keyboard under the command "Rectang."

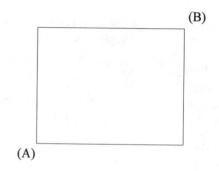

In the illustration at the right, two diagonal points are picked to define the rectangle. The rectangle drawn is in the form of a single polyline entity.

Command: **Rectang**
First corner: *(Mark a point at "A")*
Other corner: *(Mark a point at "B")*

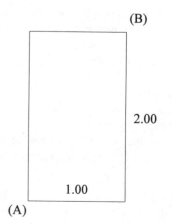

A rectangle may also be constructed by marking one point and entering a relative cooordinate for the other corner. In the prompt sequence below, a relative coordinate value of 1.00,2.00 means to make the other corner 1.00 units over in the "X" direction and 2.00 units in the "Y" direction. The "@" symbol resets the previous point at "A" to zero.

Command: **Rectang**
First corner: *(Mark a point at "A")*
Other corner: **@1.00,2.00** *(Identifying the other corner at "B")*

Using the Solid Command

The Solid command allows the user to create a filled in area of quadrilateral or triangular shapes. Two endpoints or intersections are picked as a starting edge of the solid. Two additional endpoints or intersections complete the opposite edge of the solid. Study the following prompt sequence and the illustration at the right for creating a solid.

Command: **Solid**
First point: *(Select the intersection at "A")*
Second point: *(Select the intersection at "B")*
Third point: *(Select the intersection at "C")*
Fourth point: *(Select the intersection at "D")*
Third point: *(Strike Enter to exit this command)*

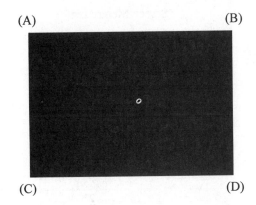

It is important how the second solid edge is selected. Instead of the third point being selected diagonally from the second point as in the illustration at the right, it was selected adjacent to the second point. This created the "hourglass" shape familiar to first-time users of the Solid command.

Command: **Solid**
First point: *(Select the intersection at "A")*
Second point: *(Select the intersection at "B")*
Third point: *(Select the intersection at "C")*
Fourth point: *(Select the intersection at "D")*
Third point: *(Strike Enter to exit this command)*

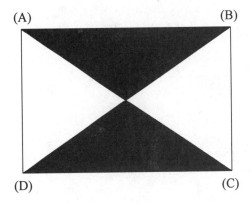

Solid edges may be continuously selected as in the illustration at the right. The key is that the third point is picked opposite or diagonally from the second point. Follow the prompt sequence below for creating this type of multiple solid.

Command: **Solid**
First point: *(Select the intersection at "A")*
Second point: *(Select the intersection at "B")*
Third point: *(Select the intersection at "C")*
Fourth point: *(Select the intersection at "D")*
Third point: *(Select the intersection at "E")*
Fourth point: *(Select the intersection at "F")*
Third point: *(Select the intersection at "G")*
Fourth point: *(Select the intersection at "H")*
Third point: *(Strike Enter to exit this command)*

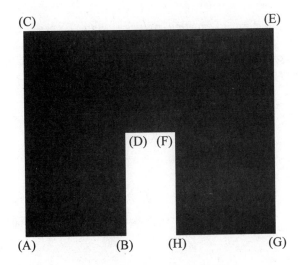

Methods of Selecting Block Commands

Screen Menu Area

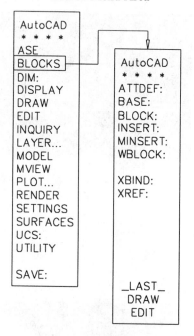

Pulldown Menu Area

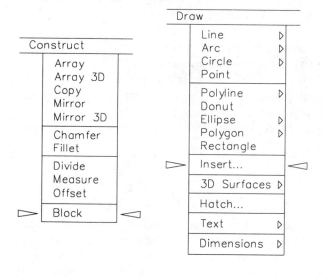

Tablet Menu Area

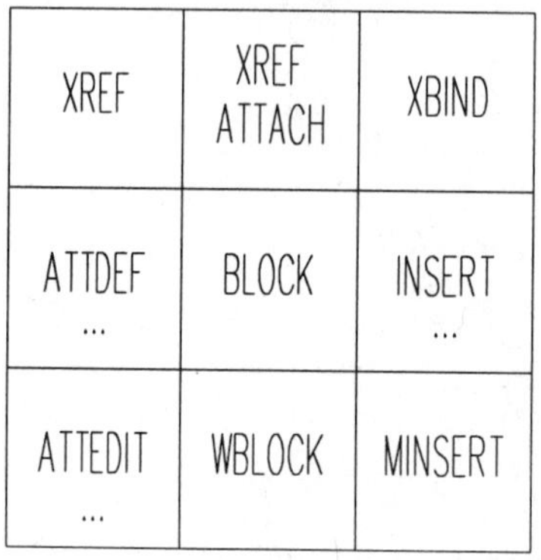

XREF	XREF ATTACH	XBIND
ATTDEF ...	BLOCK	INSERT ...
ATTEDIT ...	WBLOCK	MINSERT

Another form of creating entities is with Blocks. When creating a block, numerous entities that make up the block are all considered a single entity. The following commands will be explained in the next series of pages:

Block
Insert
Minsert
Wblock

Choose these commands from the screen menu, pulldown menu, or tablet menu areas illustrated above. The Block command can be found under the "Construct" area of the pulldown menu while insert is found under the "Draw" area of the pulldown menu.

Using the Block and Wblock Commands

Illustrated at the right is a drawing of a hex head bolt. This drawing consists of one polygon representing the hexagon, a circle indicating that the hexagon is circumscribed about the circle, and two center lines. These lines were constructed of individual segments by setting the dimension variable DIMCEN to a -0.09 value and using the Dim-Center command. Rather than copy these individual entities numerous times, a symbol will be created using the Block command and the following command prompt sequence:

Command: **Block**
Block name (or ?): **Hex-hd**
Insertion base point: **Int**
of *(Select the intersection of the two lines at "A")*
Select objects: **W** *(For Window selection mode)*
First corner: *(Mark a point at "B")*
Other corner: *(Mark a point at "C")*
Select objects: *(Strike Enter to create the block)*

When naming a block, up to 31 alphanumeric characters may be used. The insertion point is considered a point of reference. When the block is inserted into a drawing, it will be brought in relation to the insertion point.

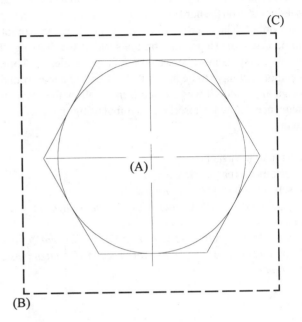

In the previous example of using the Block command, a group of entities were grouped into a single entity for insertion into a drawing. As this type of symbol was created, it is only able to be inserted into the original drawing it was created in. A more flexible command to use is the Wblock command. This command creates a symbol similar to the creation process used in the Block command. Creating the symbol using the Wblock command allows the user to insert the symbol into any drawing. Wblock is a short way of saying "Write Block." This command writes the entities to disk, which allows the symbol to be inserted into any type of drawing. Follow the prompt sequence below for creating a wblock:

Command: **Wblock**
File name: **Hex-hd**
Block name: *(Strike Enter to make the block name the same as the file name)*
Insertion base point: **Int**
of *(Select the intersection of the two lines at "A")*
Select objects: **W** *(For Window selection mode)*
First corner: *(Mark a point at "B")*
Other corner: *(Mark a point at "C")*
Select objects: *(Strike Enter to create the wblock)*

Since the wblock is written to a DOS file, a maximum of 8 characters are used in its name.

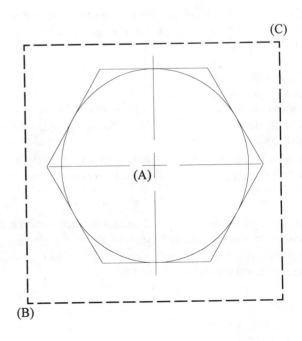

Using the Insert Command

Both Block and Wblock commands are used for creating
symbols out of individual entities. To merge these symbols
into drawings, use the Insert command. Enter the block name
of the symbol to insert. If using a wblock located in a particular
subdirectory on the hard disk, the full path must be given to get
the symbol. The insertion point is a point where the symbol is
to be inserted. Depending on the scale of the drawing, the
symbol may need to be scaled larger or smaller. Also, the
symbol may have to be inserted at a different angle from the
original position it was created from. Follow the command
sequence below for inserting a symbol using the Insert command:

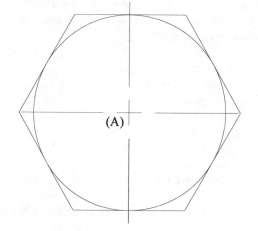

Command: **Insert**
Block name (or ?): **Hex-hd**
Insertion point: *(Mark a point at "A")*
X scale factor <1>/Corner/XYZ: *(Strike Enter to accept
default)*
Y scale factor (default=X): *(Strike Enter to accept default)*
Rotation angle <0>: *(Strike Enter to accept default and insert
the block)*

Polar coodinates may be used for insertion points to create the
gang of bolt heads illustrated at the right.

Command: **Insert**
Block name (or ?): **Hex-hd**
Insertion point: *(Mark a point at "A")*
X scale factor <1>/Corner/XYZ: *(Strike Enter to accept
default)*
Y scale factor (default=X): *(Strike Enter to accept default)*
Rotation angle <0>: *(Strike Enter to accept default and insert
the Hex-hd)*

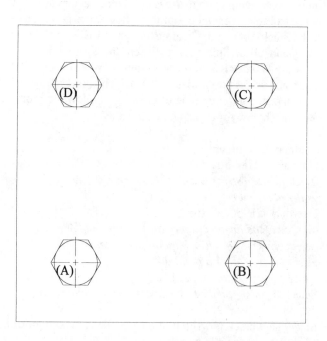

Command: **Insert**
Block name (or ?): **Hex-hd**
Insertion point: **@3<0** *(For position "B")*
X scale factor <1>/Corner/XYZ: *(Strike Enter to accept
default)*
Y scale factor (default=X): *(Strike Enter to accept default)*
Rotation angle <0>: *(Strike Enter to accept default and insert
the Hex-hd)*

Entering @3<0 places a block 3 units in the zero direction in
relation to the last inserted symbol. Insert the remaining
symbols using polar coordinate values of @3<90 and @3<180
for the insertion points at "C" and "D."

Use the DDINSERT command to dynamically insert blocks or wblocks. Select this command from the "Draw" area of the pulldown menu. Selecting "Insert..." activates the dialog boxes illustrated below. Both dialog boxes are identical except for the Options area on whether or not to specify parameters on the display screen. By default, this box is checked meaning all prompts for the insertion of symbols will occur in a similar manner as with the Insert command. If the specify parameters on screen box is not checked, the "Insertion Point," "Scale," and "Rotation" boxes activate themselves. The user must change the X, Y, and Z values to a known point on the screen. Otherwise, the symbol is inserted at point 0,0,0. Once the "Insertion Point," "Scale," and "Rotation" values have been set, selecting "OK" at the bottom of the dialog box inserts the symbol. The "Explode" box determines if the symbol is inserted as one entity or if the symbol is inserted and then exploded back to its individual entities.

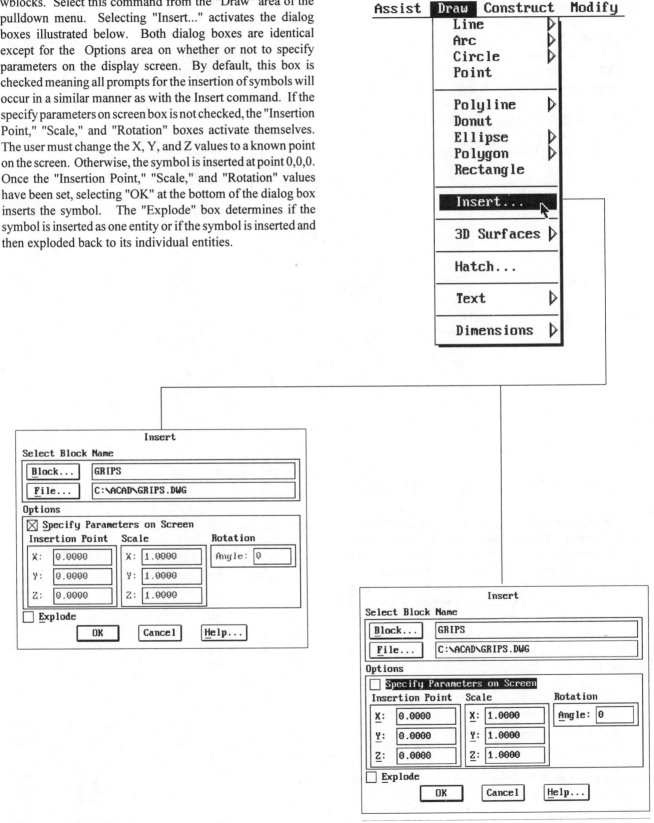

To assist the user in selecting the blocks and wblocks, two additional dialog boxes are present in DDINSERT. Illustrated at the right is the result of picking "Block...". If blocks are defined as part of the database of the current drawing, they may be selected from the dialog box at the right.

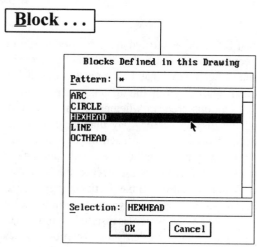

Picking "File..." from DDINSERT displays the dialog box illustrated below for inserting wblocks. This dialog box is very similar to the dialog associated with opening up a drawing file. After all, a wblock is a valid drawing file, which makes it possible to insert the file into any current drawing file. Select the desired subdirectory where the wblock is located followed by the name of the wblock. This will return the user to the insertion dialog boxes found on the previous page.

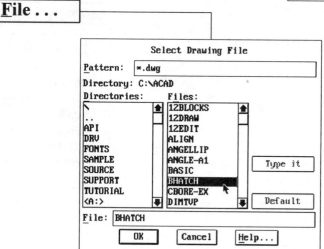

Using the Minsert Command

Use the Minsert command to produce a multiple insertion of a block or wblock. The prompts for this command are very similar to the regular Insert command regarding block name, insertion point, X and Y scale factors, and rotation angle. Minsert continues by requiring a number of rows and columns. This sets up a rectangular pattern formed by the insertion of the symbols. Following the number of rows and columns is the distance in between rows and columns. Once the distance for columns is entered, the Minsert command produces the multiple insertion.

Command: **Minsert**
Block name (or ?): **Hex-hd**
Insertion point: **1,1** *(Illustrated at the right)*
X scale factor <1>/Corner/XYZ: *(Strike Enter to accept default)*
Y scale factor (default=X): *(Strike Enter to accept default)*
Rotation angle <0>: *(Strike Enter to accept default)*
Number of rows (---) <1>: **3**
Number of columns (| | |) <1>: **2**
Unit cell or distance between rows (---): **3.00**
Distance between columns (| | |): **5.00**

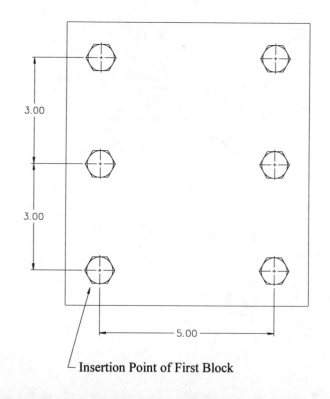

Insertion Point of First Block

Methods of Selecting Editing Commands

Screen Menu Area

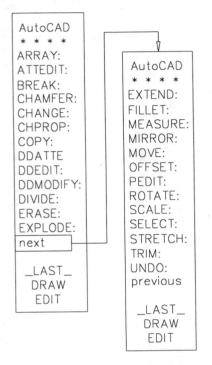

Pulldown Menu Area

Tablet Menu Area

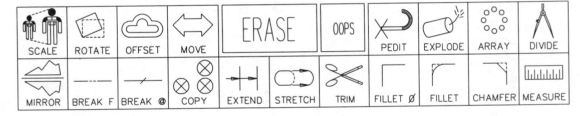

The heart of any CAD system is its ability to modify existing geometry and AutoCAD is no different. This is the function of the numerous editing command to be used by the operator.

As with all commands, the main body of editing commands may be selected from the following areas:

Screen Menu
Pulldown Menu
Tablet Menu

All commands may also be entered directly from the keyboard. The following commands may be executed by entering only its first letter:

Enter "E" for the Erase command
Enter "M" for the Move command

The following edit commands will be discussed in the next series of pages that follow:

Array	Fillet
Break	Measure
Chamfer	Mirror
Change	Move
Chprop	Offset
Copy	Pedit
Ddedit	Rotate
Divide	Scale
Erase	Stretch
Explode	Trim
Extend	

Creating Rectangular Arrays

The Array command allows the user to arrange multiple copies
of an entity or group of entities similar to the object at the right
in a rectangular or polar (circular) pattern. Suppose a rectangular pattern of the object in the lower left corner needs to be
made. The final result is to create 3 rows and 3 columns of the
object. Finally, the spacing between rows is to be 0.50 units
and the spacing between columns is 1.25 units. Study the
prompts below and the illustration to the right.

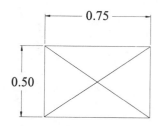

Command: **Array**
Select objects: **All**
  (This will select all entities in the original object)
Select objects: *(Strike Enter to continue)*
Rectangular or Polar array (R/P): **Rectangular**
Number of rows (---): **3**
Number of columns (||||): **3**
Unit cell distance between rows (---): **1.00**
Distance between columns (||||): **2.00**

For rectangular entities to be arrayed, a reference point in the
lower left corner of the object becomes a point where calculating the spacing between rows and columns takes place. Not
only must the spacing distance be used; the overall size of the
object plays a role in coming up with the spacing distances.
With the total height of the original object at 0.50 and a
required spacing between rows of 0.50, both object height and
spacing results in a distance of 1.00 between rows. In the same
manner, with the original length of the object at 0.75 and a
spacing of 1.25 units between columns, the total spacing
results in a distance of 2.00.

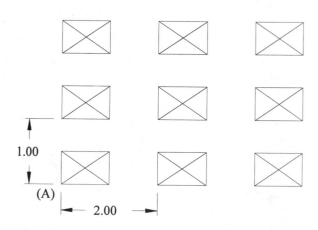

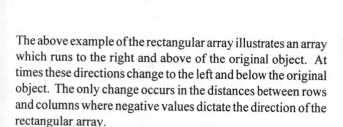

The above example of the rectangular array illustrates an array
which runs to the right and above of the original object. At
times these directions change to the left and below the original
object. The only change occurs in the distances between rows
and columns where negative values dictate the direction of the
rectangular array.

Command: **Array**
Select objects: **All**
  (This will select all entities in the original object)
Select objects: *(Strike Enter to continue)*
Rectangular or Polar array (R/P): **Rectangular**
Number of rows (---): **3**
Number of columns (||||): **2**
Unit cell distance between rows (---): **-1.50**
Distance between columns (||||): **-2.50**

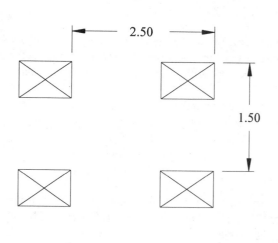

Creating Polar Arrays

Polar arrays allow the user to create multiple copies of entities in a circular or polar pattern. Follow the prompts below and illustration at the right for performing polar arrays:

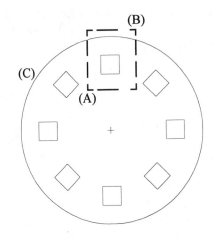

Command: **Array**
Select objects: **W** *(To select by Window mode)*
First corner: *(Select a point at "A")*
Other corner: *(Select a point at "B")*
Select objects: *(Strike Enter to continue)*
Rectangular or Polar array (R/P): **Polar**
Center point of array: **Cen**
of *(Select the edge of the large circle at "C")*
Number of items: **8**
Angle to fill (+=ccw, -=cw)<360>: *(Strike Enter to accept)*
Rotate objects as they are copied? <Y>: **Yes**

The object at the right is almost identical to the illustration above with the exception that the objects at the right were not rotated as they were copied. This is not a major problem if circles representing bolt holes are being arrayed. For rectangular or square entities, the results are illustrated at the right. Depending on how the squares or rectangles were created, they appear offset from where they should be.

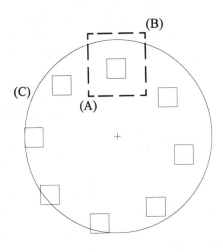

Command: **Array**
Select objects: **W** *(To select by Window mode)*
First corner: *(Select a point at "A")*
Other corner: *(Select a point at "B")*
Select objects: *(Strike Enter to continue)*
Rectangular or Polar array (R/P): **Polar**
Center point of array: **Cen**
of *(Select the edge of the large circle at "C")*
Number of items: **8**
Angle to fill (+=ccw, -=cw)<360>: *(Strike Enter to accept)*
Rotate objects as they are copied? <Y>: **No**

To array rectangular or square objects in a polar pattern without rotating the entities, the square or rectangle is first converted into a block with an insertion point located in the center of the square. Now all squares lie an equal distance from their common center.

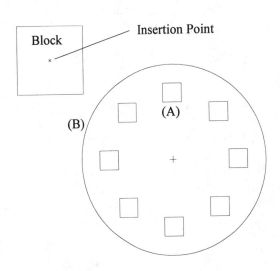

Command: **Array**
Select objects: *(Select the square block at "A")*
Select objects: *(Strike Enter to continue)*
Rectangular or Polar array (R/P): **Polar**
Center point of array: **Cen**
of *(Select the edge of the large circle at "B")*
Number of items: **8**
Angle to fill (+=ccw, -=cw)<360>: *(Strike Enter to accept)*
Rotate objects as they are copied? <Y>: **No**

Forming Bolt Holes - Method #1

Multiple copies of entities such as bolt holes are easily duplicated in circular patterns using the Array command along with the following prompts:

Command: **Array**
Select objects: *(Select both small circles illustrated at the right)*
Rectangular or Polar array (R/P): **P**
Center point of array: **Int**
of *(Select the intersection at "A")*
Number of items: **8**
Angle to fill (+=CCW, -=CW)<360>: *(Strike Enter for default)*
Rotate objects as they are copied? <Y>: *(Strike Enter for default)*

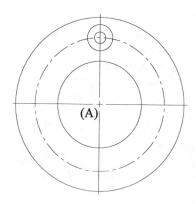

As the circles are copied in the circular pattern, center lines need to be updated to the new bolt hole positions. Again the Array command is used to copy and duplicate one center line over a 45 degree angle to fill in the counter clockwise direction.

Command: **Array**
Select objects: *(Select the vertical center line at "A")*
Rectangular or Polar array (R/P): **P**
Center point of array: **Int**
of *(Select the intersection at "B")*
Number of items: **2**
Angle to fill (+=CCW, -=CW)<360>: **45**
Rotate objects as they are copied? <Y>: *(Strike Enter for default)*

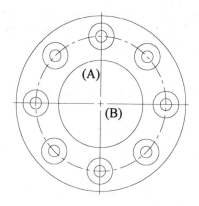

Now use the Array command to copy and duplicate the last center line and mark the remaining bolt hole circles.

Command: **Array**
Select objects: **L** *(This should select the last line)*
Rectangular or Polar array (R/P): **P**
Center point of array: **Int**
of *(Select the intersection at "A")*
Number of items: **4**
Angle to fill (+=CCW, -=CW)<360>: *(Strike Enter for default)*
Rotate objects as they are copied? <Y>: *(Strike Enter for default)*

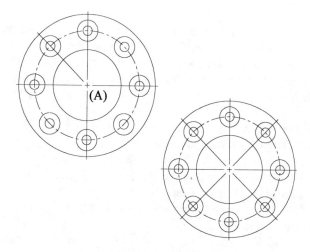

Forming Bolt Holes - Method #2

From the previous example, we have seen how easily the Array command can be used for making multiple copies of entities equally spaced around an entire circle. What if the entities are copied only partially around a circle, such as the bolt holes in the illustration at the right. The Array command is used here to copy the bolt holes in 40 degree increments. Follow the next step below for performing the operation illustrated at the right.

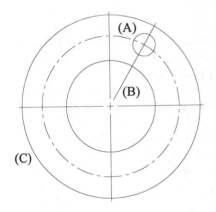

First use the Array command to rotate and copy the vertical center line at a -30 degree angle to fill. The -30 degrees will copy and rotate the center line in the clockwise direction. Next place the circle at the intersection of the center lines using the Circle command. Use the Array command, select the circle and center line, and copy the selected entities at 40 degree increments using the prompt sequence below.

Command: **Array**
Select objects: *(Select the small circle at "A" and line at "B")*
Rectangular or Polar array (R/P): **P**
Center point of array: **Cen**
of *(Select the large circle anywhere near "C")*
Number of items: *(Strike Enter to continue with this command)*
Angle to fill (+=CCW, -=CW): **-160**
Angle between items: **-40** *(To copy 40 degrees clockwise)*
Rotate objects as they are copied? <Y>: *(Strike Enter for default)*

The result is illustrated at the right. This method of identifying bolt holes shows how the number can be controlled by specifying the total angle to fill and the angle between items. In both angle specifications, a negative value is entered to force the array to be performed in the clockwise direction.

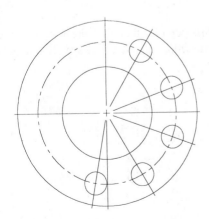

Forming Bolt Holes - Method #3

The object illustrated at the right is similar to the previous example. A new series of holes are to be placed 20 degrees away from each other using the Array command.

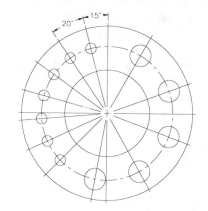

Begin placing the holes by laying out one center line using the Array command with an angle of 15 degrees to fill. Add one circle using the Circle command at the intersection of the center lines. Use the prompts below to add the remaining holes:

Command: **Array**
Select objects: *(Select the small circle at "A" and line at "B")*
Rectangular or Polar array (R/P): **P**
Center point of array: **Cen**
of *(Select the large circle anywhere near "C")*
Number of items: *(Strike Enter to continue with this command)*
Angle to fill (+=CCW, -=CW)<360>: **120**
Angle between items: **20**
Rotate objects as they are copied? <Y>: *(Strike Enter for default)*

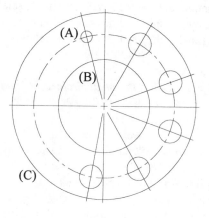

Since the direction of rotation for the array is in the counter clock-wise direction, all angles are specified in positive values.

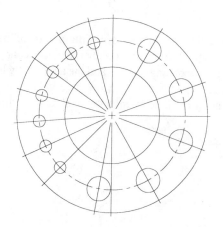

Using the Break Command

Use the Break command to partially delete a segment of an entity. Follow the command sequence below and the illustration at the right for using the Break command.

Command: **Break**
Select objects: *(Select the line at "A")*
Enter second point (or F for first point): *(Select the line at "B")*

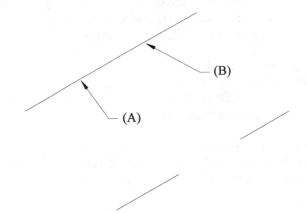

To select key entities to break using Osnap options, utilize the "First" option of the Break command. This option resets the command allowing the user to select an object to break followed by two different points that identify the break. Follow the command sequence below and illustration at the right for using the "First" option of the Break command:

Command: **Break**
Select objects: *(Select the line illustrated at the right)*
Enter second point (or F for first point): **First**
Enter first point: **Int**
of *(Select the intersection of the two lines at "A")*
Enter second point: **Endp**
of *(Select the endpoint of the line at "B")*

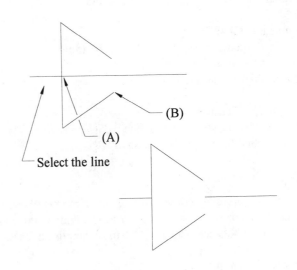

Breaking circles is always accomplished in the counter clockwise direction. Study the command sequence below and the illustration at the right for breaking circles.

Command: **Break**
Select objects: *(Select the circle illustrated at the right)*
Enter second point (or F for first point): **First**
Enter first point: *(Select the point in either illustration at the right at "A")*
Enter second point: *(Select the point in either illustration at the right at "B")*

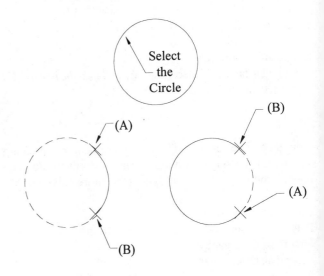

Using the Chamfer Command

The Chamfer command produces an inclined surface at an edge of two intersecting line segments. Distances determine how far from the corner the chamfer is made. One method of producing chamfers is illustrated at the right.

Command: **Chamfer**
Polyline/Distances/<Select first line>: **Distances**
First chamfer distance: **0.50**
Second chamfer distance: **0.50**

Command: **Chamfer**
Polyline/Distances/<Select first line>: *(Select the line at "A")*
Select second line: *(Select the line at "B")*

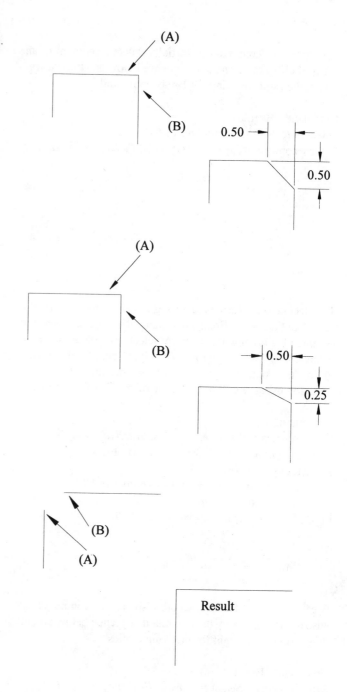

If the distances from the corner of the object are different, a beveled surface is formed illustrated at the right.

Command: **Chamfer**
Polyline/Distances/<Select first line>: **Distances**
First chamfer distance: **0.50**
Second chamfer distance: **0.25**

Command: **Chamfer**
Polyline/Distances/<Select first line>: *(Select the line at "A")*
Select second line: *(Select the line at "B")*

With non-intersecting corners, the Chamfer command could be used to connect both lines. The Chamfer command distances are both set to a value of 0 to accomplish this task.

Command: **Chamfer**
Polyline/Distances/<Select first line>: **Distances**
First chamfer distance: **0**
Second chamfer distance: **0**

Command: **Chamfer**
Polyline/Distances/<Select first line>: *(Select the line at "A")*
Select second line: *(Select the line at "B")*

Since a polyline consists of numerous segments representing a single entity, using the Chamfer command with the Polyline option produces corners throughout the entire polyline.

Command: **Chamfer**
Polyline/Distances/<Select first line>: **Distances**
First chamfer distance: **0.50**
Second chamfer distance: **0.50**

Command: **Chamfer**
Polyline/Distances/<Select first line>: **Polyline**
Select Polyline: *(Select the polyline)*

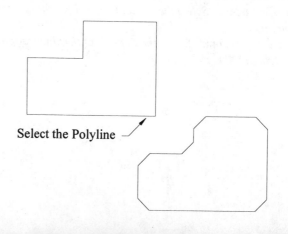

Select the Polyline

Using the Change Command

Using the Change command allows the characteristics of an entity to be modified. The example illustrated at the right shows how a line might be extended to a point without using the Extend command.

Command: **Change**
Select objects: *(Select the endpoint of the line at "A")*
Properties/<Change point>: *(Select the endpoint of the line at "B")*

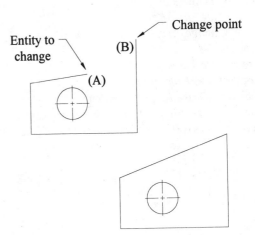

The Change command can also be used to modify the radius of a circle.

Command: **Change**
Select objects: *(Select the large circle illustrated at the right)*
Properties/<Change point>: *(Strike Enter to continue)*
Circle radius <1.00>: **0.50**

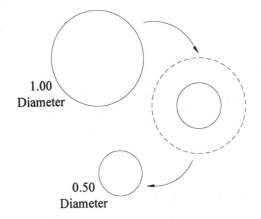

The insertion point of a block may be modified using the Change command.

Command: **Change**
Select objects: *(Select the block at the right)*
Properties/<Change point>: *(Strike Enter to continue)*
New insertion point: *(Mark a point at "A")*

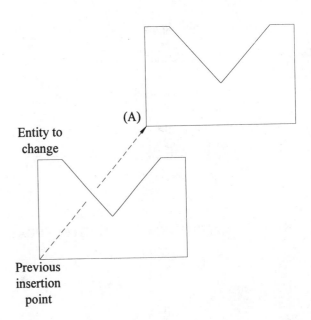

Use the Change command to modify the properties of an entity. These properties include the color, elevation, layer, linetype, and thickness of an entity.

Command: **Change**
Select objects: **W**
First corner: *(Mark a point at "A")*
Other corner: *(Mark a point at "B")*
Select objects: *(Strike Enter to continue)*
Properties/<Change point>: **P** *(To change entity property)*
Change what property (Color/Elev/LAyer/Ltype/Thickness)?

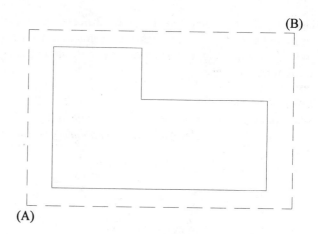

(B)

(A)

Follow the prompts at the right and the examples below for using the Change command to modify a text entity.

Command: **Change**
Select objects: *(Select the text called "MECHANICAL")*
Select objects: *(Strike Enter to continue)*
Properties/<Change point>: *(Strike Enter to continue)*
New insertion point: *(Select a new point or strike Enter to continue)*
New style or RETURN for no change: *(Enter a new text style or strike Enter to continue)*
New height <0.20>: *(Enter a new text height or strike Enter to continue)*
New rotation angle <0>: *(Enter a new text rotation angle or strike Enter to continue)*
New text <MECHANICAL>: *(Enter new text or strike Enter to continue)*

Given the following text entity:

MECHANICAL

New style: **Italicc**

MECHANICAL ⟶ *MECHANICAL*

New height <0.20>: **0.40**

MECHANICAL ⟶ MECHANICAL

New rotation angle <0>: **30**

MECHANICAL ⟶ MECHANICAL

New text <MECHANICAL>: **ARCHITECTURAL**

MECHANICAL ⟶ ARCHITECTURAL

Using the Chprop Command

The Chprop command is similar to the Change command except that the Chprop command concentrates on changing the properties of an entity. You cannont identify a new change point of a line or circle radius or block insertion point using this command. You can, however, change the color, layer, linetype, and thickenss of an entity. Use the following command sequence for the Chprop command:

Command: **Chprop**
Select objects: *(Mark a point at "A")*
Other corner: *(Mark a point at "B")*
Select objects: *(Strike Enter to continue)*
Change what property (Color/LAyer/LType/Thickness): *(Select Color, LAyer, LType, or Thickness to change)*

In the sequence above, notice that selecting a blank part of the screen at "A" for the prompt "Select objects" places the user in automatic window selection mode when selecting the other corner at "B." If "B" were the first point and "A" the other corner, automatic crossing selection mode is invoked.

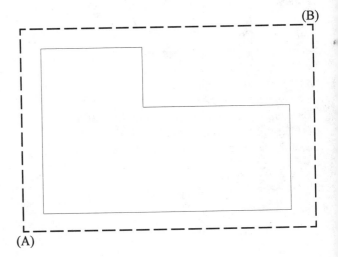

(B)

(A)

Using the DDCHPROP Dialog Box

Using the DDCHPROP command on an entity brings up a dialog box illustrated below. Use this dialog box the same way as using the normal Chprop command; namely to change an entity's color, layer, linetype, or thickness. Selecting "Color" from the Change Properties dialog box brings up another dialog box displaying the current color supported by the monitor. In the same way, an additional dialog box is displayed when selecting "Layer." A third dialog box is displayed when selecting the "Linetype" box. This dialog box will show the current linetypes loaded into the drawing file. If a desired linetype is not displayed in this dialog box, it must first be loaded by using the Linetype command along with the Load option.

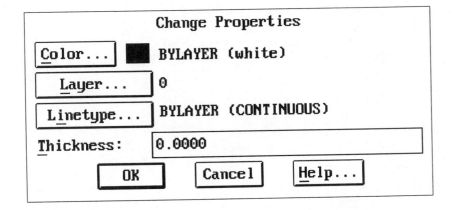

Using the Copy Command

The Copy command is used to duplicate an entity or group of entities. In the illustration at the right, the Window mode is used to select all entities to copy. Point "C" is used as the base point or displacement, or where you want to copy the entities from. Point "D" is used as the second point of displacement, or where you want to the copy the entities to.

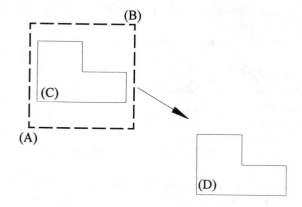

Command: **Copy**
Select objects: **W** *(To invoke the Window mode)*
First corner: *(Select near point "A")*
Other corner: *(Select near point "B")*
Select objects: *(Strike Enter to continue)*
<Base point or displacement>/Multiple: **Endp**
of *(Select the endpoint of the line at "C")*
Second point of displacement: *(Select a point near "D")*

The Copy command is also used to duplicate numerous entities while staying inside the command. Illustrated at the right are a group of entities copied using the Multiple option of the Copy command. Follow the command sequence below for using the Multiple option of the Copy command:

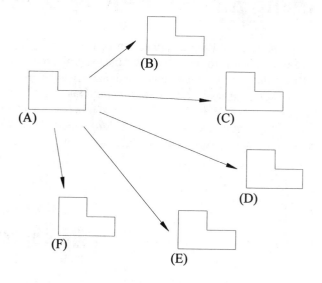

Command: **Copy**
Select objects: *(Select all entities that make up object "A")*
Select objects: *(Strike Enter to continue)*
<Base point or displacement>/Multiple: **Multiple**
Base point: **Endp**
of *(Select the endpoint of the line at "A")*
Second point of displacement: *(Select a point near "B")*
Second point of displacement: *(Select a point near "C")*
Second point of displacement: *(Select a point near "D")*
Second point of displacement: *(Select a point near "E")*
Second point of displacement: *(Select a point near "F")*
Second point of displacement: *(Strike Enter to exit this command)*

Using the DDEDIT Dialog Box

For misspelled words in a drawing, use the DDEDIT dialog box to dynamically edit the text.

Command: **DDEDIT**
<Select a TEXT or ATTDEF object>/Undo: *(Select the text)*

CHANFER

Study the next three illustrations to follow the editing of the text displayed at the right.

```
┌─────────────────────────────────────────────┐
│                 Edit  Text                    │
│                                               │
│      Text: │CHANFER                     │     │
│                                               │
│            ┌──────────┐   ┌──────────┐        │
│            │    OK    │   │  Cancel  │        │
│            └──────────┘   └──────────┘        │
└─────────────────────────────────────────────┘
```

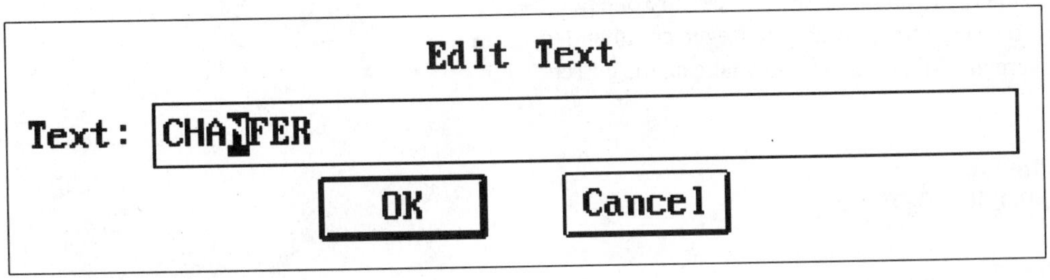

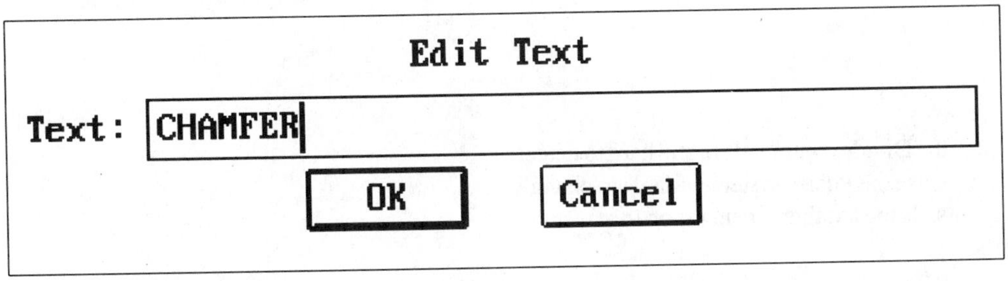

```
┌─────────────────────────────────────────────┐
│                 Edit  Text                    │
│                                               │
│      Text: │CHA FER                     │     │
│                                               │
│            ┌──────────┐   ┌──────────┐        │
│            │    OK    │   │  Cancel  │        │
│            └──────────┘   └──────────┘        │
└─────────────────────────────────────────────┘
```

```
┌─────────────────────────────────────────────┐
│                 Edit  Text                    │
│                                               │
│      Text: │CHAMFER│                    │     │
│                                               │
│            ┌──────────┐   ┌──────────┐        │
│            │    OK    │   │  Cancel  │        │
│            └──────────┘   └──────────┘        │
└─────────────────────────────────────────────┘
```

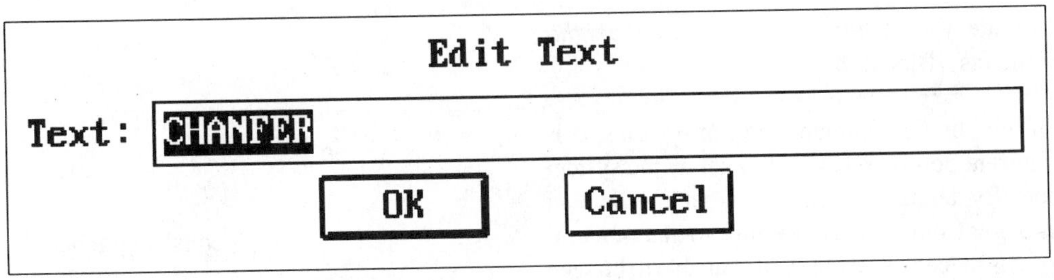

Once the text is properly edited, select the "OK" button or strike the Enter key to exit the dialog box. This will return the following prompt:

<Select a TEXT or ATTDEF object>/Undo: *(Select another text entity or strike Enter to exit this command)*

CHAMFER

Using the Divide Command

The Divide command will take an entity such as a line or arc and divide it equally depending on the number of segments desired. The Divide command accomplishes this by placing a point entity which serves as the place a division takes place. It is important to note that as a point is placed along an entity during the division process, the entity is not automatically broken at the location of the point. Rather the point is commonly used along with the Osnap-Node option to construct from. Follow the prompts below for using the Divide command.

Command: **Divide**
Select object to divide: *(Select line "A" illustrated at the right)*
<Number of segments>/Block: **8**

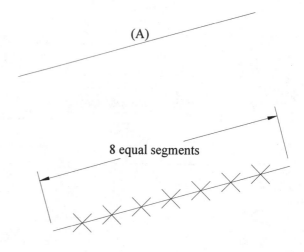

8 equal segments

If the results of the Divide command are not obvious, it is because of the current point mode which is set by the system variable Pdmode. By default, this variable is set to a value of 0, which places a point similar in appearance to that of a grid dot. When the dot is placed on an entity, it is unable to be seen easily. For this reason, use the Pdmode variable to change the point value from 0 to 3. This will produce a point similar to an "X" and thus make the point visible. Since only one point may be displayed at any one time, use the Regen command to regenerate the screen. All points will now take on the current Pdmode value.

Command: **Pdmode**
New value for PDMODE <0>: **3**

Command: **Regen**

If the points used in the Divide command are still hard to detect because of their size, use the Pdsize system variable to change the size of the points. Issue another regeneration to affect the size of all points.

Command: **Pdsize**
New value for PDSIZE <0>: **0.50**

Command: **Regen**

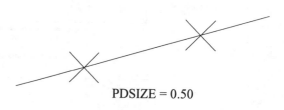

PDSIZE = 0.50

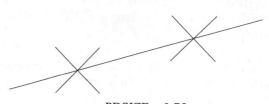

PDSIZE = 0.75

Using the Erase Command

Use the Erase command to delete entities from the database of a drawing. Entities to be erased may be selected individually or through one of the many selection set modes. The examples at the right illustrate the use of the Window and Crossing options of deleting the circle and center marker.

Erase - Window

When erasing entities by Window, be sure the entities to be erased are completely enclosed by the window as in the illuustration at the right.

Command: **Erase**
Select objects: **W** *(To invoke the window option)*
First corner: *(Mark a point at "A")*
Other corner: *(Mark a point at "B")*
Select objects: *(Strike Enter to execute the Erase command)*

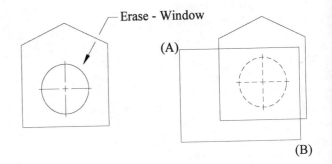

Erase - Crossing

Erasing entities by a crossing box is similar to the window box mode; however, any entity that touches the crossing box or is completely enclosed by the crossing box is selected.

Command: **Erase**
Select objects: **C** *(To invoke the crossing option)*
First corner: *(Mark a point at "A")*
Other corner: *(Mark a point at "B")*
Select objects: *(Strike Enter to execute the Erase command)*

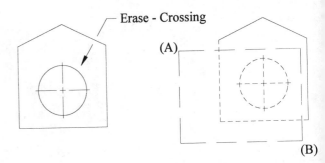

The illustration at the right displays the entities selected by the Window option to erase. However, the group of entities on the far right and left have mistakenly been selected and do not need to be affected by the Erase command. These objects will be removed from the current selection set of entities by the Remove option of select objects.

Command: **Erase**
Select objects: **W** *(To invoke the window option)*
First corner: *(Mark a point at "A")*
Other corner: *(Mark a point at "B")*

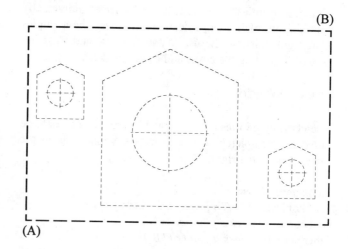

Instead of performing the Erase command on the entities in the previous illustration, issue the Remove option to unhighlight the group of entities illustrated at the right.

Select objects: **R** *(To invoke the remove option)*
Remove objects: *(Select the group of entities marked by "A")*
Remove objects: *(Select the group of entities marked by "B")*
Remove objects: *(Strike Enter to execute the Erase command)*

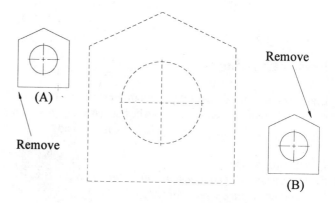

By using the Remove option of select objects, the user remains in the command instead of cancelling the command for picking the wrong entities and starting over.

The middle group of entities
are erased

Using the Explode Command

Using the Explode command on a polyline, associative dimension, or block separates the single into its individual parts. Illustrated at the right is a polyline which is considered a single entity. Using the Explode command and selecting the polyline breaks the polyline into four individual entities. If these entities were on a particular layer at a particular color, the exploded entities would return to layer 0 with the color white. If the polyline had been constructed with a certain thickness, exploding the polyline would not only separate the polyline into individual segments but would also return the individual segments to a width of 0.

Command: **Explode**
Select objects: *(Select the polyline illustrated at the right)*
The Undo command will restore it.

Polyline
Before Explode - 1 entity
After Explode - 4 entities

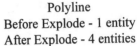

An associative dimension consists of extension lines, dimension lines, dimension text, and arrowheads, all considered a single entity. This type of dimension is covered further in Unit 5. Using the Explode command on an associative dimension breaks the dimension down into individual extension lines, dimension lines, arrowheads, and dimension text.

Command: **Explode**
Select objects: *(Select the associative dimension illustrated at the right)*
The Undo command will restore it.

Associative Dimension
Before Explode - 1 entity
After Explode - 7 entities

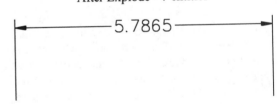

5.7865

Using the Explode command on a block breaks the block down into individual segments. If the block had yet another block nested in it, an additional Explode operation would need to be performed to break this entity into individual entities.

Command: **Explode**
Select objects: *(Select the block illustrated at the right)*
The Undo command will restore it.

Block
Before Explode - 1 entity
After Explode - 15 entities

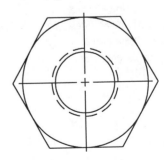

Using the Extend Command

Use the Extend command to extend entities to a specified boundary edge. In the example illustrated at the right, select the large circle "A" as the boundary edge. After striking the Enter key to continue with the command, select the arc at "B," line at "C," and arc at "D" to extend these entities to the circle. If the wrong end of an entity is selected, use the undo feature, which is an option of the command to undo the change and repeat the procedure at the correct end of the entity.

Command: **Extend**
Select boundary edge(s)...
Select objects: *(Select the large circle at "A")*
Select objects: *(Strike Enter to continue)*
<Select object to extend>/Undo: *(Select the arc at "B")*
<Select object to extend>/Undo: *(Select the line at "C")*
<Select object to extend>/Undo: *(Select the arc at "D")*
<Select object to extend>/Undo: *(Strike Enter to exit this command)*

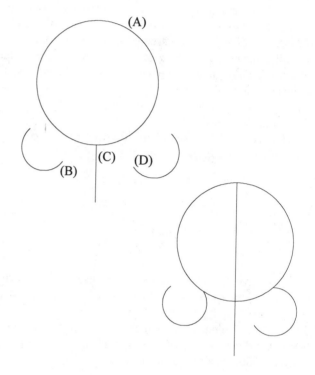

To extend multiple entities such as the 5 line segments at the right, select the line at "A" as the boundary edge and issue the fence option to select the ends of the line segments to extend. The fence option allows the user to define a crossing line or series of lines used to select multiple entities to extend.

Command: **Extend**
Select boundary edge(s)...
Select objects: *(Select the line at "A")*
Select objects: *(Strike Enter to continue)*
<Select object to extend>/Undo: **F** *(To invoke the fence option)*
Undo/<Endpoint of line>: *(Mark a point at "B")*
Undo/<Endpoint of line>: *(Mark a point at "C")*
Undo/<Endpoint of line>: *(Strike Enter to end the Fence and execute the Extend command)*
<Select object to extend>/Undo: *(Strike Enter to exit this command)*

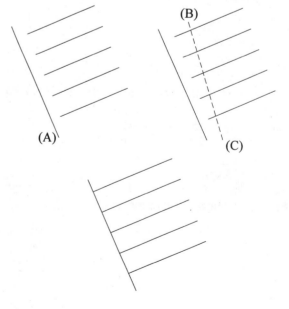

Using the Fillet Command

Fillets are corners of a part that are rounded off at a specified radius value and automatically trimmed. If the radius value is too large for the specified entities, an error message states this. Illustrated at the right is one method of producing a fillet at a designated corner.

Command: **Fillet**
Polyline/Radius/<Select first object>: **Radius**
Enter fillet radius <0.0000>: **0.50**

Command: **Fillet**
Polyline/Radius/<Select first object>: *(Select line "A")*
Select second object: *(Select line "B")*

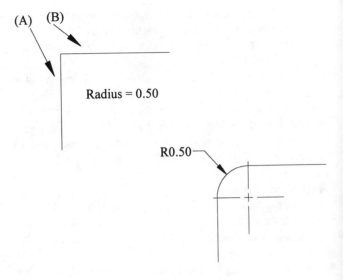

A Fillet radius of 0 produces a corner out of two non-intersecting entities.

Command: **Fillet**
Polyline/Radius/<Select first object>: **Radius**
Enter fillet radius <0.5000>: **0**

Command: **Fillet**
Polyline/Radius/<Select first object>: *(Select line "A")*
Select second object: *(Select line "B")*

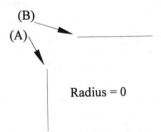

Use the Fillet command on a polyline entity to produce rounded edges at all corners of the polyline in a single operation.

Command: **Fillet**
Polyline/Radius/<Select first object>: **Radius**
Enter fillet radius <0.0000>: **0.25**

Command: **Fillet**
Polyline/Radius/<Select first object>: **Polyline**
Select 2D polyline: *(Select the pline illustrated at the right)*

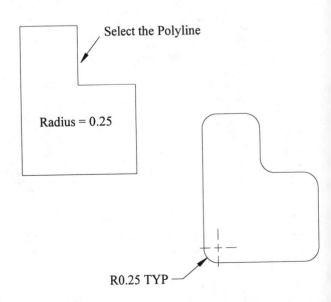

Using the Measure Command

The Measure command will take an entity such as a line or arc and measure along it depending on the length of the segment. The Measure command, similar to the Divide command, accomplishes this by placing a point entity at a specified distance given in the Measure command. It is important to note that as a point is placed along an entity during the measuring process, the entity is not automatically broken at the location of the point. Rather the point is commonly used along with the Osnap-Node option to construct from. Follow the prompts below for using the Measure command.

Command: **Measure**
Select object to measure: *(Select the designated end illustrated at the right)*
<Segment length>/Block: **0.50**

If the results of the Measure command are not obvious, it is because of the current point mode which is set by the system variable Pdmode. By default, this variable is set to a value of 0 which places a point similar in appearance to that of a grid dot. When the dot is placed on an entity, it is unable to be seen easily. For this reason, use the Pdmode variable to change the point value from 0 to 3. This will produce a point similar to an "X" and thus make the point visible. Since only one point may be displayed at any one time, use the Regen command to regenerate the screen. All points will now take on the current Pdmode value.

Command: **Pdmode**
New value for PDMODE <0>: **3**

Command: **Regen**

It is important to note that the measuring starts at the endpoint closest to the point you used to select the entity.

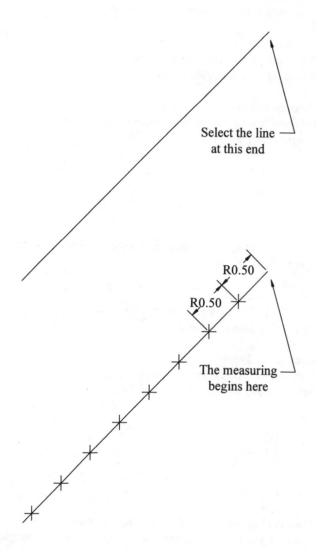

Select the line at this end

R0.50

R0.50

The measuring begins here

Using the Mirror Command - Method #1

Use the Mirror command to create a mirrored copy of an entity or group of entities. When performing a mirror, the operator has the option of deleting the original entity, which would be the same as flipping the entity, or keeping the original entity along with the mirror image, which would be the same as flipping and copying. Follow the prompts below along with the illustration at the right for using the Mirror command:

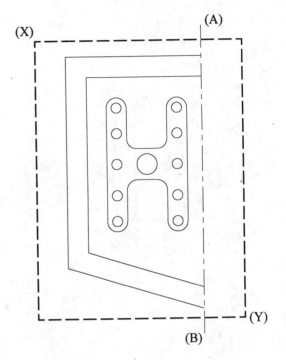

Command: **Mirror**
Select objects: **W** *(To invoke the Window option)*
First corner: *(Select a point near "X")*
Other corner: *(Select a point near "Y")*
Select objects: *(Strike the Enter key to continue)*
First point of mirror line: **Endp**
of *(Select the endpoint of the center line at "A")*
Second point: **Endp**
of *(Select the endpoint of the center line at "B")*
Delete old objects? <N> **No**

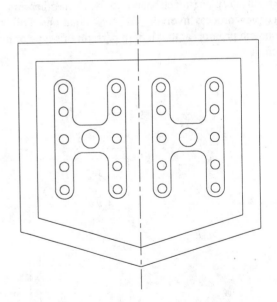

Since the original object was required to be retained by the mirror operation, the image result is illustrated at the right. The Mirror command works well when symmetry is required.

Using the Mirror Command - Method #2

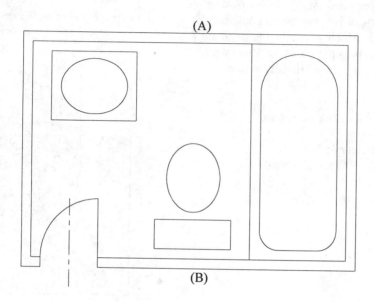

The illustration above is an interesting application of the Mirror command. Required is to have all items that make up the bathroom plan flipped but not copied to the other side. This is a typical process involving "What if" situations. Follow the command prompts at the right to perform this type of mirror operation.

Command: **Mirror**
Select objects: **All** *(This will select all entities illustrated above)*
Select objects: *(Strike the Enter key to continue)*
First point of mirror line: **Mid**
of *(Select the midpoint of the line at "A")*
Second point: **Per**
to *(Select line "B," which is perpendicular to point "A")*
Delete old objects? <N> **Yes**

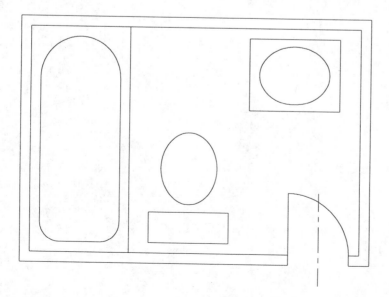

Using the Mirror Command - Method #3

Situations sometimes involve mirroring text as in the illustration at the right. A system variable called Mirrtext controls this occurance and by default is set to a value of "1" or "On." Follow the prompts below to see the results of the Mirror command on text.

Command: **Mirror**
Select objects: **W** *(To invoke the Window option)*
First corner: *(Select a point near "X")*
Other corner: *(Select a point near "Y")*
Select objects: *(Strike the Enter key to continue)*
First point of mirror line: **Endp**
of *(Select the endpoint of the center line at "A")*
Second point: **Endp**
of *(Select the endpoint of the center line at "B")*
Delete old objects? <N> **No**

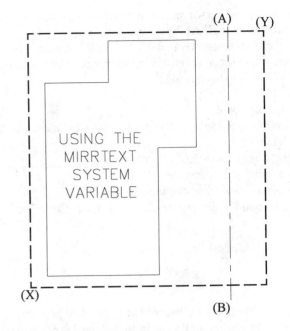

Notice that with Mirrtext turned "On," text is mirrored, which lends itself unreadable. To mirror text and have it right reading, set the Mirrtext system variable to a value of "0" or "Off" as in the following prompt sequence.

Command: **Mirrtext**
New value for MIRRTEXT <1>: **0**

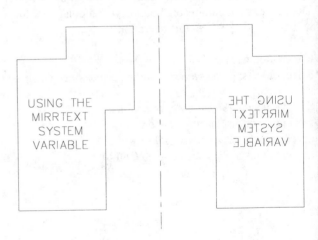

Using the Mirror command with the Mirrtext system variable set to a value of "0" results in text being able to be read similar to the illustration at the right. Follow the prompt sequence below to accomplish this.

Command: **Mirror**
Select objects: **Previous**
Select objects: *(Strike the Enter key to continue)*
First point of mirror line: **Endp**
of *(Select the endpoint of the center line at "A")*
Second point: **Endp**
of *(Select the endpoint of the center line at "B")*
Delete old objects? <N> **No**

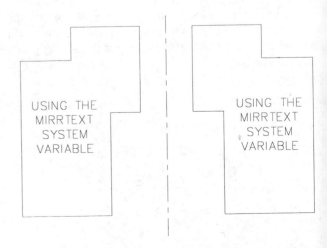

Using the Move Command

The Move command repositions an entity or group of entities at a new location. Once the entities to move are selected, a base point or displacement is found, or where the object is to move from. Next a second point of displacement is needed, or where the object is to be moved to.

Command: **Move**
Select objects: *(Select all dashed entities illustrated at the right)*
Select objects: *(Strike Enter to continue)*
Base point or displacement: **Endp**
of *(Select the endpoint of the line at "A")*
Second point of displacement: *(Mark a point at "B")*

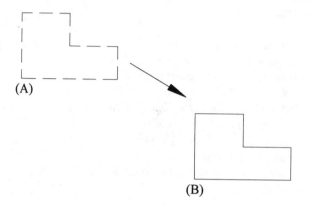

The slot illustrated at the right is incorrectly positioned; it needs to be placed 1 unit away from the left edge of the object. The Move command in combination with a polar coordinate will be used to perform this operation.

Command: **Move**
Select objects: *(Select the slot and all center lines illustrated at the right)*
Select objects: *(Strike Enter to continue)*
Base point or displacement: **Cen**
of *(Select the edge of arc "A")*
Second point of displacement: **@0.50<0**

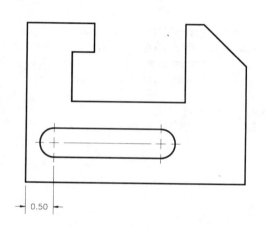

As the slot is moved into a new position using the Move command, a new horizontal dimension must be placed to reflect the correct distance from the edge of the object to the center line of the arc. Another command will be explained in the next series of pages to affect a group of entities along with the dimension.

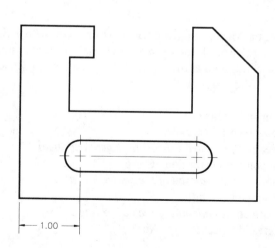

Using the Offset Command

The Offset command is commonly used for creating an entity parallel to another entity. One method of offsetting is to identify a point to offset through, called a through point. Once an entity is selected to offset, a through point is identified. The selected entity offsets to the point illustrated at the right. Follow the prompt sequence below for using this method of the Offset command.

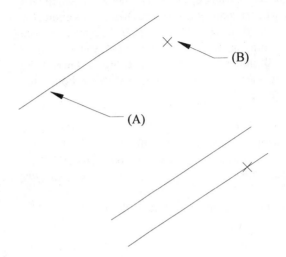

Command: **Offset**
Offset distance or Through <>: **Through**
Select object to offset: *(Select the line at "A")*
Through point: **Nod**
of *(Select the point at "B")*

Another method of offsetting is by a specified offset distance. In the illustration at the right, an offset distance of 0.50 is set. The line segment "A" is identified as the entity to offset. To complete the command, a side to offset must be identified to give the offset a direction in which to operate. Follow the command sequence below for using this method of offsetting entities.

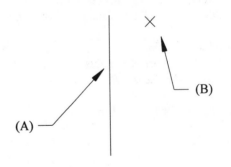

Command: **Offset**
Offset distance or Through <>: **0.50**
Select object to offset: *(Select the line at "A")*
Side to offset: *(Mark a point at "B")*
Select object to offset: *(Strike Enter to exit this command)*

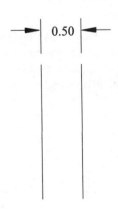

Another method of offsetting is illustrated at the right where the entities need to be duplicated at a set distance away from existing geometry. The Copy command could be used for this operation; a better command would be Offset. This allows the user to specify a distance and a side for the offset to occur. The result is an entity parallel to the original entity at a specified distance. All entities in the illustration at the right need to be offset 0.50 toward the inside of the original object. Follow the prompt sequences below.

Command: **Offset**
Offset distance or Through <0.00>: **0.50**
Select object to offset: *(Select the horizontal line at "A")*
Side to offset: *(Mark a point anywhere on the inside near "B")*

Repeat the above procedure for lines C through J.

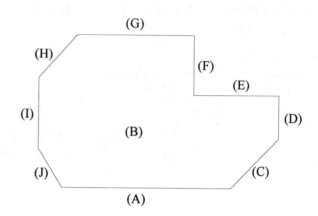

Notice that when all lines were offset, the entire original lengths of all line segments were maintained. Since all offsetting occured inside, this resulted in segments that overlap at their intersection points. In one case, at A and B, the lines didn't even meet at all. The Chamfer command may be used to edit all lines to form a sharp corner. This is accomplished by assigning a radius value of 0.00 for the chamfer distances; this happens to be the default value for the Chamfer command.

Command: **Chamfer**
Polyline/Distances/<Select first line>: *(Select the line at "A")*
Select second line: *(Select the line at "B")*

Repeat the above procedure for lines A through I

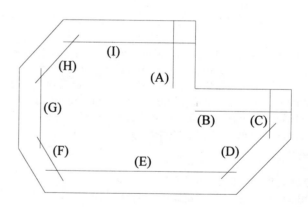

Using the Offset command along with the Chamfer command produces the result illustrated at the right. The Chamfer command must be set to a value of 0 for this special effect. The Fillet command performs the same result when set to a radius value of 0.

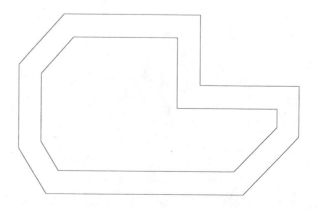

Using the Pedit Command

Editing of polylines can lead to interesting results. A few of these options will be explained over the next series of pages. Illustrated at the right is a polyline of width 0.00. The Pedit command was used to change the width of the polyline to 0.10 units. Follow the command sequence below to use the Pedit command along with the Width option.

Command: **Pedit**
Select polyline: *(Select the pline at "A")*
Close/Join/Width/Edit vertex/Fit/Spline/Decurve/Ltype gen/
Undo/eXit <X>: **W** *(To edit the Width of the polyline)*
Enter new width for all segments: **0.10**
Close/Join/Width/Edit vertex/Fit/Spline/Decurve/Ltype gen/
Undo/eXit <X>:*(Strike Enter to exit this command)*

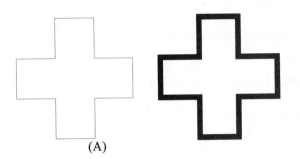

(A)

It is possible to convert regular entities into polylines. At the right, the arc segment and individual line segments may be converted into a polyline. The circle is unable to be converted unless part of the circle is broken resulting in an arc segment. Follow the prompts below for converting the line segments into a polyline.

Command: **Pedit**
Select polyline: *(Select the line at "A")*
Entity selected is not a polyline.
Do you want it to turn into one? **Yes**
Close/Join/Width/Edit vertex/Fit/Spline/Decurve/Ltype gen/
Undo/eXit <X>: **J**
Select objects: *(Select lines "B" through "D")*
3 lines added to polyline.
Close/Join/Width/Edit vertex/Fit/Spline/Decurve/Ltype gen/
Undo/eXit <X>: *(Strike Enter to exit this command)*

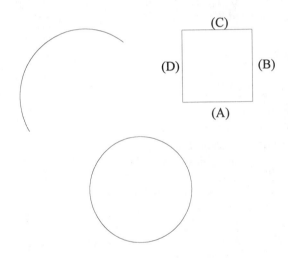

(C)
(D)   (B)
(A)

In the previous example, regular entities were selected individually before being converted into a polyline. For more complex entities, use the Window option to select numerous entities and perform the Pedit-Join operation which is faster. Follow the prompt sequence below for using this command.

Command: **Pedit**
Select polyline: *(Select the line at "A")*
Entity selected is not a polyline.
Do you want it to turn into one? **Yes**
Close/Join/Width/Edit vertex/Fit/Spline/Decurve/Ltype gen/
Undo/eXit <X>: **J**
Select objects: **W**
First corner: *(Mark a point at "B")*
Other corner: *(Mark a point at "C")*
56 lines added to polyline.
Close/Join/Width/Edit vertex/Fit/Spline/Decurve/Ltype gen/
Undo/eXit <X>: *(Strike Enter to exit this command)*

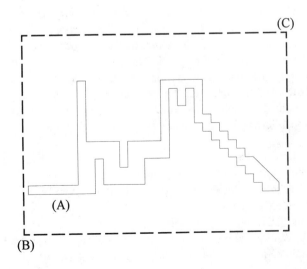

(C)
(A)
(B)

The polyline illustrated at the right will be used as an example of using the Pedit command along with various curve fitting utilities. In the examples that follow, the Spline option and Fit Curve option are illustrated. The Spline option produces a smooth fitting curve based on control points in the form of the vertices of the polyline. The Fit Curve option passes entirely through the control points producing a less desirable curve. Study the examples below that illustrate both curve options of the Pedit command.

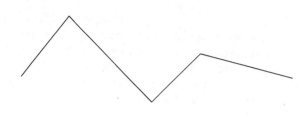

Spline Curve Generation:

Command: **Pedit**
Select polyline: *(Select the polyline at "A")*
Close/Join/Width/Edit vertex/Fit/Spline/Decurve/Ltype gen/
Undo/eXit <X>: **Spline**
Close/Join/Width/Edit vertex/Fit/Spline/Decurve/Ltype gen/
Undo/eXit <X>: *(Strike Enter to exit this command)*

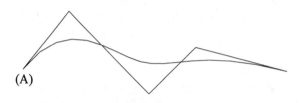

(A)

The original polyline frame is usually not visible when creating a spline and is shown only for illustrative purposes.

Fit Curve Generation:

Command: **Pedit**
Select polyline: *(Select the polyline at "B")*
Close/Join/Width/Edit vertex/Fit/Spline/Decurve/Ltype gen/
Undo/eXit <X>: **Fit**
Close/Join/Width/Edit vertex/Fit/Spline/Decurve/Ltype gen/
Undo/eXit <X>: *(Strike Enter to exit this command)*

(B)

The original polyline frame is not visible when creating a fit curve and is shown only for illustrative purposes.

The Linetype Generation option of the Pedit command controls the pattern of the linetype from polyline vertice to vertice. In example "C" illustrated at the right, the hidden linetype is generated from the first vertice to the second vertice. An entirely different pattern is formed from the second vertice to the third vertice and so on. The polyline at "C" has the linetype generated throuthout the entire polyline. In this way, the hidden linetype is smoothed throughout the polyline.

(C)

Command: **Pedit**
Select polyline: *(Select the polyline at "D")*
Close/Join/Width/Edit vertex/Fit/Spline/Decurve/Ltype gen/
Undo/eXit <X>: **Lt**
Full PLINE linetype ON/OFF <Off>: **On**
Close/Join/Width/Edit vertex/Fit/Spline/Decurve/Ltype gen/
Undo/eXit <X>: *(Strike Enter to exit this command)*

(D)

The object illustrated at the right is identical to the illustration on page 112. Also in the previous example, each individual line had to be offset to copy the lines parallel at a specified distance. Then the Chamfer command was used to clean up the corners. There is an easier way to perform this operation. First convert all individual line segments into one polyline using the Pedit command.

Command: **Pedit**
Select polyline: *(Select the line at "A")*
Entity selected is not a polyline.
Do you want it to turn into one? **Yes**
Close/Join/Width/Edit vertex/Fit/Spline/Decurve/Ltype gen/
Undo/eXit <X>: **Join**
Select objects: **W**
First corner: *(Mark a point at "X")*
Other corner: *(Mark a point at "Y")*
Select objects: *(Strike Enter to continue)*
8 lines added to polyline.
Close/Join/Width/Edit vertex/Fit/Spline/Decurve/Ltype gen/
Undo/eXit <X>: *(Strike Enter to exit this command)*

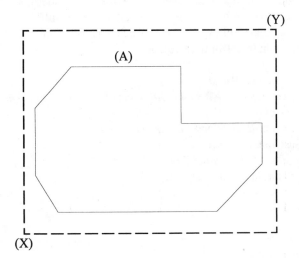

The Offset command is used to copy the shape 0.50 units on the inside. Since the object was converted into a polyline, all entities are offset at the same time. This procedure by-passes the need to use the Chamfer or Fillet commands to corner all intersections.

Command: **Offset**
Offset distance or Through <0.00>: **0.50**
Select object to offset: *(Select the polyline at "A")*
Side to offset: *(Select a point anywhere near "B")*
Select object to offset: *(Strike Enter to exit this command)*

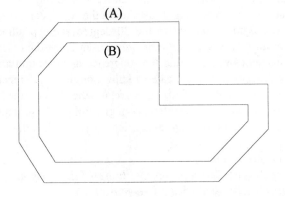

Using the Rotate Command

The Rotate command changes the orientation of an entity or group of entities by identifying a base point and a rotation angle which completes the new orientation. Illustrated at the right is an object complete with crosshatch pattern, which needs to be rotated to a 30 degree angle using point "A" as the base point. Follow the prompts below and the illustration at the right to perform the rotation.

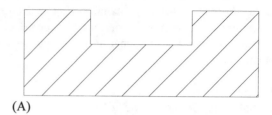

(A)

Command: **Rotate**
Select objects: **All** *(This will select all entities illustrated at the right)*
Select objects: *(Strike Enter to continue)*
Base point: **Endp**
of *(Select the endpoint of the line at "A")*
<Rotation angle>/Reference: **30**

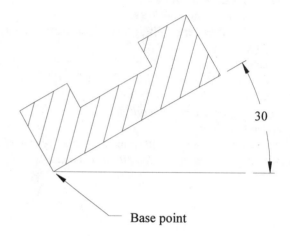

30

Base point

Rotate - Reference

At times it is necessary to rotate an object to a desired angular position. However, this must be accomplished even if the current angle of the object is not known. To maintain the accuracy of the rotation operation, the Reference option of the Rotate command is used. Illustrated at the right is an object that needs to be rotated to the 30 degree angle position. Unfortunately, we do not know the current angle the object currently lies in. Entering the Reference angle option and identifying two points creates a known angle of reference. Entering a new angle of 30 degrees rotates the object to the 30 degree position from the reference angle. Follow the prompts below and the illustrations at the right to accomplish this.

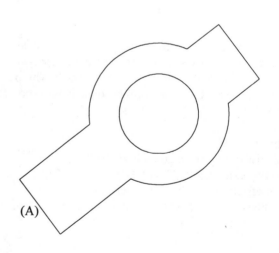

(A)

Command: **Rotate**
Select objects: *(Select the object illustrated at the right)*
Select objects: *(Strike Enter to continue)*
Base point: **Cen**
of *(Pick either the circle or two arc segments)*
<Rotation angle>/Reference: **Reference**
Reference angle <0>: **Cen**
of *(Again pick either the circle or the two arc segments)*
Second point: **Mid**
of *(Pick the line "B" to establish the reference angle)*
New angle: **30**

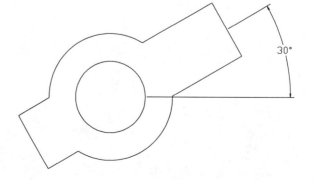

30°

Using the Scale Command

Use the scale command to change the overall size of an entity. The size may be larger or smaller in relation to the original entity or group of entities. The Scale command, similar to the Rotate command, requires a base point and scale factor to complete the command. Study the next series of examples below and illustrated at the right to perform the Scale command.

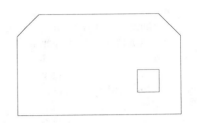

Using a base point at "A" and a scale factor of 0.50, the results of using the Scale command on a group of entities are illustrated below and to the right.

Command: **Scale**
Select objects: **All** *(This will select all dashed entities illustrated at the right)*
Select objects: *(Strike Enter to continue)*
Base point: **Endp**
of *(Select the endpoint of the line at "A")*
<Rotation angle>/Reference: **0.50**

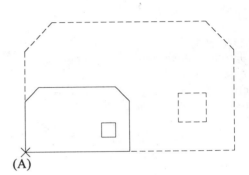

The example illustrated at the right shows the effects of identifying a new base point in the center of the object.

Command: **Scale**
Select objects: **All** *(This will select all dashed entities illustrated at the right)*
Select objects: *(Strike Enter to continue)*
Base point: (Mark a point near "A")
<Rotation angle>/Reference: **0.40**

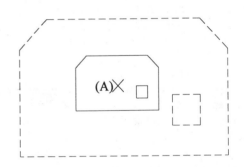

Scale - Reference

Suppose the length of a line needs to be increased to a known length; however you do not know what the exact length of the line to lengthen is. The Reference option of the Scale command can be used to identify endpoints of a line segment that act as a reference length. Entering a new length value could increase or decrease the line depending on the desired effect following the illustrations at the right and prompts below.

Command: **Scale**
Select objects: *(Select line segment "A")*
Select objects: *(Strike Enter to continue)*
Base point: **End**
of *(Pick the endpoint at "A")*
<Scale factor>/Reference: **Reference**
Reference length <1>: **End**
of *(Pick the endpoint at "A")*
Second point: **End**
of *(Pick the endpoint at "B")*

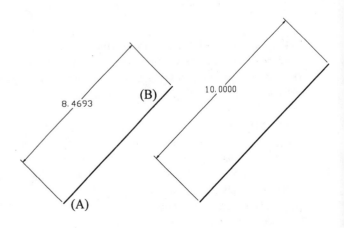

Using the Stretch Command

Use the Stretch command to move a portion of a drawing while still preserving the connections to parts of the drawing remaining in place. To perform this type of operation, the Crossing option of "Select objects" must be used. In the illustration at the right, a group of entities is selected using the crossing box. Next a base point is identified by the endpoint at "C." Finally, a second point of displacement is identified using a polar coordinate. Once the entities selected in the crossing box are stretched, the entities not only move to the new location but also mend themselves.

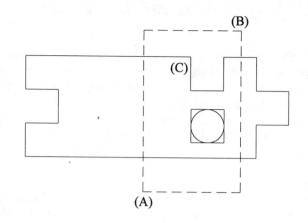

Command: **Stretch**
Select objects to stretch by window or polygon...
Select objects: **Crossing**
First corner: *(Mark a point at "A")*
Other corner: *(Mark a point at "B")*
Base point or displacement: **Endp**
of *(Select the endpoint of the line at "C")*
Second point of displacement: **@1.75<180**

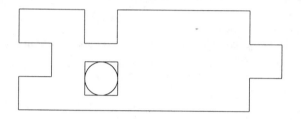

The example illustrated at the right is another example of using the Stretch command. The crossing window is employed along with a base point at "C" and a polar coordinate.

Command: **Stretch**
Select objects to stretch by window or polygon...
Select objects: **Crossing**
First corner: *(Mark a point at "A")*
Other corner: *(Mark a point at "B")*
Base point or displacement: **Endp**
of *(Select the endpoint of the line at "C")*
Second point of displacement: **@2.75<0**

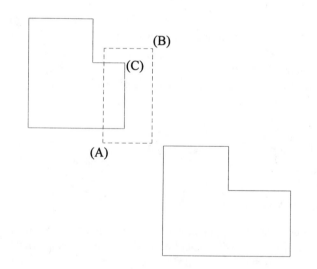

Applications of the Stretch command include the example above where a window needs to be positioned at a new location. Follow the command sequence below to stretch the window at a set distance using a polar coordinate.

Command: **Stretch**
Select objects to stretch by window or polygon...
Select objects: **Crossing**
First corner: *(Mark a point at "A")*
Other corner: *(Mark a point at "B")*
Base point or displacement: **Endp**
of *(Select the endpoint of the line at "C")*
Second point of displacement: **@10'6<0**

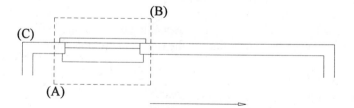

Using the Trim Command

Use the Trim command to partially delete an entity or group of entities based on a cutting edge. In the example at the right, the four dashed lines are selected as cutting edges. Next, segments of the circles are selected which trim out between the line segment cutting edges.

Command: **Trim**
Select cutting edge(s)...
Select objects: *(Select the 4 dashed lines at the right)*
Select objects: *(Strike Enter to continue)*
<Select object to trim>/Undo: *(Select the circle at "A")*
<Select object to trim>/Undo: *(Select the circle at "B")*
<Select object to trim>/Undo: *(Select the circle at "C")*
<Select object to trim>/Undo: *(Select the circle at "D")*
<Select object to trim>/Undo: *(Strike Enter to exit this command)*

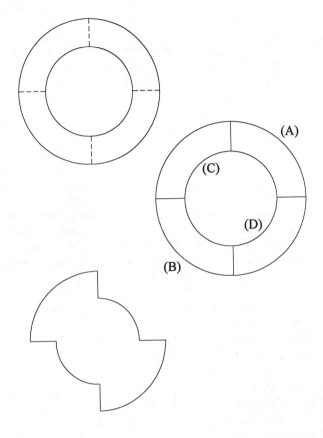

In the example at the right, even though the four dashed lines are selected as cutting edges, the middle of the cutting edge may be trimmed out at all locations identified by "A" through "D."

Command: **Trim**
Select cutting edge(s)...
Select objects: *(Select all four dashed lines at the right)*
Select objects: *(Strike Enter to continue)*
<Select object to trim>/Undo: *(Select the segment at "A")*
<Select object to trim>/Undo: *(Select the segment at "B")*
<Select object to trim>/Undo: *(Select the segment at "C")*
<Select object to trim>/Undo: *(Select the segment at "D")*
<Select object to trim>/Undo: *(Select Enter to exit this command)*

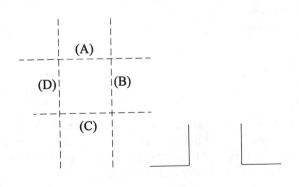

Yet another application of the Trim command uses the Fence option of "Select objects." First, invoke the Trim command and select the small circle as the cutting edge. Begin the prompt of "Select object to trim" with "Fence."

Command: **Trim**
Select cutting edge(s)...
Select objects: *(Select the small circle)*
Select objects: *(Strike Enter to continue)*
<Select object to trim>/Undo: **Fence**

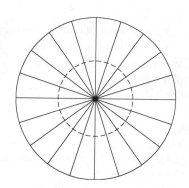

Continue with the Trim command by identifying a Fence. This consists of a series of line segments that take on a dashed appearance. This means the fence will select any entity it crosses. When completed with the construction of the desired fence illustrated at the right, strike the Enter key.

First fence point: *(Mark a point at "A" beginning the Fence)*
Undo/<Endpoint of line>: *(Mark a point at "B")*
Undo/<Endpoint of line>: *(Mark a point at "C")*
Undo/<Endpoint of line>: *(Mark a point at "D")*
Undo/<Endpoint of line>: *(Mark a point at "E")*
Undo/<Endpoint of line>: *(Mark a point at "F")*
Undo/<Endpoint of line>: *(Mark a point at "G")*
Undo/<Endpoint of line>: *(Mark a point at "H")*
Undo/<Endpoint of line>: *(Strike Enter to end the Fence and execute the Trim command)*
<Select object to trim>/Undo: *(Strike Enter to exit this command)*

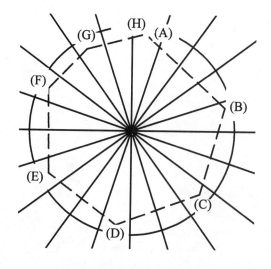

The power of the Fence option of "Select objects" is illustrated at the right. Rather than select each individual line segment inside the small circle to trim, the Fence trims all entities it touches in relation to the cutting edge.

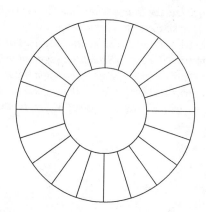

Using Entity Grips

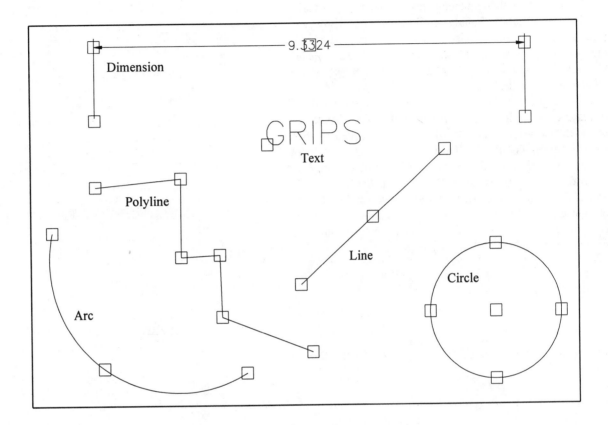

An alternate method of editing may be performed using entity grips. The grip is a small rectangle appearing at key entity positions such as the endpoints and midpoints of lines and arcs or the center and quadrants of circles. When grips are enabled, a grip pickbox is displayed at the intersection of the crosshairs illustrated in the example at the right. Once grips are selected, the entity may be stretched, moved, rotated, scaled, or mirrored. Grips are at times referred to as visual osnaps since the cursor automatically snaps to all grips displayed along an entity. Displayed above are other examples of grips on various entities.

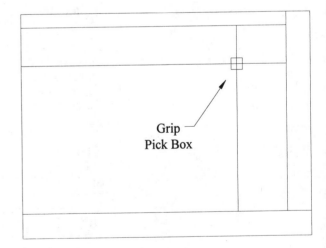

Grip
Pick Box

The DDGRIPS dialog box is available to change settings, color, and grip size. Choose this dialog box from the "Settings" area of the pulldown menu area. By default, grips are enabled; check the box "Enable Grips" to have grips displayed when selecting an entity. Also by default, a grip is placed at the insertion point when a block is selected. Check "Enable Grips Within Blocks" if you want grips to be displayed along all entities within the block. Color is applied to selected and unselected grips. Selecting "Unselected..." or "Selected..." displays a color dialog box used to change the color of selected or unselected grips. The size of grips is controlled by the system variable GRIPSIZE. By default, this value is set to 3 and may be as high as 255. Since it is difficult to determine the size differences between values of GRIPSIZE, the "Grip Size" area of the dialog box allows the user to move a slider bar left or right and visually see the size of the grip in the box at the far right. Moving the slider to the left makes the grip smaller; moving to the right makes the grip larger.

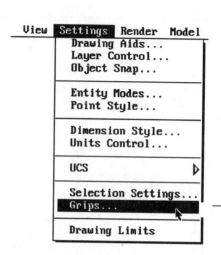

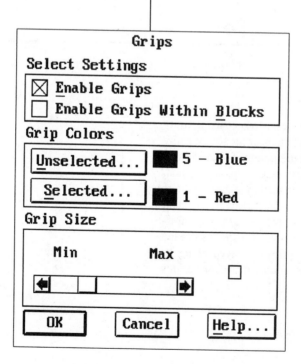

Entity Grip Modes

Illustrated at the right are the three types of grips. When an entity is first selected using the grip pickbox located at the intersection of the crosshairs, the entity highlights and the square grips are displayed; this type of grip is called a warm grip. The entire entity is subject to the many grip edit commands. When one of the grips is selected, it fills in with the current selected grip color; this type of grip is called a hot grip. Once a hot grip is selected, the following prompts appear at the command prompt:

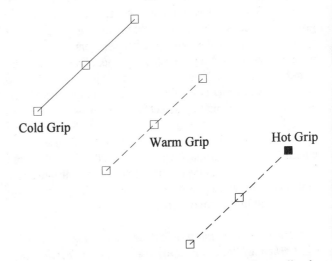

Cold Grip   Warm Grip   Hot Grip

STRETCH
<Stretch to point>/Base point/Copy/Undo/eXit:
MOVE
<Move to point>/Base point/Copy/Undo/eXit:
ROTATE
<Rotation angle>/Base point/Copy/Undo/Reference/eXit:
SCALE
<Scale factor>/Base point/Copy/Undo/Reference/eXit:
MIRROR
<Second point>/Base point/Copy/Undo/eXit:

To move from one edit command mode to another, strike the space bar. Once an editing command is completed, entering one CTRL-C removes the highlight from the entity but leaves the grip; this type of grip is considered cold. Entering a second CTRL-C removes the grips from the entity. Below are various examples illustrating each editing mode.

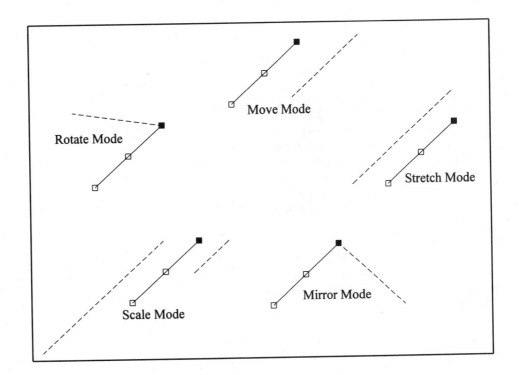

Rotate Mode   Move Mode   Stretch Mode

Scale Mode   Mirror Mode

Using the Grip - Stretch Mode

The Stretch mode of grips operates similar to the normal Stretch command. Use Stretch mode to move an entity or group of entities and have the results mend themselves similar as in the example at the right. The line segments "A" and "B" are both too long by 1 unit. To decrease these line segments by 1 unit, use the Stretch mode by selecting lines "A," "B," and "C" with the grip pickbox at the command prompt. Next, while holding down the shift key, select the warm grips "D," "E," and "F". This will make the grips hot and ready for the stretch operation. Since these hot grips are considered base points, entering a polar coordinate value of @1.00<180 will stretch the three highlighted grip entities to the left at a distance of 1 unit. To remove the entity highlight and grips, enter two CTRL-C's at the command prompt.

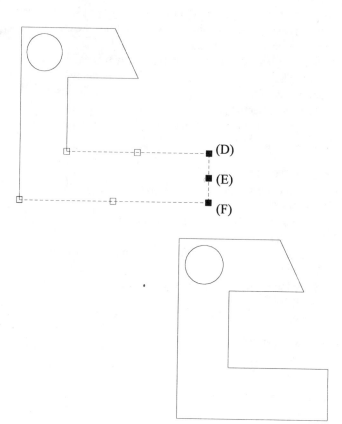

Command: *(Select the three dashed lines illustrated at the right. Then, while holding down the Shift key, select the warm grips at "D," "E," and "F" to make them hot)*

STRETCH
<Stretch to point>/Base point/Copy/Undo/eXit: **@1<180**
Command: **CTRL-C** *(To remove entity highlight)*
Command: **CTRL-C** *(To remove grips)*

Using the Grip - Scale Mode

Using the Scale mode of entity grips allows an entity to be uniformly scaled in both the X and Y directions. This means that a circle, such as the one illustrated at the right, cannot be converted into an ellipse by using different X and Y values. As the warm grip is converted into a hot grip, any cursor movement will drag the scale factor until a point is marked where the entity will be scaled to that factor. Illustrated at the right and the prompt below is an example of an absolute value to perform the scaling operation of half the circle's normal size.

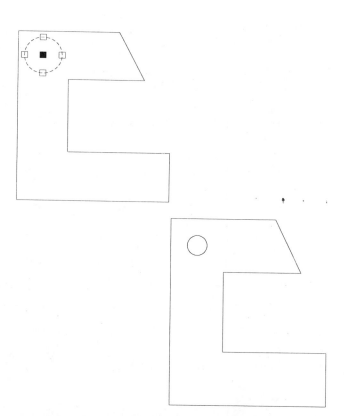

Command: *(Select the circle to enable grips, then select the warm grip at the center of the circle to make it hot. Strike the Space bar until the Scale mode appears at the bottom of the prompt line and above in the status bar)*

SCALE
<Scale factor>/Base point/Copy/Undo/Reference/eXit: **0.50**

Using the Grip - Move/Copy Mode

The multiple copy option of Move mode will be illustrated to copy the circle illustrated at the right using polar coordinates at distances 2.50 and 5.00 both in the 270 direction. This multiple copy option is actually disguised under the command options of entity grips.

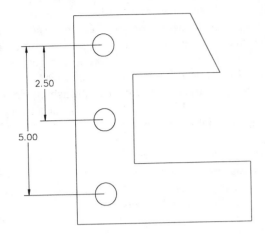

Select the circle, then the center grip of the circle to convert the warm grip into a hot grip. Use the space bar to scroll past the Stretch mode to the Move mode. Issue a copy inside of the Move option to be placed in Multiple Move mode.

Command: *(Select the circle to activate the grips at the center and quadrants; select the warm grip at the center of the circle to make it hot. Then strike the Space bar until the Move mode appears at the bottom of the prompt line and above in the status bar)*

MOVE
<Move to point>/Base point/Copy/Undo/eXit: **Copy**
MOVE (multiple)
<Move to point>/Base point/Copy/Undo/eXit: **@2.50<270**
MOVE (multiple)
<Move to point>/Base point/Copy/Undo/eXit: **@5.00<270**
MOVE (multiple)
<Move to point>/Base point/Copy/Undo/eXit: **X** *(To exit)*
Command: **CTRL-C** *(To remove entity highlight)*
Command: **CTRL-C** *(To remove grips)*

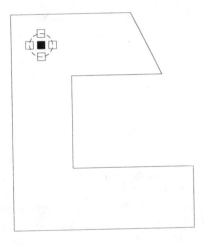

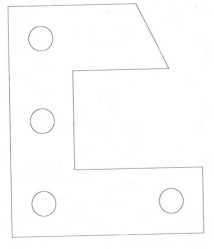

Using the Grip - Mirror Mode

Use the grip Mirror mode to flip an entity along a mirror line similar to the one used in the regular Mirror command. Follow the prompts for the Mirror option if an entity needs to be mirrored but the original does not need to be saved. This performs the mirror, but does not produce a copy of the original. If the original entity needs to be saved during the mirror operation, use the Copy option of Mirror mode. This places the user in multiple mirror mode. Locate a base point and a second point to perform the mirror.

Command: *(Select the circle at the right to enable grips, then select the warm grip at the center of the circle to make it hot. Strike the Space bar until the Mirror mode appears at the bottom of the prompt line and above in the status bar)*

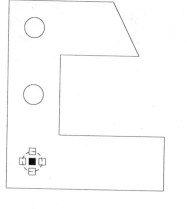

MIRROR
\<Second point>/Base point/Copy/Undo/eXit: **Copy**
MIRROR (multiple)
\<Second point>/Base point/Copy/Undo/eXit: **Base**
MIRROR (multiple)
\<Second point>/Base point/Copy/Undo/eXit: **@1<90**
MIRROR (multiple)
\<Second point>/Base point/Copy/Undo/eXit: **X**
Command: **CTRL-C** *(To remove entity highlight)*
Command: **CTRL-C** *(To remove grips)*

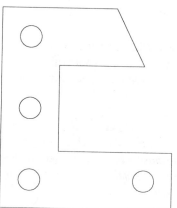

Using the Grip - Rotate Mode

Numerous grips may be selected by window or crossing boxes. At the command prompt, pick a blank part of the screen; this should place the user in Window/Crossing selection mode. Picking up or below and to the right of the previous point places the user in Window selection mode; picking up or below and to the left of the previous point places the user in Crossing selection mode. This method is used on all entities illustrated at the right. Selecting the lower left grip and using the space bar to advance to the **ROTATE** option allows all entities to be rotated at a defined angle in relation to the previously selected grip.

Command: *(Pick near "X," then near "Y" to create a window selection set and enable all grips in all entities. Select the warm grips at the lower left corner of the object to make it hot. Then strike the Space bar until the Rotate mode appears at the bottom of the prompt line and above in the status bar)*

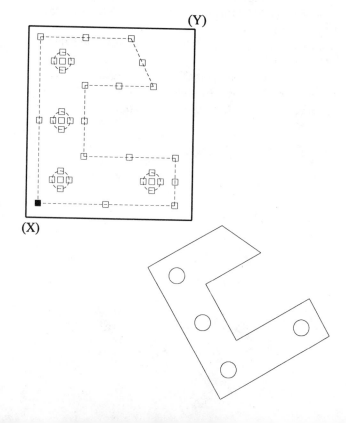

ROTATE
\<Rotation angle>/Base point/Copy/Undo/Reference/eXit: **30**
Command: **CTRL-C** *(To remove entity highlight)*
Command: **CTRL-C** *(To remove grips)*

Using the Grip-Multiple Rotate Mode

One problem of the regular Rotate command is that although an entity could be rotated, the original location could not be saved. Points had to be used to mark the original location of an entity before it was rotated. The Array command was used as a substitute to rotate and copy an entity. Now with entity grips, Rotate mode may be used to rotate and copy an entity without the use of reference points or the Array command. Illustrated at the right is a line which needs to be rotated and copied at a 40 degree angle. With a positive angle, the direction of the rotation will be counterclockwise.

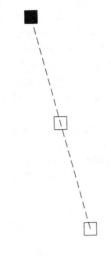

Selecting the line at the right enables grips located at the endpoints and midpoint of the line. Selecting the warm grip at the right makes it hot. This hot grip also locates the vertex of the required angle. Strike the Space bar until the Rotate mode is reached. Enter Multiple Rotate mode by entering Copy when prompted in the command sequence below. Finally enter 40 to produce a copy of the original line segment at a 40 degree angle in the counterclockwise direction.

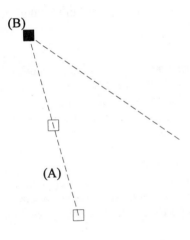

Command: *(Select line segment "A"; then select the warm grip at "B" to make it hot. Strike the Space bar until the Rotate mode appears at the bottom of the prompt line and above in the status bar)*

ROTATE
<Rotation angle>/Base point/Copy/Undo/Reference/eXit:
Copy
ROTATE (multiple)
<Rotation angle>/Base point/Copy/Undo/Reference/eXit: **30**
Command: **CTRL-C** *(To remove entity highlight)*
Command: **CTRL-C** *(To remove grips)*

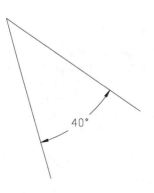

Multiple Copy Mode and Offset Snap Locations for Rotations

Any Multiple Copy modes inside of grips may be operated in a snap location mode while holding down the Shift key. Here is how it works. In the illustration at the right, the vertical center line and circle are selected using the grip pickbox. The entities highlight and the grips appear. A multiple copy of the selected entities needs to be made at an angle of 45 degrees. The grip Rotate mode is used in Multiple Copy mode.

Command: *(Select center line segment "A" and circle "B"; then select the warm grip at the center of the circle to make it hot. Strike the Space bar until the Rotate mode appears at the bottom of the prompt line and above in the status bar)*

ROTATE
<Rotation angle>/Base point/Copy/Undo/Reference/eXit: **Copy**
ROTATE (multiple)
<Rotation angle>/Base point/Copy/Undo/Reference/eXit: **Base**
New base point: **Int**
of *(Select the circle at "C" to snap to the center of the circle)*
ROTATE (multiple)
<Rotation angle>/Base point/Copy/Undo/Reference/eXit: **45**

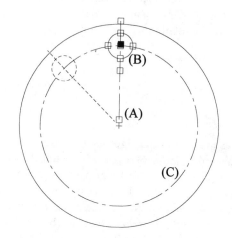

Rather than enter another angle to rotate and copy the same entities, the Shift key is held down placing the user in offset snap location mode. Moving the cursor snaps the selected entities keeping the value of the original angle of rotation, namely 45 degrees.

ROTATE (multiple)
<Rotation angle>/Base point/Copy/Undo/Reference/eXit: *(Hold down the Shift key and move the circle and center line until it snaps to the next 45 degree position illustrated at the right)*
ROTATE (multiple)
<Rotation angle>/Base point/Copy/Undo/Reference/eXit: **X**

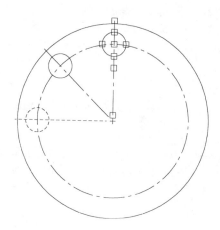

The Rotate-Copy-Snap Location mode could allow the user to create the illustration at the right without the aid of the Array command. Since all angle values are 45 degrees, continue holding down the Shift key to snap to the next 45 degree location and mark a point to place the next group of selected entities.

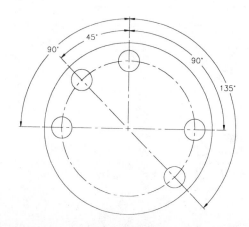

Multiple Copy Mode and Offset Snap Locations for Moving

As with the last example of using Offset Snap Locations for Rotation Mode, these same snap locations apply to Move mode. Illustrated at the right are two circles along with a common center line. The circles and center line are selected using the grip pickbox, which highlights these three entities and activates the grips. The intent is to move and copy the selected entities at 2 unit increments.

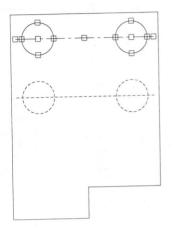

Command: *(Select the 2 circles and center line to activate the grips; select the warm grip at the midpoint of the center line to make it hot. Then strike the Space bar until the Move mode appears at the bottom of the prompt line and above in the status bar)*

MOVE
<Move to point>/Base point/Copy/Undo/eXit: **Copy**
MOVE (multiple)
<Move to point>/Base point/Copy/Undo/eXit: **@2.00<270**

In the illustration at the right, instead of remembering the previous distance and entering it to create another copy of the circles and center line, hold down the Shift key and move the cursor down to see the selected entities snap to the previous distance.

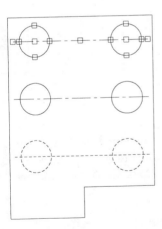

MOVE (multiple)
<Move to point>/Base point/Copy/Undo/eXit: *(Hold down the Shift key and move the cursor down to have the selected entities snap to another 2.00 unit distance)*
MOVE (multiple)
<Move to point>/Base point/Copy/Undo/eXit: **X** *(To exit)*
Command: **CTRL-C** *(To remove entity highlight)*
Command: **CTRL-C** *(To remove grips)*

The completed hole layout is illustrated at the right using the Offset Snap Location method of entity grips.

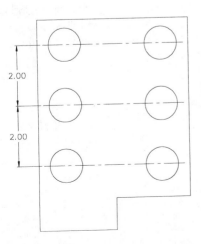

Tutorial Exercise #3
Template.Dwg

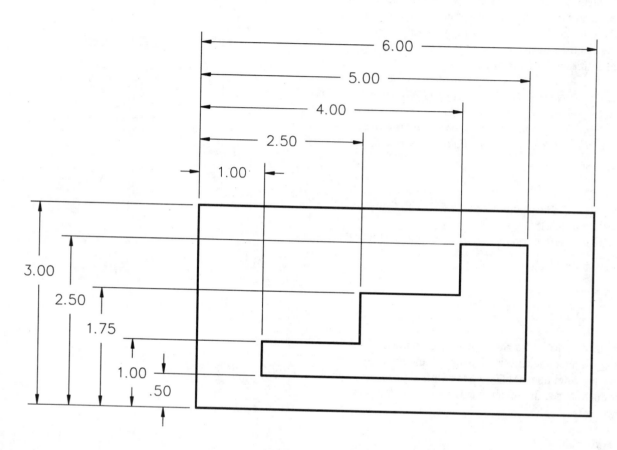

PURPOSE:
This tutorial is designed to allow the user to construct a one view drawing of the Template using the Absolute, Relative, and Polar Coordinate modes.

SYSTEM SETTINGS:
Use the current default settings for the limits of this drawing, (0,0) for the lower left corner and (12,9) for the upper right corner.
Use the Grid command and change the grid spacing from 1.0000 to 0.25 units. The grid will be used only as a guide for constructing this object. Do not turn the Snap or Ortho commands on.

LAYERS:
Create the following layers with the format :
Name-Color-Linetype
Object - White - Continuous

SUGGESTED COMMANDS:
The Line command will be used entirely for this tutorial in addition to a combination of coordinate systems. The Erase command could be used although a more elaborate method of correcting a mistake would be to use the Line-undo command to erase a previously drawn line and still stay in the Line command.

DIMENSIONING:
Dimensions may be added to this problem at a later time. Consult your instructor.

PLOTTING:
This tutorial exercise may be plotted on "A"-size paper, (8.5" x 11"). Use a plotting scale of 1=1 to produce a full size plot.

Step #1

Begin this tutorial exercise with the Line command and draw the outer perimeter of the box. Use an absolute coordinate point followed by polar coordinates using the prompt sequence below and the illustration at the right as a guide.

Command: **Line**
From point: **2,2**
To point: **@6<0**
To point: **@3<90**
To point: **@6<180**
To point: **@3<270**
To point: *(Strike Enter to exit this command)*

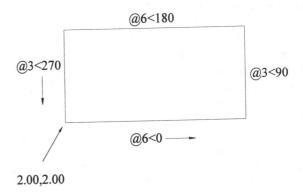

Step #2

The next step will be to draw the stair step outline of the template using the Line command again. However we first need to identify the starting point of the template. Absolute coordinates could be calculated but in more complex objects this would be difficult. A more efficient method would be to use the Line command again and at the "From point" prompt, enter a relative coordinate. This would position the new point at a specified distance from the previously used point. Use the prompt sequence below and the illustration at the right as a guide for performing this operation.

Command: **Line**
From point: **@1.00,0.50**

This relative coordinate begins a new line a distance of 1 unit in the "X" direction and 0.50 units in the "Y" direction from the last point.

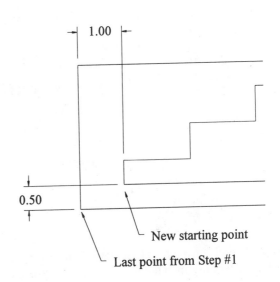

Step #3

Once the new starting point has been identified by the relative coordinate @1.00, 0.50, continue with the Line command to complete the inner part of the template by using the polar coordinate mode as in the illustration at the right.

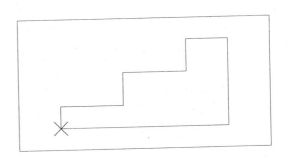

Step #4

Continue using the Line command and polar coordinate mode
to complete the inner part of the template using the prompt
sequence and the illustration below as a guide.

Command: **Line**
From point: **@1.00, 0.50**
To point: **@4.00<0**
To point: **@2.00<90**
To point: **@1.00<180**
To point: **@0.75<270**
To point: **@1.50<180**
To point: **@0.75<270**
To point **@1.50<180**
To point **@0.50<270**
To point: *(Strike Enter to exit this command)*

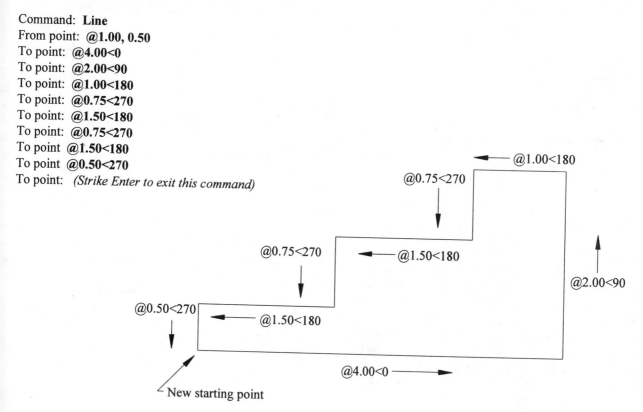

New starting point

Step #5

The completed problem is illustrated below. Dimensions may
be added at a later time upon the request of your instructor.

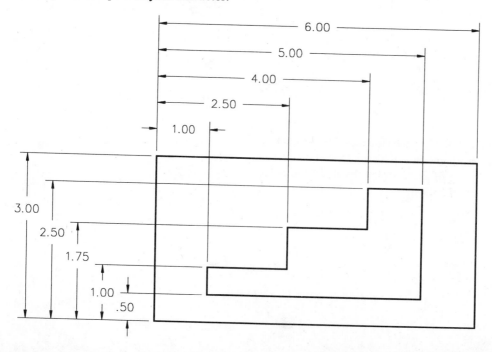

Tutorial Exercise #4
Tile.Dwg

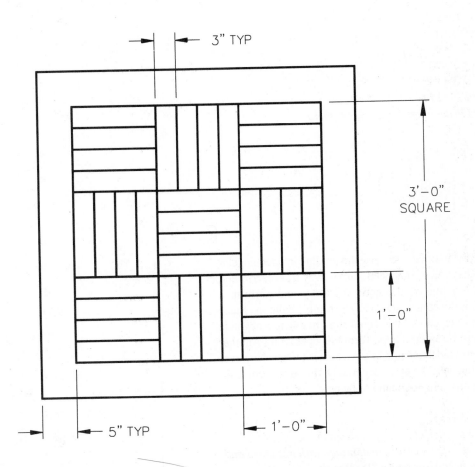

3" TYP

3'-0"
SQUARE

1'-0"

5" TYP

1'-0"

PURPOSE:
This tutorial is designed to use the Offset and Trim commands to complete the drawing of the Floor Tile.

SYSTEM SETTINGS:
Use the Units command and change the units of measure from decimal to architectural units. Keep the remaining default settings. Use the Limits command and change the limits of the drawing to (0,0) for the lower left corner and (10',8') for the upper right corner. Use the Snap command and change the value from 1.0000 to 4", (with the Grid command already set to 1.0000 the snap setting of 4" will also change the grid spacing to 4").

LAYERS:
Create the following layers with the format :
Name-Color-Linetype
Object - White - Continuous

SUGGESTED COMMANDS:
The Line command will be used to begin the Tile. The Offset command is used to copy selected line segments at a specified distance. The Trim command is then used to clean up intersecting corners. The Erase command can be used to delete entities from the drawing (Remember to use the Oops command to bring back previously erased entities deleted by mistake).

DIMENSIONING:
Dimensions may be added to this problem at a later time. Consult your instructor.

PLOTTING:
This tutorial exercise may be plotted on "B"-size paper (11" x 17"). Use a plotting scale of 1=12 to produce a scaled plot.

Step #1

Begin this exercise by using the Line command and polar coordinate mode to draw a 3'-0" square illustrated at the right.

Command: **Line**
From point: **12,12**
To point: **@3'<0**
To point: **@3'<90**
To point: **@3'<180**
To point: **@3'<270**
To point: *(Strike Enter to exit this command)*

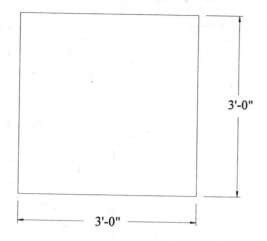

3'-0"

3'-0"

Step #2

Use the Array command to copy the top line in a rectangular pattern and have all lines spaced 3 units away from each other. Select the top line as the entity to array and perform a rectangular array consisting of 12 rows and 1 column. Since the top line selected will be copied straight down, a negative distance must be entered to perform this operation. Another popular command that could be used here is Offset. However since each line must be offset separately, the Array command is the more efficient command to be used.

Command: **Array**
Select objects: *(Select the top horizontal line at "A")*
Select objects: *(Strike Enter to continue with this command)*
Rectangular or Polar array (R/P) <R>: **R**
Number of rows (---) <1>: **12**
Number of columns (|||) <1>: **1**
Unit cell or distance between rows (---): **-3**

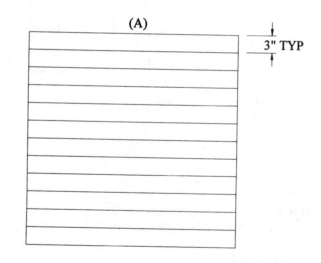

(A)

3" TYP

Step #3

Again use the Array command but this time copy the right vertical line in a rectangular pattern and have all lines spaced 3 units away from each other. Select the right line as the entity to array and perform a rectangular array consisting of 1 row and 12 columns. Since the right line selected will be copied to the left, a negative distance must be entered to perform this operation.

Command: **Array**
Select objects: *(Select the right vertical line at "A")*
Select objects: *(Strike Enter to continue with this command)*
Rectangular or Polar array (R/P) <R>: **R**
Number of rows (---) <1>: **1**
Number of columns (|||) <1>: **12**
Distance between columns (|||): **-3**

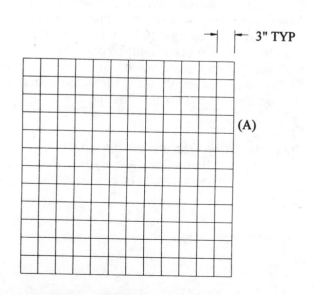

3" TYP

(A)

Step #4

Use the Trim command, select the two vertical dashed lines
illustrated at the right as cutting edges; use the illustration in
Step #5 as a guide for determining which lines to trim out.

Command: **Trim**
Select cutting edges...
Select objects: *(Select the two dashed lines at the right)*
Select objects: *(Strike Enter to continue with this command)*
<Select object to trim>/Undo: *(See Step #5)*

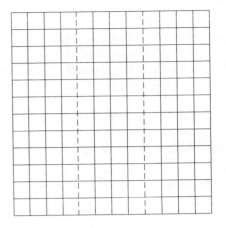

Step #5

For the last prompt of the Trim command in Step #4, select all
horizontal lines in the areas marked "A", "B", "C", and "D" at
the right. When finished selecting the entities to trim, strike the
"Enter" key to exit the command.

<Select object to trim>/Undo: *(Select all horizontal lines in
area "A")*
<Select object to trim>/Undo: *(Select all horizontal lines in
area "B")*
<Select object to trim>/Undo: *(Select all horizontal lines in
area "C")*
<Select object to trim>/Undo: *(Select all horizontal lines in
area "D")*
<Select object to trim>/Undo: *(Strike Enter to exit this com-
mand)*

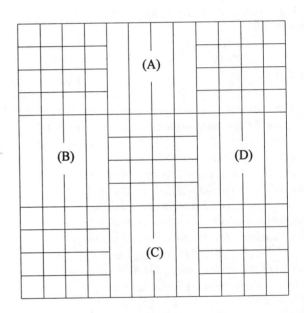

Step #6

Use the Trim command, select the two horizontal dashed lines
as cutting edges.

Command: **Trim**
Select cutting edges...
Select objects: *(Select the two dashed lines)*
Select objects: *(Strike Enter to continue with this command)*
<Select object to trim>/Undo: *(See Step #7)*

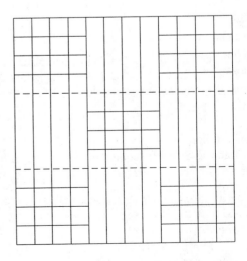

Step #7

For the last step of the Trim command back in Step #6, select all vertical lines in the areas marked "A", "B", "C", "D", and "E". When finished selecting the entities to trim, strike the "Enter" key to exit the command.

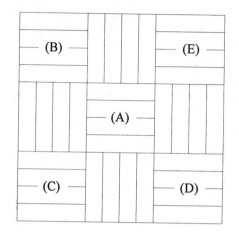

<Select object to trim>/Undo: *(Select all vertical lines in area "A")*

<Select object to trim>/Undo: *(Select all vertical lines in area "B")*

<Select object to trim>/Undo: *(Select all vertical lines in area "C")*

<Select object to trim>/Undo: *(Select all vertical lines in area "D")*

<Select object to trim>/Undo: *(Select all vertical lines in area "E")*

<Select object to trim>/Undo: *(Strike Enter to exit this command)*

Step #8

Use the Offset command again to offset the lines "A", "B", "C", and "D" five units in the directions illustrated at the right.

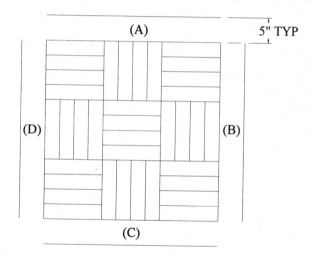

Command: **Offset**
Select offset distance or Through <Through>: **5**
Select object to offset: *(Select the line at A)*
Side to offset? *(Pick a point above the line)*
Select object to offset: *(Select the line at B)*
Side to offset? *(Pick a point right of the line)*
Select object to offset: *(Select the line at C)*
Side to offset? *(Pick a point below the line)*
Select object to offset: *(Select the line at D)*
Side to offset? *(Pick a point left of the line)*
Select object to offset: *(Strike Enter to exit this command)*

Step #9

Use the Fillet command set to a radius of 0 to place a corner at the intersection of lines "A" and "B". The radius should already be set to 0 by default so simply pick the two lines and the corner is formed.

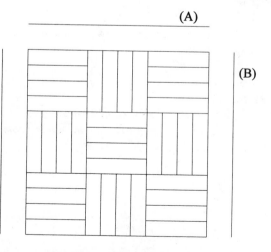

Command: **Fillet**
Polyline/Radius/<Select first object>: *(Select line "A")*
Select second object: *(Select line "B")*

Repeat this procedure for the other three corners.

Step #10

The completed tile drawing is illustrated at the right. Follow the next step to add more tiles in a rectangular pattern using the Array command.

Step #11

As an alternate step, use the Array command to copy the initial design in a rectangular pattern by row and column. Follow the prompts below to perform this operation to construct a series of tiles illustrated at the right.

Command: **Array**
Select objects: *(Select the entire tile design from Step #10)*
Select objects: *(Strike Enter to continue with this command)*
Rectangular or Polar array (R/P) <R>: **R**
Number of rows(---) <1>: **3**
Number of columns (|||) <1>: **2**
Unit cell or distance between rows (---): **3'10**
Distance between columns (|||): **3'10**

Tutorial Exercise #5
Inlay.Dwg

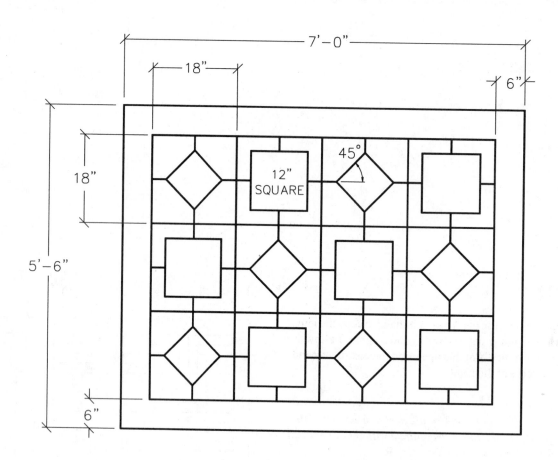

PURPOSE:
This tutorial is designed to allow the user to construct a drawing of the Inlay using the Copy command and the Multiple modifier.

SYSTEM SETTINGS:
Use the Units command and change the units of measure from decimal to architectural units. Keep the remaining default settings. Use the Limits command and change the limits of the drawing to (0,0) for the lower left corner and (15'6",9'6") for the upper right corner. Use the Snap command and change the value from 1.0000 to 3" (With the Grid command already set to 1.0000 the snap setting of 3" will also change the grid spacing to 3".)

LAYERS:
Create the following layers with the format :
Name-Color-Linetype
Object - White - Continuous

SUGGESTED COMMANDS:
Begin this tutorial by drawing a 6'-0" x 4'-6" rectangle using the Line command. Offset the edges of the rectangle by a distance of 18". Then using the 3" grid as a guide along with the Snap-On, draw the diamond and square shapes. Use the Copy command along with the Multiple modifier to copy the diamond and square shapes numerous times at the designated areas.

DIMENSIONING:
Dimensions may be added to this problem at a later time. Consult your instructor.

PLOTTING:
This tutorial exercise may be plotted on "B"-size paper (11" x 17"). Use a plotting scale of 1=12 to produce a full size plot.

Step #1

Begin this exercise by using the Line command and polar coordinate mode to draw a rectangle 6'-0" by 4'-6" illustrated at the right.

Command: **Line**
From point: **12,12**
To point: **@6'<0**
To point: **@4'6<90**
To point: **@6'<180**
To point: **@4'6<270**
To point: *(Strike Enter to exit this command)*

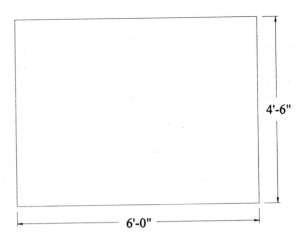

Step #2

Use the Array command to copy the top line in a rectangular pattern and have all lines spaced 18 units away from each other. Select the top horizontal line at "A" as the entity to array. Since this line will be copied straight down, a value of -18 for the spacing between rows will perform this operation. Repeat this command to copy the right vertical line at "B" 3 times to the left at a distance of 18 units. Enter a value of -18 units for the spacing in between columns since the copying is performed in the left direction as illustrated at the right.

Command: **Array**
Select objects: *(Select the top horizontal line at "A")*
Select objects: *(Strike Enter to continue with this command)*
Rectangular or Polar array (R/P) <R>: **R**
Number of rows (---) <1>: **3**
Number of columns (|||) <1>: **1**
Unit cell or distance between rows (---): **-18**

Repeat the Rectangular Array command above for the vertical line at "B." Use 1 for the number of rows, 4 for the number of columns, and -18 as the distance between columns.

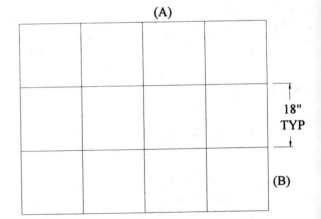

Step #3

Begin drawing one 12" x 12" diamond figure in the position illustrated at the right. Use the Zoom command along with the Window option to magnify the area around the position of the diamond figure.

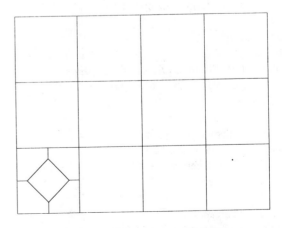

Step #4

Be sure the Grid and Snap values are set to 3"; the snap should already be turned On. Then use the Line command to draw the four lines illustrated at the right. Use "A", "B", "C", and "D" as the starting points for the four lines.

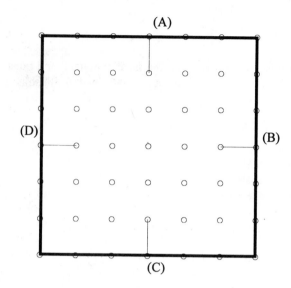

Step #5

Use the Line command to draw the diamond shaped figure illustrated at the right.

Command: **Line**
From point: **Endp**
of *(Select the endpoint of the line at "A")*
To point: **Endp**
of *(Select the endpoint of the line at "B")*
To point: **Endp**
of *(Select the endpoint of the line at "C")*
To point: **Endp**
of *(Select the endpoint of the line at "D")*
To point: **C**

When finished, perform a Zoom-Previous operation; turn the snap Off using the F9 function key.

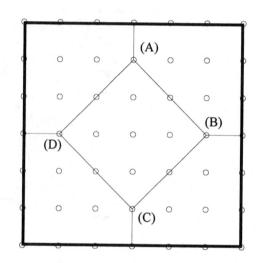

Step #6

Use the Copy command and the Multiple option to repeat the diamond shaped pattern at the right. Use the Osnap-Intersec option to assist you during this operation.

Command: **Copy**
Select objects: *(Select the highlighted lines at the right)*
Select objects: *(Strike the Enter key to continue)*
<Base point or displacement>/ Multiple: **M**
Base point: **Int**
of *(Select the intersection at "A")*
Second point of displacement: **Int**
of *(Select the intersection at "B")*
Second point of displacement: **Int**
of *(Select the intersection at "C")*
Second point of displacement: **Int**
of *(Select the intersection at "D")*
(Repeat the above procedure for "E" and "F")

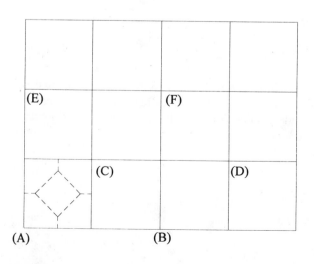

Step #7

The diamond pattern of the Inlay floor tile should be similar to the illustration at the right.

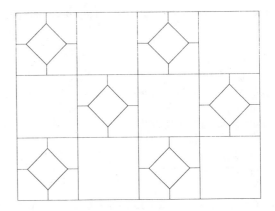

Step #8

Begin drawing one 12" x 12" square figure in the position illustrated at the right. Use the Zoom-Window command to magnify the area around the position of the square figure.

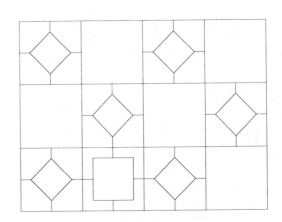

Step #9

Be sure the Grid and Snap values are set to 3" and turn the snap back On using the F9 function key. Then use the Line command to draw the four lines illustrated at the right. Use "A", "B", "C", and "D" as the starting points for the four lines.

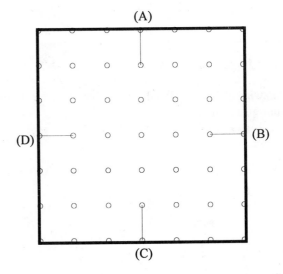

Step #10

Use the Line command to draw the square shaped object illustrated at the right.

Command: **Line**
From point: **Endp**
of *(Select the endpoint of the line at "A")*
To point: **@6<0** *(To "B")*
To point: **@12<270** *(To "C")*
To point: **@12<180** *(To "D")*
To point: **@12<90** *(To "E")*
To point: **C** *(Back to "A")*

When finished, perform a Zoom-Previous operation; turn the snap Off using the F9 function key.

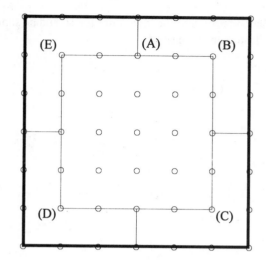

Step #11

Use the Copy command and the Multiple option to repeat the square shaped pattern at the right. Use the Osnap-Intersec option to assist you during this operation.

Command: **Copy**
Select objects: *(Select the highlighted lines at the right)*
Select objects: *(Strike the Enter key to continue)*
<Base point or displacement>/ Multiple: **M**
Base point: **Int**
of *(Select the intersection at "A")*
Second point of displacement: **Int**
of *(Select the intersection at "B")*
Second point of displacement: **Int**
of *(Select the intersection at "C")*
Second point of displacement: **Int**
of *(Select the intersection at "D")*
(Repeat the above procedure for "E" and "F")

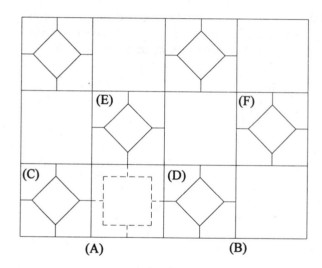

Step #12

The drawing of the Inlay should appear similar to the illustration at right.

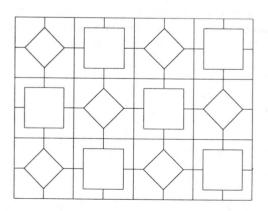

Step #13

Use the Offset command to copy the outline of the Inlay outward at a distance of 6 units.

Command: **Offset**
Select offset distance or Through <Through>: **6**
Select object to offset: *(Select line "A")*
Side to offset? *(Pick a point above the line)*
Select object to offset: *(Select line "B")*
Side to offset? *(Pick a point left of the line)*
Select object to offset: *(Select line "C")*
Side to offset? *(Pick a point below the line)*
Select object to offset: *(Select line "D")*
Side to offset? *(Pick a point right of the line)*
Select object to offset: *(Strike Enter to exit this command)*

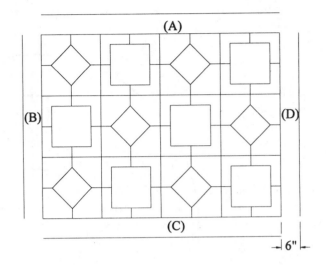

Step #14

Use the Fillet command set to a radius of "0" to place a corner at the intersection of lines "A" and "B". The radius should already be set to "0" by default so simply pick the two lines and the corner is formed.

Command: **Fillet**
Polyline/Radius/<Select firrst object>: *(select line "A")*
Select second object: *(Select line "B")*

Repeat this procedure for the other three corners.

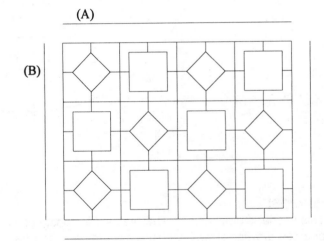

Step #15

The completed problem is illustrated at the right. Dimensions may be added upon the request of your instructor.

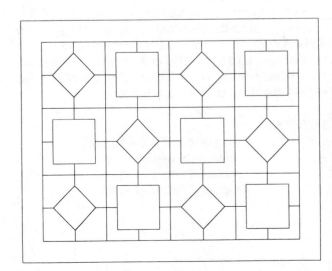

Tutorial Exercise #6
Clutch.Dwg

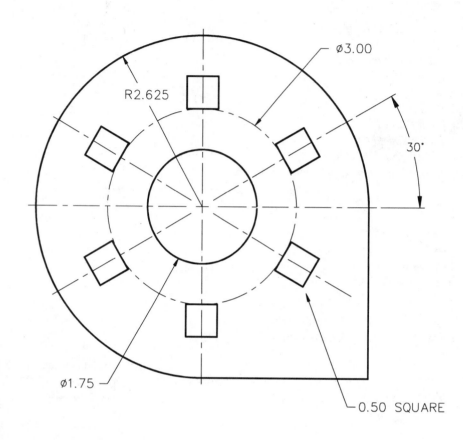

PURPOSE:
This tutorial is designed to allow the user to construct a one view drawing of the Clutch using coordinates and the Array command.

SYSTEM SETTINGS:
Use the current default settings for the limits of this drawing, (0,0) for the lower left corner and (12,9) for the upper right corner. Use the Grid command and change the grid spacing from 1.0000 to 0.25 units. The grid will be used only as a guide for constructing this object. Do not turn the Snap or Ortho commands on.

LAYERS:
Create the following layers with the format :
Name-Color-Linetype
Object - White - Continuous
Center - Yellow - Center

SUGGESTED COMMANDS:
Draw the basic shape of the object using the Line and Circle commands. Lay out a center line circle, draw one square shape, and use Array to create a multiple copy of the square in a circular pattern.

DIMENSIONING:
Dimensions may be added to this problem at a later time. Consult your instructor.

PLOTTING:
This tutorial exercise may be plotted on "A"-size paper (8.5" x 11"). Use a plotting scale of 1=1 to produce a full size plot.

Step #1

Begin drawing the clutch by placing a circle with the center at absolute coordinate (6.00,5.00) and radius of 2.625 units. Check to see that the current layer is "Object." Next, prepare to place a center mark at the center of the circle by changing the dimension variable, Dimcen from a value of 0.09 to -.09. The negative value will extend the center lines past the extremities of the circle.   Then use the Dim-Center option, identify the circle, and place the center mark. Follow the prompts carefully below.

Command: **Circle**
3P/2P/TTR/,Center point>: **6.00, 5.00**
Diameter/<Radius>: **2.625**

Command: **Dim**
Dim: **Dimcen**
Current value <0.09> New value: **-.09**

Dim: **Center**
Select arc or circle: *(Select anywhere along the circle)*
Dim: **Exit** *(This returns you to the Command: prompt)*

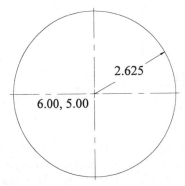

Step #2

Draw two lines at a distance of 2.625 using the polar coordinate mode and using the Osnap-Intersec option. Begin the line at point "A" illustrated at the right.

Command: **Line**
From point: **Int**
of *(Select the intersection of the line and circle at "A")*
To point: **@2.625<0**
To point: **@2.625<90**
To point: *(Strike Enter to exit this command)*

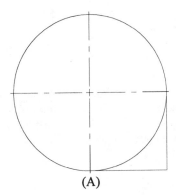

(A)

Step #3

Use the Trim command to partially delete one-fourth of the circle. Use the dashed lines illustrated at the right as the cutting edges and select the circle at "A" as the entity to trim.

Command: **Trim**
Select cutting edge(s)...
Select objects: *(Select the two dashed lines at the right)*
Select objects: *(Strike Enter to continue)*
<Select object to trim>/Undo: *(Select the circle at "A")*
<Select object to trim>/Undo: *(Strike Enter to exit this command)*

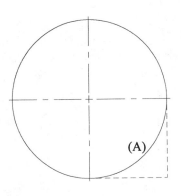

(A)

Step #4

Use the Erase command to delete the bottom vertical center line at "A." This center line will be placed back in its original position at a later step.

Command: **Erase**
Select objects: *(Select the bottom vertical center line at "A")*
Select objects: *(Strike Enter to erase and exit this command)*

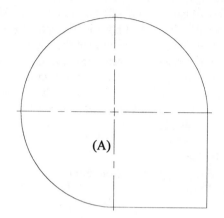

Step #5

Place a circle of 3.00 diameter from absolute coordinate (6.00, 5.00). Since this circle needs to be converted from an object line to a center line, use the Chprop command and the Layer option and change the circle to the Center layer (Be sure to have the Center layer previously created using the Layer command.)

Command: **Circle**
3P/2P/TTR/Center point>: **6.00, 5.00**
Diameter/<Radius>: **D**
Diameter: **3.00**

Command: **Chprop**
Select objects: *(Select the 3.00 diameter circle)*
Select objects: *(Strike Enter to continue with the command)*
Change what property (Color/LAyer/LType/Thickness): **LA**
New layer <object>: **Center**

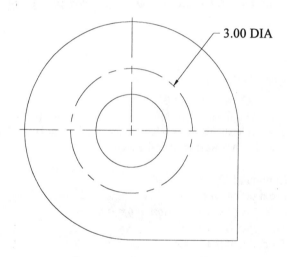

3.00 DIA

Step #6

Use the Zoom command with the Window option to magnify the upper portion of the clutch for constructing a square in the next step.

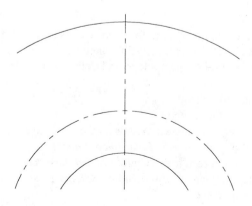

Step #7

Begin constructing the 0.50 unit square illustrated at the right. Use the Line command, begin at the intersection of the vertical and circular center lines, and use polar coordinates to assist in this operation.

Command: **Line**
From point: **Int**
of *(Select the intersection of the line and circle at "A")*
To point: **@0.25<0**
To point: **@0.50<90**
To point: **@0.50<180**
To point: **@0.50< 270**
To point: **C** *(This will close the square and exit the command)*

Perform a Zoom-Previous operation to return back to the original display.

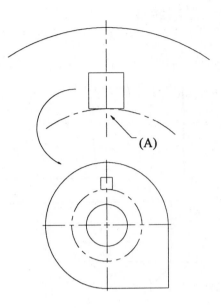

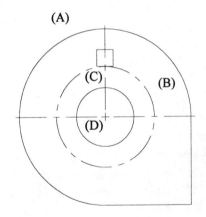

Step #8

Use the Array command to copy the square and vertical center line in a circular pattern 6 times. Follow the prompts below to perform this operation and complete the clutch.

Command: **Array**
Select objects: **W**
First corner: *(Select a point approximately at "A")*
Other corner: *(Select a point approximately at "B")*
Select objects: *(Select the vertical center line at "C")*
Select objects: *(Strike Enter to continue)*
Rectangular or Polar array (R/P) <R>: **P**
Center point of array: **Int**
of *(Select the intersection of the center mark at "D")*
Number of items: **6**
Angle to fill (+=CCW, -=CW) <360>: *(Strike Enter)*
Rotate objects as they are copied?<Y> *(Strike Enter)*

Step #9

The completed problem is illustrated at the right. Dimensions may be added at a later date.

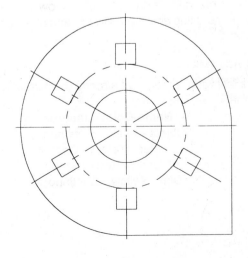

Tutorial Exercise #7
Lug.Dwg

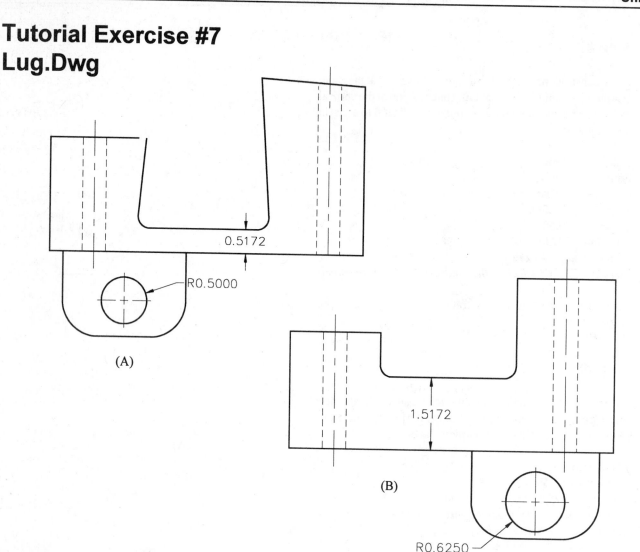

0.5172

R0.5000

(A)

1.5172

(B)

R0.6250

PURPOSE:
This tutorial is designed to use entity grips to edit the drawing of the Lug illustrated above at "A" until it appears like the illustration at "B."

SYSTEM SETTINGS:
Since this drawing is provided on diskette, open an existing drawing file called "Lug". Follow the steps in this tutorial for using entity grips to edit and make changes to the Lug.

LAYERS:
The following layers have already been created with the format:

Name-Color-Linetype
Object - White - Continuous
Center - Yellow - Center
Hidden - Red - Hidden
Dim - Yellow - Continuous

SUGGESTED COMMANDS:
Begin this tutorial by selecting the Settings area of the pulldown menu area. Next select "Grips..." and check the box "Enable Grips" to have grips present when an entity is selected. Then use the Stretch, Move, Rotate, Scale, and Mirror modes of entity grips to make changes to the existing drawing file.

DIMENSIONING:
Dimensions may be completed in this tutorial at a later time. Consult your instructor.

PLOTTING:
This tutorial exercise may be plotted on "B"-size paper (17" x 11"). Use a plotting scale of 1=1 to produce a full size plot.

Step #1

Before beginning the tutorial, be sure that entity grip mode is enabled by selecting "Settings" from the pulldown menu area. Next select "Grips...," which activates the DDGRIPS dialog box. Use the dialog box to examine that grips are enabled with the presence of the check placed in the appropriate box.

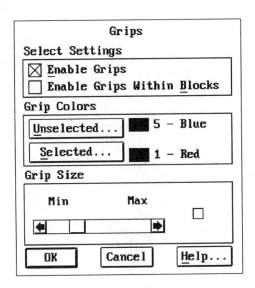

Step #2

Begin by turning ortho Off. Next, at the command prompt, use the grip cursor to select the inclined line "A" illustrated at the right. Notice the appearance of the grips at the endpoints and midpoints of the line. Continue by going on to Step #3.

Command: *(Select the inclined line "A")*

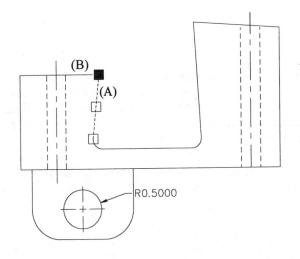

Step #3

Still at the command prompt, select the grip illustrated at the right at "A." This grip becomes the current base point for the next series of editing options. Use the Stretch option to reposition the endpoint of the highlighted line to the endpoint of the horizontal line illustrated at the right. When this operation is complete, issue two CTRL-C's at the command prompt to remove the entity highlight and grips from the display screen. Continue by going on to Step #4.

Command: *(Select the warm grip at the endpoint of the line at "A" to make it hot)*
STRETCH
<Stretch to point>/Base point/Copy/Undo/eXit: **Endp**
of *(Select the endpoint of the horizontal line at "B")*
Command: **CTRL-C** *(To remove the entity highlight)*
Command: **CTRL-C** *(To remove the grips)*

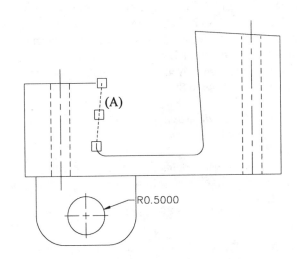

Step #4

At the command prompt, use the grip cursor to select the two inclined lines "A" and "B" illustrated at the right. Notice the appearance of the grips at the endpoints and midpoints of the line. Continue by going on to Step #5.

Command: *(Select the inclined lines "A" and "B")*

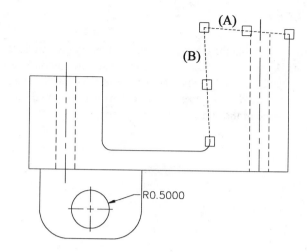

Step #5

Still at the command prompt, select the grip illustrated at the right at "A." This grip becomes the current base point for the next series of editing options. Use the Stretch option in combination with .XYZ filters to reposition the corner of the highlighted lines to form a corner. When this operation is complete, issue two CTRL-C's at the command prompt to remove the entity highlight and grips from the display screen. Continue by going on to Step #6.

Command: *(Select the warm grip at the endpoint of the line at "A" to make it hot).*
STRETCH
<Stretch to point>/Base point/Copy/Undo/eXit: **.X**
of *(Select endpoint of the line at the the grip at "B")*
need YZ: **.Y**
of *(Select the endpoint of the line at the grip at "C")*
need Z: **0**
Command: **CTRL-C** *(To remove the entity highlight)*
Command: **CTRL-C** *(To remove the grips)*

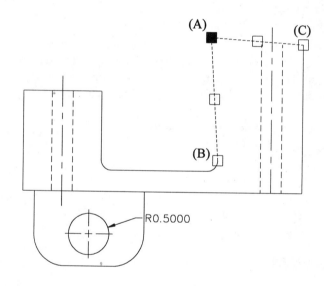

Step #6

At the command prompt, use the grip cursor to select the two vertical lines, one horizontal line, and both filleted corners illustrated at the right. Notice the appearance of the grips at the endpoints and midpoints of the lines and arcs. Continue by going on to Step #7.

Command: *(Select the two vertical lines, two arcs representing fillets, and the horizontal line)*

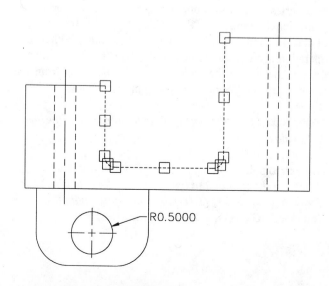

Step #7

Still at the command prompt, select the grips illustrated at the right. These grips become the current base point for the next series of editing options. Use the Stretch option in combination with a polar coordinate to stretch the entities a distance of 1 unit in the 90 degree direction. When this operation is complete, issue two CTRL-C's at the command prompt to remove the entity highlight and grips from the display screen. Continue by going on to Step #8.

Command: *(Select all warm grips illustrated at the right to make them hot. Hold down the Shift key to make multiple warm grips hot)*
STRETCH
<Stretch to point>/Base point/Copy/Undo/eXit: **@1.00<90**
Command: **CTRL-C** *(To remove the entity highlight)*
Command: **CTRL-C** *(To remove the grips)*

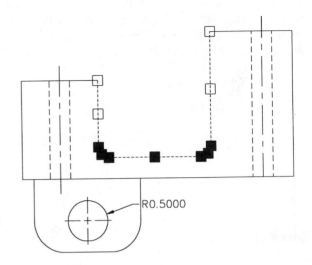

Step #8

At the command prompt, use the grip cursor to select the circle and dimension illustrated at the right. Notice the appearance of the grips at the quadrants and center of the circle and center, starting point, and text location of the radius dimension. Continue by going on to Step #9.

Command: *(Select the circle and the radius dimension)*

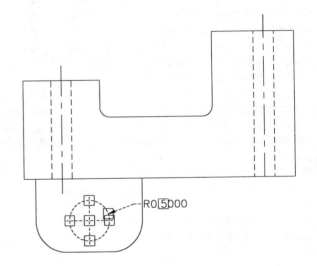

Step #9

Still at the command prompt, select the grip at the center of the circle illustrated at the right. This grip becomes the current base point for the next series of editing options. Use the Scale option to increase the size of the circle. Notice that this will also affect the value of the associative dimension. When this operation is complete, issue two CTRL-C's at the command prompt to remove the entity highlight and grips from the display screen. Continue by going on to Step #10.

Command: *(Select the warm grip at the center of the circle to make it hot. Strike the Space bar until the Scale mode appears at the bottom of the prompt line and above in the status bar)*
SCALE
<Scale factor>/Base point/Copy/Undo/Reference/eXit: **1.25**
Command: **CTRL-C** *(To remove entity highlight)*
Command: **CTRL-C** *(To remove grips)*

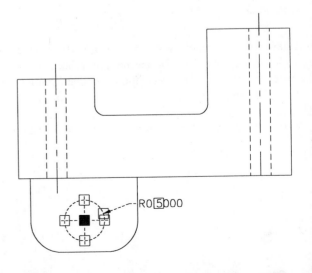

Step #10

At the command prompt, use the grip cursor to select the dimension illustrated at the right. Notice the appearance of the grips at the center, starting point, and text location of the radius dimension. The dimension will be rotated to a new angle. Continue by going on to Step #11.

Command: *(Select the dimension at the right)*

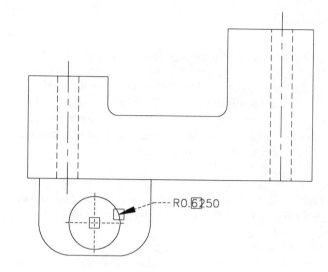

Step #11

Still at the command prompt, select the center grip illustrated at the right. This grip becomes the current base point for the next series of editing options. Use the Rotate option to rotate the dimension to a better location. When this operation is complete, issue two CTRL-C's at the command prompt to remove the entity highlight and grips from the display screen. Continue by going on to Step #12.

Command: *(Select the warm grip at the center of the circle to make it hot. Strike the Space bar until the Rotate mode appears at the bottom of the prompt line and above in the status bar)*
ROTATE
<Rotation angle>/Base point/Copy/Undo/Reference/eXit: **-45**
Command: **CTRL-C** *(To remove entity highlight)*
Command: **CTRL-C** *(To remove grips)*

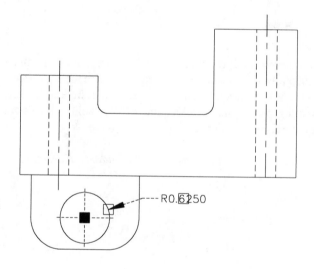

Step #12

At the command prompt, use the grip cursor to select the center line illustrated at the right. Notice the appearance of the grips at the endpoints and midpoints of the center line. Continue by going on to Step #13.

Command: *(Select the center line illustrated at the right)*

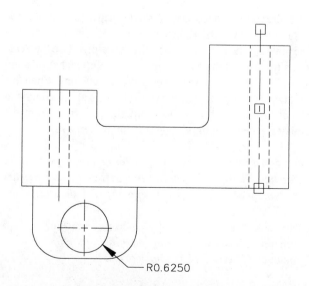

Step #13

Still at the command prompt, select the grip illustrated at the right at "A." This grip becomes the current base point for the next series of editing options. Use the Stretch option to extend the center line from a new base point at "B" to the intersection at "C." When this operation is complete, issue two CTRL-C's at the command prompt to remove the entity highlight and grips from the display screen. Continue by going on to Step #14.

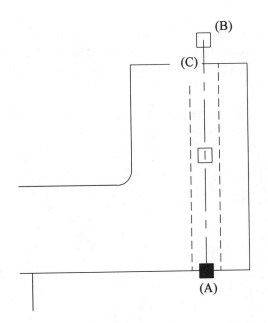

Command: *(Select the warm grip at the bottom of the center line to make it hot).*
STRETCH
<Stretch to point>/Base point/Copy/Undo/eXit: **Base**
Base point: *(Select the grip at "B")*
<Stretch to point>/Base point/Copy/Undo/eXit: **Int**
of *(Select the intersection of the center line and horizontal line at "C")*
Command: **CTRL-C** *(To remove entity highlight)*
Command: **CTRL-C** *(To remove grips)*

Step #14

At the command prompt, use the grip cursor to select the two hidden lines and center line illustrated at the right. Notice the appearance of the cold grips at the endpoints and midpoints of the lines. Continue by going on to Step #15.

Command: *(Select the two hidden lines and the center line)*

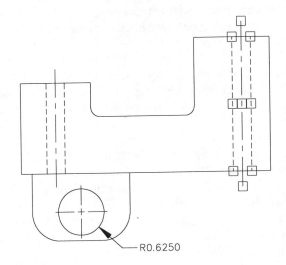

R0.6250

Step #15

Still at the command prompt, select the three grips at the midpoints of the center and hidden lines illustrated at the right. These grips become the current base point for the next series of editing options. Use the Move option to center the hidden and center lines at the middle of the horizontal line. When this operation is complete, issue two CTRL-C's at the command prompt to remove the entity highlight and grips from the display screen. Continue by going on to Step #16.

Command: *(Use the Shift key to select the three warm grips at the endpoints of the three lines to make them hot. Strike the Space bar until Move mode appears at the bottom of the prompt line and above in the status bar)*
MOVE
<Move to point>/Base point/Copy/Undo/eXit: **Base**
Base point: **Int**
of *(Select the intersection of the lines at "A")*
<Move to point>/Base point/Copy/Undo/eXit: **Mid**
of *(Select the midpoint of the horizontal line at "B")*
Command: **CTRL-C** *(To remove entity highlight)*
Command: **CTRL-C** *(To remove grips)*

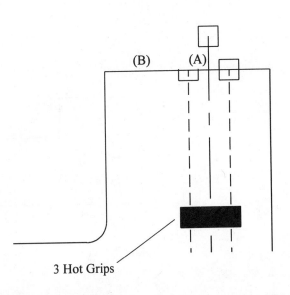

3 Hot Grips

Step #16

At the command prompt, use the grip cursor to select all of the entities illustrated at the right. Notice the appearance of the grips at the various key points of these entities. Continue by going on to Step #17.

Command: *(Select all entities illustrated at the right. This can be easily accomplished using the automatic window mode and marking points at "X" and "Y")*

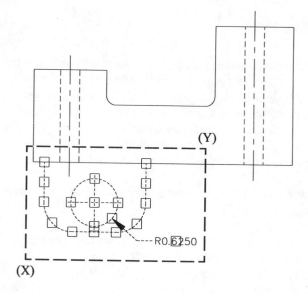

Step #17

Convert all warm grips to hot grips illustrated at the right. Use the Shift key to assist with this operation. Strike the Space bar until Mirror mode appears at the bottom of the prompt line and at the top of the status line).

Command: *(Use the Shift key to select all warm grips at the right to make them hot. Strike the Space bar until Mirror mode appears at the bottom of the prompt line and above in the status bar)*
** MIRROR **
<Second point>/Base point/Copy/Undo/eXit: **Base**
New base point: **Mid**
of *(Select the midpoint of the horizontal line at "A")*
** MIRROR **
<Second point>/Base point/Copy/Undo/eXit: **@1<90**
Command: **CTRL-C** *(To remove entity highlight)*
Command: **CTRL-C** *(To remove grips)*

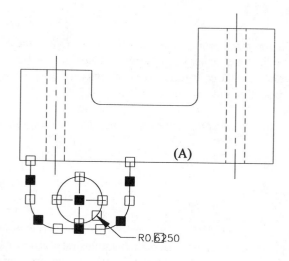

The completed object is illustrated below.

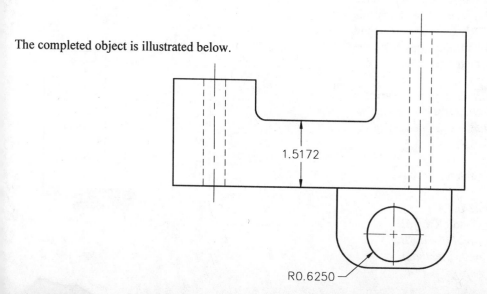

Questions for Unit 2

Draw Command Questions

2-1. To begin drawing a new line from the last known point, enter
   - (A) 0,0
   - (B) @
   - (C) Next
   - (D) Only A and B.
   - (E) Only A and C.

2-2. By default, using the POINT command displays the point as a
   - (A) circle.
   - (B) "plus" sign.
   - (C) square.
   - (D) dot.
   - (E) arc.

2-3. The special character string that causes text to be underlined is
   - (A) #U
   - (B) %%U
   - (C) ##U
   - (D) **U
   - (E) %U

2-4. The "U" option of the LINE command means
   - (A) Unite
   - (B) Unit
   - (C) Underline
   - (D) Undo
   - (E) Union

2-5. The command used to create a solid circle is
   - (A) Circfill
   - (B) Donut
   - (C) Scircle
   - (D) Arcfill
   - (E) Fillcirc

2-6. An ellipse entity is constructed as a
   - (A) series of dots.
   - (B) closed polyline.
   - (C) series of short polyline segments.
   - (D) Both B and C.
   - (E) series of points.

2-7. The special character string that causes text to be overscored is
   - (A) #O
   - (B) **O
   - (C) ##O
   - (D) %%O
   - (E) %O

2-8. The text option that prompts the user for two endpoints and then automatically computes the text height so the text is positioned between the two points is
   - (A) Aligned
   - (B) Fully Centered
   - (C) Right Justified
   - (D) Aligned
   - (E) Fit

2-9. The text option that prompts the user for two endpoints and the text height so the text is positioned between the two points is
   - (A) Centered
   - (B) Fully Centered
   - (C) Right Justified
   - (D) Aligned
   - (E) Fit

2-10. The "C" option of the LINE command stands for
   - (A) Close
   - (B) Continue
   - (C) Create
   - (D) Cling
   - (E) None of the above.

2-11. The following are all options of the PLINE command except
   - (A) Undo
   - (B) Arc
   - (C) Width
   - (D) Halfwidth
   - (E) Ltype

2-12. When using the Arc command with the "S,C,A" option, the letter "C" stands for
   - (A) circle.
   - (B) circumference.
   - (C) chord.
   - (D) center.
   - (E) None of the above.

2-13. The special character string that causes text to take on the degree symbol is
(A) #D
(B) %D
(C) ##D
(D) **D
(E) %%D

2-14. The special character string that causes text to take on the diameter symbol is
(A) %%C
(B) #C
(C) ##C
(D) **C
(E) %C

2-15. The special character string that causes text to take on the plus and minus tolerance symbol is
(A) #P
(B) %%P
(C) ##P
(D) **P
(E) %P

2-16. The Draw command that prompts the user for "First corner:" and "Other corner:" is
(A) Pline
(B) Zoom-Window
(C) Rectang
(D) Stretch-Crossing
(E) Both B and D.

2-17. Absolute coordinates are measured from the
(A) last point entered.
(B) last blip displayed.
(C) lower left corner of the screen.
(D) lower left corner as set by the LIMITS command.
(E) origin (0,0).

2-18. The option of the Line command that returns to the previous point entered, deletes the line segment created, and remains in the Line command is
(A) Close
(B) Redo
(C) Erase
(D) Continue
(E) Undo

2-19. In the figure at the right, the absolute coordinates of Point "B" are
(A) 5.00,6.00
(B) 6.00,5.00
(C) 10.00,12.00
(D) 12.00,10.00
(E) -10.00,12.00

2-20. In the figure at the right, the polar coordinates of Point "E" from Point "F" are
(A) @2.00<360
(B) @2.00<0
(C) @2.00<90
(D) @2.00<180
(E) @2.00<270

2-21. In the figure at the right, the relative coordinates of Point "C" from Point "D" are
(A) @-7.00,1.00
(B) @1.00,7.00
(C) @-7.00,-1.00
(D) @7.00,-1.00
(E) 7.00,1.00

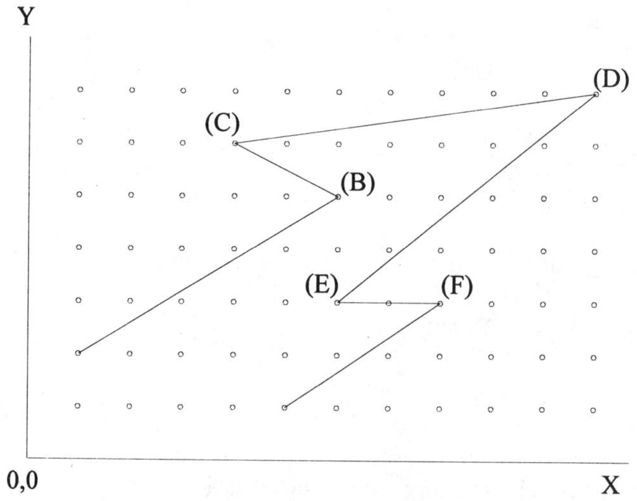

Grid Spacing = 1 Unit

2-22. In the figure at the right, the polar coordinates of Point "C" from Point "D" are
   (A) @2.00<360
   (B) @2.00<270
   (C) @2.00<180
   (D) @2.00<90
   (E) @2.00<0

2-23. In the figure at the right, the absolute coordinates of the center of circle "H" are
   (A) 8.00,3.00
   (B) 3.00,8.00
   (C) 16.00,6.00
   (D) 6.00,16.00
   (E) 14.00,4.00

2-24. In the figure at the right, the relative coordinates of Point "B" from Point "F" are
   (A) @8.00,-5.00
   (B) @-5.00,8.00
   (C) 8.00,-5.00
   (D) @-10.00,16.00
   (E) @16.00,-10.00

2-25. In the figure at the right, the polar coordinates of Point "E" from Point "D" are
   (A) @8.00<360
   (B) @8.00<0
   (C) @8.00<90
   (D) @8.00<180
   (E) @8.00<270

2-26. In the figure at the right, the polar coordinates of Point "A" from Point "H" are
   (A) @24.00<360
   (B) @24.00<0
   (C) @24.00<90
   (D) @24.00<180
   (E) @24.00<270

2-27. In the figure at the right, the absolute coordinates of the center of circle "I" are
   (A) 20.00,12.00
   (B) 12.00,20.00
   (C) 5.00,3.00
   (D) 3.00,5.00
   (E) 8.00,16.00

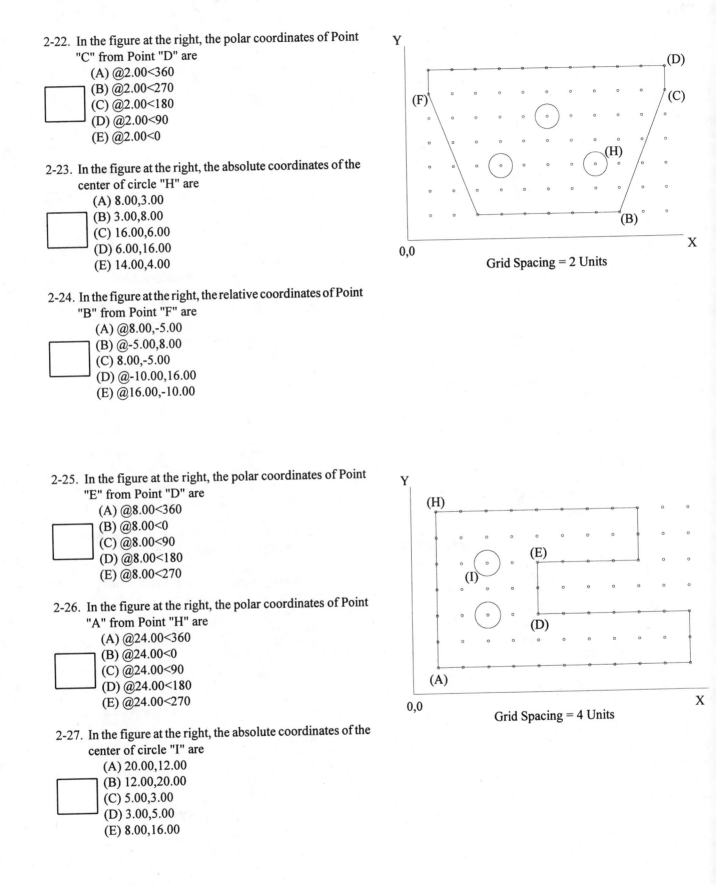

Grid Spacing = 2 Units

Grid Spacing = 4 Units

Block Command Questions

2-28. The Insert command can be used to
  (A) delete a block.
  (B) create a block.
  (C) save a block.
  (D) insert any drawing file.
  (E) redefine a block's insertion point.

2-29. When using the Wblock command to create a new block, one should respond to the block name prompt
  (A) with a block name.
  (B) with an equal sign.
  (C) with an asterisk.
  (D) by striking the Enter key to continue.
  (E) with a file name.

2-30. A Wblock is
  (A) a drawing file.
  (B) a group of nested blocks.
  (C) a symbol library.
  (D) an entity file.
  (E) a block created with the window option.

2-31. A new drawing file may be created from an existing block by using the
  (A) Wblock command.
  (B) Blocke command.
  (C) Block= command.
  (D) Writeblk command.
  (E) Insert command.

2-32. To insert a block called "PART" and have the block converted into individual entities, type
  (A) "/PART" for the block name.
  (B) "*PART" for the block name.
  (C) "?PART" for the block name.
  (D) "PART" for the block name.
  (E) "/PART" for the block name.

2-33. The command used to write all or just part of a drawing out to a new drawing file is
  (A) Wblock
  (B) Block
  (C) Dxfout
  (D) File
  (E) Blockout

2-34. The command used to return a block back to its original entities is
  (A) Explode
  (B) Break
  (C) Change
  (D) Undo
  (E) Remake

2-35. Blocks inserted using the Minsert command
  (A) may be exploded.
  (B) may not be exploded.
  (C) can be arranged by rows and columns similar to the Array command.
  (D) can be arranged in a circular pattern similar to the Array command.
  (E) Both B and C.

2-36. Blocks may be brought into a drawing using
  (A) Insert
  (B) Minsert
  (C) Join
  (D) All of the above.
  (E) Only A and B.

2-37. All of the following are valid modes of the DDINSERT dialog box except
  (A) selecting the name of a block.
  (B) selecting the file name of a wblock.
  (C) exploding the block.
  (D) specifying the insertion point of the block.
  (E) specifying the layer name of the block.

2-38. The command used to group selected entities into one entity with a distinct name is
  (A) Merge
  (B) Block
  (C) Save
  (D) Symbol
  (E) Insert

2-39. The command used to merge two drawings into one is
  (A) Insert
  (B) Merge
  (C) Join
  (D) Block
  (E) Wblock

Edit Command Questions

2-40. The command that allows you to make multiple copies of existing objects in a rectangular or circular patterns is called
(A) Repeat
(B) Array
(C) Minsert
(D) Multiple
(E) Ditto

2-41. The command used to modify the linetype of an entity is called
(A) Edit
(B) Change
(C) Linetype
(D) Ltype
(E) Line

2-42. For selecting multiple entities to trim, the selection set option to use is
(A) Window
(B) Crossing
(C) Window Polygon
(D) Fence
(E) Crossing Polygon

2-43. The command used to extend a line a known distance and direction without specifying a boundary edge is
(A) Trim
(B) Extend
(C) Modify
(D) Change
(E) Relimit

2-44. The command that prompts the user for a "Base point" is
(A) Osnap
(B) Change
(C) Trim
(D) Rotate
(E) Extend

2-45. The object selection mode most commonly used by the Stretch command is
(A) Last
(B) Window
(C) Crossing
(D) Box
(E) Single

2-46. All of the following entities can be filleted or chamfered except
(A) Blocks
(B) Arcs
(C) Rectangles
(D) Lines
(E) Polylines

2-47. All of the following text properties can be modified using the Change command except
(A) Insertion point
(B) Text height
(C) Rotation angle
(D) Text style
(E) Width factor

2-48. When breaking a circle, the break will occur in the
(A) clockwise direction.
(B) counterclockwise direction.
(C) horizontal direction.
(D) Only A and B.
(E) None of the above.

2-49. The following are valid prompts for a Polar Array except
(A) +=CCW
(B) center point of array
(C) distance between columns
(D) -=CW
(E) number of items

2-50. The Divide command
(A) divides an entity into equal parts.
(B) places markers along the entity at the divide points.
(C) uses the current point style set by Pdmmode.
(D) uses the current point size set by Pdsize.
(E) All of the above.

2-51. All of the following are valid settings of the DDGRIPS dialog box except
(A) Grip Modes
(B) Grips Enabled/Disabled
(C) Grip Colors
(D) Enabling Grips within Blocks
(E) Grip Size

2-52. All of the following are valid grip modes except
(A) Copy
(B) Stretch
(C) Rotate
(D) Break
(E) Move

Problems for Unit 2

Directions for Problems 2–53 through 2–58:
Supply the appropriate absolute, relative, and/or polar coordinates for these figures in the matrix below each object.

Problem 2–53

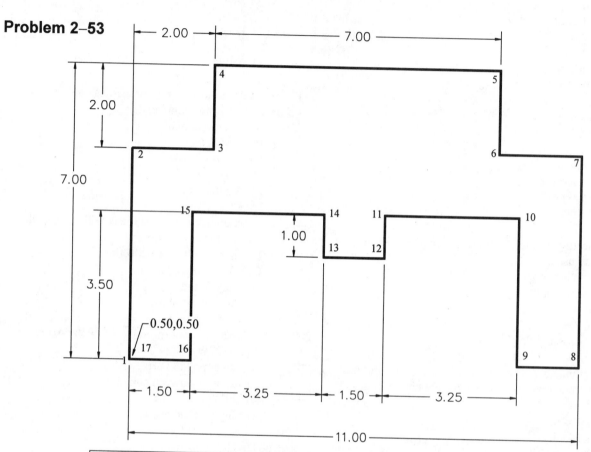

	Absolute	Relative	Polar
From Pt (1)	0.50,0.50	0.50,0.50	0.50,0.50
To Pt (2)			
To Pt (3)			
To Pt (4)			
To Pt (5)			
To Pt (6)			
To Pt (7)			
To Pt (8)			
To Pt (9)			
To Pt (10)			
To Pt (11)			
To Pt (12)			
To Pt (13)			
To Pt (14)			
To Pt (15)			
To Pt (16)			
To Pt (17)			
To Pt	Enter	Enter	Enter

Problem 2–54

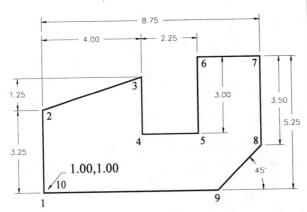

	Absolute	Relative
From Pt (1)	1.00,1.00	1.00,1.00
To Pt (2)		
To Pt (3)		
To Pt (4)		
To Pt (5)		
To Pt (6)		
To Pt (7)		
To Pt (8)		
To Pt (9)		
To Pt (10)		
To Pt	Enter	Enter

Problem 2–55

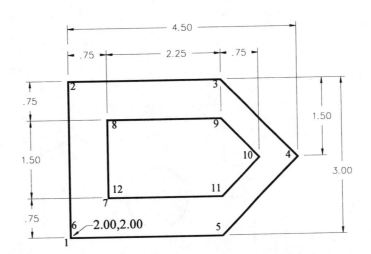

	Absolute	Relative
From Pt (1)	2.00,2.00	2.00,2.00
To Pt (2)		
To Pt (3)		
To Pt (4)		
To Pt (5)		
To Pt (6)		
To Pt		
From Pt (7)		
To Pt (8)		
To Pt (9)		
To Pt (10)		
To Pt (11)		
To Pt (12)		
To Pt	Enter	Enter

Problem 2-56

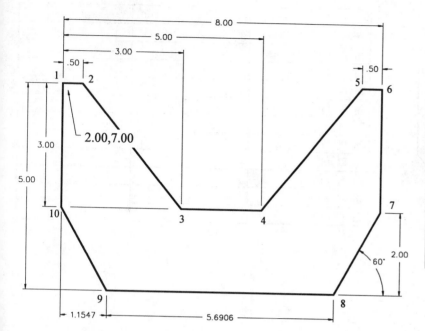

	Absolute
From Pt (1)	2.00,7.00
To Pt (2)	
To Pt (3)	
To Pt (4)	
To Pt (5)	
To Pt (6)	
To Pt (8)	
To Pt (7)	
To Pt (9)	
To Pt (10)	
To Pt	Enter

Problem 2-57

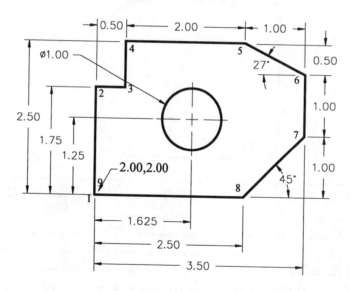

	Absolute	Relative
From Pt (1)	2.00,2.00	2.00,2.00
To Pt (2)		
To Pt (3)		
To Pt (4)		
To Pt (5)		
To Pt (6)		
To Pt (7)		
To Pt (8)		
To Pt (9)		
To Pt	Enter	Enter
Center Pt (10)		

Problem 2–58

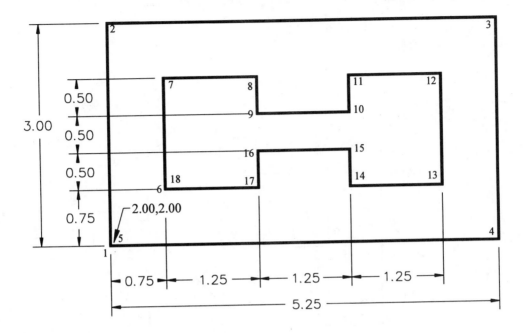

	Absolute	Relative	Polar
From Pt (1)	2.00,2.00	2.00,2.00	2.00,2.00
To Pt (2)			
To Pt (3)			
To Pt (4)			
To Pt (5)			
To Pt	Enter	Enter	Enter
From Pt (6)			
To Pt (7)			
To Pt (8)			
To Pt (9)			
To Pt (10)			
To Pt (11)			
To Pt (12)			
To Pt (13)			
To Pt (14)			
To Pt (15)			
To Pt (16)			
To Pt (17)			
To Pt (18)			
To Pt	Enter	Enter	Enter

Directions for Problems 2-59 through 2-70:
Construct one-view drawings of the following figures using existing AutoCAD commands.

Problem 2-59

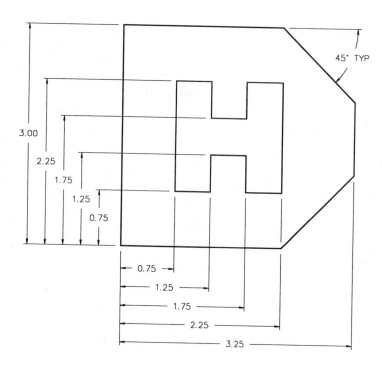

Problem 2-60

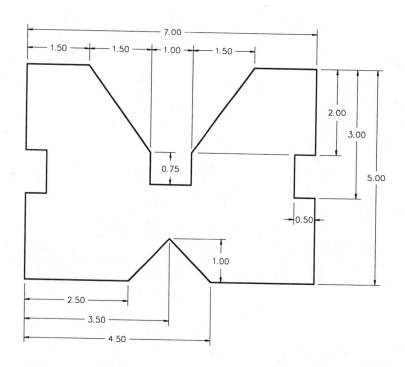

Problem 2-61

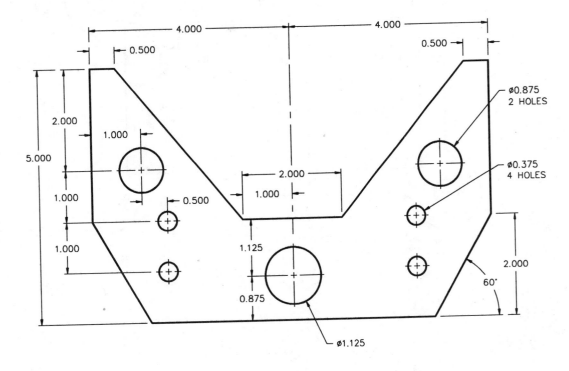

Problem 2-62

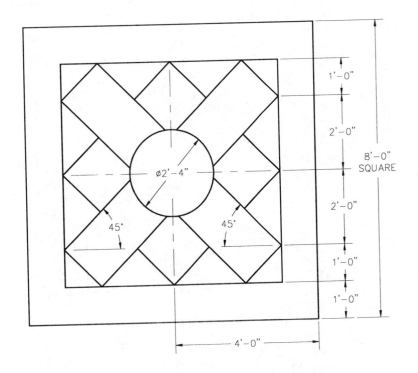

Problem 2-63

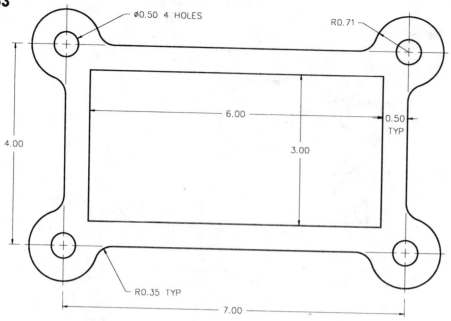

Ø0.50 4 HOLES

R0.71

6.00

0.50 TYP

4.00

3.00

R0.35 TYP

7.00

0.125 GASKET THICKNESS

Problem 2-64

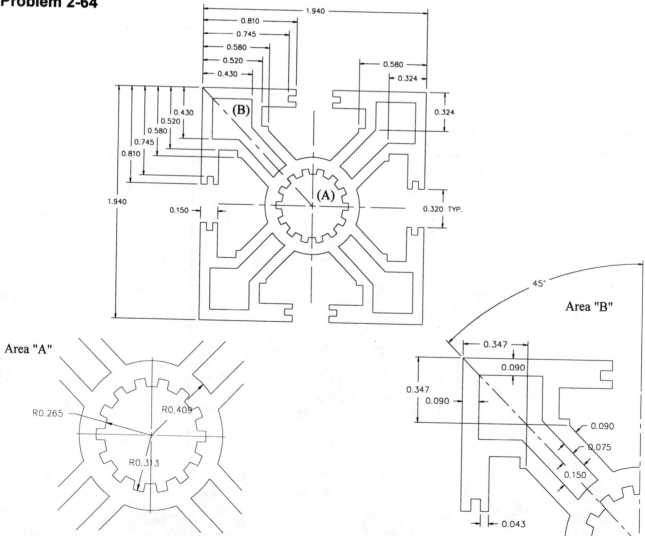

1.940

0.810

0.745

0.580

0.520

0.430

0.580

0.324

0.430

(B)

0.520

0.580

0.745

0.810

0.324

1.940

0.150

(A)

0.320 TYP.

45°

Area "B"

0.347

0.090

0.347

0.090

0.090

0.075

0.150

0.043

Area "A"

R0.265

R0.409

R0.313

Problem 2–65

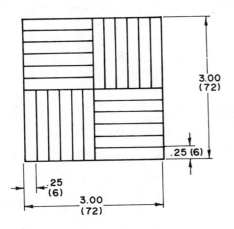

Problem 2–66

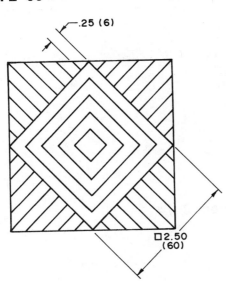

Problem 2–67

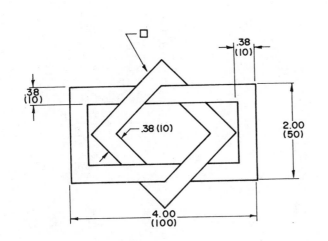

Problem 2–68

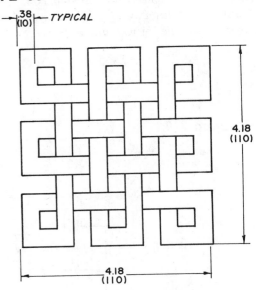

Problem 2–69

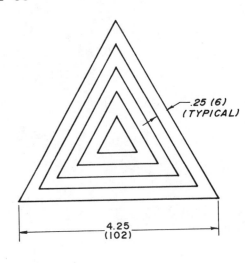

Problem 2–70

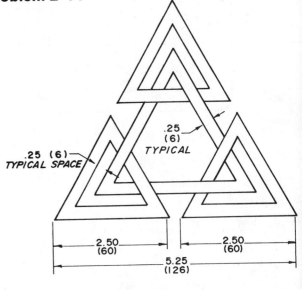

Directions for Problems 2-71 through 2-74:

Construct one-view drawings of the following figures using a grid spacing of 0.25 units. Do not dimension these drawings unless otherwise specified by your instructor.

Problem 2–71

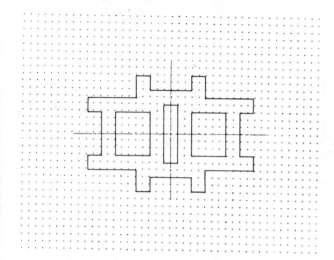

Problem 2–73

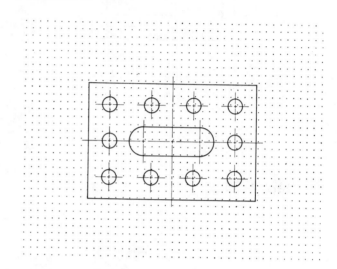

Problem 2–72

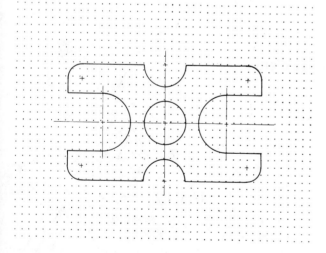

Problem 2–74

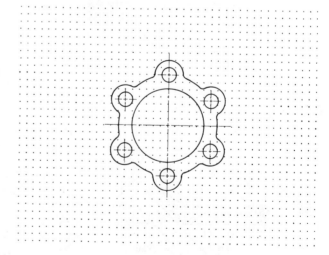

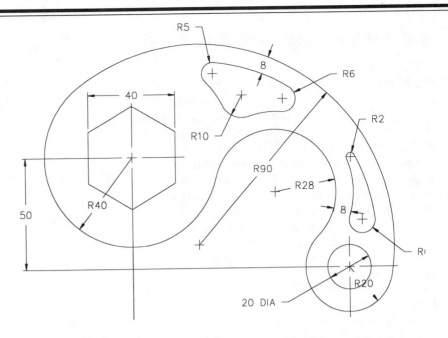

UNIT 3

Geometric Constructions

Contents

Many of the elements that go into the creation of a design revolve around geometric shapes. Applying and manipulating these shapes is the next step toward a successful design. Important manual drafting tools to assist in the construction of geometric shapes range from the T-square and drafting machine for drawing parallel and perpendicular lines to the compass and dividers for drawing circles and arcs and for setting off distances. The computer offers the designer superior control and accuracy when dealing with geometric constructions. Much of the focus of this unit will be on how to manipulate the numerous object snap modes supplied by AutoCAD through construction examples and practical applications in the form of a tutorial. Numerous other problems are supplied at the end of this chapter to challenge the designer for a complete and correct solution.

Bisecting Lines and Arcs

Illustrated at the right is an arc intersected by a line. The purpose of this problem is to locate the midpoint of the line and the arc. The Divide command will be used to accomplish this. However, if the command is used to find the midpoint, it may not be visible. This is because the Divide command places a point depending on the amount of divisions asked for. The appearance of the point is controlled by the system variable Pdmode. There is also a system variable to control the size of the point called Pdsize. Follow the steps below to change the point appearance and size before using the Divide command.

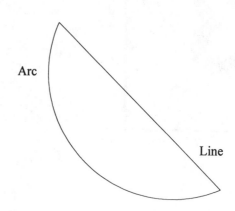

Command: **Setvar**
Variable name or ?: **Pdmode**
New value for pdmode <0>: **2**

Command: **Setvar**
Variable name or ? <PDMODE>: **Pdsize**
New value for pdsize <0>: **.25**

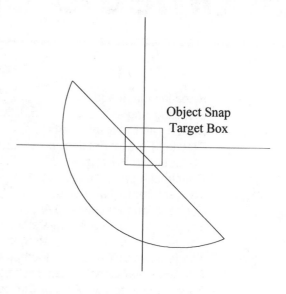

Next, use the Point command along with the Osnap-Midpoint option to locate the midpoint of the line and arc using the following steps:

Command: **Point**
Point: **Mid**
of *(Select the line or arc at any convenient location)*

The illustration at the right shows the midpoint locations of the line and arc. The current point is controlled by the system variable Pdmode. The style of point, namely the "plus," reflects the current Pdmode value of 2.

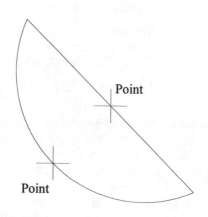

Bisecting Angles

Illustrated at the right are two lines forming an acute angle. (An acute angle measures any value less than 90 degrees.) The purpose of this problem is to bisect or divide the angle equally in two parts. A construction line will be drawn from the endpoints of the angle. Then, a line will be drawn from the vertex of the angle to the midpoint of the construction line using the Line command and the Osnap-Midpoint option.

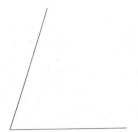

First, draw the construction line.

Command: **Line**
From point: **Endp**
of *(Select the endpoint of the line at "A")*
To point: **Endp**
of *(Select the endpoint of the line at "B")*
To point: *(Strike Enter to exit this command)*

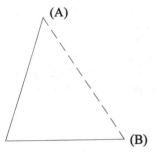

Next, use the Line command to draw a line from the vertex of the angle to the midpoint of the construction line. Use the Osnap-Endpoint and Osnap-Midpoint options to assist you in this operation.

Command: **Line**
From point: **Endp**
of *(Select the endpoint of the line at "A")*
To point: **Mid**
of *(Select the midpoint of the construction line at "B")*
To point: *(Strike Enter to exit this command)*

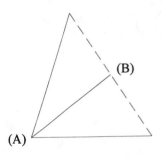

Use the Erase command to delete the construction line, leaving the angle bisected or divided into two equal angles.

Dividing an Entity into Parts

Illustrated at the right is an inclined line. The purpose of this problem is to divide the line into an equal number of parts. This proved to be a tedious task using manual drafting methods but thanks to the AutoCAD Divide command, this operation is much easier to perform. The Divide command instructs the user to supply the number of divisions and performs the division by placing a point along the entity to be divided. The size and shape are controlled by the system variables Pdsize and Pdmode, respectively. Be sure the Pdmode variable is set to a value that will produce a visible point. Otherwise, the results of the Divide command will not be obvious.

Command: **Setvar**
Variable name or ?: **Pdmode**
New value for pdmode <0>: **2**

Command: **Setvar**
Variable name or ? <PDMODE>: **Pdsize**
New value for pdsize <0>: **.25**

Next, use the Divide command, select the inclined line as the entity to divide, enter a value for the number of segments, and the command divides the entity by a series of points.

Command: **Divide**
Select object to divide: *(Select the inclined line)*
<Number of segments>/Block: **9**

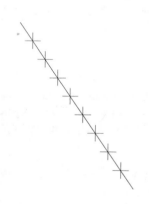

A practical application of the Divide command may be in the area of screw threads where a number of threads per inch is needed to form the profile of the thread.

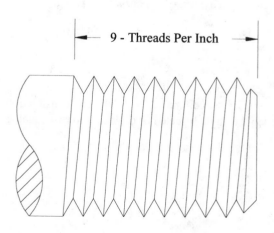

9 - Threads Per Inch

Drawing Parallel Entities

Another fundamentally important operation regarding geometric constructions is the ability to construct entities parallel to each other. Illustrated at the right is an inclined line. The purpose of this problem is to create a matching entity parallel to the original line at a set distance. The Offset command is used to accomplish this.

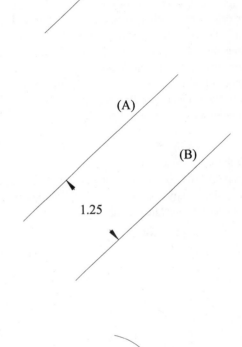

Use the Offset command, set the distance to offset, and select the entity to offset and the side of the offset. The result will be similar to the illustration at the right.

Command: **Offset**
Offset distance or Through<Through>: **1.25**
Select object to offset: *(Select the line at "A")*
Side to offset? *(Select anywhere near "B")*
Select object to offset: *(Strike Enter to exit this command)*

The Offset command will also produce concentric circles or arcs. Illustrated at the right is an arc. The purpose of this problem is to create an additional arc at a distance of 1.25 units away from the original arc.

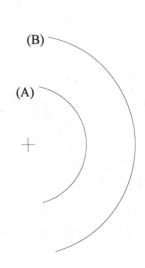

Use the Offset command, set the distance to offset, select the entity (this time the arc) to offset, and the side of the offset. The result will be similar to the illustration at the right.

Command: **Offset**
Offset distance or Through<Through>: **1.25**
Select object to offset: *(Select the arc at "A")*
Side to offset? *(Select anywhere near "B")*
Select object to offset: *(Strike Enter to exit this command)*

Constructing Hexagons

The Hexagon is an important geometric shape commonly used for such items as the plan view of a bolt or screw type of fastener. The illustrations at the right show two types of hexagons; one drawn in relation to its flat edges, and the other drawn in relation to its corners. The Polygon command is used for drawing either example. Simply supply the number of the sides and whether the figure is inscribed or circumscribed about a circle and the radius of the circle and a polygon is drawn. One interesting characteristic of polygons is that they are constructed using a series of polylines, making the polygon one entity.

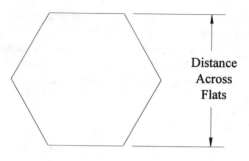

Distance Across Flats

One other note concerning polygons is that the examples on this page illustrate constructing a hexagon. Using the Polygon command, any size figure with up to 1024 sides may be constructed; not just hexagons.

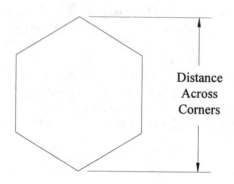

Distance Across Corners

Illustrated at the right is an example of an inscribed hexagon, or a figure constructed inside of a circle. The Polygon command is used to create this type of shape.

Command: **Polygon**
Number of sides: **6**
Edge/<Center of Polygon>: *(Select the center at "A")*
Inscribed in circle/Circumscribed about circle (I/C): **I**
Radius of circle: *(Enter a numerical value)*

Inscribed Polygon

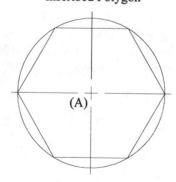

Yet another use of the Polygon is illustrated at the right when the figure needs to be drawn outside of a circle. The Circumscribed option would be used for this example.

Command: **Polygon**
Number of sides: **6**
Edge/<Center of Polygon>: *(Select the center at "A")*
Inscribed in circle/Circumscribed about circle (I/C): **C**
Radius of circle: *(Enter a numerical value)*

Circumscribed Polygon

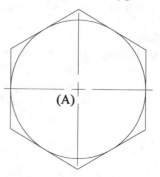

Constructing an Ellipse

Main parts of an ellipse are the major diameter or axis and minor diameter or axis illustrated at the right. Numerous construction arcs and lines were needed to construct the ellipse using manual methods. The Ellipse command prompts the user for the center of the ellipse, the endpoint of one axis, and the endpoint of the other axis. The ellipse is drawn as a series of polylines representing one entity.

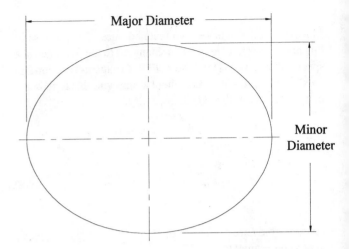

In the example at the right, a polar coordinate can be used to identify both axes of the ellipse following the prompts below.

Command: **Ellipse**
<Axis endpoint 1>/ Center: **C**
Center of ellipse: *(Select a point at "A")*
Axis endpoint: **@4<0** *(Toward a point at "B")*
<Other axis distance>/ Rotation: **@3<0** *(Toward a point at "C")*

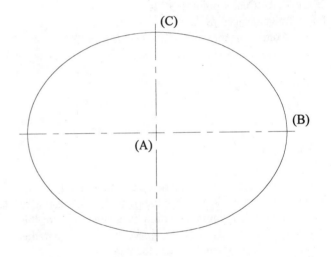

The object at the right is an example of how to outline the view with an ellipse at "A" and how to use the Offset command to offset the ellipse in the direction inside of the object at "B."

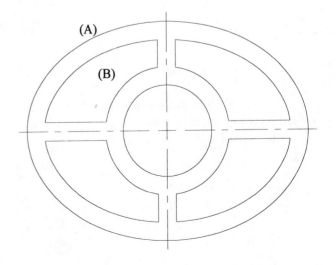

Tangent Arc to Two Lines

Illustrated at the right are two inclined lines. The purpose of this problem is to connect an arc tangent to the two lines at a specified radius. The Circle-TTR (Tangent-Tangent-Radius) command will be used here along with the Trim command to clean up the excess geometry.

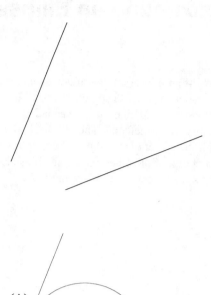

First, use the Circle-TTR command to construct an arc tangent to both lines.

Command: **Circle**
3P/2P/TTR/<Center point>: **TTR**
Enter Tangent spec: *(Select the line at "A")*
Enter second Tangent spec: *(Select the line at "B")*
Radius: *(Enter a desired radius value)*

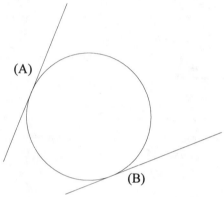

Use the Trim command to clean up the lines and arc. The completed result is illustrated at the right. It is interesting to note that the Fillet command could have been used for this procedure. Not only will the curve be drawn, but this command will automatically trim the lines.

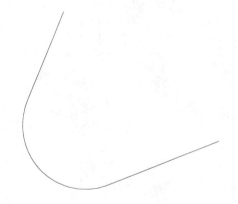

The object at the right is an example of a typical application where this procedure might be used.

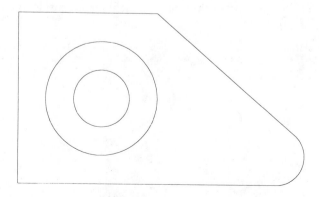

Tangent Arc to a Line and Arc

Illustrated at the right is an arc and an inclined line. The purpose of this problem is to connect an additional arc tangent to the original arc and line at a specified radius. The Circle-TTR command will be used here along with the Trim command to clean up the excess geometry.

First, use the Circle-TTR command to construct an arc tangent to the arc and inclined line.

Command: **Circle**
3P/2P/TTR/<Center point>: **TTR**
Enter Tangent spec: *(Select the arc at "A")*
Enter second Tangent spec: *(Select the line at "B")*
Radius: *(Enter a desired radius value)*

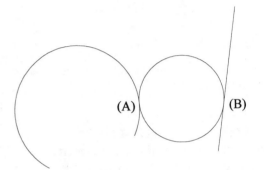

(A)   (B)

Use the Trim command to clean up the arc and line. The completed result is illustrated at the right.

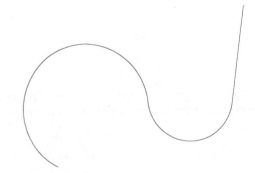

The object at the right is an example of a typical application where this procedure might be used.

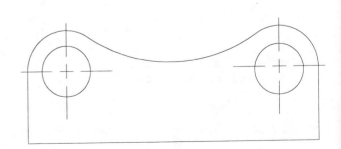

Tangent Arc to Two Arcs - Method #1

Illustrated at the right are two arcs. The purpose of this problem is to connect a third arc tangent to the original two at a specified radius. The Circle-TTR command will be used here along with the Trim command to clean up the excess geometry.

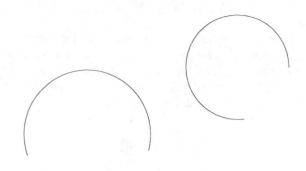

Use the Circle-TTR command to construct an arc tangent to the two original arcs.

Command: **Circle**
3P/2P/TTR/<Center point>: **TTR**
Enter Tangent spec: *(Select the first arc at "A")*
Enter second Tangent spec: *(Select the second arc at "B")*
Radius: *(Enter a desired value)*

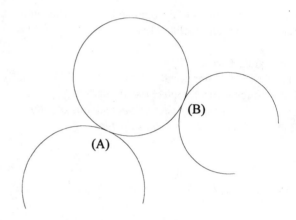

Use the Trim command to clean up the two arcs using the circle as a cutting edge. The completed result is illustrated at the right.

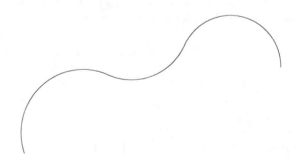

The object at the right is an example of a typical application where this procedure might be used.

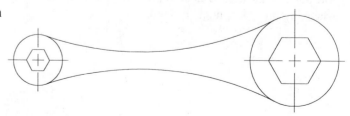

Tangent Arc to Two Arcs - Method #2

Illustrated at the right are two arcs. The purpose of this problem is to connect an additional arc tangent to and enclosing both arcs at a specified radius. The Circle-TTR command will be used here along with the Trim command.

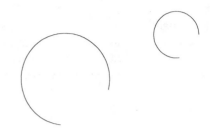

First, use the Circle-TTR command to construct an arc tangent to and enclosing both arcs.

Command: **Circle**
3P/2P/TTR/<Center point>: **TTR**
Enter Tangent spec: *(Select the arc at "A")*
Enter second Tangent spec: *(Select the arc at "B")*
Radius: *(Enter a desired radius value)*

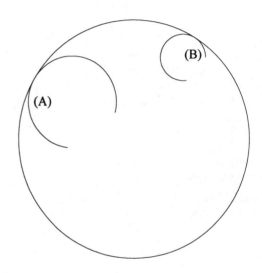

Use the Trim command to clean up all arcs. The completed result is illustrated at the right.

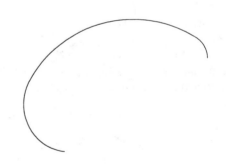

The object at the right is an example of a typical application where this procedure might be used.

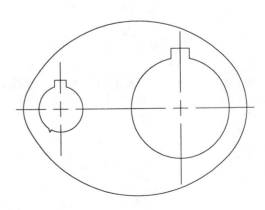

Tangent Arc to Two Arcs - Method #3

Illustrated at the right are two arcs. The purpose of this problem is to connect an additional arc tangent to one arc and enclosing the other. The Circle-TTR command will be used here along with the Trim command to clean up unnecessary geometry.

First, use the Circle-TTR command to construct an arc tangent to the two arcs. Study the illustration at the right and the prompts below to understand the proper pick points for this operation.

Command: **Circle**
3P/2P/TTR/<Center point>: **TTR**
Enter Tangent spec: *(Select the arc at "A")*
Enter second Tangent spec: *(Select the line at "B")*
Radius: *(Enter a desired radius value)*

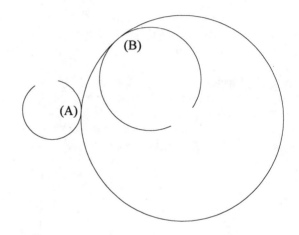

Use the Trim command to clean up the arcs. The completed result is illustrated at the right.

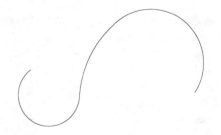

The object at the right is an example of a typical application where this procedure might be used.

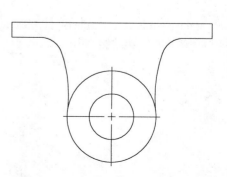

Perpendicular Construction Techniques

Illustrated at the right is an inclined line and a point. The purpose of this problem is to construct a line from a point perpendicular to another line. The Line command will be used for this operation in addition to the Osnap-Node and Osnap-Perpend options.

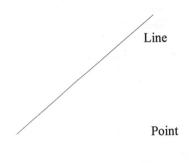

Use the Line command and the Osnap-Node option to snap to the point illustrated at the right.

Command: **Line**
From point: **Node**
of *(Select the point at "A")*

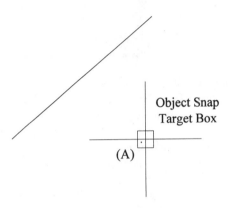

Continue the Line command and respond to the prompt, "To point" by using the Osnap-Perpend option and selecting anywhere along the inclined line.

To point: **Perpend**
of *(Select the line at "A")*
To point: *(Strike Enter to exit this command)*

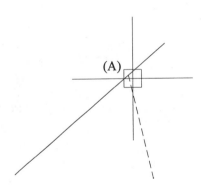

The completed solution is illustrated at the right.

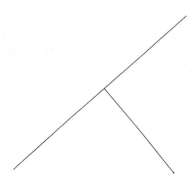

Tangent Line to Two Arcs or Circles

Illustrated at the right are two circles. The purpose of this problem is to connect the two circles with two tangent lines. This can be accomplished using the Line command and the Osnap-Tangent option.

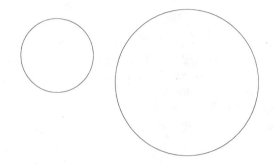

Use the Line command to connect two lines tangent to the circles. The following procedure is used for the first line. Use the same procedure for the second.

Command: **Line**
From point: **Tan**
to *(Select the circle near "A")*
To point: **Tan**
to *(Select the circle near "B")*
To point: *(Strike Enter to exit this command)*

When using the Tangent option, the rubberband cursor is not present when drawing the beginning of the line. This is due to calculations required when identifying the second point.

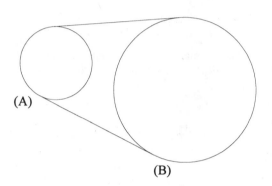

(A)

(B)

Use the Trim command to clean up the circles so the appearance of the object is similar to the illustration at the right.

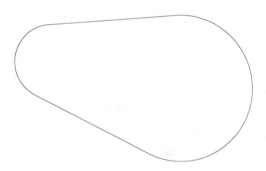

The object at the right is an example of a typical application where drawing lines tangent to circles might be used.

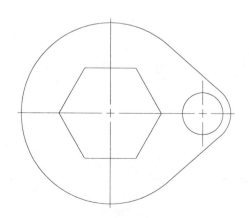

Quadrant vs Tangent Osnap Option

Various examples have been given on previous pages concerning drawing lines tangent to two circles, two arcs, or any combination of the two. The object at the right illustrates the use of the Osnap-Tangent option when used along with the Line command.

Command: **Line**
From point: **Tan**
to *(Select the arc at "A")*
To point: **Tan**
to *(Select the arc at "B")*
To point: *(Strike Enter to exit this command)*

Note that the angle of the line formed by points "A" and "B" is neither horizontal or vertical. The object at the right is a typical example of the capabilities of the Osnap option.

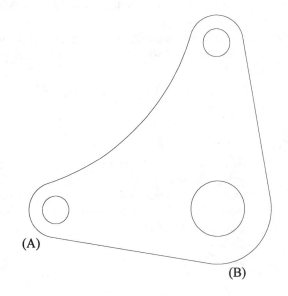

(A)

(B)

The object illustrated at the right is a modification of the drawing above with the inclined tangent lines changing to horizontal and vertical tangent lines. This example is cited to warn the user that two Osnap options are available to perform tangencies, namely Osnap-Tangent and Osnap-Quadrant. However, it is up to the user to evaluate under what conditions the Osnap options are to be used. At the right, the Osnap-Tangent or Osnap-Quadrant option could be used to draw the lines tangent to the arcs. The Quadrant option could be used only since the lines to be drawn are perfectly horizontal or vertical. Usually it is impossible to know this ahead of time, and in this case, the Osnap-Tangent option should be used whenever possible.

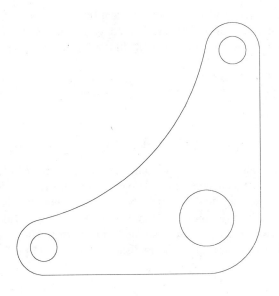

The slot at the right is an excellent example of when to use the Osnap-Quadrant option. Slots have two semi-circles connected together by two horizontal or vertical lines enabling the user to use the Quadrant option. If the slot is positioned at an odd angle, simply construct it and use the Rotate command to position it.

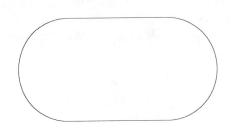

Ogee or Reverse Curve Construction

An ogee curve connects two parallel lines with a smooth flowing curve that reverses itself in symmetrical form. To begin constructing an ogee curve to line segments "AB" and "CD," first draw line "BC," which connects both parallel line segments.

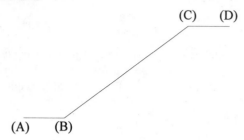

Use the Divide command to divide line segment "BC" into four equal parts. Construct vertical lines from "B" and "C." Complete this step by constructing line segment "XY," which is perpendicular to line "BC." Do not worry about where line "XY" is located at this time.

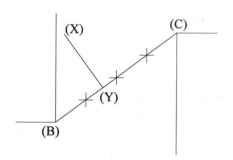

Move line "XY" to the location identified by the point at the right. Complete this step by copying line "XY" to the location identified by point "Z" illustrated at the right.

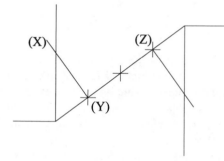

Construct two circles with centers located at points "X" and "Y" illustrated at the right. Use the Osnap-Intersec mode to accurately locate the centers. Note: if an intersection is not found from the previous step, use the Extend command to find the intersection and continue with this step. The radii of both circles are equivalent to distances "XB" and "YC."

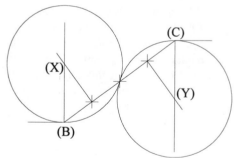

Use the Trim comand to trim away any excess arc segments to form the ogee curve. This forms the frame of the ogee for the construction of objects such as the wrench illustrated at the right.

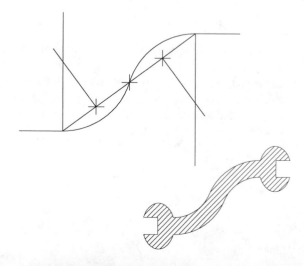

Tutorial Exercise #8
Pattern1.Dwg

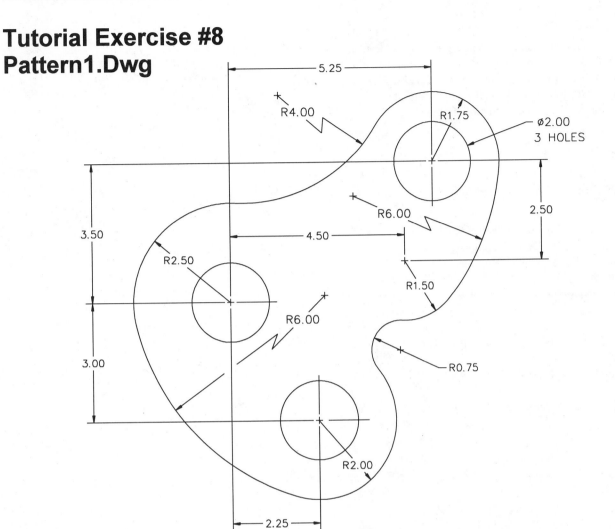

PURPOSE:
This tutorial is designed to use geometric commands to construct a one-view drawing of Pattern1. Refer to the following for special system settings and suggested command sequences.

SYSTEM SETTINGS:
Begin a new drawing called "Pattern1." Use the Units command to change the number of decimal places past the zero from 4 to 2. Keep the remaining default unit values. Using the Limits command, keep 0,0 for the lower left corner and change the upper right corner from 12,9 to 21.00,16.00. Use the Grid command and change the grid spacing from 1.00 to 0.50 units. Do not turn the snap or ortho On.

LAYERS:
Create the following layers with the format:

Name-Color-Linetype
Object - White - Continuous
Center - Yellow - Center
Dim - Yellow - Continuous

SUGGESTED COMMANDS:
Begin constructing this object by first laying out four points which will be used as centers for circles. Use the Circle-TTR command to construct tangent arcs to the circles already drawn. Use the Trim command to clean up and partially delete circles to obtain the outline of the pattern. Then, add the 2.00 diameter holes followed by the center markers using the Dim-Cen command.

DIMENSIONING:
This drawing may be dimensioned at a later date. Consult your instructor before continuing.

PLOTTING:
This tutorial exercise may be plotted on "C"-size paper (18" x 24"). Plot Pattern1 at a scale of full size, or 1=1.

Step #1

Use the Pdmode command to change the point style to a value of 2. This will form a "plus sign" when using the Point command. Locate one point at absolute coordinate 7.50,7.50. Then, use the Copy command and the dimensions at the right as a guide for duplicating the remaining points.

Command: **Pdmode**
New value for Pdmode <0>: **2**

Command: **Point**
Point: **7.50,7.50** *(Locates the point at "A")*

Command: **Copy**
Select objects: **L** *(This should select the point)*
Select objects: *(Strike Enter to continue)*
<Base point or displacement>/Multiple: **M**
Base point: **Node**
of *(Select the point at "A")*
Second point of displacement: **@2.25,-3.00** *(Locates point "B")*
Second point of displacement: **@4.50,1.00** *(Locates point "C")*
Second point of displacement: **@5.25,3.50** *(Locates point "D")*
Second point of displacement: *(Strike Enter to exit this command)*

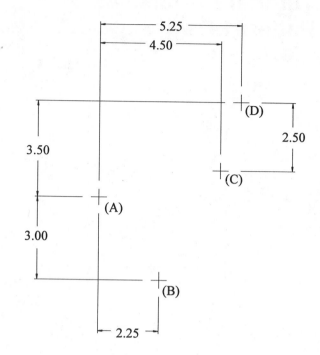

Step #2

Use the Circle command to place four circles of different sizes from points located at "A," "B," "C," and "D." When you have completed drawing the four circles, use the Erase command to erase points "A," "B," "C," and "D."

Command: **Circle**
3P/2P/TTR/<Center point>: **Node**
of *(Select the point at "A")*
Diameter/<Radius>: **2.50**

Command: **Circle**
3P/2P/TTR/<Center point>: **Node**
of *(Select the point at "B")*
Diameter/<Radius>: **2.00**

Command: **Circle**
3P/2P/TTR/<Center point>: **Node**
of *(Select the point at "C")*
Diameter/<Radius>: **1.50**

Command: **Circle**
3P/2P/TTR/<Center point>: **Node**
of *(Select the point at "D")*
Diameter/<Radius>: **1.75**

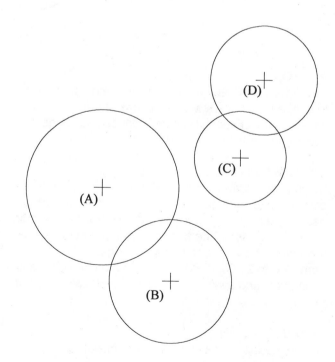

Step #3

Use the Circle-TTR command to construct a 4.00-radius circle tangent to the two dashed circles at the right. Then, use the Trim command to trim away part of circle "C."

Command: **Circle**
3P/2P/TTR/<Center point>: **TTR**
Enter Tangent spec: *(Select the dashed circle at "A")*
Enter second Tangent spec: *(Select the dashed circle at "B")*
Radius: **4.00**

Command: **Trim**
Select cutting edges...
Select objects: *(Select the two dashed circles at the right)*
Select objects: *(Strike Enter to continue)*
<Select object to trim>Undo: *(Select the large circle at "C")*
<Select object to trim>Undo: *(Strike Enter to exit this command)*

A built-in undo is provided if a mistake is made during the trimming process.

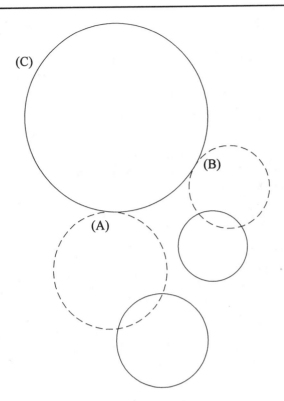

Step #4

Use the Circle-TTR command to construct a 6.00-radius circle tangent to the two dashed circles at the right. Then, use the Trim command to trim away part of circle "C."

Command: **Circle**
3P/2P/TTR/<Center point>: **TTR**
Enter Tangent spec: *(Select the dashed circle at "A")*
Enter second Tangent spec: *(Select the dashed circle at "B")*
Radius: **6.00**

Command: **Trim**
Select cutting edges...
Select objects: *(Select the two dashed circles at the right)*
Select objects: *(Strike Enter to continue)*
<Select object to trim>Undo: *(Select the large circle at "C")*
<Select object to trim>Undo: *(Strike Enter to exit this command)*

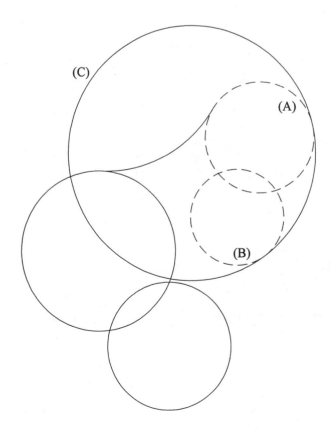

Step #5

Use the Circle-TTR command to construct a 6.00-radius circle tangent to the two dashed circles at the right. Then, use the Trim command to trim away part of circle "C."

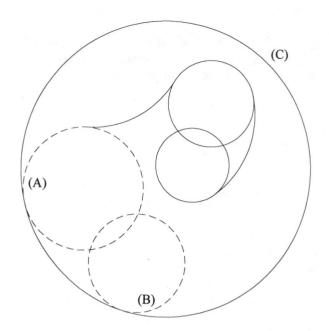

Command: **Circle**
3P/2P/TTR/<Center point>: **TTR**
Enter Tangent spec: *(Select the dashed circle at "A")*
Enter second Tangent spec: *(Select the dashed circle at "B")*
Radius: **6.00**

Command: **Trim**
Select cutting edges...
Select objects: *(Select the two dashed circles at the right)*
Select objects: *(Strike Enter to continue)*
<Select object to trim>Undo: *(Select the large circle at "C")*
<Select object to trim>Undo: *(Strike Enter to exit this command)*

Step #6

Use the Circle-TTR command to construct a 0.75-radius circle tangent to the two dashed circles at the right. Then, use the Trim command to trim away part of circle "C."

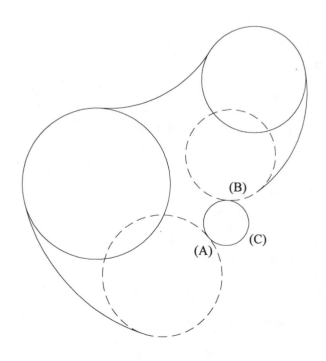

Command: **Circle**
3P/2P/TTR/<Center point>: **TTR**
Enter Tangent spec: *(Select the dashed circle at "A")*
Enter second Tangent spec: *(Select the dashed circle at "B")*
Radius: **0.75**

Command: **Trim**
Select cutting edges...
Select objects: *(Select the two dashed circles at the right)*
Select objects: *(Strike Enter to continue)*
<Select object to trim>Undo: *(Select the circle at "C")*
<Select object to trim>Undo: *(Strike Enter to exit this command)*

Step #7

Use the Trim command, select all dashed arcs at the right as cutting edges, and trim away the circular segments to form the outline of the pattern1 drawing.

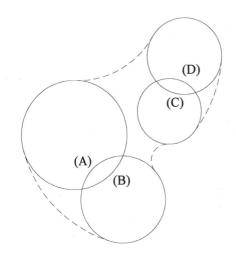

Command: **Trim**
Select cutting edges...
Select objects: *(Select the four dashed arcs at the right)*
Select objects: *(Strike Enter to continue)*
<Select object to trim>Undo: *(Select the circle at "A")*
<Select object to trim>Undo: *(Select the circle at "B")*
<Select object to trim>Undo: *(Select the circle at "C")*
<Select object to trim>Undo: *(Select the circle at "D")*
<Select object to trim>Undo: *(Strike Enter to exit this command)*

Step #8

Your drawing should be similar to the illustration at the right.

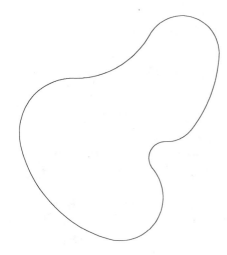

Step #9

Use the Circle command to place a circle of 2.00-unit diameter at the center of arc "A". Then, use the Copy command to duplicate the circle at the center of arcs "B" and "C".

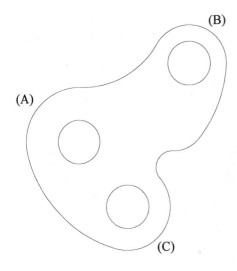

Command: **Circle**
3P/2P/TTR/<Center point>: **Cen**
of *(Select the arc at "A")*
Diameter/<Radius>: **D**
Diameter: **2.00**

Command: **Copy**
Select objects: **L**
Select objects: *(Strike Enter to continue)*
<Base point or displacement>/Multiple: **M**
Base point: **Cen**
of *(Select the arc at "A")*
Second point of displacement: **Cen**
of *(Select the arc at "B")*
Second point of displacement: **Cen**
of *(Select the arc at "C")*
Second point of displacement: *(Strike Enter to exit this command)*

Step #10

Use the Layer command to set the new current layer to Center. Enter the dimensioning area of AutoCAD and change the variable Dimcen from a value of 0.09 to -0.12. This will place the center marker identifying the centers of circles and arcs when the Dim-Cen command is used.

Command: **Layer**
?/Make/Set/New/ON/OFF/Color/Ltype/Freeze/Thaw/LOck/
Unlock: **Set**
New current layer <0>: **Center**
?/Make/Set/New/ON/OFF/Color/Ltype/Freeze/Thaw/LOck/
Unlock: (*Strike Enter to exit this command*)

Command: **Dim**
Dim: **Dimcen**
Current value <0.09> New value: **-0.12**

Dim: **Center**
Select arc or circle: (*Select the arc at "A"*)

Dim: **Center**
Select arc or circle: (*Select the arc at "B"*)

Dim: **Center**
Select arc or circle: (*Select the arc at "C"*)

Dim: **Center**
Select arc or circle: (*Select the arc at "D"*)

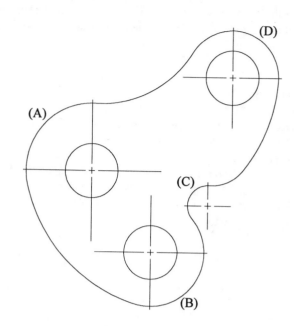

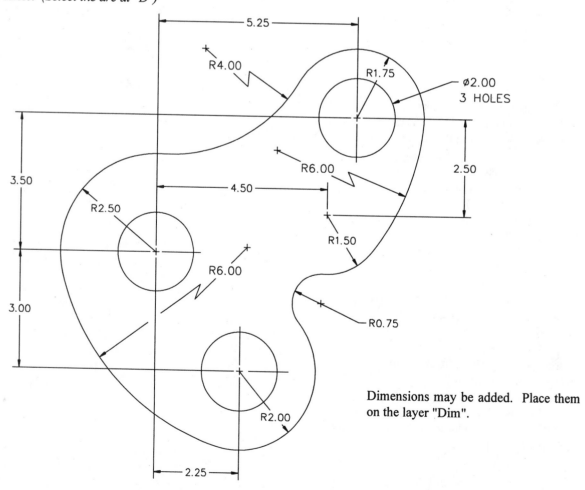

Dimensions may be added. Place them on the layer "Dim".

Tutorial Exercise #9
Gear-arm.Dwg

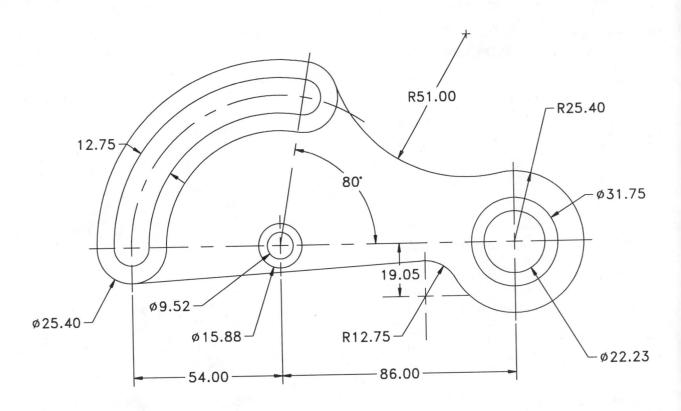

PURPOSE:
This tutorial is designed to use geometric commands to construct a one-view drawing of the Gear-arm. Follow the special system settings since this drawing is in metric.

SYSTEM SETTINGS:
Begin a new drawing called "Gear-arm." Use the Units command to change the number of decimal places past the zero from 4 to 2. Keep the remaining default unit values. Using the Limits command, keep 0,0 for the lower left corner and change the upper right corner from 12,9 to 265.00,200.00. Use the Grid command and change the grid spacing from 1.00 to 10.00 units. Do not turn the snap or ortho On. Since a layer called "Center" must be created to display center lines, use the Ltscale command and change the default value of 1.00 to 25.40. This will make the long and short dashes of the center lines appear on the display screen.

LAYERS:
Create the following layers with the format:
Name-Color-Linetype
Object - White - Continuous
Center - Yellow - Center
Dim - Yellow - Continuous

SUGGESTED COMMANDS:
The object consists of a combination of circles and arcs along with tangent lines and arcs. Use the Point command to identify and lay out the centers of all circles for construction purposes. Use the Arc command to construct a series of arcs for the left side of the Gear-arm. The Trim command will be used to trim circles, lines, and arcs to form the basic shape. Also, use the Circle-TTR command for tangent arcs to existing geometry. Since this object is metric, commands such as Ltscale and Dim-Dimscale need to be set to the metric-inch equivalent of 25.4 units. This value may be adjusted for better results.

DIMENSIONING:
This drawing may be dimensioned at a later date. Consult your instructor before continuing.

PLOTTING:
This tutorial exercise may be plotted on "A"-size paper (8.5" x 11"). Plot the Gear-arm as a metric drawing and a scale of full size, or 1=1.

Step #1

Begin the gear-arm by drawing two circles of diameters 9.52 and 15.88 using the Circle command and coordinate 112.00,90.00 as the center of both circles.

Command: **Circle**
3P/2P/TTR/<Center point>: **112.00,90.00** *(Point "A")*
Diameter/<Radius>: **D**
Diameter: **9.52**

Command: **Circle**
3P/2P/TTR/<Center point>: **@**
Diameter/<Radius>: **D**
Diameter: **15.88**

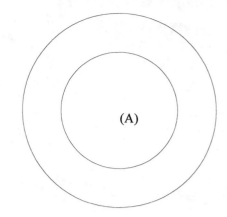

Step #2

Use the Pdmode command to change the point style to a value of 2. This will form a "plus" when using the Point command. Use the Pdsize command to change the point size to a value of 3 units. Use the Point command and the Osnap-Center option to place a point at the center of the two circles.

Command: **Pdmode**
New value for Pdmode <0>: **2**

Command: **Pdsize**
New value for Pdsize <0>: **3**

Command: **Point**
Point: **Cen**
of *(Select the large circle at "A")*

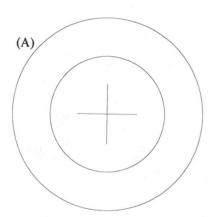

Step #3

Use the Copy command to duplicate the point using a polar coordinate distance of 86 units in the 0-degree direction.

Command: **Copy**
Select objects: **L** *(This will select the point)*
Select objects: *(Strike Enter to continue)*
<Base point or displacement>/Multiple: **Cen**
of *(Select the center of the large circle at "A")*
Second point of displacement: **@86<0**

Step #4

Use the Circle command to place three circles of different
sizes from the same point at "A".

Command: **Circle**
3P/2P/TTR/<Center point>: **Node**
of *(Select the point at "A")*
Diameter/<Radius>: **25.40**

Command: **Circle**
3P/2P/TTR/<Center point>: **@**
of *(Select the point at "A")*
Diameter/<Radius>: **D**
Diameter: **31.75**

Command: **Circle**
3P/2P/TTR/<Center point>: **@**
of *(Select the point at "A")*
Diameter/<Radius>: **D**
Diameter: **22.23**

Step #5

Use the Copy command to duplicate point "A" using a polar
coordinate distance of 54 units in the 180-degree direction.

Command: **Copy**
Select objects: *(Select the point at "A")*
Select objects: *(Strike Enter to continue)*
<Base point or displacement>/Multiple: **Cen**
of *(Select the center of the large circle at "A")*
Second point of displacement: **@54<180**

Step #6

Use the Circle command to place two circles of different
sizes at point "A". These circles will be converted to arcs in
later steps.

Command: **Circle**
3P/2P/TTR/<Center point>: **Node**
of *(Select the point at "A")*
Diameter/<Radius>: **D**
Diameter: **25.40**

Command: **Circle**
3P/2P/TTR/<Center point>: **@**
of *(Select the point at "A")*
Diameter/<Radius>: **D**
Diameter: **12.75**

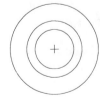

Step #7

Use the Copy command to duplicate point "A" using a polar coordinate distance of 54 units in the 80-degree direction.

Command: **Copy**
Select objects: *(Select the point at "A")*
Select objects: *(Strike Enter to continue)*
<Base point or displacement>/Multiple: **Cen**
of *(Select the center of the large circle at "A")*
Second point of displacement: **@54<80**

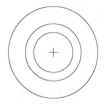

Step #8

Use the Line command to draw a line using a polar coordinate distance of 70 and a direction of 80 degrees. Start the line at point "A". Then use the Circle command to place two circles of different sizes at point "B". These circles will be converted to arcs in later steps.

Command: **Line**
From point: **Cen**
of *(Select the center of the large circle at "A")*
To point: **@70<80**
To point: *(Strike Enter to exit this command)*

Command: **Circle**
3P/2P/TTR/<Center point>: **Node**
of *(Select the point at "B")*
Diameter/<Radius>: **D**
Diameter: **25.40**

Command: **Circle**
3P/2P/TTR/<Center point>: **@**
of *(Select the point at "B")*
Diameter/<Radius>: **D**
Diameter: **12.75**

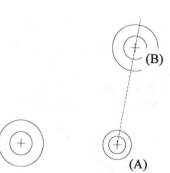
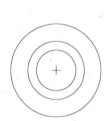

Step #9

Use the Arc command and draw an arc using point "A" as the center, point "B" as the start point, and point "C" as the endpoint.

Command: **Arc**
Center/<Start point>: **C**
Center: **Cen**
of *(Select the center of the circle at "A")*
Start point: **Int**
of *(Select the intersection of the line and circle at "B")*
Angle/Length of chord/<End point>: **Qua**
of *(Select the quadrant of the circle at "C")*

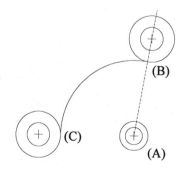
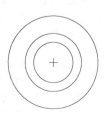

Step #10

Use the Arc command and draw an arc using point "A" as
the center, point "B" as the start point, and point "C" as
the endpoint.

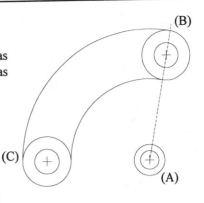
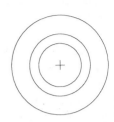

Command: **Arc**
Center/<Start point>: **C**
Center: **Cen**
of *(Select the center of the circle at "A")*
Start point: **Int**
of *(Select the intersection of the line and circle at "B")*
Angle/Length of chord/<End point>: **Qua**
of *(Select the quadrant of the circle at "C")*

Step #11

Use the Trim command, select the two dashed arcs at the
right as cutting edges, and trim the two circles at points
"C" and "D."

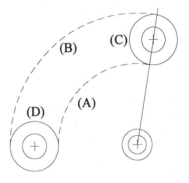
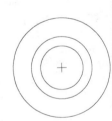

Command: **Trim**
Select cutting edge(s)...
Select objects: *(Select the two dashed arcs "A" and "B")*
Select objects: *(Strike Enter to continue)*
<Select object to trim>Undo: *(Select the circle at "C")*
<Select object to trim>Undo: *(Select the circle at "D")*
<Select object to trim>Undo: *(Strike Enter to exit this
command)*

A built-in undo is provided if a mistake is made by
trimming the wrong entity.

Step #12

Your drawing should be similar to the illustration at the
right.

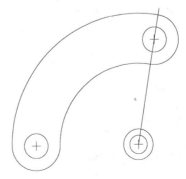
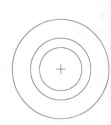

Step #13

Use the Arc command and draw an arc using point "A" as the center, point "B" as the start point, and point "C" as the endpoint.

Command: **Arc**
Center/<Start point>: **C**
Center: **Cen**
of *(Select the center of the circle at "A")*
Start point: **Int**
of *(Select the intersection of the line and circle at "B")*
Angle/Length of chord/<End point>: **Qua**
of *(Select the quadrant of the circle at "C")*

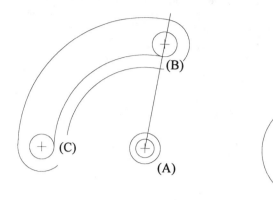

Step #14

Use the Arc command and draw an arc using point "A" as the center, point "B" as the start point, and point "C" as the endpoint.

Command: **Arc**
Center/<Start point>: **C**
Center: **Cen**
of *(Select the center of the circle at "A")*
Start point: **Int**
of *(Select the intersection of the line and circle at "B")*
Angle/Length of chord/<End point>: **Qua**
of *(Select the quadrant of the circle at "C")*

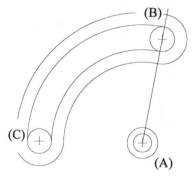

Step #15

Use the Trim command, select the two dashed arcs at the right as cutting edges, and trim the two circles at points "C" and "D".

Command: **Trim**
Select cutting edge(s)...
Select objects: *(Select the two dashed arcs "A" and "B")*
Select objects: *(Strike Enter to continue)*
<Select object to trim>Undo: *(Select the circle at "C")*
<Select object to trim>Undo: *(Select the circle at "D")*
<Select object to trim>Undo: *(Strike Enter to exit this command)*

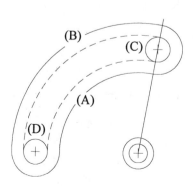

Step #16

Your drawing should be similar to the illustration at the right. Always perform periodic screen redraws using the Redraw command to clean up the display screen.

Command: **Redraw**

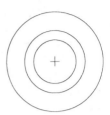

Step #17

Use the Line command and draw a line from the quadrant of the small circle to the quadrant of the large circle. This line is used only for construction purposes.

Command: **Line**
From point: **Qua**
of *(Select the circle at "A")*
To point: **Qua**
of *(Select the circle at "B")*
To point: *(Strike Enter to exit this command)*

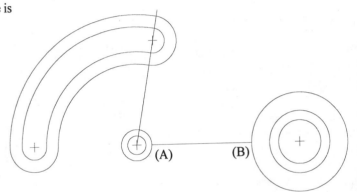

Step #18

Use the Move command to move the dashed line down the distance 19.05 units using the polar coordinate mode. Then, use the Offset command to offset the dashed circle the distance 12.75 units. The intersection of these two entities will be used to draw a 12.75 radius circle.

Command: **Move**
Select objects: *(Select the dashed line at the right)*
Select objects: *(Strike Enter to continue)*
Base point or displacement: **Endp**
of *(Select the endpoint of the line at "A")*
Second point of displacement: **@19.05<270**

Command: **Offset**
Offset distance or Through<Through>: **12.75**
Select object to offset: *(Select the circle at "B")*
Side to offset? *(Select a blank part of the screen at "C")*
Select object to offset: *(Strike Enter to exit this command)*

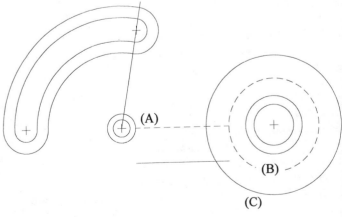

Step #19

Use the Circle command to draw a circle with a radius of 12.75. Use the center of the circle as the intersection of the large dashed circle and the dashed horizontal line illustrated at the right. Use the Erase command to erase the dashed circle and dashed line.

Command: **Circle**
3P/2P/TTR/<Center point>: **Int**
of *(Select the intersection of the line and circle at "A")*
Diameter/<Radius>: **12.75**

Command: **Erase**
Select objects: *(Select the dashed circle and dashed line)*
Select objects: *(Strike Enter to execute this command)*

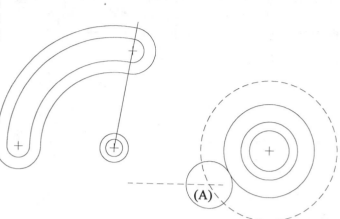

Step #20

Use the Line command to draw a line from a point tangent to the arc at "A" to a point tangent to the circle at "B". Use the Osnap-Tan option to accomplish this.

Command: **Line**
From point: **Tan**
to *(Select the arc at "A")*
To point: **Tan**
to *(Select the circle at "B")*
To point: *(Strike Enter to exit this command)*

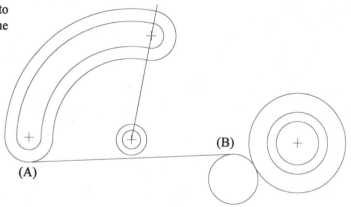

Step #21

Use the Trim command, select the dashed line and dashed circle illustrated at the right as cutting edges, and trim the circle at "A".)

Command: **Trim**
Select cutting edge(s)...
Select objects: *(Select the dashed line at the right)*
Select objects: *(Select the dashed circle at the right)*
Select objects: *(Strike Enter to continue)*
<Select object to trim>Undo: *(Select the circle at "A")*
<Select object to trim>Undo: *(Strike Enter to exit this command)*

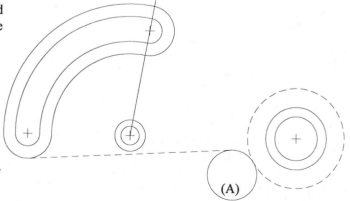

Step #22

Use the Circle-TTR command to draw a circle tangent to the arc at "A" and tangent to the circle at "B" with a radius of 51. When using the TTR option in the Circle command, the Osnap-Tan option is automatically invoked.

Command: **Circle**
3P/2P/TTR/<Center point>: **TTR**
Enter Tangent spec: *(Select the arc at "A")*
Enter second Tangent spec: *(Select the circle at "B")*
Radius: **51**

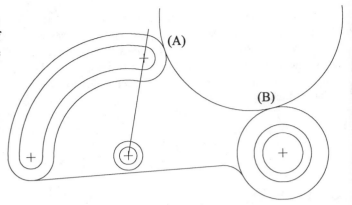

Step #23

Use the Trim command, select the dashed arc and dashed circle illustrated at the right as cutting edges, and trim the large 51 radius circle at "A".

Command: **Trim**
Select cutting edge(s)...
Select objects: *(Select the dashed arc at the right)*
Select objects: *(Select the dashed circle at the right)*
Select objects: *(Strike Enter to continue)*
<Select object to trim>Undo: *(Select the circle at "A")*
<Select object to trim>Undo: *(Strike Enter to exit this command)*

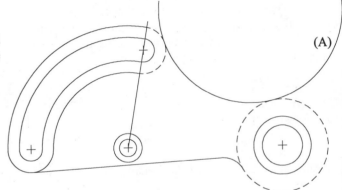

Step #24

Use the Trim command, select the two dashed arcs illustrated at the right as cutting edges, and trim the circle at "C". Use the Erase command to delete all four points used to construct the circles.

Command: **Trim**
Select cutting edge(s)...
Select objects: *(Select the dashed arc "A")*
Select objects: *(Select the dashed arc "B")*
Select objects: *(Strike Enter to continue)*
<Select object to trim>Undo: *(Select the circle at "C")*
<Select object to trim>Undo: *(Strike Enter to exit this command)*

Command: **Erase**
Select objects: *(Select the four points illustrated at the right)*
Select objects: *(Strike Enter to execute this command)*

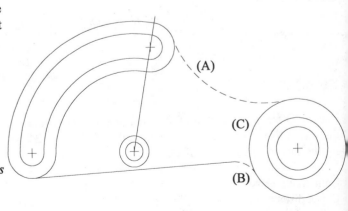

Step #25

Your display should appear similar to the illustration at the right. Notice the absence of the points that were erased in the previous step. Standard center lines will be placed to mark the center of all circles in the next series of steps. Use the Layer command to set the new current layer to Center. Change Ltscale to 25.40.

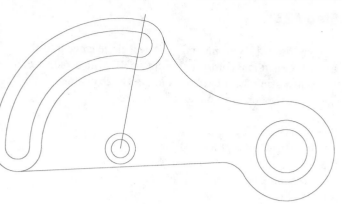

Command: **Ltscale**
New scale factor <1.0000>: **25.40**

Step #26

To place center lines for all circles and arcs, two dimension variables need to be changed. The dimension variable, Dimscale, is set to a value of 1. Since this is a metric drawing, the Dimscale variable needs to be changed to a value of 25.4. This will increase all variables by this value which is necessary because we are drawing in metric units. Also, the Dimcen variable needs to be changed from a value of 0.09 to -0.09. The negative value will extend the center lines past the edge of the circle when using the Dim-Cen command.

Command: **Dim**
Dim: **Dimscale**
Current value <1> New value: **25.4**

Dim: **Dimcen**
Current value <0.09> New value: **-0.09**

Dim: **Center**
Select arc or circle: *(Select the arc at "A")*
Dim: **Center**
Select arc or circle: *(Select the circle at "B")*
Dim: **Center**
Select arc or circle: *(Select the arc at "C")*
Dim: **Center**
Select arc or circle: *(Select the arc at "D")*
Dim: **Exit** *(To return to the "Command" prompt)*

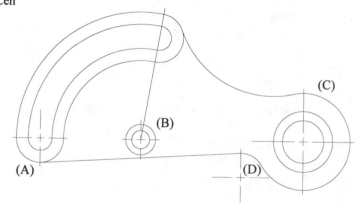

Step #27

Use the Offset command to offset the inside arc at "A" the distance 6.375 units. Indicate a point in the vicinity of "B" for the side to perform the offset.

Command: **Offset**
Offset distance or Through <Through>: **6.375**
Select object to offset: *(Select the arc at "A")*
Side to offset? *(Select a point in the vicinity of "B")*
Select object to offset: *(Strike Enter to exit this command)*

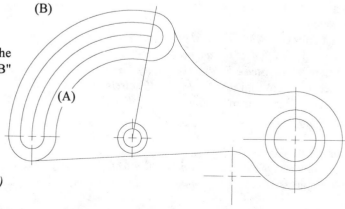

Step #28

Use the Chprop command to change the arc at "A" and the line at "B" to the Center layer. This layer should have been created earlier for the arc and line to change to the proper color and linetype.

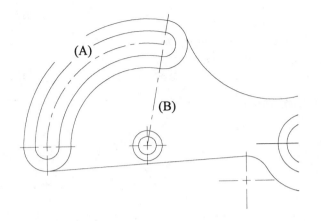

Command: **Chprop**
Select objects: *(Select the middle arc at "A")*
Select objects: *(Select the line at "B")*
Select objects: *(Strike Enter to continue)*
Change what property(Color/LAyer/LType/Thickness): **LA**
New layer <0>: **Center**
Change what property(Color/LAyer/LType/Thickness): *(Strike Enter to exit this command)*

Step #29

Use the Extend command to extend the center line arc to intersect with the circular arc at "A"; see the illustration at the right. Next, reset the dimension variable Dimcen from -0.09 to a new value of 0.09. This will change the center point to a plus without the center lines extending beyond the arc when using the Dim-Cen command. The finished object may be dimensioned as an optional step.

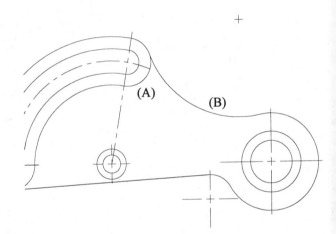

Command: **Extend**
Select boundary edge(s)...
Select objects: *(Select the arc at "A")*
Select objects: *(Strike Enter to continue)*
Select object to extend: *(Select the center line arc)*
Select object to extend: *(Strike Enter to exit this command)*

Command: **Dim**
Dim: **Dimcen**
Current value <-0.09> New value: **0.09**

Dim: **Center**
Select arc or circle: *(Select the arc at "B")*

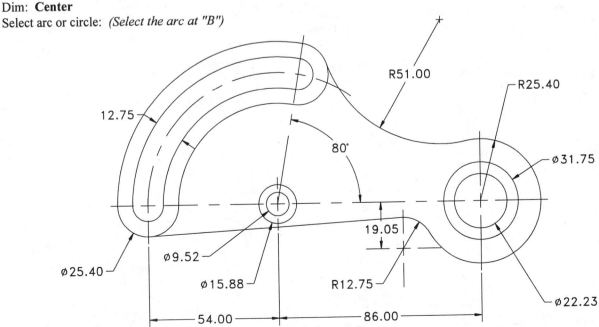

Problems for Unit 3

Directions for Problems 3–1 through 3–21:
Construct these geometric construction figures using existing AutoCAD commands.

Problem 3-1

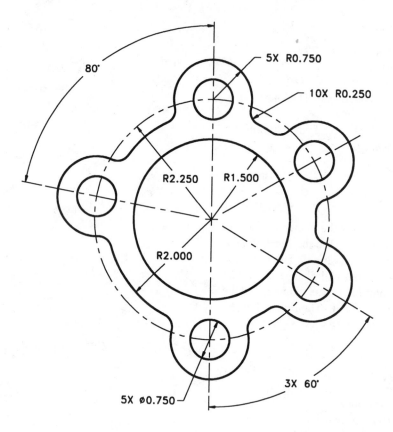

Problem 3-2

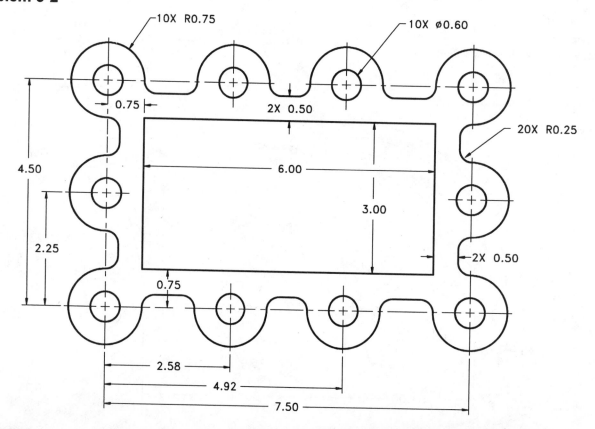

Problem 3-3

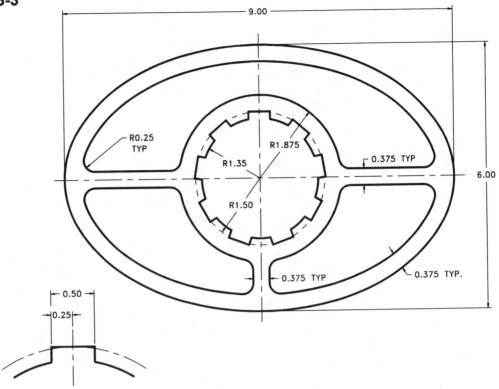

Problem 3-4

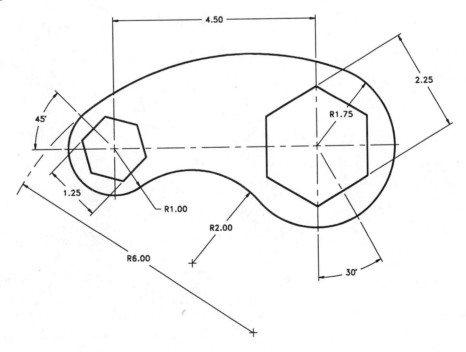

Problem 3-5

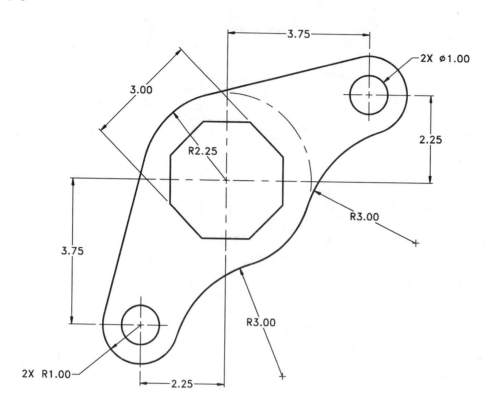

Problem 3-6

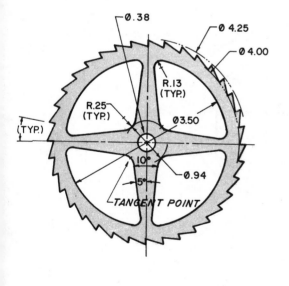

Problem 3-7

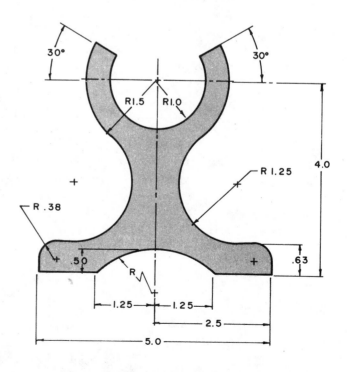

Problem 3-8

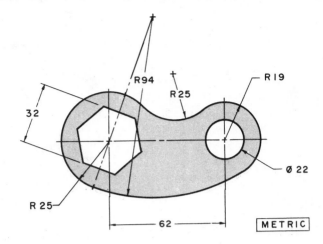

Problem 3-9

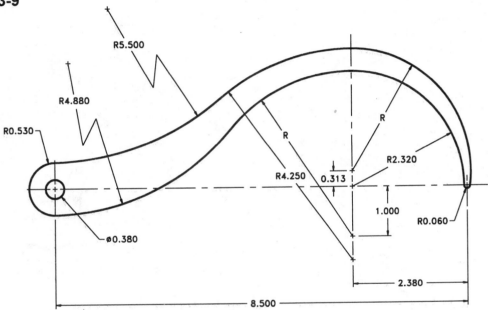

Problem 3-10

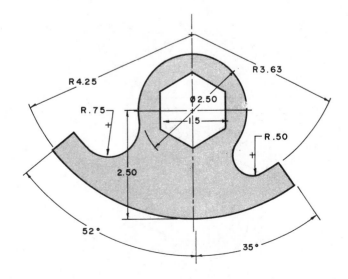

Problem 3-11

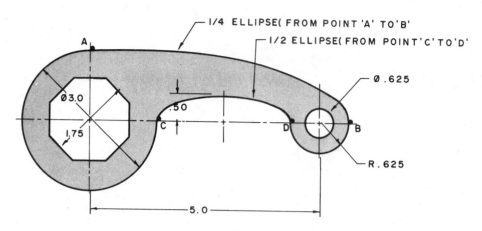

1/4 ELLIPSE(FROM POINT 'A' TO 'B'

1/2 ELLIPSE(FROM POINT 'C' TO 'D'

Ø.625

Ø3.0

.50

1.75

A

C

D

B

R.625

5.0

Problem 3-12

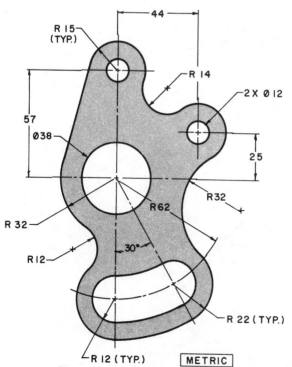

ELLIPSE
MINOR DIA.= 100
MAJOR DIA.= 150

45°

R6 (TYP.)

Ø44

Ø50

8

4

6

12

R38

12

38

METRIC

Problem 3-13

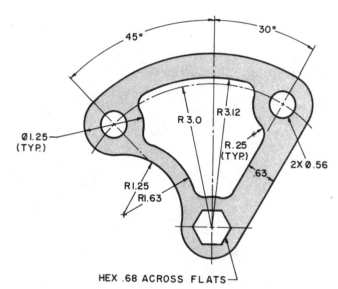

45°

30°

Ø1.25
(TYP.)

R 3.0

R3.12

R.25
(TYP.)

.63

2X Ø.56

R1.25

R1.63

HEX .68 ACROSS FLATS

Problem 3-14

R 15
(TYP.)

R 14

2X Ø 12

44

57

Ø38

25

R 32

R62

R32

R12

30°

R 22 (TYP.)

R 12 (TYP.)

METRIC

Problem 3-15

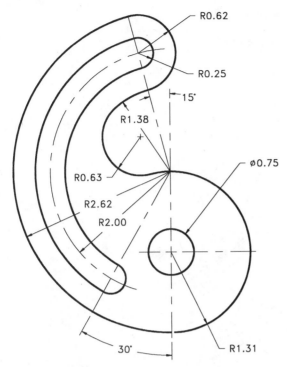

R0.62
R0.25
15°
R1.38
Ø0.75
R0.63
R2.62
R2.00
30°
R1.31

Problem 3-17

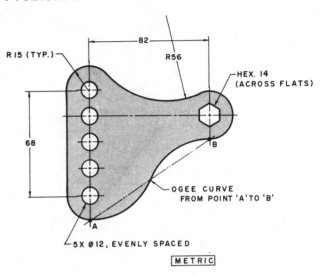

R 15 (TYP.)
82
R56
HEX. 14
(ACROSS FLATS)
68
B
OGEE CURVE
FROM POINT 'A' TO 'B'
A
5X Ø12, EVENLY SPACED

METRIC

Problem 3-16

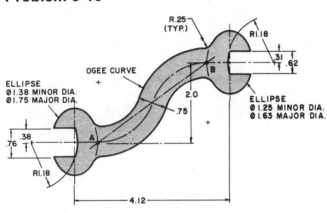

R.25
(TYP.)
R1.18
OGEE CURVE
.31
.62
B
2.0
ELLIPSE
Ø1.38 MINOR DIA.
Ø1.75 MAJOR DIA.
.75
ELLIPSE
Ø1.25 MINOR DIA.
Ø1.63 MAJOR DIA.
.38
.76
A
R1.18
4.12

Problem 3-18

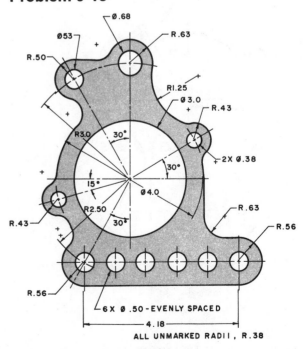

Ø.68
Ø53
R.63
R.50
R1.25
Ø3.0
R.43
R3.0
30°
2X Ø.38
30°
R.43
15°
Ø4.0
R.63
R2.50
30°
R.56
R.56
6X Ø .50-EVENLY SPACED
4.18
ALL UNMARKED RADII, R.38

Problem 3-19

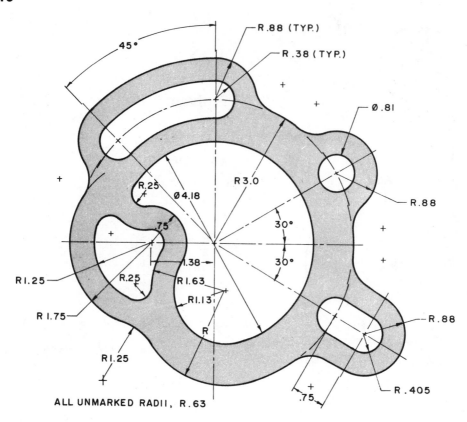

R.88 (TYP.)
R.38 (TYP.)
45°
Ø.81
R3.0
R.25
Ø4.18
30°
R.88
.75
30°
1.38
R1.25
R.25
R1.63
R1.13
R.88
R1.75
R
R.405
R1.25
.75

ALL UNMARKED RADII, R.63

Problem 3-20

Problem 3-21

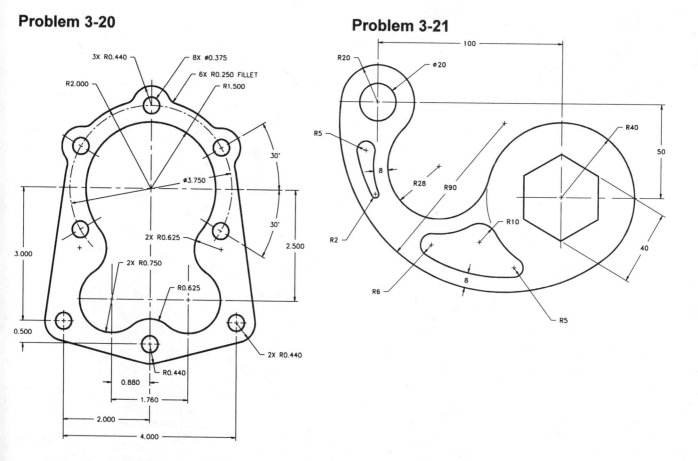

Problem 3-20

3X R0.440
8X Ø0.375
6X R0.250 FILLET
R2.000
R1.500
30°
Ø3.750
30°
2X R0.625
2.500
3.000
2X R0.750
0.625
0.500
0.880
2X R0.440
1.760
R0.440
2.000
2X R0.440
4.000

Problem 3-21

100
R20
Ø20
R5
R40
50
8
R28
R90
R2
R10
R6
40
8
R5

UNIT
4

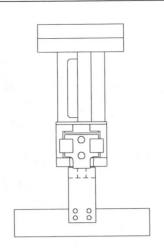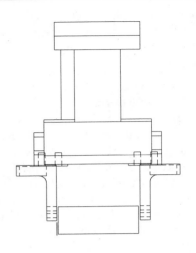

Shape Description/ Multi-View Projection

Contents

Before any object is made in production, some type of drawing needs to be made. This is not just any drawing; rather an engineering drawing consisting of overall sizes of the object and various views of the object is organized on the computer screen. This chapter introduces the topic of shape description or how many views are really needed to describe an object. The art of multi-view projection includes methods of constructing one view, two view, and three view drawings using AutoCAD commands. Linetypes are explained as a method of communicating hidden features located in different views of a drawing.

Shape Description

Before performing engineering drawings, an analysis of the object being drawn must first be made. This takes the form of describing the object by views or how an observer looks at the object. In the illustration at the right of the simple wedge, it is no surprise that this object can be viewed at almost any angle to get a better idea of its basic shape. However, some standard method of determining how and where to view the object must be exercised. This is to standardize how all objects are to be viewed in addition to limiting confusion that is usually associated with complex multi-view drawings.

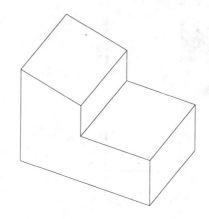

Even though the simple wedge is easy to understand because it is currently being displayed in picture or isometric form, it would be difficult to produce this object since it is unclear what the size of the front and top views is. In this way, the picture of the object is separated into six primary ways or directions to view an object and they are illustrated at the right. The Front view begins the shape discription followed by the Top view and Right Side view. Continuing on, the Left Side view, Back view, and Bottom view complete the primary ways to view an object.

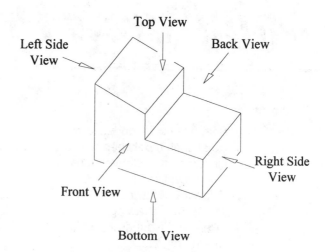

Now that the primary ways of viewing an object have been established, the views need to be organized in some type of manner to promote clarity and have the views reference themselves. Imagine the simple wedge positioned in a clear, transparent glass box similar to the illustration at the right. With the entire object at the center of the box, the sides of the box represent the ways to view the object. Images of the simple wedge are projected onto the glass surfaces of the box.

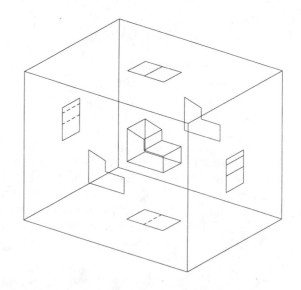

With the views projected onto the sides of the glass box, we must now prepare the views to be placed on a 2-dimensional drawing screen. To accomplish this, the glass box, which is hinged, is unfolded as in the illustration at the right. All folds occur from the front view, which remains stationary.

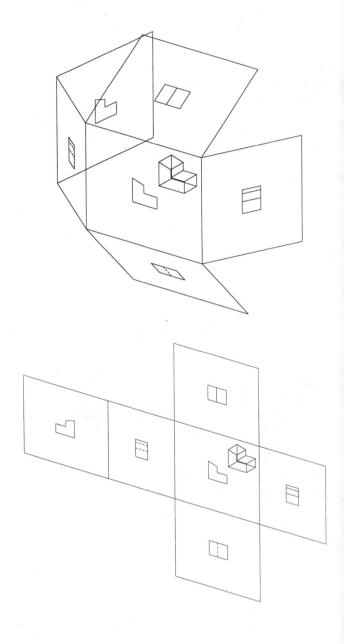

The illustration at the right shows the views in their proper alignment to one another. However this illustration is still in a pictorial view. These views need to be placed flat before continuing.

As the glass box is completely unfolded and laid flat, the result is illustrated at the right. The Front view becomes the main view with other views being placed in relation to the Front. Above the Front view is the Top view. To the right of the Front is the Right Side view. To the left of the Front is the Left Side view followed by the Back view. Underneath the Front view is the Bottom view. This becomes the standard method of laying out the necessary views needed to describe an object. But are all views necessary? Upon closer inspection, we find that except for being a mirror image, the Front and Back views are identical. The Top and Bottom views appear similar as do the Right and Left Side views. One very important rule to follow in multi-view objects is to only select those views that accurately describe the object and discard the remaining views.

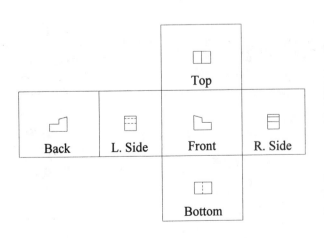

The complete multi-view drawing of the simple wedge is illustrated at the right. Only the Front, Top, and Right Side views were needed to describe this object. Important information to remember when laying out views of an object: the Front view is usually the most important view and holds the basic shape of the object being described. Directly above the Front view is the Top view and to the right of the Front is the Right Side view. All three views are separated by a space of various sizes. This space is commonly called a dimension space because it becomes a good area to place dimensions describing the size of the object. The space also acts as a separator between views; without it, the views would touch on one another, which would be difficult to read and interpret. The minimum distance of this space is usually 1.00 units.

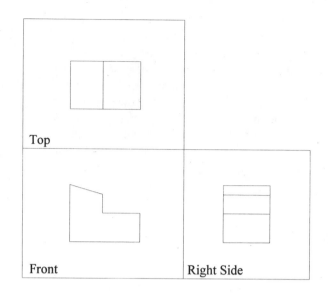

Relationships between Views

Some very interesting and important relationships are set up when placing the three views in the configuration illustrated at the right. Notice that since the Top view is directly above the Front view, both views share the same Width dimension. The Front and Right Side views share the same Height. The relationship of Depth on the Top and Right Side views can be explained by constructing a 45 degree projector line at "A" and projecting the lines over and down or vice versa to get the Depth. Yet another principle illustrated by this example is that of projecting lines up, over, and across and down to create views. Editing commands such as Erase and Trim are then used to clean up unnecessary lines

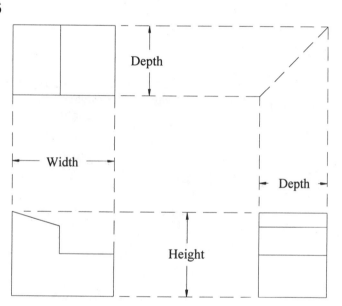

With the three views identified at the right, the only other step, and it is an important one, is to annotate the drawing, or add dimensions to the views. With dimensions, the object drawn up in multi-view projection can now be produced. Even though this has been a simple example, the methods of multi-view projection work even for the most difficult and complex of objects.

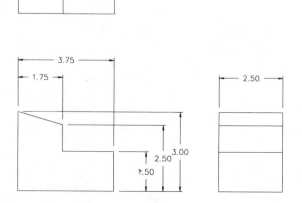

Linetypes and Conventions

At heart of any engineering drawing is the ability to assign different types of lines as a method of conveying some type of meaning to the drawing. All lines of a drawing when plotted out are dark; border and title block lines are the thickest lines of a drawing. Object lines outline visible features of a drawing and are made thick and dark (but not as thick as a border line). To identify features that invisible in an adjacent view, a hidden line is used. This line is a series of dashes 0.12 units in length with a spacing of 0.06 units. Center lines are used to identify the centers of circular features such as holes. They are also used to show that a hidden feature in one view is circular in another. The center line consists of a series of long and short dashes. The short dash measures approximately 0.12 units while the long dash may vary from 0.75 to 1.50 units. A gap of 0.06 is placed in between dashes. Study the examples of these lines in the illustration at the right.

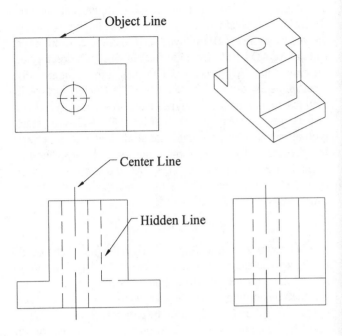

The object shown at the right is a good illustration of the use of phantom lines in a drawing. Phantom lines are especially useful where motion is applied. The arm on the right is shown using standard object lines consisting of the continuous linetype. To show that the arm rotates about a center pivot, the identical arm is duplicated using the Mirror command and all lines of this new element converted to phantom lines using the Change command. Notice that the smaller segments are not shown as phantom lines due to their short sizes.

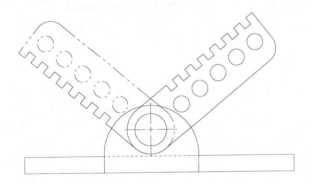

When using manual drafting techniques for drawing hidden and center lines, the drafter first had to develop a skill of drawing a series of dashes as individual line segments along with spacing each dash equally. This was very tedious work but with practice and experience, the method proved acceptable in the field of engineering drawings. Since CAD has come along, the operator has the option of assigning a linetype to the drawing. This means that if the current linetype is called Hidden, the series of dashes and spaces will automatically be drawn once a line, circle, arc, or any drawing entity is drawn. Illustrated at the right were the standard linetypes supplied with AutoCAD ever since the software was introduced in 1982.

DASHED, _ _ _ _ _ _ _ _ _ _ _ _ _ _ _ _ _

HIDDEN, _

CENTER, _ _ _ _ _ _ _ _ _ _ _ _ _ _ _ _ _ _ _

PHANTOM, _ _ _ _ _ _ _ _ _ _ _ _ _ _ _ _ _ _

DOT, ...

DASHDOT, _ . _ . _ . _ . _ . _ . _ . _ . _ . _

BORDER, _ _ . _ _ . _ _ . _ _ . _ _ . _ _

DIVIDE, _ . . _ . . _ . . _ . . _ . . _ . . _

The linetypes illustrated at the right are the current linetype definitions supplied with AutoCAD. These linetypes may be viewed by using the DOS "Type" command to type out a file called "ACAD.LIN" located in the \Support subdirectory of the main AutoCAD directory. The lines at the right are only a partial listing of the complete file. The Center linetype is identical to the standard linetype on the previous page. There also exist two more types of center lines; Center2 is the same type of center linetype except that all dashes and spaces are half the size of the orginal Center linetype. CenterX2 has the orginal linetype doubled in size. All current linetypes of AutoCAD have three possible linetypes for the operator to choose from.

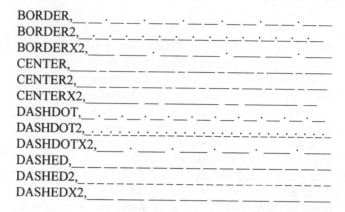

Use of linetypes in a drawing is crucial to the interpretation of the views and the final design before the object is actually made. Sometimes the linetype appears too long; in other cases the linetype does not appear at all even though using the List command on the entity will show the proper layer and linetype. The Ltscale command is used to manipulate the size of all linetypes loaded into a drawing. By default, all linetypes are assigned a scale factor of 1.00. This means that the actual dashes and/or spaces of the linetype are multiplied by this factor. The views illustrated show linetypes that use the default value of 1.00 from the Ltscale command.

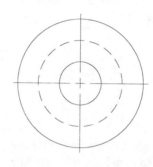

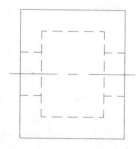

If a linetype appears too long, use the Ltscale command and set a new value to less than 1.00. If a linetype appears too short, use the Ltscale command and set a new value to greater than 1.00. The same views illustrated at the right show the effects of the Ltscale command set to a new value of 0.75. Notice the center in the right side view has one more series of dashes in its appearance than the same object illustrated above. The value 0.75 is the new multiplier that affects all dashes and spaces defined in the linetype.

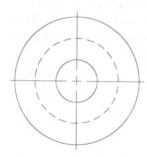

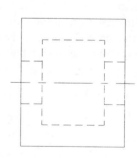

In this third example using the Ltscale command, a new value of 0.50 has been used to shorten the linetypes even more. Now even the center marks identifying the circles in the front view have been changed into center lines. One other important note to realize when using the Ltscale command: the new value, whether larger or smaller that 1.00, affects all linetypes visible on the display screen. In other words it is not possible to affect a hidden line set to a certain scale without affecting a center line with the same scale.

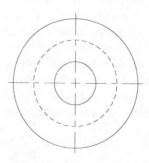

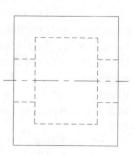

One View Drawings

An important rule to remember concerning multi-view drawings is draw only enough views to accurately describe the object. In the drawing of the gasket at the right, a front and side view are shown. However, the side view is so narrow that it is difficult to interpret the hidden lines drawn inside. A better approach would be to leave out the side view and construct a one view drawing consisting of just the front view.

Begin the one view drawing of the gasket by first laying out center lines marking the centers of all circles and arcs. A layer containing center lines could be used to show all lines as center lines.

Use the Circle command to lay out all circles representing the bolt holes of the gasket. The Offset command could be used to form the large rectangle on the inside of the gasket. If lines of the rectangle extend past each other, use the Fillet command set to a value of "0." Selecting two lines of the rectangle will form a corner. Repeat this procedure for any other lines that do not form exact corners.

Use the Trim command to begin forming the outside arcs of the gasket.

Use the Fillet command set to the desired radius to form a smooth transition from the arcs to the outer rectangle.

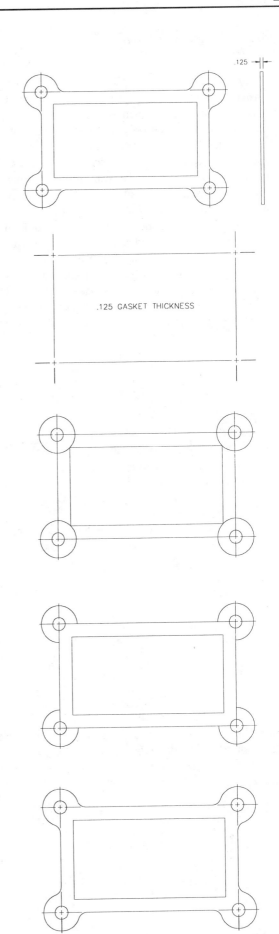

.125

.125 GASKET THICKNESS

Two View Drawings

Before attempting any drawing, some thought needs to be made in determining how many views need to be drawn. Only a minimum number of views are needed to describe an object. Drawing extra views is not only time consuming, but may result in two identical views with mistakes in each view. The operator must interpret which is the correct set of views. The illustration at the right is a three view multi-view drawing of a coupler. The front view is identified by the circles and circular hidden circle. Except for their rotation angles, the top and right side views are identical. In this example or for other symetrical objects, only two views are needed to accurately describe the object being drawn. The top view has been deleted to leave the front and right side views. The side view could have easily been deleted in favor of leaving the front and top views. This decision is up to the designer depending on sheet size and which views are best suited for the particular application.

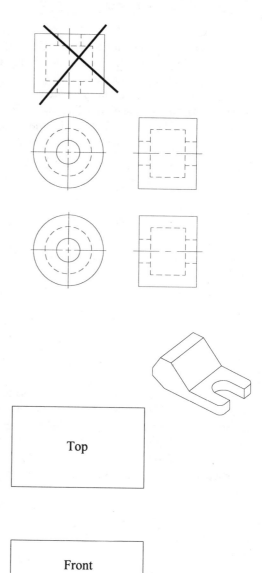

To illustrate how AutoCAD is used as the vehicle for creating a two view engineering drawing, study the pictorial drawing illustrated at the right to get an idea of how the drawing will appear. Begin the two view drawing by using the Line command to lay out the front and side views. The width of the top view may be found by projecting lines up from the front since both views share the same width. Provide a space of 1.50 units in between views to act as a separator and allow for dimensions at a later time.

Top

Front

Begin adding visible details to the views such as circles, filleted corners, and angles. Use various editing commands such as trim, extend, and offset to clean up unnecessary geometry.

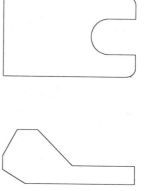

From the front view, project corners up into the top view. These corners will form visible edges in the top view. Use the same projection technique to project features from the top view into the front view.

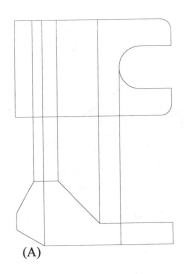

(A)

Use the Trim command to delete any geometry that appears in the 1.50 dimension space. The views now must conform to engineering standards by determining which lines are visible and which are invisible. The corner at "A" represents an area hidden in the top view. Use the Change command to convert the line in the top view from the continuous linetype to the hidden linetype. In the same manner, the slot visible in the top view is hidden in the front view. Again use Change to convert the continuous line in the front view to the hidden linetype. Since the slot in the top view represents a circular feature, use the Dim-Cen command to place a center marker at the center of the semi-circle. To show in the front view that the hidden line represents a circular feature, add one center line consisting of one short dash and two short dashes. If the slot in the top view was square instead of circular, center lines would not be necessary.

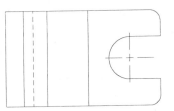

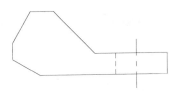

Use the spaces provided to properly add dimensions to the drawing. Once the dimension spaces are filled with numbers, use outside areas to call out distances. Placing dimensions will be discussed in a later chapter.

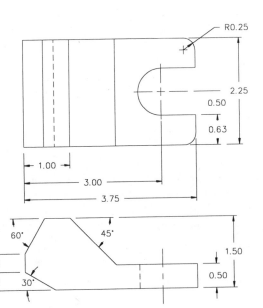

Three View Drawings

If two views are not enough to describe an object, a three view drawing of the object is drawn. This consists of front, top, and right side views. A three view drawing of the guide block illustrated in pictorial format will be the focus of this segment. Notice the broken section exposing the spotfacing operation above a drill hole. Begin this drawing by laying out all views using overall dimensions of width, depth, and height. The Line command along with Offset are popular commands used to accomplish this. Provide a space in between views to accommodate dimensions at a later time.

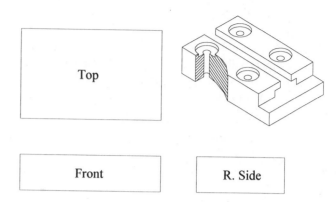

Begin drawing features in the views they appear visible in. Since the spotface holes appear above, draw these in the top view. The notch appears in the front view; draw it there. A slot is visible in the right side view and is drawn there.

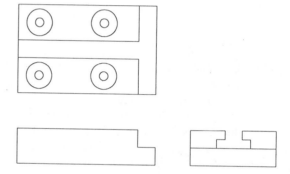

As in two view drawings, all features are projected down from top to front view. To project depth measurements from top to right side views, construct a 45 degree line at "A."

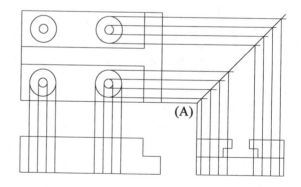

Use the 45 degree line to project the slot from the right side to the top view. Project the height of the slot from the right side to the front view. Use the Change command to convert continuous lines to hidden lines where features appear invisible such as the holes in the front and right side views.

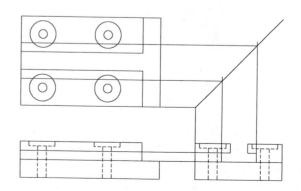

Use the Change command to change the remaining lines from continuous to hidden. Erase any construction lines including the 45 degree projection line.

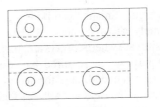

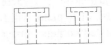

Begin adding center lines to label circular features. The Dim-Cen command is used where the circles are visible. Where features are hidden but represent circular features, the single center line consisting of one short dash and two long dashes is used. In the illustration below, dimensions remain the final step in completing the engineering drawing before being checked and shipped off for production.

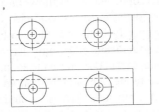

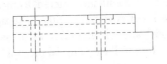

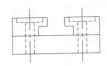

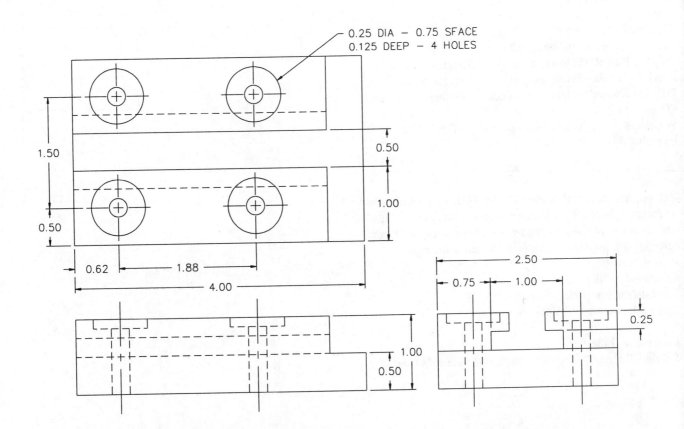

Fillets and Rounds

Numerous objects require highly finished and polished surfaces consisting of extremely sharp corners. Fillets and Rounds represent the opposite case where corners are rounded off either for ornamental purposes or required by design. Generally a fillet consists of a rounded edge formed in the corner of an object illustrated at the right at "A." A round is formed at an outside corner similar to "B." Fillets and rounds are primarily used where objects are cast or made from poured metal. The metal will form easier around a pattern that has rounded corners versus sharp corners which usually break away. Some drawings have so many fillets and rounds that a note is used to convey the size of all similar to "All Fillets and Rounds 0.125 Radius."

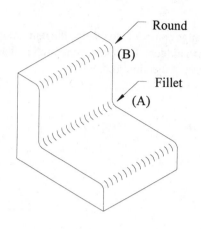

AutoCAD provides the Fillet command which allows the user to enter a radius followed by the selection of two lines. The result will be a fillet of the specified radius to the two lines selected. The two lines are also automatically trimmed leaving the radius drawn from the endpoint of one line to the endpoint of the other line. Illustrated to the right and the bottom are examples of using the Fillet command. Because sometimes the command is used over and over again, the Multiple automatically repeats the next command entered from the keyboard, in this case the Fillet command. Since Multiple continually repeats the command, use the CTRL-C keys to cancel the command and return to the command prompt.

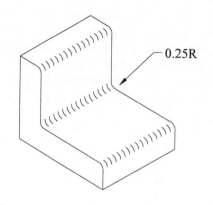

Command: **Multiple**
Fillet
Polyline/Radius/<Select two objects>: **R**
Enter fillet radius <0.0000>: **0.25**
Polyline/Radius/<Select two objects>: *(Select at "A" and "B")*
Polyline/Radius/<Select two objects>: *(Select at "B" and "C")*
Polyline/Radius/<Select two objects>: *(Select at "C" and "D")*
Polyline/Radius/<Select two objects>: *(Type CTRL C to cancel)*

Yet another powerful feature of the Fillet command is to connect two lines at their intersections or corner the two lines. This is accomplished by setting the fillet radius to "0" and selecting the two lines. This is illustrated at the right.

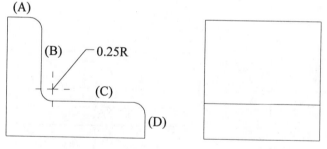

Command: **Fillet**
Polyline/Radius/<Select two objects>: **R**
Enter fillet radius <0.2500>: **0**

Command: **Fillet**
Polyline/Radius/<Select two objects>: *(Select at "A" and "B")*

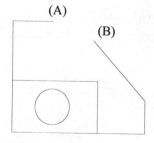
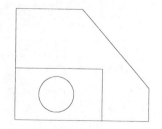

Chamfers

Chamfers represent yet another way finish a sharp corner of an object. As fillets and rounds result from a pattern-making operation and remain unfinished, a chamfer is a machining operation which may even result in a polishing operation. The illustration at the right is one example of an object which has been chamfered along its top edge.

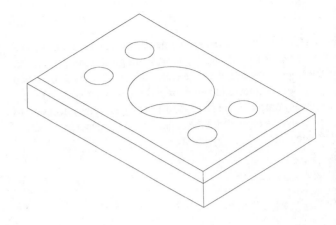

As with the Fillet command, AutoCAD also provides a Chamfer command designed to draw an angle across a sharp corner given two chamfer distances. The most popular chamfer involves a 45 degree angle which is illustrated at the right. Even though this command does not allow the user to specify an angle, the operator may control the angle by the distances entered. In the example at the right, by specifying the same numeric value for both chamfer distances, a 45 degree chamfer will automatically be formed. As long as both distances are the same, a 45 degree chamfer will always be drawn. Study the illustration at the right and prompts below:

Command: **Chamfer**
Polyline/Distances/<Select first line>: **D**
Enter first chamfer distance <0.0000>: **0.15**
Enter second chamfer distance <0.1500>: *(Strike Enter to accept)*

Command: **Chamfer**
Polyline/Distances/<Select first line>: *(Select the line at "A")*
Select second line: *(Select the line at "B")*

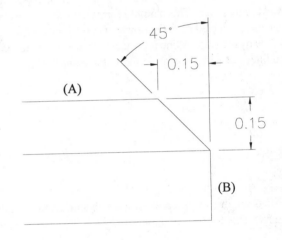

The illustration at the right is similar to the 45 degree chamfer above with the exception that the angle is different. This results from two different chamfer distances outlined in the prompts below. This type of edge is commonly called a bevel.

Command: **Chamfer**
Polyline/Distances/<Select first line>: **D**
Enter first chamfer distance <0.0000>: **0.30**
Enter second chamfer distance <0.3000>: **0.15**

Command: **Chamfer**
Polyline/Distances/<Select first line>: *(Select the line at "A")*
Select second line: *(Select the line at "B")*

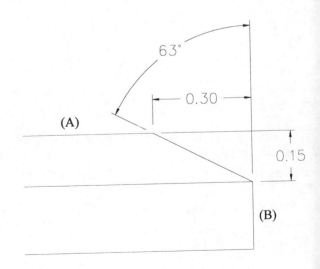

Runouts

Where flat surfaces become tangent to cylinders, there must be some method of accurately representing this using fillets. In the object illustrated at the right, the front view shows two cylinders connected to each other by a tangent slab. The top view is complete; the front view has all geometry necessary to describe the object with the exception of the exact intersection of the slab with the cylinder.

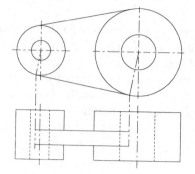

This illustration displays the correct method for finding intersections or runouts; areas where surfaces intersect others and blend in, disappear, or simply runout. A point of intersection is found at "A" in the top view with the intersecting slab and the cylinder. This actually forms a 90 degree angle with the line projected from the center of the cylinder and the angle made by the slab. A line is projected from "A" in the top view to intersect with the slab found in the front view.

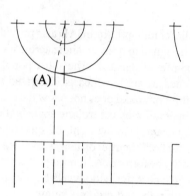

Fillets are drawn to represent the slab and cylinder intersections. This forms the runout.

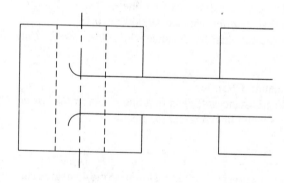

The resulting two view drawing complete with runouts is illustrated at the right.

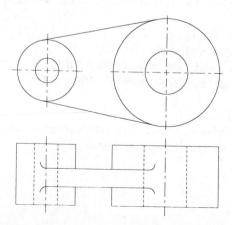

Tutorial Exercise #10
Shifter.Dwg

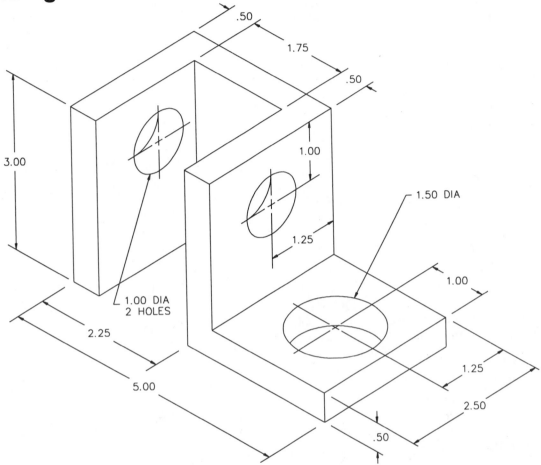

PURPOSE:
This tutorial is designed to allow the user to construct a three view drawing of the Shifter.

SYSTEM SETTINGS:
Use the Units command and change the number of decimal places from 4 to 2. Keep the remaining default settings. Use the default settings for the screen limits (0,0 for the lower left corner and 12,9 for the upper right corner). The grid and snap do not need to be set to any certain values.

LAYERS:
Create the following layers with the format:
Name-Color-Linetype
Object - Green - Continuous
Hid - Red - Hidden
Cen - Yellow - Center
Dim-Yellow-Continuous

SUGGESTED COMMANDS:
The primary commands used during this tutorial are Offset and Trim. The Offset command is used for laying out all views before using the Trim command to clean up excess lines. Since different linetypes represent certain parts of a drawing, the Change command is used to convert to the desired linetype needed as set by the Layer command. Once all visible details are identified in the primary views, project the visible features into the other views using the Line command. A 45 degree inclined line is constructed to project lines from the top view to the right side view and vice versa.

DIMENSIONING:
Dimensions may be added to this problem at a later time. Consult your instructor.

PLOTTING:
This tutorial exercise may be plotted on "B"-size paper (11"x17"). Use a plotting scale of 1=1 to produce a full size plot.

Step #1

Begin the orthographic drawing of the Shifter by constructing a right angle consisting of one horizontal and one vertical line. The corner formed by the two lines will be used to orient the front view.

Command: **Line**
From point: **1,1**
To point: **@11<0**
To point: *(Strike Enter to exit this command)*

Command: **Line**
From point: **1,1**
To point: **@8<90**
To point: *(Strike Enter to exit this command)*

Step #2

Begin the layout of the primary views by using the Offset command to offset the vertical line at "A" the distance of 5.00 units, which represents the length of the Shifter.

Command: **Offset**
Offset distance or Through <Through>: **5.00**
Select object to offset: *(Select the vertical line at "A")*
Side to offset? *(Pick a point anywhere near "B")*
Select object to offset: *(Strike Enter to exit this command)*

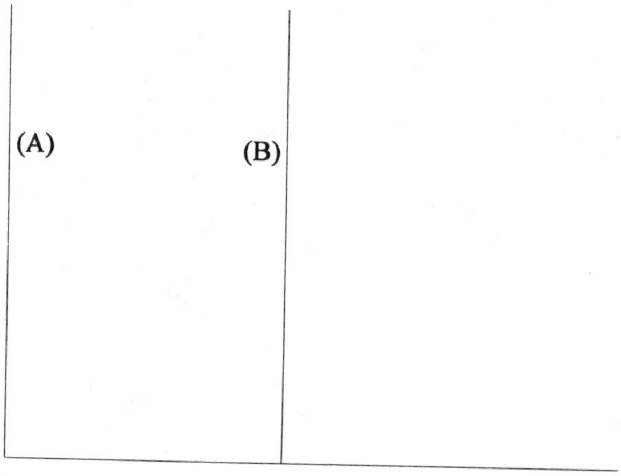

Step #3

Use the Offset command to offset the horizontal line at "A" the distance of 3.00 units, which represents the height of the Shifter.

Command: **Offset**
Offset distance or Through <5.00>: **3.00**
Select object to offset: *(Select the horizontal line at "A")*
Side to offset? *(Pick a point anywhere near "B")*
Select object to offset: *(Strike Enter to exit this command)*

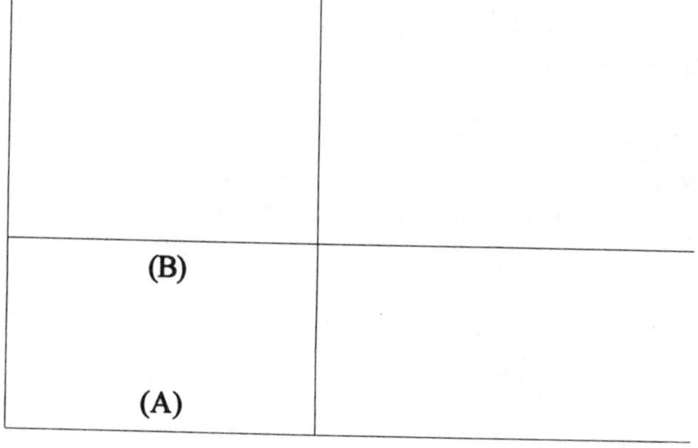

Step #4

Begin laying out dimension spaces which will act as separators between views and allow for the placement of dimensions once the Shifter is completed. A spacing of 1.50 units will be more than adequate for this purpose. Again use the Offset command to accomplish this.

Command: **Offset**
Offset distance or Through <3.00>: **1.50**
Select object to offset: *(Select the vertical line at "A")*
Side to offset? *(Pick a point anywhere near "B")*
Select object to offset: *(Select the horizontal line at "C")*
Side to offset? *(Pick a point anywhere near "D")*
Select object to offset: *(Strike Enter to exit this command)*

Step #5

Use the Offset command to lay out the depth of the Shifter at a distance of 2.50 units.

Command: **Offset**
Offset distance or Through <1.50>: **2.50**
Select object to offset: *(Select the vertical line at "A")*
Side to offset? *(Pick a point anywhere near "B")*
Select object to offset: *(Select the horizontal line at "C")*
Side to offset? *(Pick a point anywhere near "D")*
Select object to offset: *(Strike Enter to exit this command)*

Step #6

Use the Trim command to trim away excess construction lines used when laying out the primary views of the Shifter.

Command: **Trim**
Select cutting edge(s)...
Select objects: *(Select the lines at "A" and "B")*
Select objects: *(Strike Enter to continue)*
<Select object to trim>/Undo: *(Select the line at "C")*
<Select object to trim>/Undo: *(Select the line at "D")*
<Select object to trim>/Undo: *(Select the line at "E")*
<Select object to trim>/Undo: *(Select the line at "F")*
<Select object to trim>/Undo: *(Strike Enter to exit this command)*

Step #7

Use the Trim command again to complete trimming away excess construction lines used when laying out the primary views of the Shifter.

Command: **Trim**
Select cutting edge(s)...
Select objects: *(Select the lines at "A" and "B")*
Select objects: *(Strike Enter to continue)*
<Select object to trim>/Undo: *(Select the line at "C")*
<Select object to trim>/Undo: *(Select the line at "D")*
<Select object to trim>/Undo: *(Select the line at "E")*
<Select object to trim>/Undo: *(Select the line at "F")*
<Select object to trim>/Undo: *(Strike Enter to exit this command)*

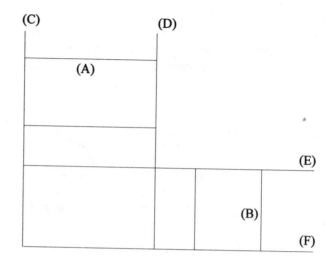

Step #8

Your display should appear similar to the illustration at the right with the layout of the front, top, and right-side views. Begin adding details to all views through methods of projection. Use the Offset command to offset lines "A", "C", and "E" a distance of 0.50.

Command: **Offset**
Offset distance or Through<2.50>: **0.50**
Select object to offset: *(Select the vertical line at "A")*
Side to offset? *(Pick a point anywhere near "B")*
Select object to offset: *(Select the horizontal line at "C")*
Side to offset? *(Pick a point anywhere near "D")*
Select object to offset: *(Select the horizontal line at "E")*
Side to offset? *(Pick a point anywhere near "F")*
Select object to offset: *(Strike Enter to exit this command)*

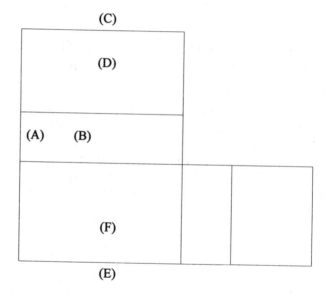

Step #9

Use the Offset command and offset the vertical line at "A" the distance of 1.75.

Command: **Offset**
Offset distance or Through<0.50>: **1.75**
Select object to offset: *(Select the vertical line at "A")*
Side to offset? *(Pick a point anywhere near "B")*
Select object to offset: *(Strike Enter to exit this command)*

Step #10

Use the Offset command to offset the vertical line at "A" a distance of 0.50.

Command: **Offset**
Offset distance or Through<1.75>: **0.50**
Select object to offset: *(Select the vertical line at "A")*
Side to offset? *(Pick a point anywhere near "B")*
Select object to offset: *(Strike Enter to exit this command)*

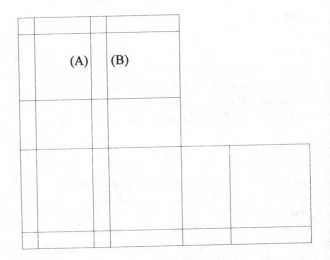

Step #11

Your display should appear similar to the illustration at the right. Next, use the Trim command to partially delete the line segments located in the spaces in between views.

Command: **Trim**
Select cutting edge(s)...
Select objects: *(Select the lines at "A," "B," "C," and "D")*
Select objects: *(Strike Enter to continue)*
<Select object to trim>/Undo: *(Select the line at "E")*
<Select object to trim>/Undo: *(Select the line at "F")*
<Select object to trim>/Undo: *(Select the line at "G")*
<Select object to trim>/Undo: *(Select the line at "H")*
<Select object to trim>/Undo: *(Select the line at "I")*
<Select object to trim>/Undo: *(Select the line at "J")*
<Select object to trim>/Undo: *(Select the line at "K")*
<Select object to trim>/Undo: *(Select the line at "L")*
<Select object to trim>/Undo: *(Strike Enter to exit this command)*

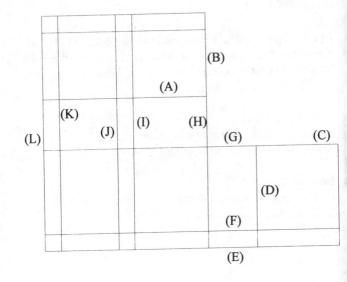

Step #12

Use the Zoom-Window command to magnify the top view similar to the illustration at the right. Then use the trim command to trim the excess entity labeled "B."

Command: **Trim**
Select cutting edge(s)...
Select objects: *(Select the vertical line at "A")*
Select objects: *(Strike Enter to continue)*
<Select object to trim>/Undo: *(Select the line at "B")*
<Select object to trim>/Undo: *(Strike Enter to exit this command)*

Step #13

Zoom back to the original display using the Zoom-Previous command. Use the Zoom-Window command to magnify the display to show the front view illustrated at the right. Use the Trim command to clean-up the excess lines in the front view using the illustration at the right as a guide.

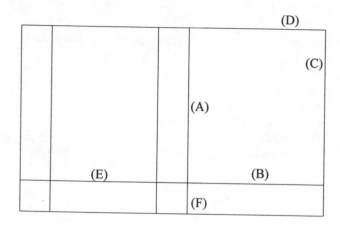

Command: **Trim**
Select cutting edge(s)...
Select objects: *(Select the lines at "A" and "B")*
Select objects: *(Strike Enter to continue)*
<Select object to trim>/Undo: *(Select the line at "C")*
<Select object to trim>/Undo: *(Select the line at "D")*
<Select object to trim>/Undo: *(Select the line at "E")*
<Select object to trim>/Undo: *(Select the short line at "F")*
<Select object to trim>/Undo: *(Strike Enter to exit this command)*

Step #14

Use the Zoom-Previous command to zoom back to the previous screen display containing the three views. Use the Fillet command to create a corner by selecting lines "A", "B", and "C" illustrated at the right.

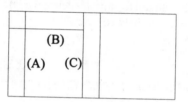

Command: **Fillet**
Polyline/Radius/<Select first object>: *(Select line "A")*
Select second object: *(Select line "B")*

Command: **Fillet**
Polyline/Radius/<Select first object>: *(Select line "B")*
Select second object: *(Select line "C")*

Step #15

Use the Trim command to partially delete the horizontal line at "C" and form the upside-down "U" shape illustrated at the right.

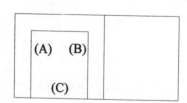

Command: **Trim**
Select cutting edge(s)...
Select objects: *(Select the vertical lines at "A" and "B")*
Select objects: *(Strike Enter to continue)*
<Select object to trim>/Undo: *(Select the horizontal line at "C")*
<Select object to trim>/Undo: *(Strike Enter to exit this command)*

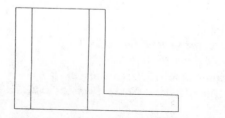

Step #16

Begin placing a circle in the top view representing the 1.50 diameter drill hole. Before accomplishing this, place a point of reference in the upper right corner of the top view. This point will then be referenced in the next step to accurately place the circle. In order to see the reference, use the Pdmode system variable to change the shape of the point to cross. Place the point using the Osnap-Endpoint command.

Command: **Pdmode**
New value for PDMODE <0>: **3**

Command: **Point**
Point: **Int**
of *(Select the intersection of the two lines at "A")*

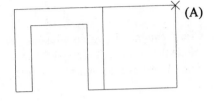
(A)

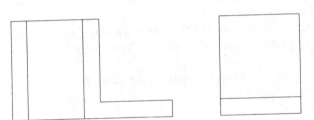

Step #17

Since the point was the last entity placed, immediately use the Circle command. Use coordinates to locate the center of the circle the distance @-1.00,-1.25 away from the reference point.

Command: **Circle**
3P/2P/TTR/<Center point>: **@-1.00,-1.25**
Diameter/<Radius>: **D**
Diameter: **1.50**

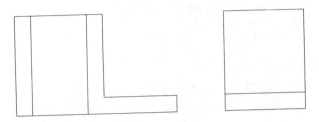

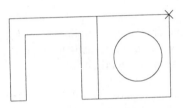

Step #18

Use the Point command again to place a reference point in the upper right corner of the right side view. This point will be used to help identify the center point of the 1.00 diameter drill holes. The current Pdmode of "3" will keep the shape of the point looking like a cross.

Command: **Point**
Point: **Int**
of *(Select the intersection of the right side view at "A")*

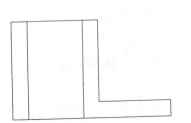

(A)

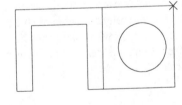

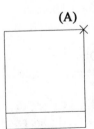

Step #19

Immediately use the Circle command and coordinates to locate the center of the circle the distance @-1.25,-1.00 away from the last reference point. Use the Erase command to delete the points.

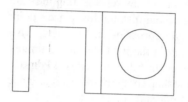

Command: **Circle**
3P/2P/TTR/<Center point>: **@-1.25,-1.00**
Diameter/<Radius>: **D**
Diameter: **1.00**

Command: **Erase**
Select objects: *(Select both points)*
Select objects: *(Strike Enter to execute the erase command)*

Step #20

Use the Line command and draw projection lines from both circles into the front view. Use the Osnap-Quadrant and Osnap-Perpend modes to accomplish this.

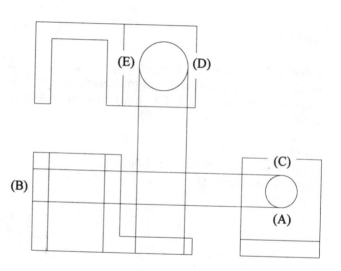

Command: **Line**
From point: **Qua**
of *(Select the quadrant of the circle at "A")*
To point: **Per**
to *(Select the line at "B" to obtain the perpendicular point)*
To point: *(Strike Enter to exit this command)*

Repeat the above procedure for the quadrants "C", "D", and "E".

Step #21

Place center marks at the centers of both circles as illustrated at the right. Before proceding with this operation, a system variable needs to be set to a certain value to achieve the desired results. Set the Dimcen system variable to a value of -0.12. This will not only place the center mark when using the Dim-Cen command but will also extend the center line a short distance outside of both circles.

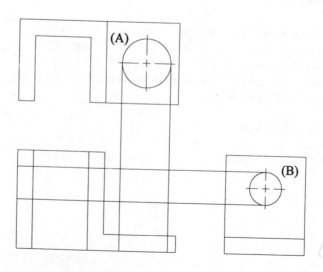

Command: **Dim**
Dim: **Dimcen**
Current value <0.09> New value: **-0.12**

Dim: **Cen**
Select arc or circle: *(Select the circle at "A")*

Dim: **Cen**
Select the arc or circle: *(Select the circle at "B")*

Dim: **Exit**

Step #22

Project two lines from the endpoints of both center marks using the Osnap-Endpoint command. Turn ortho On to assist with this operation. The lines will be converted to center lines at a later step. Draw the lines 0.50 units past the front view as illustrated at the right.

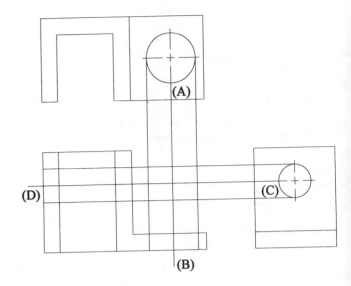

Command: **Line**
From point: **End**
of *(Select the endpoint of the center mark in the top view at "A")*
To point: *(Mark a point just below the front view at "B")*
To point: *(Strike Enter to exit this command)*

Command: **Line**
From point: **End**
of *(Select the endpoint of the center mark in the side view at "C")*
To point: *(Mark a point to the left of the front view at "D")*
To point: *(Strike Enter to exit this command)*

Step #23

Use the Trim command and the illustration at the right to trim away excess lines.

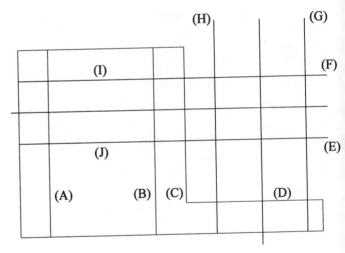

Command: **Trim**
Select cutting edge(s)...
Select objects: *(Select the lines at "A," "B," "C," and "D")*
Select objects: *(Strike Enter to continue)*
<Select object to trim>/Undo: *(Select the line at "E")*
<Select object to trim>/Undo: *(Select the line at "F")*
<Select object to trim>/Undo: *(Select the line at "G")*
<Select object to trim>/Undo: *(Select the line at "H")*
<Select object to trim>/Undo: *(Select the line at "I")*
<Select object to trim>/Undo: *(Select the line at "J")*
<Select object to trim>/Undo: *(Strike Enter to exit this command)*

Step #24

Use the Change command to change the six lines illustrated at the right from their current layer assignment of "0" to a new layer assignment of "Hid" which will change the lines from object to hidden lines.

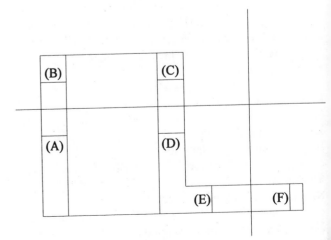

Command: **Change**
Select objects: *(Select all six short lines labeled "A" to "F")*
Select objects: *(Strike Enter to continue this command)*
Properties/<Change point>: **P**
Change what property (Color/Elev/LAyer/LType/Thickness)? **LA**
New layer <0>: **Hid**
Change what property (Color/Elev/LAyer/LType/Thickness)?
(Strike Enter to exit this command)

Step #25

Use the Break command to partially delete the lines illustrated at the right before converting them to center lines. Remember, centerlines extend past the object lines when identifying hidden drill holes; it would be inappropriate to use the Trim command for this step.

Command: **Break**
Select object: *(Select the horizontal line at "A")*
Enter second point (or F for first point): *(Select the line at "B")*

Command: **Break**
Select object: *(Select the vertical line at "C")*
Enter second point (or F for first point): *(Select the line at "D")*

Command: **Break**
Select object: *(Select the horizontal line at "E")*
Enter second point (or F for first point): **@**

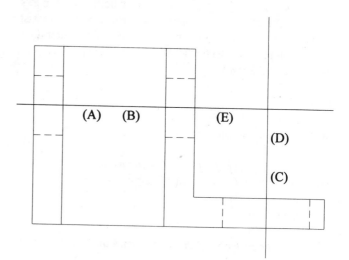

Step #26

The purpose of the "@" symbol in the step above is to break a line into two segments without showing the break. The "@" symbol means "the last known point" which completes the Break command by satisfying the "Second point" prompt. To prove this, use the Erase command to delete the segments no longer needed.

Command: **Erase**
Select objects: *(Carefully select the lines at "A" and "B")*
Select objects: *(Strike Enter to perform the erase)*

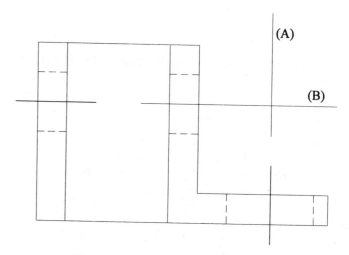

Step #27

Use the Change command to change the three lines illustrated at the right from their current layer assignment of "0" to a new layer assignment of "Cen" which will change the lines from object to center lines.

Command: **Change**
Select objects: *(Select all three short lines labeled "A" to "C")*
Select objects: *(Strike Enter to continue this command)*
Properties/<Change point>: **P**
Change what property (Color/Elev/LAyer/LType/Thickness)?**LA**
New layer <0>: **Cen**
Change what property (Color/Elev/LAyer/LType/Thickness)?
(Strike Enter to exit this command)

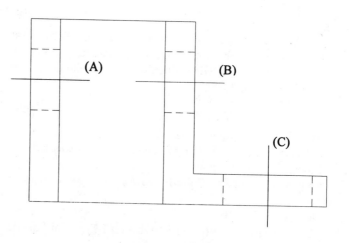

Step #28

A 45 degree angle needs to be constructed in order to begin projecting features from the top view to the right side view and then back again. This angle is formed by extending the bottom edge of the top view to intersect with the left edge of the side view. Use the Fillet command to accomplish this. Then draw the 45 degree line; the length of this line is not important.

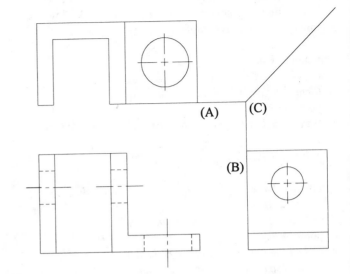

Command: **Fillet**
Polyline/Radius/<Select first object>: *(Select line "A")*
Select second object: *(Select line "B")*

Command: **Line**
From point: **Int**
of *(Select the intersection at "C")*
To point: **@4<45**
To point: *(Strike Enter to exit this command)*

Step #29

Draw lines from points "A", "B", and "C" to intersect with the 45 degree angle projector. Be sure Ortho mode is on to draw horizontal lines. Use the Osnap-Intersec, Endpoint, and Quadrant modes to assist in this operation.

Command: **Line**
From point: **Int**
of *(Select the intersection of the corner at "A")*
To point: *(Draw a line just past the 45 degree angle)*
To point: *(Strike Enter to exit this command)*

Command: **Line**
From point: **End**
of *(Select the endpoint of the center line at "B")*
To point: *(Draw a line just past the 45 degree angle)*
To point: *(Strike Enter to exit this command)*

Command: **Line**
From point: **Qua**
of *(Select the quadrant of the circle at "C")*
To point: *(Draw a line just past the 45 degree angle)*
To point: *(Strike Enter to exit this command)*

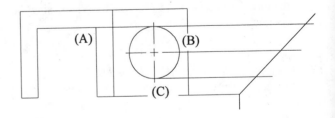

Step #30

Draw lines from the intersection of the 45 degree angle projection line to points in the right side view. Follow the commands below to accomplish this.

Command: **Line**
From point: **Int**
of *(Select the intersection of the angle at "A")*
To point: **Per**
to *(Select the bottom horizontal line of the right side view at "B")*
To point: *(Strike Enter to exit this command)*

Command: **Line**
From point: **Int**
of *(Select the intersection of the angle at "C")*
To point: **Per**
to *(Select the bottom line of the right side view at "B")*
To point: *(Strike Enter to exit this command)*

Command: **Line**
From point: **Int**
of *(Select the intersection of the angle at "D")*
To point: *(Select a point below the bottom of the side view at "E")*
To point: *(Strike Enter to exit this command)*

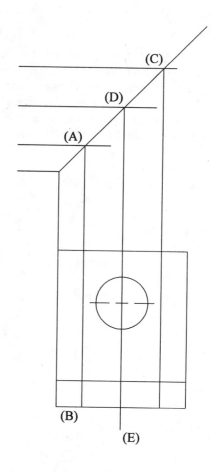

Step #31

Erase the three projection lines from the top view using the Erase command.

Command: **Erase**
Select objects: *(Select lines "A", "B", and "C")*
Select objects: *(Strike Enter to perform this command)*

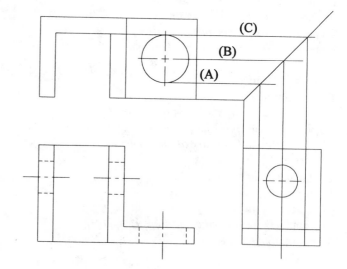

Step #32

Use the Trim command to trim away unnecessary geometry using the illustration at the right as a guide. The hidden hole and slot will be formed in the side view through this operation.

Command: **Trim**
Select cutting edges...
Select objects: *(Select the horizontal line at "A")*
Select objects: *(Strike Enter to continue)*
<Select object to trim>/Undo: *(Select the line at "B")*
<Select object to trim>/Undo: *(Strike Enter to exit this command)*

Repeat the above procedure for the line illustrated at the right using the horizontal line "C" as the cutting edge and the vertical line "D" as the line to trim.

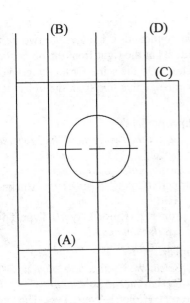

Step #33

Use the Break command to split the vertical line at "A" into two separate entities. This will be accomplished by typing @ in response to the prompt "Enter second point". This will split the line in two without noticing the break.

Command: **Break**
Select object: *(Select the vertical line at "A")*
Enter second point (or F for first point): **@**

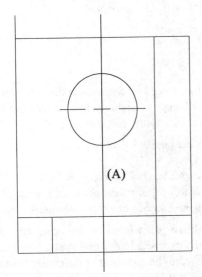

Step #34

Use the Erase command to delete the top half of the line broken previously in Step #33. This will leave a short line segment that will be changed or converted into a center line marking the center of a hidden hole. Redraw the screen to refresh the vertical center lines at the hole.

Command: **Erase**
Select objects: *(Select the vertical line at "A")*
Select objects: *(Strike Enter to execute this command)*

Command: **Redraw**

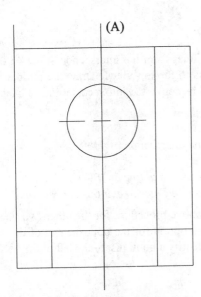

Step #35

Use the Chprop command to convert the two vertical lines labeled "A" and "B" at the right from the object layer to the layer named "Hid." Do the same for the longer vertical line labeled "C" but change this line from the object layer to the layer named "Cen."

Command: **Chprop**
Select objects: *(Select the two vertical lines labeled "A" and "B")*
Select objects: *(Strike Enter to continue)*
Change what property(Color/LAyer/LType/Thickness)? **LA**
New layer <0>: **Hid**
Change what property (Color/LAyer/LType/Thickness)? *(Strike Enter to exit this command)*

Command: **Chprop**
Select objects: *(Select the vertical line labeled "C")*
Select objects: *(Strike Enter to continue)*
Change what property(Color/LAyer/LType/Thickness)? **LA**
New layer <0>: **Cen**
Change what property (Color/LAyer/LType/Thickness)? *(Strike Enter to exit this command)*

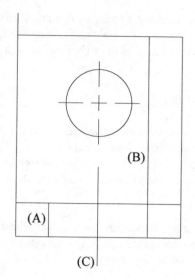

Step #36

Draw three lines from key features on the side view to intersect with the 45 degree angle projector. Use Osnap options whenever possible. Ortho mode must be On.

Command: **Line**
From point: **Qua**
of *(Select the quadrant of the circle at "A")*
To point: *(Identify a point past the 45-degree angle)*
To point: *(Strike Enter to exit this command)*

Repeat the above procedure for the other two projection lines. Use the Osnap-Endpoint option and begin the second projector line from the endpoint of the center marker at "B." Begin the third projector line from the quadrant of the circle at "C."

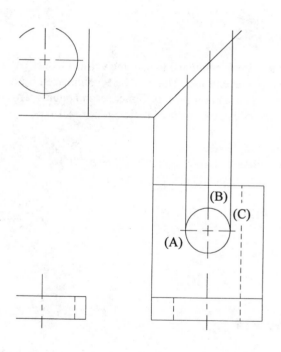

Step #37

Draw three lines from the intersection at the 45 degree projector across to the top view. Draw the middle line longer since it will be converted into a center line at a later step.

Command: **Line**
From point: **Int**
of *(Select the intersection at "A")*
To point: **Per**
to *(Select the vertical line at "B")*
To point: *(Strike Enter to exit this command)*

Repeat the above procedure for the other two lines. Do not use an Osnap option for the opposite end of the middle line but rather identify a point just to the left of the vertical line at "B."

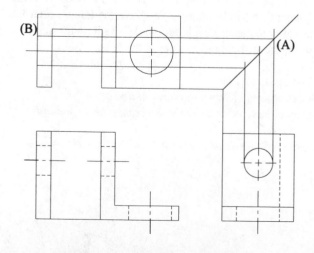

Step #38

Use the Erase command to erase the three vertical projector lines from the side view.

Command: **Erase**
Select objects: *(Select the vertical lines labeled "A", "B", and "C")*
Select objects: *(Strike Enter to execute this command)*

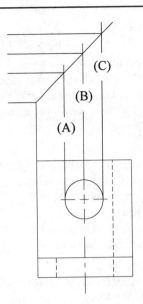

Step #39

Use the Trim command to trim away excess lines in the top view as illustrated at the right. Perform a redraw when completed.

Command: **Trim**
Select cutting edges...
Select objects: *(Select the vertical lines labeled "A", "B", and "C")*
Select objects: *(Strike Enter to continue)*
<Select object to trim>/Undo: *(Select the line at "D")*
<Select object to trim>/Undo: *(Select the line at "E")*
<Select object to trim>/Undo: *(Select the line at "F")*
<Select object to trim>/Undo: *(Select the line at "G")*
<Select object to trim>/Undo: *(Strike Enter to exit this command)*

Command: **Redraw**

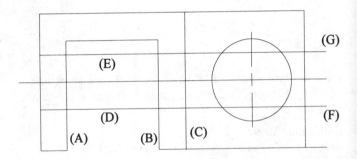

Step #40

Use the Break command to partially delete the horizontal line segment from "A" to "B". Use the Break command with the @ option to break the line into two segments at "C". Use the Erase command to delete the trailing line segment at "D". Redraw the screen.

Command: **Break**
Select object: *(Select the horizontal line at "A")*
Enter second point (or F for first): *(Select the line at "B")*

Command: **Break**
Select object: *(Select the horizontal line at "C")*
Enter second point (or F for first): **@**

Command: **Erase**
Select objects: *(Select the horizontal line segment at "D")*
Select objects: *(Strike Enter to execute this command)*

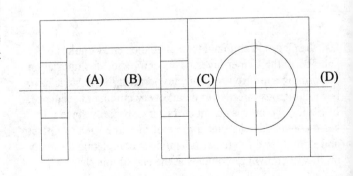

Step #41

Use the Chprop command to change the line segments labeled "A," "B," "C," and "D" to the new layer named "Hid," which will display hidden lines. Change the line segments labeled "E" and "F" to the new layer named "Cen," which will display center lines.

Command: **Chprop**
Select objects: *(Select the four lines labeled "A" to "D")*
Select objects: *(Strike Enter to continue)*
Change what property (Color/LAyer/LType/Thickness)? **LA**
New layer <0>: **Hid**
Change what property (Color/LAyer/LType/Thickness)? *(Strike Enter to exit this command)*

Command: **Chprop**
Select objects: *(Select the two lines labeled "E" and "F")*
Select objects: *(Strike Enter to continue)*
Change what property (Color/LAyer/LType/Thickness)? **LA**
New layer <0>: **Cen**
Change what property (Color/LAyer/LType/Thickness)? *(Strike Enter to exit this command)*

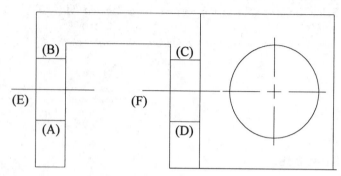

Step #42

Use the Erase command to delete the 45 degree angle line. Use the Fillet command to create corners in the top view and right side view.

Command: **Erase**
Select objects: *(Select the inclined line at "A")*
Select objects: *(Strike Enter to execute this command)*

Command: **Fillet**
Polyline/Radius/<Select first object>: *(Select line "B")*
Select second object: *(Select line "C")*

Command: **Fillet**
Polyline/Radius/<Select first object>: *(Select line "D")*
Select second object: *(Select line "E")*

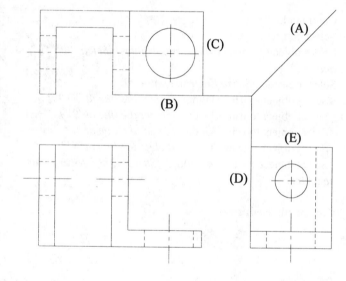

Use the Chprop command to change the center marks at "A" and "B" to the Center layer. This completes this tutorial on performing a multi-view projection drawing. The steps have been numerous in order to detail every command sequence. In reality, the process is much faster, especially since a few basic commands were used most of the time such as Offset and Trim. Use this tutorial as a guide in completing the many multi-view drawing problems at the end of this chapter.

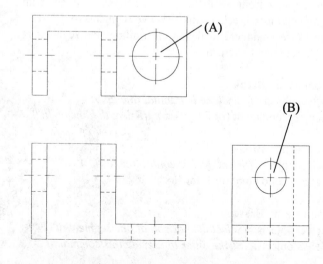

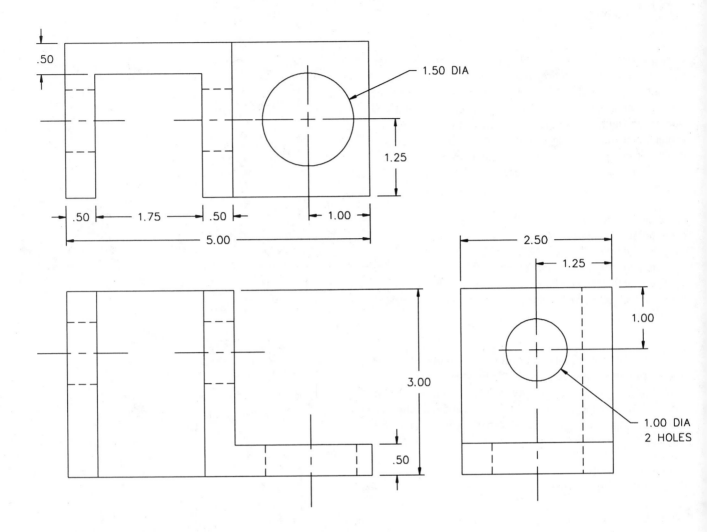

In keeping with the layers created, all object lines are to be changed from the layer they were created on to the object layer using the Chprop command.

Command: **Chprop**
Select objects: *(Select all twenty-four object lines)*
Select objects: *(Strike Enter to continue)*
Change what property(Color/LAyer/LType/Thickness)? **LA**
New layer <0>: **Object**
Change what property (Color/LAyer/LType/Thickness)?
    (Strike Enter to exit this command)

The final process in completing a multi-view drawing is to place dimensions to define size and locate features. This topic will be discussed in a later chapter.

Tutorial Exercise #11
XYZ.Dwg

It is possible to perform multi-view projections from one view to another without drawing construction lines and then trimming them to size. XYZ filters provide the means of accomplishing this along with a little practice. In the illustration at the right, the problem is to draw a circle using the center point of the circle exactly at the center of the rectangle. To complicate the issue, no other command or setting can be used to perform this operation. First, construct the rectangle.

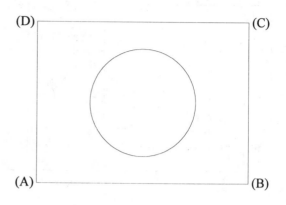

Command: **Line**
From point: **2,2** *(Point "A")*
To point: **@8<0** *(To Point "B")*
To point: **@6<90** *(To Point "C")*
To point: **@8<180** *(To Point "D")*
To point: **C** *(Back to Point "A" to close and exit the command)*

Use the Circle command and begin with the normal "Center point" prompt sequence. Begin finding the center of the rectangle by first filtering out the midpoint of the X value at "A" followed by the Y value at "B." Answer the prompt for Z by entering a value of 0. Complete the Circle command by supplying the radius value of 2.

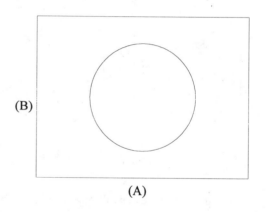

Command: **Circle**
3P/2P/TTR/<Center point>: **.X**
of **Mid**
of *(Select the horizontal line at "A")*
(Need YZ): **.Y**
of **Mid**
of *(Select the vertical line at "B")*
(Need Z): **0**
Diameter/<Radius>: **2**

To review, filtering out the midpoint of the horizontal line at "A" saved the point to satisfy the center point of the circle. However, a circle needs at least a second point; this is the reason for the prompt "Need YZ." A YZ point was needed to find the center point. Instead of selecting any point, the Y point was filtered out at the midpoint of the vertical line at "B." Now, the X and Y values were saved to satisfy the center of the circle. However, since AutoCAD now exists in a 3-dimensional (3D) database, a prompt "Need Z" appears. Since the entire drawing is located at an elevation of Ø, entering a value of 0 satisfies the centerpoint of the circle and marks a point. This may seem very tedious and too much trouble; with a little practice, however, filters become another drawing aid to work in your favor.

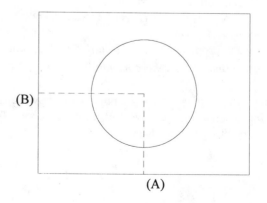

Tutorial Exercise #12
Gage.Dwg

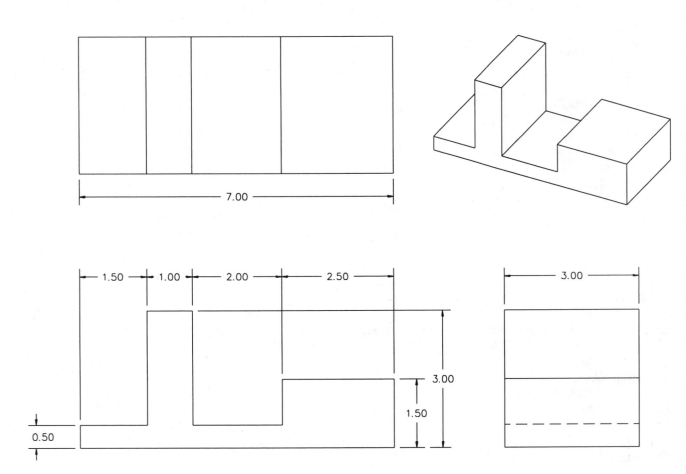

PURPOSE:

This tutorial is designed to allow the user to construct a three view drawing of the Gage with the aid of XYZ filters.

SYSTEM SETTINGS:

Begin a new drawing called "Gage." Use the Units command to change the number of decimal places past the zero from 4 to 2. Keep the remaining default unit values. Using the Limits command, keep 0,0 for the lower left corner and change the upper right corner from 12,9 to 15.50,9.50. Use the Grid command and change the grid spacing from 1.00 to 0.25 units. Do not turn the snap or ortho On.

LAYERS:

Create the following layers with the format:

Name-Color-Linetype
Hidden-Red-Hidden

SUGGESTED COMMANDS:

Begin this tutorial by laying out the three primary views using the Line and Offset commands. Use the Trim command to clean up any excess line segments. As an alternate method used for projection, use XYZ filters in combination with Osnap options to add features in other views.

DIMENSIONING:

Dimensions may be added to this problem at a later time. Consult your instructor.

PLOTTING:

This tutorial exercise may be plotted on "B"-size paper (11" x 17"). Use a plotting scale of 1=1 to produce a full size plot.

Step #1

Begin the multi-view drawing of the gage by constructing
the front view using absolute and polar coordinates. Start
the front view at coordinate 1.50,1.00.

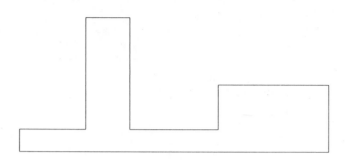

Command: **Line**
From point: **1.50,1.00**
To point: **@7.00<0**
To point: **@1.50<90**
To point: **@2.50<180**
To point: **@1.00<270**
To point: **@2.00<180**
To point: **@2.50<90**
To point: **@1.00<180**
To point: **@2.50<270**
To point: **@1.50<180**
To point: **C**

Step #2

Begin the construction of the top view by locating the lower
left corner at coordinate 1.50,5.50.

Command: **Line**
From point: **1.50,5.50**
To point: **@7.00<0**
To point: **@3.00<90**
To point: **@7.00<180**
To point: **C**

Step #3

Begin the construction of the right side view by locating the
lower left corner at coordinate 10.50,1.00.

Command: **Line**
From point: **10.50,1.00**
To point: **@3.00<0**
To point: **@3.00<90**
To point: **@3.00<180**
To point: **C**

Step #4

Your display should appear similar to the illustration at the right with the placement of the top view above the front view and the right side view directly to the right of the front. Filtering methods will now be used to complete the missing lines in the top and right side views.

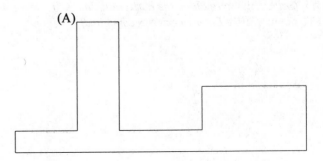

Step #5

Zoom into the drawing so the front and top views appear similar to the illustration at the right. Follow the steps below for the proper use of XYZ filters when used in conjunction with the Line command.

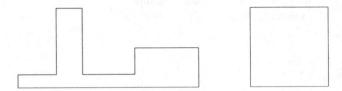

Command: **Line**
From point: **.X**
of **Int**
of *(Select the intersection of the lines at "A")*
(Need YZ): **Nea**
to *(Select anywhere along the horizontal line "B")*
To point: **Per**
to *(Select anywhere along the horizontal line "C")*
To point: *(Strike Enter to exit this command)*

The X value identified by selecting the intersection on the front view was saved for later use by projecting the value to the horizontal line of the top view and completing the Line command with the Osnap-Perpendicular option.

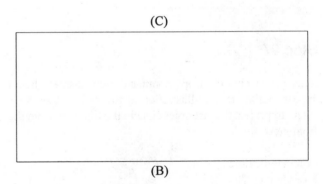

Step #6

Your display should appear similar to the illustration at the right. Rather than use filters for the next line, simply duplicate the last line drawn using the Copy command.

Command: **Copy**
Select objects: **L**
Select objects: *(Strike Enter to continue)*
<Base point or displacement>/Multiple: **Endp**
of *(Select the endpoint of the line at "A")*
Second point of displacement: **Endp**
of *(Select the endpoint of the line at "B")*

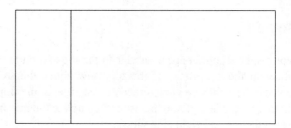

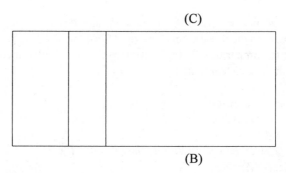

Step #7

After performing the Copy command, your display should appear similar to the illustration at the right. Use XYZ filters to project the last object line in the top view from the front view.

Command: **Line**
From point: **.X**
of **Int**
of *(Select the intersection of the lines at "A")*
(Need YZ): **Nea**
to *(Select anywhere along the horizontal line "B")*
To point: **Per**
to *(Select anywhere along the horizontal line "C")*
To point: *(Strike Enter to exit this command)*

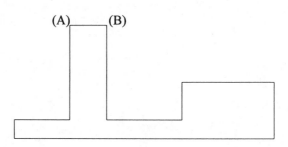

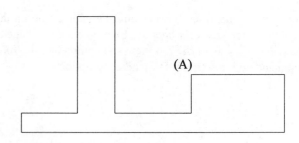

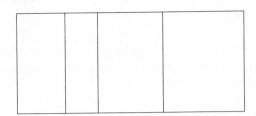

Step #8

The complete front and top views are illustrated at the right. Use the same procedure with XYZ filters for completing the right side view.

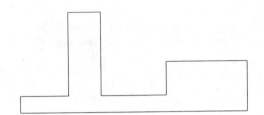

Step #9

Begin using XYZ filters to add missing lines to the right side view. First, use the Zoom-Window option to magnify the area illustrated at the right.

Command: **Zoom**
All/Center/Dynamic/Extents/Left/Previous/Vmax/Window/
<Scale(X/XP)>: **W**
First corner: *(Select a point on the screen at "A")*
Second corner: *(Select a point on the screen at "B")*

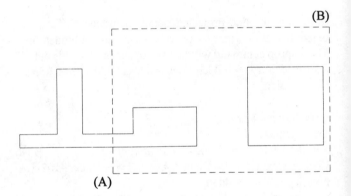

Step #10

Instead of filtering or saving the X coordinate, the same procedure can be used for saving the Y coordinate for later use.

Command: **Line**
From point: **.Y**
of **Int**
of *(Select the intersection of the lines at "A")*
(Need XZ): **Nea**
to *(Select anywhere along the vertical line "B")*
To point: **Per**
to *(Select anywhere along the vertical line "C")*
To point: *(Strike Enter to exit this command)*

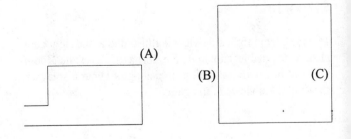

Step #11

Filter out the Y coordinate value of the intersecting area of the front view at the right to project the coordinate to the right side view.

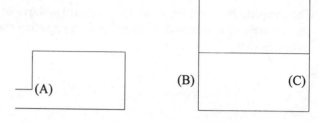

Command: **Line**
From point: **.Y**
of **Int**
of *(Select the intersection of the lines at "A")*
(Need XZ): **Nea**
to *(Select anywhere along the vertical line "B")*
To point: **Per**
to *(Select anywhere along the vertical line "C")*
To point: *(Strike Enter to exit this command)*

Step #12

Since the last line projected details a hidden surface, the object line at the right needs to be changed to a hidden line. The Chprop command will be used to accomplish this only if a layer identifying hidden lines, such as Hidden, has already been created.

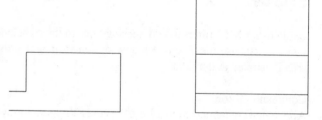

Command: **Chprop**
Select objects: **L**
Select objects: *(Strike Enter to continue)*
Change what property(Color/LAyer/LType/Thickness)? **LA**
New layer<0>: **Hidden**
Change what property(Color/LAyer/LType/Thickness)?
(Strike Enter to exit this command)

Step #13

The completed right side view is illustrated at the right complete with visible and invisible surfaces. Use the Zoom-Previous or Zoom-All options to demagnify your display and show all three views of the gage.

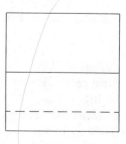

As a general rule of thumb, an X coordinate was filtered out and retrieved later for projecting lines from the front to the top view or vice versa. A Y coordinate was filtered out and retrieved later for projecting lines from the front to the right side view and vice versa. Filters may also be used for projecting such features as holes and slots between views. The key is to lock onto a significant part of an entity using one of the many options provided by the Osnap option.

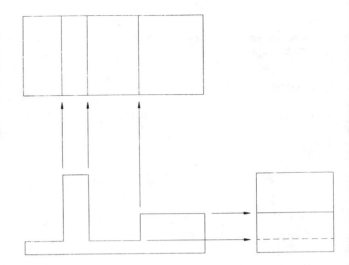

With the completed three-view drawing illustrated below, the next step would be to add dimensions to the views for manufacturing purposes. This topic will be discussed in a later chapter.

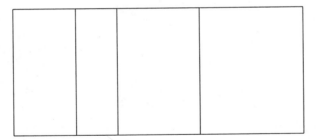

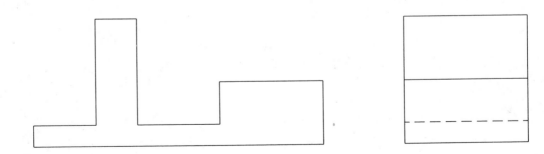

Questions for Unit 4

Directions for Questions 4-1 through 4-4:
Based on the isometric drawings, identify which set of orthographic projections correctly depict the object. Place your answer in the box provided below.

Question 4-1

Answer

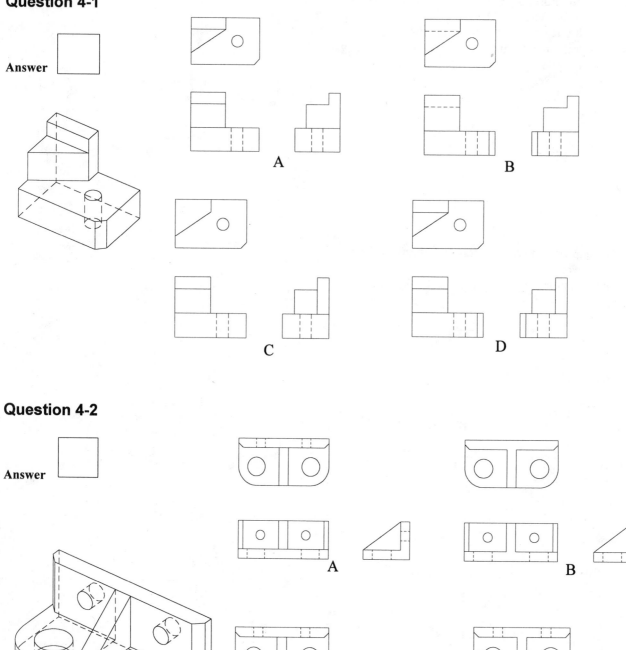

Question 4-2

Answer

Question 4-3

Answer

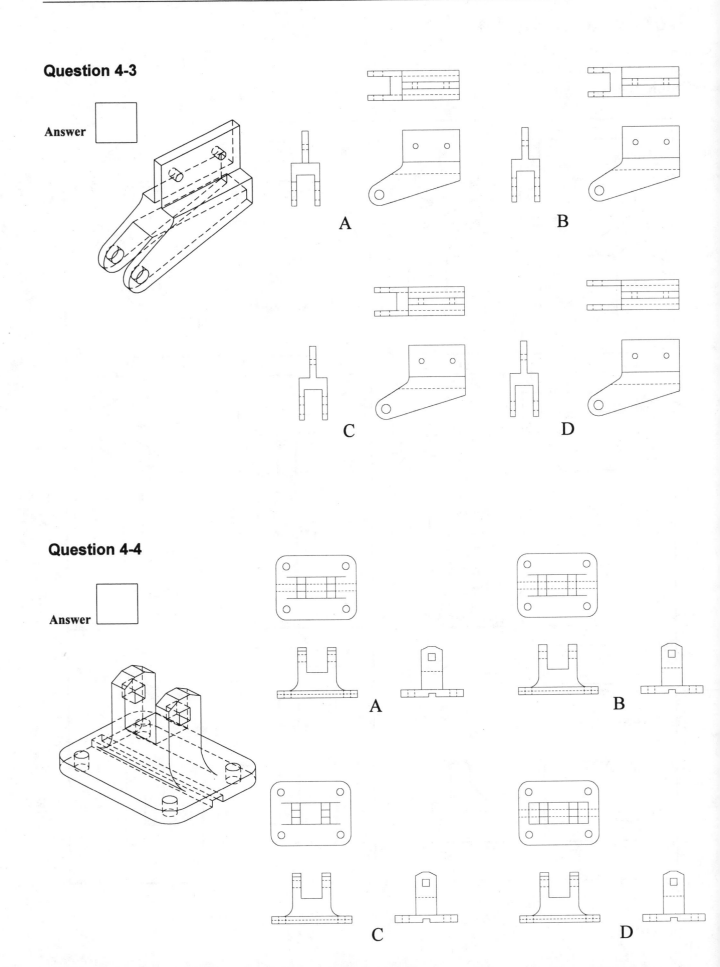

A

B

C

D

Question 4-4

Answer

A

B

C

D

Problems for Unit 4

Directions for Problems 4-5 and 4-6:
Find the missing lines in these problems and sketch the correct solution.

Problem 4-5

Problem 4-6

Directions for Problems 4-7 through 4-11:

Construct a multi-view drawing by sketching the front, top, and right side views using the grid below as a guide.

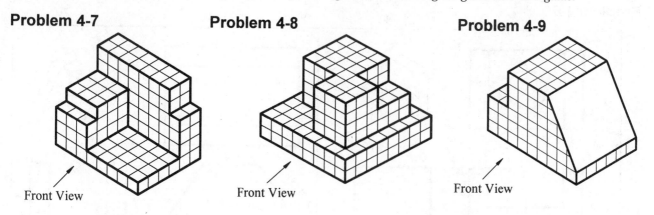

Problem 4-7

Front View

Problem 4-8

Front View

Problem 4-9

Front View

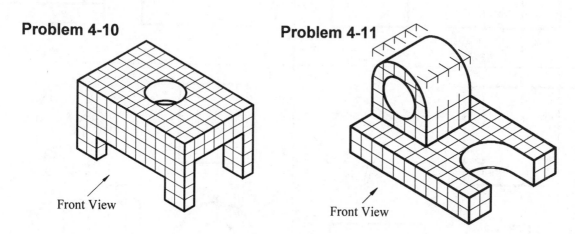

Problem 4-10

Front View

Problem 4-11

Front View

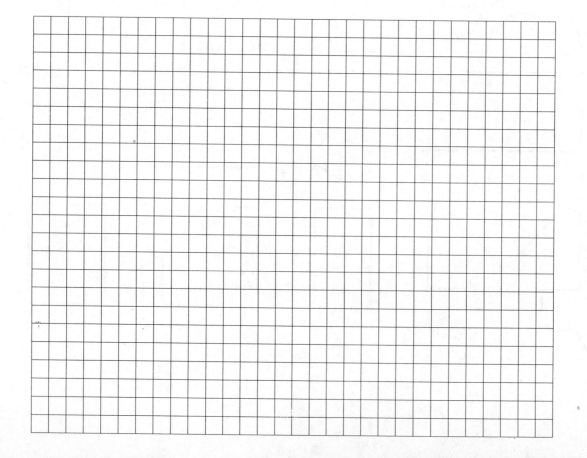

Directions for Problems 4–12 through 4–21
Construct a multi-view drawing of each object. Be sure to construct only those views that accurately describe the object.

Problem 4–12

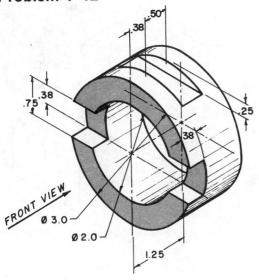

Problem 4–15

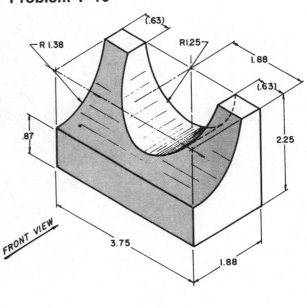

Problem 4–13

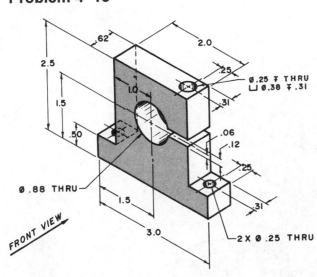

Problem 4–16

Problem 4–14

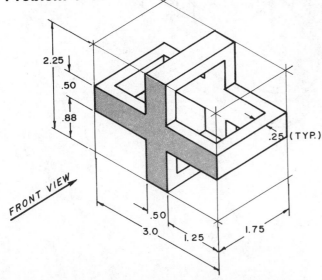

Problem 4–17

Problem 4–18

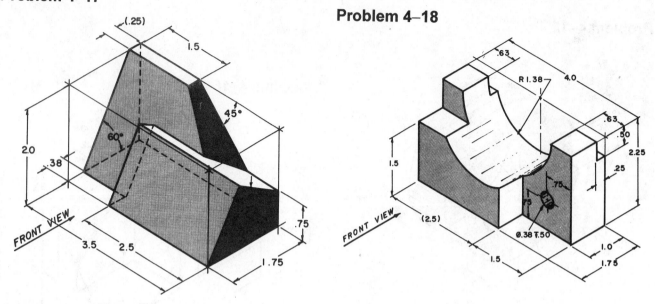

Problem 4–19

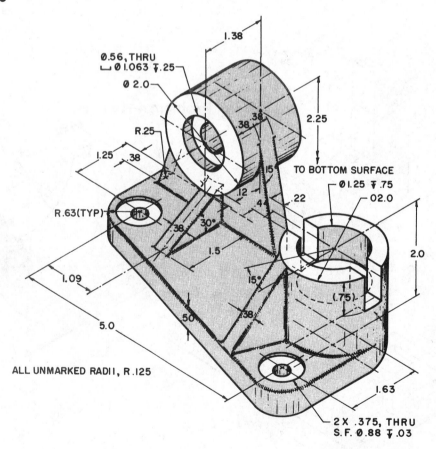

R.63(TYP)

Ø.56, THRU
⌴ Ø1.063 ⊼.25
Ø 2.0

R.25

1.25   .38

1.09

5.0

ALL UNMARKED RADII, R.125

1.38

.38

2.25

TO BOTTOM SURFACE
Ø1.25 ⊼.75
O2.0

2.0

(.75)

15°

.38

1.5

.50

15°

.22

.4

.12

38

30°

38

1.63

2X .375, THRU
S.F. Ø.88 ⊼.03

Problem 4–20

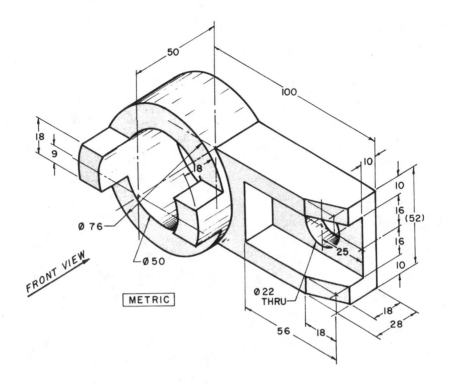

FRONT VIEW

METRIC

Problem 4–21

Directions for Problems 4–22 through 4–33
Construct a 3-view drawing of each object.

Problem 4–22

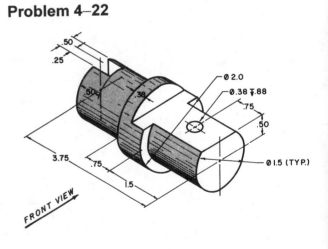

FRONT VIEW

METRIC

Problem 4–23

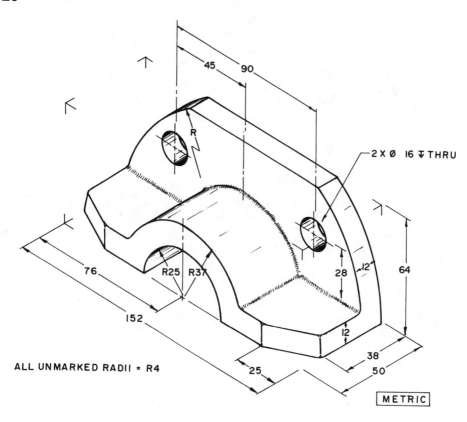

ALL UNMARKED RADII = R4

METRIC

Problem 4–24

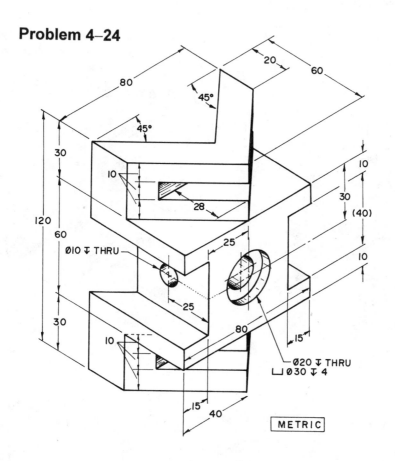

METRIC

Problem 4–25

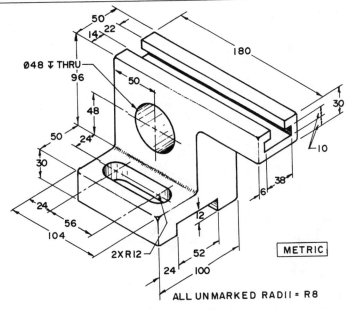

Ø48 ⤓ THRU

96

50

48

50 24

30

24

56

104

2×R12

52

24 100

12

50

14 22

180

30

10

38

6

METRIC

ALL UNMARKED RADII = R8

Problem 4–26

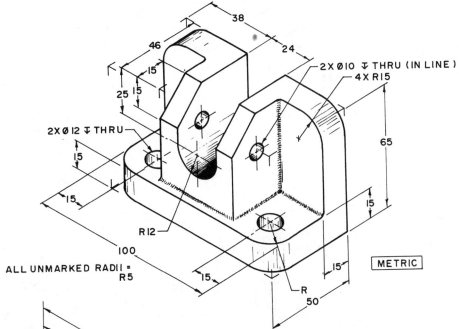

38

46

24

15

25 15

2×Ø10 ⤓ THRU (IN LINE)

4×R15

2×Ø12 ⤓ THRU

15

65

15

R12

15

100

ALL UNMARKED RADII = R5

15

R

50

METRIC

Problem 4–27

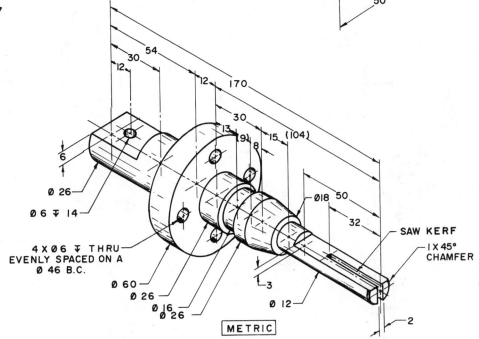

54

30

12

12 170

30

13 (9) 8

15 (104)

6

Ø 26

Ø 6 ⤓ 14

4 × Ø6 ⤓ THRU
EVENLY SPACED ON A
Ø 46 B.C.

Ø 60

Ø 26

Ø 16
Ø 26

Ø18

Ø 12

50

32

SAW KERF

1 × 45°
CHAMFER

3

2

METRIC

Problem 4–28

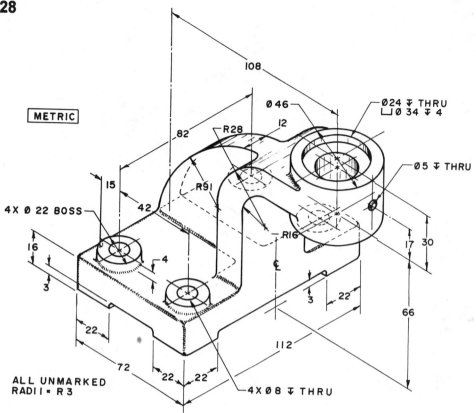

METRIC

Ø 46
R28
12
Ø24 ⇩ THRU
⊔ Ø 34 ⇩ 4
108
82
R91
15
R16
42
Ø5 ⇩ THRU
4X Ø 22 BOSS
16
4
17 30
3
℄
3 22
66
22
112
72
22 22
4XØ8 ⇩ THRU

ALL UNMARKED
RADII = R3

Problem 4–29

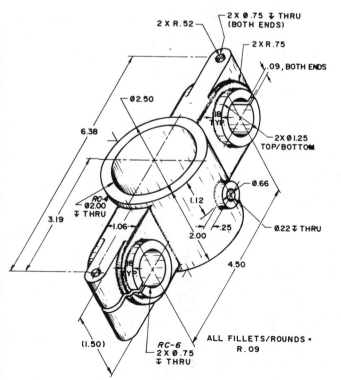

2 X R.52
2 X Ø.75 ⇩ THRU
(BOTH ENDS)
2 X R.75
.09, BOTH ENDS
Ø2.50
6.38
R
TYP
2X Ø1.25
TOP/BOTTOM
RC-4
Ø2.00
⇩ THRU
Ø.66
1.12
3.19
1.06
.25
2.00
Ø.22 ⇩ THRU
4.50
R
TYP
(1.50)
RC-6
2 X Ø.75
⇩ THRU
ALL FILLETS/ROUNDS =
R.09

Problem 4–30

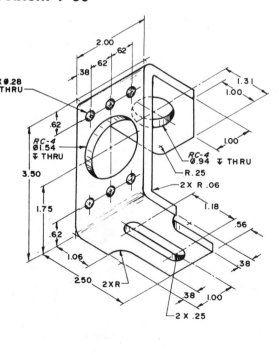

2.00
.62
6X Ø.28
⇩ THRU
.38 .62
.62
1.31
1.00
RC-4
Ø1.54
⇩ THRU
3.50
RC-4
Ø.94 ⇩ THRU
R.25
1.00
2 X R .06
1.75
1.18
.62
.56
1.06
2.50 2XR
38
38 1.00
2 X .25

Problem 4–31

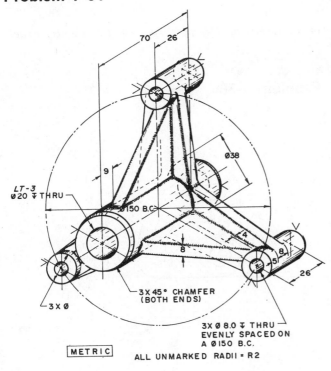

LT-3
Ø20 ⊥ THRU

70   26

Ø38

9

Ø150 B.C.

4

8   8

5

26

3X Ø

3X 45° CHAMFER
(BOTH ENDS)

3X Ø 8.0 ⊥ THRU
EVENLY SPACED ON
A Ø150 B.C.

METRIC

ALL UNMARKED RADII = R2

Problem 4–33

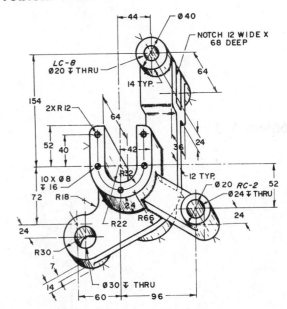

44   Ø 40

NOTCH 12 WIDE X
68 DEEP

LC-8
Ø20 ⊥ THRU

64

14 TYP.

154

2X R 12

64

52   40

42

36   24

10 X Ø 8
⊥ 16

R32

12 TYP.

Ø 20 RC-2   52
Ø24 ⊥ THRU

72   R18

24

24

R66

R30

R22

Ø30 ⊥ THRU

7

14

60   96

Problem 4–32

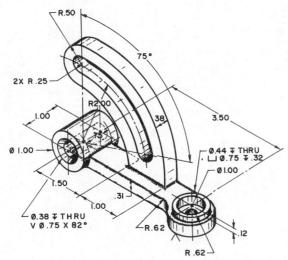

R.50

75°

2X R.25

R2.00

1.00

38   3.50

Ø 1.00

0.44 ⊥ THRU
⌴ Ø.75 ⊥ .32

Ø1.00

1.50

.31

1.00

Ø.38 ⊥ THRU
V Ø.75 X 82°

R.62

.12

R .62

ALL UNMARKED RADII = R.12

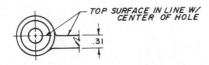

TOP SURFACE IN LINE W/
CENTER OF HOLE

.31

Directions for Problems 4–34 through 4–37
Using the background grid as a guide, reproduce each problem on a CAD system. Use a grid spacing of 0.25 units for each problem.

Problem 4–34

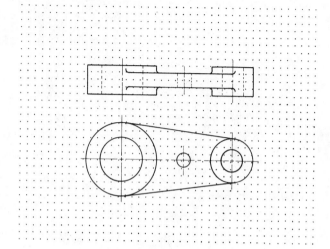

Problem 4–36

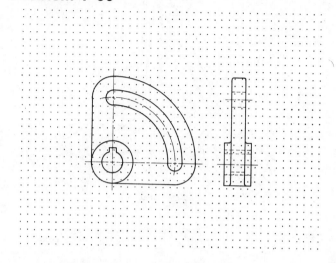

Problem 4–35

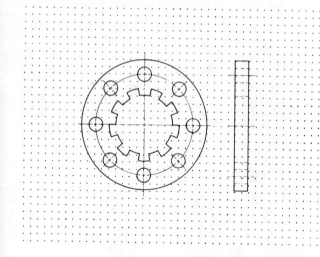

Problem 4–37

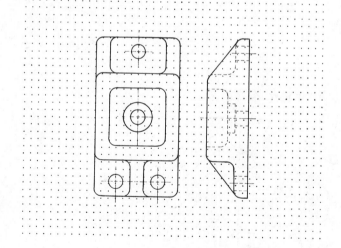

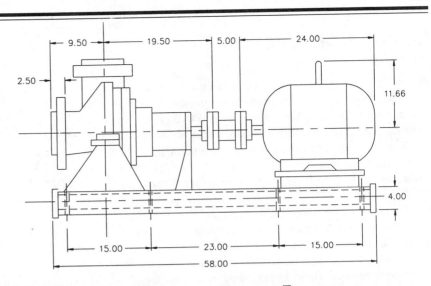

UNIT 5

Dimensioning Techniques

Contents

Once views have been laid out, a design is not ready for the production line until numbers describing how wide, tall, or deep the object is are added to the drawing. However, these numbers must be added in a certain organized fashion; otherwise, the drawing becomes difficult to read. This may lead to confusion and the possible production of a part that is incorrect by the original design. This unit will focus on the basics of dimensioning and includes a few rules for proper dimensioning practices with numerous examples. Dimensioning techniques using

AutoCAD will also be discussed in great detail including linear dimensioning, radius and diameter dimensioning, dimensioning angles, leader line usage, and a complete listing of all dimension variables with an explanation of their purpose. Special topics include dimensioning isometric drawings and ordinate or datum dimensioning. Three tutorials follow the main body of text; all three tutorials are complete regarding geometry. To complete them, dimensions need to be added or edited in some cases.

Dimension Basics

Before discussing the components of a dimension, remember that object lines (at "A") continue to be the thickest lines of a drawing with the exception of border or title blocks. To promote contrasting lines, dimensions become visible, yet thin lines. The heart of a dimension is the dimension line (at "B"), which is easily identified by the location of arrow terminators at both ends (at "D"). In mechanical cases, the dimension line is broken in the middle, which provides an excellent location for the dimension text (at "E"). For architectural applications, dimension text is usually placed above an unbroken dimension line. The extremities of the dimension lines are limited by placing lines that act as stops for the arrow terminators. These lines, called extension lines (at "C"), begin close to the object without touching the object. Extension lines will be highlighted in greater detail in the pages that follow. For placing diameter and radius dimensions, a leader line consisting of an inclined line with a short horizontal shoulder is used (at "F"). Other applications of leader lines are for adding notes to drawings.

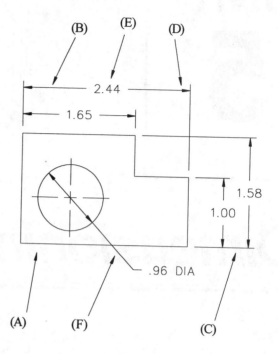

When placing dimensions in a drawing, it is recommended to provide a spacing of at least 0.38 units between the first dimension line and object being dimensioned (at "A"). If placing stacked or baseline dimensions, provide a minimum spacing of at least 0.25 units between the first and second dimension line at "B" or any other dimension lines placed thereafter. This will prevent dimensions from being placed too close to each other.

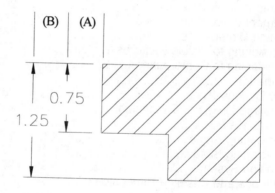

It is recommended that extensions never touch the object being dimensioned and begin approximately between 0.03 and 0.06 units away from the object at "A." As dimension lines are added, extension lines should extend no further than 0.12 beyond the arrow or any other terminator (at "B"). The height of dimension text is usually 0.125 units (at "C"). This value also applies to notes placed on objects with leader lines. Certain standards may require a taller lettering height. Become familiar with office practices that may deviate from these recommended values.

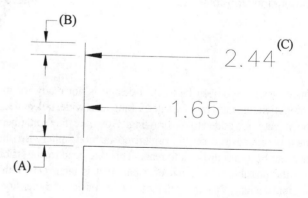

Placement of Dimensions

When placing multiple dimensions on one side of an object, place the shorter dimension closest to the object followed by the next larger dimension. When placing multiple horizontal and vertical dimensions involving extension lines that cross other extension lines, do not place gaps in the extension lines at their intersection points.

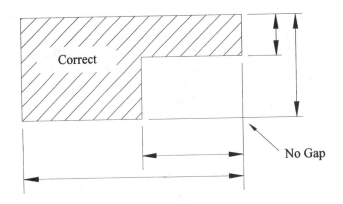

As it is acceptable for extension lines to cross each other, it is considered unacceptable practice for extension lines to cross dimension lines as in the example at the right. The shorter dimension is placed closest to the object followed by the next larger dimension.

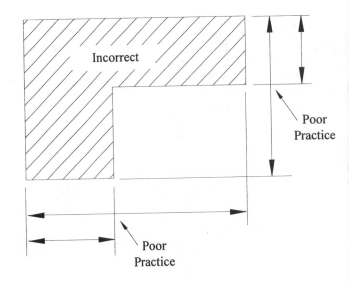

It is considered poor practice to dimension on the inside of an object when there is sufficient room to place dimensions on the outside. There may be exceptions to this rule, however. It is also considered poor practice to cross dimension lines since this may render the drawing confusing and possibly result in the inaccurate interpretation of the drawing.

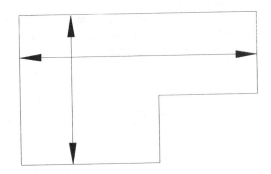

Placement of Extension Lines

As two extension lines may intersect without providing a gap, so also may extension lines and object lines intersect with each other without the need for a gap in between them. This is the same rule practiced when using center lines that extend beyond the object without gapping.

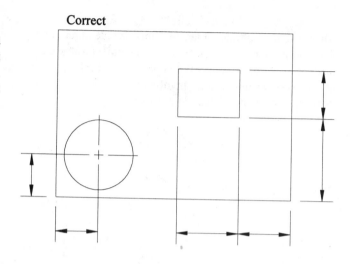

Correct

In the example at the right, the gaps in the extension lines may appear acceptable; however, in a very complex drawing, gaps in extension lines would render a drawing confusing. Draw extension lines as continuous lines without providing breaks in the lines.

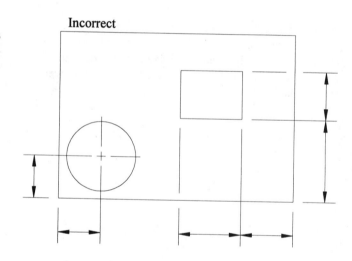

Incorrect

In the same manner, when center lines are used as extension lines for dimension purposes, no gap is provided at the intersection of the center line and the object. As with extension lines, the center line should extend no further than 0.125 units past the arrow terminator.

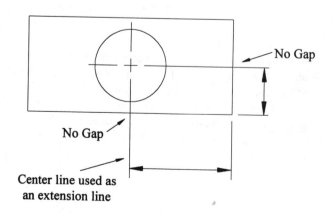

No Gap

No Gap

Center line used as
an extension line

Grouping Dimensions

To promote ease of reading and interpretation, it is considered good practice to group dimensions whenever possible as in the example at the right. This promotes good organizational skills and techniques in addition to making the drawing and dimensions easier to read. As in previous examples, always place the shorter dimensions closest to the drawing followed by any larger or overall dimensions.

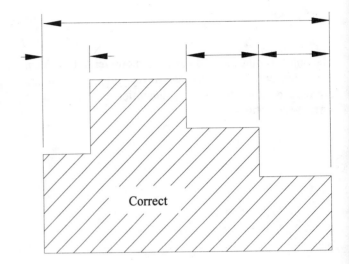

Avoid placing dimensions to an object line substituting for an extension line as in the illustration at the right. The drawing is more difficult to follow with the dimensions being placed at different levels instead of being grouped. It must be pointed out at this time, however, that there may be cases where even this practice of dimensioning is unavoidable.

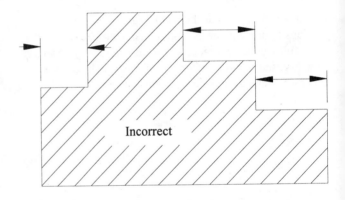

For tight spaces, arrange dimensions as in the illustration at the right. Extra care needs to be exercised to follow proper dimension rules without sacrificing clarity. AutoCAD dimension variables Dimtix and Dimsoxd may aid in the placing of dimensions in small spaces.

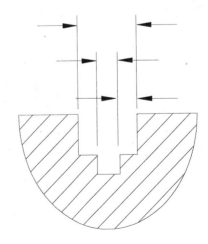

Dimensioning to Visible Features

The object is dimensioned correctly, however, the problem is that hidden lines are used to dimension to. As there are always exceptions, try to avoid dimensioning to any hidden surfaces or features.

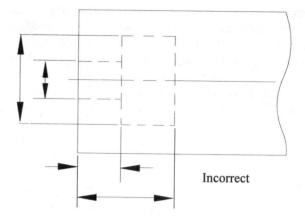

Incorrect

The object illustrated at the right is almost identical to the previous figure with the exception that it has been converted into a full section. Surfaces that were previously hidden are now exposed. This example illustrates a better way to dimension details that were previously invisible.

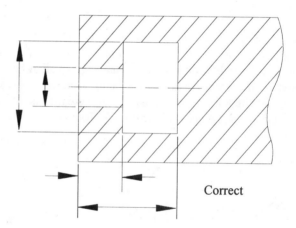

Correct

Dimensioning to Center Lines

Center lines are used to identify the center of circular features as in "A". The AutoCAD dimension variable Dimcen may be used to control the size of the center marker and whether the center marker extends beyond the largest circle. Center lines can also be used to indicate an axis of symmetry as in "B". Here, the center line consisting of a short dash flanked by two long dashes signifies the feature is circular in shape and form. Center lines may take the place of extension lines when placing dimensions in drawings.

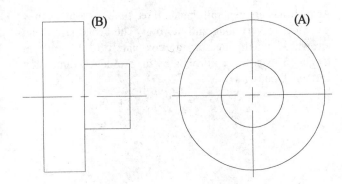

The illustration at the right represents the top view of a "U"-shaped object with two holes placed at the base of the "U". It also represents the correct way of utilizing center lines as extension lines when dimensioning to holes. What makes this example correct is the rule of always dimensioning to visible features. The example at the right uses center lines to dimension to holes that appear as circles. This is in direct contrast to the next example.

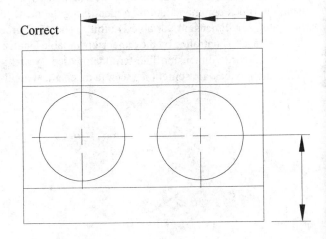

This illustration represents the front view of the "U"-shaped object. The hidden lines display the circular holes passing through the object along with center lines. Center lines are being used as extension lines for dimensioning purposes; however, it is considered poor practice to dimension to hidden features or surfaces. Always attempt to dimension to a view where the features are visible before dimensioning to hidden areas.

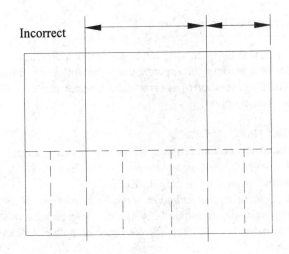

Arrowheads

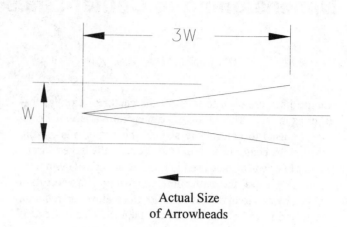

Actual Size
of Arrowheads

Arrowheads are generally made three times as long as they are wide, or very long and narrow. The actual size of an arrowhead would measure approximately 0.125 units in length. This size is controlled by the AutoCAD dimension variable Dimasz.

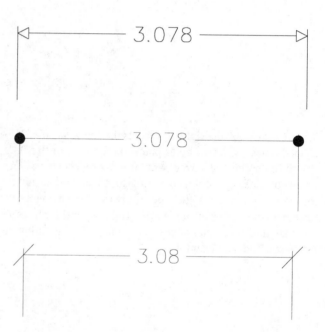

Dimension line terminators may take the form of shapes other than filled-in arrowheads. Open arrowheads and filled-in dots are controlled by first defining the shape as a symbol using the Block command and then identifying the name in the AutoCAD dimension variable Dimblk. The 45-degree slash or "tick" is controlled by the dimension variable Dimtsz. This is a favorite dimension line terminator used by architects although they are sometimes seen in mechanical applications.

Linear Dimensions

Horizontal Dimensioning

This linear dimensioning mode generates a dimension line that is horizontal in appearance. The following prompts illustrate generation of a horizontal dimension using the Dim-Horizontal option:

Dim: **Hor**
First extension line origin or RETURN to select: **Endp**
of *(Select the endpoint of the horizontal line at "A")*
Second extension line origin: **Endp**
of *(Select the other endpoint of the horizontal line at "B")*
Dimension line location (Text/Angle): *(Select a point at "C")*
Dimension text <2.00>: *(Strike Enter to accept the default value)*

Vertical Dimensioning

This linear dimensioning mode generates a dimension line that is vertical in appearance. The following prompts illustrate generation of a horizontal dimension using the Dim-Vertical option:

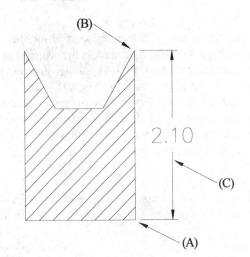

Dim: **Ver**
First extension line origin or RETURN to select: **Endp**
of *(Select the endpoint of the vertical line at "A")*
Second extension line origin: **Endp**
of *(Select the other endpoint of the vertical line at "B")*
Dimension line location (Text/Angle): *(Select a point at "C")*
Dimension text <2.10>: *(Strike Enter to accept the default value)*

Aligned Dimensioning

This linear dimensioning mode generates a dimension line that is parallel to the distance specified by the two extension line origins. The following prompts illustrate generation of an aligned dimension using the Dim-Aligned option:

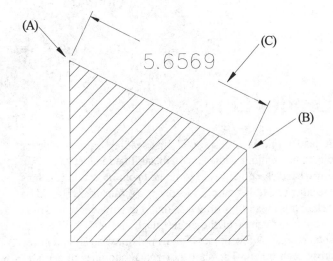

Dim: **Ali**
First extension line origin or RETURN to select: **Endp**
of *(Select the endpoint of the line at "A")*
Second extension line origin: **Endp**
of *(Select the other endpoint of the line at "B")*
Dimension line location (Text/Angle): *(Select a point at "C")*
Dimension text <5.6569>: *(Strike Enter to accept the default value)*

Rotated Dimensioning

This linear dimensioning mode generates a dimension line that is rotated at a specified angle. The following prompts illustrate generation of a rotated dimension using the Dim-Rotated option:

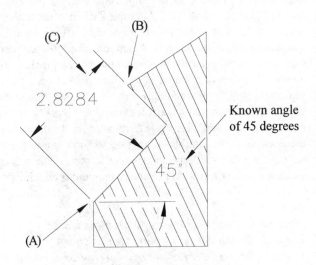

Dim: **Rot**
Dimension line angle <0>: **45**
First extension line origin or RETURN to select: **Endp**
of *(Select the endpoint of the line at "A")*
Second extension line origin: **Endp**
of *(Select the other endpoint of the line at "B")*
Dimension line location (Text/Angle): *(Select a point at "C")*
Dimension text <2.8284>: *(Strike Enter to accept the default value)*

Radius and Diameter Dimensioning

Arcs and circles are to be dimensioned in the view where their true shape is visible. The mark in the center of the circle or arc indicate its center point. The dimension text may be placed either inside or outside of the circle or arc depending on the current values of two dimension variables, namely Dimtix and Dimtofl. Both of these variables will be discussed in detail at a later time. The prompts for Diameter and Radius dimensions are as follow:

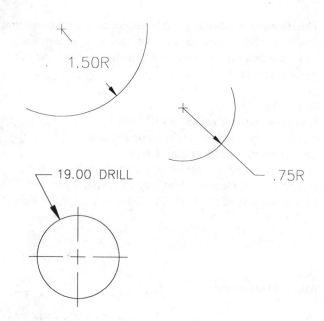

Dim: **Dia**
Select arc or circle: *(Select the edge of an arc or circle)*
Dimension text <value>: *(Strike Enter to accept the default)*

Dim: **Rad**
Select arc or circle: *(Select the edge of an arc or circle)*
Dimension text <value>: *(Strike Enter to accept the default)*

Leader Lines

A leader line is a thin, solid line leading from a note or dimension ending with an arrowhead illustrated at "A". The arrowhead should always terminate on an object line such as the edge of a hole or arc. A leader to a circle or arc should be radial; this means it is drawn so that if extended it would pass through the center of the circle illustrated at "B". Leaders should cross as few object lines as possible and should never cross each other. The short horizontal shoulder of a leader should meet the dimension illustrated at "A". It is considered poor practice to underline the dimension with the horizontal shoulder illustrated at "C". Example "C" also illustrates a leader not lined up with the center or radial. This may affect the appearance of the leader. Again, check for the standard office practices to ensure this example is acceptable. Yet another function of a leader is to attach notes to a drawing illustrated at "D". Notice the two notes attached to the view have different terminators, arrows, and dots. It is considered good practice to adopt only one terminator for the duration of the drawing. The Dimblk dimension variable may be used for defining different terminators such as dots. The prompt sequence for the AutoCAD Leader command is as follows:

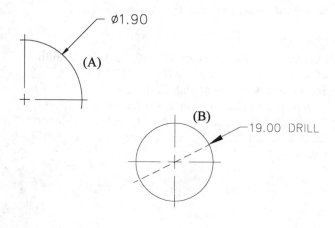

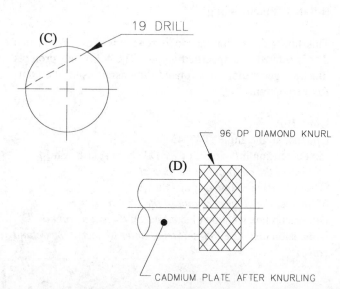

Dim: **Lea**
From point: *(Select a starting point for the leader)*
To point: *(Select an ending point for the leader)*
To point: *(Strike Enter to place the short horizontal shoulder)*
Dimension text < >: *(Enter the desired text for the leader)*

Dimensioning Angles

Dimensioning angles requires two lines forming the angle in addition to the location of the vertex of the angle. Other important information needed before dimensioning an angle includes the dimension arc location, the dimension text or what the angle actually measures, and the location of the dimension text. Before going any further, understand where the curved arc for the angular dimension is derived from. In illustration "A" at the right, the dimension arc is struck from an imaginary center or vertex of the arc. The following prompts are taken from the AutoCAD Angular dimension command:

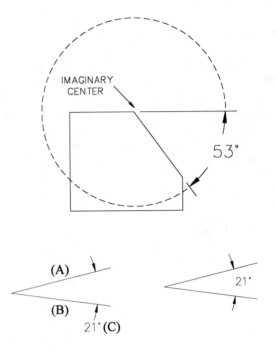

Dim: **Ang**
Select arc, circle, line or RETURN: *(Select the line at "A")*
Second line: *(Select the line at "B")*
Dimension line arc location (Text/Angle): *(Select a point at "C")*
Dimension text <21>: *(Strike Enter to accept the default value)*
Enter text location (or RETURN): *(Strike Enter to place text in the center of the dimension arc)*

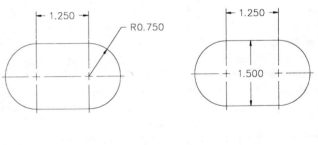

Dimensioning Slots

For slots, first select the view where the slot is visible. Two methods of dimensioning the slot are illustrated at the right. A slot may be called out by locating the center-to-center distance of the two semi-circles followed by a radius dimension to one of the semi-circles; which radius dimension selected depends on the available room to dimension. A second method involves the same center-to-center distance followed by an overall distance designating the width of the slot. This dimension happens to be the same as the diameter of the semi-circles. It is also considered good practice to place this dimension inside of the slot. A more complex example involves slots formed by curves and angles. Here, the radius of the circular center arc is called out. Angles reference each other for accuracy. As in the previous example, the overall width of the slot is dimensioned which happens to be the diameter of the semi-circles at opposite ends of the slot.

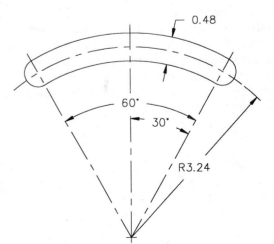

Dimensioning Systems

The Unidirectional System

When placing dimensions using AutoCAD, a typical example is illustrated at the right. Here, all text items are right-reading or horizontal. This goes for all vertical, aligned, angular, and diameter dimensions. When dimension text can be read right-reading, this is the Unidirectional Dimensioning System. By default, all AutoCAD dimension variables are set to dimension in the Unidirectional System.

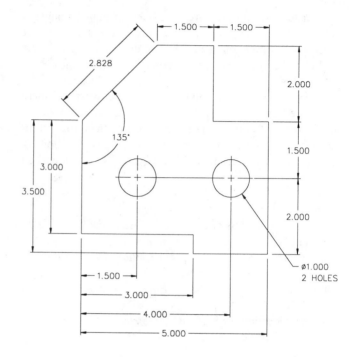

The Aligned System

The identical object from the previous example is illustrated at the right. Notice that all horizontal dimensions have the text positioned in the horizontal direction as in the previous example. However, vertical and aligned dimension text is rotated or aligned with the direction being dimensioned. This is the most notable feature of the Aligned Dimensioning System. Text along vertical dimensions is rotated in such a way that the drawing must be read from the right. Angular dimensions remain unaffected in the Aligned System; however, aligned dimension text is rotated parallel with the feature being dimensioned. Dimtih and Dimtoh are the two dimension variables that control whether the dimension text is horizontal. Both are currently in an Off mode. Once switched On, text for vertical dimensions will appear similar to the example at the right.

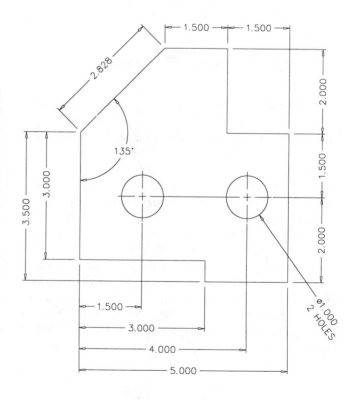

Continuous and Baseline Dimensions

The power of grouping dimensions for ease of reading has already been explained. The illustration at the right shows yet another feature while dimensioning in AutoCAD; namely, the practice of using Continuous dimensions. With one dimension already placed, the Continuous subcommand of DIM: is selected, which prompts the user for the second extension line location. Picking the second extension line location strings the dimensions next to each other or continues the dimension.

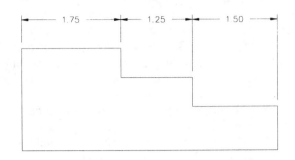

Yet another aid in grouping dimensions is using the Baseline mode, which is also located in the DIM: area of AutoCAD. Continuous dimensions place dimensions next to each other; Baseline dimensions establish a base or starting point for the first dimension. Any dimensions that follow in Baseline mode are calculated from the common base point already established. This is a very popular mode to use when one end of an object acts as a reference edge. As dimensions are placed in Baseline mode, the AutoCAD dimension variable Dimdli controls the spacing of the dimensions away from each other. This variable has a default spacing of 0.38 units.

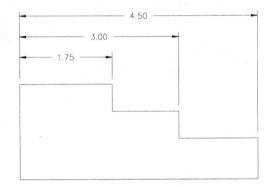

Tolerances

Interchangeability of parts requires replacement parts to fit in an assembly of an object no matter where the replacement part comes from. In the example at the right, the "U"-shaped channel of 2.000 units in width is to accept a mating part of 1.995 units. Under normal situations, there is no problem with this drawing or callout. However, what if the person cannot make the channel piece exactly at 2.000? What if he is close and the final product measures 1.997? Again, the mating part will have no problems fitting in the 1.997 slot. What if the mating part is not made exactly 1.995 units but is instead 1.997? You see the problem. As easy as it is to attach a dimension to a drawing, some thought needs to go into the possibility that based on the numbers, maybe the part cannot be easily made. Instead of locking dimensions in using one number, a range of numbers would allow the production individual the flexibility to vary in any direction and still have the parts fit together. This is the purpose of converting some basic dimensions to tolerances.

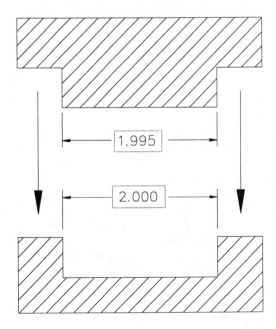

The example at the right shows the same two mating parts; this time two sets of numbers for each part are assigned. For the lower part, the machinist must make the part anywhere in between 2.002 and 1.998, which creates a range of 0.004 units of variance. The upper mating part must be made within 1.997 and 1.993 units in order for the two to fit correctly. The range for the upper part is also 0.004 units. As we will soon see, no matter if upper or lower numbers are used, the parts will always fit together. If the bottom part is made to 2.002 and the top part is made to 1.993, the parts will fit. If the bottom part is made to 1.998 and the upper part is made to 1.997, the parts will fit. In any case or combination, if the dimensions are followed exactly as stated by the tolerances, the pieces will always fit. If the bottom part is made to 1.998 and the upper part is made to 1.999, the upper piece is rejected. The method of assigning upper and lower values to dimensions is called limit dimensioning. Here, the larger value in all cases is placed above the smaller value. This is also called a clearance fit since any combinations of numbers may be used and the parts will still fit together. The AutoCAD dimension variable Dimlim controls the display of limit dimensions. Variables Dimtp and Dimtm assign positive and negative ranges for the tolerance. These will be discussed later in this chapter.

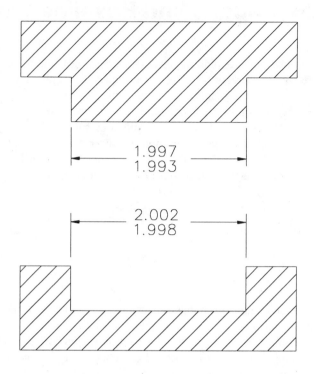

The object at the right has a different tolerance value assigned to it. The basic dimension is 2.000; in order for this part to be accepted, the width of this object may go as high as 2.002 units or as low as 2.001 units giving a range of 0.003 units by which the part may vary. This type of tolerance is called a plus/minus dimension with an upper limit of 0.002 difference from the lower limit of -0.001. The AutoCAD dimension Dimtol controls the display of plus/minus dimensions. Variables Dimtp and Dimtm assign positive and negative ranges for the tolerance. These will be discussed later in this chapter.

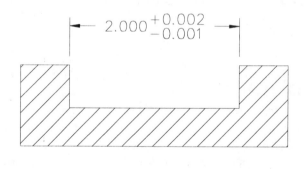

Illustrated at the right is yet another way to display tolerances. It is very similar to the plus/minus method except that both upper and lower limits are the same; for this reason, this method is called plus and minus dimensions. The basic dimension is still 2.000 with upper and lower tolerance limits of 0.002 units. The dimension variable Dimtol again controls the display of plus and minus dimensions. When both the Dimtp and Dimtm variables have the same values, the result is both tolerances listed together with the plus and minus symbol (±) placed in front of the tolerance. All variables will be discussed later in this chapter.

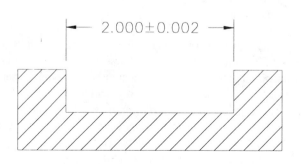

Repetitive Dimensions

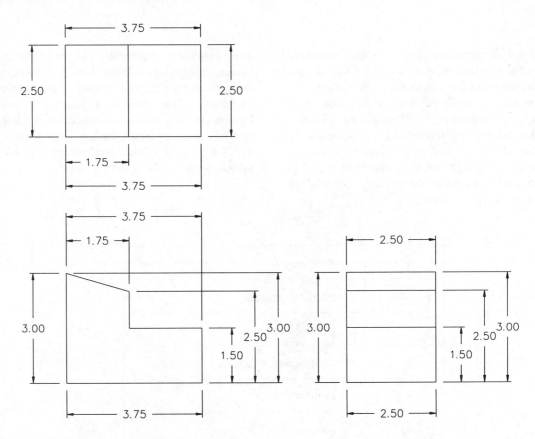

Throughout this chapter, numerous methods of dimensioning have been discussed supported by many examples. Just as it was important in multi-view projection to draw only those views that accurately described the object, so also is it important to dimension these views. In this manner, the actual production of the part may start. However, care needs to be taken when placing dimensions; dimensioning takes planning to better place a dimension. The problem with the illustration above is that even though the views are correct and the dimensions call out the overall sizes of the object, there are too many cases where dimensions are duplicated. Once a feature has been dimensioned, such as the overall width of 3.75 units, this number does not need to be placed in the top view. This is the purpose of understanding the relationship between views or what dimensions the views have in common with each other. Adding unnecessary dimensions also makes the drawing very busy and cluttered in addition to being very confusing to read. Compare the illustration above with the illustration at the right, which shows just those dimensions needed to describe the size of the object. Do not be concerned that the top view has no dimensions; the designer should interpret the width of the top as 3.75 units from the front view and the depth as 2.50 from the right side view.

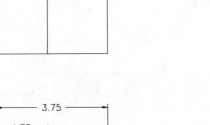

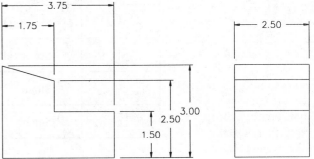

General Dimensioning Symbols

In todays global economy, the transfer of documents in the form of drawings is becoming a standard form of doing business. As drawings are shared with a subsidiary of an overseas company, two different forms of language may be needed to interpret the dimensions of the drawing. This in the past has led to confusion of interpreting a drawing and as a result, a system of dimensioning symbols has been developed. It is the hope that the recognition of a symbol will be easier to interpret than adding a note on the particular feature being dimensioned in a different language.

The illustration below shows some of the more popular dimensioning symbols in use today on drawings. Notice how the symbols are designed to make as clear and consistent an interpretation of the dimension as possible. As an example, the Deep or Depth symbol displays an arrow pointing down. This symbol is used to identify the how far into a part the depth of a drill hole goes. Other symbols such as the Arc Length symbol identifies the length of an arc.

⌒	Arc Length
X.XX	Basic Dimension
▷	Conical Taper
⊔	Counterbore or Spotface
⌄	Countersink
↧	Deep or Depth
⌀	Diameter
X.XX	Dimension Not to Scale
2X	Number of Times - Places
R	Radius
(X.XX)	Reference Dimension
S⌀	Spherical Diameter
SR	Spherical Radius
◺	Slope
☐	Square

The size of all general dimensioning symbols is illustrated below. All values are in relation to "h" or the relative height of the dimension numerals. As an example, with a dimension numeral height of 0.18, study the illustration at the right for finding the size of the counterbore dimension symbol. Since the height of the counterbore symbol is defined by "h," simply substitute the dimension numeral height of 0.18 for the height of the counterbore symbol. Since the width of the symbol is "2h," multiply 0.18 by 2 to obtain a value of 0.36. These symbols may be created and saved as blocks for insertion into drawings. They may also be part of an existing text font mapped to a particular key on the keyboard. Striking that key brings up the dimension symbol.

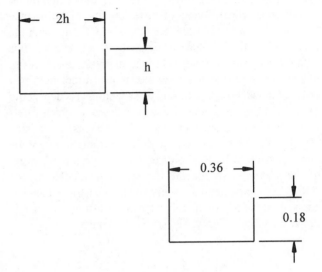

Applying the Basic Dimension Symbol

A basic dimension is identified by drawing a rectangular box around the dimension numeral. The basic dimension represents the theoretical distance of the feature such as the distance from the center hole to the left hole as 4.00 units with the distance from the left hole to the right hole as 8.00 units. It is a known fact that it is impossible to maintain tolerances to obtain an exact value of 4.00 or 8.00 units. A rectangle may be constructed; however, this may be a tedious task depending on the number of decimal places past the zero which determines the size of the rectangle. The dimension variable "DIMGAP," which controls the distance from the end of the dimension line to the dimension text, also controls the drawing of the rectangle around the dimension text. Setting the "DIMGAP" variable to a negative value constructs the rectangle; the dimension numerals at the right illustrate a rectangle drawn by setting "DIMGAP" set to -0.20 units.

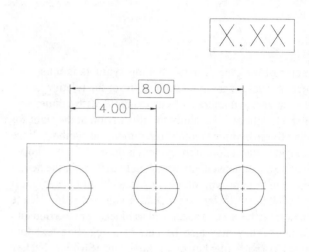

Applying the Reference Dimension Symbol

Illustrated at the right are a series of continuous dimensions strung together in one line. The dimension numeral 3.00 units is enclosed in parenthesis and is referred to as a reference dimension. Reference dimensions are not required drawing dimensions; they are placed for information purposes. The parenthesis located on the keyboard are used along with the dimension numeral to create a reference dimension.

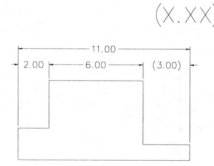

Applying the Not to Scale Dimension Symbol

At times it is necessary to place a dimension which is not to scale as in the example illustrated at the right. To distinguish regular dimensions from a dimension which is not to scale, this dimension is identified in the drawing by drawing a line under the dimension. To perform this in an AutoCAD drawing, enter the following text string for the dimension value: %%u5.00%%u. The "%%u" toggles on underline text mode followed by the dimension text of 5.00 units. To toggle underline off, complete the text string with another "%%u."

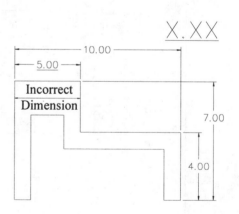

Applying the Square Dimension Symbol

Use this dimension symbol to identify the cross section of an object as a square. This will limit the need for two dimensions of the same value to be placed identifying a square. The square dimension symbol should be placed before the dimension numeral as in the illustration at the right.

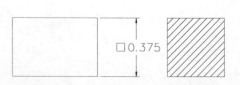

Applying the Spherical Diameter Dimension Symbol

The spherical illustrated at the right is in the form of a dome-shaped feature. Use the diameter symbol preceded by the letter "S" for spherical.

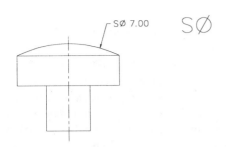

Applying the Spherical Radius Dimension Symbol

The illustration at the right is similar to the previous spherical example except the spherical is identified by a radius. Place the letter "S" infront of the "R" symbol.

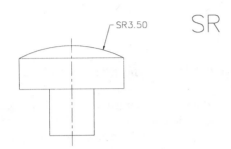

Applying the Diameter Dimension Symbol

This symbol is represented by a circle with a diagonal line through it. This diameter symbol always precedes the dimension numeral with no space in between. This symbol is automatically placed before the dimension numeral when using the AutoCAD Diameter dimension command. All holes represented by the diameter symbol are understood to pass completely through a part as in the illustration at the right.

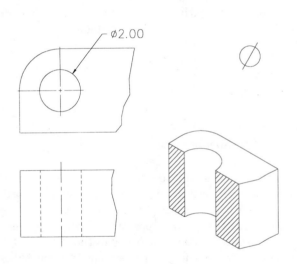

Applying the Deep Dimension Symbol

When a hole does not pass completely through a part as in the illustration at the right, the depth or deep dimension symbol is used. The diameter of the hole is first placed followed by the depth symbol and the distance the hole is to be drilled. Below this example is yet another method of identifying holes although it is considered dated.

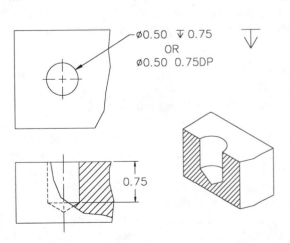

Applying the Deep Dimension Symbol

Illustrated at the right is yet another application of the depth dimension symbol. First the diameter of the hole is identified. Next, a counterdrill diameter with angle of the counterdrill is given. The depth of the counterdrill completes the dimension. This operation is used when a flat-head screw needs to be recessed below the surface of a part.

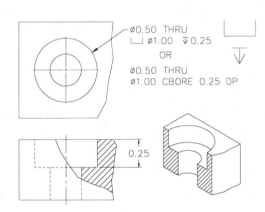

Applying the Counterbore Dimension Symbol

A counterbore is an enlarged portion of a previously drilled hole. The purpose of a counterbore is to receive the head of such screws as socket-head or fillister-head screws. The counterbore example at the right first identifies the diameter of the thru hole. On the next line comes the counterbore symbol and its diameter. The final specification is the depth of the counterbore identified by the Depth dimension symbol. Below this example is yet another method of identifying counterbores although it is considered dated.

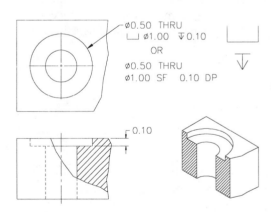

Applying the Spotface Dimension Symbol

A spotface is similar to the counterbore except that it is usually made quite shallower than the counterbore. The purpose of the spotface is to seat a washer from moving around along the surface of a part. The counterbore symbol is used to identify a spotface as in the example at the right. The diameter of the thru hole is first given. On the next line, the counterbore symbol followed by diameter is given. Finally the depth of the counterbore is identified by the distance and the Depth dimension symbol. Below this example is yet another method of identifying spotfaces although it is considered dated.

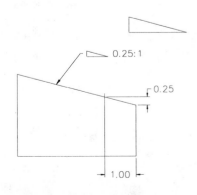

Applying the Slope Dimension Symbol

The slope dimension symbol applies to an inclined surface and the amount of rise in the surface given by two dimensions as in the illustration at the right. This rise is indicated by the change in height per unit distance along a base line. In the example at the right, the slope symbol is placed followed by the change in height and the change in unit distance. This makes the slope dimension a ratio of the height and unit distance.

Applying the Countersink Dimension Symbol

A countersink is a V-shaped conical taper at one end of a hole. The purpose is to accept a flat-head screw and make it flush with the top surface of part. In the example at the right, three holes of 0.50 diameter are first identified. On the next line, the countersink symbol followed by the number of degrees in the countersink is specified. Below this example is yet another method of identifying countersinks although it is considered dated.

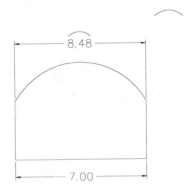

Applying the Arc Dimension Symbol

Illustrated at the right is the difference between dimensioning the chord of an arc and the actual distance of the arc. The 7.00 unit dimension in addition to defining the width of the part also specifies the distance of the chord of an arc. The 8.48 unit dimension is the distance of the arc and is specified by a small arc symbol above the distance. This is accomplished by drawing a small arc and moving it above the arc distance. Although AutoCAD does not dimension the length of an arc, it does allow the user to use the DDMODIFY command to get the distance of an arc. This value is then entered at the dimension distance prompt.

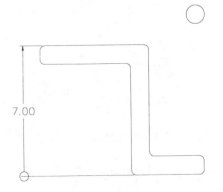

Applying the Origin Dimension Symbol

The origin dimension symbol is used to indicate the origin of the dimension. The small circle is substituted for an arrowhead as the termination of the dimension line. This is accomplished in AutoCAD by first defining the circle as a block and assigning the name of the block to the "DIMBLK1" dimension variable. Next the "DIMSAH" dimension variable is turned on. This variable controls separate arrowheads and will place the circle at the intersection of the first extension line and dimension line and an arrowhead at the intersection of the second extension line and dimension line.

Applying the Conical Taper Dimension Symbol

Similar to the slope dimension symbol, the conical taper symbol is used when the amount of taper per unit of length is desired. The conical taper symbol is placed before the value of the taper. The value of the taper is based on a ratio of 1.00 units in length to a Delta diameter illustrated at the right. The Delta diameter is based on the large shaft diameter minus the diameter taken at the area where the 1.00 unit length is located. This value becomes the ratio to 1.00 unit of length.

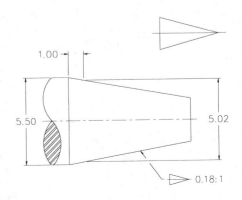

Location of Uniformly Spaced Features

Illustrated at the right is an example of a plate consisting of 8 holes equally spaced at 45 degre angles about a 7.30 diameter bolt hole circle. Rather than dimension each individual angle, the dimension symbol "X" is used for the number of times a feature is repeated. Notice at the right that "8X" is used to call out the number of times the angle and diameter of the holes repeat.

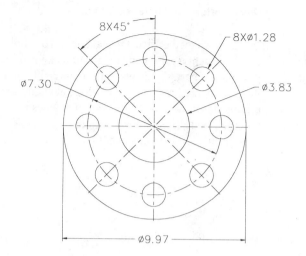

Location of Uniformly Spaced Features

The example at the right is similar to the previous illustration. Five 72 degree angles are used to locate five slots each with a width of 1.50 units.

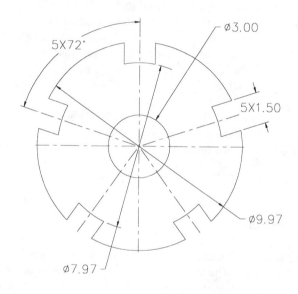

Location of Uniformly Spaced Features

The example at the right shows a different application to the location of features that are uniformly spaced. Six holes of 1.00 diameter are located along a bolt circle of 7.30 diameter. All six holes are spaced at 30 degree angles. Since the holes are not laid out in a full circle, only five angles are required to locate the six holes. The symbol "5X is used to identify the number of angles while the symbol "6X" is used to identify the number of holes.

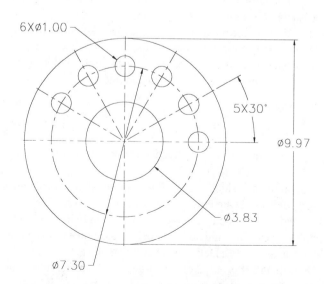

Character Mapping for Dimension Symbols

A text font has been supplied with the enclosed diskette; this file is called "ANSI_SYM.SHX". Rather than create the symbols as individual drawings, save them as Wblocks, insert them into a drawing, and move the symbols into position, this special text font contains the following dimension symbols: Square, Depth, Diameter, Conical Taper, Countersink, Counterbore, and Slope. The following regular keyboard characters have been replaced by the dimension symbols: ~, !, @, $, ^, &, and *. This means to bring up the Square dimension symbol, simply hold down the shift key and type the exclamation point from the keyboard. The Square dimension symbol will appear on the screen. The ANSI_SYM.SHX font was derived from the original SIMPLEX.SHP font by replacing the seven characters previously identified with the

new definitions containing the dimension symbols. Assign this font to a new text style and begin dimensioning with these symbols. See the following breakdown and the illustration below for matching the dimension symbols with certain keys based on the ANSI_SYM.SHX font:

> Shift + ~ = Square
> Shift + ! = Depth
> Shift + @ = Diameter
> Shift + $ = Conical Taper
> Shift + ^ = Countersink
> Shift + & = Counterbore
> Shift + * = Slope

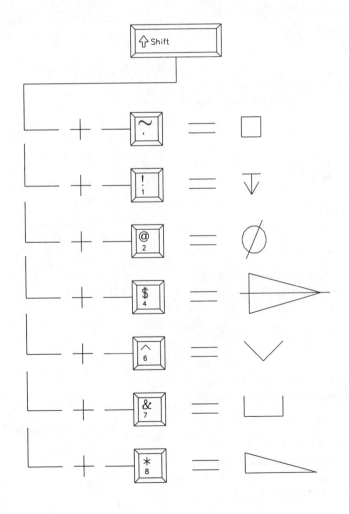

Generating Dimension Symbols

Illustrated at the right is a typical counterbore operation and the correct dimension layout. The Leader command was used to begin the diameter dimension. The counterbore and depth symbols were generated using the Dtext command. In any case, the first step to using the dimension symbols on the previous page is to first create a new text style of which any name may be used; for purposes of this illustration, a new text style will be created called "ANSI" using the supplied text font "ANSI_SYM.SHX". Follow the prompts below for the creation of this new text style:

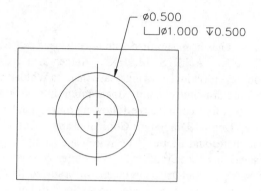

Command: **Style**
Text style name (or ?) <STANDARD>: **ANSI**
Font file <TXT>: **ANSI_SYM**
(Strike Enter to accept the remaining defaults)
ANSI is now the current text style.

It is important to point out that the dimension text height controlled by the DIMTXT variable must be the same as the text placed with the Dtext command.

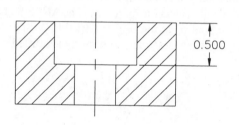

Use the Dim-Leader command to begin dimensioning the counterbore using the dimension symbols. Follow the prompt sequence for the proper key-ins:

Command: **Dim**
Dim: **Leader**
From point: *(Select the circle at "A")*
To point: *(Mark a point at "B")*
To point: *(Strike Enter to continue)*
Dimension text <>: **@0.500**
Dim: **Exit**

Notice placing the "@" in front of the 0.500 diameter dimension draws the diameter symbol as defined by the ANSI_SYM.SHX font.

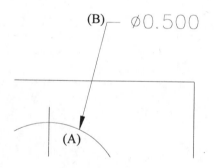

Place the remaining symbology using the Dtext command. Identify a point near "A" as the starting point for the text.

Command: **Dtext**
Justify/Style/<Start point>: *(Mark a point near "A")*
Height <0.2000>: **0.18** *(or the same height as the Dimtxt variable)*
Rotation angle <0>: *(Strike Enter to accept this default)*
Text: ***@1.000 !0.500**

Notice that typing "*" displays the counterbore symbol, typing "@" displays the diameter symbol, and typing the "!" displays the depth symbol.

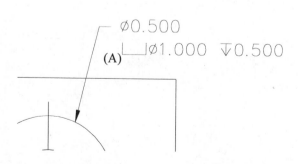

Dimension Variables

Dimension variables control the manner in which dimensions are placed in an AutoCAD drawing. These variables range from specifying a new size for arrowheads or text to turning On or Off variables that control things like if dimension text is placed horizontal or vertical. Illustrated at the right is a partial list of variables with default values. Each variable will be explained in detail in the next series of pages that follow.

DIMTIH	On
DIMTIX	Off
DIMTM	0.00
DIMTOFL	Off
DIMTOH	Off
DIMTOL	Off
DIMTP	0.00
DIMTSZ	0.00
DIMTVP	0.00
DIMTXT	0.18
DIMZIN	0

DIMALT
Alternate Dimension Units
DIMALTF
Alternate Dimension Unit Scale Factor
DIMALTD
Alternate Dimension Unit Decimal Places

All three variables operate together in controlling an alternate unit display. If Dimalt is On, as at "B" in the illustration at the right, a second dimension text string is placed next to the existing dimension text. The alternate text string is placed in brackets to separate it from the original string. The value of the alternate dimension is calculated from the current value held in the Dimaltf variable. This value is multiplied by the current dimension text string to arrive at the alternate string. Illustrated at "C" is the Dimaltd which controls the number of decimal places held in the alternate dimension string. The prompt for this dimension variable is:

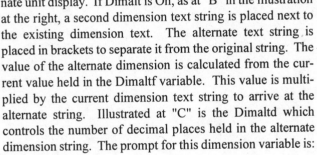

Dim: DIMALT
Current value <Off> New value: *(Enter On or accept default)*

Dim: DIMALTF
Current value <25.4000> New value: *(Enter a numeric value)*

Dim: DIMALTD
Current value <2> New value: *(Enter a numeric value)*

DIMAPOST
The Default Suffix for Alternate Text

When Dimalt is On and Dimapost is set to the text string, "mm," the alternate dimension text value, will have the Dimapost string added to its end. To change this value back to a null response, type a "." for the new value. The prompt for this dimension variable is:

Dim: DIMAPOST
Current value <> New value: *(Enter a numeric value)*

DIMASO
The Associative Dimensioning Control Variable

Use this variable to turn associative dimensioning On or Off. If On, all entities that make up the dimension will be considered one entity. If Off, all dimension components such as arrowheads, extension lines, dimension lines, and dimension text will be considered single entities. This is the same effect as using the Explode command on an associative dimension. With Dimaso set to Off, commands that normally affect dimensions that are associative will have no effect. The prompt for this dimension variable is:

Dim: DIMASO
Current value <On> New value: *(Enter Off or keep the default)*

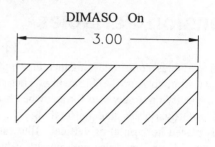

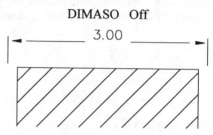

DIMASZ
Dimension Arrow Size

Use this variable to control the size of the arrow terminator. This arrow is solid which is controlled by the Fill command. With Fill turned On, the arrow is filled-in as illustrated at the right. With Fill turned Off, just the outline of the arrow is displayed without being filled-in. The prompt for this dimension variable is:

Dim: DIMASZ
Current value <0.18> New value: *(Enter a numeric value)*

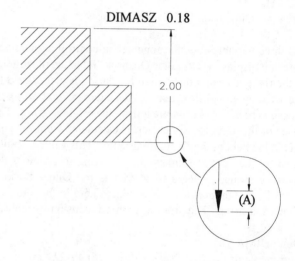

DIMBLK
New Dimension Line Terminator Name

This variable controls the shape of the terminator at the two ends of the dimension line. Dimblk waits for the name of a Block that has already been defined in the drawing data base such as the "Dot" at the right. Once the dot has been called out in the Dimblk variable, all dimension lines will be terminated at the extension lines by dots at both ends. There is some skill involved in creating the block of the arrow. Generally the block is constructed inside a 1-unit by 1-unit grid area. A short horizontal shoulder is drawn to complete the block; this shoulder will be added to the dimension line preventing a gap between the end of the dimension line and the new terminator. To disable this terminator and return to standard arrows, reply to the new value of the variable with a period, "." The prompt for this dimension variable is:

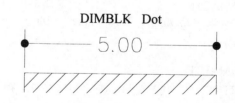

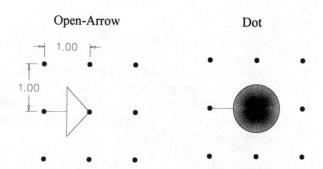

Dim: DIMBLK
Current value < > New value: *(Enter the name of a block)*

DIMBLK1
First Dimension Terminator Block Name
DIMBLK2
Second Dimension Terminator Block Name
DIMSAH
Separate Dimension Terminators

Some applications require two different dimension line terminators. All of the above variables work in combination with each other when specifying two different dimension line terminators. When a Block called "Dot" is assigned to Dimblk1 and open "Arrow" is assigned to Dimblk2, the user may place a dot at the first extension line origin and an open arrow at the second extension line origin. This is only possible if two different blocks have been defined and called out in Dimblk1 and Dimblk2 in addition to having Dimsah set to On. The prompts for Dimblk1 and Dimblk2 are identical to Dimblk. The prompt for Dimsah is:

Dim: DIMSAH
Current value <Off> New value: *(Enter On or keep the default)*

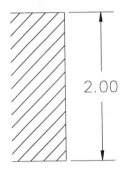

```
DIMBLK1  Dot
DIMBLK2  Arrow
DIMSAH   Off
```

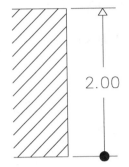

```
DIMBLK1  Dot
DIMBLK2  Arrow
DIMSAH   On
```

DIMCEN
Center Mark Size for Circles and Arcs

This dimension variable controls the size of the center mark that is placed when using the Dim-Cen command. Selecting a circle or arc places a mark similar to "A" when the value of Dimcen is 0.09. A new value of -0.09 for Dimcen places the center mark at the center in addition to drawing extender lines beyond the perimeter of the circle. The prompt for this dimension variable is:

Dim: DIMCEN
Current value <0.09> New value: *(Enter a numeric value)*

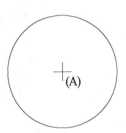

DIMCEN 0.12

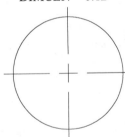

DIMCEN -0.12

DIMCLRE
Color for Extension Lines
DIMCLRD
Color for Dimension Lines
DIMCLRT
Color for Dimension Text

These three variables control the color of the following dimension components, extension lines, dimension lines and arrowheads, and dimension text. It may seem that these variables merely give the operator control of colorfully displaying dimensions. However, a more practical use of these variables is to assign colors to the different dimension components to control line quality during plotting by assigning different pen weights. The prompt for all three variables is:

Dim: DIMCLRE
Current value < > New value: *(Enter the name of a color)*

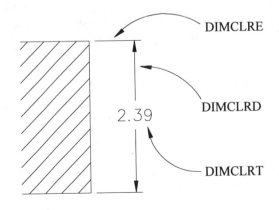

DIMDLE

Dimension Line Extension

Some professions prefer the dimension line be drawn past the current arrow terminator. Setting the variable Dimdle to a value of 0.20 extends the dimension line past the terminator at that distance. The prompt for this dimension variable is:

Dim: DIMDLE
Current value <0.00> New value: *(Enter a numeric value)*

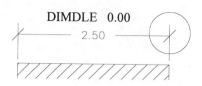

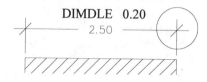

DIMDLI

Dimension Line Increment for Continuation

This variable controls the spacing when multiple dimensions are placed away from each other similar to the illustration at the right. The default value of Dimdli (0.38) satisfies the minimum value on the spacing of dimensions especially while in Baseline mode. The prompt for this dimension variable is:

Dim: DIMDLI
Current value <0.38> New value: *(Enter a numeric value)*

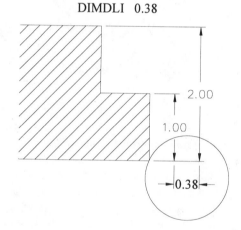

DIMEXE

Extension Above Dimension Line

Use this variable to control how far the extension line extends past the dimension line. In keeping with dimension basics, this variable may be changed from the default value of 0.18 to a new value of 0.12. The prompt for this dimension variable is:

Dim: DIMEXE
Current value <0.18> New value: *(Enter a numeric value)*

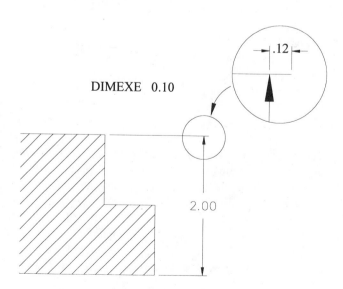

DIMEXO
Extension Line Origin Offset

This variable controls how far away from the object the extension will start (at "A"). The prompt for this dimension variable is:

Dim: DIMEXO
Current value <0.0625> New value: *(Enter a numeric value)*

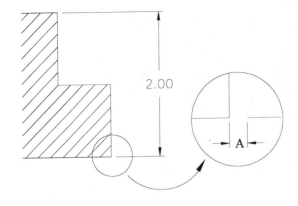

DIMLFAC
Linear Unit Scale Factor

This variable acts as a multiplier for all linear dimensions including radius and diameter dimensions. As a dimension is calculated by AutoCAD, the current Dimlfac value is multiplied by the dimension to arrive at a new dimension value. With a value of 1.00, all dimensions are taken as default. With a value of 2.00, all dimension values are first multiplied by 2.00 before being placed. In the same manner, a Dimlfac value of 0.50 would cut all dimensions in half. The prompt for this dimension variable is:

Dim: DIMLFAC
Current value <1.00> New value: *(Enter a numeric value)*

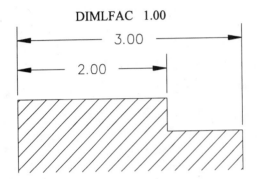

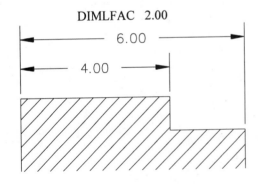

DIMGAP
Dimension Line Gap Increment

This variable maintains the area around dimension text by providing the gap between the text and the dimension line ends. The prompt for this dimension variable is:

Dim: DIMGAP
Current value <0.09> New value: *(Enter a numeric value)*

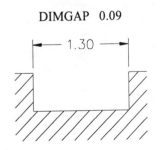

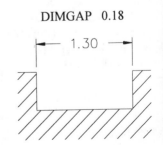

DIMPOST
Default Suffix for Dimension Text

This variable adds a character string immediately after all dimension values except for angular dimensions. The text string may be disabled by entering a single period "." at the "New value:" prompt. The prompt for this dimension variable is:

Dim: DIMPOST
Current value <0> New value: *(Enter a numeric value)*

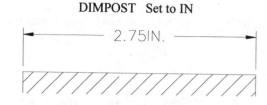

DIMRND
Dimension Round-Off Value

This variable will round-off all dimension distances based on the current rounding value. If the Dimrnd value is set to 0.25, all distances will be rounded to the nearest 0.25 unit. This variable does not affect angular dimensions. The prompt for this dimension variable is:

Dim: DIMRND
Current value <0> New value: *(Enter a numeric value)*

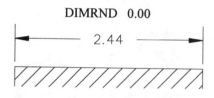

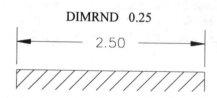

DIMSCALE
Overall Dimension Scale Factor

This variable acts as a multiplier and globally affects all current dimension variables that are specified by sizes or distances. This means that if the current Dimscale value is 1.00, variables such as Dimtxt set to 0.18 will remain unchanged. If the Dimscale value is changed to 2.00, the dimension text visible on the display screen will be changed from 0.18 to 0.36. The prompt for this dimension variable is:

Dim: DIMSCALE
Current value <1.0000> New value: *(Enter a numeric value)*

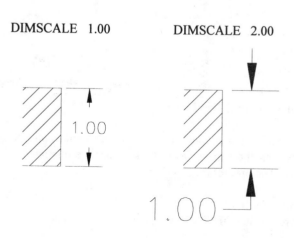

DIMSE1
Suppress the First Extension Line
DIMSE2
Suppress the Second Extension Line

These two variables control the display of extension lines. They are useful when dimensioning to an object line and to avoid placing the extension line on top of the object line. Dimse1, if turned On, suppresses the first extension. This is another way of saying that if Dimse1 is On, all first extension lines from that point on will be turned off. The same is true for Dimse2, which will turn off or suppress the second extension line if set to On. Study the many examples illustrated at the right. The prompt for this dimension variable is:

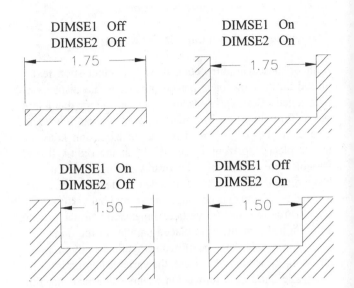

Dim: DIMSE1
Current value <Off> New value: *(Enter On or keep the default)*

Dim: DIMSE2
Current value <Off> New value: *(Enter On of keep the default)*

DIMTAD
Place Text above the Dimension Line

This variable controls whether dimension text will be placed inside of the dimension line or placed above the dimension. The default value is Off, which means the dimension line will break to allow text to be placed in between. If turned On, text will be placed above the dimension line as in the illustration at the right. This variable is very important to architectural office practices although some mechanical applications use this variable On. The prompt for this dimension variable is:

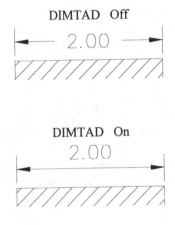

Dim: DIMTAD
Current value <Off> New value: *(Enter On or keep the default)*

DIMTIX
Place Text Inside Extension Lines

This variable controls whether text is placed outside of extension lines or forced in between extension lines. By default, the variable is turned Off; depending on the current value of Dimasz and Dimtxt, AutoCAD will calculate whether the dimension will fit or be placed outside of extension lines. Turning this variable On forces the text to be placed inside the extension lines even if the text is so large that it goes beyond the extensions. The prompt for this dimension variable is:

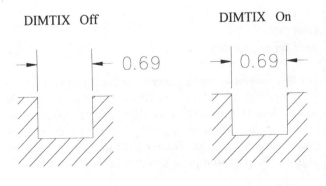

Dim: DIMTIX
Current value <Off> New value: *(Enter On or keep the default)*

DIMTIH
Text Inside Extension Lines Is Horizontal
DIMTOH
Text Outside Extension Lines Is Horizontal

Both of these variables control whether dimension text is
placed horizontally or is placed parallel to the distance dimensioned. Both variables are either On or Off; the default
setting for both is On. This means that if text can fit inside of
extension lines or is placed outside of extension lines, the
text is placed horizontally as in "A" at the right. If both
variables are turned Off, the results are illustrated at "B."
Here, dimension text placed outside of extension lines, such
as the 0.50 distance, is no longer horizontal but vertical. The
same is true for the 1.00 vertical dimension. Notice that the
2.06 aligned dimension is placed parallel to the dimension
line. One variable may be turned On while the other remains
Off as in "C" at the right. Here, the variable Dimtih is turned
Off while Dimtoh is turned On. Dimension text that falls
outside of extension lines will be placed horizontally as in the
0.50 dimension while dimension text that falls inside extension lines will not be horizontal as with the 1.00 and 2.06
dimensions. The prompts for these dimension variables are:

Dim: DIMTIH
Current value <On> New value:

Dim: DIMTOH
Current value <On> New value:

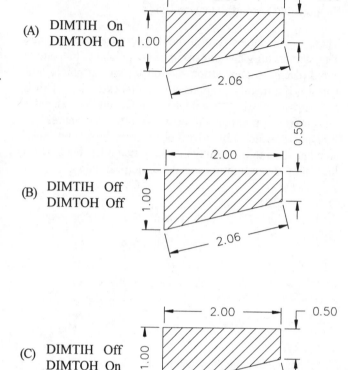

DIMTOFL
Force the Dimension Line Inside the Extension Lines

This variable forces the dimension line to be drawn inside of
extension lines even if the text is placed outside of the
extension lines. The prompt for this dimension variable is:

Dim: DIMTOFL
Current value <Off> New value: *(Enter On or keep the
default)*

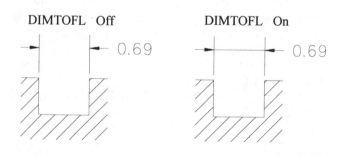

DIMTSZ
Tick Terminator Size

Use this variable to convert arrow terminators into 45 degree
slashes or "ticks." With the Dimtsz value 0.00, arrowheads
will be drawn. Setting the variable to a value of 0.12 overrides the current arrow setting and places ticks at the ends of
the dimension line. Set Dimtsz back to 0.00 to return to
arrowheads. The prompt for this dimension variable is:

Dim: DIMTSZ
Current value <0.00> New value: *(Enter a numeric value)*

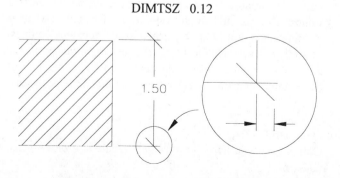

DIMSOXD
Suppress Dimension Line Outside of Extension Lines

This variable controls the dimension line for dimension text placed outside of extension lines. The default value is Off, which means the dimension line with arrowheads will be drawn whether the text is placed inside or outside. With this variable turned On, the dimension line is suppressed or turned off leaving the text and extension lines.
The prompt for this dimension variable is:

Dim: DIMSOXD
Current value <Off> New value: *(Enter Off or keep the default)*

DIMTIX On
DIMSOXD Off

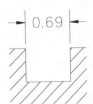

DIMTIX On
DIMSOXD On

DIMTOL
Generate Tolerance Dimensions
DIMLIM
Generate Limit Dimensions
DIMTP
Positive Tolerance Value
DIMTM
Minus Tolerance Value

These variables work together in converting normal dimensions to limit or tolerance dimensions. Dimtp and Dimtm require numeric values; Dimtol and Dimlim are On/Off switches. An example of a tolerance dimension is illustrated at "A" complete with "plus" and "minus" values. An example of a limit dimension is illustrated at "B." Here the upper limit is placed over the lower limit. As both Dimtol and Dimlim can be switched On or Off, only one variable may remain On. This means if Dimlim is On and Dimtol is changed from Off to On, the Dimlim variable automatically switches itself Off. With Dimtol On in example "A" and different values entered for Dimtp and Dimtm, the base dimension of 4.000 is placed along with the upper limit, (Dimtp) and lower limit, (Dimtm). In example "B," Dimlim is On with Dimtp and Dimtm the same values. The limit dimension takes the basic dimension size (4.000), adds the Dimtp value, subtracts the Dimtm value, and places the upper limit and lower limit as the new dimension. If Dimtp and Dimtm have values assigned but both Dimtol and Dimlim are Off, the result is the basic size at "C." If Dimtp and Dimtm are the same size and Dimtol is On, a plus/minus dimension is placed as in "D." The prompts for these variables are:

Dim: DIMTOL
Current value <Off> New value:

Dim: DIMLIM
Current value <Off> New value:

Dim: DIMTP
Current value <0.00> New value:

Dim: DIMTM
Current value <0.00> New value:

If...
DIMTP = 0.005
DIMTM = 0.003

And...
DIMTOL is On
DIMLIM is Off

Then...

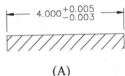

(A)

If...
DIMTP = 0.003
DIMTM = 0.003

And...
DIMTOL is Off
DIMLIM is On

Then...

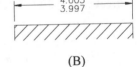

(B)

If...
DIMTP = 0.005
DIMTM = 0.005

And...
DIMTOL is Off
DIMLIM is Off

Then...

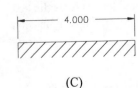

(C)

If...
DIMTP = 0.005
DIMTM = 0.005

And...
DIMTOL is On
DIMLIM is Off

Then...

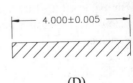

(D)

DIMTFAC
Dimension Tolerance Text Size Factor

This variable acts as a multiplier and affects the height of tolerance text. Dimtfac will multiply its current value by the value set in Dimtxt. This variable is designed to work when Dimtm and Dimtp are not equal and Dimtol and Dimlim are On, but not at the same time. The prompt for this dimension variable is:

Dim: DIMTFAC
Current value <1.00> New value: *(Enter a numeric value)*

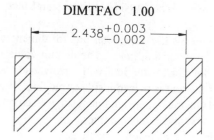

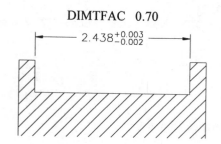

DIMTXT

This variable controls the height of text added to dimensions. To control the text font, use the Style command. The prompt for this dimension variable is:

Dim: DIMTXT
Current value <0.18> New value: *(Enter a numeric value)*

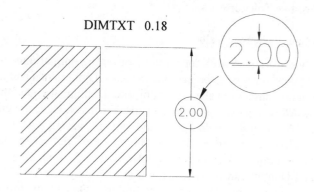

DIMTVP
Vertical Dimension Text Position

This variable allows you to control the position of the dimension text either above or below the dimension line. With Dimtad Off and if Dimtvp is set to 1.00 as in the illustration at the right, AutoCAD will multiply this value by the current Dimtxt value and place the text above the dimension line. If Dimtvp is set to -1.00, the negative multiplier places the text below the dimension line. The prompt for this dimension variable is:

Dim: DIMTVP
Current value <0> New value: *(Enter a numeric value)*

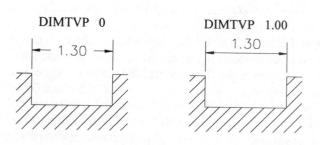

DIMZIN
Zero Inches/Feet Control

This variable controls the display of feet and inches in a drawing depending on what the current Dimzin value is set to. This variable may also suppress the leading zero or trailing zeros of decimal units depending on the setting. See the table at the right for the effects of Dimzin on different types of units. The prompt for this dimension variable is:

Dim: DIMZIN
Current value <0> New value: *(Enter a supported numeric value)*

DIMZIN Value				
0	3/8"	5"	2'	0.7500
1	0'-0 3/8"	0'-5"	2'-0"	
2	0'-0 3/8"	0'-5"	2'	
3	3/8"	5"	2'-0"	
4				.7500
8				0.75
12				.75

Diameter and radius dimensioning can be affected by the Dimtix and Dimtofl system variables depending on the results desired. With both variables Off in example "A," the dimension text is placed on the outside of the circle resembling a leader line. This dimension, however, remains associative as long as Dimaso is On. If Dimtix is turned On as in example "B," the dimension text will be forced inside of the circle. Turning Dimtix Off and Dimtofl On places the dimension text outside of the circle and forces the dimension line to be drawn through the entire circle.

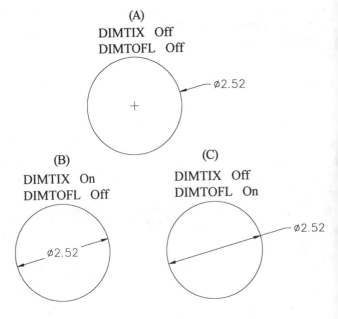

The effects of Dimtix and Dimtofl on diameter dimensions are identical to radius dimensions.

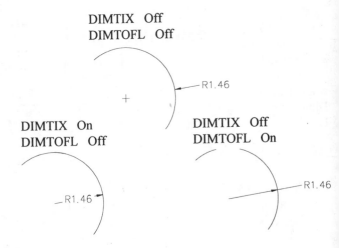

Using the DDIM Command

The Dimension Styles and Variables dialog box is displayed when selecting "Dimension Style" from the "Settings" pull-down menu area. This dialog box may also be brought up by entering "DDIM" at the command prompt. Dimension styles are a collection of dimension variables organized under a user defined name. Dimension styles may be called up or "Restored" to activate certain dimension variables. The DDIM dialog box consists of subdialog boxes each controlling specific dimension variables. Some dimension variables may appear in more than one subdialog box. After a set of variables is changed or set to different values, these changes may be assigned a new dimension style name to be added to the main dialog box illustrated below. Each subdialog box and the dimension variables affected are explained and displayed in the next series of pages.

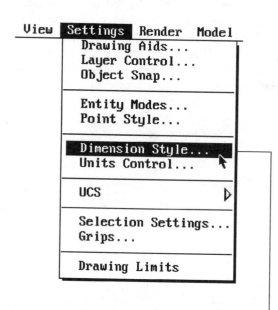

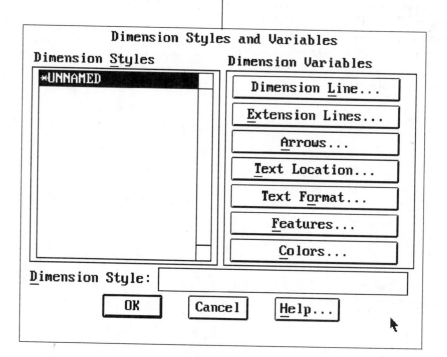

DDIM - Dimension Line

Selecting the "Dimension Line" option of the DDim command displays a subdialog box affecting the following dimension variables:

Feature Scaling:
DIMSCALE = 1.0000

Dimension Line Color:
DIMCLRD = Bylayer

Force Interior Lines:
DIMTOFL = ON *(If checked)*
DIMTOFL = OFF *(If not checked)*

Text Gap:
DIMGAP = 0.0900

Baseline Increment:
DIMDLI = 0.3800

Dimension **L**ine...

```
                Dimension Line
  Style: *UNNAMED
  Feature Scaling:          1.00000
  [ ] Use Paper Space Scaling
  Dimension Line Color:  BYBLOCK  [ ]
  Dimension Line
  ┌──────────────────────────────────┐
  │ [ ]  Force Interior Lines        │
  │ [ ]  Basic Dimension             │
  │ Text Gap:            0.0900       │
  │ Baseline Increment: 0.3800       │
  └──────────────────────────────────┘
    [ OK ]    [ Cancel ]   [ Help... ]
```

Illustrated at the right are examples of the results if "Basic Dimension" is selected. By default, all dimensions will display as a standard dimension. If, however, "Basic Dimension" is checked, the result will be a dimension with a box drawn around the text of the dimension. This happens to be the dimensioning standard denoting a dimension without a tolerance value assigned.

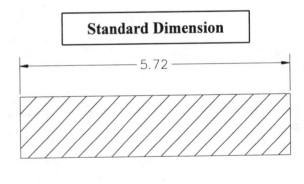

Standard Dimension

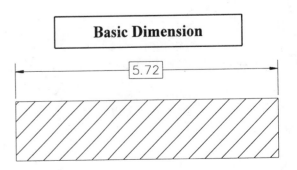

Basic Dimension

DDIM - Extension Lines

Selecting the "Extension Lines" option of the DDim command
displays a subdialog box affecting the following dimension
variables:

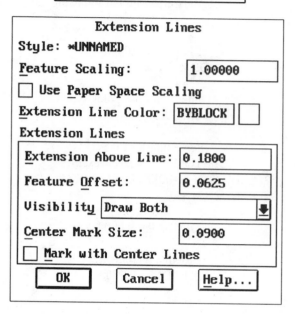

Feature Scaling:
DIMSCALE = 1.0000

Extension Line Color:
DIMCLRE = Bylayer

Extension Above Line:
DIMEXE = 0.1800

Feature Offset:
DIMEXO = 0.625

Visibility (Draw Both Extension Lines):
DIMSE1 = OFF
DIMSE2 = OFF

Center Mark Size:
DIMCEN = 0.0900

Mark with Center Lines:
DIMCEN = -0.0900 *(If checked)*
DIMCEN = 0.0900 *(If not checked)*

To draw a center mark with center lines when using the Center,
Diameter, and Radius commands, check the "Mark with
Center Lines" box. The value assigned to the "Center Mark
Size" is stored as a negative value in the DIMCEN dimension
variable.

Selecting the "Visibility" option of the "Extension Lines"
subdialog box affects the following dimension variables:

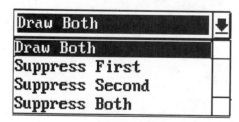

Draw Both:
DIMSE1 = OFF
DIMSE2 = OFF

Suppress First:
DIMSE1 = ON
DIMSE2 = OFF

Suppress Second:
DIMSE1 = OFF
DIMSE2 = ON

Suppress Both:
DIMSE1 = ON
DIMSE2 = ON

DDIM - Arrows

Selecting the "Arrows" option of the DDim command displays a subdialog box affecting the following dimension variables:

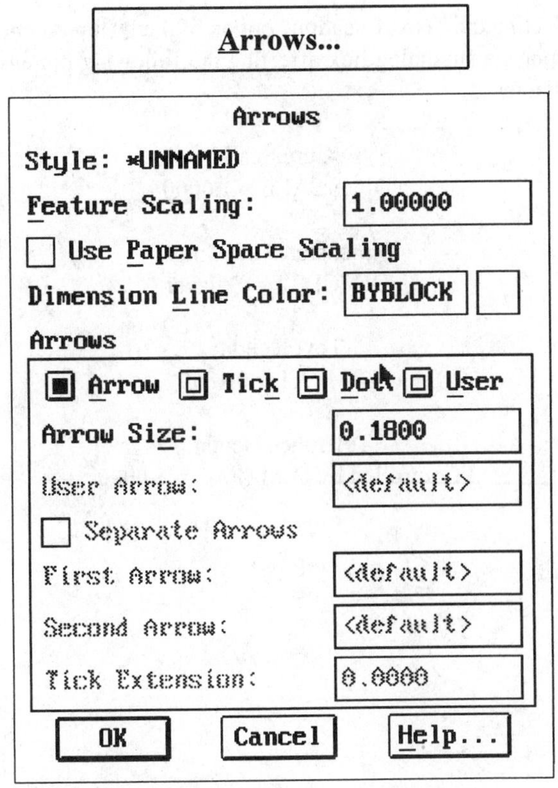

Feature Scaling:
DIMSCALE = 1.0000

Dimension Line Color:
DIMCLRD = Bylayer

Arrow:
DIMASZ = 0.1800

Tick:
DIMTSZ = Value

Dot:
DIMBLK = Dot

If "Tick" is selected, the "Tick Extension" box activates. The tick extension is used for determining how far the dimension line extends past the tick mark. The following dimension variable controls this condition:

DIMDLE = 0.0000

Selecting "Separate Arrows" affects the following dimension variables:

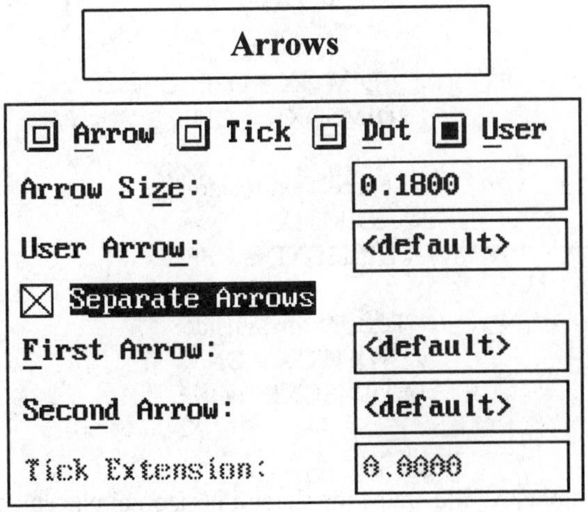

User:
DIMBLK = Name

User Arrow:
DIMBLK = Name

Separate Arrows:
DIMSAH = ON *(If checked)*
DIMSAH = OFF *(If not checked)*

First Arrow:
DIMBLK1 = Name

Second Arrow:
DIMBLK2 = Name

User defined arrow blocks are designed to display dimension line terminators other than normal arrows, ticks, or dots.

DDIM - Text Location

Selecting the "Text Location" option of the DDim command displays a subdialog box affecting the following dimension variables:

Feature Scaling:
DIMSCALE = 1.0000

Dimension Text Color:
DIMCLRT = Bylayer

Text Height:
DIMTXT = 0.1800

Tolerance Height:
Controlled by DIMTFAC = 1.0000

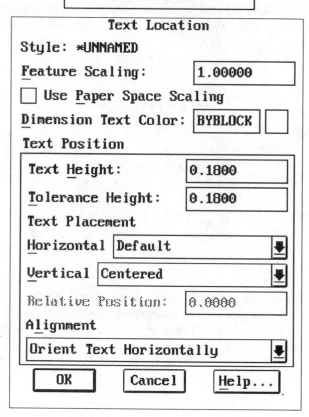

Selecting "Horizontal" brings up another subdialog box affecting the following dimension variables:

Default:
DIMTIX = OFF
DIMSOXD = OFF

Force Text Inside:
DIMTIX = ON
DIMSOXD = OFF

Text, Arrows Inside:
DIMTIX = ON
DIMSOXD = ON

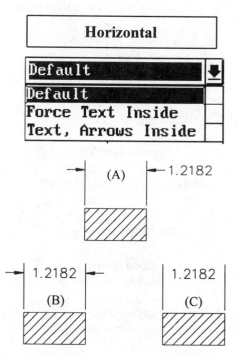

Example "A" illustrates the default horizontal placement of dimension text. Example "B" forces the dimension text to be placed inside of the extension lines. Example "C" forces the dimension text inside of the extension lines and suppresses the outside dimension line and arrowheads.

Selecting "Vertical" brings up another subdialog box affecting the following dimension variables:

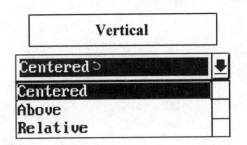

Centered:
DIMTAD = OFF

Above:
DIMTAD = ON

Relative:
DIMTVP = Value

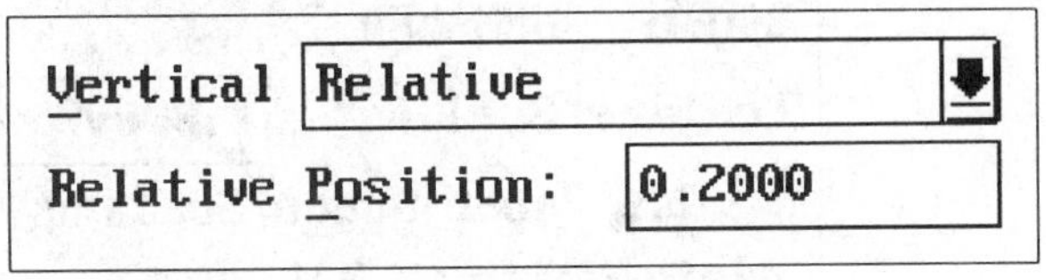

Selecting the "Relative" mode for dimension text activates the "Relative Position" box. The relative position value is divided by the text height and then sent to the dimension variable DIMTVP.

Example "A" illustrates the result of the default vertical dimension text placement. Example "B" turns the dimension variable DIMTAD On, which places the dimension text above the dimension line. Example "C" illustrates the effects of the "Relative" vertical text position. A value of -0.20 assigned to the "Relative Position" box moves the dimension text below the dimension line.

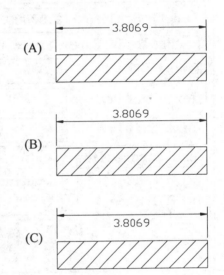

Selecting "Alignment" brings up another subdialog box affecting the following dimension variables:

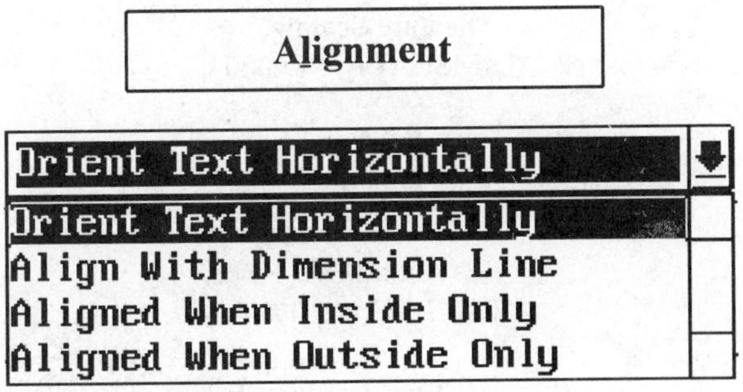

Orient Text Horizontally:
DIMTIH = ON
DIMTOH = ON

Align With Dimension Line:
DIMTIH = OFF
DIMTOH = OFF

Aligned When Inside Only:
DIMTIH = ON
DIMTOH = OFF

Aligned When Outside Only:
DIMTIH = OFF
DIMTOH = ON

DDIM - Text Format

```
                              ┌─────────────────────────┐
                              │      Text Format...      │
                              └─────────────────────────┘
```

```
┌────────────────────────────────────────────────────────────────┐
│                          Text Format                            │
│                                                                  │
│  Style: *UNNAMED                      Tolerances                 │
│  Feature Scaling:  ┌──────────┐      ┌─────────────────────────┐ │
│                    │ 1.00000  │      │ ■ None                  │ │
│                    └──────────┘      │ □ Variance              │ │
│  □ Use Paper Space Scaling           │ □ Limits                │ │
│  Basic Units                         │ Upper Value: ┌────────┐ │ │
│  ┌──────────────────────────────┐    │              │ 0.0000 │ │ │
│  │ Length Scaling: ┌──────────┐ │    │              └────────┘ │ │
│  │                 │ 1.00000  │ │    │ Lower Value: ┌────────┐ │ │
│  │                 └──────────┘ │    │              │ 0.0000 │ │ │
│  │ □ Scale In Paper Space Only  │    │              └────────┘ │ │
│  │ Round Off:   ┌──────────┐    │    └─────────────────────────┘ │
│  │              │ 0.0000   │    │    Alternate Units             │
│  │              └──────────┘    │    ┌─────────────────────────┐ │
│  │ Text Prefix: ┌──────────┐    │    │ □ Show Alternate Units? │ │
│  │              └──────────┘    │    │ Decimal Places: ┌─────┐ │ │
│  │ Text Suffix: ┌──────────┐    │    │                 │ 2   │ │ │
│  │              └──────────┘    │    │                 └─────┘ │ │
│  └──────────────────────────────┘    │ Scaling: ┌───────────┐  │ │
│  Zero Suppression                    │          │ 25.40000  │  │ │
│  ┌──────────────────────────────┐    │          └───────────┘  │ │
│  │ ⊠ 0 Feet    □ Leading        │    │ Suffix:  ┌───────────┐  │ │
│  │ ⊠ 0 Inches  □ Trailing       │    │          └───────────┘  │ │
│  └──────────────────────────────┘    └─────────────────────────┘ │
│        ┌──────┐   ┌────────┐   ┌─────────┐                       │
│        │  OK  │   │ Cancel │   │ Help... │                       │
│        └──────┘   └────────┘   └─────────┘                       │
└────────────────────────────────────────────────────────────────┘
```

Selecting the "Text Format" option of the DDim command displays a subdialog box affecting the following dimension variables:

Feature Scaling:
DIMSCALE = 1.0000

Length Scaling:
DIMLFAC = 1.0000

Round Off:
DIMRND = 0.0000

Text Prefix:
DIMPOST = Name

Text Suffix:
DIMPOST = Name

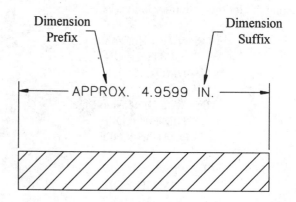

Selecting "Tolerances" brings up another subdialog box affecting the following dimension variables:

Tolerances (None):
DIMLIM = OFF
DIMTOL = OFF

Tolerances (Variance):
DIMLIM = OFF
DIMTOL = ON

Tolerances (Limits):
DIMLIM = ON
DIMTOL = OFF

Upper Value:
DIMTP = 0.0000

Lower Value:
DIMTM = 0.0000

```
┌─────────────────────────────────┐
│           Tolerances            │
└─────────────────────────────────┘
┌──────────────────────────────────────────┐
│  □  None                                   │
│  ■  Variance                               │
│  □  Limits                                 │
│  Upper Value:      ┌──────────┐            │
│                    │ 0.0000   │            │
│                    └──────────┘            │
│  Lower Value:      ┌──────────┐            │
│                    │ 0.0000   │            │
│                    └──────────┘            │
└──────────────────────────────────────────┘
```

Selecting "Alternate Units" brings up another subdialog box
affecting the following dimension variables:

Show Alternate Units:
DIMALT = ON *(If checked)*
DIMALT = OFF *(If not checked)*

Decimal Places:
DIMALTD = 2

Scaling:
DIMALTF = 25.4000

Suffix:
DIMAPOST = Name

```
┌─────────────────────────────────┐
│        Alternate Units          │
└─────────────────────────────────┘
┌──────────────────────────────────────────┐
│  ⊠  Show Alternate Units?                  │
│  Decimal Places:   ┌──────────┐            │
│                    │ 2        │            │
│                    └──────────┘            │
│  Scaling:          ┌──────────┐            │
│                    │ 25.40000 │            │
│                    └──────────┘            │
│  Suffix:           ┌──────────┐            │
│                    │          │            │
│                    └──────────┘            │
└──────────────────────────────────────────┘
```

The "Zero Suppression" area of the "Text Format" subdialog box affects the following dimension variables:

0 Feet:
DIMZIN = 3

0 Inches:
DIMZIN = 2

Leading:
DIMZIN = 5

Trailing:
DIMZIN = 9

Combinations of "Zero Suppression" may be achieved by checking one or more options. By default, "0 Feet" and "0 Inches" are checked resulting in the DIMZIN dimension variable being set to 0.

```
┌──────────────────────────────────┐
│          Zero Suppression        │
└──────────────────────────────────┘
┌──────────────────────────────────┐
│  ⊠ 0 Feet      ☐ Leading         │
│  ⊠ 0 Inches    ☐ Trailing        │
└──────────────────────────────────┘
```

DDIM - Colors

Selecting the "Colors" option of the DDim command displays a subdialog box affecting the following dimension variables:

Feature Scaling:
DIMSCALE = 1.0000

Dimension Line Color:
DIMCLRD = Bylayer

Extension Line Color:
DIMCLRE = Bylayer

Dimension Text Color:
DIMCLRT = Bylayer

```
┌──────────────────────────────────┐
│             Colors...            │
└──────────────────────────────────┘
┌──────────────────────────────────────┐
│                Colors                  │
│  Style: *UNNAMED                       │
│  Feature Scaling:       │1.00000 │     │
│  ☐ Use Paper Space Scaling             │
│  Dimension Line Color:  │BYBLOCK│ □    │
│  Extension Line Color:  │BYBLOCK│ □    │
│  Dimension Text Color:  │BYBLOCK│ □    │
│   ┌────┐  ┌──────┐  ┌──────┐           │
│   │ OK │  │Cancel│  │Help..│           │
│   └────┘  └──────┘  └──────┘           │
└──────────────────────────────────────┘
```

DDIM - Features

```
┌─────────────────────────────────────┐
│          Features...                 │
└─────────────────────────────────────┘
```

```
┌──────────────────────────────────────────────────────────────────┐
│                            Features                                │
│                                                                    │
│   Style: *UNNAMED                   Extension Lines                │
│   Feature Scaling:    [1.00000]     Extension Above Line: [0.1800] │
│   ☐ Use Paper Space Scaling         Feature Offset:       [0.0625] │
│   Dimension Line                                                   │
│   ┌────────────────────────────┐    Visibility [Draw Both      ▼]  │
│   │ ☐ Force Interior Lines      │   Center Mark Size:       [0.0900]│
│   │ ☐ Basic Dimension           │   ☐ Mark with Center Lines        │
│   │ Text Gap:        [0.0900]   │   Text Position                   │
│   │ Baseline Increment: [0.3800]│   ┌──────────────────────────────┐│
│   └────────────────────────────┘    │ Text Height:       [0.1800]  ││
│   Arrows                             │ Tolerance Height:  [0.1800]  ││
│   ┌────────────────────────────┐    │ Text Placement               ││
│   │ ■ Arrow ☐ Tick ☐ Dot ☐ User│   │ Horizontal [Default       ▼] ││
│   │ Arrow Size:      [0.1800]   │   │ Vertical [Centered        ▼] ││
│   │ User Arrow:      [<default>]│   │ Relative Position: [0.0000]  ││
│   │ ☐ Separate Arrows           │   │ Alignment                    ││
│   │ First Arrow:     [<default>]│   │ [Orient Text Horizontally ▼] ││
│   │ Second Arrow:    [<default>]│   └──────────────────────────────┘│
│   │ Tick Extension:  [0.0000]   │    [ OK ]  [ Cancel ]  [Help...]  │
│   └────────────────────────────┘                                   │
└──────────────────────────────────────────────────────────────────┘
```

Using the Features option of the DDim command displays a large subdialog box covering the following DDim command options:

Dimension Line Options
Arrows Option
Extension Lines Option
Text Position Option

This dialog box is usually used to modify or examine all of the settings that relate to the Dimension Line, Extension Line, Arrows, Text Location, and Text Format subdialog boxes. For details on each of the individual subdialog boxes, pick the major headings back at the overall dimension style dialog box.

Creating a New Dimension Style

By default, the current dimension style is unnamed. To save and restore dimension styles, use the Dimension Style edit box of the main Dimension Styles and Variables dialog box illustrated below. The values of all dimension variables will be stored in the current dimension style name with the exception of DIMASO and DIMSHO which must be entered at the command line and not the DDIM dialog box.

Dimension styles are always displayed in alphabetical order. As a displayed style is selected to be the current dimension style, it stays highlighted until another dimension style is selected to be the current style.

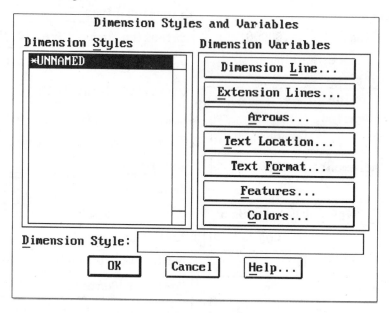

To save a dimension style, enter the name of the style in the Dimension Style edit box. As this is done, the name appears in the listing of dimension styles and is highlighted signifying it is the current dimension style.

This new dimension style is a complete copy of the previous style which, in the example below, was the unnamed style. To verify this, a message appears in the the lower left corner of the dialog box consisting of the following:

New style STANDARD created from UNNAMED

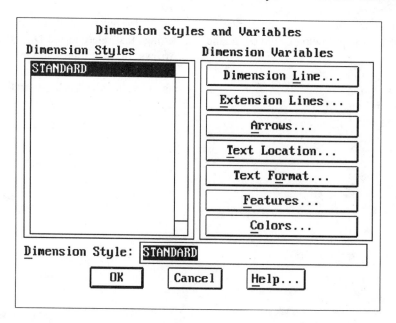

Dimensioning Isometric Drawings

In past versions of AutoCAD, it was possible to use the Aligned dimensioning mode to make the dimension line parallel with the surface being dimensioned. Arrowheads were also drawn parallel; however, extension lines were drawn perpendicular to the dimension lines and not at an isometric angle. The results of this type of dimensioning technique are illustrated at the right. One of the only ways to simulate isometric dimensions was to turn off the extension lines, manually draw new extension lines at isometric angles, and move the dimension to a new location because of the position of the extension lines.

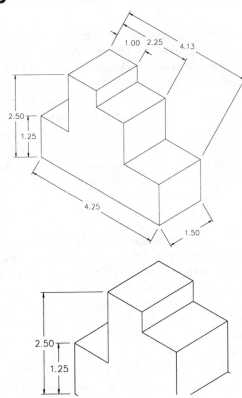

The Oblique command allows the user to enter an obliquing angle which will rotate the extension lines and reposition the dimension line.

Dim: **Obl**
Select objects: *(Select the 2.50 and 1.25 dimensions at the right)*
Select objects: *(Strike Enter to continue with this command)*
Enter obliquing angle (Return for none): **150**

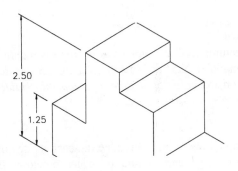

Both dimensions at the right were repositioned using the Oblique command. Notice that the extension lines and dimension line were affected; however, the text remains in the horizontal position as defined by the Dimtih and Dimtoh dimension variables. The Tedit and Trotate commands will allow text to be rotated at an angle; however, the text will not be in true isometric form.

The Oblique command was used to rotate the dimension at "A" at an obliquing angle of 210 degrees. An obliquing angle of -30 degrees was used to rotate the dimension at "B" and the dimensions at "C" required an obliquing angle of 90. This represents proper isometric dimensions except for the orientation of the text.

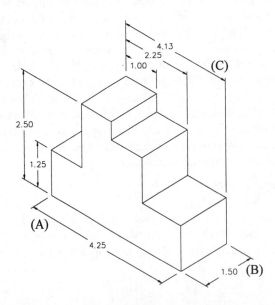

Ordinate Dimensioning

The plate at the right consists of numerous drill holes with a few slots in addition to numerous 90-degree angle cuts along the perimeter. This object is not considered difficult to draw or make since it mainly consists of drill holes. However, conventional dimensioning techniques make the plate appear complex since a dimension is required for the location of every hole and slot in both the X and Y directions. Add to that standard dimension components such as extension lines, dimension lines, and arrowheads and it is easy to get lost in the complexity of the dimensions even on this simple object.

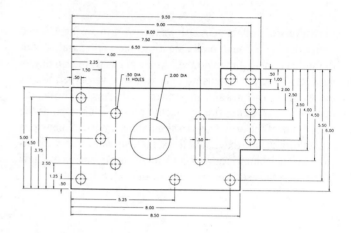

A better dimensioning method to use is illustrated at the right, called Ordinate or Datum dimensioning. Here, no dimension lines or arrowheads are drawn; instead, one extension line is constructed from the selected feature to a location specified by the user. A dimension is added to identify this feature in either the X or Y directions. It is important to understand that all dimension calculations occur in relation to the current User Coordinate System (UCS), or the current 0,0 origin. In the example at the right, with the 0,0 origin located in the lower left corner of the plate, all dimensions in the horizontal and vertical directions are calculated in relation to this 0,0 location. Holes and slots are called out using the Dim-Diameter option. The following illustrates a typical ordinate dimensioning prompt sequence:

Command: **Dim**
Dim: **Ord**
Select feature: *(Select a feature using an Osnap option)*
Leader endpoint (Xdatum/Ydatum): *(Locate an outside point)*
Dimension text < >: *(Strike Enter to accept the default value)*

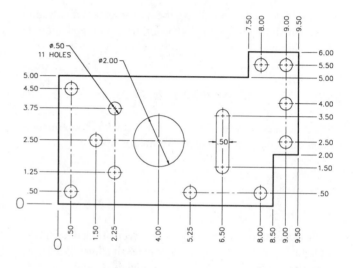

To illustrate how to place ordinate dimensions, see the example at the right and the prompt sequence below. Before placing any dimensions, a new UCS must be moved to a convenient location on the object using the UCS-Origin option. All ordinate dimensions will reference this new origin since it is located at coordinate 0,0. Enter the dimensioning area by picking from a menu or typing "Dim". Once in dimensioning mode, select or type Ordinate to enter ordinate dimensioning. Select the Quadrant of the arc at "A" as the feature. For the leader endpoint, pick a point at "B". Be sure Ortho mode is On. It is also helpful to snap to a convenient grid point for this and other dimensions along this direction. Follow the prompt sequence below:

Command: **Dim**
Dim: **Ord**
Select feature: *(Select the Quadrant of the arc at "A")*
Leader endpoint (Xdatum/Ydatum): *(Select a point at "B")*
Dimension text <1.50>: *(Strike Enter to accept this value)*

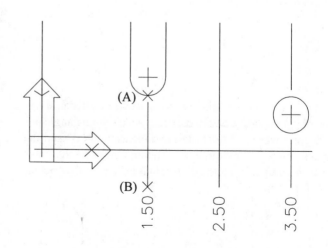

With the previous example highlighting horizontal ordinate dimensions, placing vertical ordinate dimensions is identical. With the location of the UCS still located in the lower left corner of the object, select the feature at "A" using either the Endpoint or Quadrant modes. Pick a point at "B" in a convenient location on the drawing. Accept the default value and the dimension is placed. Again, it is helpful if Ortho is On and a grid dot is snapped to.

Command: **Dim**
Dim: **Ord**
Select feature: *(Select the Endpoint of the line or arc at "A")*
Leader endpoint (Xdatum/Ydatum): *(Select a point at "B")*
Dimension text <3.00>: *(Strike Enter to accept this value)*

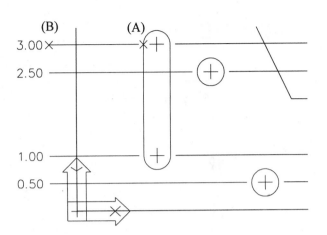

When spaces are tight to dimension to, two points not parallel to the X or Y axis will result in a "jog" being drawn. The jog always occurs in the middle of the extension line and always at a 90 degree angle to the extension line. It is still helpful to snap to a grid dot when performing this operation; however, be sure Ortho is Off.

Command: **Dim**
Dim: **Ord**
Select feature: *(Select the Endpoint of the line at "A")*
Leader endpoint (Xdatum/Ydatum): *(Select the point at "B")*
Dimension text <2.00>: *(Strike Enter to accept this default value)*

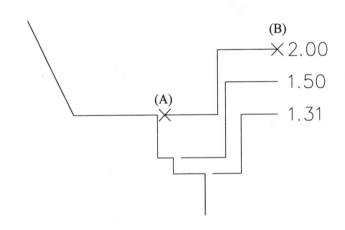

Ordinate dimensioning provides a neat and easy way of organizing dimensions for drawings where geometry leads to applications involving numerical control. Only two points are required to place the dimension which references the current location of the UCS. Certain dimension variables have no effect on ordinate dimensions. Dimension text along the X direction will never be drawn horizontally regardless of what Dimtih and Dimtoh are set to. However, dimension text may be placed above the extension line controlled by Dimtad or at specified distances above or below the extension line controlled by Dimtvp. The gap between the dimension text and the end of the extension line can be manipulated with Dimgap. The spacing between the selected feature and the beginning of the extension line is controlled by Dimexo.

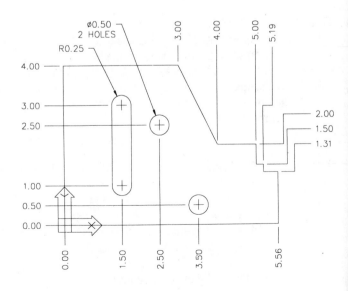

Tutorial Exercise #13
Dimex.Dwg

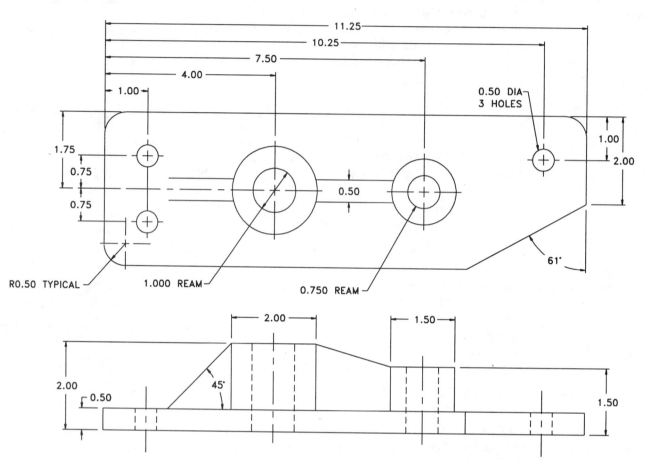

PURPOSE:
The purpose of this tutorial is to draw the two view object illustrated above and place dimensions on the drawing. This drawing is available on diskette under the file name "Dimex.Dwg."

SYSTEM SETTINGS:
Either copy "Dimex.Dwg" from the diskette provided or begin a new drawing called "Dimex." Units and Limits of the drawing should be already set. Use the Grid command and change the grid spacing from 1.00 units to 0.25 units to aid in the placement of dimension lines. Do not turn the snap or ortho modes on.

LAYERS:
The drawing file "Dimex.Dwg" has the following layers already created for this tutorial.

Name-Color-Linetype
Object - Magenta - Continuous
Hidden-Red-Hidden
Center-Yellow-Center
Dim - Yellow - Continuous

SUGGESTED COMMANDS:
All commands for this tutorial deal with dimensioning; all dimensioning commands and options begin at the prompt, "Dim:." The following options of the Dim: command will be used: Horizontal, Vertical, Continuous, Baseline, Center, Leader, Radius, Diameter, and Angular. All dimension options may be picked from the digitizing pad, screen menu, or entered from the keyboard. When entering dimension options from the keyboard, the first three letters are all that is required, (Horizontal=Hor). Use the Zoom command to get a closer look at details and features that are being dimensioned.

DIMENSIONING:
Follow the tutorial for manipulating and setting dimension variables.

PLOTTING:
This tutorial exercise may be plotted on "B"-size paper (11 x 17"). Use a plotting scale of 0.75=1 to produce a scaled size plot.

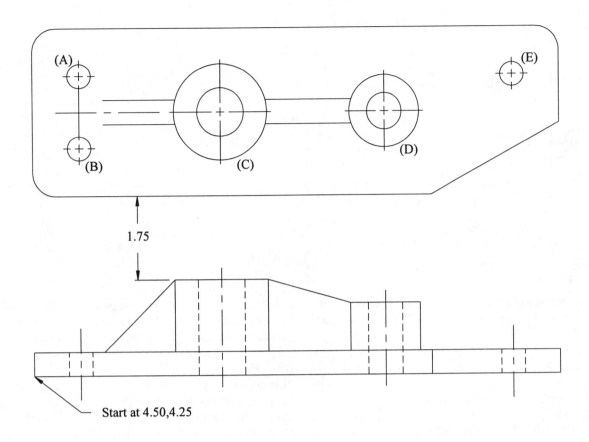

Step #1

Before beginning to assign dimensions, construct the front and top views. Starting the front view at absolute coordinate (2.50,1.50) and spacing the views a distance 1.75 units away from each other will ensure the dimensions will fit on the defined limits. Next use the DIM: command, enter the dimensioning mode, and use the Center option to place center marks at the center of all circles. Be sure the dimension variable, DIMCEN is set to a value of -0.09. This value will place the markers at the circle center in addition to the lines that extend to the outside of the circle.

Command: **Dim**
Dim: **Dimcen**
Current value <0.09) New value: **-0.09**
Dim: **Center**
Select arc or circle: *(Select the circle at "A")*
Dim: **Center**
Select arc or circle: *(Repeat for circles "B", "C", "D", and "E")*

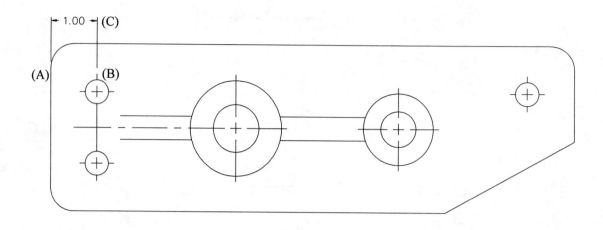

Step #2

Change the following Dimension variables by entering the changes in at the keyboard or by using the DDIM dialog box:

Dim: **Dimtxt**
Current value <0.18> New value: **0.12**
Dim: **Dimasz**
Current value <0.18> New value: **0.12**
Dim: **Dimexo**
Current value <0.06) New value: **0.12**
Dim: **Dimexe**
Current value <0.18> New value: **0.07**
Dim: **Dimsho**
Current value <Off> New value: **On**

Use the Dim-Horizontal mode and place the 1.00 dimension as illustrated above. All dimensioning options can be entered by the first three letters. This will be used throughout this tutorial.

Dim: **Hor**
First extension line origin or RETURN to select: **Endp**
of *(Select the endpoint of the line at "A")*
Second extension line origin: **Endp**
of *(Select the endpoint of the line at "B")*
Dimension line location (Text/Angle): *(Select a point at "C")*
Dimension text <1.00>: *(Strike Enter to accept this default value)*

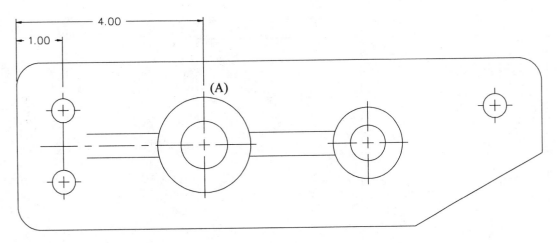

Step #3

The first extension line location of the last dimension, 1.00, will be used to establish a point of reference for the next 4 dimensions. This is accomplished with the DIM-Baseline option. Again, only the first three letters will be used to begin the command.

Dim: **Bas**
Second extension line origin or RETURN to select: **Endp**
of *(Select the endpoint of the line at "A")*
Dimension text <4.00>: *(Strike Enter to accept this default value)*

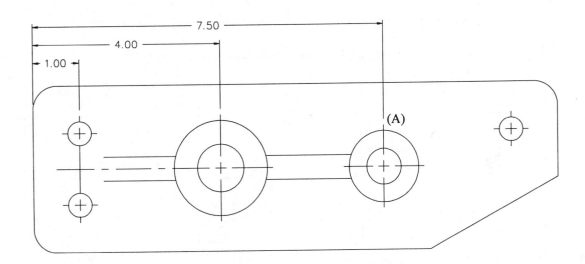

Step #4

Repeat the Dim-Baseline option for the 7.50 dimension illustrated above.

Dim: **Bas**
Second extension line origin or RETURN to select: **Endp**
of *(Select the endpoint of the line at "A")*
Dimension text <7.50>: *(Strike Enter to accept this default value)*

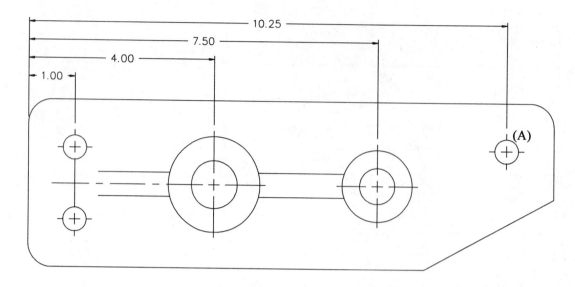

Step #5

Repeat the Dim-Baseline option for the 10.25 dimension illustrated above.

Dim: **Bas**
Second extension line origin or RETURN to select: **Endp**
of *(Select the endpoint of the line at "A")*
Dimension text <10.25>: *(Strike Enter to accept this default value)*

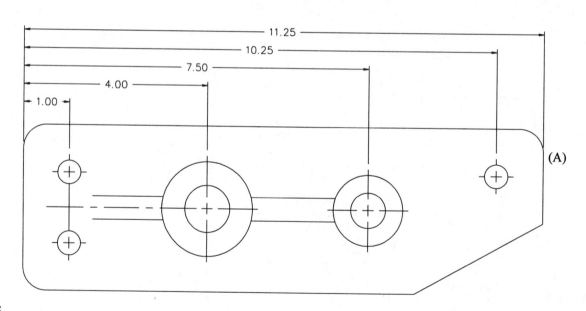

Step #6

Repeat the Dim-Baseline option for the 11.25 dimension illustrated above.

Dim: **Bas**
Second extension line origin or RETURN to select: **Endp**
of *(Select the endpoint of the line at "A")*
Dimension text <11.25>: *(Strike Enter to accept this default value)*

Step #7

Use the Zoom command to get a closer look at the left side of the top view as illustrated to the right. Then use the Dim-Vertical mode to place the 0.75 dimension.

Dim: **Ver**
First extension line origin or RETURN to select: **Endp**
of *(Select the endpoint of the line at "A")*
Second extension line origin: **Endp**
of *(Select the endpoint of the line at "B")*
Dimension line location (Text/Angle): *(Select a point at "C")*
Dimension text <0.75>: *(Strike Enter to accept this default value)*

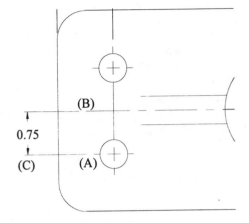

Step #8

The Dim-Continuous mode is used to place the next dimension in line or along side the previous dimension. When using the Continuous mode, the placement of the dimension line is remembered from the previous dimension.

Dim: **Con**
Second extension line origin or RETURN to select: **Endp**
of *(Select the endpoint of the line at "A")*
Dimension text <0.75>: *(Strike Enter to accept this default value)*

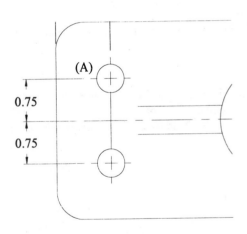

Step #9

Use the Dim-Vertical mode to place the 1.75 dimension illustrated at the right.

Dim: **Ver**
First extension line origin or RETURN to select: **Endp**
of *(Select the endpoint of the line at "A")*
Second extension line origin: **Endp**
of *(Select the endpoint of the line at "B")*
Dimension line location (Text/Angle): *(Select a point at "C")*
Dimension text <1.75>: *(Strike Enter to accept this default value)*

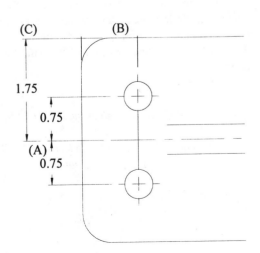

Step #10

Use the Zoom command to get a closer look at the right side of the top view as illustrated to the right. Then use the Dim-Vertical mode to place the 1.00 dimension.

Dim: **Ver**
First extension line origin or RETURN to select: **Endp**
of *(Select the endpoint of the line at "A")*
Second extension line origin: **Endp**
of *(Select the endpoint of the line at "B")*
Dimension line location (Text/Angle): *(Select a point at "C")*
Dimension text <1.00>: *(Strike Enter to accept this default value)*

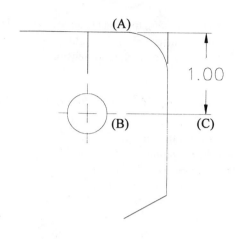

Step #11

Repeat the Dim-Baseline option for the 2.00 dimension illustrated at the right.

Dim: **Bas**
Second extension line origin or RETURN to select: **Endp**
of *(Select the endpoint of the line at "A")*
Dimension text <2.00>: *(Strike Enter to accept this default value)*

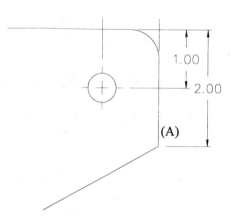

Step #12

Use the Dim-Angular mode to place the 61 degree dimension illustrated to the right. Follow the system prompts carefully when performing angular dimensioning. If the results are not satisfactory, type "U" at the "Dim:" prompt to erase the last dimension and try again.

Dim: **Ang**
Select arc, circle, line or RETURN: *(Select the line at "A")*
Second line: *(Select the line at "B")*
Dimension line arc location (Text/Angle): *(Select near "C")*
Dimension text <61>: *(Strike Enter to accept this default value)*
Enter text location (or RETURN): *(Strike the Enter key)*

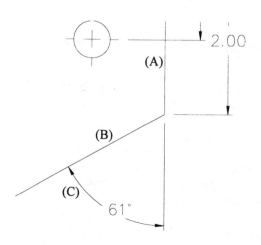

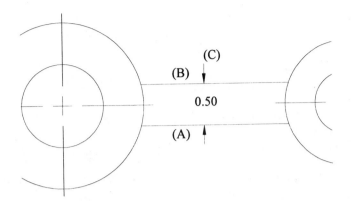

Step #13

The vertical dimension, 0.50, will be placed using the Dim-Vertical mode. To prevent extension lines from being placed on top of object lines, the dimension variables Dimse1 and Dimse2 need to be turned On. This will turn off or suppress the extension lines for this dimension. Also, to place the dimension in between the arrowheads, set the dimension variable, Dimtix, from Off to On.

Dim: **Dimse1**
Current value <Off> New value: **On**
Dim: **Dimse2**
Current value <Off> New value: **On**
Dim: **Dimtix**
Current value <Off> New value: **On**

Dim: **Ver**
First extension line origin or RETURN to select: **Nea**
to *(Select the line at "A")*
Second extension line origin: **Per**
to *(Select the line at "B")*
Dimension line location (Text/Angle): *(Select a point at "C")*
Dimension text <0.50>: *(Strike Enter to accept this default value)*

Step #14

The Dim-Radius will be used to place the 0.50 dimension illustrated at the right. Before performing this operation, reset the last dimension variables, Dimse1, Dimse2, and Dimtix, back to their original values. Since other corners of this drawing have the same radius value, the note, "TYPICAL," is typed in for new dimension text along with the 0.50 value. Follow the prompts at the below to place this dimension.

Dim: **Dimse1**
Current value <On> New value: **Off**
Dim: **Dimse2**
Current value <On> New value: **Off**
Dim: **Dimtix**
Current value <On> New value: **Off**
Dim: **Dimtofl**
Current value <Off> New value: **On**
Dim: **Rad**
Select arc or circle: *(Select the arc at "A")*
Dimension text <0.50>: **R0.50 TYPICAL**
Enter leader length for text: *(Select a point at "B")*
Dim: **Center**
Select arc or circle: *(Select the arc at "A" to place a center mark)*

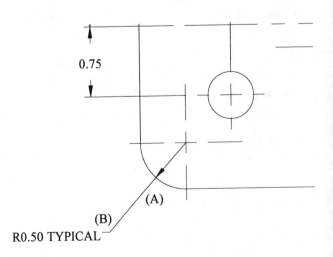

Step #15

Use the Dim-Diameter mode to place the diameter dimension illustrated to the right. The term "Ream" is a precision drilling operation and needs to be called-out in note form. Reams usually carry decimal places of 3 or more places. Therefore, when the dimension text is displayed in the diameter prompt, a new value of 1.000 REAM needs to be entered. Also, since center marks were placed back in Step #1, and since the diameter mode automatically places a center mark, the dimension variable, Dimcen, needs to be changed from -0.09 to 0. The zero will prevent a center mark from being placed. Follow the prompts below to place the diameter dimension.

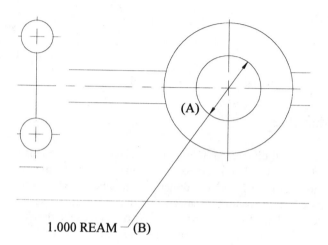

Dim: **Dimcen**
Current value <-0.09> New value: **0**
Dim: **Dia**
Select circle or arc: *(Select the circle at "A")*
Dimension text <1.00>: **1.000 REAM**
Enter leader length for text: *(Select near "B")*

Step #16

Use the Dim-Diameter mode to place the diameter dimension illustrated to the right. Again, as in the previous dimension, substitute the new dimension text, 0.750 REAM for the default value <0.75>. Leave the dimension variable Dimcen set to zero to prevent the placing of double center marks. The dimension variable Dimtofl may be turned back Off at this time.

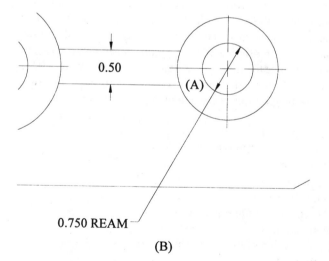

Dim: **Dia**
Select circle or arc: *(Select the circle at "A")*
Dimension text <0.75>: **0.750 REAM**
Enter leader length for text: *(Select near "B")*

Dim: **Dimtofl**
Current value <On> New value: **Off**

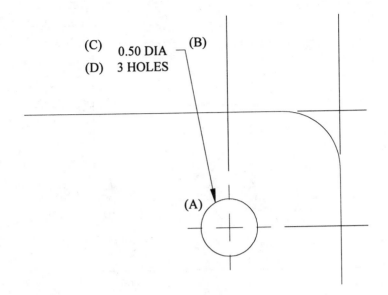

Step #17

The Dim-Leader mode will now be used to place a value and note describing 3 holes that share the save diameter value of 0.50. When placing a leader dimension, use the Osnap-Nearest mode when prompted for "Select circle or arc". This will select a point on the circle for the location of the tip of the arrowhead. The prompt continues with the familiar "To point" phrase. Follow the commands at the bottom to place this dimension. Once the dimension is placed, copy the text down a short distance and use the Change command to change the text from 0.50 DIA to 3 HOLES.

Dim: **Lea**
Leader start: **Nea**
of *(Select the circle at "A")*
To point: *(Select a point at "B")*
To point: *(Strike Enter to continue)*
Dimension text <0.75>: **.50 DIA**
Dim: **Exit**

Command: **Copy**
Select objects: **L**
Select objects: *(Strike Enter to continue)*
<Base point or displacement>/Multiple:*(Select at "C")*
Second point of displacement: *(Select a point at "D")*

Command: **Change**
Select objects: *(Select the text at "D")*
Select objects: *(Strike Enter to continue)*
Properties/<Change point>: *(Strike Enter to accept default)*
Enter text insertion point: *(Strike Enter to accept default)*
New style or RETURN for no change: *(Strike Enter to accept default)*
New Height <0.12>: *(Strike Enter to accept default)*
New rotation angle <0>: *(Strike Enter to accept default)*
New text <0.50 DIA>: **3 HOLES**

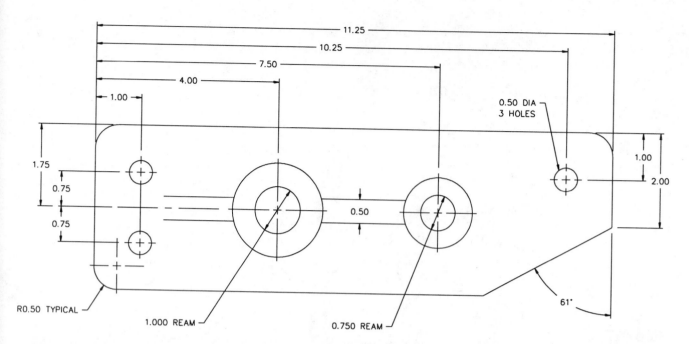

The completed top view including dimensions is illustrated above. Use this example to check that all dimensions have been placed and all features (holes, fillets, etc.) have been properly identified.

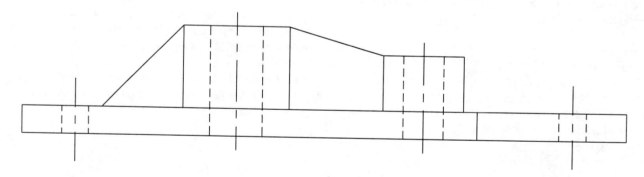

The front view will now be the focus of the next series of dimensioning steps. Again use the Zoom command whenever dimensioning to smaller surfaces.

Step #18

Use the Dim-Horizontal mode to place the 2.00 dimension calling out the size of the cylinder as illustrated at the right.

Dim: **Hor**
First extension line origin or RETURN to select: **Endp**
of *(Select the endpoint of the line at "A")*
Second extension line origin: **Endp**
of *(Select the endpoint of the line at "B")*
Dimension line location (Text/Angle): *(Select a point at "C")*
Dimension text <2.00>: *(Strike Enter to accept this default value)*

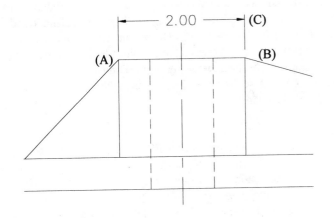

Step #19

Use the Dim-Horizontal mode to place the 1.50 dimension calling out the size of the smaller cylinder as illustrated at the right.

Dim: **Hor**
First extension line origin or RETURN to select: **Endp**
of *(Select the endpoint of the line at "A")*
Second extension line origin: **Endp**
of *(Select the endpoint of the line at "B")*
Dimension line location (Text/Angle): *(Select a point at "C")*
Dimension text <1.50>: *(Strike Enter to accept this default value)*

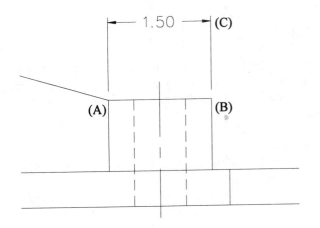

Step #20

Use the Zoom command to magnify the right end of the front view as illustrated at the right. Next use the Dim-Vertical mode and place the 1.50 dimension. Follow the prompts below.

Dim: **Ver**
First extension line origin or RETURN to select: **Endp**
of *(Select the endpoint of the line at "A")*
Second extension line origin: **Endp**
of *(Select the endpoint of the line at "B")*
Dimension line location (Text/Angle): *(Select a point at "C")*
Dimension text <1.50>: *(Strike Enter to accept this default value)*

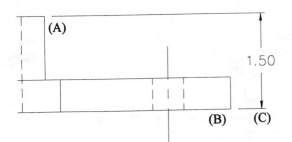

Step #21

Use the Zoom command and magnify the left end of the front view as illustrated to the right. Use the Dim-Vertical mode and place the 0.50 dimension. Do not be concerned that the dimension is not in between the arrowheads. Remember, this is controlled by the dimension variable Dimtix. If you want the text to be inside the arrowheads, change the values of Dimtix from Off to On.

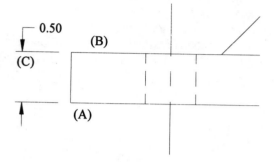

Dim: **Ver**
First extension line origin or RETURN to select: **Endp**
of *(Select the endpoint of the line at "A")*
Second extension line origin: **Endp**
of *(Select the endpoint of the line at "B")*
Dimension line location (Text/Angle): *(Select a point at "C")*
Dimension text <0.50>: *(Strike Enter to accept this default value)*

Step #22

Use the Dim-Baseline mode to place the 2.00 dimension. Remember, when in Baseline mode, the first extension line location is remembered from the first or previous dimension placed.

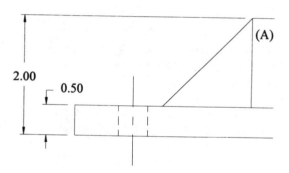

Dim: **Bas**
Second extension line origin or RETURN to select: **Endp**
of *(Select the endpoint of the line at "A")*
Dimension text <2.00>: *(Strike Enter to accept this default value)*

Step #23

Use the Dim-Angular mode to place the 45 degree dimension illustrated to the right. Follow the system prompts carefully when performing angular dimensioning. If the results are not satisfactory, type "U" at the Dim: prompt to erase the last dimension at try again.

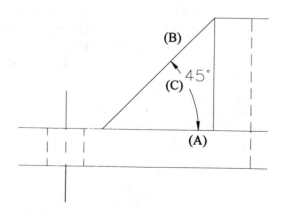

Dim: **Ang**
Select arc, circle, line, or RETURN: *(Select the line at "A")*
Second line: *(Select the line at "B")*
Dimension line arc location (Text/Angle): *(Select near "C")*
Dimension text <45>: *(Enter)*
Enter text location (or RETURN): *(Select near "C")*

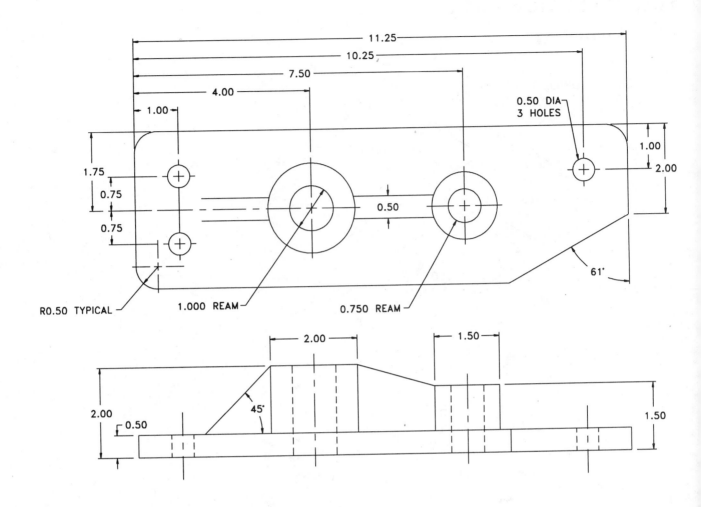

The front and top views complete with dimensions are shown
above. Use this illustration to check your final results.

Tutorial Exercise #14
Tblk-iso.Dwg

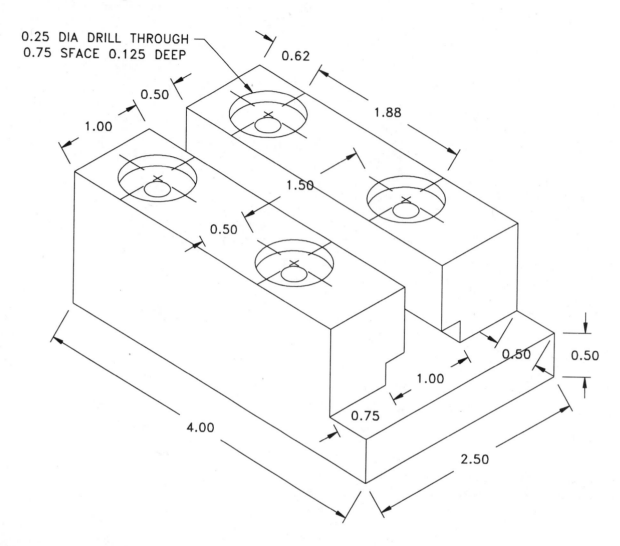

0.25 DIA DRILL THROUGH
0.75 SFACE 0.125 DEEP

0.62

0.50

1.00

1.88

1.50

0.50

0.50

1.00

0.50

0.50

4.00

0.75

2.50

PURPOSE:
The purpose of this tutorial is to align dimensions on an isometric drawing to oblique dimensions.

SYSTEM SETTINGS:
This drawing is already constructed and dimensioned up to a certain point. Enter AutoCAD and edit an existing drawing called "TBLK-ISO." Follow the steps in this tutorial for converting the dimensions to proper isometric mode.

LAYERS:
All layers have already been created:
Name-Color-Linetype
Object - White - Continuous
Dim - Yellow - Continuous

SUGGESTED COMMANDS:
All commands for this tutorial deal with the Oblique subcommand of Dim:.

DIMENSIONING:
Follow the tutorial for manipulating and setting dimension variables.

PLOTTING:
This tutorial exercise may be plotted on "A"-size paper (8.5 x 11"). Use a plotting scale of 1=1 to produce a scaled size plot.

Step #1

Before converting all dimensions to isometric form, create a new dimension style called "EXT-OFF". This style has the two variables, Dimse1 and Dimse2 turned On which will suppress or not show the extension lines of a dimension. This will be used later for one dimension. Then use the DIM-OBL command to rotate the dimension for isometric purposes.

Command: **Dim**
Dim: **Save**
?/Name for new dimension style: **Standard**

Dim: **Dimse1**
Current value <Off> New value: **On**

Dim: **Dimse2**
Current value <Off> New value: **On**

Dim: **Save**
?/Name for new dimension style: **Ext-off**

Dim: **Restore**
New dimension style to restore: **Standard**

Dim: **Obl**
Select objects: *(Select dimensions "A" through "G")*
Select objects: *(Strike Enter to continue)*
Enter obliquing angle (RETURN for none): **150**

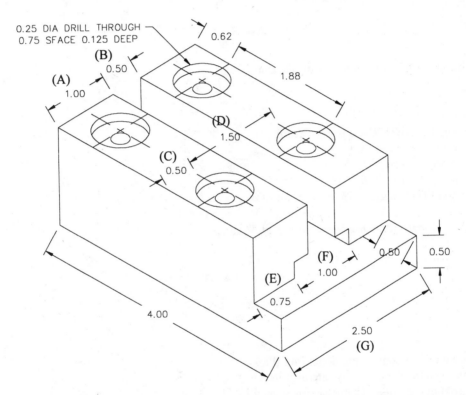

Step #2

Use the Dim-Obl command to convert the three dimensions illustrated at the right to an isometric form at an obliquing angle of 30 degrees.

Dim: **Obl**
Select objects: *(Select dimensions "A" through "D")*
Select objects: *(Strike Enter to continue with this command)*
Enter obliquing angle (RETURN for none): **210**

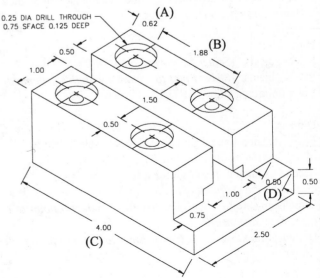

Step #3

Convert the dimension illustrated at the right to an isometric dimension and an obliquing angle of 30 degrees using the Dim-Obl command.

Dim: **Obl**
Select objects: *(Select the dimension at "A")*
Select objects: *(Strike Enter to continue)*
Enter obliquing angle (RETURN for none): **30**

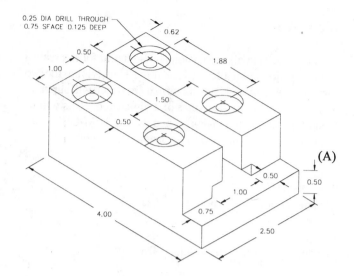

Step #4

Place the 1.00 dimension illustrated at the right using the Dim-Aligned command.

Dim: **Ali**
First extension line origin or RETURN to select: **Endp**
of *(Select the endpoint of the line at "A")*
Second extension line origin: **Endp**
of *(Select the endpoint of the line at "B")*
Dimension line location (Text/Angle): *(Select a point at a convenient distance)*
Dimension text <1.00>: *(Strike Enter to accept this value)*

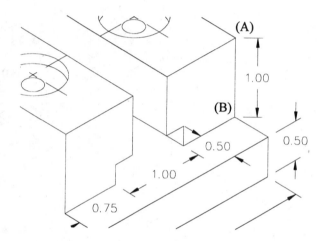

Step #5

Use the Dim-Obl command to convert the dimension illustrated at the right to an isometric dimension by entering a value of 30 degrees for the obliquing angle. Then place an aligned dimension using the Dim-Ali command.

Dim: **Obl**
Select objects: *(Select the dimension at "A")*
Select objects: *(Strike Enter to continue with this command)*
Enter obliquing angle (RETURN for none): **30**

Dim: **Ali**
First extension line origin or RETURN to select: **Endp**
of *(Select the endpoint of the line at "B")*
Second extension line origin: **Endp**
of *(Select the endpoint of the line at "C")*
Dimension line location (Text/Angle): *(Select a point at a convenient distance)*
Dimension text <0.75>: *(Strike Enter to accept this default value)*

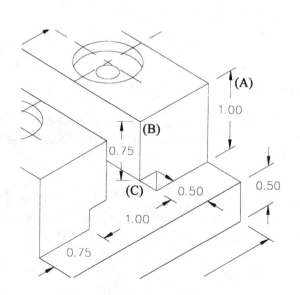

Step #6

In this final step, use the Dim-Obl command to convert the dimension at "A" to an isometric form. Then, restore the dimension style, "Ext-off", and use the Dim-Update command to update the dimension at "B". This dimension style has both Dimse1 and Dimse2 turned on which will turn off the extension lines of the 0.50 dimension and prevent lines from plotting over existing object lines.

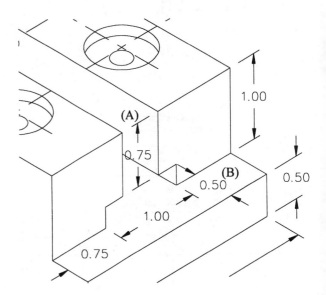

Dim: **Obl**
Select objects: *(Select the dimension at "A")*
Select objects: *(Strike Enter to continue with this command)*
Enter obliquing angle (RETURN for none): **30**

Dim: **Restore**
?/Enter dimension style name or RETURN to select dimension: **Ext-off**

Dim: **Upd**
Select objects: *(Select the dimension at "B")*
Select objects: *(Strike Enter to update the dimension to the new dimension style)*

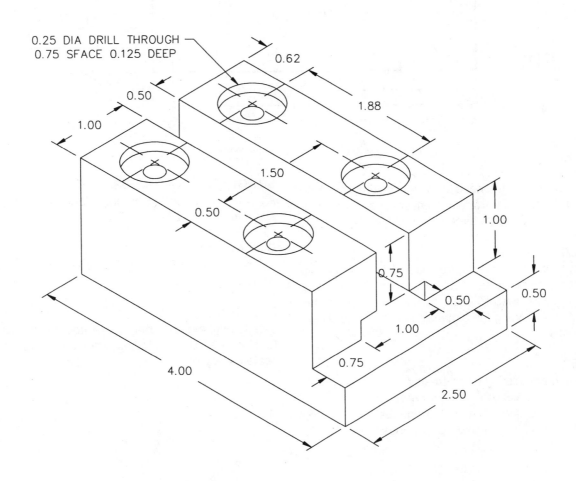

Tutorial Exercise #15
Bas-plat.Dwg

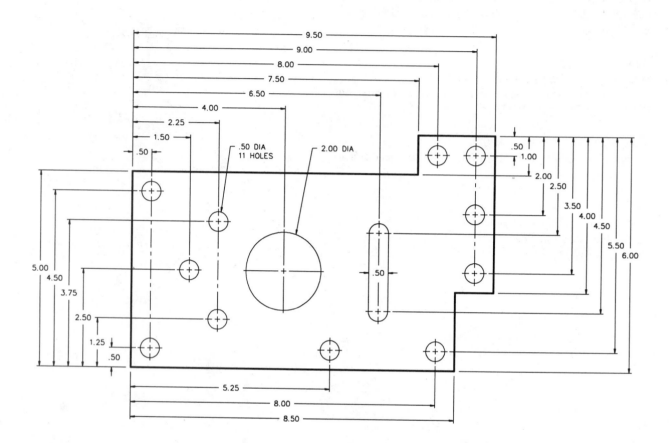

PURPOSE:
The purpose of this tutorial is to convert the drawing of the Bas-plat (Base plate) from the conventional dimensioning style to the ordinate dimensioning style.

SYSTEM SETTINGS:
This drawing is already constructed and dimensioned up to a certain point. Enter AutoCAD and edit an existing drawing called "BAS-PLAT." Follow the steps in this tutorial for converting the dimensions to proper ordinate dimensions.

LAYERS:
All layers have already been created:
Name-Color-Linetype
Object - White - Continuous
Dim - Yellow - Continuous

SUGGESTED COMMANDS:
All commands for this tutorial deal with the Ordinate subcommand of Dim:.

DIMENSIONING:
Follow the tutorial for manipulating and setting dimension variables.

PLOTTING:
This tutorial exercise may be plotted on "A"-size paper (8.5 x 11"). Use a plotting scale of 1=1 to produce a scaled size plot.

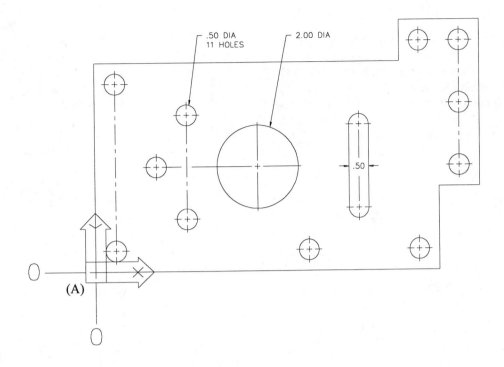

Step #1

All ordinate dimensions make reference to the current 0,0 location identified by the position of the user coordinate system . Since this icon is located in the lower left corner of the display screen by default, the coordinate system must be moved to a point on the object where all ordinate dimensions will be referenced from. First use the UCS command to define a new coordinate system with the orgin at the lower left corner of the object. Then use the UCSICON command to force the icon to display at the new origin.

Command: **UCS**
Origin/ZAxis/3point/Entity/View/X/Y/Z/Prev/Restore/Save/
Del/?/<World>: **Origin**
Origin point <0,0,0>: **Int**
of *(Select the intersection at "A")*

Command: **UCSICON**
ON/OFF/All/Noorgin/ORigin <ON>: **OR**

Step #2

Use the Zoom-Center command to magnify the screen similar to the illustration at the right.

Command: **Zoom**
All/Center/Dynamic/Extents/Left/Previous/Vmax/Window/
<Scale(X/XP)>: **C**
Center point: **0,0**
Magnification or Height <11.23>: **4**

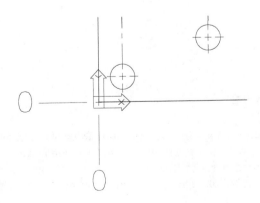

Step #3

Before continuing with this next step, be sure that the grid is turned On. Use the F7 function key to accomplish this. Next, begin to place the first ordinate dimension using the Dim-Ord command. Use the Osnap-Quadrant mode to select the circle as the feature. With the Snap and Ortho turned on, mode two grid dots to identify the leader endpoint. AutoCAD will determine if the dimension is Xdatum or Ydatum.

Command: **Dim**
Dim: **Ord**
Select feature: **Qua**
of *(Select the quadrant of the circle at "A")*
Leader endpoint (Xdatum/Ydatum): *(Pick a point two grid dots below the edge of the object at "B")*
Dimension text<0.50>: *(Strike Enter to accept this default value)*

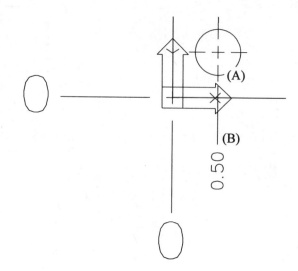

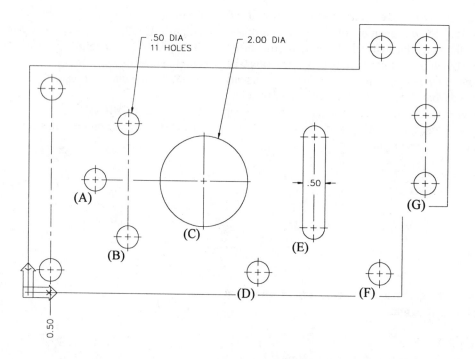

Step #4

Perform a Zoom-Previous operation to display the overall drawing of the bas-plat similar to the illustration above. Repeat the procedure back in Step #3 to place ordinate dimensions at locations "A" through "G". Be sure Ortho mode is on and that the leader location is two grid dots below the bottom edge of the object. The Osnap-Quadrant mode should be used on each circle and arc to satisfy the prompt, "Select feature".

Step #5

Continue placing ordinate dimensions similar to the procedure used in Step #3. Use the Osnap-Endpoint mode to select features at "A" and "B" in the illustration at the right. Have Ortho mode On and identify the leader endpoint two grid dots below the bottom edge of the object.

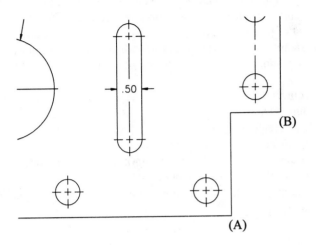

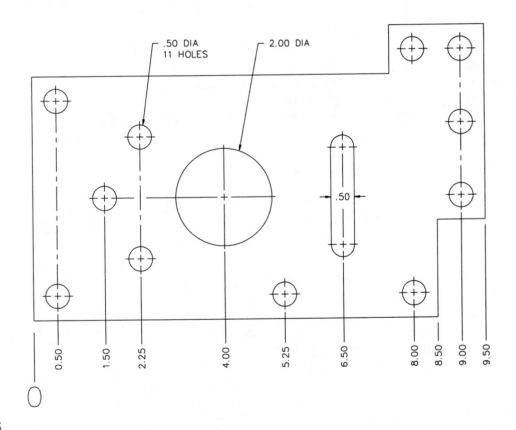

Step #6

Your display should appear similar to the illustration above. Notice the user coordinate system icon has disappeared. The Ucsicon command may be used to turn off the display of the icon while still keeping the 0,0 origin at the lower left corner of the object.

Command: **Ucsicon**
ON/OFF/All/Noorgin/ORigin <ON>: **Off**

Step #7

With Snap On and Ortho On, begin placing the first vertical ordinate dimension; the procedure and prompts are identical to that of placing a horizontal ordinate dimension. Follow the example at the right and the prompts below for performing this operation.

Command: **Dim**
Dim: **Ord**
Select feature: **Qua**
of *(Select the quadrant of the circle at "A")*
Leader endpoint (Xdatum/Ydatum): *(Pick a point two grid dots to the left of the object at "B")*
Dimension text<0.50>: *(Strike Enter to accept this default value)*

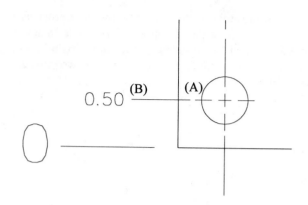

Step #8

Follow the procedure in the previous step to complete the vertical ordinate dimensions along this edge of the object. Use the Osnap-Quadrant mode for "A" through "D" and Osnap-Endpoint for "E". Again have Ortho On and Snap On. For the leader endpoint count two grid dots to the left of the object and place the dimensions.

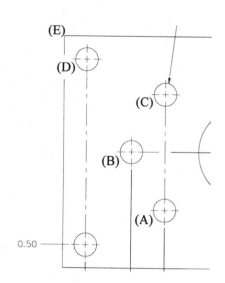

Step #9

Your display should appear similar to the illustration at the right.

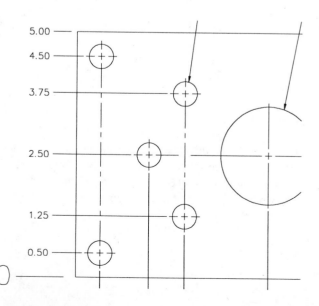

Step #10

Magnify the right portion of the object using the Zoom-Window command. Use ordinate dimensions and a combination of Osnap-Endpoint and Quadrant modes to place vertical ordinate dimensions from "A" to "I".

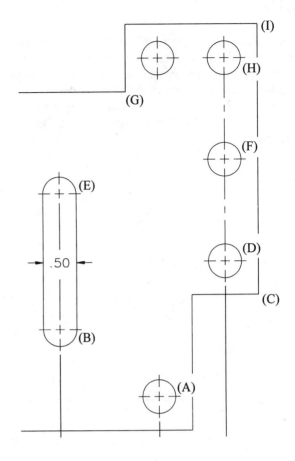

Step #11

Your display should appear similar to the illustration at the right.

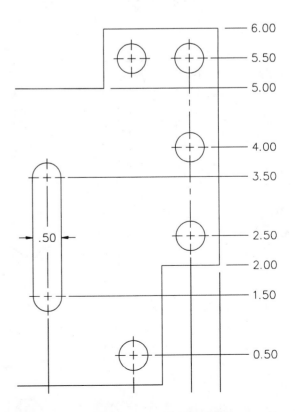

Step #12

Complete the dimensioning by placing two horizontal ordinate dimensions at the locations illustrated at the right. As in the past, have Ortho On, Snap On, and use a combination of the Osnap-Quadrant and Endpoint modes to select the features. Place the leader endpoint two grid dots above the top line of the object. Use the Zoom-All option to return the entire object to your display.

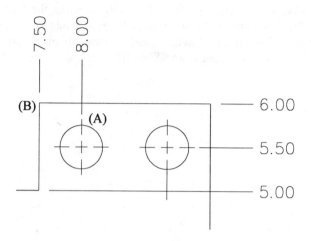

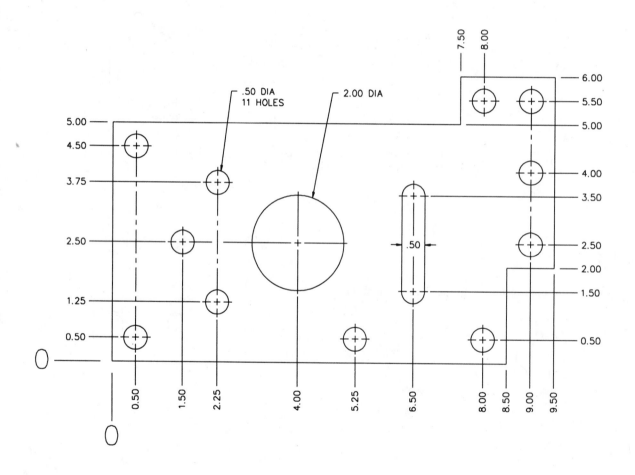

Questions for Unit 5

5-1. The Continue dimensioning subcommand always
(A) offsets each added dimension.
(B) references the variable DIMDLI.
(C) prompts for the first extension origin.
(D) is used as part of an angular dimension.
(E) requires an existing linear dimension.

5-2. To change the sizes of all dimension variables, use
(A) DIMTSZ
(B) DIMSCALE
(C) DIMLFAC
(D) DIMRND
(E) DIMASZ

5-3. The Dimension Variable that controls associative dimensioning is
(A) DIMASC
(B) DIMASO
(C) DIMSHO
(D) DIMAOC
(E) DIMSHX

5-4. The sequence of commands used to change the size of the dimension line arrowhead is
(A) DIM - DIMASZ
(B) DIM - DIMSIZE
(C) DIM - ARROW
(D) DIM - ASIZE
(E) DIMSIZE

5-5. The Dimension Variable DIMEXE controls the
(A) extension line increment.
(B) extension above the dimension line.
(C) executable dimensions.
(D) extension line origin offset.
(E) extension line color.

5-6. The dimensioning subcommand that lists all dimension variables with their current values is called
(A) DIMVAR
(B) STATUS
(C) VARIABLE
(D) DIMSTAT
(E) LIST

5-7. The dimension subcommand that allows the user to change the dimension text for an existing associative dimension is called
(A) DIMTEXT
(B) NEWTEXT
(C) UPDTEXT
(D) DIMTXT
(E) DIM1

5-8. All of the following are valid dimension subcommands except
(A) CREATE
(B) UPDATE
(C) NEWTEXT
(D) HOMETEXT
(E) CENTER

5-9. The Dimension Variable that controls the dimension text being placed above the dimension line is
(A) DIMTABOVE
(B) DIMTXABO
(C) DIMTAD
(D) DIMTAB
(E) DIMTXTAB

5-10. By default, all dimension text is
(A) aligned horizontally.
(B) aligned vertically.
(C) aligned to the entity being dimensioned.
(D) rotated to the entity being dimensioned.
(E) None of the above.

5-11. When using the Baseline Dimensioning option, the variable that controls the offset distance for each new dimension is
(A) DIMTAD
(B) DIMTOH
(C) DIMTIH
(D) DIMDLE
(E) DIMDLI

5-12. Dimension line terminators may take the form of
(A) Arrows
(B) Ticks
(C) Dots
(D) All of the above.
(E) Only A and B.

5-13. The Dimension Variable used to force dimension text inside of the extension lines is
(A) DIMTEX
(B) DIMFTEX
(C) DIMEXT
(D) DIMEXO
(E) DIMTIX

5-14. To change the default dimension display of the value from 0.50 to .50, set the following Dimension Variable
(A) DIMZIN to 0
(B) DIMZIN to 1
(C) DIMZIN to 2
(D) DIMZIN to 3
(E) DIMZIN to 4

5-15. To affect the scale factor of all linear dimension distances, use the dimension variable
- (A) DIMEXO
- (B) DIMSCALE
- (C) DIMEXE
- (D) DIMSE1
- (E) DIMLFAC

5-16. To turn off extension lines so they are not displayed when placing a dimension, use
- (A) DIMEXO
- (B) DIMSE1
- (C) DIMSCALE
- (D) DIMSE2
- (E) Both B and D

5-17. When using associative dimensions, the layer name automatically created is called
- (A) DEFPOINTS
- (B) DIMASO
- (C) DIM
- (D) Both A and C.
- (E) None of the above.

5-18. The dimension variable, DIMTAD controls
- (A) text and dimensions.
- (B) text above dimension line.
- (C) time spent dimensioning.
- (D) text after dimension line.
- (E) None of the above.

5-19. The dimension variable DIMBLK controls
- (A) dimension block name.
- (B) arrow block name.
- (C) tick block name.
- (D) dot block name.
- (E) dimension block size.

5-20. The Dimension Variable that controls the tick size is
- (A) DIMTIC
- (B) DIMASZ
- (C) DIMSIZ
- (D) DIMTSZ
- (E) DIMTIK

5-21. The Dimension Variable DIMEXE controls
- (A) executable dimensions.
- (B) extension line increment.
- (C) extension above the dimension line.
- (D) extension line origin offset.
- (E) None of the above.

5-22. The Dimension Variable DIMEXO controls
- (A) extension line origin.
- (B) extension line On/Off.
- (C) exit On/Off.
- (D) extension line offset.
- (E) None of the above.

5-23. To control the color of the dimension line in an associative dimension, use the
- (A) DIMCLRD variable.
- (B) DIMCLRT variable.
- (C) DIMCOLOR variable.
- (D) DIMCLRE variable.
- (E) None of the above.

5-24. To place an associative diameter dimension with the dimension line, dimension text, and arrowheads all inside of the circle, set
- (A) DIMTIX-ON and DIMTOFL-ON
- (B) DIMTIX-ON and DIMSOXD-OFF
- (C) DIMTIX-OFF and DIMTOFL-ON
- (D) DIMTIX-OFF and DIMTOFL-OFF
- (E) None of the above.

5-25. To place an associative diameter dimension with the dimension line, dimension text and arrowheads all outside of the circle resembling a leader line, set
- (A) DIMTIX-ON and DIMTOFL-ON
- (B) DIMTIX-ON and DIMTOFL-OFF
- (C) DIMTIX-OFF and DIMTOFL-ON
- (D) DIMTIX-OFF and DIMTOFL-OFF
- (E) None of the above.

5-26. To place an associative radius dimension with dimension text outside of the arc but with the dimension line and arrowheads inside of the arc, set
- (A) DIMTIX-ON and DIMTOFL-ON
- (B) DIMTIX-ON and DIMTOFL-OFF
- (C) DIMTIX-OFF and DIMTOFL-ON
- (D) DIMTIX-OFF and DIMTOFL-OFF
- (E) None of the above.

5-27. To control the color of an arrowhead in an associative dimension, use the
- (A) DIMCLRD variable.
- (B) DIMCLRL variable.
- (C) DIMCLRA variable.
- (D) DIMASZ variable.
- (E) DIMCLRE variable.

5-28. The dimension subcommand used to change the justification or rotate the text of an associative dimension is
- (A) DTEXT
- (B) TEDIT
- (C) TROTATE
- (D) ROTATE
- (E) Both B and C.

5-29. To create a rectangular box around dimension text representing a basic or reference dimension, set
- (A) DIMGAP to zero.
- (B) DIMBOX to a positive value.
- (C) DIMBOX to a negative value.
- (D) DIMGAP to a positive value.
- (E) DIMGAP to a negative value.

Problems for Unit 5

Problems 5–30 through 5–48

1. Use the grid and a spacing of 0.50 units to determine all dimensions.
2. Reproduce the views shown, and fully dimension the drawings.

Problem 5-30

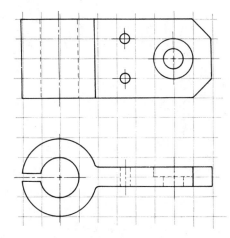

Problem 5-33

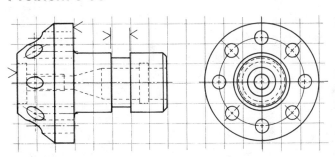

Problem 5-31

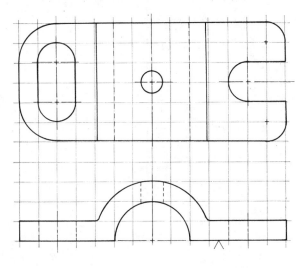

Problem 5-34

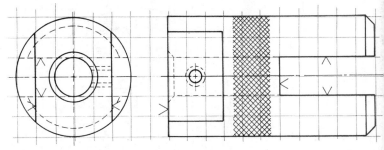

Problem 5-35

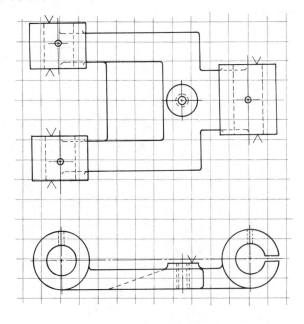

Problem 5-32

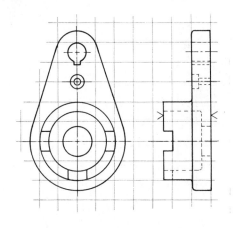

Problem 5–36

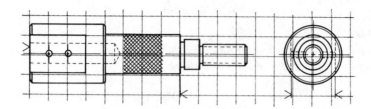

Problem 5–37

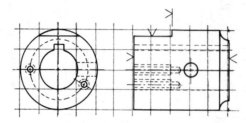

Problem 5–38

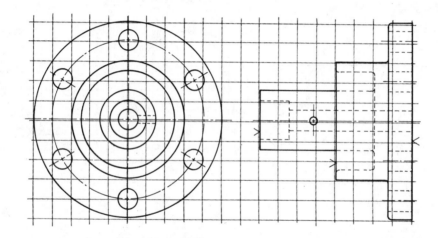

Problem 5–39

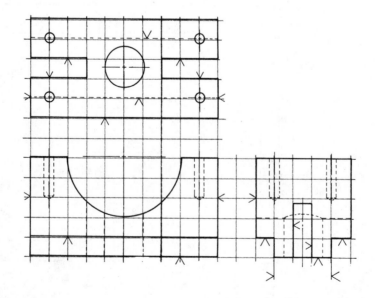

Problem 5–40

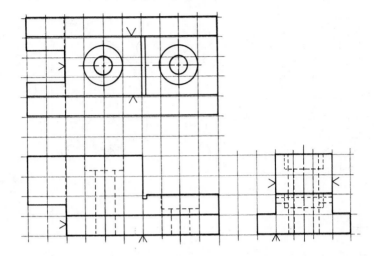

Problem 5–41

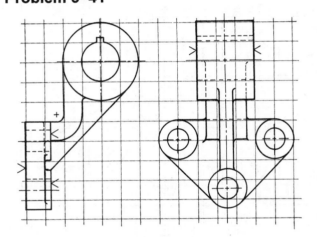

Problem 5–43

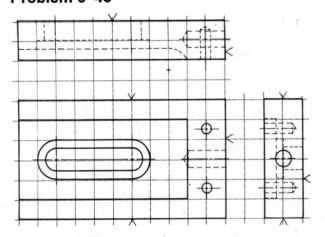

Problem 5–42

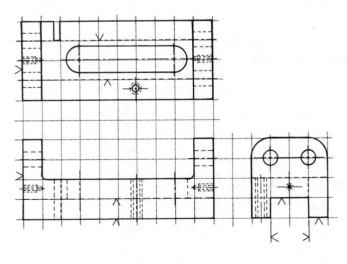

Problem 5–44

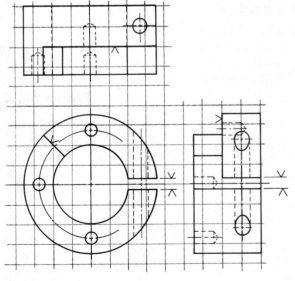

Problem 5-45

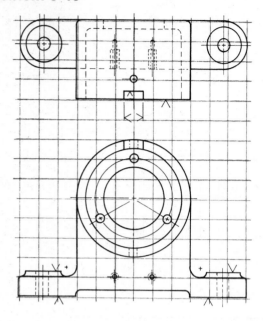

Problem 5-46

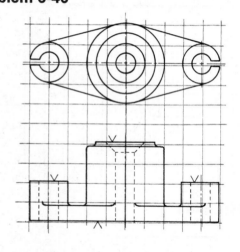

Problem 5-47

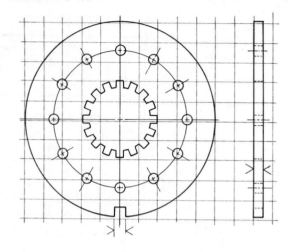

Problem 5-48

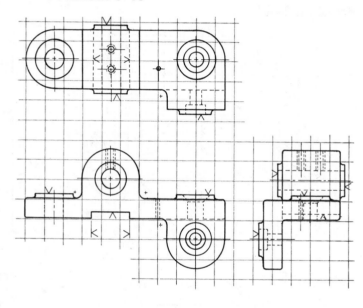

Problems 5–49 through 5–54

1. *Convert the isometric drawings provided into orthographic drawings, showing as many views as necessary to communicate the design.*
2. *Fully dimension your drawings.*

Problem 5-49

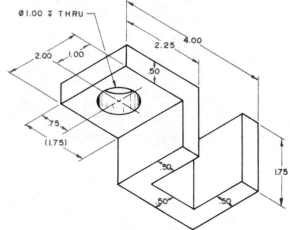

Ø1.00 ⊥ THRU

2.25   4.00

2.00   1.00

.50

.75

(1.75)

.50   .50

1.75

.50

Problem 5–50

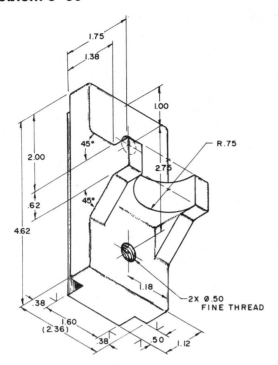

Problem 5–52

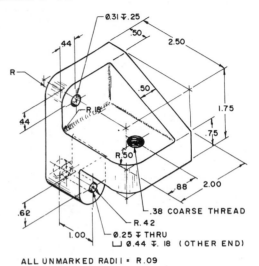

ALL UNMARKED RADII = R.09

Problem 5–51

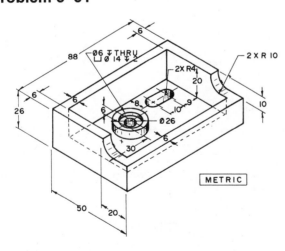

METRIC

Problem 5–53

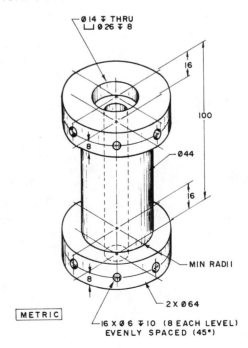

METRIC

Problem 5-54

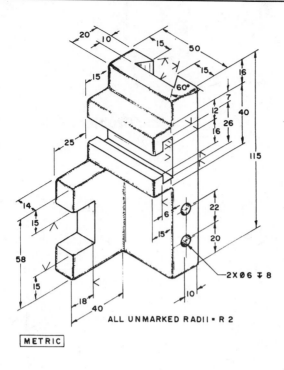

ALL UNMARKED RADII = R 2

METRIC

Directions for Problem 5–55
Use ordinate dimensioning techniques to dimension this drawing.

Problem 5–55

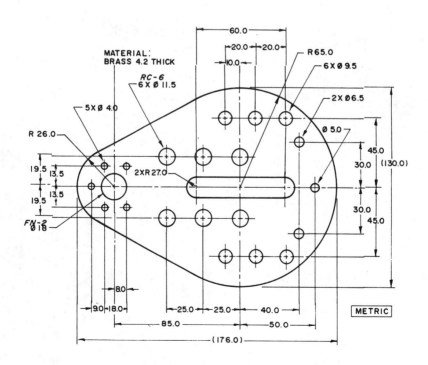

METRIC

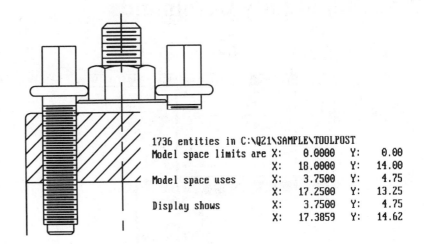

1736 entities in C:\Q21\SAMPLE\TOOLPOST
Model space limits are X: 0.0000   Y: 0.00
                       X: 18.0000   Y: 14.00
Model space uses   X: 3.7500   Y: 4.75
                       X: 17.2500   Y: 13.25
Display shows   X: 3.7500   Y: 4.75
                       X: 17.3859   Y: 14.62

Analyzing
2-D Drawings

Contents

Completed drawings are usually plotted out and checked with scales for accuracy purposes. Depending on the thickness of pen used to perform the plot and the scale used, a range of accuracy or tolerance is assigned. A proper computer-aided design system is equipped with a series of commands to calculate distances and angles of selected entities. Surface areas may be performed on complex geometric shapes.

The next series of pages highlights all Inquiry commands and how they are used to display useful information on an entity or group of entities. The DDMODIFY command is also explained in great detail on what type of entity control it supplies to the user. Use this information in Unit 6 to become more comfortable with all Inquiry commands and the DDMODIFY dialog box.

Choosing Inquiry Commands

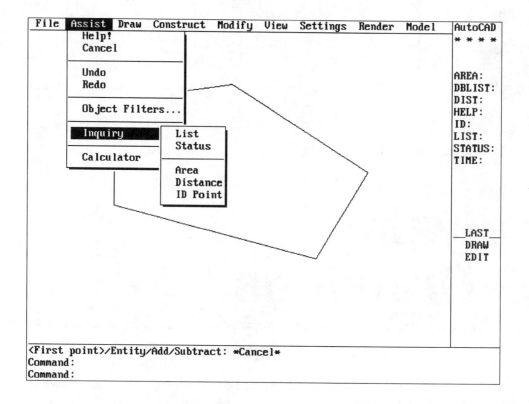

AutoCAD's Inquiry commands may be selected from the digitizing tablet, may be keyed in at the keyboard, or may be selected from the pulldown menu bar or side bar screen menus as illustrated above. The following is a listing of the Inquiry commands with a short description of each:

AREA - used to calculate the surface area given a series of points or by selecting a polyline or circle. Multiple entities may be added or subtracted to calculate the area with holes and cutouts.

DBLIST - provides a listing of all entities that make up the current drawing file.

DIST - calculates the distance between two points. Also provides the delta X,Y,Z coordinate values, the angle in the X-Y plane, and the angle from the X-Y plane.

HELP - provides online help for any command. May be entered at the keyboard or selected from a dialog box.

ID - displays the X,Y,Z absolute coordinate of a selected point.

LIST - displays key information depending on the entity selected.

STATUS - displays important information on the current drawing.

TIME - displays the time spent in the drawing editor.

Finding the Area of an Enclosed Shape

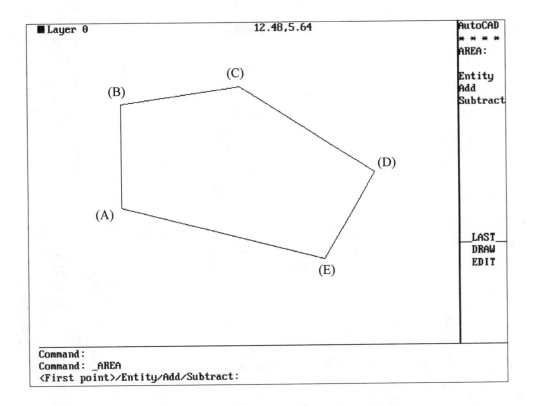

The Area command is used to calculate the area through the selection of a series of points. Select the endpoints of all vertices of the object illustrated above with the Osnap-End-point option. Once the first point is selected along with the remaining points in either a clockwise or counter-clockwise pattern, the command prompt "Next point:" is followed by the "Enter" key in order to calculate the area of the shape. Along with the area is a calculation of the perimeter. Use the illustrations above and to the right to gain a better understanding of the prompt sequence used for finding the area by identifying a series of points.

```
Command: AREA

<First point>/Entity/Add/Subtract: ENDP
of (Select Point "A")
Next point: ENDP
of (Select Point "B")
Next point: ENDP
of (Select Point "C")
Next point: ENDP
of (Select Point "D")
Next point: ENDP
of (Select Point "E")
Next point: ENDP
of (Select Point "A")
Next point:(Strike Enter to calculate the area)

Area = 25.00, Perimeter = 20.02
```

Finding the Area of an Enclosed Polyline or a Circle

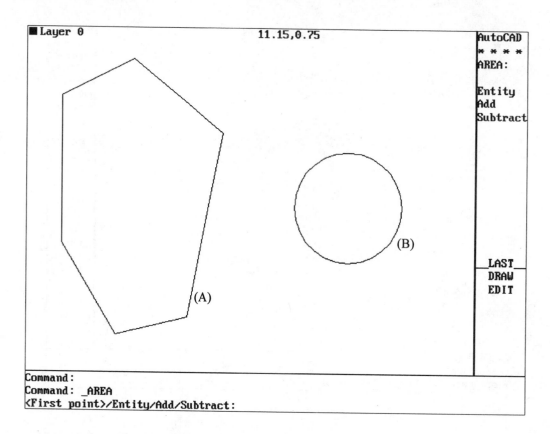

On the previous page, an example was given of finding the area of an enclosed shape using the Area command and identifying the corners and intersections of the enclosed area by a series of points. For a complex area, this could be a very tedious operation. As a result, the Area command has a built-in Entity option which will calculate the area and perimeter on a polyline and the area and circumference on a circle. Study the illustrations above and to the right for these operations.

Finding the area of a polyline can only be accomplished if one of the following is satisfied:

- The entity must have already been constructed using the Pline command.
- The entity must have already been converted into a polyline using the Pedit command if originally constructed out of individual entities.

```
_AREA
<First point>/Entity/Add/Subtract: _ENTITY
Select circle or polyline: (Select polyline "A")
Area = 24.88, Perimeter = 19.51

Command: _AREA
<First point>/Entity/Add/Subtract: _ENTITY
Select circle or polyline: (Select circle "B")
Area = 7.07, Circumference = 9.42
```

Finding the Area of a Surface by Subtraction

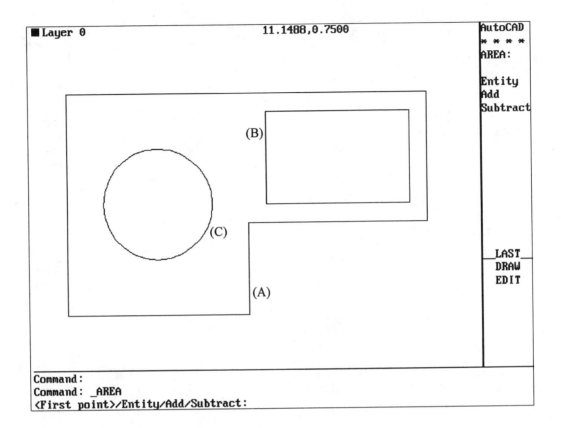

```
■Layer 0                        11.1488,0.7500              AutoCAD
                                                            * * * *
                                                            AREA:

                                                            Entity
                                                            Add
                                                            Subtract
          (B)

              (C)
                                                            LAST
                                                            DRAW
                                                            EDIT
          (A)

Command:
Command: _AREA
<First point>/Entity/Add/Subtract:
```

All previous examples of finding the area have dealt with either individual entities in the form of circles and polylines or identification of key intersections of an enclosed shape consisting of individual line segments. The true power of the Area command is illustrated in the examples above and to the right. The object above consists of an irregular profile with a circular and rectangular cutout. The steps used to calculate the total surface area are to first calculate the area of the outline and then subtract the entities inside of the outline. This is easily accomplished using the Area command in combination with the Pedit command. All individual entities with the exception of circles must first be converted into polylines using the Pedit command. Next, the overall area is found and added to the database using the Add mode of the Area command. "Add" mode is exited and the inner shapes are removed using the Subtract mode of the Area command. Remember, all shapes must be in the form of a circle or polyline. This means the inner shape at "B" must also be converted into a polyline using the Pedit command before calculating the area. Once the final entity is subtracted from the outline, the total area is displayed based on the current number of decimal places set by the Units command. Care must be taken when selecting the entities to subtract. If an entity is selected twice, it is subtracted twice and may yield an inaccurate area in the final calculation. Before performing this type of area, first redraw your screen to get rid

```
Command: _AREA
<First point>/Entity/Add/Subtract: _ADD
<First point>/Entity/Subtract: _ENTITY
(ADD mode) Select circle or polyline: (Select "A")
Area = 47.5000, Perimeter = 32.0000
Total area = 47.5000

(ADD mode) Select circle or polyline: (Enter)

<First point>/Entity/Subtract: _SUBTRACT
<First point>/Entity/Add: _ENTITY
(SUBTRACT mode) Select circle or polyline:("B")
Area = 10.0000, Perimeter = 13.0000
Total area = 37.5000

(SUBTRACT mode) Select circle or polyline:("C")
Area = 7.0686, Circumference = 9.4248
Total area = 30.4314

(SUBTRACT mode) Select circle or polyline:(Enter)

<First point>/Entity/Add: (Strike Enter)
```

of any unnecessary blips on the screen. Next use the blips as guides during the subtraction process to see what entities have been selected.

Using the Dblist (Database List) Command

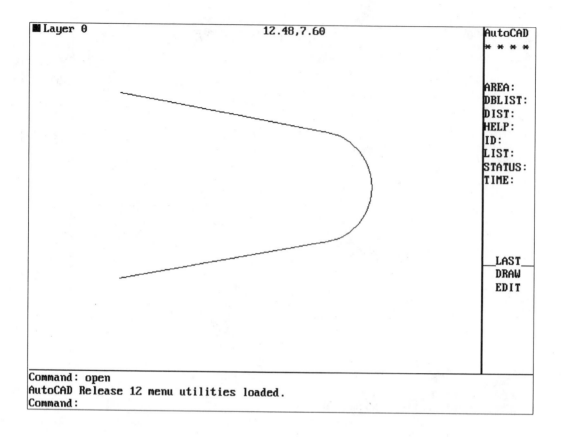

The Dblist commands simply lists all of the entities of a particular drawing that make up its database. Depending on the size of the drawing, the list of entities may be quite extensive. As a result, all entities continue to scroll over and over the text screen until the last entity lists ending the string of entities.

Depending on the number of entities in a drawing file, sometimes all of this information will not fit on the display screen. Once the screen is filled with information, strike the Enter key to continue listing more entity information.

```
    ARC        Layer: 0
               Space: Model space
 center point, X=      8.00  Y=      5.00  Z=      0.00
 radius      1.50
   start angle    280
   end angle      80

    LINE       Layer: 0
               Space: Model space
  from point, X=     2.50  Y=      7.50  Z=      0.00
    to point, X=     8.26  Y=      6.48  Z=      0.00
Length =      5.85,   Angle in X-Y Plane =      350
       Delta X =       5.76,  Delta Y =      -1.02, Delta Z =      0.00

    LINE       Layer: 0
               Space: Model space
  from point, X=     2.50  Y=      2.50  Z=      0.00
    to point, X=     8.26  Y=      3.52  Z=      0.00
Length =      5.85,   Angle in X-Y Plane =      10
       Delta X =       5.76,  Delta Y =       1.02, Delta Z =      0.00
```

Using the Dist (Distance) Command

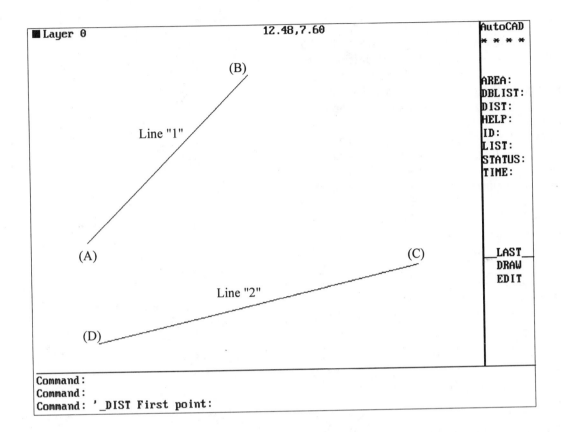

In its simplest form, the Dist command calculates the linear distance between two points on an entity whether it be the distance of a line, the distance between two points, or the distance from the quadrant of one circle to the quadrant of another circle. However, the linear distance is not the only information given when using this command. The following additional information is supplied when using the Dist command: the angle in the X-Y plane; the angle from the X-Y plane; the delta X, Y, and Z coordinate values.

The angle in the X-Y plane is given in the current angular mode set by the Units command. This mode is explained further on the next page. The angle from the X-Y plane is also given which is used mainly for finding angle information of 3D entities. The delta X, Y, and Z coordinate is a relative coordinate value taken from the first point identified by the Dist command to the second point. Using "Line 1" above as an example, "A" is identified as the first point and "B" the second. The relative distance from "A" to "B" is 5.00 units along the X axis, 4.50 units along the Y axis, and 0.00 units along the Z axis.

Care needs to be taken with the interpretation of angles using the Dist command. Angles are calculated to the nearest decimal place set by the Units command. "Line 1" above displays a 42 degree angle. However, if angle accuracy is set to 4 decimal places, an angle value of 41.9072 is displayed.

```
                        Line "1"
Command: _DIST First point: ENDP
of  Second point: ENDP
of
Distance = 6.73,  Angle in X-Y Plane = 42,  Angle from X-Y Plane = 0
Delta X = 5.00,  Delta Y = 4.50,   Delta Z = 0.00
```

```
                        Line "2"
Command: _DIST First point: ENDP
of  Second point: ENDP
of
Distance = 5.20,  Angle in X-Y Plane = 193,  Angle from X-Y Plane = 0
Delta X = -5.08,  Delta Y = -1.15,   Delta Z = 0.00
```

```
Command: _DIST First point: ENDP
of  Second point: ENDP
of
Distance = 6.73,  Angle in X-Y Plane = 41.9072,  Angle from X-Y Plane = 0.0000
Delta X = 5.00,  Delta Y = 4.50,   Delta Z = 0.00

Command: _DIST First point: ENDP
of  Second point: ENDP
of
Distance = 5.20,  Angle in X-Y Plane = 192.7500,  Angle from X-Y Plane = 0.0000
Delta X = -5.08,  Delta Y = -1.15,   Delta Z = 0.00
```

Interpretation of Angles

On the previous page, it was already pointed out that the Dist command yields information regarding distance, delta X,Y coordinate values, and angle information. Of particular interest is the angle in the X-Y plane formed between two points. In the illustration at the right, picking the endpoint of the line segment at "A" as the first point followed by the endpoint of the line segment at "B" as the second point displays an angle of 42 degrees. This angle is formed from an imaginary horizontal line drawn from the endpoint of the line segment at "A" in the zero direction. The 42 degrees are calculated from this baseline to the inclined line segment using the following prompt sequence of the Dist command:

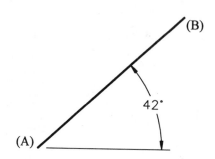

Command: **Dist**
First point: **Endp**
of *(Select the endpoint of the line at "A")*
Second point: **Endp**
of *(Select the endpoint of the line at "B")*

Angle in X-Y Plane = 42 degrees

Care needs to be taken with using the Dist command to find an angle on an identical line segment illustrated to the right as with the example above. However, notice that the two points for identifying the angle are selected differently. Using the Dist command, the endpoint of the line segment at "B" is selected as the first point followed by the endpoint of the segment at "A" for the second point. A new angle in the X-Y plane of 222 degrees is formed. In the illustration at the right, the angle is calculated by constructing a horizontal line from the endpoint at "B" the new first point of the Dist command. This horizonal line is also drawn in the zero direction. Notice the relationship of the line segment to the horizontal baseline. Be careful identifying the endpoints of line segments when extracting angle information. As you can see, a line segment may have two possible angles depending on the sequence in which the first and second points are selected.

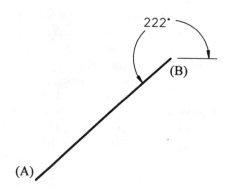

Command: **Dist**
First point: **Endp**
of *(Select the endpoint of the line at "B")*
Second point: **Endp**
of *(Select the endpoint of the line at "A")*

Angle in X-Y Plane = 222 degrees

Using the ID (Identify) Command

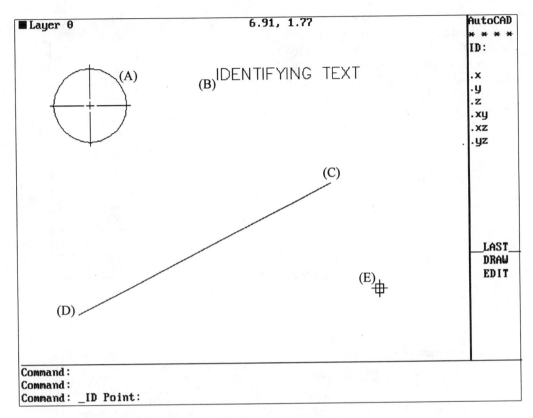

The ID command is probably one of the more straightforward of the Inquiry commands. ID stands for "Identify" and allows the user to obtain the current absolute coordinate listing of an entity. This coordinate listing is in the form of X, Y, and Z values although for 2-dimensional objects, the Z value will always be zero. When using the ID command, it is imperative that one or more of the supported Osnap modes is used in order to obtain the most accurate coordinate value. This accuracy is dependent on the current number of decimal places set by the Units command. In the example above and with the command sequence illustrated at the right, the following absolute coordinate values were found using the command sequence:

Command: **ID**
Point: *(Select a point using one of the Osnap modes)*

As a result, the coordinate value of the center of the circle at "A" was found by using ID and the Osnap-Center mode; the coordinate value of the starting point of text string "B" was found using ID and the Osnap-Insert mode; the coordinate value of the endpoint of line segment "C" was found using ID and the Osnap-Endpoint mode; the coordinate value of the midpoint of the line segment at "D" was found by using ID and

```
Command: ID
Point: CEN (Select circle "A")
of  X = 2.00      Y = 7.00        Z = 0.00

Command: ID
Point: INS (Select text at "B")
of  X = 5.54      Y = 7.67        Z = 0.00

Command: ID
Point: END (Select line at "C")
of  X = 8.63      Y = 4.83        Z = 0.00

Command: ID
Point: MID (Select line at "D")
of  X = 5.13      Y = 3.08        Z = 0.00

Command: ID
Point: NOD (Select point at "E")
of  X = 9.98      Y = 1.98        Z = 0.00
```

the Osnap-Midpoint mode; and the coordinate value of the current position of point "E" was found by using ID and the Osnap-Node mode.

Using the DDMODIFY Dialog Box

The DDMODIFY command allows the user to select an entity and display its properties in a dialog box on the screen. Choose this command by selecting "Entity" from the "Modify" section of the pulldown menu area. Entering DDMODIFY at the command prompt is yet another way to access this command. A typical DDMODIFY dialog box consists of the following:

Properties:
Color
Linetype
Layer
Thickness

Depending upon the entity selected, the DDMODIFY dialog box also displays important information on the entity. Since this information is different for each type of entity, it will be discussed in the next series of pages. Below and on the next page is a brief description of the additional dialog boxes directly related to DDMODIFY, namely Color..., Linetype..., Layer..., and Thickness.

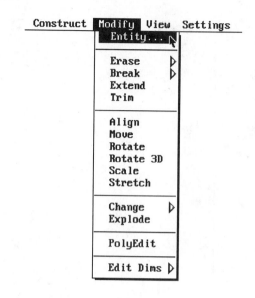

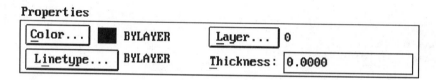

Changing the Color of an Entity

Under the "Properties" section of the DDMODIFY dialog box is the Color option. Use this property to change the color of the entity selected. Choosing the color property brings up an additional dialog box displaying the current number of colors supported by the monitor configured to use AutoCAD. Picking a different color changes the selected entity to that color. This is a quick way to change color on the fly. However, changing colors this way may affect the original color of an entity set by the Layer command. It is for this reason that color must be controlled by changing to a new layer.

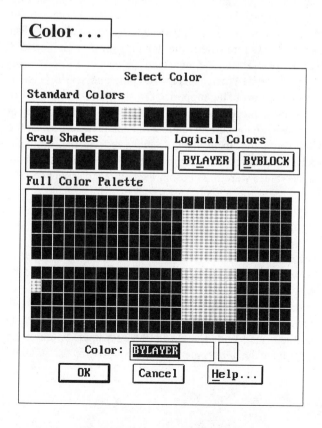

Changing the Linetype of an Entity

The "Properties" section of the DDMODIFY dialog box also has an option for modifying the linetype of a selected entity; this new dialog box is illustrated at the right. Choosing the linetype property brings up an additional dialog box displaying the current linetypes loaded into the drawing. By default, only the continuous linetype displays in this dialog box. As linetypes are loaded using the Linetype command, they will appear in this dialog box. As with color, picking a different linetype changes the selected entity to that linetype. Also as with color, changing linetype this way may affect the original linetype of an entity set by the Layer command. It is for this reason that linetypes must be controlled by changing to a new layer.

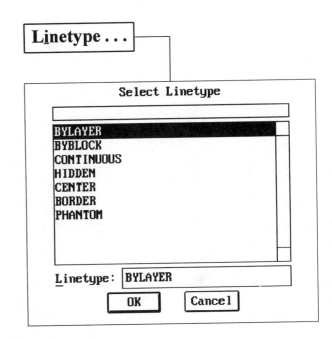

Changing the Layer of an Entity

The "Properties" section of the DDMODIFY dialog box also has an option for modifying the layer of a selected entity illustrated below. By default, only layer 0 is created when entering a new drawing. As layers are created, they will appear in the "Select Layer" dialog box.

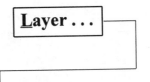

Select Layer

Current Layer: 0

Layer Name	State	Color	Linetype
0	On	white	CONTINUOUS
BORDER	On	cyan	BORDER
CENTER	On	green	CENTER
DIM	On	green	CONTINUOUS
HIDDEN	On	red	HIDDEN
OBJECT	On	yellow	CONTINUOUS
PHANTOM	On	magenta	PHANTOM
SECTION	On	blue	CONTINUOUS

Set Layer Name: 0

OK   Cancel

Listing the Properties of Arcs

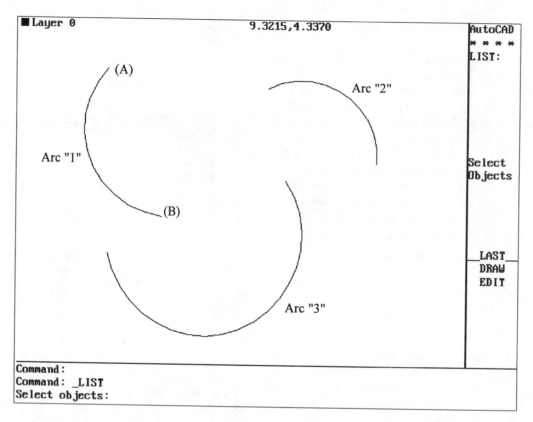

```
■ Layer 0                          9.3215,4.3370          AutoCAD
                                                          * * * *
                                                          LIST:
              (A)

                              _____              Arc "2"
                          /            \

  Arc "1"                                            Select
                                                     Objects


                 (B)



                                                     LAST_
                                                     DRAW
                                                     EDIT
                              Arc "3"



  Command:
  Command: _LIST
  Select objects:
```

This segment of the Inquiry commands begins a study of the List command and all of the properties displayed on the particular entity listed. Using the List command on an arc provides the user with the following information: the name of the entity being listed, (Arc); the layer the entity was drawn on; whether the entity occupies model or paper space; the center point of the arc; the radius of the arc; the starting angle of the arc in degrees; the ending angle of the arc in degrees.

The center point lists the current X, Y, and Z absolute coordinate values. Number of decimal places is determined by the current value set by the Units command.

The starting angle and ending angle of an arc are determined by the original construction of the arc in the counterclockwise direction. In the illustration above, listing "Arc 1" displays a starting angle at "A" of 136 degrees. Moving in a counterclockwise direction, the end angle at "B" measures 265 degrees.

Using the List command, the total included angle of an arc is not calculated nor is the length of the arc.

```
                      Arc "1"
        ARC         Layer: 0
                    Space: Model space
  center point, X=  4.2065  Y=  6.3587  Z=  0.0000
  radius    2.3677
    start angle    136
    end angle      265
```

```
                      Arc "2"
        ARC         Layer: 0
                    Space: Model space
  center point, X=  7.9286  Y=  5.6429  Z=  0.0000
  radius    2.0763
    start angle    356
    end angle      117
```

```
                      Arc "3"
        ARC         Layer: 0
                    Space: Model space
  center point, X=  5.2000  Y=  3.5000  Z=  0.0000
  radius    2.7459
    start angle    190
    end angle       33
```

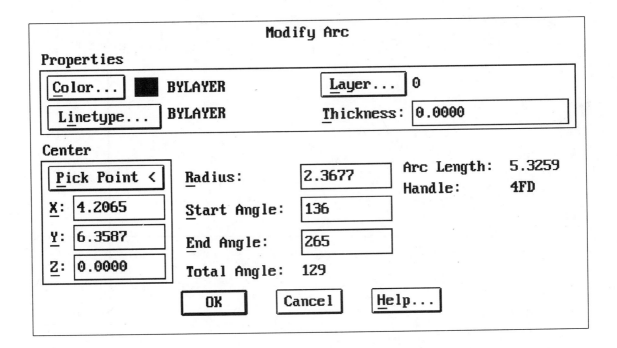

Superior entity control is enhanced through the use of the DDMODIFY command which displays the dialog box illustrated above. At first glance, the DDList dialog box allows the user to control the following entity properties: Color; Linetype; Layer Name; Thickness.

When selecting the Color, Linetype, and Layer Name Properties, an additional dialog box appears allowing the user to graphically change the entity property desired. In addition to the above uses, the DDMODIFY dialog box allows for the following: displays the center of the arc selected; displays the radius of the arc; displays the starting angle of the arc; displays the ending angle of the arc; displays the total included angle of the arc; displays the length of the arc.

All of the above parameters of the arc that are placed inside of a box may be changed in this dialog box and have the entity update itself after exiting the dialog box. The illustrations at the right explain the starting angle, ending angle, total angle, and total length of the arc. The DDMODIFY command on an arc provides a more powerful listing of the entity compared with the List command on the previous page.

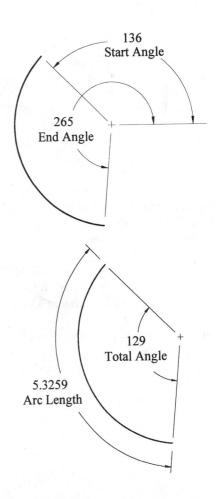

Listing the Properties of an Arc converted into a Polyline

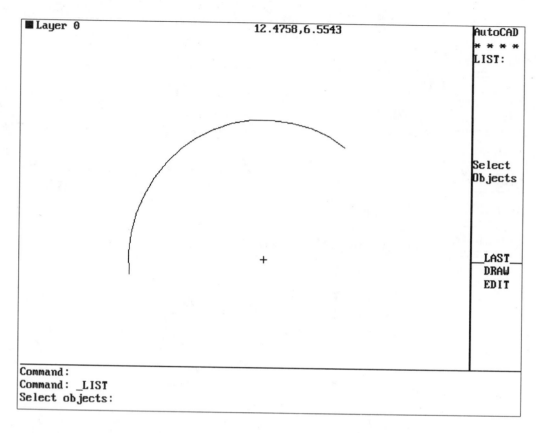

Using the List command on a polyarc entity lists the absolute coordinate values of each vertex of the polyline along with the following information: The name of the entity being listed, (Polyline); The layer the polyarc was drawn on; Whether the polyarc occupies model or paper space; The specific vertex being listed in absolute coordinates; The center point of the polyarc segment; The radius of the polyarc; The starting angle of the polyarc; The ending angle of the polyarc.

Depending on the number of vertices, sometimes this information will not fit on the display screen all at once. Once the screen is filled with information, strike the Enter key to continue listing more vertex information.

Once the last screen is displayed, the List command calculates the area occupied by a closed or open polyarc. If the polyarc is open, the total length of the polyarc segment is given.

Using the Explode command on a polyarc separates the entity into individual arc segments.

```
          VERTEX    Layer: 0
                    Space: Model space
        at point, X=   9.0000  Y=   6.0000  Z=   0.0000
starting width    0.0000
 ending width     0.0000
      bulge    0.6510
     center X=   6.7746  Y=   2.9221  Z=   0.0000
     radius    3.7981
 start angle     54
 end angle      186

          VERTEX    Layer: 0
                    Space: Model space
        at point, X=   3.0000  Y=   2.5000  Z=   0.0000
starting width    0.0000
 ending width     0.0000
      bulge    1.5362

       END SEQUENCE  Layer: 0
                    Space: Model space
      area    11.3095
      length   8.7668
```

```
                          Modify Polyline
Properties
  [Color...]  ■  BYLAYER          [Layer...]  0
  [Linetype...]  BYLAYER          Thickness: [0.0000]

Polyline Type: 2D polyline        Entity Handle: None
Vertex Listing        Fit/Smooth      Mesh          Polyline
[Vertex:1] [Next]     ■ None          M: □ Closed   □ Closed
                      □ Quadratic     N: □ Closed   □ LT Gen
 X: 9.0000            □ Cubic         U: [    ]
 Y: 6.0000            □ Bezier        V: [    ]
 Z: 0.0000            □ Curve Fit
           [  OK  ]  [ Cancel ]  [ Help... ]
```

Using the DDMODIFY command on a polyarc lists the usual entity properties such as color, linetype, layer name, and entity thickness. This command also lists the following information: the X, Y, and Z coordinates of each polyarc vertex; the type of curve fitted to the polyarc; whether the polyarc is closed or open; whether to generate linetype scaling per vertex.

The coordinates of all polyline vertices are identified above in the DDMODIFY dialog box. Selecting "Next" lists the next set of coordinates. This may not be as important a function on a polyarc as on a multiple vertex polyline since the polyarc only has two vertices, one at the beginning of the polyarc and the other at the end of the polyarc. Also, the methods of fitting a curve have no affects on a polyarc.

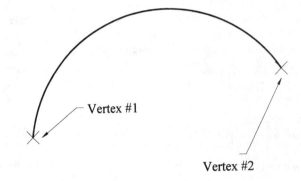

Vertex #1

Vertex #2

Listing the Properties of an Associative Dimension

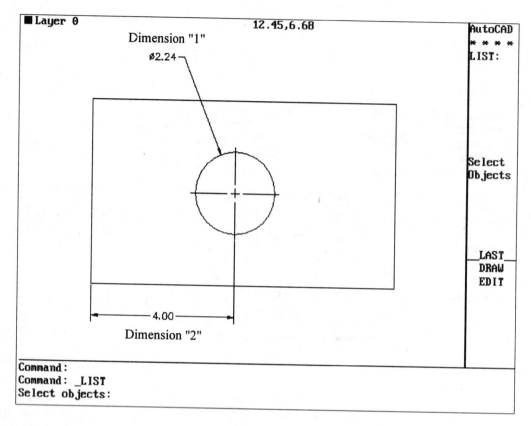

Using the List command on a diameter dimension entity lists the following information: the specific entity being listed (Dimension); the layer the dimension was drawn on: whether the dimension occupies model or paper space; the first defining point of the leader; the second defining point of the leader; the text position (in absolute coordinates); the default text; the current dimension style.

Using the List command on a linear dimension entity lists the following information: the specific entity being listed (Dimension); the layer the dimension was drawn on; whether the dimension occupies model or paper space; the first extension line defining point; the second extension defining point; the text position (in absolute coordinates); the default text; the current dimension style.

```
                    Dimension "1"

              DIMENSION  Layer: 0
                         Space: Model space
   type: diameter
            defining point: X=     6.41  Y=    3.57  Z=    0.00
   defining point:      X=     5.59  Y=    5.65  Z=    0.00
   leader length    2.83
   default text position: X=     3.91  Y=    8.27  Z=    0.00
   default text
   dimension style: STANDARD
```

```
                    Dimension "2"

              DIMENSION  Layer: 0
                         Space: Model space
   type: horizontal
   1st extension  defining point: X=     2.00  Y=    2.11  Z=    0.00
   2nd extension  defining point: X=     6.00  Y=    3.37  Z=    0.00
   dimension line defining point: X=     6.00  Y=    1.25  Z=    0.00
   default text position: X=    4.00  Y=    1.25  Z=    0.00
   default text
   dimension style: STANDARD
```

```
┌─────────────────────────────────────────────────────────────┐
│                    Modify Dimension                           │
│  Properties                                                   │
│  ┌──────────────────────────────────────────────────────┐    │
│  │ ┌─────────┐  ▉▉▉  BYLAYER    ┌─────────┐ 0            │    │
│  │ │ Color...│                  │ Layer...│              │    │
│  │ └─────────┘                  └─────────┘              │    │
│  │ ┌───────────┐  BYLAYER    Thickness: ┌──────────┐     │    │
│  │ │ Linetype..│                        │ 0.00     │     │    │
│  │ └───────────┘                        └──────────┘     │    │
│  └──────────────────────────────────────────────────────┘    │
│  Dimension Type:  Linear      Dimension Style:  STANDARD      │
│  Dimension Text:  Default Text  Handle:         None          │
│            ┌──────┐   ┌────────┐   ┌────────┐                 │
│            │  OK  │   │ Cancel │   │ Help...│                 │
│            └──────┘   └────────┘   └────────┘                 │
└─────────────────────────────────────────────────────────────┘
```

Using the DDMODIFY command on a linear dimension lists the following information:

- The dimension type.
- The dimension text.
- The current dimension style.

Compared with the effect DDMODIFY has on other entities, this dialog box is similar to using the List command on a dimension.

No special tools are available to change or modify an associative dimension except for the following standard items displayed at the top of the dialog box:

- Modifying the Color of the associative dimension.
- Modifying the Linetype of the associative dimension.
- Modifying the Layer of the associative dimension.
- Modifying the Thickness of an associative dimension.

Listing the Properties of a Block

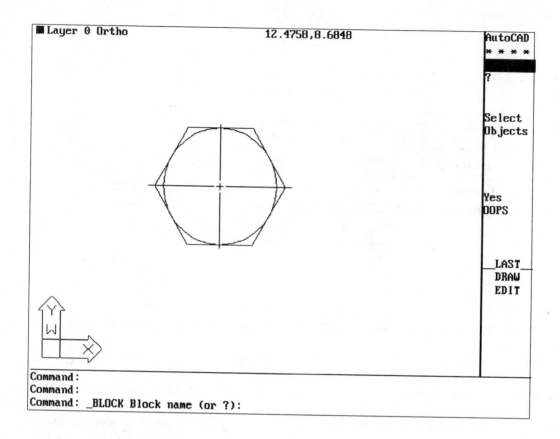

```
Layer 0 Ortho                    12.4758,8.6848                    AutoCAD
                                                                   * * * *

                                                                   ?

                                                                   Select
                                                                   Objects


                                                                   Yes
                                                                   OOPS


                                                                   LAST
                                                                   DRAW
                                                                   EDIT

Command:
Command:
Command: _BLOCK Block name (or ?):
```

Using the List command on a block provides the user with the following information: the name of the entity being listed, (Block Reference); the layer the block was inserted on; whether the block occupies model or paper space; the name of the block, which in this case is "Bolthead"; the insertion point of the block in X, Y, and Z coordinates; the X, Y, and Z scale factors of the block; the rotation angle of the block.

Number of decimal places for all coordinate values and scale factors are determined by the current value set by the Units command.

The name of the block is not limited to 8 characters; 31 characters may be used in the naming of a block. However it is still best to keep the names short and simple, which will speed up entry of block names from the keyboard.

Illustrated at the right is an example of the "Bolthead" block. Point "A" identifies the insertion point of the block or where the block is inserted in relation to. All scaling and rotation values are referenced by the insertion point. The insertion point of a block is also used for reference where multiple block insertions at specific distances are required.

```
BLOCK REFERENCE  Layer: 0
               Space: Model space
     BOLTHEAD
     at point, X=  5.7756  Y=   4.7283  Z=   0.0000
         X scale factor    1.0000
         Y scale factor    1.0000
rotation angle       0
         Z scale factor    1.0000
```

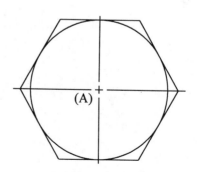

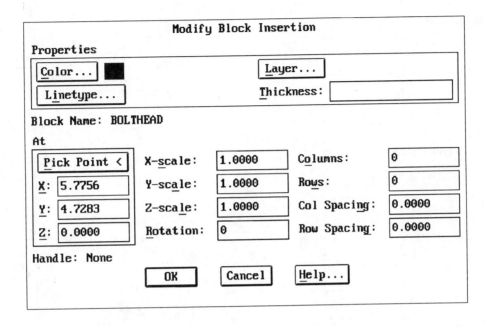

Control of blocks is enhanced through the use of the DDMODIFY command which displays the dialog box illustrated above. Using this dialog box allows the user to dynamically change each of the following items: the insertion point of the block; the X, Y, and Z scale values of the block; the rotation angle of the block.

The dialog box in the first illustration above also controls multiple inserts of a block by allowing the user to change each of the following items: the number of columns; the number of rows; a value for the spacing in between columns; a value for the spacing in between rows.

Using DDMODIFY to control multiple block insertions is illustrated in the second illustration above. With the insertion point of the block at the center of the bolt head, 2 columns and 3 rows are specified. Both spacings in between columns and rows are 2.50 units. This creates the arrangement of bolt heads at the right.

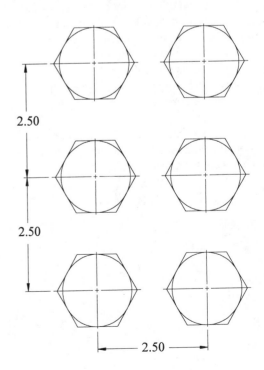

Listing the Properties of a Circle

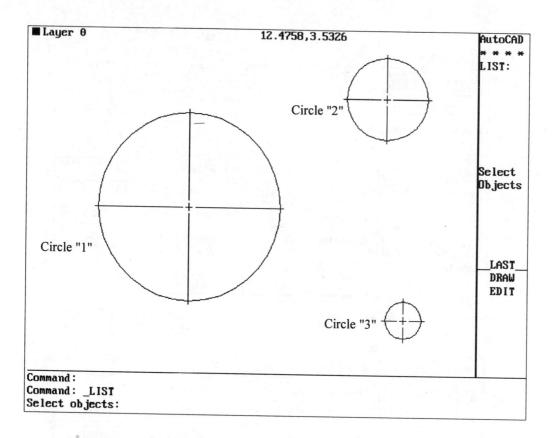

Using the List command on a circle displays the following information:

- The name of the entity being listed, (Circle).
- The layer the circle was drawn on.
- Whether the circle occupies model or paper space.
- The center point of the circle.
- The radius of the circle.
- The circumference of the circle.
- The area of the circle.

As with all circle properties such as center point, radius, circumference, and area, the number of decimal places of accuracy is set by the Units command.

```
                    Circle "1"

         CIRCLE    Layer: 0
                   Space: Model space
      center point, X=  4.5000  Y=  4.5000  Z=  0.0000
         radius    2.5495
   circumference   16.0190
         area    20.4204
```

```
                    Circle "2"

         CIRCLE    Layer: 0
                   Space: Model space
      center point, X= 10.0000  Y=  7.5000  Z=  0.0000
         radius    1.1180
   circumference    7.0248
         area     3.9270
```

```
                    Circle "3"

         CIRCLE    Layer: 0
                   Space: Model space
      center point, X= 10.5000  Y=  1.5000  Z=  0.0000
         radius    0.5000
   circumference    3.1416
         area     0.7854
```

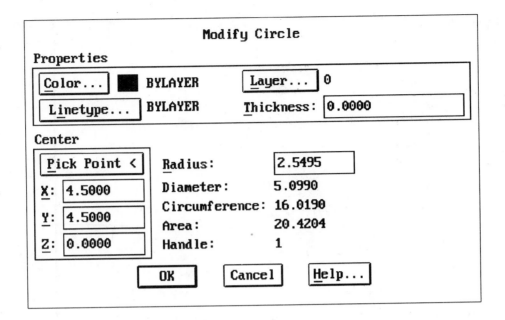

Modify Circle

Properties

Color... ■ BYLAYER   Layer... 0

Linetype... BYLAYER   Thickness: 0.0000

Center

Pick Point <   Radius: 2.5495

X: 4.5000   Diameter: 5.0990
                Circumference: 16.0190
Y: 4.5000   Area: 20.4204
Z: 0.0000   Handle: 1

OK   Cancel   Help...

Using the DDMODIFY command on a circle lists the identical properties to change as with such entities as lines and arcs. Color, linetype, and layer name all bring up additional dialog boxes to make it easier to edit the entity listed.

In addition to the above properties, the following additional parameters are listed of the circle: center point of the circle; radius of the circle; diameter of the circle; circumference of the circle; area of the circle.

A new center point may be selected by selecting the box "Pick point" or by entering a new coordinate value in the appropriate X, Y, and/or Z boxes. Also, the radius of the circle may be changed by editing the value in the radius box.

The diameter, circumference, and area values are all listed but may not be changed from this dialog box. Notice, however, that when changing the circle radius, the diameter, circumference, and area values update themselves based on the new radius.

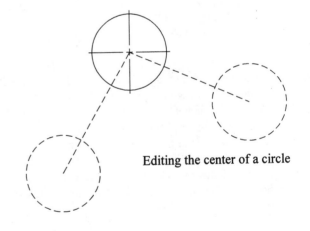

Editing the center of a circle

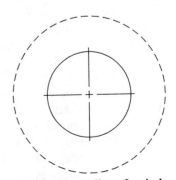

Editing the radius of a circle

Listing the Properties of a Donut

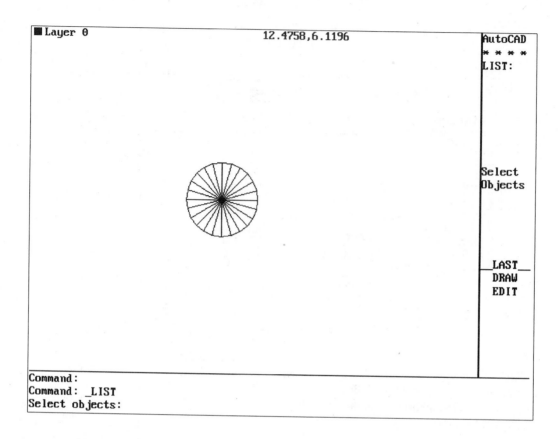

Using the List command on a donut entity lists the absolute coordinate values of each vertex since the donut is actually constructed as a polyline with the following information: the name of the entity being listed (Polyline); the layer the donut was drawn on; whether the donut occupies model or paper space; the specific vertex being listed in absolute coordinates; the center point of the polyarc segment; the radius of the polyarc; the starting angle of the polyarc; the ending angle of the polyarc.

Depending on the number of vertices, sometimes this information will not fit on the display screen all at once. Once the screen is filled with information, strike the Enter key to continue listing the coordinates of the next vertex.

Since the donut is a closed polyline entity, the List command calculates the area and perimeter. If the donut is open, the total area and length of the donut segment is given. Numerous donuts consume an extreme amount of drawing time especially when the screen display is regenerated. It is for this purpose to use the Fill command and turn off the filled-in part of a donut. They will display faster on the screen.

```
  ending width     1.0000
         bulge     1.0000
        center X=   5.3458   Y=    4.7169  Z=   0.0000
        radius     0.5000
   start angle      180
     end angle        0

              VERTEX    Layer: 0
                    Space: Model space
           at point, X=   5.8458   Y=    4.7169  Z=   0.0000
 starting width     1.0000
   ending width     1.0000
         bulge     1.0000
        center X=   5.3458   Y=    4.7169  Z=   0.0000
        radius     0.5000
   start angle        0
     end angle      180

          END SEQUENCE  Layer: 0
                    Space: Model space
         area      0.7854
    perimeter      3.1416
```

Open Donut

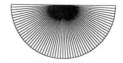

```
                        Modify Polyline
 Properties
  ┌─────────┐  ████  BYLAYER        ┌─────────┐ 0
  │ Color...│                       │ Layer...│
  └─────────┘                       └─────────┘
  ┌───────────┐  BYLAYER           Thickness: │0.0000│
  │ Linetype..│
  └───────────┘

 Polyline Type: 2D polyline        Entity Handle: None

 Vertex Listing      Fit/Smooth    Mesh          Polyline
 ┌──────────────┐  ┌────────────┐  M:  ☐ Closed  ┌─────────────┐
 │Vertex:1 ┌────┐│  │■ None      │  N:  ☐ Closed  │☒ Closed     │
 │         │Next││  │☐ Quadratic │  U: ┌──┐       │☐ LT Gen     │
 │         └────┘│  │☐ Cubic     │     └──┘       └─────────────┘
 │X: 4.8458     │  │☐ Bezier    │  V: ┌──┐
 │Y: 4.7169     │  │☐ Curve Fit │     └──┘
 │Z: 0.0000     │  └────────────┘
 └──────────────┘
              ┌────┐   ┌────────┐   ┌────────┐
              │ OK │   │ Cancel │   │ Help...│
              └────┘   └────────┘   └────────┘
```

Using the DDMODIFY command on a donut lists the usual entity properties such as color, linetype, layer name, and entity thickness. This command also lists the following information: the X, Y, and Z coordinates of each donut vertex; the type of curve fitted to the donut; whether the donut is closed or open; whether to generate linetype scaling per vertex

The coordinates of all donut vertices are identified above in the DDMODIFY dialog box. Selecting "Next" lists the next set of coordinates.

By default, a donut has normal curve generation which displays the donut as a filled in circular polyline. Selecting Quadratic, Cubic, and Fit Curve displays the results similar to the upper illustration at the right. As a result, any special curve generations of a donut should not be used.

Using the Explode command on a closed donut separates the entity into two individual arc segments as in the illustration at the right. The center dividing line is not present when the donut is exploded and is used only for illustrative purposes. Using the Explode command on an open donut converts the entity into a single arc segment.

Quadratic, Cubic, and Curve Fit on a Donut

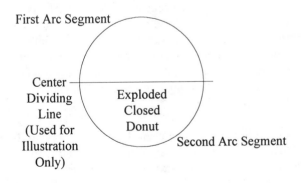

First Arc Segment

Center Dividing Line (Used for Illustration Only)

Exploded Closed Donut

Second Arc Segment

Exploded Open Donut

Listing the Properties of an Ellipse

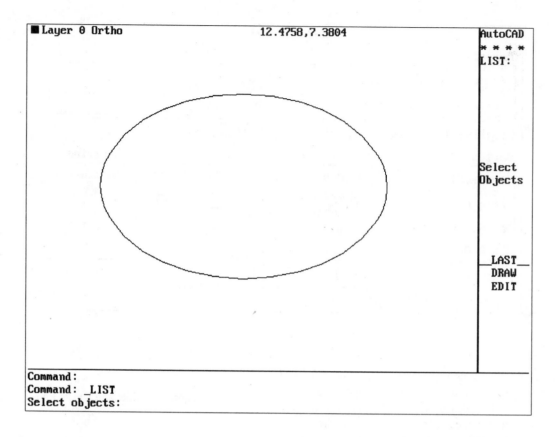

```
■Layer 0 Ortho                    12.4758,7.3804            AutoCAD
                                                            * * * *
                                                            LIST:


                                                            Select
                                                            Objects



                                                            LAST
                                                            DRAW
                                                            EDIT



Command:
Command: _LIST
Select objects:
```

An ellipse is an entity consisting of numerous arcs all converted into one polyline entity. Using the List command on an ellipse displays the following information: the name of the entity being listed (Polyline); the layer the ellipse was drawn on; whether the ellipse occupies model or paper space; the specific vertex of the ellipse being listed in absolute coordinates; the origin of the vertex; the starting width of the polyline; the ending width of the polyline; the bulge value of the vertex; the center point of the radius; the radius of the arc segment; the starting angle of the arc segment; the ending angle of the arc segment.

As the ellipse is selected using List, all vertices are listed. When the text screen fills up with information, simply strike the Enter key to list the next series of vertices.

In addition to the above information, the List command automatically calculates the area and perimeter of the ellipse.

Using the Explode command on an ellipse breaks the ellipse into individual arc entity segments.

```
        bulge    0.1071
       center X=  7.2497  Y=   5.7582  Z=   0.0000
       radius    2.9787
  start angle     302
    end angle     326
Curve direction:   32

            VERTEX    Layer: 0
                    Space: Model space
        at point, X=   9.7326  Y=   4.1126  Z=   0.0000
 starting width    0.0000
   ending width    0.0000
        bulge    0.1474
       center X=  8.3937  Y=   5.0000  Z=   0.0000
       radius    1.6063
  start angle     326
    end angle       0

          END SEQUENCE  Layer: 0
                    Space: Model space
        area    31.4079
   perimeter    20.6859
```

```
                        Modify Polyline
   Properties
   ┌─────────┐  ███                    ┌─────────┐
   │ Color...│      BYLAYER            │ Layer...│  0
   └─────────┘                         └─────────┘
   ┌──────────┐                                   ┌──────────┐
   │ Linetype.│.    BYLAYER            Thickness: │ 0.0000   │
   └──────────┘                                   └──────────┘

   Polyline Type: 2D polyline         Entity Handle: None
   Vertex Listing      Fit/Smooth     Mesh              Polyline
   ┌────────────────┐ ┌────────────┐ ┌──────────────┐ ┌──────────────┐
   │Vertex:1 ┌─────┐│ │■ None      │ │M:  □ Closed  │ │⊠ Closed      │
   │         │Next ││ │□ Quadratic │ │N:  □ Closed  │ │⊠ LT Gen      │
   │         └─────┘│ │□ Cubic     │ │U: ┌──┐        │ │              │
   │X: 10.0000      │ │□ Bezier    │ │   └──┘        │ │              │
   │Y: 5.0000       │ │□ Curve Fit │ │V: ┌──┐        │ │              │
   │Z: 0.0000       │ │            │ │   └──┘        │ │              │
   └────────────────┘ └────────────┘ └──────────────┘ └──────────────┘
              ┌──────┐   ┌──────────┐   ┌──────────┐
              │  OK  │   │  Cancel  │   │  Help... │
              └──────┘   └──────────┘   └──────────┘
```

Using the DDMODIFY command on an ellipse lists the usual entity properties such as color, linetype, layer name, and entity thickness. Using this command on an ellipse may not allow for dramatic changes such as on arcs and circles. This dialog box does list the following: the X, Y, and Z coordinates of each vertex; the type of curve fitted to the ellipse; whether the polyline is closed or open; whether to generate linetype scaling per vertex.

Since an ellipse consists of individual arc segments joined into one entity, the location of each vertex is identified at the endpoints of these arc segments similar to the illustration at the right. The coordinates of these vertices are identified above in the DDMODIFY dialog box. Selecting "Next" lists the next set of coordinates.

The LT Gen option stands for generate a linetype. If LT Gen is unselected in the DDMODIFY dialog box, or is set to "Off," the linetype is applied to each individual vertex. In the example at the right, with the Ltscale command set to 0.70 only the top and bottom arc segments show with the proper linetype. This is because the other arc segments of the ellipse are too short to support the linetype at the current Ltscale value.

A better result would be to check the LT Gen box in DDMODIFY. This applies the linetype throughout the entire polyline and not per vertex.

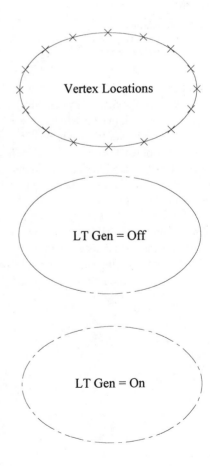

Vertex Locations

LT Gen = Off

LT Gen = On

Listing the Properties of a Line

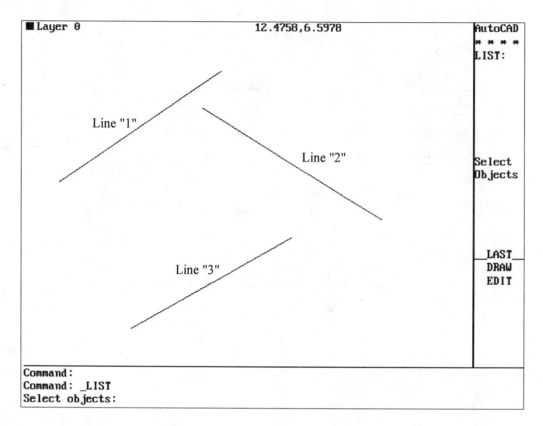

Using the List command on a line entity lists the following information: the specific specific entity being listed (Line); the layer the line segment was drawn on; whether the line occupies model or paper space; the starting endpoint of the line segment; the ending endpoint of the line segment; the length of the line segment; the angle of the line segment in the X-Y plane; the relative coordinate value from the starting point to the ending point of the line segment.

Study the illustrations at the right and above to isolate the information on the individual line segments. Number of decimal places for the beginning and ending points of the line in addition to the length of the line are governed by the Units command.

```
                    Line "1"

        LINE      Layer: 0
                  Space: Model space
     from point, X=   1.0000  Y=   5.0000  Z=   0.0000
       to point, X=   5.5000  Y=   8.0000  Z=   0.0000
   Length =   5.4083, Angle in X-Y Plane =      34
          Delta X =   4.5000, Delta Y =   3.0000, Delta Z =   0.0000
```

```
                    Line "2"

        LINE      Layer: 0
                  Space: Model space
     from point, X=   5.0000  Y=   7.0000  Z=   0.0000
       to point, X=  10.0000  Y=   4.0000  Z=   0.0000
   Length =   5.8310, Angle in X-Y Plane =     329
          Delta X =   5.0000, Delta Y =  -3.0000, Delta Z =   0.0000
```

```
                    Line "3"

        LINE      Layer: 0
                  Space: Model space
     from point, X=   3.0000  Y=   1.0000  Z=   0.0000
       to point, X=   7.5000  Y=   3.5000  Z=   0.0000
   Length =   5.1478, Angle in X-Y Plane =      29
          Delta X =   4.5000, Delta Y =   2.5000, Delta Z =   0.0000
```

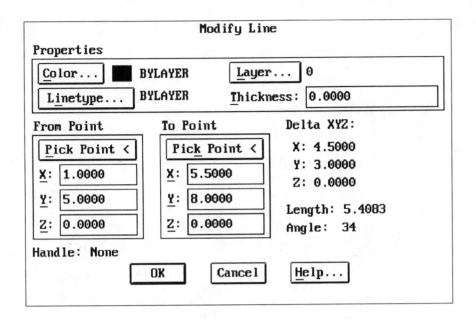

Using the DDMODIFY command on a line segment lists the following information: the X, Y, and Z coordinates of the starting of the line; the X, Y, and Z coordinates of the end of the line; the Delta XYZ coordinate value of the line; the total length of the line segment; the angle the line segment makes in the X-Y plane.

The X, Y, and Z values in the dialog box above may be changed to affect the beginning or end of the line segment. When any of these values change, the Delta XYZ, Length, and Angle values update themselves to the new values of the line segment.

The top illustration at the right shows how the length and angle are calculated. Angles will always be calculated in the counterclockwise direction depending on the current setting in the Units command.

The Delta XYZ value at the lower right shows the horizontal and vertical distances needed to go from the beginning of the line segment to the end of the line segment. These values may be negative at times if the Delta directions are to the left or down.

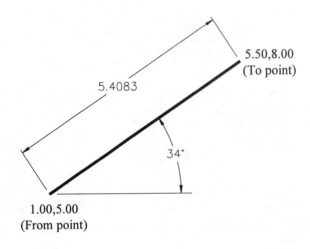

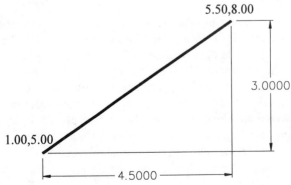

Listing the Properties of a Multiple Block Insertion

```
■ Layer 0                           12.1495,5.0978              AutoCAD
                                                               * * * *
                                                               LIST:


       I        I        I        I        I



                                                               Select
       I        I        I        I        I         Objects



       I        I        I        I        I
                                                               LAST
                                                               DRAW
                                                               EDIT
       I        I        I        I        I



 Command:
 Command: _LIST
 Select objects:
```

Using the List command on a block which has been inserted numerous times using the Minsert command provides the user with the following information: the name of the entity being listed (Block Reference); the layer the block was inserted on; whether the block occupies model or paper space; the name of the block, which in this case is "I-Beam"; the insertion point of the block in X, Y, and Z coordinates; the X, Y, and Z scale factors of the block; the rotation angle of the block; the number of columns along with the spacing; the number of rows and the spacing.

It is always very important to consider the insertion point of the block before performing the Minsert command. For the I-Beam example illustrated above, both the column and row spacing relate directly to the block insertion point. This means that the column and row spacing begin from the insertion point of the block and does not necessarily reflect the actual spacing between the blocks. For the I-Beam example, the insertion point for this particular block is in the lower left corner at "A" illustrated at the right. This point will be used on the next page to change the column and row spacing of this multiple inserted block.

```
        BLOCK REFERENCE  Layer: 0
                     Space: Model space
        I-BEAM
         at point, X=   0.8133  Y=    0.9651  Z=   0.0000
         X scale factor    1.0000
         Y scale factor    1.0000
   rotation angle       0
         Z scale factor    1.0000
     # columns 5
 column spacing    2.2500
       # rows 4
   row spacing    2.0000
```

(A)

```
                         Modify Block Insertion
 Properties
 ┌──────────┐  ██         ┌──────────┐
 │ Color... │             │ Layer... │
 └──────────┘             └──────────┘
     ┌──────────────┐          Thickness: ┌──────────────────┐
     │ Linetype...  │                     │                  │
     └──────────────┘                     └──────────────────┘
 Block Name: I-BEAM
 At
 ┌──────────────┐  X-scale: ┌────────┐  Columns:    ┌────────┐
 │ Pick Point < │           │ 1.0000 │              │ 2      │
 └──────────────┘           └────────┘              └────────┘
 X: ┌────────┐    Y-scale: ┌────────┐  Rows:       ┌────────┐
    │ 0.8133 │             │ 1.0000 │              │ 3      │
    └────────┘             └────────┘              └────────┘
 Y: ┌────────┐    Z-scale: ┌────────┐  Col Spacing: ┌────────┐
    │ 0.9651 │             │ 1.0000 │               │ 5.0000 │
    └────────┘             └────────┘               └────────┘
 Z: ┌────────┐    Rotation: ┌────────┐  Row Spacing: ┌────────┐
    │ 0.0000 │              │ 0      │               │ 3.0000 │
    └────────┘              └────────┘               └────────┘
 Handle: None
              ┌──────┐   ┌────────┐   ┌────────┐
              │  OK  │   │ Cancel │   │ Help...│
              └──────┘   └────────┘   └────────┘
```

Control of blocks inserted using the Minsert command are enhanced through the use of the DDMODIFY command, which displays the dialog box illustrated above. Using this dialog box allows the user to dynamically change each of the following items: the insertion point of the block; the X, Y, and Z scale values of the block; the rotation angle of the block.

This dialog box also controls multiple inserts of a block by allowing the user to change each of the following items: the number of columns; the number of rows; a value for the spacing in between columns; a value for the spacing in between rows.

Using DDMODIFY to control multiple block insertions is illustrated above. With the insertion point of the block at the center of the bolt head, 2 columns and 3 rows are specified. Spacing between columns is 5.0000 units and between rows is 3.0000 units. The results are illustrated at the right.

Of interest is the use of the insertion point as the point of reference where all values for column and row spacing are calculated from. With the lower left corner of the column used as the insertion point, notice the column and row spacing span from the insertion point of one block to the insertion point of the other block.

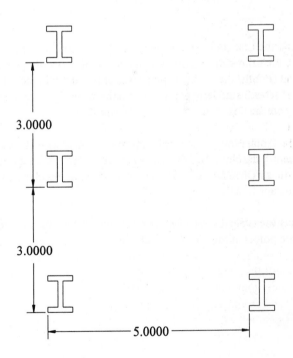

Listing the Properties of a Point

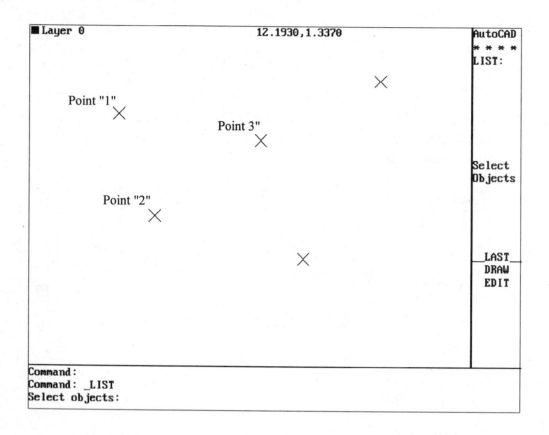

Using the List command on an point entity provides the user with the following information: the name of the entity being listed (Point); the layer the point was drawn on; whether the point occupies model or paper space; the current point "Handle"; the coordinates of the point in X, Y, and Z values.

As with all points, the size and appearance are dictated by the system variables PDSIZE and PDMODE. In the example above, a PDMODE of 3 has been specified forming all points in the appearance of an "X."

Study the typical screens at the right and see how they relate to the points in the illustration above.

```
                        Point "1"

     POINT       Layer: 0
                 Space: Model space
       Handle = 1
     at point, X=   2.5482  Y=   6.9289  Z=   0.0000
```

```
                        Point "2"

     POINT       Layer: 0
                 Space: Model space
       Handle = 3
     at point, X=   3.5675  Y=   4.1313  Z=   0.0000
```

```
                        Point "3"

     POINT       Layer: 0
                 Space: Model space
       Handle = 2
     at point, X=   6.5819  Y=   6.2133  Z=   0.0000
```

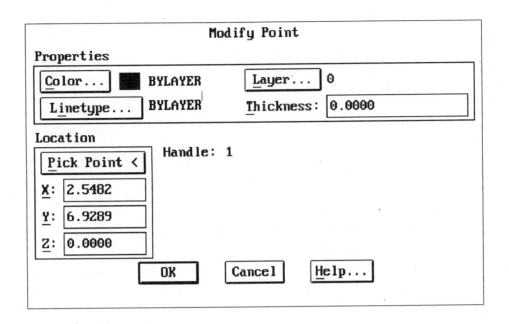

Using the DDMODIFY command on a point lists the usual entity properties such as color, linetype, layer name, and entity thickness. This command also lists the following information: the X, Y, and Z coordinates of the point; the current Handle assigned to the point.

Notice in the above illustration the appearance of the X, Y, and Z coordinate values located in edit boxes. A new value may be entered in one of these boxes, which will change the location of the point.

If the absolute coordinates of a point to move are not known, the "Pick Point<" button may be used to locate the point with the current pointing device such as a digitizing puck or mouse. Object snap modes are usually used to locate the point on an entity using the "Pick Point<" button.

Listing the Properties of a Polyline

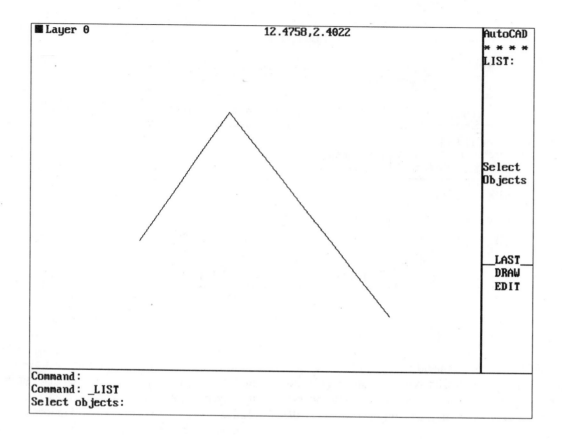

Using the List command on a polyline entity lists the absolute coordinate values of each vertex of the polyline along with the following information: the name of the entity being listed (Polyline); the layer the polyline was drawn on; whether the polyline occupies model or paper space; the specific vertex of the polyline being listed in absolute coordinates; the absolute coordinate value at the vertex; the starting width of the vertex; the ending width of the vertex.

Depending on the number of vertices, sometimes this information will not fit on the display screen all at once. Once the screen is filled with information, strike the Enter key to continue listing more vertex information.

Once the last screen is displayed, the List command calculates the area occupied by a closed or open polyline. For an open polyline, the total length of the polyline is given. If the polyline was closed, the perimeter of the polyline would be calculated.

Using the Explode command on a polyline separates the polyline into individual line segments.

```
        VERTEX    Layer: 0
                  Space: Model space
            at point, X=   3.0000  Y=   3.5000  Z=   0.0000
starting width    0.0000
 ending width     0.0000

        VERTEX    Layer: 0
                  Space: Model space
            at point, X=   5.5000  Y=   7.0000  Z=   0.0000
starting width    0.0000
 ending width     0.0000

        VERTEX    Layer: 0
                  Space: Model space
            at point, X=  10.0000  Y=   1.5000  Z=   0.0000
starting width    0.0000
 ending width     0.0000

        END SEQUENCE  Layer: 0
                      Space: Model space
        area   14.7500
      length   11.4075
```

```
                        Modify Polyline
 Properties
 ┌─────────┐  ■  BYLAYER          ┌─────────┐  0
 │ Color...│                      │ Layer...│
 └─────────┘                      └─────────┘
   ┌───────────┐  BYLAYER         Thickness: ┌──────────────────────┐
   │ Linetype..│                             │ 0.0000               │
   └───────────┘                             └──────────────────────┘

 Polyline Type: 2D polyline        Entity Handle: None
 Vertex Listing      Fit/Smooth    Mesh            Polyline
 ┌─────────────────┐ ┌──────────┐  M:  ☐ Closed   ┌──────────────┐
 │ Vertex:1 ┌────┐ │ │ ■ None   │                 │ ☐ Closed     │
 │          │Next│ │ │ ☐ Quadratic N: ☐ Closed    │ ☒ LT Gen     │
 │          └────┘ │ │ ☐ Cubic  │  U: ┌──┐        │              │
 │ X: 3.0000       │ │ ☐ Bezier │     └──┘        │              │
 │ Y: 3.5000       │ │ ☐ Curve Fit V: ┌──┐        │              │
 │ Z: 0.0000       │ │          │     └──┘        │              │
 └─────────────────┘ └──────────┘                 └──────────────┘
              ┌──────┐  ┌────────┐  ┌────────┐
              │  OK  │  │ Cancel │  │ Help...│
              └──────┘  └────────┘  └────────┘
```

Using the DDMODIFY command on a polyline lists the usual entity properties such as color, linetype, layer name, and entity thickness. Using this command lists the following information: the X, Y, and Z coordinates of each polyline vertex; the type of curve fitted to the polyline; whether the polyline is closed or open; whether to generate linetype scaling per vertex.

The coordinates of all polyline vertices are identified above in the DDMODIFY dialog box. Selecting "Next" lists the next set of coordinates.

By default, a polyline has normal curve generation which means is is absent of any curves. Selecting Quadratic, Cubic, and Fit Curve displays the results similar to the illustration at the right.

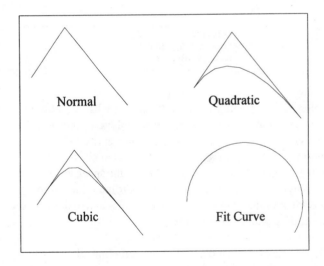

The LT Gen option stands for generate a linetype. If LT Gen is unselected in the DDMODIFY dialog box, or is set to "Off," the linetype is applied to each individual vertex. In the example at the right, with the Ltscale command set to 1.10 one leg of the polyline has a single center line while the other leg has three center lines. Checking the LT Gen box above in DDMODIFY turns on LT Gen with the results at the far right. Here, the first leg of the polyline has two center lines that continue into the second leg.

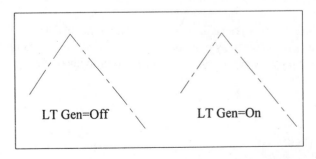

Listing the Properties of a Polygon

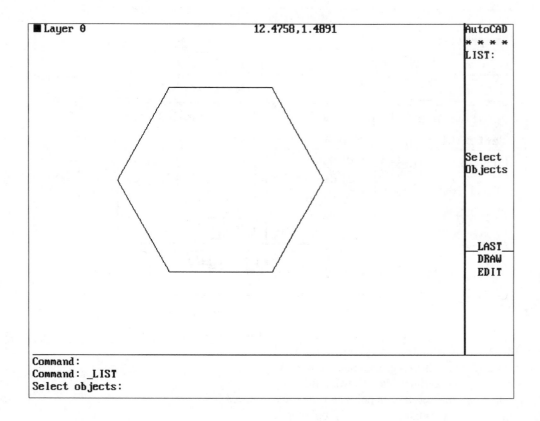

```
 Layer 0                    12.4758,1.4891              AutoCAD
                                                        * * * *
                                                        LIST:



                                                        Select
                                                        Objects



                                                        LAST
                                                        DRAW
                                                        EDIT




Command:
Command: _LIST
Select objects:
```

Using the List command on a polygon entity lists the absolute coordinate values of each vertex of the polygon along with the following information: the name of the entity being listed (Polyline); the layer the polygon was drawn on; whether the polygon occupies model or paper space; the specific vertex of the polygon being listed in absolute coordinates; the absolute coordinate value at the vertex; the starting width of the vertex; the ending width of the vertex.

Depending on the number of vertices, sometimes this information will not fit on the display screen all at once. Once the screen is filled with information, strike the Enter key to continue listing more vertex information.

Once the last screen is displayed, the List command calculates the area occupied by polygon. Using the Explode command on a polyline separates the polyline into individual line segments.

```
          VERTEX    Layer: 0
                    Space: Model space
             at point, X=   2.4783  Y=   4.9325  Z=   0.0000
starting width     0.0000
  ending width     0.0000

          VERTEX    Layer: 0
                    Space: Model space
             at point, X=   3.9783  Y=   2.3345  Z=   0.0000
starting width     0.0000
  ending width     0.0000

          VERTEX    Layer: 0
                    Space: Model space
             at point, X=   6.9783  Y=   2.3345  Z=   0.0000
starting width     0.0000
  ending width     0.0000

          END SEQUENCE  Layer: 0
                    Space: Model space
        area    23.3827
   perimeter    18.0000
```

```
                          Modify Polyline
 Properties
   Color...   ▓   BYLAYER        Layer...   0

   Linetype...   BYLAYER        Thickness:  0.0000

 Polyline Type: 2D polyline      Entity Handle: None
 Vertex Listing    Fit/Smooth    Mesh          Polyline
  Vertex:1  Next   ■ None        M:   □ Closed  ⊠ Closed
                   □ Quadratic   N:   □ Closed  □ LT Gen
  X: 8.4783        □ Cubic       U:
  Y: 4.9325        □ Bezier
                                 V:
  Z: 0.0000        □ Curve Fit
                  OK     Cancel    Help...
```

Using the DDMODIFY command on a polygon lists the usual entity properties such as color, linetype, layer name, and entity thickness. Using this command lists the following information: the X, Y, and Z coordinates of each polygon vertex; the type of curve fitted to the polygon; whether the polygon is closed or open; whether to generate linetype scaling per vertex.

Illustrated at the right are examples of an inscribed polygon (inside of a circle) and a circumscribed polygon (around a circle).

The coordinates of all polygon vertices are identified above in the DDMODIFY dialog box. Selecting "Next" lists the next set of coordinates. By default, a polygon has normal curve generation which means it is absent of any curves. Selecting Quadratic, Cubic, and Fit Curve displays the results similar to the illustration at the right.

The LT Gen option stands for generate a linetype. If LT Gen is unselected in the DDMODIFY dialog box, or is set to "Off," the linetype is applied to each individual vertex. In the example at the right, with the Ltscale command set to 1.10 one leg of the polyline has a single center line while the other leg has three center lines. Checking the LT Gen box above in DDMODIFY turns on LT Gen with the results at the far right. Here, the first leg of the polyline has two center lines that continue into the second leg.

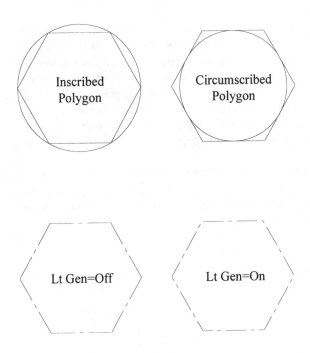

Inscribed Polygon

Circumscribed Polygon

Lt Gen=Off

Lt Gen=On

Listing the Properties of Text

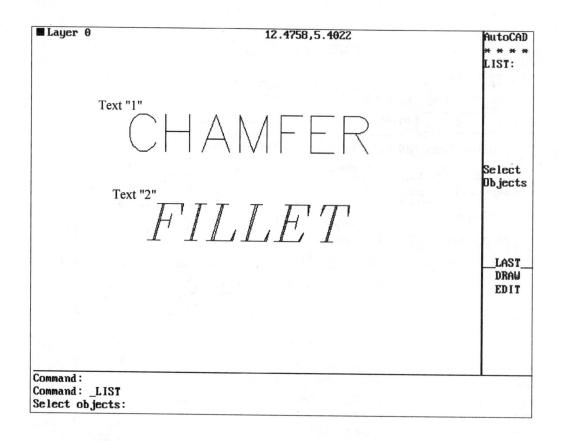

Using the List command on text displays the following information: the name of the entity being listed (Text); the layer the entity was drawn on; whether the entity occupies model or paper space; the text style the entity was drawn in; the font file the entity was drawn in; the starting point of the text; the text height; the actual text that was entered from the keyboard; the text rotation angle; the current width factor of the text; the current obliquing angle for the text; the type of text generation.

In the examples above and to the right, the width scale factor, obliquing angle, and generation of text are all controlled by the Style command when the text font was originally associated with a text style.

```
                    Text "1"
        TEXT        Layer: 0
                    Space: Model space
     Style = STANDARD    Font file = SIMPLEX
     center point, X=   6.0000  Y=   6.0000  Z=   0.0000
     height    1.0000
        text CHAMFER
  rotation angle        0
     width scale factor      1.0000
  obliquing angle       0
 generation normal
```

```
                    Text "2"
        TEXT        Layer: 0
                    Space: Model space
     Style = ITALICC    Font file = ITALICC
     start point, X=   3.1711  Y=   3.5867  Z=   0.0000
     height    1.0000
        text FILLET
  rotation angle        0
     width scale factor      1.0000
  obliquing angle       0
 generation normal
```

```
                        Modify Text
 Properties
  [ Color... ]  ■  BYLAYER          [ Layer... ]  0
  [ Linetype... ]  BYLAYER          Thickness: [ 0.0000        ]

 Text:  [ CHAMFER                                              ]
 Origin
  [  Pick Point <  ]  Height:   [ 1.0000 ]  Justify: [ Center    ▼ ]
  X: [ 6.0000 ]  Rotation:  [ 0      ]  Style:   [ STANDARD   ▼ ]
  Y: [ 6.0000 ]  Width Factor: [ 1.0000 ]  □ Upside Down
  Z: [ 0.0000 ]  Obliquing: [ 0      ]  □ Backward

 Handle: None
            [   OK   ]  [ Cancel ]  [ Help... ]
```

Using the DDMODIFY command on a text entity provides for superior control of text. All listings appearing in a box may be dynamically changed to affect the final form of the text entity. In addition to the usual entity properties that may be changed, such as color, linetype, layer name, and entity thickness, the following may be changed using this dialog box: the actual text may be edited in a way similar to the DDEDIT command; selecting a new text origin point; entering a new text height; entering a new rotation angle for the text; entering a new width factor for the text; entering an obliquing angle to make text inclined; selecting a new justification position for the text; selecting a new text style.

Keep in mind that all of the above changes only apply to the text entity selected and do not globally affect all text entities.

Selecting the current text justification position opens up an additional dialog box illustrated at the right. All valid text justification positions appear allowing the user to scroll up or down to select a new justification position.

Selecting the current text style opens another dialog box illustrated at the right. Use this dialog box to select additional text styles. These styles are defined using the Style command.

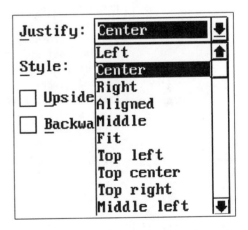

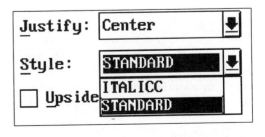

Using the Status Command

Once inside of a large drawing, it becomes difficult sometimes to keep track of various settings that have been changed from their default values to different values required by the drawing such as current number of entities, current color, linetype, and layer along with such settings as grid and snap spacing. To obtain a listing of these important settings contained in the current drawing file, use the Status command. Once the Status command is invoked, the graphics screen changes to a text screen displaying all of the information illustrated at the right.

```
1736 entities in C:\QZ1\SAMPLE\TOOLPOST
Model space limits are X:    0.0000  Y:     0.0000  (Off)
                       X:   18.0000  Y:    14.0000
Model space uses       X:    3.7500  Y:     4.7500
                       X:   17.2500  Y:    13.2500
Display shows          X:    3.7500  Y:     4.7500
                       X:   17.3859  Y:    14.6259
Insertion base is      X:    0.0000  Y:     0.0000  Z:    0.0000
Snap resolution is     X:    1.0000  Y:     1.0000
Grid spacing is        X:    0.0000  Y:     0.0000

Current space:         Model space
Current layer:         0
Current color:         BYLAYER -- 7 (white)
Current linetype:      CONTINUOUS
Current elevation:     0.0000  thickness:    0.0000
Fill on  Grid off  Ortho off  Qtext off  Snap off  Tablet off
Object snap modes:     None
Free disk: 8177664 bytes
Virtual memory allocated to program: 3624K
Amount of program in physical memory/Total (virtual) program size: 67%
-- Press RETURN for more --
Total conventional memory: 292K      Total extended memory: 7424K
Swap file size: 388K bytes
```

Using the Time Command

```
Command: _TIME
Current time:          22 Jun 1992 at 21:25:58.340
Drawing created:       21 Aug 1990 at 15:24:58.640
Drawing last updated:  14 Apr 1992 at 20:45:47.990
Time in drawing editor: 0 days 00:25:50.700
Elapsed timer:          0 days 00:25:50.700
Next automatic save in: 0 days 01:56:56.160
Timer on.
Display/ON/OFF/Reset:
```

The Time command provides the operator with the following information:

Current Time:
All dates and times are set by the DOS Date and Time commands. If using the Time command, it is important to make sure these DOS commands are properly set in order to display the desired results.

Drawing Created:
This date and time value is set whenever using the New command for creating a new drawing file. This value is also set to the current date and time whenever a Wblock is created or a drawing is saved under a different name using the Save command.

Drawing Last Updated:
This data consists of the date and time the current drawing was last updated. This value updates itself whenever using the Save command or the End command.

Time in Drawing Editor:
This represents the total time spent editing the drawing. The timer is always updating itself and cannot be reset to a new or different value.

Elapsed Timer:
This timer runs while AutoCAD is in operation and can be turned on or off or reset by the user.

Next Automatic Save In:
This timer displays when the next automatic save will occur. This value is controlled by the system variable "SAVETIME". If this system variable is set to zero, the automatic save utility is disabled. If the timer is set to a nonzero value, the timer displays when the next automatic save will take place. The increment for automatic saving is in minutes.

Tutorial Exercise #16
Extrude.Dwg

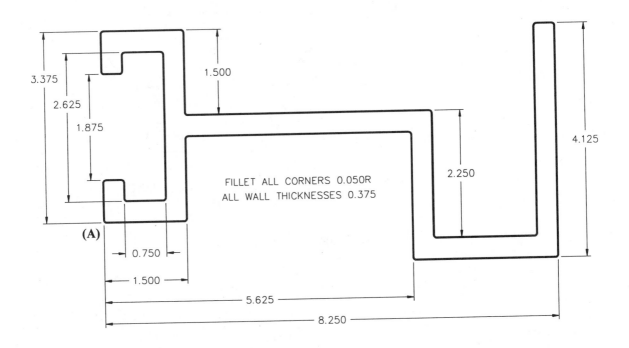

FILLET ALL CORNERS 0.050R
ALL WALL THICKNESSES 0.375

PURPOSE:
This tutorial is designed to show the user various methods in constructing the extruded pattern above. The surface area of the extrusion will also be found using the Area command.

SYSTEM SETTINGS:
Keep the default drawing limits at 0.0000,0.0000 for the lower left corner and 12.0000,9.0000 for the upper right corner. Use the Units command and change the number of decimal places past the zero from 4 units to 3 units.

LAYERS:
No special layers need be created for this drawing.

SUGGESTED COMMANDS:
Begin drawing the extrusion with point "A" illustrated above at absolute coordinate 2.000,3.000. Use either of the following methods to construct the extrusion:
- Using a series of absolute, relative, and polar coordinates to construct the profile of the extrusion.
- Constructing a few lines; then using the Offset command followed by the Trim command to construct the extrusion profile.

The Fillet command is used to create the 0.050 radius rounds at all corners of the extrusion. Before calculating the area of the extrusion, convert and join all entities into one single polyline. This will allow the Area command to be used in a more productive way.

DIMENSIONING:
This drawing does not need to be dimensioned in order to answer any Inquiry command question.

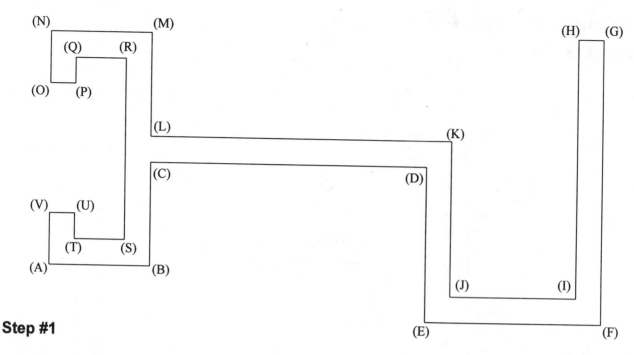

Step #1

One method of constructing the extrusion above is to use the measurements on the previous page to calculate a series of polar coordinate system distances.

Command: **Line**
From point: **2.000,3.000** *(Starting at "A")*
To point: **@1.500<0** *(To "B")*
To point: **@1.500<90** *(To "C")*
To point: **@4.125<0** *(To "D")*
To point: **@2.250<270** *(To "E")*
To point: **@2.625<0** *(To "F")*
To point: **@4.125<90** *(To "G")*
To point: **@0.375<180** *(To "H")*
To point: **@3.750<270** *(To "I")*
To point: **@1.875<180** *(To "J")*
To point: **@2.250<90** *(To "K")*
To point: **@4.500<180** *(To "L")*

To point: **@1.500<90** *(To "M")*
To point: **@1.500<180** *(To "N")*
To point: **@0.750<270** *(To "O")*
To point: **@0.375<0** *(To "P")*
To point: **@0.375<90** *(To "Q")*
To point: **@0.750<0** *(To "R")*
To point: **@2.625<270** *(To "S")*
To point: **@0.750<180** *(To "T")*
To point: **@0.375<90** *(To "U")*
To point: **@0.375<180** *(To "V")*
To point: **Close** *(Back to "A")*

An alternate step of constructing the extrusion is to place a horizontal and a vertical line. These lines are then offset using the Offset command and the measurements on the previous page. This method may prove to be quicker than the coordinate method since dimensions are simply read instead of having to be calculated. Once all lines are offset to form a rough image of the extrusion, the trim command is used to clean up all corners and intersections.

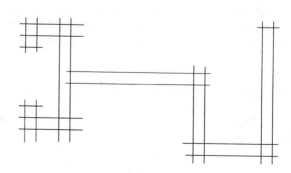

Step #2

All corners come together at 90 degree intersections. From the original dimensions of the extrusion, a note calls out that all corners be rounded off with a 0.500 radius. This is easily accomplished using the Fillet command. However since so many corners need to be filleted, the risk is high of forgetting to fillet one or more corners. All corners may be filleted at one time only if the entire extrusion consists of one polyline. First the Pedit command will be used to perform this conversion followed by the Fillet command.

Command: **Pedit**
Select polyline: *(Select the entity labeled "A")*
Entity selected is not a polyline.
Do you want to turn it into one? <Y>: *(Strike Enter to continue)*
Close/Join/Width/Edit vertex/Fit/Spline/Decurve/Ltype gen/ Undo/eXit <X>: **Join**
Select objects: **W** *(To window in all entities)*
First corner: **0.000,0.000**
Other corner: **12.000,9.000**
Select objects: *(Strike Enter to continue)*
21 segments added to polyline
Open/Join/Width/Edit vertex/Fit/Spline/Decurve/Ltype gen/ Undo/eXit <X>: *(Strike Enter to exit this command)*

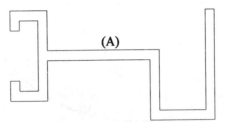

(A)

Step #3

With the entire extrusion converted into a polyline, use the Fillet command, set a radius of 0.050, and use the polyline option of the Fillet command to fillet all corners of the extrusion at once.

Command: **Fillet**
Polyline/Radius/<Select first object>: **Radius**
Enter fillet radius <0.0000>: **0.050**

Command: **Fillet**
Polyline/Radius/<Select first object>: **Polyline**
Select 2D polyline: *(Select the polyline at the right)*
22 lines were filleted

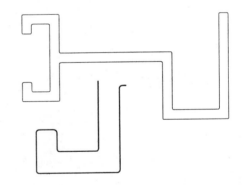

Step #4

Question #1
The total surface area of the Extrusion is_____.

Use the Area command to calculate the surface area of the extrusion. This is easily accomplished since the extrusion has already been converted into a polyline.

Command: **Area**
<First point>/Entity/Add/Subtract: **Entity**
Select circle or polyline: *(Select any part of the extrusion)*
Area = 7.170,   Perimeter = 38.528

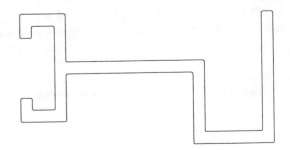

Total surface area of the Extrusion is 7.170

Tutorial Exercise #17
C-Lever.Dwg

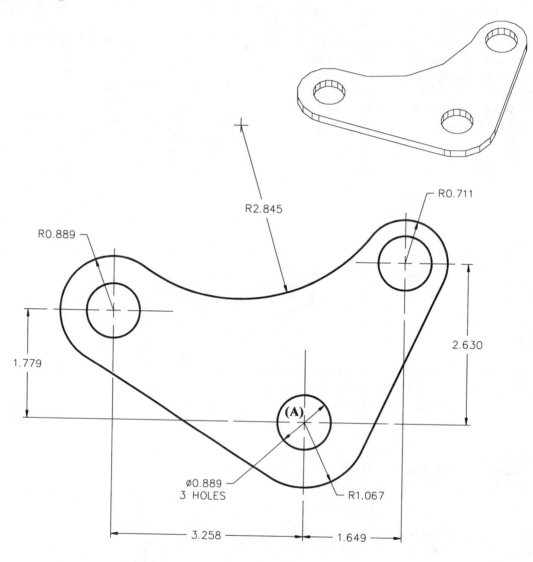

PURPOSE:
This tutorial is designed to show the user various methods in constructing the C-Lever object above. Numerous questions will be asked about the object requiring the use of a majority of Inquiry commands.

SYSTEM SETTINGS:
Keep the default drawing limits at 0.0000,0.0000 for the lower left corner and 12.0000,9.0000 for the upper right corner. Use the Units command and change the number of decimal places past the zero from 4 units to 3 units.

LAYERS:
No special layers need be created for this drawing.

SUGGESTED COMMANDS:
Begin drawing the C-Lever with point "A" illustrated above at absolute coordinate 7.000,3.375. Begin laying out all circles. Then draw tangent lines and arcs. Use the Trim command to clean up unnecessary entities. To prepare to answer the Area command question, convert the profile of the C-Lever into a polyline using the Pedit command. Other questions pertaining to distances, angles, and point identifications follow.

DIMENSIONING:
This drawing does not need to be dimensioned in order to answer any Inquiry command question.

Step #1

Construct one circle of 0.889 diameter with the center of the circle at absolute coordinate 7.000,3.375. Construct the remaining circles of the same diameter by using the Copy command with the multiple option. Use of the "@" symbol for the base point in the copy command identifies the last known point which in this case is the center of the first circle drawn at coordinate 7.000,3.375

Command: **Circle**
3P/2P/TTR/<Center point>: **7.000,3.375**
Diameter/<Radius>: **Diameter**
Diameter: **0.889**

Command: **Copy**
Select objects: **Last**
Select objects: *(Strike Enter to continue)*
<Base point or displacement>/Multiple: **Multiple**
Base point: **@**
Second point of displacement: **@1.649,2.630**
Second point of displacement: **@-3.258,1.779**
Second point of displacement: *(Strike Enter to exit this command)*

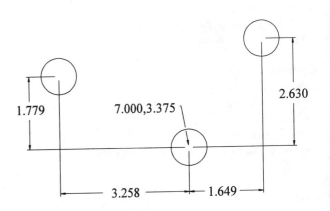

Step #2

Construct three more circles. Even though these entities actually represent arcs, circles will be drawn now and trimmed later to form the arcs.

Command: **Circle**
3P/2P/TTR/<Center point>: **Cen**
of *(Select the edge of circle "A")*
Diameter/<Radius>: **1.067**

Command: **Circle**
3P/2P/TTR/<Center point>: **Cen**
of *(Select the edge of circle "B")*
Diameter/<Radius>: **0.889**

Command: **Circle**
3P/2P/TTR/<Center point>: **Cen**
of *(Select the edge of circle "C")*
Diameter/<Radius>: **0.711**

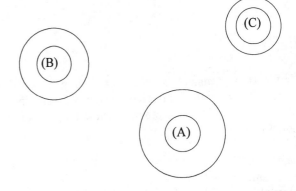

Step #3

Construct lines tangent to the three outer circles illustrated at the right.

Command: **Line**
From point: **Tan**
to *(Select the outer circle near "A")*
To point: **Tan**
to *(Select the outer circle near "B")*
To point: *(Strike Enter to exit this command)*

Command: **Line**
From point: **Tan**
to *(Select the outer circle near "C")*
To point: **Tan**
to *(Select the outer circle near "D")*
To point: *(Strike Enter to exit this command)*

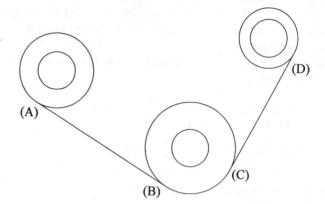

Step #4

Construct a circle tangent to the two circles illustrated at the right using the Circle command with the Tangent-Tangent-Radius option (TTR).

Command: **Circle**
3P/2P/TTR/<Center point>: **TTR**
Enter Tangent spec: *(Select the outer circle near "A")*
Enter second Tangent spec: *(Select the outer circle near "B")*
Radius: **2.845**

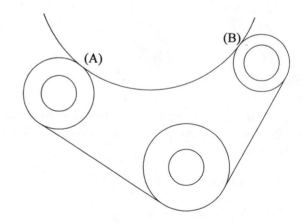

Step #5

Use the Trim command to clean up and form the finished object. Select all of the entities represented by dashed lines as cutting edges. Follow the prompts below for selecting the entities to trim.

Command: **Trim**
Select cutting edge(s)...
Select objects: *(Select all dashed entities illustrated at the right)*
Select objects: *(Strike Enter to continue)*
<Select object to trim>/Undo: *(Select the circle at "A")*
<Select object to trim>/Undo: *(Select the circle at "B")*
<Select object to trim>/Undo: *(Select the circle at "C")*
<Select object to trim>/Undo: *(Select the circle at "D")*
<Select object to trim>/Undo: *(Strike Enter to exit this command)*

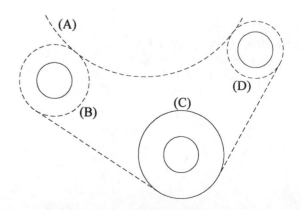

Checking the Accuracy of C-Lever.Dwg

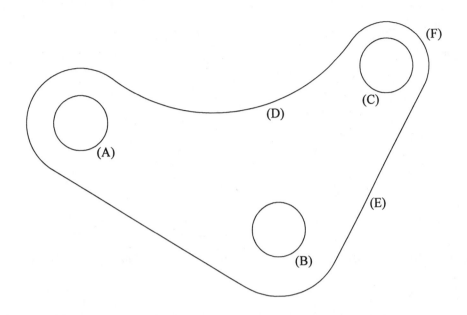

Once the C-Lever has been constructed, answer the questions below to determine the accuracy of the drawing. Use the illustration above to assist in answering the questions.

1. The total area of the C-Lever with all three holes removed is_____

2. The total distance from the center of circle "A" to the center of circle "B" is_____

3. The angle formed in the X-Y plane from the center of circle "C" to the center of circle "B" is

   _____

4. The delta X-Y distance from the center of circle "C" to the center of circle "A" is _____

5. The absolute coordinate value of the center of arc "D" is_____

6. The total length of line "E" is_____

7. The total length of arc "F" is_____

A solution for each question follows complete with the method used to arrive at the answer. Apply these methods to any type of drawing with similar needs.

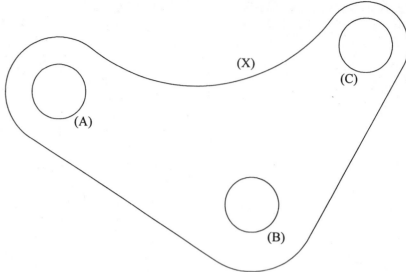

Question #1

The total area of the C-Lever with all three holes removed is_____.

The Area command will be used to first calculate the total area of the object and then subtract all three holes. However before using the Area command, all entities representing the outline of the object must be converted into a polyline. The Pedit command with the Join option is used to best accomplish this.

Command: **Pedit**
Select polyline: *(Select the entity labeled "X")*
Entity selected is not a polyline.
Do you want to turn it into one? <Y>: *(Strike Enter to continue)*
Close/Join/Width/Edit vertex/Fit/Spline/Decurve/Ltype gen/ Undo/eXit <X>: **Join**
Select objects: **W**
First corner: **0.000,0.000**
Other corner: **12.000,9.000**
Select objects: *(Strike Enter to continue)*
5 segments added to polyline
Open/Join/Width/Edit vertex/Fit/Spline/Decurve/Ltype gen/ Undo/eXit <X>: *(Strike Enter to exit this command)*

All outer entities now consist as one polyline entity. Notice that when selecting all entities including the three circles, only the entities connected formed the polyline. Since the circles were independent of the outer entities, they were not included in the creation of the polyline. An alternative method of selecting entities to join into a polyline is to select each entity individually.

Now the Area command may be successfully used to calculate the area of the object with the holes removed.

Command: **Area**
<First point>/Entity/Add/Subtract: **Add**
<First point>/Entity/Subtract: **Entity**
(ADD mode) Select circle or polyline: *(Select the edge of the object near "X")*
Area = 15.611,   Perimeter = 17.771
Total area = 15.611
(ADD mode) Select circle or polyline: *(Strike Enter to continue)*
<First point>/Entity/Subtract: **Subtract**
<First point>/Entity/Add: **Entity**
(SUBTRACT mode) Select circle or polyline: *(Select circle"A")*
Area = 0.621,   Circumference = 2.793
Total area = 14.991
(SUBTRACT mode) Select circle or polyline: *(Select circle"B")*
Area = 0.621,   Circumference = 2.793
Total area = 14.370
(SUBTRACT mode) Select circle or polyline: *(Select circle"C")*
Area = 0.621,   Circumference = 2.793
Total area = 13.749
(SUBTRACT mode) Select circle or polyline: *(Strike Enter to continue)*
<First point>/Entity/Add: *(Strike Enter to exit this command)*

The total area of the C-Lever with all three holes removed is 13.749

Question #2
The total distance from the center of circle "A" to the center of circle "B" is_____.

Use the Dist (Distance) command to calculate the distance from the center of circle "A" to the center of circle "B". Be sure to use the Osnap-Center mode. Notice that additional information is given when using the Dist command. For the purpose of this question, we will only be looking for the distance.

Command: **Dist**
First point: **Cen**
of *(Select the edge of circle "A")*
Second point: **Cen**
of *(Select the edge of circle "B")*
Distance = 3.712

The total distance from the center of circle "A" to the center of circle "B" is 3.712

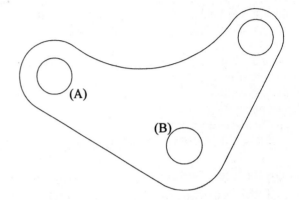

Question #3
The angle formed in the X-Y plane from the center of circle "C" to the center of circle "B" is_____.

Use the Dist (Distance) command to calculate the angle from the center of circle "C" to the center of circle "B". Be sure to use the Osnap-Center mode. Notice that additional information is given when using the Dist command. For the purpose of this question, we will only be looking for the angle in the X-Y plane.

Command: **Dist**
First point: **Cen**
of *(Select the edge of circle "C")*
Second point: **Cen**
of *(Select the edge of circle "B")*
Angle in X-Y Plane = 238

The angle formed in the X-Y plane from the center of circle "C" to the center of circle "B" is 238 degrees

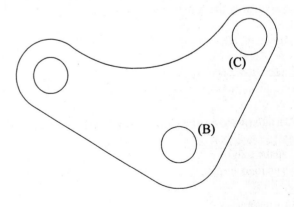

Question #4
The delta X-Y distance from the center of circle "C" to the center of circle "A" is_____.

Use the Dist (Distance) command to calculate the delta X-Y distance from the center of circle "C" to the center of circle "A". Be sure to use the Osnap-Center mode. Notice that additional information is given when using the Dist command. For the purpose of this question, we will only be looking for the delta X-Y distance. The Dist command will display the relative X, Y, and Z distances. Since this is a 2 dimensional problem, only the X and Y values will be used.

Command: **Dist**
First point: **Cen**
of *(Select the edge of circle "C")*
Second point: **Cen**
of *(Select the edge of circle "A")*
Delta X = -4.907, Delta Y = -0.851, Delta Z = 0.000

The delta X-Y distance from the center of circle "C" to the center of circle "A" is -4.907, -0.851

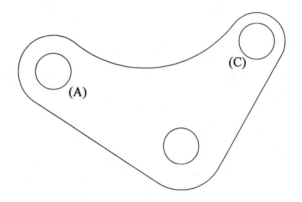

Question #5
The absolute coordinate value of the center of arc "D" is_____.

The ID command is used to get the current absolute coordinate information on a desired point. This command will display the X, Y, and Z coordinate values. Since this is a 2 dimensional problem, only the X and Y values will be used.

Command: **ID**
Point: **Cen**
of *(Select the edge of arc "D")*
X = 5.869, Y = 8.223, Z = 0.000

The absolute coordinate value of the center of arc "D" is 5.869,8.223

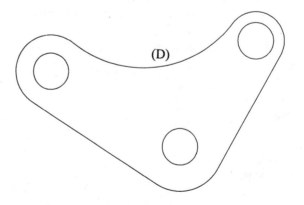

Question #6
The total length of line "E" is_____.

Use the Dist (Distance) command to find the total length of line "E". Be sure to use the Osnap-Endpoint mode. Notice that additional information is given when using the Dist command. For the purpose of this question, we will only be looking for the distance.

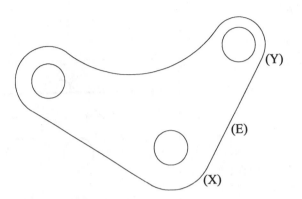

Command: **Dist**
First point: **Endp**
of *(Select the endpoint of the line at "X")*
Second point: **Endp**
of *(Select the endpoint of the line at "Y")*
Distance = 3.084

The total length of line "E" is **3.084**

Question #7
The total length of arc "F" is_____.

The List command is used to calculate the lengths of arcs; however a little preparation is needed before preforming this operation. If arc "F" is selected at the right, notice the entire outline is selected since it is a polyline. Use the Explode command to break the outline back into individual entities. Again use the List command on arc "F". The following information is displayed: Entity name, Layer entity is found on, Model Space or Paper Space, Center point of the arc, Radius of the arc, Starting and ending angles of the arc. However, the length of the arc is not given. Convert the arc into a polyline using the Pedit command and then use List. Along with the vertices of the polyarc, the area and length are given.

Command: **Explode**
Select objects: *(Select the dashed polyline anywhere)*
Select objects: *(Strike Enter to execute this command)*

Command: **Pedit**
Select polyline: *(Select arc "F")*
Entity selected is not a polyline.
Do you want to turn it into one? <Y>: *(Strike Enter to continue)*
Close/Join/Width/Edit vertex/Fit/Spline/Decurve/Ltype gen/ Undo/eXit <X>: *(Strike Enter to exit this command)*

Command: **List**
Select objects: *(Select arc "F")*
Select objects: *(Strike Enter to continue)*
Length = 2.071

The total length of arc "F" is **2.071**

An alternate command to use to find the total length of an arc is DDMODIFY. This command displays a dialog box listing various properties and information regarding the selected arc. DDMODIFY will yield the total length of an arc segment without first converting the arc segment into a polyline.

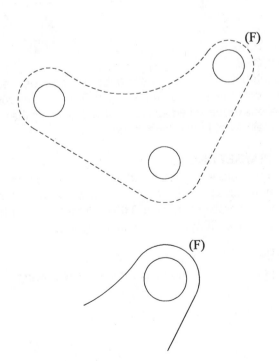

Tutorial Exercise #18
Fitting.Dwg

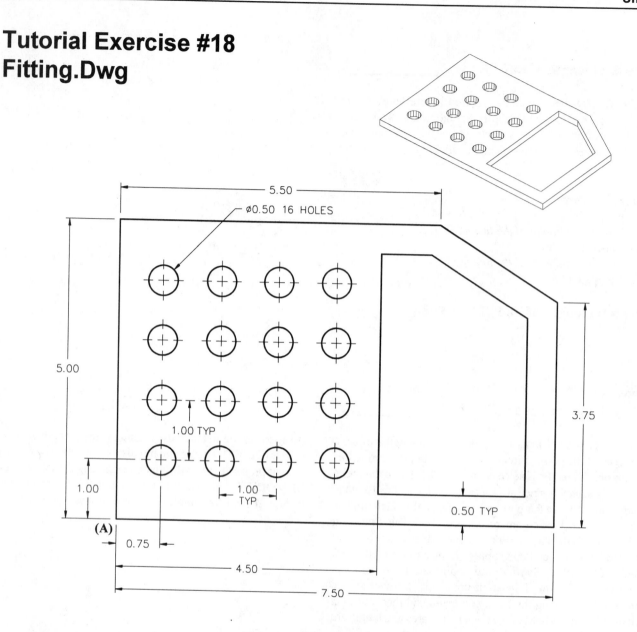

5.50

Ø0.50 16 HOLES

5.00

1.00 TYP

1.00
TYP

3.75

0.50 TYP

(A)

1.00

0.75

4.50

7.50

PURPOSE:
This tutorial is designed to show the user various methods in constructing the Fitting object above. Numerous questions will be asked about the object requiring the use of a majority of Inquiry commands.

SYSTEM SETTINGS:
Use the Units command and change the number of decimal places past the zero from 4 units to 2 units. Keep the default drawing limits at 0.00,0.00 for the lower left corner and 12.00,9.00 for the upper right corner.

LAYERS:
No special layers need be created for this drawing.

SUGGESTED COMMANDS:
Begin drawing the Fitting with point "A" illustrated above at absolute coordinate 2.24,1.91. Begin by laying out the profile of the Fitting. Locate one circle and use the Array command to produce 4 rows and columns of the circle. Use a series of Offset, Trim, and Fillet commands to construct the 5 sided figure on the inside of the Fitting profile. To prepare for the Area question, convert the outer and inner profiles into a polyline using the Pedit command. Other questions pertaining to distances, angles, and point identifications follow.

DIMENSIONING:
This drawing does not need to be dimensioned in order to answer any Inquiry command question.

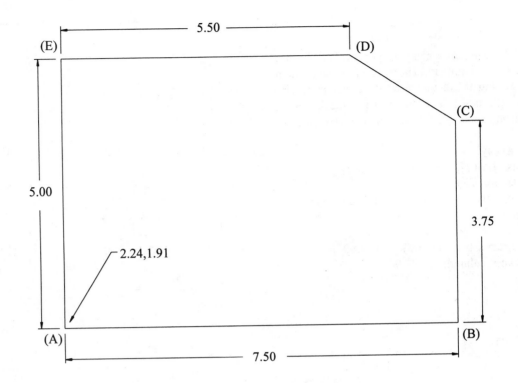

Step #1

Use the Line command to construct the outline of the Fitting using a combination of absolute, polar, and relative coordinates. Begin the lower left corner of the Fitting at absolute coordinate 2.24,1.91. Do not use the Close option of the Line command for constructing the last side of the Fitting.

Command: **Line**
From point: **2.24,1.91** *(Starting at "A")*
To point: **@7.50<0** *(To "B")*
To point: **@3.75<90** *(To "C")*
To point: **@-2.00,1.25** *(To "D")*
To point: **@5.50<180** *(To "E")*
To point: **@5.00<270** *(Back to "A")*
To point: *(Strike Enter to exit this command)*

Step #2

Construct a circle of 0.50 diameter using the Circle command. Since the last known point is at "A" from use of the previous Line command, this point is referenced using the "@" symbol followed by a coordinate value for the center point. This identifies the center of the circle from the last known point.

Command: **Circle**
3P/2P/TTR/<Center point>: **@0.75,1.00**
Diameter/<Radius>: **Diameter**
Diameter: **0.50**

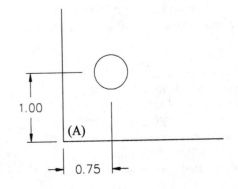

Step #3

Create multiple copies in a rectangular pattern of the last circle by using the Array command. There are 4 rows and columns each with a spacing of 1.00 units from the center of one circle to the center of the other. Since the array is to the right and up from the existing circle, all 1.00 spacing units are positive.

Command: **Array**
Select objects: **Last** *(This should select the circle)*
Select objects: *(Strike Enter to continue)*
Rectangular or Polar array (R/P)<R>: **Rectangular**
Number of rows (---) <1>: **4**
Number of columns (|||) <1>: **4**
Unit cell or distance between rows (---): **1.00**
Distance between columns (|||): **1.00**

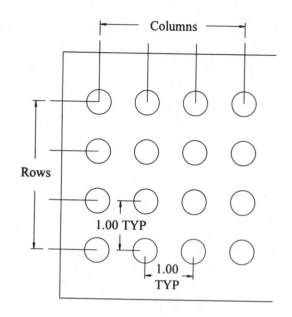

Step #4

Use the Offset command to copy the line at "A" parallel at a distance of 4.50 in the direction of "B."

Command: **Offset**
Offset distance or Through <Through>: **4.50**
Select object to offset: *(Select line "A")*
Side to offset? *(Pick a point anywhere near "B")*
Select object to offset: *(Strike Enter to exit this command)*

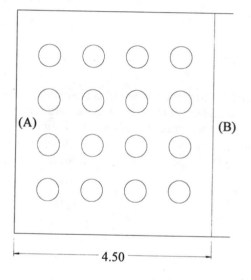

Step #5

Use the Offset command to copy the lines at "A," "B," "C," and "D" parallel at a distance of 0.50 in the direction of "E."

Command: **Offset**
Offset distance or Through <4.50>: **0.50**
Select object to offset: *(Select line "A")*
Side to offset? *(Pick a point anywhere near "E")*
Select object to offset: *(Select line "B")*
Side to offset? *(Pick a point anywhere near "E")*
Select object to offset: *(Select line "C")*
Side to offset? *(Pick a point anywhere near "E")*
Select object to offset: *(Select line "D")*
Side to offset? *(Pick a point anywhere near "E")*
Select object to offset: *(Strike Enter to exit this command)*

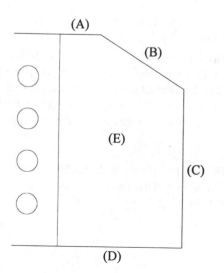

Step #6

Use the Trim command to partially delete the horizontal and vertical segments labeled "A," "B," "C," and "D" at the right. Select the three dashed entities at the right as cutting edges.

Command: **Trim**
Select cutting edge(s)...
Select objects: *(Select the three dashed entities at the right)*
Select objects: *(Strike Enter to continue)*
<Select object to trim>/Undo: *(Select the short vertical line segment at "A")*
<Select object to trim>/Undo: *(Select the horizontal line segment at "B")*
<Select object to trim>/Undo: *(Select the short vertical line segment at "C")*
<Select object to trim>/Undo: *(Select the horizontal line segment at "D")*
<Select object to trim>/Undo: *(Strike Enter to exit this command)*

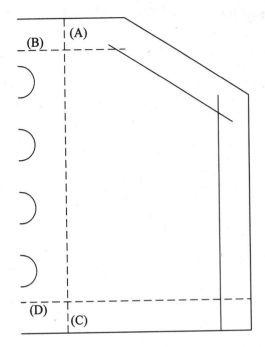

Step #7

Use the Fillet command to create corners at the intersections of the four line segments illustrated at the right. The Fillet radius by default is set to "0" to accomplish this unless it has been set to a positive value. The Trim command could also be used to accomplish this step.

Command: **Fillet**
Polyline/Radius/<Select first object>: *(Select line "A")*
Select second object: *(Select line "B")*

Command: **Fillet**
Polyline/Radius/<Select first object>: *(Select line "B")*
Select second object: *(Select line "C")*

Command: **Fillet**
Polyline/Radius/<Select first object>: *(Select line "C")*
Select second object: *(Select line "D")*

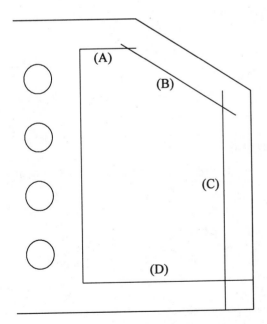

Checking the Accuracy of Fitting.Dwg

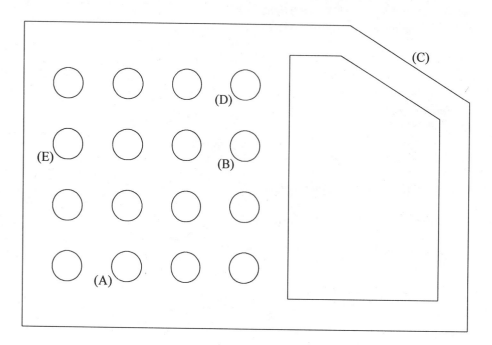

Once the Fitting has been constructed, answer the questions below to determine the accuracy of the drawing. Use the illustration above to assist in answering the questions.

1. The total area of the Fitting with the inner slot and all holes removed is_____

2. The total distance from the center of circle "A" to the center of circle "B" is_____

3. The total length of line "C" is_____

4. The angle formed in the X-Y plane from the center of circle "D" to the center of circle "E" is

_____

5. The absolute coordinate value of the center of circle "D" is_____

A solution for each question follows complete with the method used to arrive at the answer. Apply these methods to different objects with similar needs.

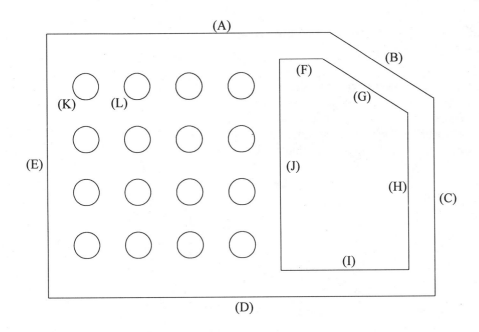

Question #1
The total area of the Fitting with the inner slot and all holes removed is_____.

The Area command will be used to first calculate the total area of the object and then subtract the slot and all holes. However before using the Area command, all entities representing the outline of the Fitting and slot must be converted into a polyline. The Pedit command with the Join option is used to best accomplish this. Refer to the illustration above for the following:

Command: **Pedit**
Select polyline: *(Select the entity labeled "A")*
Entity selected is not a polyline.
Do you want to turn it into one? <Y>: *(Strike Enter to continue)*
Close/Join/Width/Edit vertex/Fit/Spline/Decurve/Ltype gen/ Undo/eXit <X>: **Join**
Select objects: *(Select lines "B", "C", "D", and "E")*
Select objects: *(Strike Enter to continue)*
4 segments added to polyline
Open/Join/Width/Edit vertex/Fit/Spline/Decurve/Ltype gen/ Undo/eXit <X>: *(Strike Enter to exit this command)*

All outer entities now consist as one polyline entity. Repeat the above Pedit procedure for converting entities "F", "G", "H", "I", and "J" into one polyline. Now the Area command may be successfully used to calculate the area of the object with theslot and all holes removed.

Command: **Area**
<First point>/Entity/Add/Subtract: **Add**
<First point>/Entity/Subtract: **Entity**
(ADD mode) Select circle or polyline: *(Select the edge of the Fitting near "A")*
Area = 36.25,   Perimeter = 24.11
Total area = 36.25
(ADD mode) Select circle or polyline: *(Strike Enter to continue)*
<First point>/Entity/Subtract: **Subtract**
<First point>/Entity/Add: **Entity**
(SUBTRACT mode) Select circle or polyline: *(Select the slot near "F")*
Area = 9.16,   Perimeter = 12.27
Total area = 27.09
(SUBTRACT mode) Select circle or polyline: *(Select circle"K")*
Area = 0.20,   Circumference = 1.57
Total area = 26.90
(SUBTRACT mode) Select circle or polyline: *(Select circle"L")*
Area = 0.20,   Circumference = 1.57
Total area = 26.70
(SUBTRACT mode) Select circle or polyline: *(Carefully select the remaining 14 circles individually)*
Total area = 23.95
(SUBTRACT mode) Select circle or polyline: *(Strike Enter to continue)*
<First point>/Entity/Add: *(Strike Enter to exit this command)*

The total area of the Fitting with the slot and all holes removed is <u>23.95</u>

Question #2
The total distance from the center of circle "A" to the center of circle "B" is_____.

Use the Dist (Distance) command to calculate the distance from the center of circle "A" to the center of circle "B". Be sure to use the Osnap-Center mode. Notice that additional information is given when using the Dist command. For the purpose of this question, we will only be looking for the distance.

Command: **Dist**
First point: **Cen**
of *(Select the edge of circle "A")*
Second point: **Cen**
of *(Select the edge of circle "B")*
Distance = 2.83

The total distance from the center of circle "A" to the center of circle "B" is **2.83**

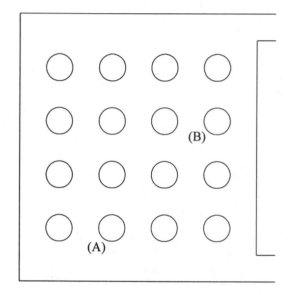

Question #3
The total length of line "C" is_____.

Use the Dist (Distance) command to find the total length of line "C". Be sure to use the Osnap-Endpoint mode. Notice that additional information is given when using the Dist command. For the purpose of this question, we will only be looking for the distance. The List command could also be used to perform this operation. However since the outline of the Fitting consists of one continuous polyline segment, the polylines would have to be converted into individual line segments using the Explode command before using the list command on segment "C".

Command: **Dist**
First point: **Endp**
of *(Select the endpoint of the line at "X")*
Second point: **Endp**
of *(Select the endpoint of the line at "Y")*
Distance = 2.36

The total length of line "C" is **2.36**

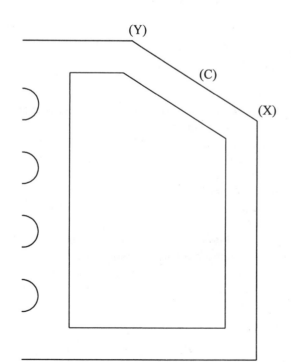

Question #4
The angle formed in the X-Y plane from the center of circle "D" to the center of circle "E" is_____.

Use the Dist (Distance) command to calculate the angle from the center of circle "D" to the center of circle "E". Be sure to use the Osnap-Center mode. Notice that additional information is given when using the Dist command. For the purpose of this question, we will only be looking for the angle in the X-Y plane.

Command: **Dist**
First point: **Cen**
of *(Select the edge of circle "D")*
Second point: **Cen**
of *(Select the edge of circle "E")*
Angle in X-Y Plane = 198

The angle formed in the X-Y plane from the center of circle "D" to the center of circle "E" is 198 degrees

Question #5
The absolute coordinate value of the center of circle "D" is_____.

The ID command is used to get the current absolute coordinate information on a desired point. This command will display the X, Y, and Z coordinate values. Since this is a 2 dimensional problem, only the X and Y values will be used.

Command: **ID**
Point: **Cen**
of *(Select the edge of circle "D")*
X = 5.99, Y = 5.91, Z = 0.000

The absolute coordinate value of the center of circle "D" is 5.99,5.91

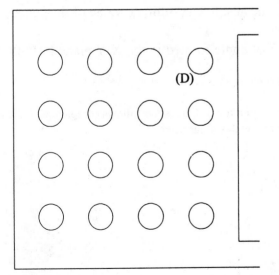

Additional Questions for Fitting.Dwg

Suppose the Fitting is rotated at a -10 degree angle using "X" as the center of rotation. Follow the command sequence below to perform this operation.

Command: **Rotate**
Select objects: **All**
Select objects: *(Strike Enter to continue)*
Base point: **Endp**
of *(Select the inclined line near "X")*
<Rotation angle>/Reference: **-10**

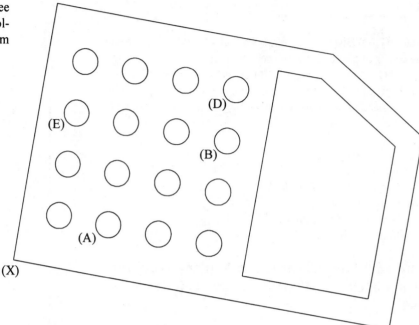

Once the Fitting has been rotated, answer the questions below to determine the accuracy of the drawing. Use the illustration above to assist in answering the questions.

1. The absolute coordinate value of the center of circle "D" is_____

2. The absolute coordinate value of the center of circle "A" is_____

3. The absolute coordinate value of the center of circle "B" is_____

4. The angle formed in the X-Y plane from the center of circle "D" to the center of circle "E" is_____

A solution for each question follows complete with the method used to arrive at the answer. Apply these methods to different objects with similar needs.

Question #1
The absolute coordinate value of the center of circle "D" is_____.

The ID command is used to get the current absolute coordinate information on a desired point. This command will display the X, Y, and Z coordinate values. Since this is a 2 dimensional problem, only the X and Y values will be used.

Command: **ID**
Point: **Cen**
of *(Select the edge of circle "D")*
X = 6.63, Y = 5.20, Z = 0.000

The absolute coordinate value of the center of circle "D" is **6.63,5.20**

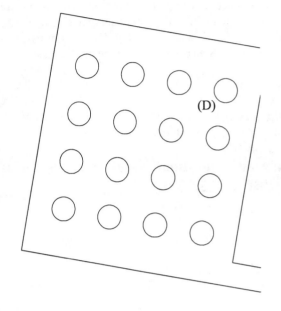

Question #2
The absolute coordinate value of the center of circle "A" is_____.

The ID command is used to get the current absolute coordinate information on point "A". This command will display the X, Y, and Z coordinate values. Since this is a 2 dimensional problem, only the X and Y values will be used.

Command: **ID**
Point: **Cen**
of *(Select the edge of circle "A")*
X = 4.14, Y = 2.59, Z = 0.000

The absolute coordinate value of the center of circle "A" is **4.14,2.59**

Question #3
The absolute coordinate value of the center of circle "B" is_____.

The ID command is used to get the current absolute coordinate information on point "B". This command will display the X, Y, and Z coordinate values. Since this is a 2 dimensional problem, only the X and Y values will be used.

Command: **ID**
Point: **Cen**
of *(Select the edge of circle "B")*
X = 6.45, Y = 4.21, Z = 0.000

The absolute coordinate value of the center of circle "B" is 6.45,4.21

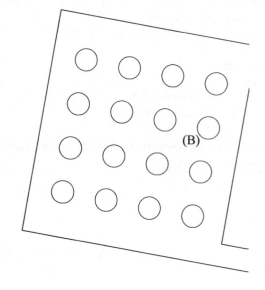

Question #4

The angle formed in the X-Y plane from the center of circle "D" to the center of circle "E" is_____.

Use the Dist (Distance) command to calculate the angle from the center of circle "D" to the center of circle "E". Be sure to use the Osnap-Center mode. Notice that additional information is given when using the Dist command. For the purpose of this question, we will only be looking for the angle in the X-Y plane.

Command: **Dist**
First point: **Cen**
of *(Select the edge of circle "D")*
Second point: **Cen**
of *(Select the edge of circle "E")*
Angle in X-Y Plane = 188

The angle formed in the X-Y plane from the center of circle "D" to the center of circle "E" is 188 degrees

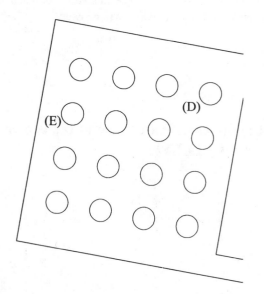

Questions for Unit 6

6-1. The command used to identify the coordinates of a specific point is
- (A) Id
- (B) Find
- (C) Inquire
- (D) Status
- (E) Analysis

6-2. The ID command lists
- (A) only an X and Y coordinate value.
- (B) the type of entity selected.
- (C) an X, Y, and Z coordinate value.
- (D) the layer name for the selected entity.
- (E) Both B and C.

6-3. The command which lets you access information about an entity is
- (A) DDMODIFY
- (B) DDATTE
- (C) List
- (D) Both A and B.
- (E) Both A and C.

6-4. To obtain the length of an arc
- (A) use the List command on the arc.
- (B) use the Id command on the arc.
- (C) convert the arc into a polyline, then use List.
- (D) use the Dblist command.
- (E) Only A and B.

6-5. All of the following are listed when executing the Status command except
- (A) current limits of the display screen.
- (B) Grid On/Off.
- (C) current layer, color, and linetype.
- (D) Ortho On/Off.
- (E) current Object snap modes.

6-6. Using the List command on a closed polyline displays the
- (A) area.
- (B) perimeter.
- (C) total polyline length.
- (D) Only A and B.
- (E) Only A and C.

6-7. Using the List command on an open polyline displays the
- (A) area.
- (B) perimeter.
- (C) total polyline length.
- (D) Only A and B.
- (E) Only A and C.

6-8. Using DDMODIFY on a circle lists the
- (A) center of the circle.
- (B) radius of the circle.
- (C) diameter of the circle.
- (D) area of the circle.
- (E) All of the above.

6-9. The command used to verify the name of the current drawing is
- (A) Status
- (B) Inquiry
- (C) List
- (D) Find
- (E) ?

6-10. The command used to get a complete listing of all entities in the current drawing file is
- (A) Status
- (B) List
- (C) Dist
- (D) Inquiry
- (E) Dblist

Problems for Unit 6

Problems 1 through 27 consist of objects to be drawn using basic draw and edit commands. Seven questions follow each problem. Inquiry commands are to be used to answer the questions by selecting the correct choice from the list of possible answers available.

These problems are designed to provide the individual with an idea as to the level of accuracy of their drawing. As a result, layers do not have to be made unless it helps with the construction process of the object. Also, each problem does not have to be dimensioned unless requested by the instructor.

Problem 6-1
Angleblk.Dwg

Directions for Angleblk.Dwg
Use the Units command to set the units to decimal. Set the number of digits to the right of the decimal point from 4 to 2. Be sure the system of angle measure is set to decimal degrees and the number of decimal places for the display of angles is zero (0). Keep the remaining default unit values. Keep the default settings for the drawing limits.

Begin this drawing by locating the lower left corner of Angleblk identified by "X" at coordinate (2.35,3.17).

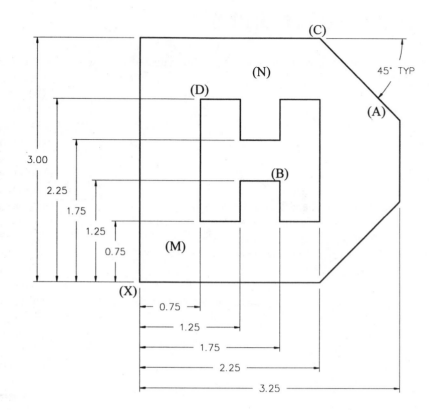

Refer to the drawing of the Angleblk above to answer the following questions:

1. The total surface area of the Angleblk with the inner "H" shape removed is
   - (A) 6.66
   - (B) 6.77
   - (C) 6.89
   - (D) 7.00
   - (E) 7.11

2. The total area of the inner "H" shape is
   - (A) 1.75
   - (B) 1.81
   - (C) 1.87
   - (D) 1.93
   - (E) 1.99

3. The total length of line "A" is
   - (A) 1.29
   - (B) 1.35
   - (C) 1.41
   - (D) 1.47
   - (E) 1.53

4. The absolute coordinate value of the endpoint of the line at "B" is
   - (A) 4.04,4.42
   - (B) 4.10,4.42
   - (C) 4.16,4.42
   - (D) 4.22,4.42
   - (E) 4.28,4.42

5. The absolute coordinate value of the endpoint of the line at "C" is
   - (A) 4.60,6.11
   - (B) 4.60,6.17
   - (C) 4.60,6.23
   - (D) 4.60,6.29
   - (E) 4.60,6.35

6. Use the Stretch command and extend the inner "H" shape a distance of 0.37 units in the 180 direction. Use "M" as the first corner of the crossing window and "N" as the other corner. Use the endpoint of "D" as the base point of the stretching operation. The new surface area of Angleblk with the inner "H" shape removed is closest to
   - (A) 6.63
   - (B) 6.69
   - (C) 6.75
   - (D) 6.81
   - (E) 6.87

7. Use the Scale command with the endpoint of the line at "D" as the base point. Reduce the size of just the inner "H" using a scale factor of 0.77. The new surface area of Angleblk with the inner "H" removed is
   - (A) 7.48
   - (B) 7.54
   - (C) 7.60
   - (D) 7.66
   - (E) 7.72

Problem 6-2
Lever1.Dwg

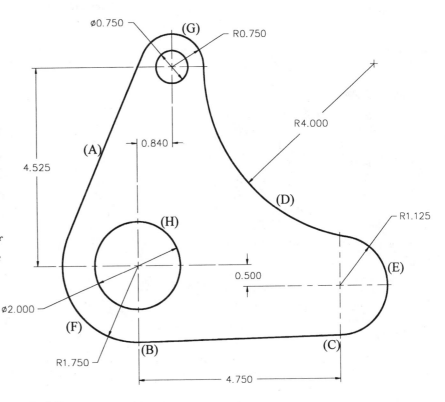

Directions for Lever1.Dwg

Use the Units command to set the units to decimal. Set the number of digits to the right of the decimal point from 4 to 3. Be sure the system of angle measure is set to decimal degrees and the number of decimal places for the display of angles is zero (0). Keep the remaining default unit values.

Begin this drawing by locating the center of the 2.000 diameter circle at coordinate (4.500,3.250).

Refer to the drawing of Lever1 above to answer the following questions:

1. The total length of line segment "A" is
   - (A) 4.483
   - (B) 4.492
   - (C) 4.499
   - (D) 4.504
   - (E) 4.515

2. The absolute coordinate value of the center of the 4.00 radius arc "D" is
   - (A) 10.090,7.773
   - (B) 10.090,7.782
   - (C) 10.090,7.791
   - (D) 10.090,7.800
   - (E) 10.090,7.806

3. The length of the 1.125 radius arc segment "E" is
   - (A) 3.299
   - (B) 3.308
   - (C) 3.319
   - (D) 3.329
   - (E) 3.337

4. The distance from the center of the 1.750 radius arc "F" to the center of the 0.750 radius arc "G" is
   - (A) 4.572
   - (B) 4.584
   - (C) 4.596
   - (D) 4.602
   - (E) 4.610

5. The total area of Lever1 with the two holes removed is
   - (A) 26.271
   - (B) 26.282
   - (C) 26.290
   - (D) 26.298
   - (E) 26.307

6. The circumference of the 2.000 diameter circle "H" is
   - (A) 6.269
   - (B) 6.276
   - (C) 6.283
   - (D) 6.289
   - (E) 6.294

7. Use the Scale command to reduce Lever1 in size by a scale factor of 0.333. Use the center of the 2.000 diameter hole as the base point. The absolute coordinate value of the center of the 0.750 arc "G" is
   - (A) 4.780,4.757
   - (B) 4.780,4.767
   - (C) 4.780,4.777
   - (D) 4.785,4.757
   - (E) 4.793,4.777

Problem 6-3
Plate1.Dwg

Directions for Plate1.Dwg

Use the Units command and set to decimal units. Set the number of digits to the right of the decimal point from 4 to 3. Be sure the system of angle measure is set to decimal degrees and the number of decimal places for the display of angles is zero (0). Accept the defaults for the remaining unit prompts. Use the Limits command and set the upper right corner of the screen area to a value of 36.000,24.000.

Begin this drawing by placing the center of the 4.000 diameter arc with keyway at coordinate (16.000,13.000).

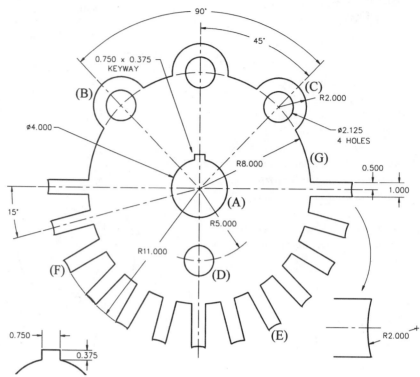

Refer to the drawing of Plate1 above to answer the following questions:

1. The distance from the center of the 2.000 radius arc "B" to the center of the 2.000 radius arc "C" is
   - (A) 10.286
   - (B) 10.293
   - (C) 11.300
   - (D) 11.307
   - (E) 11.314

2. The absolute coordinate value of the center of arc "C" is
   - (A) 21.657,18.657
   - (B) 21.657,18.664
   - (C) 21.657,18.671
   - (D) 21.657,18.678
   - (E) 21.657,18.685

3. The angle formed in the X-Y plane from the center of the 2.000 radius arc "C" to the center of the 2.125 diameter hole "D" is
   - (A) 242 degrees
   - (B) 244 degrees
   - (C) 246 degrees
   - (D) 248 degrees
   - (E) 250 degrees

4. The total length of arc "E" is
   - (A) 0.999
   - (B) 1.005
   - (C) 1.011
   - (D) 1.017
   - (E) 1.023

5. The total length of arc "C" is
   - (A) 6.766
   - (B) 6.772
   - (C) 6.778
   - (D) 6.784
   - (E) 6.790

6. The total area of Plate1 with all holes including keyway removed is
   - (A) 232.259
   - (B) 232.265
   - (C) 232.271
   - (D) 232.277
   - (E) 232.283

7. The distance from the center of the 2.000 radius arc "B" to the center of the 2.000 radius arc "E" is
   - (A) 20.732
   - (B) 20.738
   - (C) 20.744
   - (D) 20.750
   - (E) 20.756

Problem 6-4
Hanger.Dwg

Directions for Hanger.Dwg
Use the Units command set to decimal. Set the number of digits to the right of the decimal point from 4 to 3. Be sure the system of angle measure is set to decimal degrees and the number of decimal places for the display of angles is zero (0). Accept the defaults for the remaining unit prompts. Use the Limits command and set the upper right corner of the screen area to a value of 250.000,150.000.

Begin drawing the Hanger by locating the center of the 40.000 radius arc at coordinate (55.000,85.000).

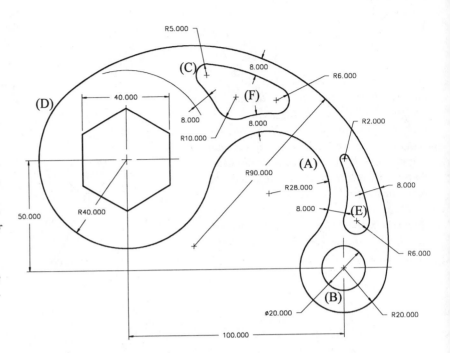

Refer to the drawing of the Hanger above to answer the following questions:

1. The area of the outer profile of the Hanger is closest to
   (A) 9970.567
   (B) 9965.567
   (C) 9975.567
   (D) 9975.005
   (E) 9980.347

2. The area of the Hanger with the polygon, circle, and irregular shapes removed is closest to
   (A) 7304.089
   (B) 7305.000
   (C) 7303.890
   (D) 7304.000
   (E) 7306.098

3. The absolute coordinate value of the center of the 28.000 radius arc "A" is closest to
   (A) 120.000,69.000
   (B) 121.082,68.964
   (C) 121.520,68.237
   (D) 121.082,69.000
   (E) 121.082,66.937

4. The absolute coordinate value of the center of the 20.000 diameter circle "B" is closest to
   (A) 154.000,35.000
   (B) 156.000,36.000
   (C) 156.147,35.256
   (D) 155.000,35.000
   (E) 156.000,37.000

5. The angle formed in the X-Y plane from the center of the 5.000 radius arc "C" to the center of the 40.000 radius arc "D" is
   (A) 219 degrees
   (B) 221 degrees
   (C) 223 degrees
   (D) 225 degrees
   (E) 227 degrees

6. The total area of irregular shape "E" is
   (A) 271.613
   (B) 271.723
   (C) 271.784
   (D) 271.801
   (E) 271.822

7. The total area of irregular shape "F" is
   (A) 698.511
   (B) 698.621
   (C) 699.817
   (D) 699.856
   (E) 699.891

Problem 6-5
Gasket1.Dwg

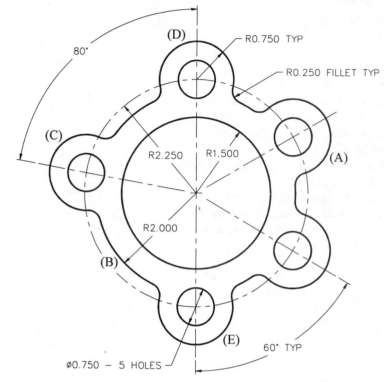

R0.750 TYP
R0.250 FILLET TYP
(D)
80°
(C)
R2.250 R1.500 (A)
R2.000
(B)
(E)
60° TYP
⌀0.750 − 5 HOLES

Directions for Gasket1.Dwg
Use the Units command to set the units to decimal. Set the number of digits to the right of the decimal point from 4 to 3. Be sure the system of angle measure is set to decimal degrees and the number of decimal places for the display of angles is zero (0). Keep the remaining default unit values.

Begin this drawing by locating the center of the 1.500 radius circle at coordinate (5.750,4.750).

Refer to the drawing of Gasket1 above to answer the following questions:

1. The total area of Gasket1 with the all holes removed is closest to
   - (A) 9.918
   - (B) 9.921
   - (C) 9.924
   - (D) 9.927
   - (E) 9.930

2. The absolute coordinate value of the center of the 0.750 radius arc "A" is closest to
   - (A) 7.669,5.875
   - (B) 7.669,5.870
   - (C) 7.666,5.875
   - (D) 7.699,5.875
   - (E) 7.699,5.975

3. The length of arc segment "B" is closest to
   - (A) 1.698
   - (B) 1.704
   - (C) 1.710
   - (D) 1.716
   - (E) 1.722

4. The angle formed in the X-Y plane from the center of the arc "C" to the center of arc "D" is closest to
   - (A) 30 degrees
   - (B) 35 degrees
   - (C) 40 degrees
   - (D) 45 degrees
   - (E) 50 degrees

5. The total length of the 0.750 radius arc "C" is
   - (A) 2.674
   - (B) 2.680
   - (C) 2.686
   - (D) 2.692
   - (E) 2.698

Use the Move command to reposition Gasket1 at a distance of 1.832 in the -45 degree direction. Use the center of the 1.500 radius circle as the base point of the move. After performing the operation, answer the following questions:

6. The absolute coordinate value of the center of the 1.500 radius circle is
   - (A) 7.045,3.437
   - (B) 7.045,3.443
   - (C) 7.045,3.449
   - (D) 7.045,3.455
   - (E) 7.045,3.461

7. The absolute coordinate value of the center of the 0.750 radius arc "C" is
   - (A) 4.806,3.845
   - (B) 4.812,3.845
   - (C) 4.818,3.845
   - (D) 4.824,3.845
   - (E) 4.830,3.845

Problem 6-6
Gasket2.Dwg

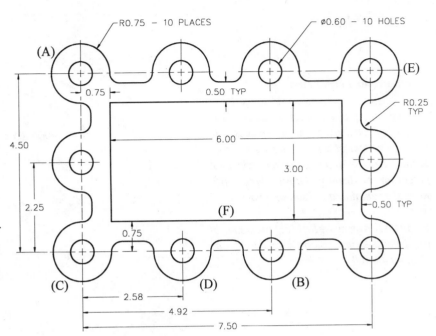

Directions for Gasket2.Dwg

Use the Units command to set the units to decimal. Set the number of digits to the right of the decimal point from 4 to 2. Be sure the system of angle measure is set to decimal degrees and the number of decimal places for the display of angles is zero (0). Keep the remaining default unit values.

Begin this drawing by locating the center of the 6.00 x 3.00 rectangle at coordinate (6.00, 4.75).

Refer to the drawing of Gasket2 above to answer the following questions:

1. The total surface area of Gasket2 with the rectangle and all ten holes removed is closest to
   - (A) 21.46
   - (B) 21.48
   - (C) 21.50
   - (D) 21.52
   - (E) 21.54

2. The distance from the center of arc "A" to the center of arc "B" is closest to
   - (A) 6.63
   - (B) 6.67
   - (C) 6.71
   - (D) 6.75
   - (E) 6.79

3. The length of arc segment "C" is closest to
   - (A) 3.44
   - (B) 3.47
   - (C) 3.50
   - (D) 3.53
   - (E) 3.56

4. The absolute coordinate value of the center of the 0.75 radius arc "D" is closest to
   - (A) 4.83,2.50
   - (B) 4.83,2.47
   - (C) 4.83,2.53
   - (D) 4.80,2.50
   - (E) 4.83,2.56

5. The angle formed in the X-Y plane from the center of the 0.75 radius arc "D" to the center of the 0.75 radius arc "A" is
   - (A) 116 degrees
   - (B) 118 degrees
   - (C) 120 degrees
   - (D) 122 degrees
   - (E) 124 degrees

6. The delta X,Y distance from the intersection at "E" to the midpoint of the line at "F" is
   - (A) -4.50,3.65
   - (B) -4.50,-3.65
   - (C) -4.50,-3.70
   - (D) -4.50,3.75
   - (E) -4.50,-3.75

7. Use the Scale command to reduce the size of the inner rectangle. Use the midpoint of the line at "F" as the base point. Use a scale factor of 0.83 units. The new total surface area with the rectangle and all ten holes removed is
   - (A) 26.99
   - (B) 27.04
   - (C) 27.09
   - (D) 27.14
   - (E) 27.19

Problem 6-7
Lever2.Dwg

Directions for Lever2.Dwg
Use the Units command to set the units to decimal. Keep the number of digits to the right of the decimal point at 4 places. Be sure the system of angle measure is set to decimal degrees and the number of decimal places for the display of angles is zero (0). Keep the remaining default unit values.

Begin this drawing by locating the center of the 1.0000 diameter circle at coordinate (2.2500,4.0000).

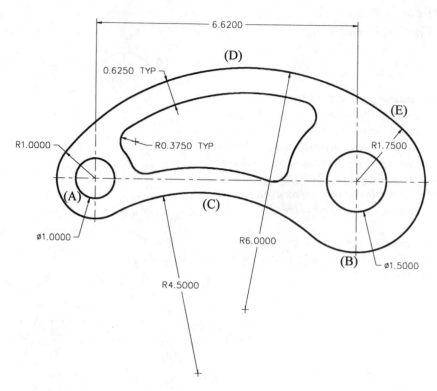

Refer to the drawing of Lever2 above to answer the following questions:

1. The total area of Lever2 with the inner irregular shape and both holes removed is
   - (A) 17.6813
   - (B) 17.6819
   - (C) 17.6825
   - (D) 17.6831
   - (E) 17.6837

2. The absolute coordinate value of the center of the 4.5000 radius arc "C" is
   - (A) 4.8944,-0.0822
   - (B) 4.8944,-0.8226
   - (C) 4.8950,-0.8232
   - (D) 4.8956,-0.8238
   - (E) 4.8962,-0.8244

3. The absolute coordinate value of the center of the 6.0000 radius arc "D" is
   - (A) 6.0828,0.7893
   - (B) 6.0834,0.7899
   - (C) 6.0840,0.7905
   - (D) 6.0846,0.7911
   - (E) 6.0852,0.7917

4. The total length of arc "C" is
   - (A) 5.3583
   - (B) 5.3589
   - (C) 5.3595
   - (D) 5.3601
   - (E) 5.3607

5. The distance from the center of the 1.0000 diameter circle "A" to the intersection of the circle and center line at "B" is closest to
   - (A) 6.8456
   - (B) 6.8462
   - (C) 6.8474
   - (D) 6.8480
   - (E) 6.8486

6. The angle formed in the X-Y plane from the upper quadrant of arc "D" to the center of the 1.5000 circle is
   - (A) 313 degrees
   - (B) 315 degrees
   - (C) 317 degrees
   - (D) 319 degrees
   - (E) 321 degrees

7. The delta X,Y distance from the upper quadrant of arc "C" to the center of the 1.0000 hole "A" is
   - (A) -2.6444,0.3220
   - (B) -2.6444,0.3226
   - (C) -2.6444,0.3232
   - (D) -2.6444,0.3238
   - (E) -2.6444,0.3244

Problem 6-8
Flange1.Dwg

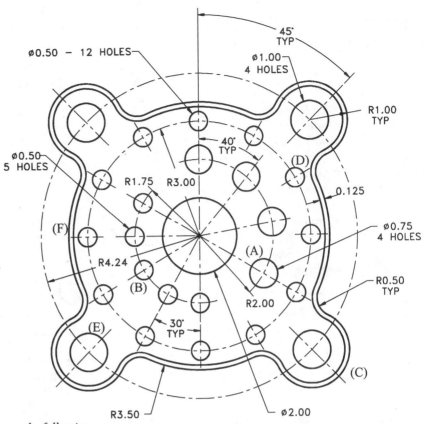

Directions for Flange1.Dwg
Use the Units command to set the units to decimal. Set the number of digits to the right of the decimal point from 4 to 2. Be sure the system of angle measure is set to decimal degrees and the number of decimal places for the display of angles is zero (0). Keep the remaining default unit values.

Begin this drawing by locating the center of the 2.00 diameter circle at coordinate (6.00,5.50).

Refer to the drawing of Flange1 above to answer the following questions:

1. The total area of the 0.125 strip around the perimeter of Flange1 is
   - (A) 4.10
   - (B) 4.12
   - (C) 4.14
   - (D) 4.16
   - (E) 4.18

2. The absolute coordinate value of the center of the 0.75 diameter circle "A" is
   - (A) 7.73,4.45
   - (B) 7.73,4.50
   - (C) 7.73,4.55
   - (D) 7.75,4.55
   - (E) 7.75,4.60

3. The absolute coordinate value of the center of the 0.50 diameter circle "B" is closest to
   - (A) 4.45,4.55
   - (B) 4.48,4.55
   - (C) 4.48,4.62
   - (D) 4.48,4.69
   - (E) 4.50,4.56

4. The length of the 1.00 radius arc "C" is
   - (A) 3.82
   - (B) 3.85
   - (C) 3.88
   - (D) 3.91
   - (E) 3.97

5. The total surface area of the inner part of Flange1 with all holes removed is closest to
   - (A) 35.12
   - (B) 35.18
   - (C) 35.24
   - (D) 35.30
   - (E) 35.36

6. The total length of outer arc "F" is
   - (A) 2.91
   - (B) 2.97
   - (C) 3.03
   - (D) 3.09
   - (E) 3.15

7. The angle formed in the X-Y plane from the center of the 0.50 hole "D" to the center of the 0.50 hole "B" is
   - (A) 204 degrees
   - (B) 206 degrees
   - (C) 208 degrees
   - (D) 210 degrees
   - (E) 212 degrees

Problem 6-9
Wedge.Dwg

Directions for Wedge.Dwg
Use the Units command to set the units to decimal. Set the number of digits to the right of the decimal point from 4 to 2. Be sure the system of angle measure is set to decimal degrees and the number of decimal places for the display of angles is zero (0). Keep all remaining default unit values.

Begin constructing the Wedge with vertex "A" located at coordinate (30,30).

Segment Lengths
AB=73
BC=34
CD=17
DE=93
EF=47
FG=20

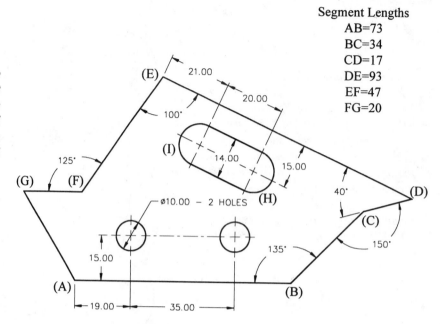

Refer to the drawing of Wedge above to answer the following questions:

1. The total area of the Wedge with the 2 holes and slot removed is
   - (A) 4367.97
   - (B) 4368.54
   - (C) 4370.12
   - (D) 4371.83
   - (E) 4374.91

2. The distance from the intersection of vertex "E" to the intersection of vertex "G" is closest to
   - (A) 60.72
   - (B) 60.74
   - (C) 60.80
   - (D) 60.85
   - (E) 60.87

3. The distance from the intersection of vertex "D" to the intersection of vertex "G" is closest to
   - (A) 131.00
   - (B) 131.12
   - (C) 131.24
   - (D) 131.36
   - (E) 131.48

4. The length of arc "H" is closest to
   - (A) 21.00
   - (B) 21.50
   - (C) 21.99
   - (D) 22.50
   - (E) 22.99

5. The overall height of the Wedge from the base of line "AB" to the peak at "E" is
   - (A) 60.72
   - (B) 65.87
   - (C) 67.75
   - (D) 69.08
   - (E) 71.98

6. The distance from the intersection of vertex "A" to the center of arc "I" is
   - (A) 61.09
   - (B) 61.67
   - (C) 61.98
   - (D) 62.93
   - (E) 63.02

7. The length of line "AG" is closest to
   - (A) 31.92
   - (B) 32.47
   - (C) 33.62
   - (D) 34.22
   - (E) 35.33

Problem 6-10
Pattern1.Dwg

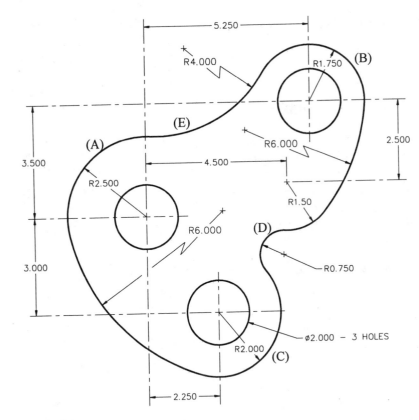

Directions for Pattern1.Dwg
Use the Units command to set the units to decimal. Set the number of digits to the right of the decimal point from 4 to 3. Be sure the system of angle measure is set to decimal degrees and the number of decimal places for the display of angles is zero (0). Keep all remaining default unit values. Use the Limits command and set the upper right corner of the limits to 16.000,12.000.

Begin this drawing by locating the center of the 2.500 radius arc at coordinate (4.250,5.750).

Refer to the drawing of Pattern1 above to answer the following questions:

1. The total surface area of Pattern1 with the 3 holes removed is
    - (A) 47.340
    - (B) 47.346
    - (C) 47.386
    - (D) 47.486
    - (E) 47.586

2. The distance from the center of the 2.500 radius arc "A" to the center of the 1.750 radius arc "B" is
    - (A) 6.310
    - (B) 6.315
    - (C) 6.210
    - (D) 6.321
    - (E) 6.305

3. The absolute coordinate value of the center of the 4.000 radius arc at "E" is
    - (A) 4.580,12.241
    - (B) 4.589,12.249
    - (C) 4.589,12.237
    - (D) 4.480,12.237
    - (E) 4.589,12.241

4. The perimeter of the outline of Pattern1 is
    - (A) 31.741
    - (B) 31.747
    - (C) 31.753
    - (D) 31.759
    - (E) 31.765

5. The total length of arc "A" is
    - (A) 4.633
    - (B) 4.639
    - (C) 4.645
    - (D) 4.651
    - (E) 4.657

6. The angle formed in the X-Y plane from the center of the 2.500 radius arc "A" to the center of the 2.000 radius arc "C" is
    - (A) 301 degrees
    - (B) 303 degrees
    - (C) 305 degrees
    - (D) 307 degrees
    - (E) 309 degrees

7. Use the Mirror command to flip but not duplicate Pattern1. Use the center of the 2.000 radius arc "C" as the first point of the mirror line. Use polar coordinate @1.000<90 as the second point. The new absolute coordinate value of the center of the 0.750 radius arc "D" is
    - (A) 4.378,4.504
    - (B) 4.382,4.504
    - (C) 4.386,4.504
    - (D) 4.390,4.504
    - (E) 4.394,4.504

Problem 6-11
Bracket1.Dwg

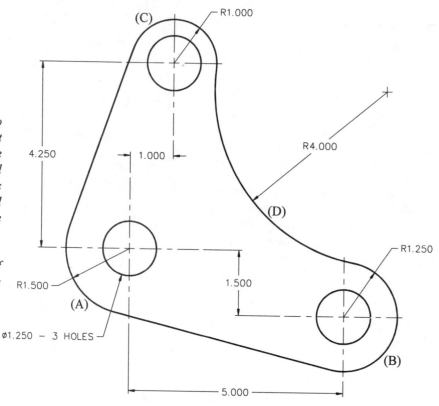

Directions for Bracket1.Dwg
Use the Units command to set the units to decimal. Set the number of digits to the right of the decimal point from 4 to 3. Be sure the system of angle measure is set to decimal degrees and the number of decimal places for the display of angles is zero (0). Keep all remaining default unit values. Keep the default values for the limits.

Begin this drawing by locating the center of the 1.500 radius arc "A" at coordinate (4.000,3.500).

Refer to the drawing of Bracket1 above to answer the following questions:

1. The distance from the center of the 1.500 radius arc "A" to the center of the 1.250 radius arc "B" is
    (A) 5.205
    (B) 5.210
    (C) 5.215
    (D) 5.220
    (E) 5.228

2. The distance from the center of the 1.500 radius arc "A" to the center of the 1.000 radius arc "C" is
    (A) 4.366
    (B) 4.370
    (C) 4.374
    (D) 4.378
    (E) 4.382

3. The distance from the center of the 1.250 radius arc "B" to the center of the 1.000 radius arc "C" is
    (A) 6.990
    (B) 6.995
    (C) 6.998
    (D) 7.000
    (E) 7.004

4. The length of arc "B" is
    (A) 3.994
    (B) 4.000
    (C) 4.006
    (D) 4.012
    (E) 4.018

5. The absolute coordinate value of the center of the 4.000 radius arc "D" is
    (A) 9.965,7.112
    (B) 9.965,7.250
    (C) 9.960,7.161
    (D) 9.965,7.161
    (E) 9.995,1.161

6. The total area of Bracket1 with all three 1.250 diameter holes removed is
    (A) 27.179
    (B) 27.187
    (C) 27.193
    (D) 27.198
    (E) 28.003

7. The angle formed in the X-Y plane from the center of the 1.250 radius arc "B" to the center of the 1.000 radius arc "C" is
    (A) 121 degrees
    (B) 123 degrees
    (C) 125 degrees
    (D) 127 degrees
    (E) 129 degrees

Problem 6-12
Lever3.Dwg

Directions for Lever3.Dwg
Begin the construction of Lever3 illustrated at the right by keeping the default units set to decimal but changing the number of decimal places past the zero from 4 to 2. Be sure the system of angle measure is set to decimal degrees and the number of decimal places for the display of angles is zero (0). Keep the remaining default unit values. Use the Limits command to change the drawing limits to 0.00,0.00 for the lower left corner and 15.00,12.00 for the upper right corner.

Begin this drawing by placing the center of the regular hexagon and 2.25 radius arc at coordinate (6.25,6.50).

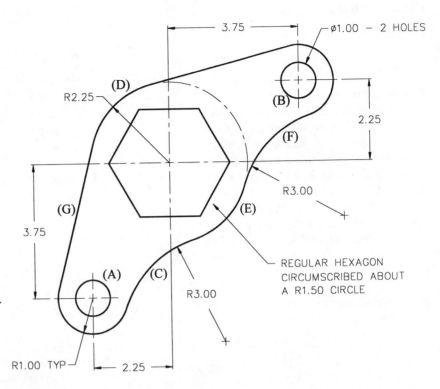

Refer to the drawing of Lever3 above to answer the following questions:

1. The distance from the center of the 1.00 diameter hole "A" to the center of the other 1.00 diameter hole "B" is
   - (A) 8.39
   - (B) 8.42
   - (C) 8.46
   - (D) 8.49
   - (E) 8.52

2. The absolute coordinate value of the center of the 3.00 radius arc "C" is
   - (A) 7.79,1.48
   - (B) 7.79,1.51
   - (C) 7.79,1.54
   - (D) 7.76,1.51
   - (E) 7.76,1.48

3. The total length of line "G" is
   - (A) 4.14
   - (B) 4.19
   - (C) 4.24
   - (D) 4.29
   - (E) 4.34

4. The length of the 2.25 radius arc segment "E" is
   - (A) 2.10
   - (B) 2.13
   - (C) 2.16
   - (D) 2.19
   - (E) 2.22

5. The angle formed in the X-Y plane from the center of the 1.00 diameter circle "B" to the center of the 2.25 radius arc "D" is closest to
   - (A) 203 degrees
   - (B) 205 degrees
   - (C) 207 degrees
   - (D) 209 degrees
   - (E) 211 degrees

6. The total surface area of Lever3 with the hexagon and both 1.00 diameter holes removed is
   - (A) 20.97
   - (B) 21.00
   - (C) 21.03
   - (D) 21.06
   - (E) 21.09

7. The total length of arc "D" is
   - (A) 2.31
   - (B) 2.36
   - (C) 2.41
   - (D) 2.46
   - (E) 2.51

Problem 6-13
Housing1.Dwg

Directions for Housing1.Dwg

Begin the construction of Housing1 illustrated at the right by keeping the default units set to decimal but change the number of decimal places past the zero from 4 to 3. Be sure the system of angle measure is set to decimal degrees and the number of decimal places for the display of angles is zero (0). Keep all remaining default unit values. Keep the default drawing limits set to 0.000,0.000 for the lower left corner and 12.000,9.000 for the upper right corner.

Place the center of the 1.500 radius circular center line at coordinate (6.500,5.250).

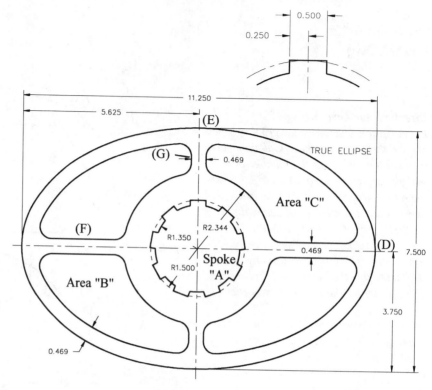

ALL FILLETS AND ROUNDS R0.375

Refer to the drawing of Housing1 above to answer the following questions:

1. The perimeter of Spoke "A" is closest to
   - (A) 11.564
   - (B) 11.570
   - (C) 11.576
   - (D) 11.582
   - (E) 11.588

2. The perimeter of Area "B" is closest to
   - (A) 12.513
   - (B) 12.519
   - (C) 12.525
   - (D) 12.531
   - (E) 12.537

3. The total area of Area "C" is closest to
   - (A) 7.901
   - (B) 7.907
   - (C) 7.913
   - (D) 7.919
   - (E) 7.927

4. The absolute coordinate value of the intersection of the ellipse and center line at "D" is
   - (A) 12.125,5.244
   - (B) 12.125,5.250
   - (C) 12.125,5.256
   - (D) 12.125,5.262
   - (E) 12.125,5.268

5. The total surface area of Housing1 with the spoke and all slots removed is
   - (A) 27.095
   - (B) 28.101
   - (C) 28.107
   - (D) 28.113
   - (E) 28.119

6. The distance from the midpoint of the horizontal line segment at "F" to the midpoint of the vertical line segment at "G" is
   - (A) 4.235
   - (B) 4.241
   - (C) 4.247
   - (D) 4.253
   - (E) 4.259

7. Increase Spoke "A" in size using the Scale command. Use the center of the 1.500 radius arc as the base point. Use a scale factor of 1.115 units. The new total area of Housing1 with the spoke and all slots removed is closest to
   - (A) 26.536
   - (B) 26.542
   - (C) 26.548
   - (D) 26.554
   - (E) 26.560

Problem 6-14
Cam1.Dwg

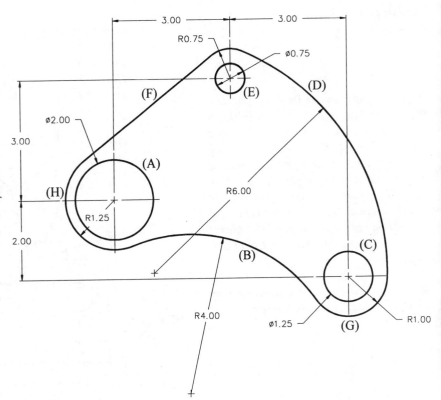

Directions for Cam1.Dwg

Start a new drawing called Cam1. Keep the default settings of decimal units but change the number of decimal places past the zero from 4 to 2. Be sure the system of angle measure is set to decimal degrees and the number of decimal places for the display of angles is zero (0). Keep all remaining default unit values. Keep the default limit settings at 0.00,0.00 by 12.00,9.00.

Begin this drawing by constructing the center of the 2.00 unit diameter circle at coordinate (3.00,4.00).

Refer to the drawing of Cam1 above to answer the following questions:

1. The absolute coordinate of the center of the 4.00 radius arc "B" is
   - (A) 4.92,0.89
   - (B) 4.92,-0.89
   - (C) 4.95,-0.91
   - (D) 4.92,-0.85
   - (E) -4.95,-0.85

2. The angle formed in the X-Y plane from the center of the 2.00 diameter circle "A" to the center of the 1.25 diameter circle "C" is
   - (A) 334 degrees
   - (B) 336 degrees
   - (C) 338 degrees
   - (D) 340 degrees
   - (E) 342 degrees

3. The length of arc "D" is
   - (A) 7.20
   - (B) 7.23
   - (C) 7.26
   - (D) 7.29
   - (E) 7.32

4. The total area of Cam1 with all three holes removed is
   - (A) 26.78
   - (B) 26.81
   - (C) 26.84
   - (D) 26.87
   - (E) 26.90

5. The total length of line "F" is
   - (A) 4.05
   - (B) 4.09
   - (C) 4.13
   - (D) 4.17
   - (E) 4.21

6. The delta X distance from the quadrant of the 1.00 radius arc at "G" to the quadrant of the 1.25 radius arc at "H" is
   - (A) -7.13
   - (B) -7.16
   - (C) -7.19
   - (D) -7.22
   - (E) -7.25

7. Use the Rotate command to re-align Cam1. Use the center of the 2.00 diameter circle "A" as the base point of the rotation. Rotate Cam1 from this point at a -10 degree angle. The absolute coordinate value of the center of the 0.75 diameter hole "E" is
   - (A) 6.48,6.37
   - (B) 6.48,6.40
   - (C) 6.48,6.43
   - (D) 6.51,6.46
   - (E) 6.54,6.49

Problem 6-15
Pattern4.Dwg

Directions for Pattern4.Dwg
Start a new drawing called Pattern4. Even though this is a metric drawing, no special limits need be set. Keep the default setting of decimal units but change the number of decimal places past the zero from 4 to 0, (zero). Be sure the system of angle measure is set to decimal degrees and the number of decimal places for the display of angles is zero (0). Keep all remaining default unit values.

Begin this drawing by constructing Pattern4 with vertex "A" at absolute coordinate (50,30).

Segment Lengths
AB = 94
BC = 40
CD = 35
DE = 57
EF = 82
FG = 61
GH = 38
HJ = 85
JK = 53

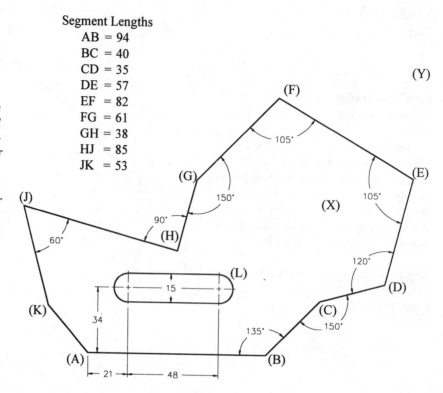

Refer to the drawing of Pattern4 above to answer the following questions:

1. The total distance from the intersection of vertex "K" to the intersection of vertex "A" is
   (A) 33
   (B) 34
   (C) 35
   (D) 36
   (E) 37

2. The total area of Pattern4 with the slot removed is
   (A) 14493
   (B) 14500
   (C) 14529
   (D) 14539
   (E) 14620

3. The perimeter of the outline of Pattern4 is
   (A) 570
   (B) 578
   (C) 586
   (D) 594
   (E) 602

4. The distance from the intersection of vertex "A" to the intersection of vertex "E" is
   (A) 186
   (B) 190
   (C) 194
   (D) 198
   (E) 202

5. The absolute coordinate value of the intersection at vertex "G" is
   (A) 104,117
   (B) 105,118
   (C) 106,119
   (D) 107,120
   (E) 108,121

6. The total length of arc "L" is
   (A) 20
   (B) 21
   (C) 22
   (D) 23
   (E) 24

7. Stretch the portion of Pattern4 around the vicinity of angle "E." Use "X" as the first corner of the crossing box. Use "Y" as the other corner. Use the endpoint of "E" as the base point of the stretching operation. For the new point, enter a polar coordinate value of 26 units in the 40 degree direction. The new degree value of the angle formed at vertex "E" is
   (A) 78 degrees
   (B) 79 degrees
   (C) 80 degrees
   (D) 81 degrees
   (E) 82 degrees

Problem 6-16
Rotor.Dwg

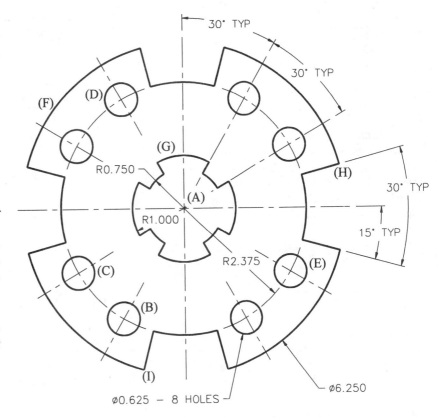

Directions for Rotor.Dwg
Start a new drawing called Rotor. Keep the default setting of decimal units but change the number of decimal places past the zero from 4 to 3. Be sure the system of angle measure is set to decimal degrees and the number of decimal places for the display of angles is zero (0). Keep all remaining default unit values. Keep the default limit settings at 0.000,0.000 for the lower left corner and 12.000,9.000 for the upper right corner.

Begin this drawing by constructing the center of the 6.250 unit diameter circle at "A" at coordinate (5.500,5.000).

Refer to the drawing of the Rotor above to answer the following questions:

1. The absolute coordinate value of the center of the 0.625 diameter circle "B" is closest to
   (A) 4.294,2.943
   (B) 4.300,2.943
   (C) 4.306,2.943
   (D) 4.312,2.943
   (E) 4.318,2.943

2. The total area of the Rotor with all 8 holes and the center slot removed is
   (A) 21.206
   (B) 21.210
   (C) 21.214
   (D) 21.218
   (E) 21.222

3. The total length of arc "F" is
   (A) 3.260
   (B) 3.264
   (C) 3.268
   (D) 3.272
   (E) 3.276

4. The distance from the center of the 0.625 circle "C" to the center of the 0.625 circle "D" is
   (A) 3.355
   (B) 3.359
   (C) 3.363
   (D) 3.367
   (E) 3.371

5. The angle formed in the X-Y plane from the center of the 0.625 circle "B" to the center of the 0.625 circle "E" is
   (A) 11 degrees
   (B) 13 degrees
   (C) 15 degrees
   (D) 17 degrees
   (E) 19 degrees

6. The delta X,Y distance from the intersection at "H" to the intersection at "I" is
   (A) -3.827,-3.827
   (B) -3.827,3.827
   (C) -3.834,3.820
   (D) -3.841,3.813
   (E) -3.848,-3.806

7. Use the Scale command to increase the size of just the center slot "G." Use the center of arc "F" as the base point. Use a scale factor of 1.500 units. The new surface area of the Rotor with the center slot and all 8 holes removed is
   (A) 17.861
   (B) 17.868
   (C) 17.875
   (D) 17.882
   (E) 17.889

Problem 6-17
Template.Dwg

Directions for Template.Dwg
Use the Units command to set the units to decimal. Set the number of digits to the right of the decimal point from 4 to 3. Change the system of degrees from "Decimal" to "Degrees/Minutes/Seconds." Keep the default values for the limits of the drawing.

Begin this drawing by locating Point "A" at coordinate (4.500,2.750).

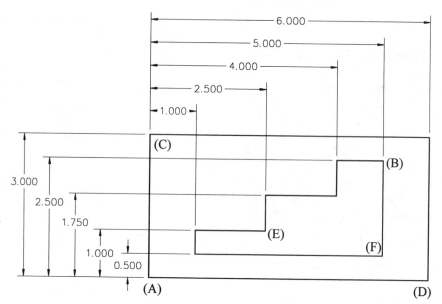

Refer to the drawing of the Template above to answer the following questions:

1. The angle formed in the X-Y plane from the intersection at "A" to the intersection at "B" is closest to
   (A) 26d25'21"
   (B) 26d28'3"
   (C) 26d30'10"
   (D) 26d32'37"
   (E) 26d33'54"

2. The distance from the intersection at "C" to the intersection at "D" is
   (A) 6.704
   (B) 6.708
   (C) 6.712
   (D) 6.716
   (E) 6.720

3. The angle formed in the X-Y plane from the intersection at "E" to the intersection at "F" is closest to
   (A) 347d31'10"
   (B) 347d47'32"
   (C) 348d1'53"
   (D) 348d20'20"
   (E) 348d41'24"

4. The total area of the Template with the step pattern removed is
   (A) 13.367
   (B) 13.371
   (C) 13.375
   (D) 13.379
   (E) 13.383

Using the Scale command to reduce the size of the Template. Use Point "A" as the base point and 0.950 as the scale factor and answer the following questions:

5. The perimeter of the inside step pattern is
   (A) 11.384
   (B) 11.388
   (C) 11.392
   (D) 11.396
   (E) 11.400

6. The new total area of the Template with the step pattern removed is
   (A) 12.059
   (B) 12.063
   (C) 12.067
   (D) 12.071
   (E) 12.075

7. The absolute coordinate value of the endpoint of the line at "B" is
   (A) 9.200,5.100
   (B) 9.250,5.125
   (C) 9.250,5.130
   (D) 9.260,5.135
   (E) 9.260,5.140

Problem 6-18
S-Cam.Dwg

Directions for S-Cam.Dwg
Begin the construction of S-Cam illustrated at the right by keeping the default units set to decimal but changing the number of decimal places past the zero from 4 to 3. Be sure the system of angle measure is set to decimal degrees and the number of decimal places for the display of angles is zero (0). Keep all remaining default unit values. Keep the default drawing limits set to 0.000,0.000 for the lower left corner and 12.000,9.000 for the upper right corner.

Begin the drawing by placing the center of the 1.750 diameter circle at coordinate (2.500,3.750).

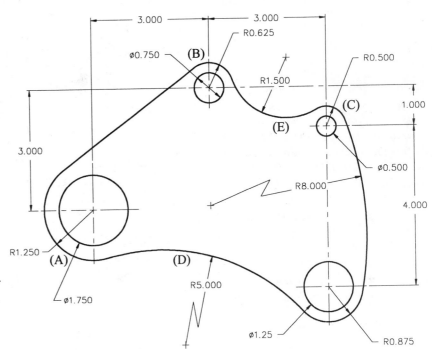

Refer to the drawing of the S-Cam above to answer the following questions:

1. The total surface area of the S-Cam with all 4 holes removed is
   (A) 27.654
   (B) 27.660
   (C) 27.666
   (D) 27.672
   (E) 28.678

2. The distance from the center of the 1.250 radius arc "A" to the center of the 0.625 radius arc "B" is
   (A) 4.237
   (B) 4.243
   (C) 4.249
   (D) 4.255
   (E) 4.261

3. The absolute coordinate value of the center of arc "D" is
   (A) 4.208,-2.258
   (B) 4.208,-2.262
   (C) 4.208,-2.266
   (D) 4.208,-2.270
   (E) 4.208,-2.274

4. The total length of arc "D" is
   (A) 5.480
   (B) 5.486
   (C) 5.492
   (D) 5.498
   (E) 6.004

5. The angle formed in the X-Y plane from the center of the 0.625 arc "B" to the center of the 0.500 arc "C" is closest to
   (A) 334 degrees
   (B) 336 degrees
   (C) 338 degrees
   (D) 340 degrees
   (E) 342 degrees

6. The delta X,Y distance from the center of the 0.500 arc "C" to the quadrant of the 5.000 arc "D" is
   (A) -4.284,-3.012
   (B) -4.288,-3.012
   (C) -4.292,-3.012
   (D) -4.296-3.012
   (E) -4.300,-3.012

7. Use the Scale command to reduce S-Cam in size. Use the center of the 1.250 radius arc "A" as the base point. Use a scale factor of 0.822 units. The new surface area of S-Cam with all four 4 removed is
   (A) 18.694
   (B) 18.698
   (C) 18.702
   (D) 18.706
   (E) 18.710

Problem 6-19
Housing2.Dwg

Directions for Housing2.Dwg

Begin the construction of Housing2 illustrated at the right by keeping the default units set to decimal but change the number of decimal places past the zero from 4 to 3. Be sure the system of angle measure is set to decimal degrees and the number of decimal places for the display of angles is zero (0). Keep all remaining default unit values. Keep the default drawing limits set to 0.000,0.000 for the lower left corner and 12.000,9.000 for the upper right corner.

Begin the drawing by placing the center of the ellipse at coordinate (5.500,4.500).

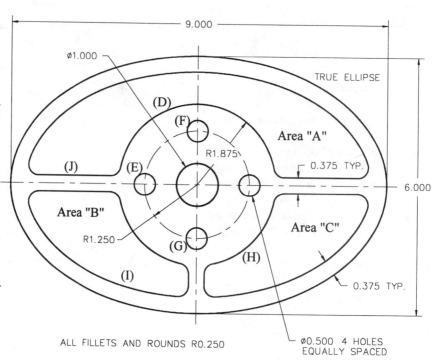

ALL FILLETS AND ROUNDS R0.250

Refer to the drawing of Housing2 above to answer the following questions:

1. The total area of Housing2 with all 5 holes and Areas "A," "B," and "C" removed is
   (A) 20.082
   (B) 20.088
   (C) 20.094
   (D) 20.100
   (E) 20.106

2. The perimeter of Area "A" is
   (A) 19.837
   (B) 19.843
   (C) 19.849
   (D) 19.855
   (E) 19.861

3. The length of arc "D" is
   (A) 5.107
   (B) 5.113
   (C) 5.119
   (D) 5.125
   (E) 5.131

4. The distance from the center of the 0.500 diameter hole "E" to the center of the 0.500 diameter hole "F" is
   (A) 1.762
   (B) 1.768
   (C) 1.774
   (D) 1.780
   (E) 1.786

5. The total length of arc "I" is
   (A) 4.374
   (B) 4.380
   (C) 4.386
   (D) 4.392
   (E) 4.398

6. The delta X,Y distance from the center of circle "G" to the midpoint of the horizontal line at "J" is
   (A) -2.943,1.438
   (B) 2.943,1.438
   (C) 2.937,-1.438
   (D) -2.931,1.431
   (E) -2.931,-1.425

7. Use the Move command to relocate Housing2 at a distance of 0.375 units in a 45 degree angle direction. The new absolute coordinate value of the center of the 0.500 diameter hole "G" is
   (A) 5.759,3.515
   (B) 5.765,3.503
   (C) 5.759,3.509
   (D) 5.765,3.509
   (E) 5.765,3.515

Problem 6-20
Pattern5.Dwg

Segment Lengths
AB = 94
BC = 40
CD = 35
DE = 57
EF = 82
FG = 61
GH = 73
HJ = 43

Directions for Pattern5.Dwg

Begin the construction of Pattern5 illus-trated at the right by keeping the default units set to decimal but change the number of decimal places past the zero from 4 to 0. Be sure the system of angle measure is set to decimal degrees and the number of decimal places for the display of angles is zero (0). Keep all remaining default unit values. Use the Limits command to set the drawing limits to 0,0 for the lower left corner and 250,200 for the upper right corner.

Begin the drawing by placing Vertex "A" at absolute coordinate (190,30).

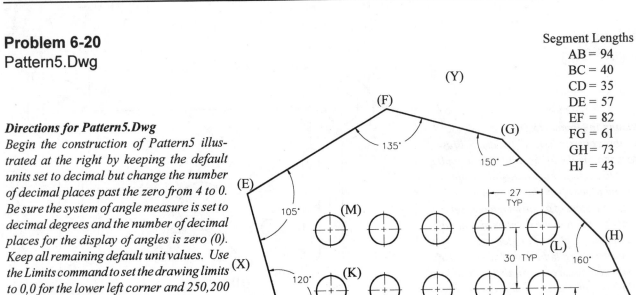

ø15 – 10 HOLES

Refer to the drawing of Pattern5 above to answer the following questions:

1. The distance from the intersection of vertex "J" to the intersection of vertex "A" is
   (A) 38
   (B) 39
   (C) 40
   (D) 41
   (E) 42

2. The perimeter of Pattern5 is
   (A) 523
   (B) 524
   (C) 525
   (D) 526
   (E) 527

3. The total area of Pattern5 with all 10 holes removed is
   (A) 16369
   (B) 16370
   (C) 16371
   (D) 16372
   (E) 16373

4. The distance from the center of the 15 diameter hole "K" to the center of the 15 diameter hole "L" is
   (A) 109
   (B) 110
   (C) 111
   (D) 112
   (E) 113

5. The angle formed in the X-Y plane from the center of the 15 diameter hole "M" to the center of the 15 diameter hole "N" is closest to
   (A) 340 degrees
   (B) 342 degrees
   (C) 344 degrees
   (D) 346 degrees
   (E) 348 degrees

6. The absolute coordinate value of the intersection at "F" is
   (A) 90,163
   (B) 92,163
   (C) 94,165
   (D) 96,167
   (E) 98,169

7. Use the Stretch command to lengthen Pattern5. Use "X" as the first point of the stretch crossing box. Use "Y" as the other corner. Pick the intersection at "F" as the base point and stretch Pattern5 a total of 23 units in the 180 direction. The new total area of Pattern5 with all 10 holes removed is
   (A) 17746
   (B) 17753
   (C) 17760
   (D) 17767
   (E) 17774

Problem 6-21
Bracket5.Dwg

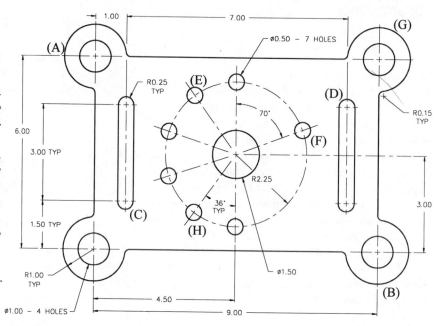

Directions for Bracket5.Dwg
Begin the construction of Bracket5 illustrated at the right by keeping the default units set to decimal but change the number of decimal places past the zero from 4 to 2. Be sure the system of angle measure is set to decimal degrees and the number of decimal places for the display of angles is zero (0). Keep all remaining default unit values. Use the Limits command to set the drawing limits to 0.00,0.00 for the lower left corner and 15.00,12.00 for the upper right corner.

Begin the drawing by placing the center of the 1.50 diameter hole at coordinate (7.50,5.75).

Refer to the drawing of Bracket5 above to answer the following questions:

1. The distance from the quadrant of the 1.00 radius arc at "A" to the quadrant of the 1.00 radius arc at "B" is
   - (A) 12.09
   - (B) 12.12
   - (C) 12.15
   - (D) 12.18
   - (E) 12.21

2. The distance from the center of the 0.25 radius arc at "C" to the center of the 0.25 radius arc at "D" is
   - (A) 7.62
   - (B) 7.65
   - (C) 7.68
   - (D) 7.71
   - (E) 7.74

3. The distance from the center of the 0.50 diameter circle at "E" to the center of the 0.50 diameter circle at "F" is
   - (A) 3.50
   - (B) 3.53
   - (C) 3.56
   - (D) 3.59
   - (E) 3.62

4. The length of arc "G" is
   - (A) 4.42
   - (B) 4.45
   - (C) 4.48
   - (D) 4.51
   - (E) 4.54

5. The total area of Bracket5 with all holes and slots removed is
   - (A) 53.72
   - (B) 53.75
   - (C) 53.78
   - (D) 53.81
   - (E) 53.84

6. The delta X,Y distance from the center of the 0.25 radius arc "C" to the center of the 0.50 circle "E" is
   - (A) 2.12,3.38
   - (B) 2.18,3.32
   - (C) 2.24,3.26
   - (D) 2.30,3.20
   - (E) 2.36,3.14

7. The angle formed in the X-Y plane from the center of the 0.25 radius arc "D" to the center of the 1.00 radius arc "G" is
   - (A) 50 degrees
   - (B) 52 degrees
   - (C) 54 degrees
   - (D) 56 degrees
   - (E) 58 degrees

Problem 6-22
Plate2A.Dwg

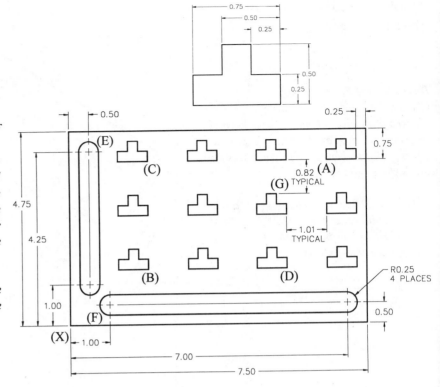

Directions for Plate2A.Dwg

Begin Plate2A by keeping the default units set to decimal but changing the number of decimal places past the zero from 4 to 2. Keep the system of angle measure set to decimal degrees but change the fractional places for display of angles to 2. Keep the remaining default unit values. Keep the default drawing limits set to 0.00,0.00 for the lower left corner and 12.00,9.00 for the upper right corner.

Begin constructing Plate2A by starting the lower left corner "X" at absolute coordinate (2.25,2.25).

Refer to the drawing of Plate2A above to answer the following questions:

1. The total perimeter of the horizontal slot is
   - (A) 13.42
   - (B) 13.47
   - (C) 13.52
   - (D) 13.57
   - (E) 13.62

2. The total perimeter of the vertical slot is
   - (A) 7.98
   - (B) 8.03
   - (C) 8.07
   - (D) 8.14
   - (E) 8.20

3. The total area of the horizonal slot is
   - (A) 3.02
   - (B) 3.08
   - (C) 3.14
   - (D) 3.20
   - (E) 3.26

4. The total area of the vertical slot is
   - (A) 1.82
   - (B) 1.88
   - (C) 1.94
   - (D) 2.00
   - (E) 2.06

5. The distance from the endpoint of the line at "C" to the endpoint of the line at "D" is
   - (A) 4.25
   - (B) 4.30
   - (C) 4.35
   - (D) 4.40
   - (E) 4.45

6. The angle formed in the X-Y plane formed from the center of arc "E" to the endpoint of the line at "B" is
   - (A) 296.90 degrees
   - (B) 296.96 degrees
   - (C) 297.02 degrees
   - (D) 297.08 degrees
   - (E) 297.14 degrees

7. The angle formed in the X-Y plane formed from the endpoint of the line at "A" to the center of arc "F" is
   - (A) 212.47 degrees
   - (B) 212.53 degrees
   - (C) 212.59 degrees
   - (D) 212.65 degrees
   - (E) 212.71 degrees

Problem 6-23
Plate2B.Dwg

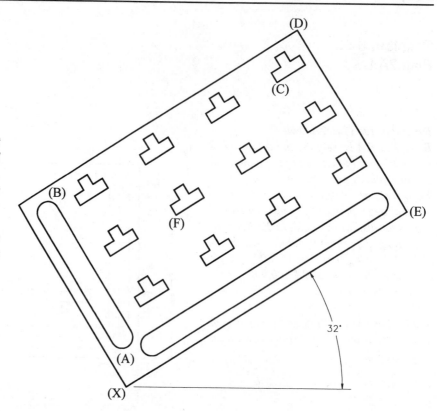

Directions for Plate2B.Dwg
Change the number of decimal places past
the zero from 2 to 3. Change from decimal
degrees to degrees/minutes/seconds.
Change the number of fractional places for
display of angles to 4.

Rotate Plate2B at a 32 degree, 0', 0" angle
using absolute coordinate (2.250,2.250) at
"X" below as the base point of rotation.
Perform a Zoom-Extents.

*Refer to the drawing of Plate2B above to answer the following
questions:*

1. The angle in the X-Y plane from the center of arc "A" to the
   center of arc "B" is
   - (A) 122d0'0"
   - (B) 122d20'0"
   - (C) 122d40'0"
   - (D) 122d40'30'
   - (E) 123d0'0"

2. The absolute coordinate value of the endpoint of the line at
   "C" is
   - (A) 5.627,9.071
   - (B) 5.631,9.075
   - (C) 5.635,9.079
   - (D) 5.639,9.083
   - (E) 5.643,9.087

3. The absolute coordinate value of the endpoint of the line at
   "D" is
   - (A) 6.077,10.237
   - (B) 6.081,10.241
   - (C) 6.085,10.245
   - (D) 6.089,10.249
   - (E) 6.093,10.253

4. The absolute coordinate value of the endpoint of the line at
   "E" is
   - (A) 8.606,6.220
   - (B) 8.610,6.224
   - (C) 8.614,6.228
   - (D) 8.618,6.232
   - (E) 8.622,6.236

5. The angle formed in the X-Y plane from the endpoint of the
   line at "F" to the center of arc "B" is
   - (A) 178d1'2"
   - (B) 178d27'37"
   - (C) 179d15'12"
   - (D) 179d39'49"
   - (E) 180d34'29"

6. The delta X,Y distance from the intersection at "D" to the
   center of the 0.250 radius arc "A" is
   - (A) -3.937,-6.878
   - (B) -3.941,-6.882
   - (C) -3.945,-6.886
   - (D) -3.949,-6.890
   - (E) -3.953,-6.894

7. Use the Scale command to reduce Plate2B in size. Use
   absolute coordinate 2.250,2.250 as the base point. Use
   0.777 as the scale factor. The area of Plate2B with both slots
   and the 12 T-shaped objects removed is
   - (A) 16.659
   - (B) 16.663
   - (C) 16.667
   - (D) 16.671
   - (E) 16.675

Problem 6-24
Ratchet.Dwg

Directions for Ratchet.Dwg

Use the Units command to change the number of decimal places past the zero from 4 to 2. Be sure the system of angle measure is set to decimal degrees and the number of decimal places for the display of angles is zero (0). Keep the remaining default unit values. Use the Limits command to set the upper right corner of the display screen to 13.00,10.00.

Begin by drawing the center of the 1.00 radius arc of the Ratchet at absolute coordinate (6.00,4.50).

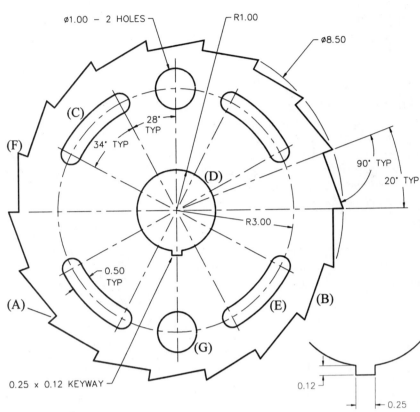

Refer to the drawing of the Ratchet above to answer the following questions:

1. The total length of the short line segment "A" is
   - (A) 0.22
   - (B) 0.24
   - (C) 0.26
   - (D) 0.28
   - (E) 0.30

2. The total length of line "B" is
   - (A) 1.06
   - (B) 1.19
   - (C) 1.33
   - (D) 1.45
   - (E) 1.57

3. The total length of arc "C" is
   - (A) 1.81
   - (B) 1.93
   - (C) 2.08
   - (D) 2.19
   - (E) 2.31

4. The perimeter of the 1.00 radius arc "D" with the 0.25 x 0.12 keyway is
   - (A) 6.52
   - (B) 6.63
   - (C) 6.77
   - (D) 6.89
   - (E) 7.02

5. The total surface area of the Ratchet with all 4 slots, the two 1.00 diameter holes, and the 1.00 radius arc with the keyway removed is
   - (A) 42.98
   - (B) 43.04
   - (C) 43.10
   - (D) 43.16
   - (E) 43.22

6. The absolute coordinate value of the endpoint at "F" is
   - (A) 2.01,5.10
   - (B) 2.01,5.63
   - (C) 2.01,5.95
   - (D) 2.20,5.95
   - (E) 2.37,5.95

7. The angle formed in the X-Y plane from the endpoint of the line at "F" to the center of the 1.00 diameter hole "G" is
   - (A) 304 degrees
   - (B) 306 degrees
   - (C) 308 degrees
   - (D) 310 degrees
   - (E) 312 degrees

Problem 6-25
Slide.Dwg

Directions for Slide.Dwg
Use the current unit settings and limits settings for this drawing. The number of decimal places past the zero should already be set to 4. Be sure the system of angle measure is set to decimal degrees and the number of decimal places for the display of angles is zero (0).

Begin by drawing the center of the 0.5000 diameter circle of the Slide at absolute coordinate (2.0000, 2.2500).

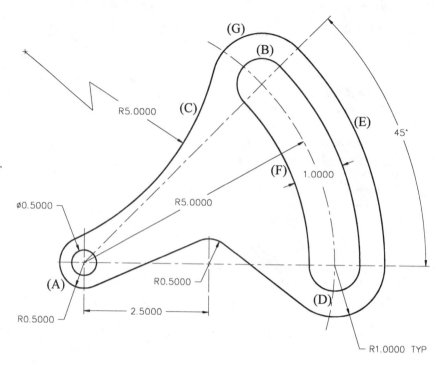

Refer to the drawing of the Slide above to answer the following questions:

1. The absolute coordinate value of the center of the 5.0000 radius arc "C" is
   (A) -0.2800,6.5603
   (B) 0.2887,-5.4393
   (C) -0.3953,7.2503
   (D) -0.2819,7.2543
   (E) -0.3953,6.2543

2. The total length of arc "E" is
   (A) 4.7264
   (B) 4.7124
   (C) 5.4302
   (D) 4.8710
   (E) 4.6711

3. The distance from the center of the 0.5000 diameter circle "A" to the center of the 5.0000 radius arc "C" is
   (A) 5.5000
   (B) 5.0043
   (C) 4.8768
   (D) 4.6691
   (E) 4.0001

4. The total area of the of the 1.0000 diameter slot is
   (A) 4.7124
   (B) 4.7863
   (C) 6.8370
   (D) 4.7625
   (E) 5.9102

5. The angle formed in the X-Y plane from the upper quadrant of the 1.0000 diameter slot at "B" to the lower quadrant of the slot at "D" is
   (A) 282 degrees
   (B) 284 degrees
   (C) 286 degrees
   (D) 288 degrees
   (E) 290 degrees

6. The total area of the Slide with the 0.5000 diameter hole and slot "F" removed is
   (A) 8.9246
   (B) 13.6370
   (C) 13.8750
   (D) 13.8333
   (E) 14.0297

7. The absolute coordinate value of the center of the 1.0000 radius arc "G" is
   (A) 5.5120,5.5621
   (B) 5.5237,5.5551
   (C) 5.5355,5.5590
   (D) 5.5355,5.6123
   (E) 5.5355,5.7855

Problem 6-26
Geneva.Dwg

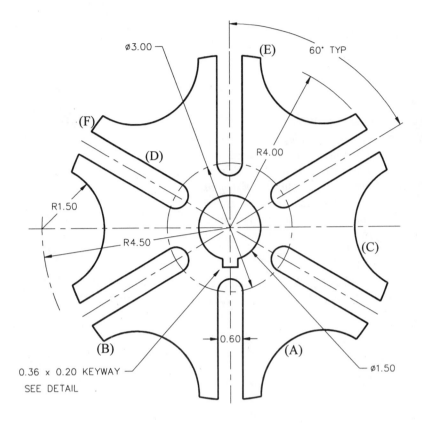

Ø3.00

(E)

60° TYP

(F)

(D)

R4.00

R1.50

R4.50

(C)

(B)

0.60

(A)

0.36 x 0.20 KEYWAY
SEE DETAIL

Ø1.50

Directions for Geneva.Dwg
Start a new drawing called Geneva. Keep the default settings of decimal units but change the number of decimal places past the zero from 4 to 2. Be sure the system of angle measure is set to decimal degrees and the number of decimal places for the display of angles is zero (0). Keep the remaining default unit values.

Begin the drawing by constructing the 1.50 diameter arc at absolute coordinate (7.50,5.50).

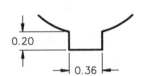

0.20

0.36

Refer to the drawing of the Geneva above to answer the following questions:

1. The total length of arc "A" is closest to
   - (A) 3.00
   - (B) 3.10
   - (C) 3.20
   - (D) 3.30
   - (E) 3.40

2. The angle formed in the X-Y plane from the intersection at "B" to the center of arc "C" is
   - (A) 11 degrees
   - (B) 13 degrees
   - (C) 15 degrees
   - (D) 17 degrees
   - (E) 19 degrees

3. The absolute coordinate value of the midpoint of line "D" is
   - (A) 5.27,7.13
   - (B) 5.27,7.17
   - (C) 5.23,7.13
   - (D) 5.31,7.13
   - (E) 5.31,7.09

4. The total area of the Geneva with the 1.50 diameter hole and keyway removed is closest to
   - (A) 27.20
   - (B) 27.30
   - (C) 27.40
   - (D) 27.50
   - (E) 27.60

5. The total distance from the midpoint of arc "F" to the center of arc "A" is
   - (A) 8.24
   - (B) 8.29
   - (C) 8.34
   - (D) 8.39
   - (E) 8.44

6. The delta X,Y distance from the intersection at "E" to the center of arc "C" is
   - (A) 3.93,-3.75
   - (B) 3.93,3.75
   - (C) 3.75,3.80
   - (D) 3.75,3.93
   - (E) 3.75,-3.93

7. Use the Scale command to reduce the Geneva in size. Use 7.50,5.50 as the base point; use a scale factor of 0.83 units. The absolute coordinate value of the intersection at "E" is
   - (A) 8.12,8.71
   - (B) 8.12,8.76
   - (C) 8.12,8.81
   - (D) 8.12,8.86
   - (E) 8.12,8.91

Problem 6-27
Rotor2.Dwg

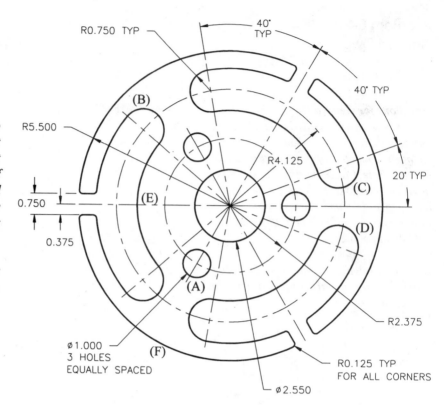

Directions for Rotor2.Dwg
Start a new drawing called Rotor2. Keep the default settings of decimal units but change the number of decimal places past the zero from 4 to 3. Be sure the system of angle measure is set to decimal degrees and the number of decimal places for the display of angles is zero (0). Keep all remaining default unit values.

Begin by constructing the 2.550 diameter circle at absolute coordinate (11.125,9.225).

Refer to the drawing of Rotor2 above to answer the following questions:

1. The absolute coordinate value of the center of hole "A" is closest to
   - (A) 9.937,7.158
   - (B) 9.937,7.163
   - (C) 9.937,7.168
   - (D) 9.942,7.168
   - (E) 9.947,7.173

2. The perimeter of Rotor2 is
   - (A) 81.850
   - (B) 81.855
   - (C) 81.860
   - (D) 81.865
   - (E) 81.870

3. The distance from the center of arc "B" to the center of arc "C" is
   - (A) 7.125
   - (B) 7.130
   - (C) 7.135
   - (D) 7.140
   - (E) 7.145

4. The absolute coordinate value of the center of arc "D" is closest to
   - (A) 15.001,7.814
   - (B) 15.001,7.819
   - (C) 15.001,7.824
   - (D) 15.006,7.829
   - (E) 15.011,7.834

5. The total length of arc "F" is
   - (A) 10.474
   - (B) 10.479
   - (C) 10.484
   - (D) 10.489
   - (E) 10.494

6. The total area of Rotor2 with all four holes removed is
   - (A) 54.902
   - (B) 54.907
   - (C) 54.912
   - (D) 54.917
   - (E) 54.922

7. Change the diameter of all three 1.000 diameter holes to 0.700 diameter. Change the diameter of the 2.550 hole to a new diameter of 1.625. The new area of Rotor2 with all four holes removed is
   - (A) 59.122
   - (B) 59.127
   - (C) 59.132
   - (D) 59.137
   - (E) 59.142

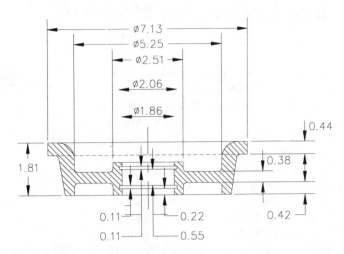

Region Modeling Techniques

Contents

Introducing the Region Modeler

As an alternate method of performing geometric constructions, a region modeler has been developed. A region model
represents a closed 2-dimensional shape that is treated as a
single entity similar to a block. As a result, such properties as
Area and Perimeter are associated with regions. In some cases,
the use of a region modeler may make geometric constructions
easier and less time consuming. The method of constructing
regions that makes it completely different from conventional
construction methods is the use of "boolean operations."
These operations of Union, Subtraction, and Intersection
allow entities to be joined together into a single entity or
subtracted from each other leaving a difference in both entities. Once a region is constructed, a separate mass property
utility is available to perform calculations such as area, perimeter, centroid, and even moments of inertia.

Illustrated at the right and below is a method of loading the
region modeler using the Appload command.

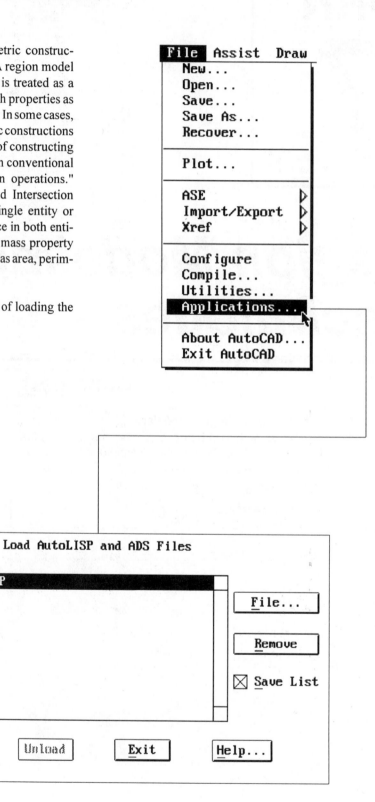

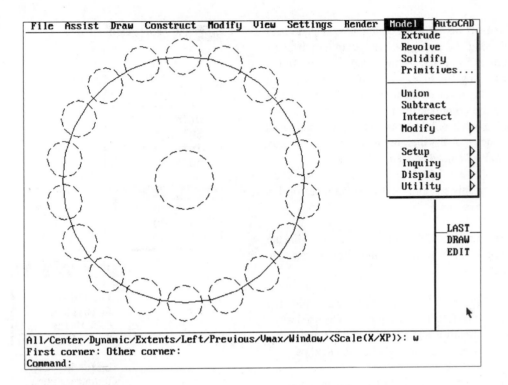

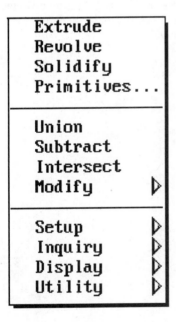

Illustrated above and to the right is one area used to obtain region modeling commands, namely the pulldown menu area. To display the dialog box illustrated at the right, pick "Model" from the pulldown menu. As this menu displays, notice that some commands appear more bolder than others. The commands identified in bold print are valid commands that can be used to create, edit, or analyze the region model.

Setting Up a Region Model

The "Setup" area of the "Model" dialog box activates the "Variables" submenu. Selecting "Variables" in turn displays the Region Modeler System Variables dialog box illustrated below. Use this dialog box to set values affecting the region modeler. This dialog box may also be displayed by entering DDSOLVAR at the command prompt from the keyboard. Areas of the dialog box in bold print are active areas supported by the region modeler. Areas that are greyed out support only the 3-dimensional Advanced Modeling Extension.

For the purposes of this section, the following areas of this dialog box will be explained:

Selecting "Units . . ." activates another dialog box allowing the individual to make changes to the current units used to create the region model.

Selecting "Hatch Patterns . . ." activates a dialog box allowing the individual to use different hatch patterns to be used when the region is automatically crosshatched. Both of these dialog boxes will be discussed in greater detail on the next page.

The "Other Parameters . . ." dialog box pertains mainly to system variables that primarily affect the 3-dimensional solid model created in the Advanced Modeling Extension.

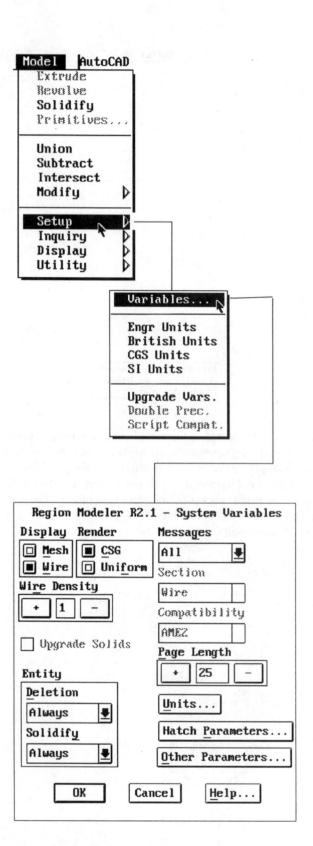

Selecting the "Units . . ." button from the Region Modeler System Variables dialog box displays the current units used when calculations are performed. Two measurements may be changed to reflect the desired units of the region model, namely Length and Area. The remaining two measurements of volume and mass are reserved for a 3-dimensional solid model constructed using the Advanced Modeling Extension. The default units for all region models are centimeters for length and square centimeters for area.

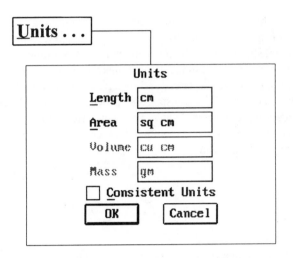

When changing from centimeters to other forms of units, enter "in" as the length unit value in place of "cm" as in the example at the right. To have the area reflect these new units, pick the check box labeled "Consistent Units" at the right. This will automatically update all units to the current value placed in the "Length" edit box.

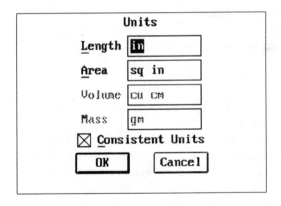

Selecting "Hatch Parameters . . ." from the "Region Modeler System Variables" dialog box displays the current hatch parameters. These parameters will be used to automatically hatch a region as it is being created. All supplied hatch patterns are supported in the region modeler. If the pattern "None" is substituted for the current pattern "U," no hatch pattern is applied to the region during the creation, editing, or analysis processes.

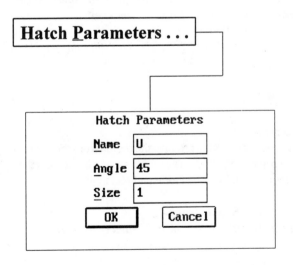

Region Modeling Construction Techniques

One of the most fundamental methods of creating regions is through the use of Constructive Solids Geometry. This is accomplished through the boolean operations of union, subtraction, and intersection. All three operations are explained below and utilize the circle and rectangle illustrated at the right. To aid in the construction process, the four lines representing the rectangle have been converted into a single polyline. The Rectang command automatically draws a rectangle as a polyline.

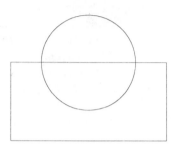
Individual Entities

Union
Use this command to join a group of entities into one complete region. Selecting the circle and rectangular polyline creates the region illustrated at the right.

Command: **Union**
Select objects: *(Select the circle and rectangle)*
Select objects: *(Strike Enter to create a union of both entities)*

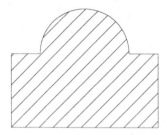
Creating a Region by Union

Subtraction
Use this command to subtract an entity or series of entities from a source object. To create the cove illustrated at the right, select the rectangle as the source object; the circle is subtracted from the rectangle to form the cove shape.

Command: **Subtract**
Source objects...
Select objects: *(Select the rectangle as the source object)*
Select objects: *(Strike Enter to continue)*
Objects to subtract from them...
Select objects: *(Select the circle. This will subtract it from the rectangle)*

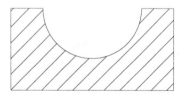
Creating a Region by Subtraction

Intersection
Use this command to produce the region from the intersection of two or more entities. The region shape illustrated at the right is common to both the circle and rectangle.

Command: **Solint**
Select objects: *(Select the circle and rectangle)*
Select objects: *(Strike Enter to create the intersection)*

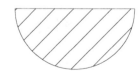
Creating a Region by Intersection

Analyzing a Region Model

Selecting "Mass Property . . ." from the "Inquiry" submenu of the "Model" pulldown menu area activates the "Mass Properties" dialog box illustrated below. This dialog displays various calculations based on a selected region. The following properties of the region selected are calculated and listed below:

Area
Perimeter
Bounding Box
Centroid
Moments of Inertia
Product of Inertia
Radii of Gyration
Principal Moments and X-Y directions about centroid

The "Mass Properties" dialog box may also be displayed by entering the command DDSOLMASSP from the keyboard. If Solmassp is entered from the keyboard and a region is selected, the mass properties are displayed in a text screen format.

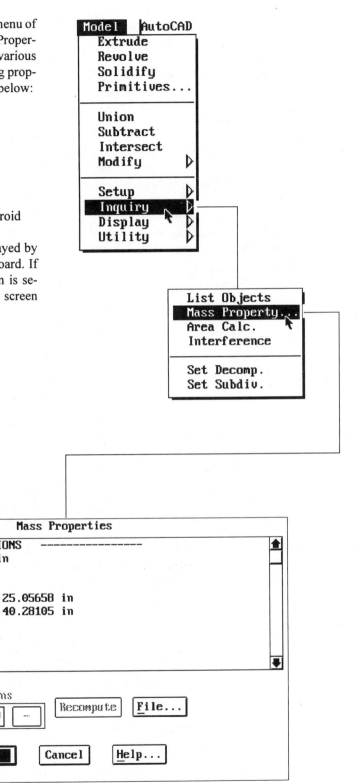

The following mass property calculations are explained below:

The Area option displays the enclosed area of the region.

The Perimeter represents the total length of the inside or outside of the region.

The Bounding Box is represented by a rectangular box that totally encloses the region. The box is identified by two diagonal points as in "A" and "B" in the illustration at the right.

The Centroid represents the center of the region's area. The location of the centroid is identified by a point entity placed on the current layer. The System Variable PDMODE controls the style of point displayed.

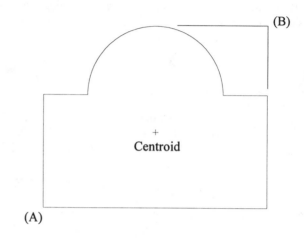

(B)

Centroid

(A)

If the "File . . ." button is selected from the "Mass Properties" dialog box, the following prompt appears:
            Write to a file? <N>:

Answering "Yes" displays the next prompt:
            File name <Default drawing name>:

After entering a file name, all mass property calculations are written out to a DOS text file similar to the illustration at the right. The text file that is created has the extension .Mpr.

File . . .

Area:	54.49189 sq cm
Perimeter:	30.81973 cm
Bounding box:	X: 46.34819 — 55.58921 cm
	Y: 0.8098652 — 8.173072 cm
Centroid:	X: 50.9687 cm
	Y: 3.874955 cm
Moments of inertia:	X: 1012.65 sq cm sq cm
	Y: 141884.2 sq cm sq cm
Product of inertia:	XY: 10762.22 sq cm sq cm
Radii of gyration:	X: 4.310859 cm
	Y: 51.02713 cm

Principal moments(sq cm sq cm) and X-Y directions about centroid:
            I: 194.4393 about [1 0]
            J: 324.771 about [0 1]

Editing a Region Model

The spoke assembly illustrated at the right was constructed as a region; it is now required to go through a design change; the 8.000 unit diameter hole in the center of the spoke needs to be changed to a new diameter of 4.000 units. Under normal circumstances, the Change command could have been used to change a circle to a new radius or diameter. However, since the 8.000 diameter circle is considered a primitive component of the total region, it is difficult to select just the circle using conventional editing commands. As a result, the Solchp command (Solid Change Property) is reserved for editing primitive shapes that belong to a region. Use this command to change the size or color of a primitive; you can also delete, copy, move, or replace a primitive.

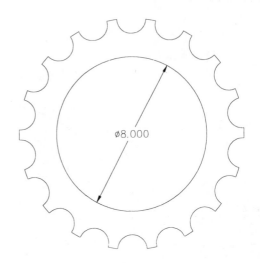

In the illustration at the right, the Solchp command is used to change the size of the large hole. After selecting the entire region followed by the desired primitive to change (the large circle at "A"), use the Size option of the command. Notice that the Motion Control System icon (MCS) appears giving you directions for such options as Move. After changing the size of the large hole, the icon disappears and the change is made.

Command: **Solchp**
Select a region: *(Select any part if the spoke illustrated at the right)*
Select primitive: *(Select the large circle at "A")*
Color/Delete/Evaluate/Instance/Move/Next/Pick/Replace/
Size/eXit <N>: **Size** *(The Motion Control System icon appears illustrated at the right)*
Radius of circle <4>: **2.000**
Color/Delete/Evaluate/Instance/Move/Next/Pick/Replace/
Size/eXit <N>: **X** *(To exit this command)*

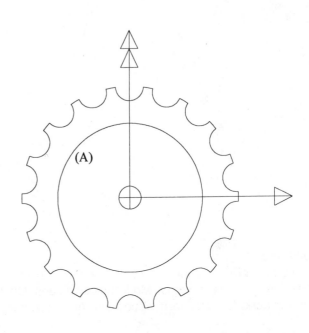

When the Solchp command is properly exited, the region updates to the latest changes such as the size of the large center circle changing from 8.000 to 4.000 units in diameter.

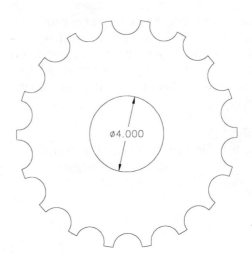

Tutorial Exercise #19
Region.Dwg

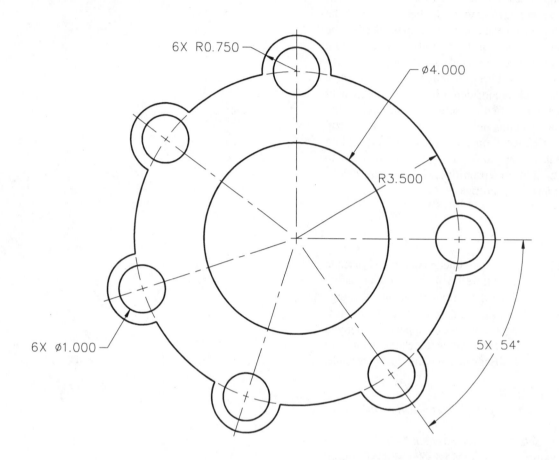

PURPOSE:
This tutorial is designed to construct a region of the Region above using Region Modeling techniques. Once the object is constructed, it will be analyzed for accuracy.

SYSTEM SETTINGS:
Begin a new drawing called "Region." Use the Units command to change the number of decimal places past the zero from 4 to 2. Keep the remaining default unit values. Keep the current limits set to 0,0 for the lower left corner and 12,9 for the upper right corner. The Grid or Snap commands do not need to be set to any certain values.

LAYERS:
Create the following layers with the format:
Name-Color-Linetype
Object - Green - Continuous

SUGGESTED COMMANDS:
Begin by locating the center of the large 4.000 diameter hole at absolute coordinate 6.000,4.500. Construct all necessary circles; use the Array command to create the circular pattern of 6 holes spaced 54 degrees away from each other. Begin joining the outer perimeter of the Region using the Union command. To create the holes as cutouts, use the Subtract command. Use the Solmassp or DDSOLMASSP commands to analyze the region. Edit the region by changing the size of the large inside hole and the six holes arranged in the circular pattern. Perform another analysis of the region.

DIMENSIONING:
This object does not have to be dimensioned.

PLOTTING:
Plot this drawing on a sheet of "B" size paper at a scale factor of 1=1.

Step #1

Construct two circles, one of radius 3.500 and the other of diameter 4.000. Use the center point of 6.000,4.500 for both circles. The "@" symbol is used for the center of the second circle to use the last known coordinate as the center.

Command: **Circle**
3P/2P/TTR/<Center point>: **6.000,4.500**
Diameter/<Radius>: **3.500**

Command: **Circle**
3P/2P/TTR/<Center point>: **@**
Diameter/<Radius>: **Diameter**
Diameter: **4.000**

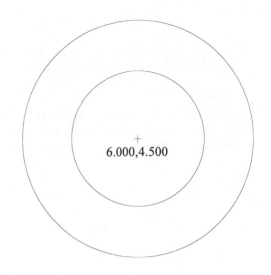

Step #2

Begin forming the lobes and holes of the region by constructing a circle of radius 0.750 and circle of diameter 1.00 from the upper quadrant of the large circle at "A" illustrated at the right.

Command: **Circle**
3P/2P/TTR/<Center point>: **Qua**
of *(Select the quadrant of the large circle at "A")*
Diameter/<Radius>: **0.750**

Command: **Circle**
3P/2P/TTR/<Center point>: **@**
Diameter/<Radius>: **Diameter**
Diameter: **1.000**

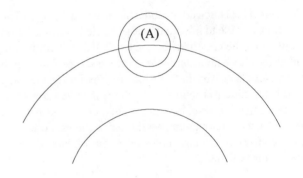

Step #3

Use the Array command to duplicate the last two circles six times in an included angle of 270 degrees. Since this is a positive angle, the direction of the array will be in the counterclockwise direction.

Command: **Array**
Select objects: *(Select the last two circles)*
Select objects: *(Strike Enter to continue)*
Rectangular or Polar array (R/P) <R>: **Polar**
Center point of array: **Cen**
of *(Select the edge of the circle at "A")*
Number of items: **6**
Angle between items (+=CCW, -=CW): **270**
Rotate object as they are copied? <Y>: *(Strike Enter to accept this default value)*

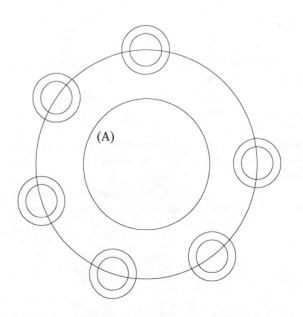

Step #4

Begin creating the region by using the Union command to join
the large circle with the 6 smaller 0.750 radius circles.

Command: **Union**
Select objects: *(Select the large dashed circle at the right)*
Select objects: *(Select the six dashed circles at the right)*
Select objects: *(Strike Enter to create a union of both entities)*

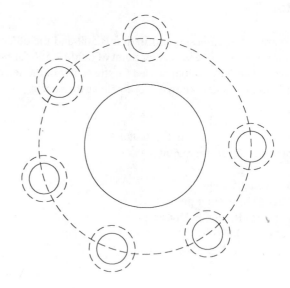

Step #5

The result of the Union command is illustrated at the right. The
smaller circles blend in with the large circle to form a region
identified by the crosshatching pattern. This pattern is controlled by the Solhpat variable. The spacing of the hatch
pattern is controlled by the Solhsize variable. Finally the angle
of the hatch pattern is controlled by the Solhangle variable. If
no hatch pattern is desirable, set the Solhpat variable to none.
Notice that the hatch pattern goes through all holes. Follow the
next step to subtract these holes from the original region to
form a new region.

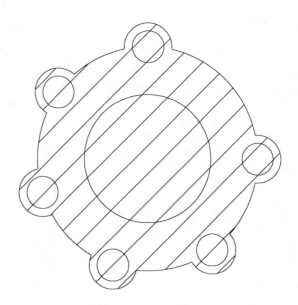

Step #6

Use the Subtract command to subtract the 7 dashed holes at the
right from the region.

Command: **Subtract**
Source objects...
Select objects: *(Select the region at "A" as the source object)*
Select objects: *(Strike Enter to continue)*
Objects to subtract from them...
Select objects: *(Select all seven dashed circles illustrated at
the right. This will subtract them from the original region)*

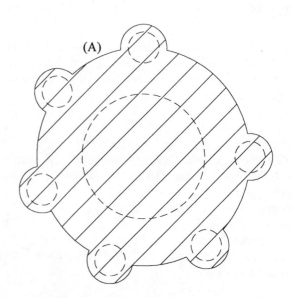

Step #7

As the holes subtract from the first region, a new region illustrated at the right is formed. The same hatch pattern, hatch size, and hatch angle are used on the new region.

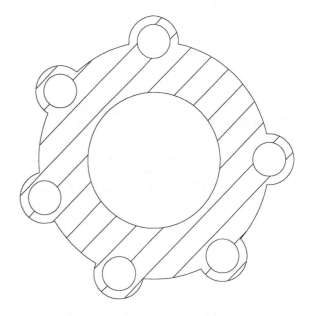

Step #8

Follow the prompts below to set the hatch pattern to none, which will display the region similar to the illustration at the right.

Command: **Solhpat**
Hatch pattern <U>: **None**

Once the hatch pattern is displayed and the Solhpat variable is set to none, only boolean operations or solid editing operations update the region to the new hatch pattern. The use of the Regen command to update the region to the current hatch pattern is ineffective. The hatch pattern assigned to the region model will disappear after completing the editing operation in Step #10 on page 445.

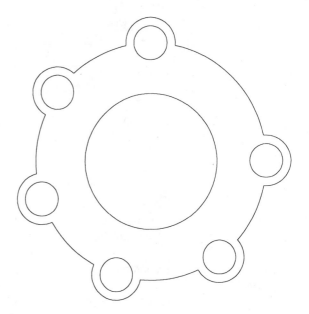

Step #9

To perform such calculations on a region, the DDSOLMASSP or Solmassp commands are used. These commands are very useful to extract the area, perimeter, bounding box, and centroid of a region. Other information such as Moments of inertia, Product of inertia, Radii of gyration, and Principal moments are also calculated. The differences in the two commands lie in the methods they display this information. The DDSOLMASSP command displays all calculations in a dialog box illustrated just below. The Solmassp command provides the same information about the region; it displays its information in text form. Once the mass properties of a region are found, the centroid is automatically marked by the current point mode. This is displayed as the "+" sign at the right signifying a Pdmode of 2.

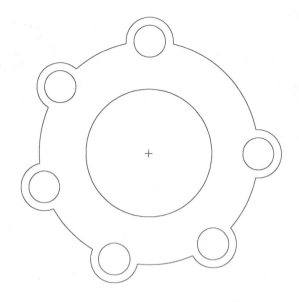

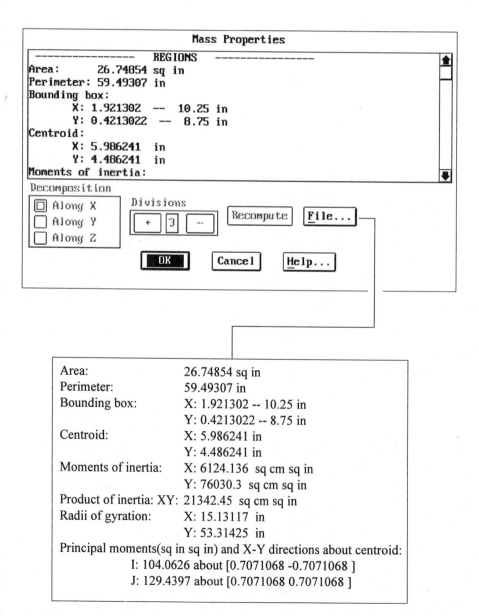

Area:	26.74854 sq in
Perimeter:	59.49307 in
Bounding box:	X: 1.921302 -- 10.25 in
	Y: 0.4213022 -- 8.75 in
Centroid:	X: 5.986241 in
	Y: 4.486241 in
Moments of inertia:	X: 6124.136 sq cm sq in
	Y: 76030.3 sq cm sq in
Product of inertia: XY:	21342.45 sq cm sq in
Radii of gyration:	X: 15.13117 in
	Y: 53.31425 in

Principal moments(sq in sq in) and X-Y directions about centroid:

I: 104.0626 about [0.7071068 -0.7071068]

J: 129.4397 about [0.7071068 0.7071068]

Step #10

A change has occurred in the original design of the region. The large 4.000 diameter hole needs to be changed to a new diameter of 5.000 units. The Solchp command along with the Size option will be used to perform this task.

Command: **Solchp**
Select a region: *(Select the region at "A")*
Select primitive: *(Select the 4.000 unit diameter hole at "B")*
Color/Delete/Evaluate/Instance/Move/Next/Pick/Replace/ Size/eXit <N>: **Size** *(The Motion Control System icon will appear)*
Radius of circle <2>: **2.500**
Color/Delete/Evaluate/Instance/Move/Next/Pick/Replace/ Size/eXit <N>: **X** *(To exit this command)*

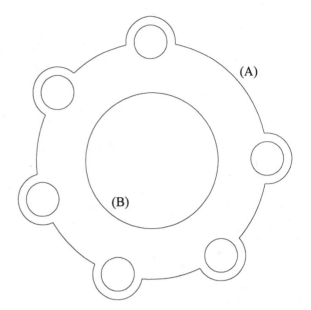

Step #11

Notice at the right that the 4.000 diameter hole is changed to the new diameter of 5.000 when the Solchp command is properly exited. Another change must be made to the region; all six 1.000 diameter holes must be changed to a new diameter of 0.750. The Solchp command is first made to change one hole. Then a copy or instance of the hole is made before it is arrayed and used to replace all existing holes. Follow the prompts below to perform this step:

Command: **Solchp**
Select a region: *(Select the region at "A")*
Select primitive: *(Select the 1.000 unit diameter hole at "B")*
Color/Delete/Evaluate/Instance/Move/Next/Pick/Replace/ Size/eXit <N>: **Size** *(The Motion Control System icon will appear)*
Radius of circle <0.5>: **0.375**
Color/Delete/Evaluate/Instance/Move/Next/Pick/Replace/ Size/eXit <N>: **Instance** *(To create a copy of the new primitive)*
Color/Delete/Evaluate/Instance/Move/Next/Pick/Replace/ Size/eXit <N>: **X** *(To exit this command)*

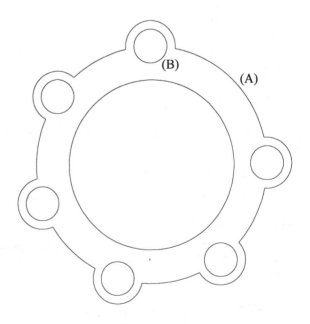

After performing the Instance option of the Solchp command, two holes now exist at location "B".

Step #12

Use the Array command to copy the last entity six times in an included angle of 270 degrees. Use the Window option to select only the copy of the hole using a selection box from "A" to "B." After performing the array, each circle represents six individual regions which will replace the existing 1.000 diameter holes. Follow the prompts below to perform this operation:

Command: **Array**
Select objects: **W**
First corner: *(Pick a point at "A")*
Other corner: *(Pick a point at "B" to select the last circle which was copied using the Instance option of Solchp)*
Select objects: *(Strike Enter to continue)*
Rectangular or Polar array (R/P) <R>: **Polar**
Center point of array: **Cen**
of *(Select the edge of the circle at "C")*
Number of items: **6**
Angle between items (+=CCW, -=CW): **270**
Rotate object as they are copied? <Y>: *(Strike Enter to accept this default value)*

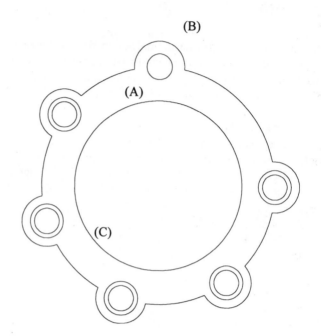

Step #13

Rather than change the individual size of each 1.000 diameter hole, the Replace option of the Solchp command will be used.

Command: **Solchp**
Select a region: *(Select the region at "A")*
Select primitive: *(Select the 1.000 unit diameter hole at "B")*
Color/Delete/Evaluate/Instance/Move/Next/Pick/Replace/
Size/eXit <N>: **Replace**
Select region to replace primitive: *(Select the inside region at "C")*
Retain detached primitive? <N>: *(Strike Enter to accept this default value)*
Color/Delete/Evaluate/Instance/Move/Next/Pick/Replace/
Size/eXit <N>: **Pick**
Select primitive: *(Select the next 1.000 diameter hole)*

Continue using the Replace option of Solchp until all 1.000 diameter holes at "A", "B", "C", and "D" are changed to the new diameter of 0.750. Exit the Solchp command to update all hole changes.

Color/Delete/Evaluate/Instance/Move/Next/Pick/Replace/
Size/eXit <N>: **X** *(To exit this command)*

Two holes still exist at "E". Use the Erase command and the Window option to window in the hole to erase it. Selecting the hole may select the entire region.

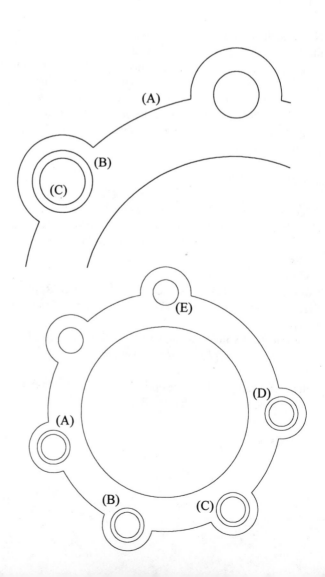

Step #14

Use the DDSOLMASSP command to perform a mass property calculation on the new region. Compare the differences in the Area, Perimeter, and Centroid in the dialog box below with the results located in the dialog box on page 444 on the region before the changes were made. Create a text file of the mass property calculations similar to the illustration below.

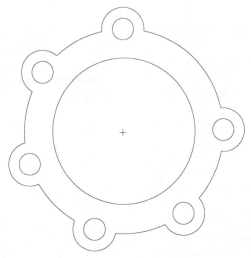

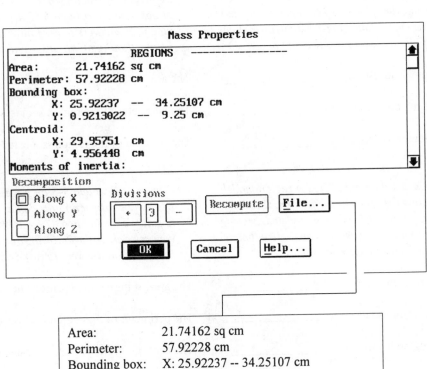

Area: 21.74162 sq cm
Perimeter: 57.92228 cm
Bounding box: X: 25.92237 -- 34.25107 cm
                Y: 0.9213022 — 9.25 cm
Centroid: X: 29.95751 cm
                Y: 4.956448 cm
Moments of inertia: X: 645.5438 sq cm sq cm
                         Y: 62286.42 sq cm sq cm
Product of inertia: XY: 5748.426 sq cm sq cm
Radii of gyration: X: 5.449001 cm
                         Y: 53.52426 cm
Principal moments(sq cm sq cm) and X-Y directions about centroid:
                I: 97.17697 about [0.7071068 -0.7071068]
                J: 125.6846 about [0.7071068 0.7071068]

Problems for Unit 7

1. *Construct Problem 3-1 on page 202 as a region model. Begin by constructing the center of the 1.50 radius circle at absolute coordinate 6.50,4.25. When completed, provide your answers to the following questions on the lines below. Round off all answers to two decimal places.*

The area of the region is closest to: _____

The perimeter of the region is closest to:_____

The centroid of the region is located at: _____

2. *Construct Problem 3-2 on page 202 as a region model. Begin by constructing the center of the 6.00 X 3.00 rect-angle at absolute coordinate 6.54,4.21. When completed, provide your answers to the following questions on the lines below. Round off all answers to two decimal places.*

The area of the region is closest to: _____

The perimeter of the region is closest to:_____

The centroid of the region is located at: _____

3. *Construct Problem 3-3 on page 203 as a region model. Begin by constructing the center of the 1.875 radius arc at absolute coordinate 6.36,4.61. When completed, provide your answers to the following questions on the lines below. Round off all answers to two decimal places.*

The area of the region is closest to: _____

The perimeter of the region is closest to:_____

The centroid of the region is located at: _____

4. *Construct Problem 3-4 on page 203 as a region model. Begin by constructing the center of the 1.00 radius arc at absolute coordinate 7.12,5.38. When completed, provide your answers to the following questions on the lines below. Round off all answers to two decimal places.*

The area of the region is closest to: _____

The perimeter of the region is closest to:_____

The centroid of the region is located at: _____

5. *Construct Problem 3-5 on page 204 as a region model. Begin by constructing the center of the 2.25 radius arc at absolute coordinate 8.40,5.90. When completed, provide your answers to the following questions on the lines below. Round off all answers to two decimal places.*

The area of the region is closest to: _____

The perimeter of the region is closest to:_____

The centroid of the region is located at: _____

6. *Construct Problem 3-6 on page 204 as a region model. Begin by constructing the center of the 0.38 diameter circle at absolute coordinate 6.67,4.19. When completed, pro-vide your answers to the following questions on the lines below. Round off all answers to two decimal places.*

The area of the region is closest to: _____

The perimeter of the region is closest to:_____

The centroid of the region is located at: _____

7. *Construct Problem 3-7 on page 204 as a region model. Begin by constructing the center of the 1.0 radius arc at absolute coordinate 6.87,8.42. When completed, provide your answers to the following questions on the lines below. Round off all answers to two decimal places.*

The area of the region is closest to: _____

The perimeter of the region is closest to:_____

The centroid of the region is located at: _____

8. *Construct Problem 3-8 on page 205 as a region model. Begin by constructing the center of the 25 radius arc at absolute coordinate 92,45. When completed, provide your answers to the following questions on the lines provided below. Round off all answers to two decimal places.*

The area of the region is closest to: _____

The perimeter of the region is closest to:_____

The centroid of the region is located at: _____

9. *Construct Problem 3-9 on page 205 as a region model. Begin by constructing the center of the 0.53 radius arc at absolute coordinate 1.156,3.594. When completed, provide your answers to the following questions on the lines below. Round off all answers to two decimal places.*

The area of the region is closest to: _____

The perimeter of the region is closest to:_____

The centroid of the region is located at: _____

10. *Construct Problem 3-10 on page 205 as a region model. Begin by constructing the center of the 2.50 diameter arc at absolute coordinate 12.70,9.29. When completed, provide your answers to the following questions on the lines below. Round off all answers to two decimal places.*

The area of the region is closest to: _____

The perimeter of the region is closest to:_____

The centroid of the region is located at: _____

11. *Construct Problem 3-11 on page 206 as a region model. Begin by constructing the center of the 3.0 diameter arc at absolute coordinate 4.74,4.19. When completed, provide your answers to the following questions on the lines below. Round off all answers to two decimal places.*

The area of the region is closest to: _____

The perimeter of the region is closest to:_____

The centroid of the region is located at: _____

12. *Construct Problem 3-12 on page 206 as a region model. Begin by constructing the center of the 50 diameter arc at absolute coordinate 100,90. When completed, provide your answers to the following questions on the lines below. Round off all answers to two decimal places.*

The area of the region is closest to: _____

The perimeter of the region is closest to:_____

The centroid of the region is located at: _____

13. *Construct Problem 3-13 on page 206 as a region model. Begin by constructing the center of the 3.12 radius arc at absolute coordinate 7.69,2.55. When completed, provide your answers to the following questions on the lines below. Round off all answers to two decimal places.*

The area of the region is closest to: _____

The perimeter of the region is closest to:_____

The centroid of the region is located at: _____

14. *Construct Problem 3-14 on page 206 as a region model. Begin by constructing the center of the 38 diameter circle at absolute coordinate 51,64. When completed, provide your answers to the following questions on the lines provided below. Round off all answers to two decimal places.*

The area of the region is closest to: _____

The perimeter of the region is closest to:_____

The centroid of the region is located at: _____

15. *Construct Problem 3-15 on page 207 as a region model. Begin by constructing the center of the 2.0 radius center arc at absolute coordinate 4.22,5.95. When completed, provide your answers to the following questions on the lines below. Round off all answers to two decimal places.*

The area of the region is closest to: _____

The perimeter of the region is closest to:_____

The centroid of the region is located at: _____

16. *Construct Problem 3-16 on page 207 as a region model. Begin by constructing the center of the leftmost 1.38 major diameter ellipse at absolute coordinate 4.49,3.28. When completed, provide your answers to the following questions on the lines below. Round off all answers to two decimal places.*

The area of the region is closest to: _____

The perimeter of the region is closest to:_____

The centroid of the region is located at: _____

17. *Construct Problem 3-17 on page 207 as a region model. Begin by constructing the center of the 14 unit hexagon at absolute coordinate 141,77. When completed, provide your answers to the following questions on the lines below. Round off all answers to two decimal places.*

The area of the region is closest to: _____

The perimeter of the region is closest to:_____

The centroid of the region is located at: _____

18. *Construct Problem 3-18 on page 207 as a region model. Begin by constructing the center of the 3.00 diameter circle at absolute coordinate 12.13,8.87. When completed, provide your answers to the following questions on the lines below. Round off all answers to two decimal places.*

The area of the region is closest to: _____

The perimeter of the region is closest to:_____

The centroid of the region is located at: _____

19. *Construct Problem 3-19 on page 208 as a region model. Begin by constructing the center of the 4.18 diameter arc at absolute coordinate 6.34,6.48. When completed, provide your answers to the following questions on the lines below. Round off all answers to two decimal places.*

The area of the region is closest to: _____

The perimeter of the region is closest to:_____

The centroid of the region is located at: _____

20. *Construct Problem 3-20 on page 208 as a region model. Begin by constructing the center of the 1.50 radius arc at 6.40,10.73. When completed, provide your answers to the following questions on the lines provided below. Round off all answers to two decimal places.*

The area of the region is closest to: _____

The perimeter of the region is closest to:_____

The centroid of the region is located at: _____

21. *Construct Problem 3-21 on page 208 as a region model. Begin by constructing the center of the 40 radius arc at absolute coordinate 150,69. When completed, provide your answers to the following questions on the lines provided below. Round off all answers to two decimal places.*

The area of the region is closest to: _____

The perimeter of the region is closest to:_____

The centroid of the region is located at: _____

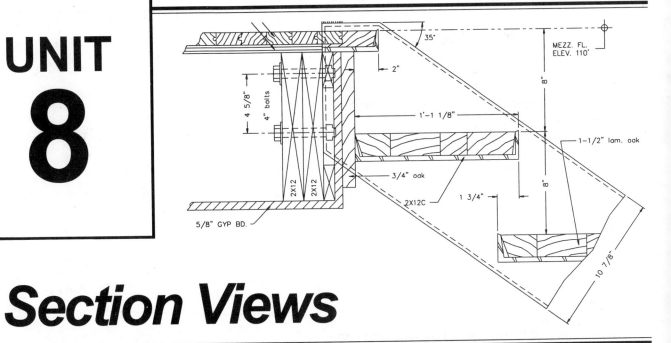

Section Views

Contents

Principles of orthographic projections remain the key method for the production of engineering drawings, whether using manual methods or CAD. As these drawings get more complicated in nature, the job of the operator or designer becomes more difficult in the interpretation of views, especially where hidden features are involved. The concept of slicing a view to expose these interior details is the purpose of performing a section. Section views then follow the same rules as orthographic or multi-view drawings except that the creation of a section makes the drawing easier to read since hidden features are converted to visible features. In this unit you will learn how sections are formed in addition to the many types of sections available to the designer. Three tutorial exercises at the end of the unit are designed to give you experience using two methods of crosshatching when using AutoCAD as a drafting tool.

Section View Basics

The illustration at the right is a pictorial representation of a typical flange consisting of eight bolt holes and counterbore hole in the center.

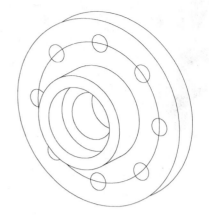

The drawings shown at the right show a typical solution to a multi-view problem complete with front and side views. The front view displaying the eight bolt holes is obvious to interpret; however, the numerous hidden lines in the side view make the drawing difficult to understand, and this is considered a relatively simple drawing. To relieve the confusion associated with a drawing too difficult to understand because of numerous hidden lines, a section is made of the part. Orthographic methods are followed up to the creation of the side view.

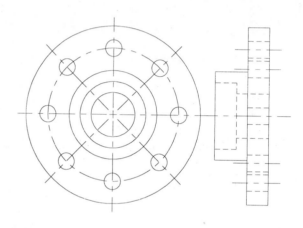

To understand section views more, see the illustration at the right. Creating a section view slices an object in such a way so as to expose what used to be hidden features and convert them into visible features. This slicing or cutting operation can be compared to that of using a glass plate or cutting plane to perform the section. In the object at the right, the glass plate cuts the object in half. It is the responsibility of the designer or CAD operator to convert one half of the object into a section and to discard the other half. Surfaces that come in contact with the glass plane are crosshatched to show where the actual cutting took place.

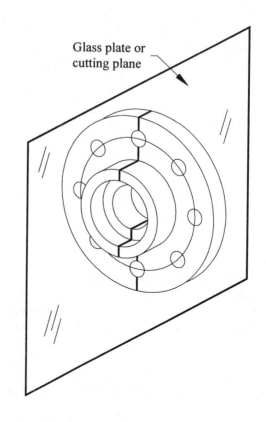

Glass plate or cutting plane

A completed section view drawing is shown at the right. Two new lines are also illustrated, a cutting plane line and section lines. The cutting plane line performs the cutting operation on the front view. In the side view, section lines show the surfaces that were cut. Notice that holes are not section lined since the cutting plane passes across the center of the hole. Notice also that hidden lines are not displayed in the side view. It is considered poor practice to merge hidden lines into a section view although there are always exceptions. The arrows of the cutting plane line tell the designer to view the section in the direction of the arrows and discard the other half.

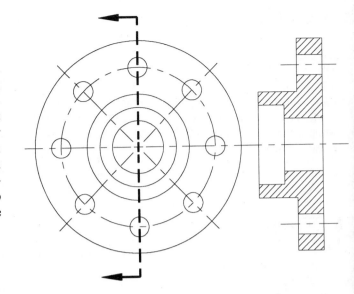

The cutting plane line consists of a very thick line at a series of dashes approximately 0.25" in length. A polyline is used to create this line of 0.05 thickness. The arrows point in the direction of sight used to create the section with the other half generally discarded. Assign this line one of the dashed linetypes; the hidden linetype is reserved for detailing invisible features in views. The section line, by contrast with the cutting plane line, is a very thin line. This line identifies the surfaces being cut by the cutting plane line. The section line is usually drawn at an angle and at a specified spacing.

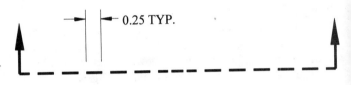

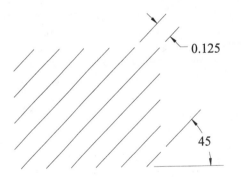

Depending on whether you are using AutoCAD Release 12 or 13, a wide variety of hatch patterns are already supplied with the software. One of these patterns, Ansi31, is displayed at the right. This is one of the more popular patterns with lines spaced 0.125 units apart from each other and at a 45 degree angle.

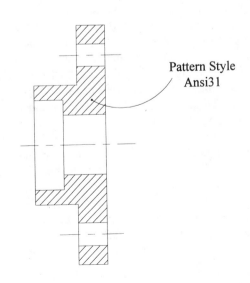

The object at the right illustrates proper section lining techniques. Much of the pain of spacing the section lines apart from each other and at angles has been eased considerably by using the computer as a tool. However, the designer must still practice proper section lining techniques at all times for clarity of the section.

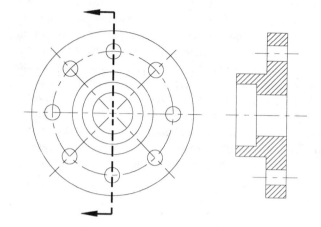

The next four examples illustrate common errors involved with section lines. At "A," section lines run the correct directions and at the same angle; however, the hidden lines have not been converted into object lines. This will confuse the more experienced designer since the presence of hidden lines in the section means more complicated invisible features. Example "B" is yet another error encountered when creating sections. Again, the section lines are properly placed; however, all surfaces representing holes have been removed, which displays the object as a series of sectioned blocks unconnected, implying four separate parts.

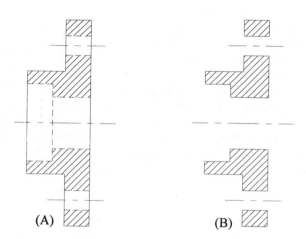

(A)   (B)

Example "C" appears to be a properly sectioned object; however, upon closer inspection, we see the angle of the cross-hatch lines in the upper half different from the same lines in the lower left half. This suggests two different parts, when in actuality, it is the same part. In example "D" at the right, all section lines run the correct direction. The problem is the lines run through areas that were not sliced by the cutting plane line. These areas at "D" represent drill and counterbore holes and are left unsectioned.

These have been identified as the most commonly made errors when crosshatching an object. Remember just a few rules to follow: section lines are present only on surfaces that are cut by the cutting plane line; section lines are drawn in one direction when crosshatching the same part; hidden lines are usually omitted when creating a section view; areas such as holes are not sectioned since the cutting line only passes across this feature.

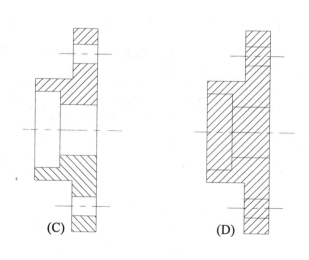

(C)   (D)

Full Sections

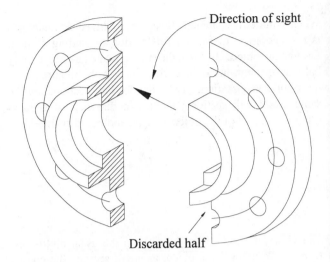

When the cutting plane line passes through the entire object, a full section is formed. In the illustration at the right, a full section would be the same as taking an object and cutting it completely in half. Depending on the needs of the designer, one half is kept, the other half is discarded. The half that is kept is section lined.

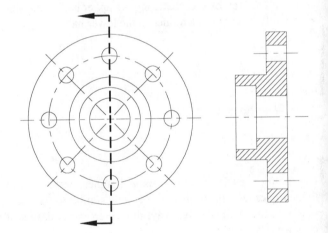

At the right is the multi-view solution to the problem above. The front view is drawn with lines projected across to form the side view. To show the side view is in section, a cutting plane line is added to the front view. This line performs the physical cut. The designer has the option of keeping either half of the object. This is the purpose of adding arrowheads to the cutting plane line. The arrowheads define the direction of sight the designer views the object to form the section. The designer must then interpret what surfaces are being cut by the cutting plane line in order to properly add crosshatching lines to the section which is located in the right side view. Hidden lines are not necessary once a section has been made.

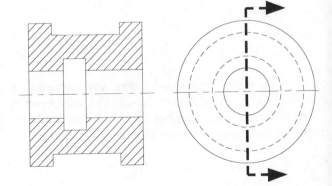

Numerous examples illustrate a cutting plane line with the direction of sight off to the the left. This does not mean that a cutting plane line cannot have a direction of sight going to the right as in the example. In this example, the section is formed from the left side view if the circular features are located in the front view.

Half Sections

When symmetrical shaped objects are involved, sometimes it is not necessary to form a full section by cutting straight through the object. Instead, the cutting plane line passes only halfway through the object which makes the illustration at the right a half section. The rules for half sections are the same as for full sections; namely, a direction of sight is established, part of the object is kept, and part discarded.

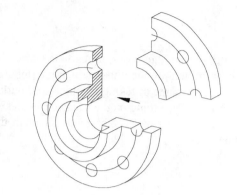

The views are laid out at the right in the usual multi-view format. To prepare the object as a half section, the cutting plane line passes halfway through the front view before being drawn off to the right. The right side view is converted into a half section by crosshatching the upper half of the side view while leaving hidden lines in the lower half.

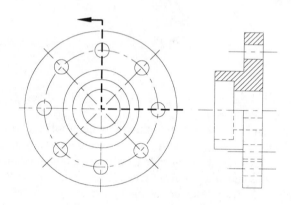

Depending on office practices, some designers prefer to omit hidden lines entirely from the side view similar to the illustration at the right. In this way, the lower half is drawn of only what is visible.

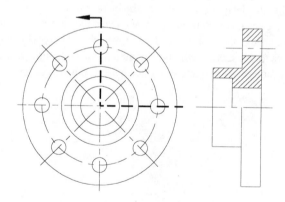

The illustration at the right shows another way of drawing the cutting plane line to conform to the right side view drawn in section. Hidden lines have been removed from the lower half; only those lines visible are displayed.

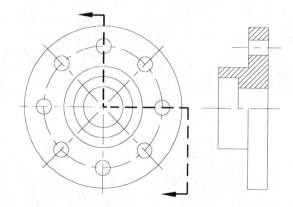

Assembly Sections

It would be unfair to give designers the impression that section views are only used for displaying internal features of individual parts. Yet another advantage of using section views is that it permits the designer to create numerous objects, assemble them, and then slice the assembly to expose internal details of all parts. This type of section is an assembly section similar to the illustration at the right. For all individual parts, notice the section lines running the same directions. This follows one of the basic rules of section views: keep section lines at the same angle for each individual part.

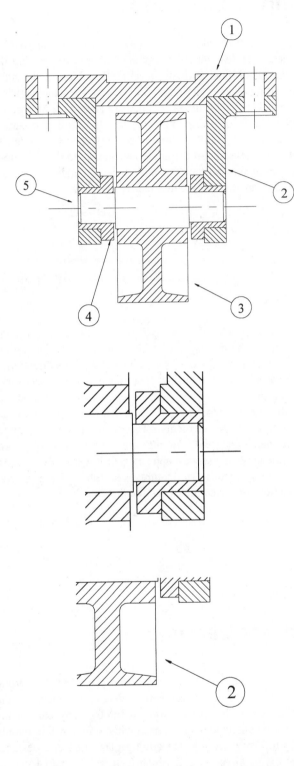

Illustrated at the right is the difference of assembly sections and individual parts that have been sectioned. For parts in an assembly that contact each other, it is considered good practice to alternate the directions of the section lines and make the assembly much more clear and distinguish the parts from each other. This can be accomplished by changing the angle of the hatch pattern or even the scale of the pattern.

To identify parts in an assembly, an identifying part number along with a circle and arrowhead line are used. The line is very similar to a leader line used to call out notes for specific parts on a drawing. The addition of the circle highlights the part number. Sometimes this type of call out is referred to as a "bubble."

In the enlarged assembly illustrated at the right, the large area in the middle is actually a shaft used to support a pulley system. With the cutting plane passing through the assembly including the shaft, it is considered good practice to refrain from crosshatching features such as shafts, screws, pins, or other types of fasteners. The overall appearance of the assembly is actually enhanced by not crosshatching these items.

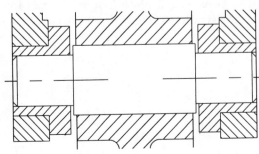

Aligned Sections

Aligned sections take into consideration the angular position of details or features of a drawing. Instead of drawing the cutting plane line vertically through the object at the right, the cutting plane is angled or aligned with the same angle the elements are at. Aligned sections are also made to produce better clarity of a drawing. At the right, with the cutting plane forming a full section of the object, it is difficult to obtain the true size of the angled elements. In the side view, the appear foreshortened or not to scale. Hidden lines were added as an attempt to better clarify the view.

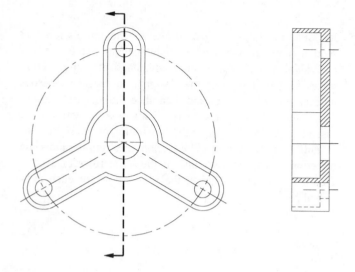

Instead of drawing the cutting plane line all the way through the object, the line is bent at the center of the object before being drawn through one of the angled legs. The direction of sight arrows on the cutting plane line not only determines which direction the view will be sectioned, but also shows another direction for rotating the angled elements so they line up with the upper elements. This rotation is usually never more than 90 degrees. As lines are projected across to form the side view, the section appears as if it were a full section. This is only because the features were rotated and projected in section for greater clarity of the drawing.

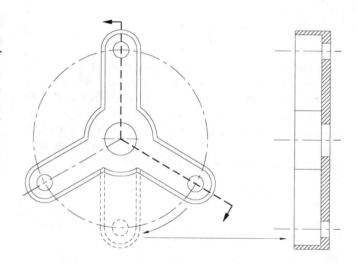

Offset Sections

Offset sections take their name from offsetting the cutting plane line to pass through certain details in a view. If the cutting plane line passes straight through any part of the object, details would be exposed while others would remain hidden. By offsetting the cutting plane line, the designer controls its direction and which features of a part it passes through. The view to section follows the basic section rules.

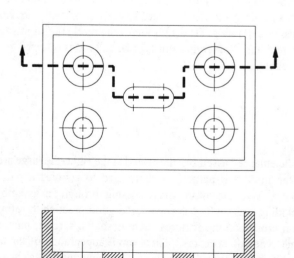

Sectioning Ribs

Contrary to section view principles, parts made out of cast iron with webs or ribs used for reinforcement do not follow basic rules of sections. In the example at the right, the front view has the cutting plane line passing through the entire view; the side view at "A" is crosshatched according to section view basics. However, it is difficult to read the thickness of the base since the crosshatching includes the base along with the web. A more efficient method is to ignore crosshatching webs as in "B." Therefore, not crosshatching the web exposes other important details such as thicknesses of bases and walls.

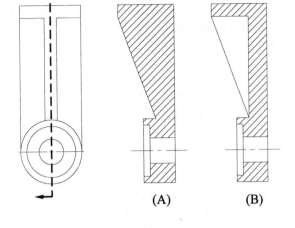

(A)　　　　(B)

The object at the right is another example of performing a full section on an area consisting of webbed or ribbed features. By not crosshatching the webbed areas, more information is available such as the thickness of the base and wall areas around the cylindrical hole. This may not be considered true projection; however, it is considered good practice.

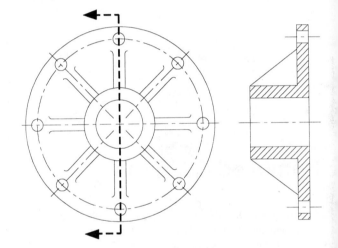

Broken Sections

At times only a partial section of an area needs to be created. For this reason, a broken section might be used. The object at "A" shows a combination of sectioned areas and conventional areas outlined by the hidden lines. When converting an area to a broken section, the designer creates a break line, crosshatches one area, and leaves the other area as in conventional drawing. Break lines may take the form of short freehanded line segments illustrated in example "B" or a series of long lines separated by break symbols as in example "C." Caution needs to be exercised when drawing the freehanded break line using the Sketch command. Be careful not to use a very small sketch increment since this will increase the size of the drawing file. The Line command can be used with Ortho-Off to produce the desired effect as in the examples at the right.

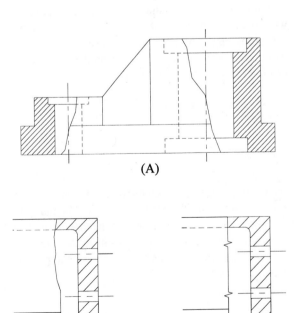

(A)

(B)　　　　(C)

Revolved Sections

Section views may be constructed as part of a view by using revolved sections. In the example at the right, the elliptical shape is constructed while it is revolved into position and then crosshatched.

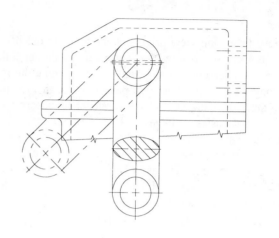

Illustrated at the right is another example of a revolved section where a cross-section of the C-clamp was cut away and revolved to display its shape.

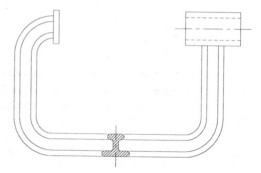

Removed Sections

Removed sections are very similar to revolved sections with the exception that instead of drawing the section somewhere inside of the object as is the case of a revolved section, the section is placed elsewhere or removed to a new location in the drawing. The cutting plane line is present with the arrows showing the direction of sight. Identifying letters are placed on the cutting plane and underneath the section to keep track of the removed sections especially when there are a number of them on the same drawing sheet.

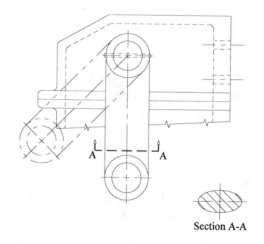

Section A-A

Another way of displaying removed sections is to use center lines as a substitute for the cutting plane line. In the example at the right, the center lines determine the three shapes of the chisel and display the basic shapes from circle to octagon to rectangle. Identification numbers are not required in this particular example.

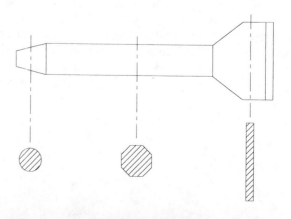

Isometric Sections

Section views may be incorporated into pictorial drawings that are illustrated at the right. The object at "A" is an example of a full isometric section with the cutting plane passing through the entire object. In keeping with basic section rules, only those surfaces sliced by the cutting plane line are crosshatched. Isometric sections make it easy to view cut edges compared to holes or slots. Illustrated at "B" is an example of an isometric drawing converted into a half section.

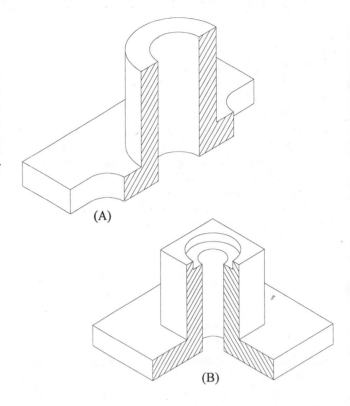

(A)

(B)

Architectural Sections

Mechanical representations of machine parts are not the only type of drawings where section views are used. Architectural drawings rely on sections to show the type of building materials that go into the construction of foundation plans, roof details, or wall sections in the example at the right. Here numerous types of crosshatching symbols are used to call out the different types of building materials such as brick veneer at "A," insulation at "B," finished flooring at "C," floor joists at "D," concrete block at "E," earth at "F," and poured concrete at "G." Section symbols provided in AutoCAD were used to crosshatch most of the building components with the exception of the floor joists and insulation.

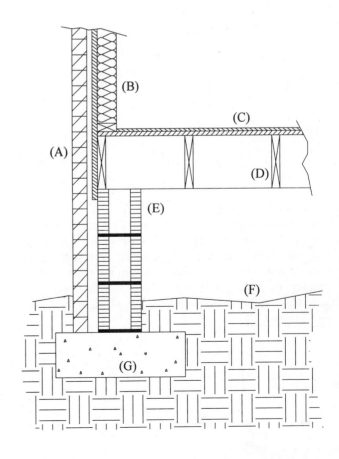

Hatching Techniques Using AutoCAD

The process of crosshatching an object using AutoCAD is considered by some to be an art or even bordering on a science. This all revolves around boundary or area to be crosshatched. Using manual methods, the designer followed a boundary with the aid of his eye along with the familiar T-square and triangle. Each line was stepped off and drawn individually. AutoCAD provides considerable help with this process since once a boundary is defined, the system crosshatches the area automatically. The problem is defining the area to crosshatch. The illustration at the right will be used to show two methods of crosshatching available to the user.

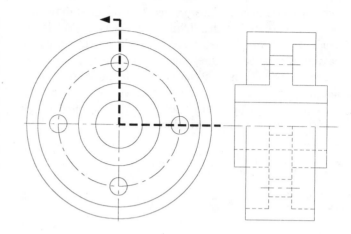

To perform a successful crosshatching operation, a boundary must first be defined similar to the illustration at the right. This example shows the areas cut by the cutting plane line before crosshatch lines are added. As this boundary is easily identified by eye, the actual construction of the view using AutoCAD may not yield the boundary at the right.

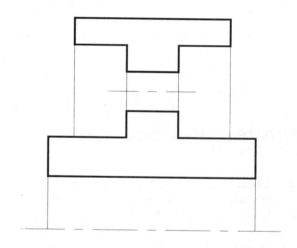

When constructing this view, horizontal lines were projected from the front view into the side view where they were sized using the Trim command. The dashed lines at the right merely represent the lines projected from the front view; in actuality, they are object lines. From this illustration, all horizontal lines are of the proper length to form a crosshatching boundary.

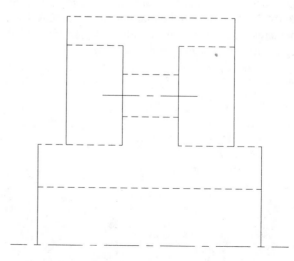

In this particular hatching example, problems will occur with the vertical being constructed as single entities labeled "A," "B," "C," "D," "E," and "F." Single entity lines are often difficult to define a boundary to crosshatch. Two methods may be used to define boundaries; using the Break command to split segments in two for boundary creation, or tracing a boundary using a polyline. Both methods are outlined in the text that follows.

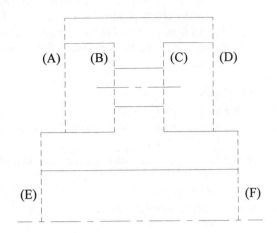

Lines may be formed into segments used to define boundaries using the Break command along with the @ option. In the illustration at the right, line "A" is the entity to be broken. Next, the first break point is selected at "B" followed by @ for the second break point. The significance of using @ is to select the previous point. The result is a break so small it is undetected by eye, yet it breaks line "A" into two segments. Follow the prompt sequence below:

Command: **Break**
Select object: *(Select line segment "A")*
Enter second point (or F for first point): **F**
Enter first point: **Int**
of *(Select the intersection of the selected line at "B")*
Enter second point: **@**

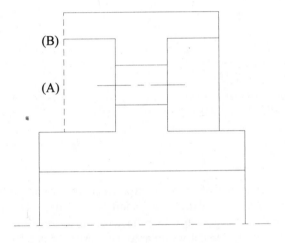

Repeat the procedure above for all points illustrated at the right. The result will be a series of broken vertical lines. When combined with the horizontal segments, they form the boundary needed to perform a proper crosshatching operation using AutoCAD.

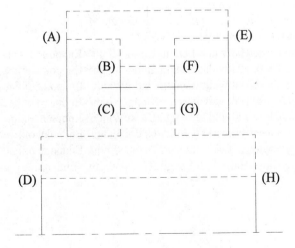

Once the boundaries are defined, use the Hatch command to place a certain crosshatching pattern. Use the boxes illustrated at the right to form standard object selection set windows for performing the hatching operation. Follow the hatching prompts below:

Command: **Hatch**
Pattern (? or name/U,style): **Ansi31**
Scale for pattern <1.0000>: *(Strike Enter to accept this default)*
Angle for pattern <0>: *(Strike Enter to accept this default)*
Select objects: **W**
First corner: *(Select a point at "A")*
Other corner: *(Select a point at "B")*
Select objects: **W**
First corner: *(Select a point at "C")*
Other corner: *(Select a point at "D")*
Select objects: *(Strike Enter to perform the hatching operation)*

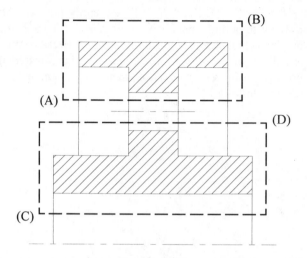

Another method of defining a crosshatching boundary is to make a separate layer called "Pline," short for polyline. Use the Pline command along with Osnap-Intersection to trace the boundary over existing geometry similar to the illustration at the right.

Command: **Osnap**
Object snap mode(s): **Int**

Command: **Pline**
From point: *(Select the intersection at "A")*
Current line-width is 0.00000
Arc/Close/Halfwidth/Length/Undo/Width/<Endpoint of line>: *("B")*
Arc/Close/Halfwidth/Length/Undo/Width/<Endpoint of line>: *("C")*
Arc/Close/Halfwidth/Length/Undo/Width/<Endpoint of line>: *("D")*
Arc/Close/Halfwidth/Length/Undo/Width/<Endpoint of line>: *("E")*
Arc/Close/Halfwidth/Length/Undo/Width/<Endpoint of line>: *("F")*
Arc/Close/Halfwidth/Length/Undo/Width/<Endpoint of line>: *("G")*
Arc/Close/Halfwidth/Length/Undo/Width/<Endpoint of line>: *("H")*
Arc/Close/Halfwidth/Length/Undo/Width/<Endpoint of line>: **Cl**

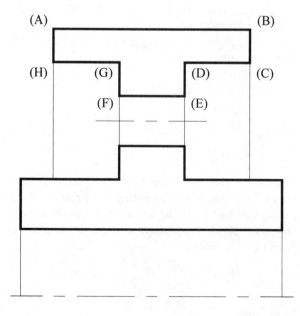

Use the Hatch command to crosshatch the boundaries outlined by the polylines. Instead of using the Window option
to group the selection set, simply select the polylines to
perform the hatching.

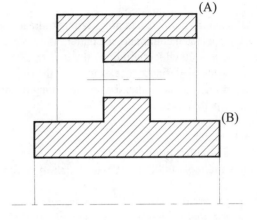

Command: **Hatch**
Pattern (? or name/U,style): **Ansi31**
Scale for pattern <1.0000>: *(Strike Enter to accept this
default)*
Angle for pattern <0>: *(Strike Enter to accept this default)*
Select objects: *(Select the polyline at "A")*
Select objects: *(Select the polyline at "B")*
Select objects: *(Strike Enter to perform the hatching
operation)*

To prevent the original geometry lines from plotting underneath the polyline, use the Erase command to delete both
polyline boundaries. A more efficient method would be to
use the Layer command and turn off or freeze the layer
named "Pline."

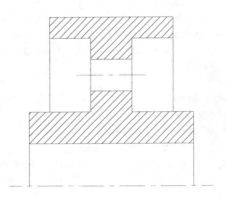

Hatching with AutoCAD is made easier only if a proper
boundary to be hatched has been defined. The two methods
of using the Break command or polyline to create a boundary
will be illustrated in the tutorial exercises at the end of this
unit.

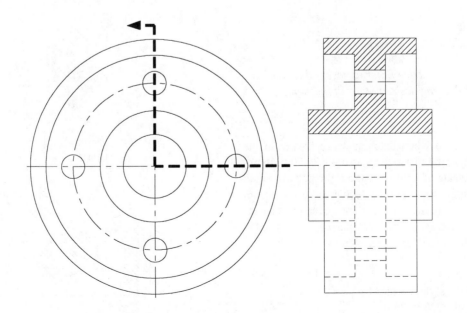

Hatch Pattern Scaling

AutoCAD has pre-defined hatching patterns already sized. When using the Hatch command, the pattern used is assigned a scale value of 1.00 which will draw the pattern exactly the way it was orginally defined. The example at the right shows the affects of the Hatch command when accepting the default value for the pattern scale.

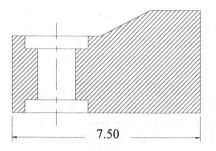

Command: **Hatch**
Pattern (? or name/U,style): **Ansi31**
Scale for pattern <1.0000>: *(Strike Enter to accept this default)*
Angle for pattern <0>: *(Strike Enter to accept this default)*
Select objects: *(Select the desired areas at the right)*

Entering a different scale value for the pattern will either increase or decrease the spacing in between crosshatch lines. At the right is an example of the Ansi31 pattern with a new scale value of 0.50.

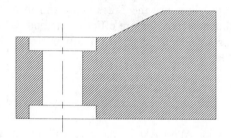

Command: **Hatch**
Pattern (? or name/U,style): **Ansi31**
Scale for pattern <1.0000>: **0.50**
Angle for pattern <0>: *(Strike Enter to accept this default)*
Select objects: *(Select the desired areas at the right)*

As the scale of a pattern can be decreased to hatch small areas, so also may the pattern be scaled up for large areas. The example at the right has a hatch scale of 2.00 which doubles all distances in between hatch lines.

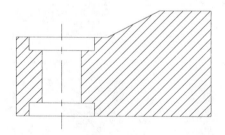

Command: **Hatch**
Pattern (? or name/U,style): **Ansi31**
Scale for pattern <1.0000>: **2.00**
Angle for pattern <0>: *(Strike Enter to accept this default)*
Select objects: *(Select the desired areas at the right)*

Care needs to be used when hatching large areas. In the example at the right, the distance measures 190.50 millimeters. If the hatch scale of 1.00 were used, the pattern would take on a filled appearance similar to the Solid command. The problem is that numerous lines are generated that increase the size of the drawing file. A value of 25.4 is used to scale hatch lines for metric drawings.

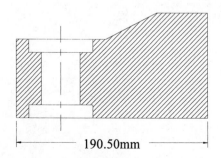

Command: **Hatch**
Pattern (? or name/U,style): **Ansi31**
Scale for pattern <1.0000>: **25.4**
Angle for pattern <0>: *(Strike Enter to accept this default)*
Select objects: *(Select the desired areas at the right)*

Hatch Pattern Angle Manipulation

As with the scale of the hatch pattern, the angle for the hatch pattern can be controlled by the designer depending on the effect the pattern has with the area being hatched. By default, the Hatch command displays a "0 degree" angle for all patterns. In the example at the right, the angle for "Ansi31" is 45 degrees. This is because the pattern was originally created at a 45 degree angle.

Command: **Hatch**
Pattern (? or name/U,style): **Ansi31**
Scale for pattern <1.0000>: *(Strike Enter to accept this default)*
Angle for pattern <0>: *(Strike Enter to accept this default)*
Select objects: *(Select the desired areas at the right)*

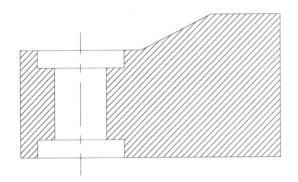

Entering any angle different from the default value of "0" will rotate the hatch pattern by that value. This means if a pattern was originally designed at a 45 degree angle like "Ansi31," entering a new angle for the pattern would begin rotating the pattern starting at the 45 degree position. In the example at the right, a new angle of 45 degrees is entered. Since the original angle was already 45 degrees, this new angle value is added to the original to obtain a vertical crosshatch pattern.

Command: **Hatch**
Pattern (? or name/U,style): **Ansi31**
Scale for pattern <1.0000>: *(Strike Enter to accept this default)*
Angle for pattern <0>: **45**
Select objects: *(Select the desired areas at the right)*

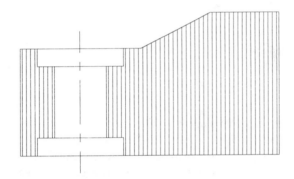

Again entering an angle other than the default rotates the pattern from the original angle to a new angle. In the example at the right, the "Ansi31" pattern was rotated by 90 degrees.

Command: **Hatch**
Pattern (? or name/U,style): **Ansi31**
Scale for pattern <1.0000>: *(Strike Enter to accept this default)*
Angle for pattern <0>: **90**
Select objects: *(Select the desired areas at the right)*

Providing different angles for patterns is useful when creating section assemblies where different parts are in contact with each other and patterns are placed at different angles making the parts easy to see.

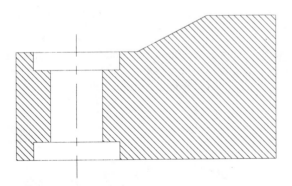

Using the BHATCH Command

Thus far, two methods have been used to crosshatch areas. Whether breaking entities or tracing polyline entities to define the area to be crosshatched, the methods tend to become tedious. A more powerful method of defining areas to be crosshatched is to use the Bhatch command, which stands for boundary hatch. The object illustrated at the right will be used to demonstrate the boundary crosshatching method.

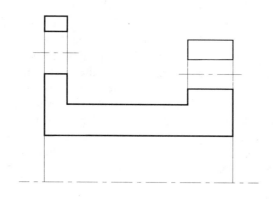

Instead of using the Break command to form exact intersections for areas to be hatched, or instead of tracing a polyline entity around all areas to crosshatch, then erasing the polyline, the Bhatch command automatically outlines the area to be hatched with a polyline entity similar to the illustration at the right.

The Bhatch command is able to outline boundaries to be crosshatched using a method of ray tracing. Selecting an internal point in the area to be crosshatched sends out tracers in all directions. As a tracer indentifies the edge of an entity, it begins drawing a polyline from the first point to the next intersection and so on until a closed polyline is formed. Illustrated at the right are 3 areas with internal points identified by the "X 's." The dashed areas identify the outline of the areas traced with a polyline.

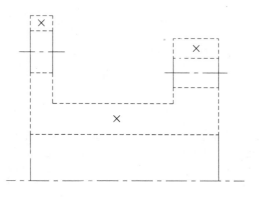

In the final step of the Bhatch command, as the crosshatch pattern is applied to the areas, the polyline trace outline is automatically deleted from the crosshatch outlines.

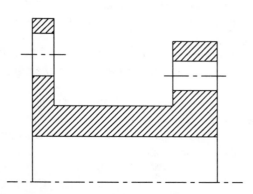

Picking "Hatch . . ." from the "Draw" pulldown menu activates the main Boundary Hatch dialog box illustrated below. Use this dialog box to pick a point identifying the area to be crosshatched, select objects using one or more of the popular selection set modes, or select "Hatch Options . . ." to activate another dialog box containing such hatch options as pattern, scale, and rotation angle. All areas greyed out in the illustration at the bottom are currently unavailable. Only when additional parameters have been satisfied will these buttons activate for use.

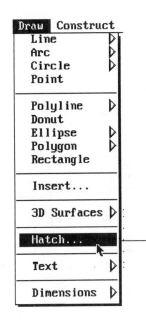

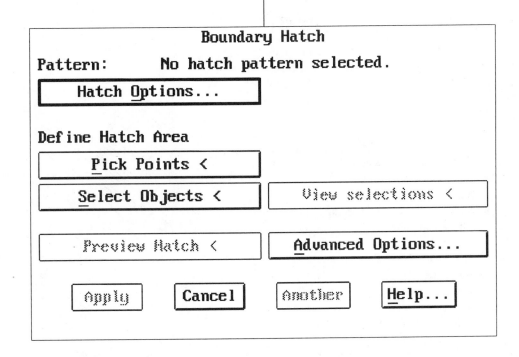

Hatch Options . . .

Hatch Options

Pattern Type

■ Stored Hatch Pattern
☐ User-Defined Pattern (U)

Pattern...

Scale: 1

Angle: 0

Spacing: 1.0000

Hatching Style

■ Normal

☐ Outer

☐ Ignore

☐ Exploded Hatch

☐ Double Hatch

Copy Existing Hatch <

OK   Cancel

Selecting the "Hatch Options . . ." button from the main
Boundary Hatch dialog box activates the dialog box illustrated
above. This dialog box holds a number of crosshatching
options such as Pattern type, Pattern scale, Pattern angle, and
Hatching style.

The radio button "Stored Hatch Pattern" refers to a series of
pre-defined hatch patterns already created. Selecting the
"User-Defined Pattern" allows the individual to create his own
hatch pattern. Selecting this radio button activates the "Spacing" edit box allowing the user to enter a spacing in between
entities that make up the hatch pattern.

Illustrated at the right are 3 hatching styles and how they affect
levels of crosshatching. The normal hatch style is the default
hatching method where upon selecting all entities with a
window, the hatching begins with the outermost boundary,
skipping the next inside boundary, hatching the next inner-
most boundary, etc. Notice the hatching pattern still exposing
the text entities for easy reading. The outermost hatch style
hatches only the outermost boundary of the object. The ignore
hatch style ignores the default hatching methods of alternating
crosshatching and hatches the entire object.

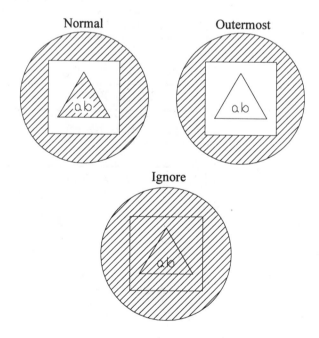

Normal   Outermost

Ignore

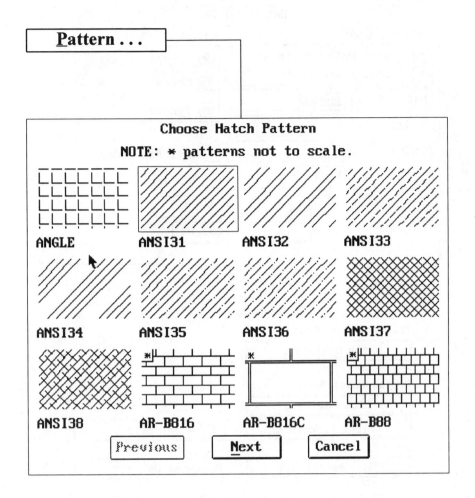

Selecting the "Pattern . . ." button from the "Hatch Options" dialog box displays a series of crosshatching patterns already created and ready for use. Select a particular pattern by picking the pattern itself. If the wrong pattern was selected, simply pick the correct one. Selecting the "Next" button takes the user to the next hatch pattern screen. Once on the second hatch pattern collection, "Previous" activates allowing the user to either go back to the previous hatch dialog box or go to the next pattern dialog box by selecting the "Next" button.

```
                    Hatch Options
Pattern Type                  Hatching Style
 ■ Stored Hatch Pattern        ■ Normal
 □ User-Defined Pattern (U)    □ Outer
 [Pattern...] [ANSI31]         □ Ignore
 Scale:    [1        ]
 Angle:    [0        ]         □ Exploded Hatch
 Spacing:  [1.0000   ]         □ Double Hatch
              [Copy Existing Hatch <]
                [OK]    [Cancel]
```

```
                    Boundary Hatch
Pattern:      ANSI31,N
           [   Hatch Options...   ]

Define Hatch Area
           [     Pick Points <    ]
           [   Select Objects <   ]   [  View selections <  ]
           [   Preview Hatch <    ]   [ Advanced Options... ]
        [Apply]  [Cancel]  [Another]  [Help...]
```

Once a hatch pattern is selected, the name is displayed back in the "Hatch Options . . ." dialog box illustrated above. If the scaling, rotation, and style setttings look favorable, select the "OK" button to go back to the main Boundary Hatch dialog box. Selecting the button "Pick Points" and marking 3 points in the illustration at the right automatically defines the boundary to be hatched without manually tracing a polyline.

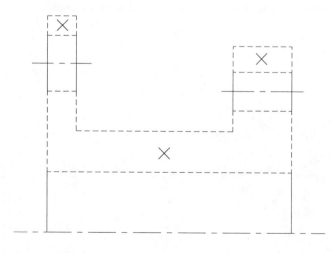

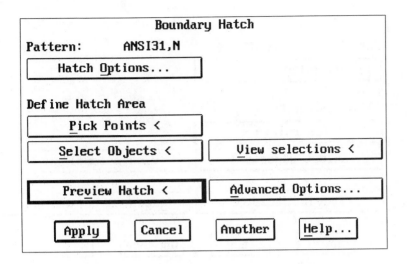

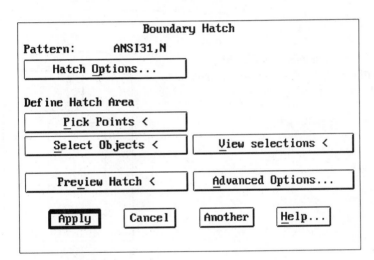

As in the previous page, once a series of boundaries are selected, the user is returned back to the main Boundary Hatch dialog box illustrated above. The user now has the option of first previewing the hatch to see if all settings and the appearance of the pattern are desirable. Pick the "Preview Hatch" button to accomplish this. The results are illustrated at the right. Technically, the pattern is still not placed on the object. Striking Enter to exit preview mode returns the user back to the main "Hatch Options" dialog box. If the hatch pattern is correct in appearance, pick the "Apply" button to place the pattern with the drawing. As the pattern is placed, the temporary polyline entity is automatically erased.

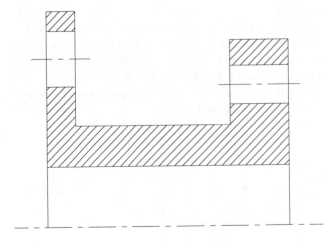

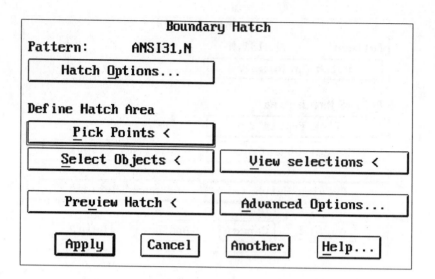

Care needs to be used when picking points to define the area to be crosshatched. In the illustration at the right, a point has been selected to crosshatch the outer area of the object.

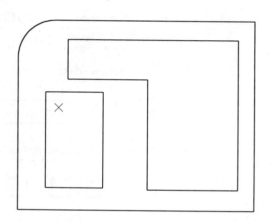

However, instead of tracing a temporary polyline around the outer perimeter of the object, the boundary hatch command identified the inner rectangle as the pattern to hatch. At this point, an error box alerts the user that the point selected is outside of the boundary. All of this occurred because the boundary point was selected too close to an inner feature; this time a rectangle. When using the Bhatch command, it is considered good practice to assist the command by selecting a point towards the outer perimeter.

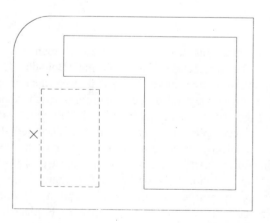

Using BHATCH to Hatch Islands

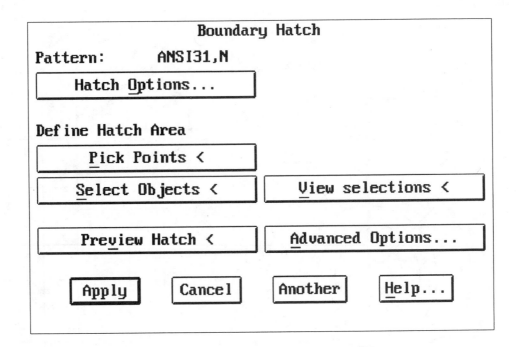

When confronted with crosshatching islands in the illustration at the right, first issue the Bhatch command, select a pattern, and begin picking points by marking a point at "A" which will define the outer perimeter of the object. While marking points, make picks inside of the two islands at "B" and "C." All areas should highlight similar to the illustration at the right signifying that a series of polylines have been traced around all entities.

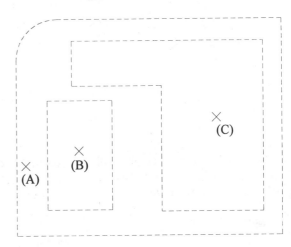

Selecting "Preview Hatch" should display the hatch pattern illustrated at the right. If changes need to be made, such as a change in the hatch scale or angle, preview allows these changes to be made. After changes, be sure to preview the pattern once again to check if the results are desirable. Selecting "Apply" places the pattern and deletes the temporary polylines used in the boundary identification process.

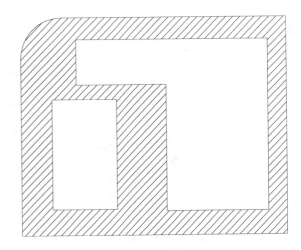

Tutorial Exercise #20
Dflange.Dwg

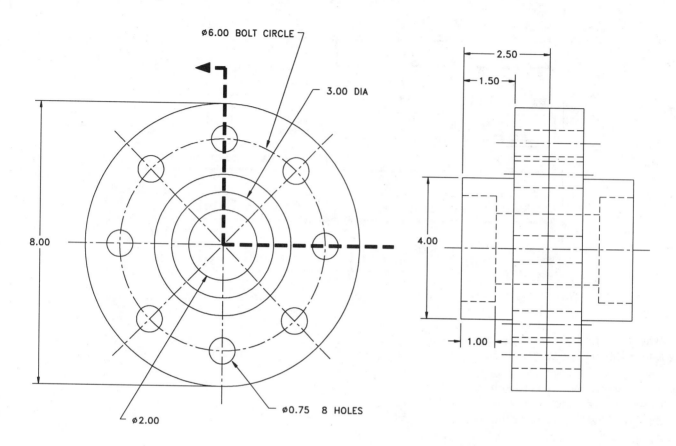

PURPOSE:

This tutorial is designed to use the Mirror, Change, and Hatch commands to convert the Dflange (Double Flange), into a half section assembly.

SYSTEM SETTINGS:

This drawing is complete except for converting the upper half into a half section. All Units, Limits, Grid, and Snap have already been set. From the drawing disk supplied with this text, call up a drawing called Dflange.

LAYERS:

The following layers have already been created:

Name-Color-Linetype
Object - White - Continuous
Cen - Yellow - Center
Hid - Red - Hidden
Xhatch - Magenta - Continuous
Cpl - Yellow - Dashed
Pline - Green - Continuous
Dim - Yellow - Continuous

SUGGESTED COMMANDS:

Begin this tutorial by converting one half of the object into a section by erasing unnecessary hidden lines. Next use the Change command and change the remaining hidden lines to the Object layer. Use the Break command to define boundaries to hatch. The selected boundaries are section lined using the Hatch command and the entire object is duplicated and copied to form a matching flange using the Mirror command. Use the Ansi31 Hatching pattern for this exercise.

DIMENSIONING:

Dimensions may be added to this tutorial at a later time. Use the Dim layer already created. Consult your instructor.

PLOTTING:

This tutorial exercise may be plotted on "B"-size paper (11" x 17"). Use a plotting scale of 1=1 to produce a scaled plot.

Step #1

Prepare the object to be mirrored by creating a selection set of the side view. Use the Select command to accomplish this by windowing all entities from "A" to "B."

Command: **Select**
Select objects: **Window**
First corner: *(Mark a point at "A"*
Other corner: *(Mark a point at "B")*
Select objects: *(Strike Enter to exit this command and create the selection set)*

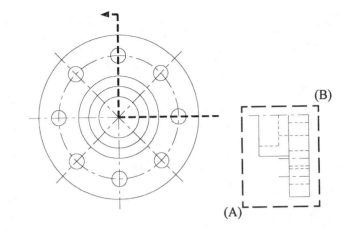

Step #2

Use the Mirror command to create a duplicate copy of the side view of the Dflange. Use the vertical line from "A" to "B" as the mirror line.

Command: **Mirror**
Select objects: **Previous**
Select objects: **Remove**
Remove objects: *(Select the vertical line at "C" to remove it from the selection set)*
Remove objects: *(Strike Enter to continue)*
First point of mirror line: **Endp**
of *(Select the endpoint of the line at "A")*
Second point: **Endp**
of *(Select the other endpoint of the line at "B")*
Delect old objects? <N>: *(Strike Enter to exit this command)*

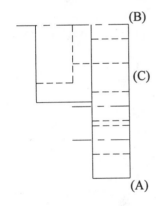

Step #3

Again use the Mirror command to copy and duplicate one-half of the bottom of the Dflange. It is this half that will be converted into a half section.

Command: **Mirror**
Select objects: **Previous**
Select objects: *(Select the vertical line at "C" to add it to the current selection set)*
Select objects: **Remove**
Remove objects: *(Select the horizontal center line at "D" to remove it from the selection set)*
Select objects: *(Strike Enter to continue)*
First point of mirror line: **Endp**
of *(Select the endpoint of the center line at "A")*
Second point: **Endp**
of *(Select the other endpoint of the center line at "B")*
Delect old objects? <N>: *(Strike Enter to exit this command)*

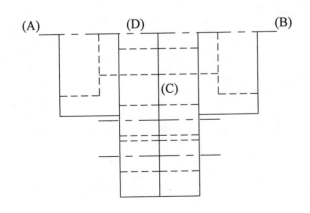

Step #4

Begin the preparation of the section by deleting any unnecessary entities such as the hidden and center lines located at "A," "B," "C," and "D" shown in the illustration at the right.

Command: **Erase**
Select objects: *(Select the 4 lines at "A,, "B," "C," and "D")*
Select objects: *(Strike the Enter key to execute this command)*

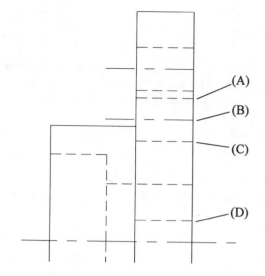

Step #5

Use the change command to convert the 5 hidden lines illustrated at the right from hidden lines to object lines. Perform the change from the layer "Hid" to the layer "Object."

Command: **Change**
Select objects: *(Select the 5 lines at "A," "B," "C," "D," and "E")*
Select objects: *(Strike Enter to continue with this command)*
Properties/ <Change point>: **P**
Change what property(Color/Elev/LAyer/LType/Thickness)? **LA**
New layer <Hid>: **Object**

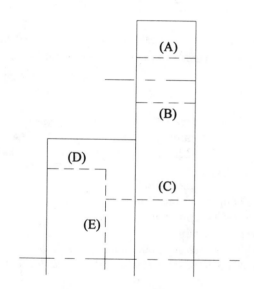

Step #6

Use the Trim command, select the horizontal line at "A" as the cutting edge, and select the vertical line at "B" as the entity to trim.

Command: **Trim**
Select cutting edge(s)...
Select objects: *(Select the horizontal line at "A")*
Select objects: *(Strike Enter to continue with this command)*
<Select object to trim>/Undo: *(Select the vertical line at "B")*
<Select object to trim>/Undo: *(Strike Enter to exit this command)*

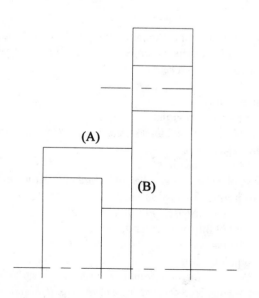

Step #7

The side view now needs to have the surfaces sliced by the cutting plane line crosshatched. Two areas labeled "A" and "B" need to be prepared for crosshatching. However, since all vertical lines are continuous and not broken into individual segments, polylines will be used to trace a boundary to be accepted by the Hatch command.

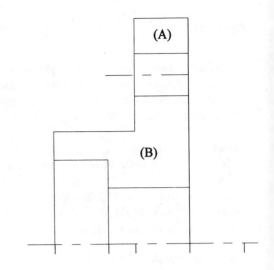

Step #8

Use the Layer command and make the layer "Pline" the new current layer. Then use the Pline command to trace a boundary using the illustration at the right as a guide. Because of the many intersections selected, use the Osnap command to lock into the intersection mode.

Command: **Layer**
?/Make/Set/New/ON/OFF/Color/Ltype/Freeze/Thaw/LOck/
Unlock: **Set**
New current layer<0>: **Pline**
?/Make/Set/New/ON/OFF/Color/Ltype/Freeze/Thaw/LOck/
Unlock: *(Strike Enter)*
Command: **Osnap**
Object snap mode(s): **Int**

Command: **Pline**
From point: *(Pick the intersection of the lines at "A")*
Arc/Close/Halfwidth/Length/Undo/Width/<Endpoint of line>:
(Pick the intersection at "B")
Arc/Close/Halfwidth/Length/Undo/Width/<Endpoint of line>:
(Pick the intersection at "C")
Arc/Close/Halfwidth/Length/Undo/Width/<Endpoint of line>:
(Pick the intersection at "D")
Arc/Close/Halfwidth/Length/Undo/Width/<Endpoint of line>:
Close

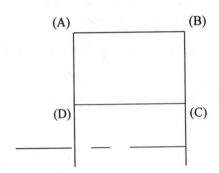

Step #9

Repeat the procedure above for tracing a polyline using the illustration at the right as a guide. Draw the polyline from the intersection of "A" to "B," "C," "D," "E," "F," "G," "H," and complete the polyline by using the Close option. Since you are still locked into the Osnap-Intersection mode, use the Osnap command again to free up any Osnap modes previously in use by entering the "None" mode.

Command: **Osnap**
Object selection mode(s): **None**

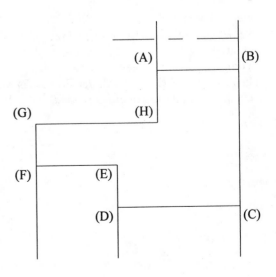

Step #10

Using the polylines constructed in the previous step as boundaries, use the Hatch command and the pattern "Ansi31" to
crosshatch the two areas at the right. Before performing this
step, change layers from "Pline" to "Xhatch."

Command: **Layer**
?/Make/Set/New/ON/OFF/Color/Ltype/Freeze/Thaw/LOck/
Unlock: **Set**
New current layer<Pline>: **Xhatch**
?/Make/Set/New/ON/OFF/Color/Ltype/Freeze/Thaw/LOck/
Unlock: *(Strike Enter to exit this command)*

Command: **Hatch**
Pattern (? or name/U,style): **Ansi31**
Scale for pattern <1.0000>: *(Strike Enter to accept this
default)*
Angle for pattern <0>: *(Strike Enter to accept this default)*
Select objects: *(Select the polyline at "A")*
Select objects: *(Select the polyline at "B")*
Select objects: *(Strike Enter to execute this command)*

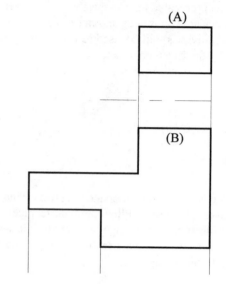

Step #11

The complete crosshatching of the upper half of the side view
is illustrated at the right. The pattern, "Ansi31" is drawn at a
45 degree angle with a spacing of approximately .125"
inbetween lines. Because the polyline used to define the
boundary is on top of existing lines, use the Layer command
to turn the "Pline" layer off.

Command: **Layer**
?/Make/Set/New/ON/OFF/Color/Ltype/Freeze/Thaw/LOck/
Unlock: **Off**
New current layer<0>: **Pline**
?/Make/Set/New/ON/OFF/Color/Ltype/Freeze/Thaw/LOck/
Unlock: *(Strike Enter to exit this command)*

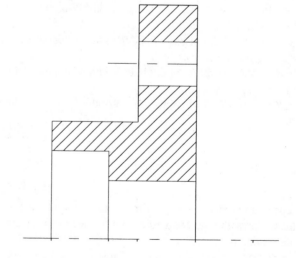

Step #12

Duplicate and flip the upper half of the side view to form the opposite half of the flange assembly using the Mirror command. Use the crossing option to select the objects illustrated at the right. Select "C" and "D" as the first and second points of the mirror lines.

Command: **Mirror**
Select objects: **C**
First corner: *(Select at "A")*
Other corner: *(Select at "B")*
Select objects: **Remove**
Remove objects: *(Select the vertical line at "E" to remove it from the selection set)*
Remove objects: *(Strike Enter to continue with this command)*
First point of mirror line: **Endp**
of *(Select the endpoint of the center line at "C")*
Second point: **Endp**
of *(Select the endpoint of the line at "D")*
Delete old objects? <N> *(Strike Enter to accept the default)*

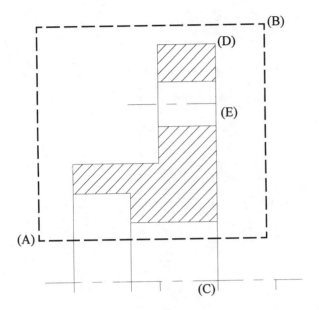

Step #13

Dimensions may be added at a later time similar to the illustration at the right. Notice the bottom halves of the side views show conventional hidden lines. This is one method of representing half sections.

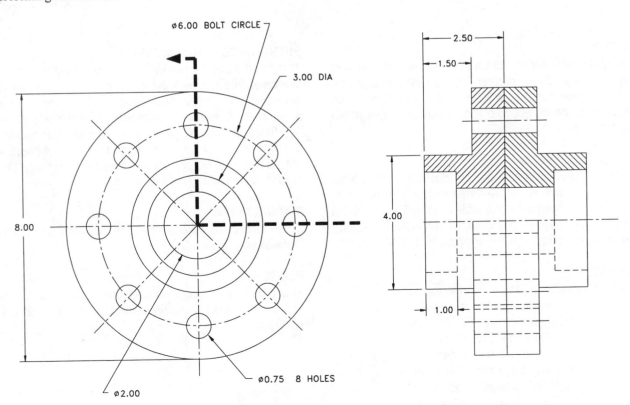

Tutorial Exercise #21
Coupler.Dwg

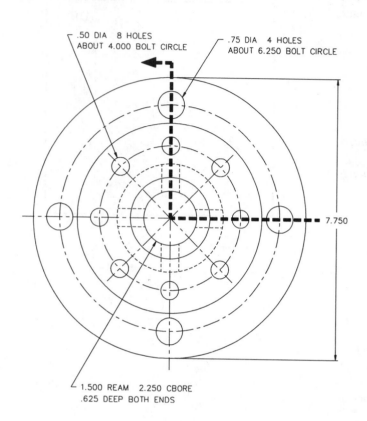

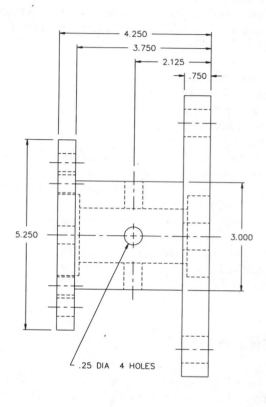

PURPOSE:
This tutorial is designed to use the Mirror, Change, and Hatch commands to convert the Coupler into a half section. The methods will be similar to the previous tutorial, with the exception of the crosshatching segment being different.

SYSTEM SETTINGS:
Use the Units command to change the number of decimal places from 4 to 2. Keep the remaining defaults settings. Use the Limits command to set the lower left corner to 0,0 and the upper right corner to 14,10. The Grid command may be used to change the spacing from 1.00 to 0.50.

LAYERS:
Create the following layers with the format:
Name-Color-Linetype
Object - White - Continuous
Cen - Yellow - Center
Hid - Red - Hidden
Xhatch - Magenta - Continuous
Cpl - Yellow - Dashed
Dim - Yellow - Continuous

SUGGESTED COMMANDS:
This tutorial begins similar in procedure to the previous exercise, Dflange, with the exception of the crosshatching segment. Instead of tracing a boundary using a polyline, the Break command will be used along with the "@" function. Boundaries will be formed when two lines are broken at an exact intersection so that the break will not be seen. This will also define the boundary to be crosshatched with the "Ansi31" pattern.

DIMENSIONING:
Dimensions may be added to this tutorial at a later time. Consult your instructor.

PLOTTING:
This tutorial exercise may be plotted on "B"-size paper (11" x 17"). Use a plotting scale of 1=1 to produce a scaled plot.

Step #1

Magnify the side view using the Zoom-Window command. Then, use the Select command to group all entities illustrated at the right into one selection set. The Remove option of the Select command will be used to remove the center line at "A" and "C" and circle at "B" from the selection set.

Command: **Select**
Select objects: **Window**
First corner: *(Mark a point at "X")*
Other corner: *(Mark a point at "Y")*
Select objects: **Remove**
Remove objects: *(Select the center lines at "A" and "C")*
Remove objects: *(Select the circle at "B")*
Remove objects: *(Select the two short horizontal and vertical lines located completely inside of the small circle)*
Remove objects: *(Strike Enter to exit this command and create the selection set)*

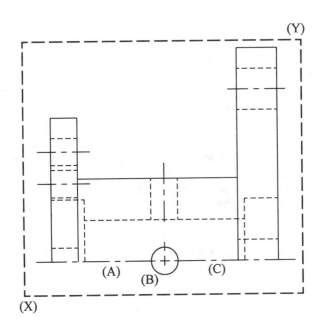

Step #2

Use the mirror command to copy and flip the upper half of the side view and form the lower half.

Command: **Mirror**
Select objects: **Previous**
Select objects: *(Strike Enter to continue)*
First point of mirror line: **Endp**
of *(Select the endpoint of the center line at "A")*
Second point: **Endp**
of *(Select the other endpoint of the center line at "B")*
Delect old objects? <N>: *(Strike Enter to exit this command)*

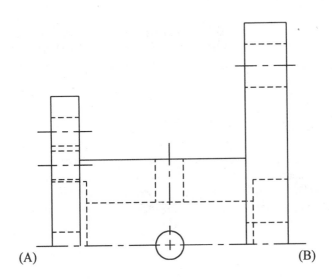

Step #3

Begin converting the upper half of the side view to a half section by using the Erase command to remove any unnecessary hidden lines and center lines from the view.

Command: **Erase**
Select objects: *(Carefully select the hidden lines labeled "A," "B," "C," and "D")*
Select objects: *(Select the center line labeled "E")*
Select objects: *(Strike Enter to execute the Erase command)*

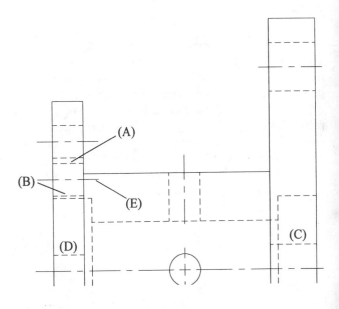

Step #4

Since the remaining hidden lines actually represent object lines when shown in section, use the Change command to convert all hidden lines labeled at the right from the "Hid" layer to the "Object" layer.

Command: **Change**
Select objects: *(Select all hidden lines labeled "A" through "K")*
Select objects: *(Strike Enter to continue with this command)*
Properties/<Change point>: **Prop**
Change what property (Color/Elev/LAyer/LType/Thickness)? **LA**
New layer <Hid>: **Object**

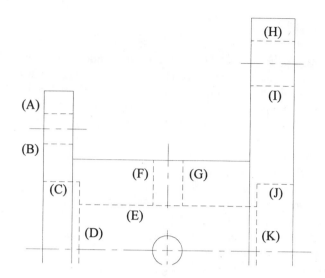

Step #5

Remove unnecessary line segments from the upper half of the converted section using the Trim command. Use the horizontal line at "A" as the cutting edge, and select the two vertical segments at "B" and "C" as the entities to trim.

Command: **Trim**
Select cutting edges...
Select objects: *(Select the horizontal line at "A")*
Select objects: *(Strike Enter to continue with this command)*
<Select object to trim>/Undo: *(Select the vertical line at "B")*
<Select object to trim>/Undo: *(Select the vertical line at "C")*
<Select object to trim>/Undo: *(Strike Enter to exit this command)*

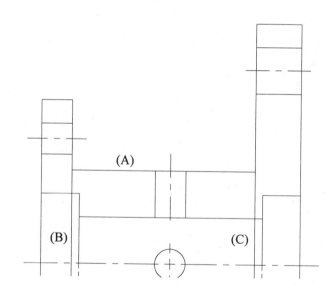

Step #6

Before beginning the crosshatching segment, four boundaries need to be defined in order to succeed with the Hatch command. The four areas to hatch are labeled to the right. One previous method used was to trace the profile of the boundary with a polyline, hatch the area outlined by the polyline, and finally delete the polyline. For this tutorial, the Break command will be featured along with the "@" option. The task will be to perform breaks at key intersections to define boundaries. The breaks, however, are to be small enough so as not to be noticed when performing a zoom of the area or a plot.

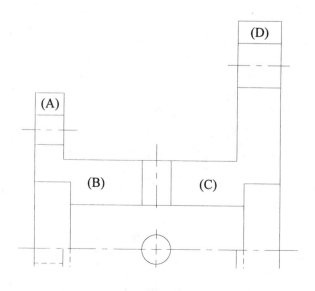

Step #7

Since all vertical and horizontal lines consist of continuous segments, the Break-@ command will split lines into numerous segments. Follow the prompts below to understand this operation:

Command: **Break**
Select object: *(Select the vertical line at "A")*
Enter second point (or F for first point): **F**
Enter first point: **Int**
of *(Select the intersection of the two lines at "A")*
Enter second point: **@**

The "@" means "last point" and performs the break at the exact same location as the first point selected. In this way, the line is separated into two segments without the separation being noticed. Use the above same procedure for intersections "B" through "H."

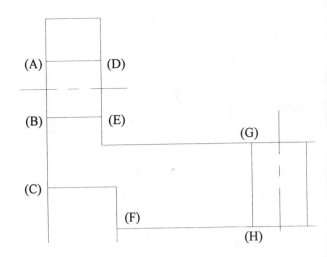

Step #8

Repeat the Break-@ command on the shape at the right to layout out the necessary boundaries to perform the hatching operation.

Command: **Break**
Select object: *(Select the horizontal line at "A")*
Enter second point (or F for first point): **F**
Enter first point: **Int**
of *(Select the intersection of the two lines at "A")*
Enter second point: **@**

Repeat the procedure above for intersections "B" through "H."

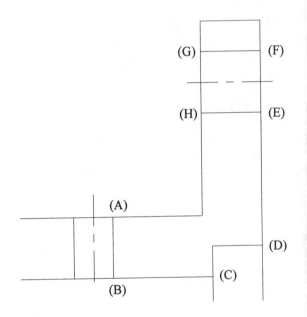

Step #9

Use the Layer command to set the current layer to "Xhatch." Once all areas have been broken identifying boundaries, use the hatch command and select each boundary to hatch.

Command: **Hatch**
Pattern (? or name/U,style): **Ansi31**
Scale for pattern <1.0000>: *(Strike Enter to accept this default)*
Angle for pattern <0>: *(Strike Enter to accept this default)*
Select objects: **W**
First corner: *(Select a point at "A")*
Other corner: *(Select a point at "B")*
Select objects: *(Use another window to select Area "C")*
Select objects: *(Use another window to select Area "D")*
Select objects: *(Use another window to select Area "E")*
Select objects: *(Strike Enter to execute this command)*

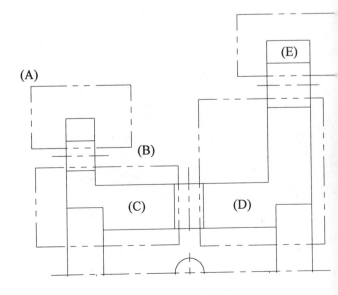

Step #10

As a type of check on progress, when using the Window option of the Hatch command and selecting the boundaries created back in Step #9 should highlight the boundaries similar to the illustration at the right. Here, the dashed lines signify the selected boundaries for the Hatch command to properly cross-hatch the area. If your display is not similar to the illustration at the right, use the Break-@ command to complete all breaks before using the Hatch command.

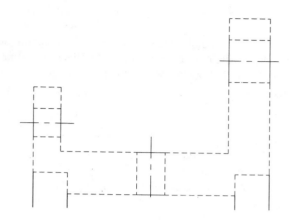

Step #11

Selecting multiple Window options at the "Select objects" prompt of the Hatch command will crosshatch all defined areas as in the illustration at the right. Using multiple selections for the same command will also prevent having to use the command four separate times. The completed drawing is also illustrated below complete with dimensions. As with other half section examples, it is the option of the operator or designer to show all hidden lines in the lower half or delete all hidden and center lines from the lower half and simply interpret the section in the upper half.

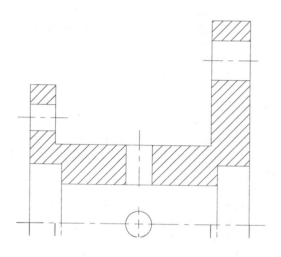

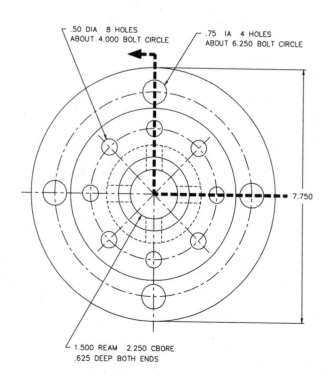

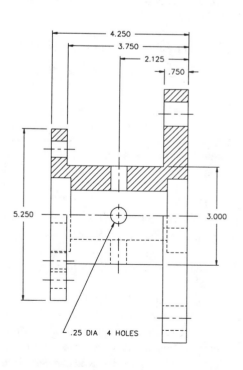

Tutorial Exercise #22
Assembly.Dwg

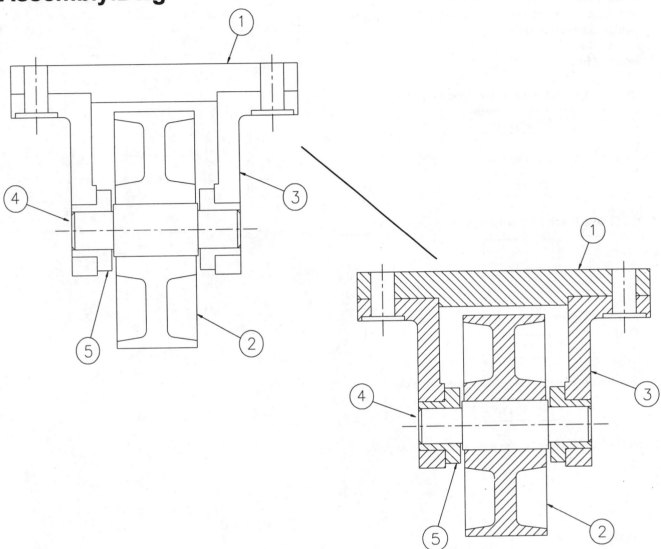

PURPOSE:
This tutoral is designed around using the Bhatch command to simplify crosshatching of an object. To illustrate this procedure, an assembly section will be made of an existing drawing file.

SYSTEM SETTINGS:
All settings have been made since this exercise is available on disk.

LAYERS:
Layers have already been created for this tutorial exercise.

Name-Color-Linetype
Object - White - Continuous
Leader - Cyan - Continuous
Section - Magenta - Continuous

SUGGESTED COMMANDS:
The Bhatch command will be used exclusively during this tutorial exercise.

DIMENSIONING:
This object does not have to be dimensioned.

PLOTTING:
Plot this drawing on a sheet of "B" size paper at a scale factor of 1=1.

Step #1

Issue the Bhatch command and select Ansi31 as the hatching pattern. Respond to the prompt "Pick Points<" by selecting the point illustrated at the right.

Command: **Bhatch**

As the Boundary Hatch dialog box appears, pick:

> **Hatch Options...**

As the Hatch Options dialog box appears, pick:

> **Pattern...**

As the Choose Hatch Pattern dialog box appears, pick:

As the Hatch Options dialog box reappears, pick:

> **OK**

As the Boundary Hatch dialog box reappears, pick:

> **Pick Points <**

Select the "X" illustrated at the right as the internal point.

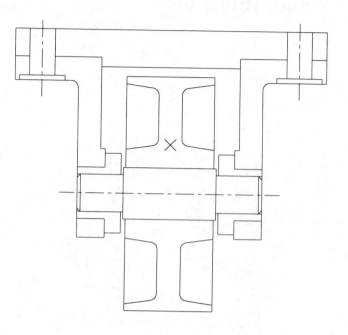

Step #2

As the area surrounding the last internal point highlights, continue selecting additional internal points illustrated by the "X's" at the right.

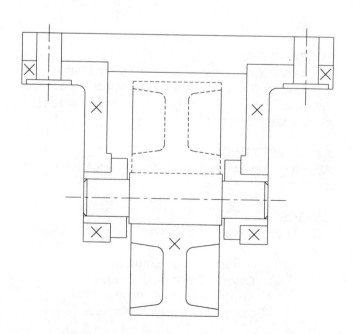

Step #3

As all areas identified by the internal points highlight, illustrated at the right with the dashed lines, strike Enter to get back to the Boundary Hatch dialog box. First preview the hatch pattern and see if the results are favorable.

After all of the areas illustrated at the right highlight, strike the Enter key to return to the Boundary Hatch dialog box. As the Boundary Hatch dialog box appears, pick:

| Preview Hatch < |

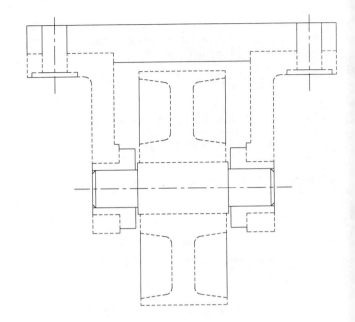

Step #4

The results of previewing the boundary hatch pattern is illustrated at the right. Follow the final step below to place the hatch pattern.

While the hatch pattern is previewed on the screen, strike Enter to return to the Boundary Hatch dialog box. If the results are favorable, pick:

| Apply |

AutoCAD outlines all areas to be hatched by tracing a polyline over that area. The polyline remains in place until the hatch is applied to the area. Upon placing the hatch pattern, the polyline outline is automatically deleted.

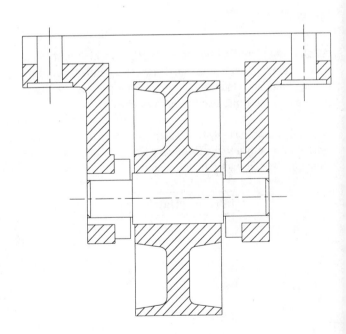

Step #5

Prepare the next series of areas to be hatched by reentering the Bhatch command. Respond to the prompt "Pick Points<" by selecting all "X's" illustrated at the right.

Command: **Bhatch**

As the Boundary Hatch dialog box appears, pick:

> **P̲ick Points <**

Select the "X's" illustrated at the right as internal points.

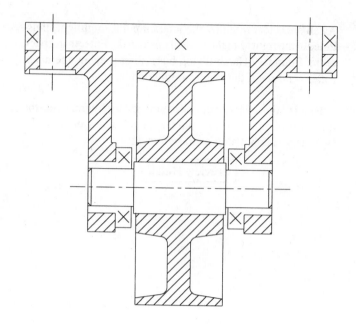

Step #6

As all areas identified by the internal points highlight, illustrated at the right with the dashed lines, strike Enter to get back to the Boundary Hatch dialog box. First preview the hatch pattern and see if the results are favorable.

After all of the areas illustrated at the right highlight, strike the Enter key to return to the Boundary Hatch dialog box. As the Boundary Hatch dialog box appears, pick:

> **Previe̲w Hatch <**

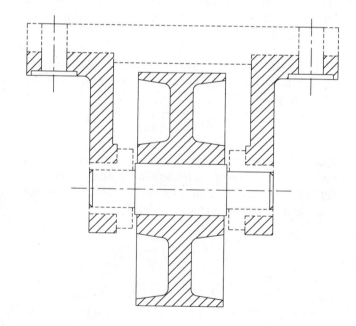

Step #7

The previewed hatch pattern is illustrated at the right. However, notice the pattern is running in the same direction as the
previous hatch pattern. Because this is an assembly section,
different parts that touch each other are more easily identified
if the hatch pattern runs in the opposite direction. Strike the
Enter key to return to the Boundary Hatch dialog box.

As the Boundary Hatch dialog box reappears, pick:

| **Hatch Options...** |

As the Hatch Options dialog box appears, change:

| **Angle: 0** |

To:

| **Angle: 90** |

To exit the Hatch Options dialog box, pick:

| **OK** |

As the Boundary Hatch dialog box reappears, pick:

| **Preview Hatch <** |

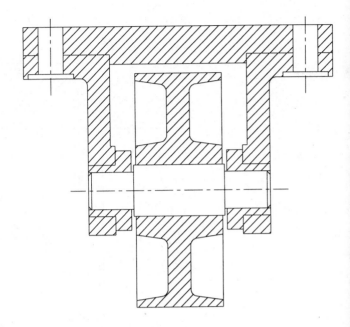

Step #8

Previewing the boundary hatch reveals the hatch pattern
rotated by 90 degrees. More importantly, it is easier to
distinguish the individual parts with the hatch patterns going
in different directions.

While the hatch pattern is previewed on the screen, strike Enter
to return to the Boundary Hatch dialog box. If the results are
favorable, pick:

| **Apply** |

The completed assembly section completely crosshatched is
illustrated at the right.

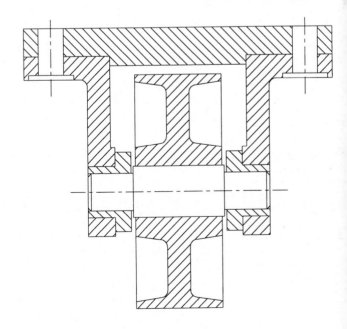

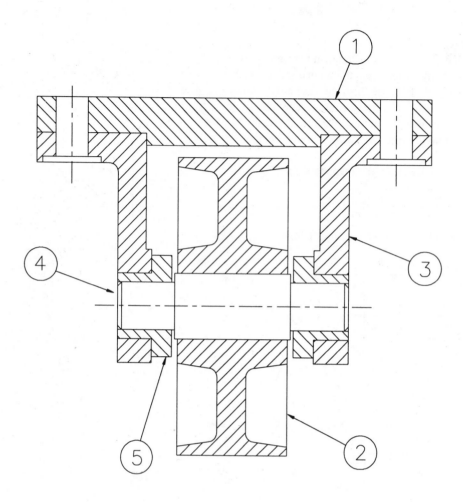

Step #9

The completed assembly including identifying bubbles is
illustrated above.

Questions for Unit 8

Directions for Question 8-1:

Based on the isometric and orthographic drawings below, identify which of the following best depicts a half section of the object. Place your answer in the box provided below.

Answer

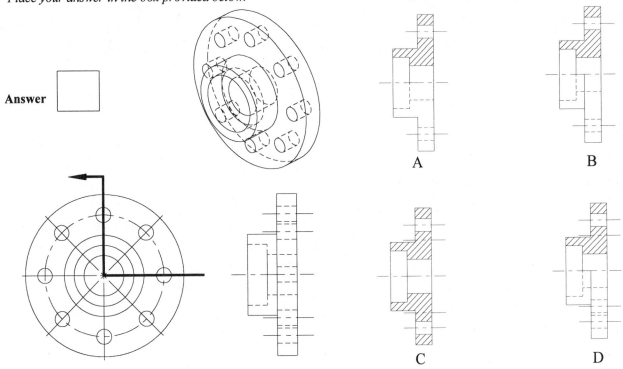

A

B

C

D

Directions for Question 8-2:

Based on the isometric drawing and orthographic drawing below, identify which of the following best depict a full section of the object. Place your answer in the box provided below.

Answer

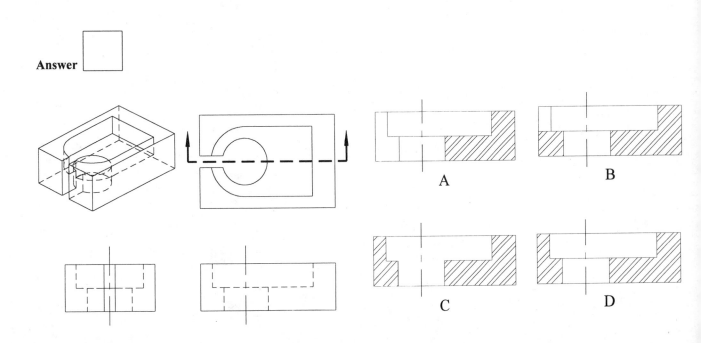

A

B

C

D

Directions for Question 8-3

Based on the isometric drawing and orthographic front view below, identify which of the following best depict a full section of the object. Place your answer in the box provided below.

Answer

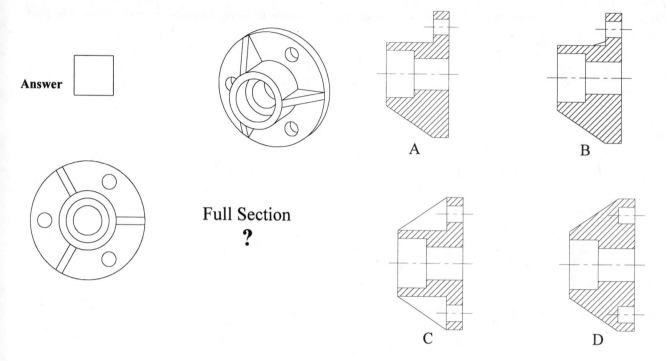

Full Section
?

A

B

C

D

Directions for Question 8-4:

Based on the isometric and orthographic drawings below, identify which of the following best depict an offset section of the object. Place your answer in the box provided below.

Answer

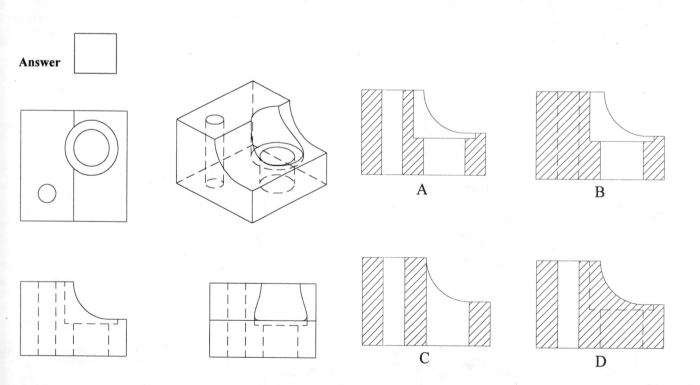

A

B

C

D

Problems for Unit 8

Problem 8-5

Center a 3 view drawing and make the front view a full section.

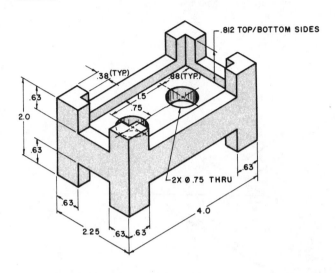

Problem 8-7

Center 2 views within the work area, and make 1 view a full section.

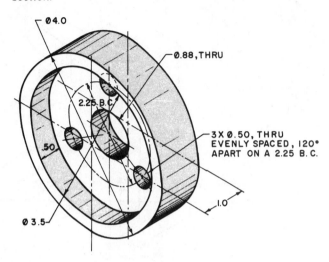

Problem 8-6

Center 2 views within the work area, and make 1 view a full section. Use correct drafting practices for the ribs.

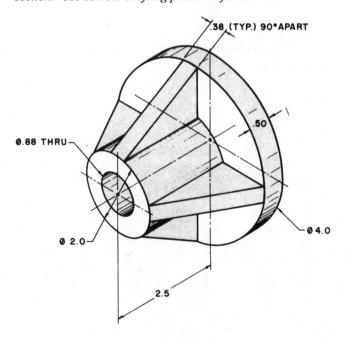

Problem 8-8

Center the front view and top view within the work area. Make 1 view a full section.

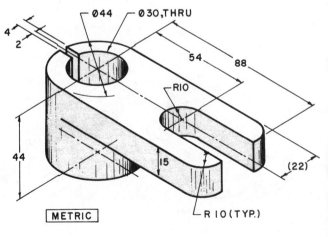

Problem 8-9

Center the required views within the work area, and add removed section A-A.

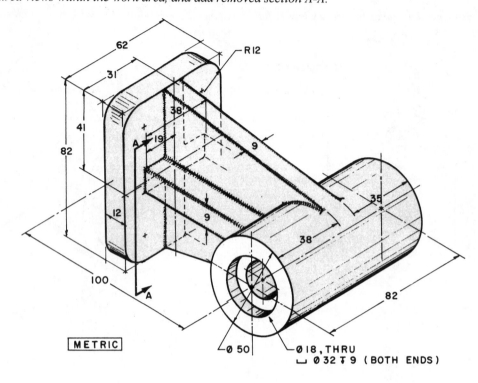

62
31
R12
38°
41
82
A 19
9
12
9
35
38
100
A
82
METRIC
Ø 50
Ø 18, THRU
⌴ Ø 32 ⊤ 9 (BOTH ENDS)

Problem 8-10

Center 3 views within the work area, and make 1 view an offset section.

Problem 8-11

Center 3 views within the work area, and make 1 view an offset section.

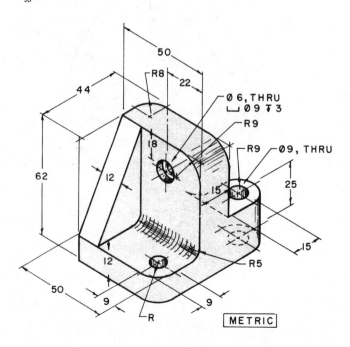

2 X .25 , THRU
⌴ Ø .50 ⊤ .125
3.5
1.25   .75
.50   2.0
1.0
R.50 (TYP.)
1.75
R.125 (TYP.)
1.38
R.50
.38
.25
.31
1.0
.31
R 22 (TYP.)
.38

50
R8   22
44
Ø 6 , THRU
⌴ Ø 9 ⊤ 3
R9
18
R9   Ø 9 , THRU
62
12
15
25
12
R5
50
9
9
15
R
METRIC

Problem 8-12

Center 3 views within the work area, and make 1 view an offset section.

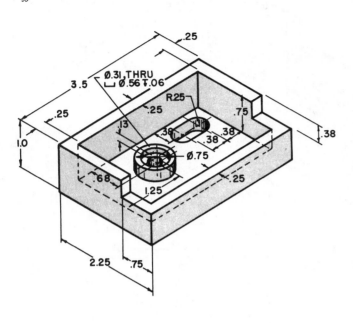

Problem 8-14

Center 2 views within the work area, and make 1 view an offset section.

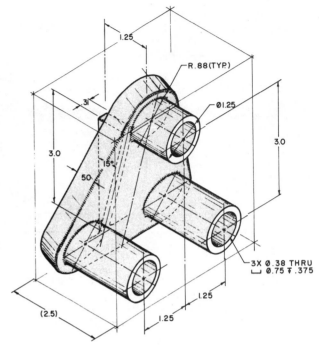

Problem 8-13

Center 3 views within the work area, and make 1 view an offset section.

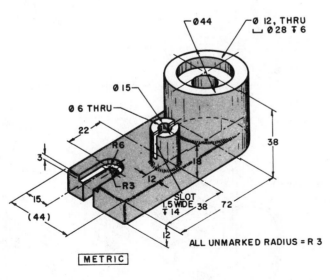

Problem 8-15

Center the front view and top view within the work area. Make 1 view a half section.

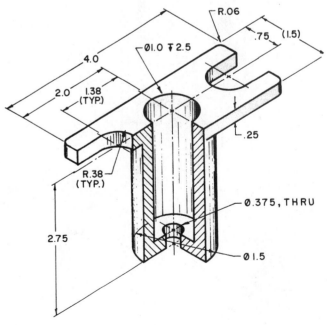

Problem 8-16
Center 2 views within the work area, and make 1 view a half section.

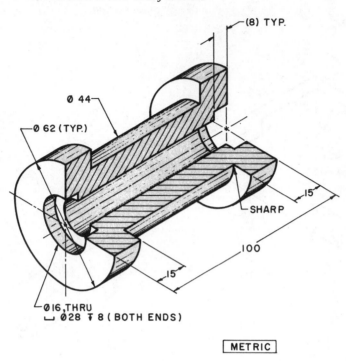

(8) TYP.

Ø 44

Ø 62 (TYP.)

SHARP

15

100

15

Ø16, THRU
⌴ Ø28 ⊤ 8 (BOTH ENDS)

METRIC

Problem 8-17
Center the 2 views within the work area, and make 1 view a half section.

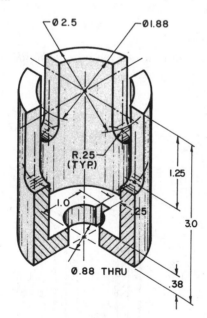

Ø 2.5   Ø1.88

R.25
(TYP.)

1.25

1.0

.25

3.0

Ø 0.88 THRU

.38

Problem 8-18
Center 2 views within the work area, and make 1 view a half section.

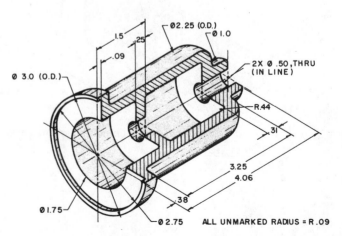

Ø2.25 (O.D.)
Ø 1.0

1.5   .25

.09

2X Ø .50, THRU
(IN LINE)

Ø 3.0 (O.D.)

R.44

.31

3.25
4.06

.38

Ø1.75

Ø 2.75   ALL UNMARKED RADIUS = R.09

Problem 8-19

Center 2 views within the work area, and make 1 view a half section.

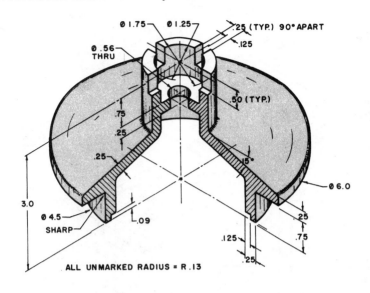

Ø 1.75 — Ø 1.25 — .25 (TYP.) 90° APART
.125
Ø .56 THRU
.75
.25
.50 (TYP.)
.25
5°
3.0
.25
Ø 6.0
Ø 4.5 SHARP
.09
.25
.125
.75
.25

ALL UNMARKED RADIUS = R.13

Problem 8-20

Center the required views within the work area, and make 1 view a broken-out section to illustrate the complicated interior area.

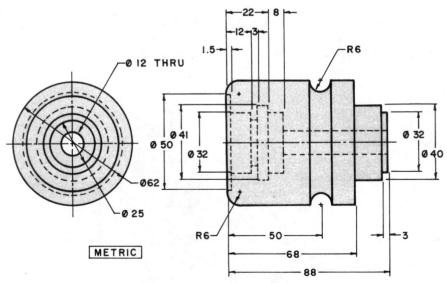

Ø 12 THRU
22
8
12
3
1.5
R6
Ø 41
Ø 50
Ø 32
Ø 32
Ø 40
Ø62
Ø 25
R6
50
3
METRIC
68
88

Problem 8-21
Center the required views within the work area, and add removed section A-A.

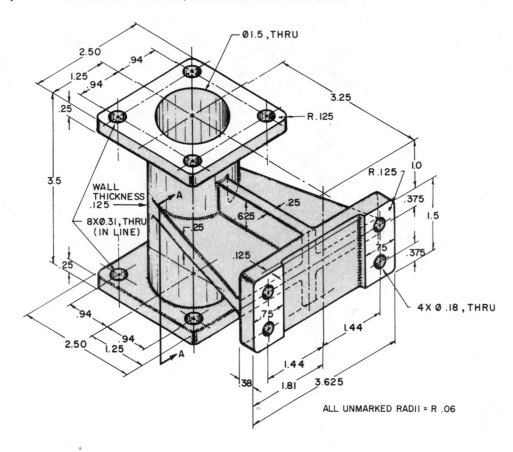

Problem 8-22
Center the required views within the work area, and add removed section A-A.

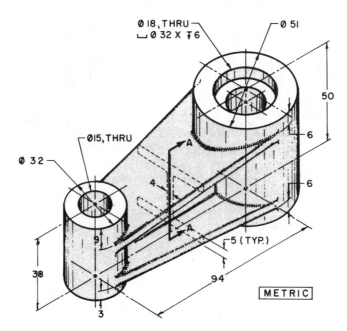

Directions for Problems 8-23 through 8-25:
Center required views within the work area. Leave a 1 inch or 25 mm space between views. Make 1 view into a section view to fully illustrate the object. Use a full half, offset, broken-out, revolved, or removed section. Consult your instructor if dimensions are to be added.

Problem 8-23

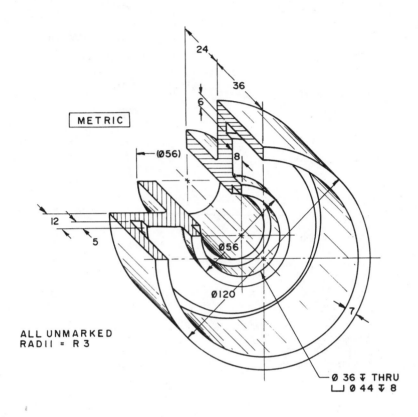

METRIC

(Ø56)

24

36

6

8

12

5

Ø56

Ø120

7

ALL UNMARKED
RADII = R 3

Ø 36 ⊤ THRU
⌴ Ø 44 ⊤ 8

Problem 8-24

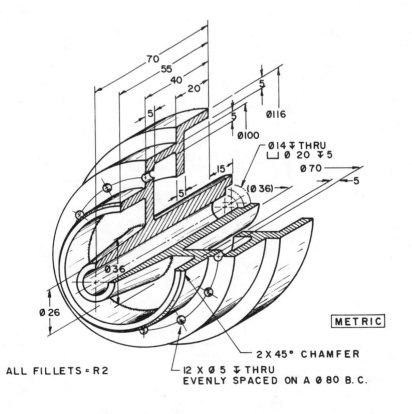

70

55

40

20

5

5

Ø116

5

Ø100

Ø14 ⊤ THRU
⌴ Ø 20 ⊤ 5

Ø 70

15

5

5

(Ø 36)

Ø 36

Ø 26

METRIC

ALL FILLETS = R2

12 X Ø 5 ⊤ THRU
EVENLY SPACED ON A Ø 80 B.C.

2 X 45° CHAMFER

Problem 8-25

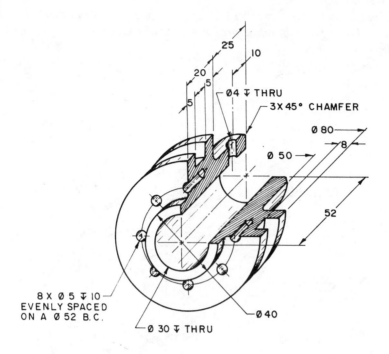

Directions for Problem 8-26:
Using the background grid as a guide, reproduce each problem on a CAD system. Add all dimensions. Use a grid spacing of
0.25 units.

Problem 8-26

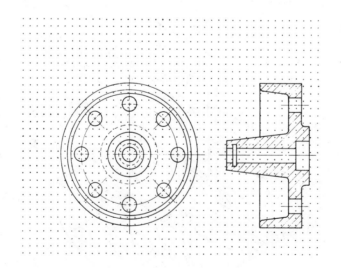

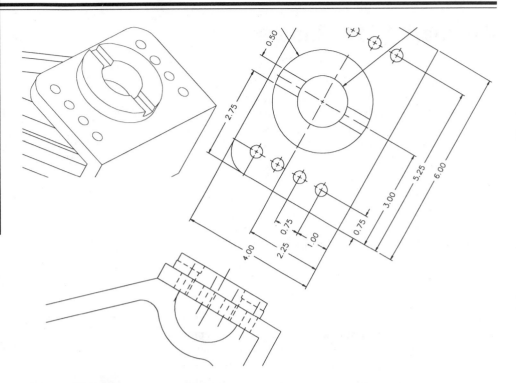

UNIT 9

Auxiliary Views

Contents

During the discussion of multi-view drawings it was pointed out the need to draw enough views of an object in order to accurately describe it. This requires a front, top, and right side view in most cases. Sometimes additional views are required such as left side, bottom, and back views to show features not visible in the three primary views. Other special views like sections are taken to expose interior details for better clarity. Sometimes all of these views are still not enough to describe the object, especially when features are located on an inclined surface. To produce a view perpendicular to this inclined surface, an auxiliary view is drawn. This unit will describe where auxiliary views are used and how they are projected from one view to another. A tutorial exercise is presented to go through the steps in the construction of an auxiliary view. Additional problems are provided at the end of this unit for further study of auxiliary views.

Auxiliary View Basics

The illustration at the right presents interesting results if constructed as a multi-view drawing or orthographic projection. Let us see how this object differs from others previously discussed in Unit 4.

The illustration at the right should be quite familiar; it represents the standard glass box with object located in the center. The purpose again of this box is to prove how orthographic views are organized and laid out. The example at the right is no different. First the primary views, front, top, and right side views are projected from the object to intersect perpendicular with the glass plane. Under normal circumstances, this procedure would satisfy most multi-view drawing cases. Remember, only those views necessary to describe the object are drawn. However, under closer inspection, we notice the object in the front view consists of an angle forming an inclined surface.

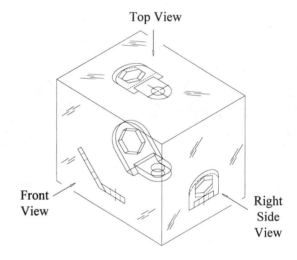

When laying out the front, top, and right side views, a problem occurs. The front view shows the basic shape of the object, the angle of the inclined surface. The top view shows true size and shape of the surface formed by the circle and arc. The right side view shows the true thickness of the hole from information found in the top view. However, there does not exist a true size and shape of the features found in the inclined surface at "A." We see the hexagonal hole going through the object in the top and right side views. These views, however, show the detail not to scale, or foreshortened. For that matter, the entire inclined surface is foreshortened in all views. This is one case where the front, top, and right side views are not enough to describe the object. An additional view, or auxiliary view, is used to display the true shape of surfaces along an incline.

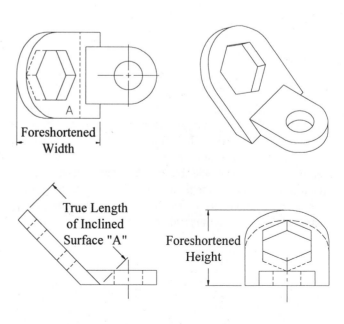

To prove the formation of an auxiliary view, lets create another glass box; this time an inclined plane is formed. This plane is always parallel to the inclined surface of the object. Instead of just projecting the front, top, and right side views, the geometry describing the features along the inclined surface is projected to the auxiliary plane similar to the illustration at the right.

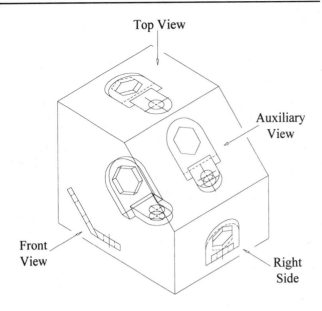

As in multi-view projection, the edges of the glass box are unfolded out using the edges of the front view plane as the pivot.

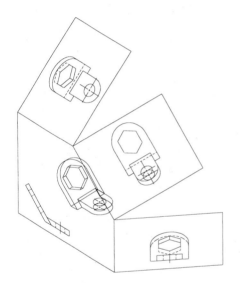

All planes are extended perpendicular to the front view where the rotation stops. The result is the organization of the multi-view drawing complete with an auxiliary viewing plane.

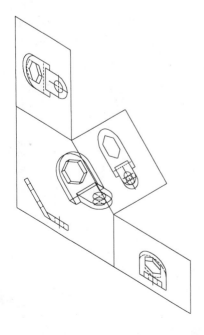

Illustrated at the right is the final layout complete with auxiliary view. This example shows the auxiliary being formed as
a result of the inclined surface being in the front view. An
auxiliary view may be made in relation to any inclined surface located in any view. Also, the illustration displays
circles and arcs in the top view which appear as ellipses in the
auxiliary view. It is usually not required to draw elliptical
shapes in one view where the feature is shown true size and
shape in another. The resulting view minus these elliptical
features is called a partial view which is used extensively in
auxiliary views. An example of the top view converted into a
partial view is displayed at "A".

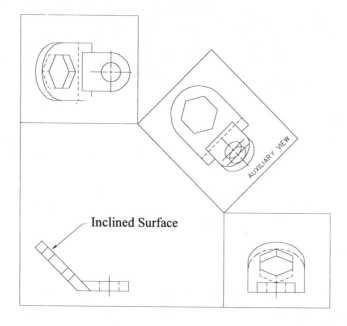

Inclined Surface

(A)

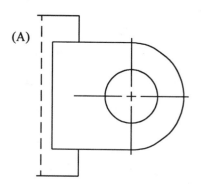

A few rules to follow when constructing auxiliary views are
displayed pictorially at the right. First, the auxiliary plane is
always constructed parallel to the inclined surface. Once this
is established, visible as well as hidden features are projected
from the incline to the auxiliary view. These projection lines
are always drawn perpendicular to the inclined surface and
the auxiliary view.

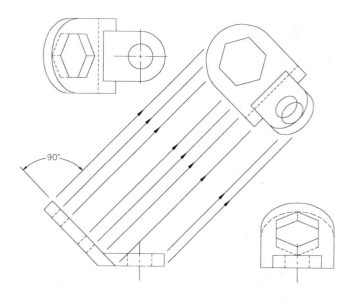

Constructing an Auxiliary View

Given at the right is a multi-view drawing consisting of front, top and right side views. The inclined surface in the front view is displayed in the top and right side views; however, the surface appears foreshortened in both adjacent views. An auxiliary view of the incline needs to be made to show its true size and shape. Currently the display screen has Grid On in addition to the position of the typical AutoCAD cursor. Follow the next series of illustrations for one suggested method for projecting to find auxiliary views.

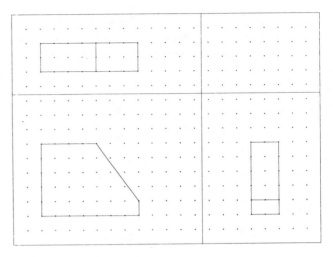

To assist with the projection process, it would help if the current grid display could be rotated parallel and perpendicular to the inclined surface. In fact, this can be accomplished using the Snap command and the prompts below:

Command: **Snap**
Snap spacing or ON/OFF/Aspect/Rotate/Style <0.50>: **Rotate**
Base point <0,0,0>: **Endp**
of *(Select the endpoint of the line at "A" as the new base point)*
Rotation angle <0>: **Endp**
of *(Select the endpoint of the line at "B" to define the rotation angle by pointing)*

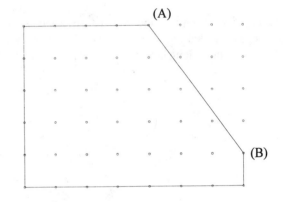

The results of rotating the grid through the Snap command are illustrated at the right. This operation has had no effect on the already existing views; however, the grid is now placed rotated in relation to the incline located in the front view. Notice the appearance of the standard AutoCAD cursor has also changed to conform to the new grid orientation. In addition to snapping to these new grid dots, lines are easily drawn perpendicular to the incline using Ortho On, which will draw lines in relation to the current cursor.

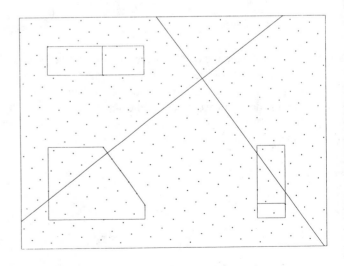

Use the Offset command to construct a reference line at a specified distance away from the incline in the front view. This reference line becomes the start for the auxiliary view.

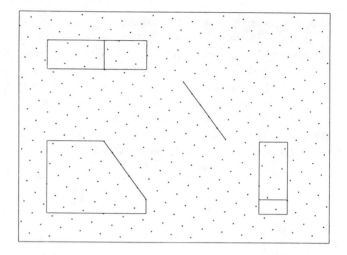

Use the Change command to extend the two endpoints of the previous line. The exact distances are not critical; however, the line should be long enough to accept projector lines from the front view.

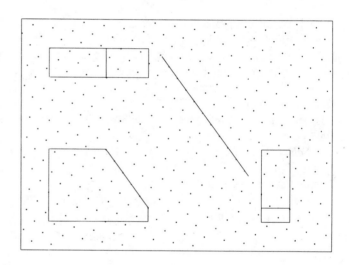

Use the Offset command to copy the auxiliary reference line the thickness of the object. This distance may be retrieved from the depth of the top or right side views since they both contain the depth measurement of the object.

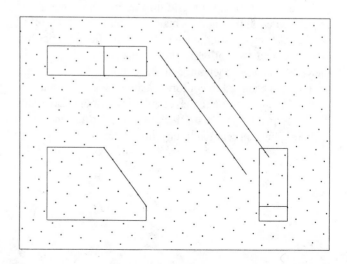

Use the Line command and connect each intersection on the front view perpendicular to the outer line on the auxiliary. Draw the lines starting with the Osnap-Intersec option and ending with the Osnap-Perpend option.

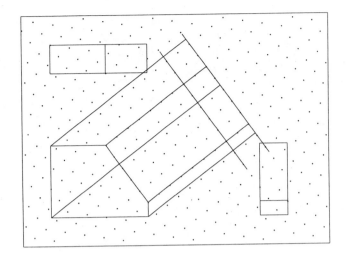

Before editing the auxiliary view, analyze the drawing to see if any corners in the front view are to be represented as hidden lines in the auxiliary view. It turns out that the lower left corner of the front view is hidden in the auxiliary view. Use the Chprop command to convert this projection line from a linetype of continuous to hidden. This is best accomplished by assigning the hidden linetype to a layer and then using Chprop to convert the line to the different layer.

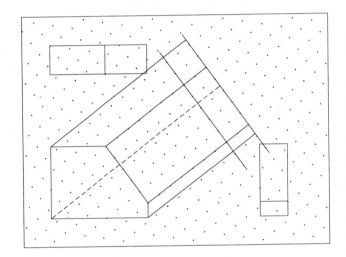

Use the Trim command to partially delete all projection lines using the inside auxiliary view line as the cutting edge.

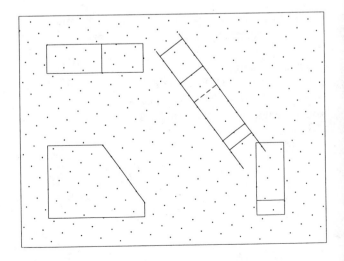

Use the Trim command to trim the corners of the auxiliary
view. A better solution would be to use the Fillet command
set to a radius value of 0. Selecting two lines would automatically corner the view.

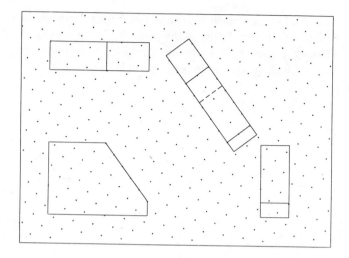

The result is a multi-view drawing complete with auxiliary
view displaying the true size and shape of the inclined
surface.

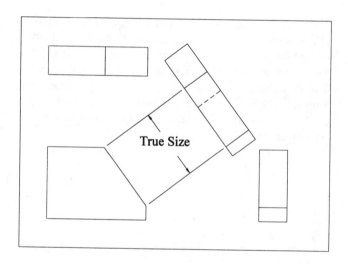

For dimensioning purposes, use the Snap command and set
the grid and cursor appearance back to normal.

Command: **Snap**
Snap spacing or ON/OFF/Aspect/Rotate/Style <0.50>: **Rotate**
Base point <Current default>: **0,0,0**
Rotation angle <Current default>: **0**

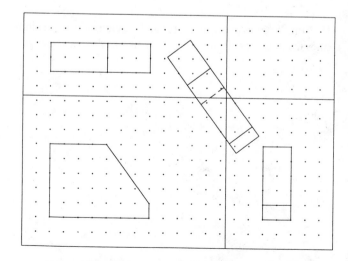

Tutorial Exercise #23
Bracket.Dwg

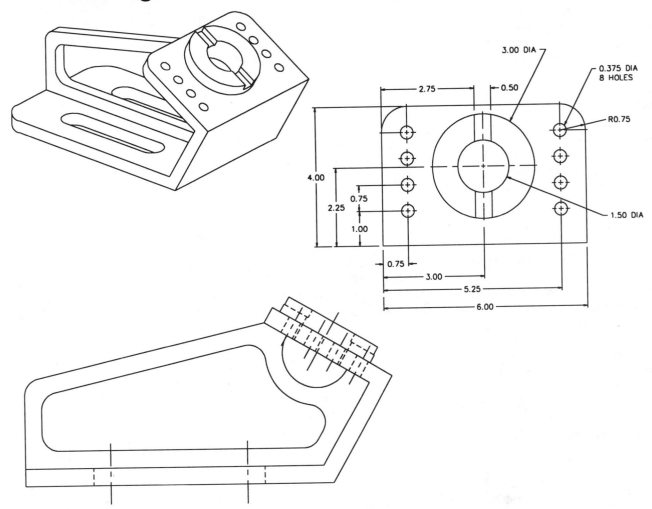

PURPOSE:
This tutorial is designed to allow the user to construct an auxiliary view of the inclined surface above in the Bracket.

SYSTEM SETTINGS:
Since this drawing is provided on diskette, edit an existing drawing called "Bracket." Follow the steps in this tutorial for the creation of an auxiliary view. All units, limits, grid, and snap values have been previously set.

LAYERS:
The following layers have already been created with the following format:

Name-Color-Linetype
Cen - Yellow - Center
Dim - Yellow - Continuous
Hid - Red - Hidden

SUGGESTED COMMANDS:
Begin this tutorial by using the Offset command to construct a series of lines parallel to the inclined surface containing the auxiliary view. Next construct lines perpendicular to the inclined surface. Use the Circle command to begin laying out features that lie in the auxiliary view. Use Array to copy the circle in a rectangular pattern. Add center lines using the Dim-Center command. Insert a predefined view called "Top." A three-view drawing consisting of Front, Top, and Auxiliary views is completed.

DIMENSIONING:
This drawing may be dimensioned at a later date. Consult your instructor before continuing.

PLOTTING:
This tutorial exercise may be plotted on "D"-size paper (24" x 36"). Use a plotting scale of 1=1 to produce a full size plot.

Step #1

Before beginning, understand that an auxiliary view will be taken from a point of view illustrated at the right. This direction of sight is always perpendicular to the inclined surface. This perpendicular direction ensures the auxiliary view of the inclined surface will be of true size and shape. Begin this tutorial by turning ortho Off for the time being. Restore a previously saved view called "Front."

Command: **Ortho**
ON/OFF <On>: **Off**

Command: **View**
?/Delete/Restore/Save/Window: **R**
View name to restore: **Front**

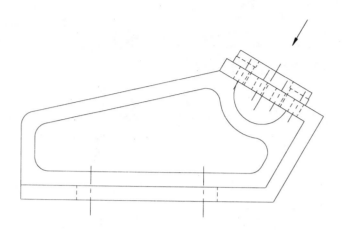

Step #2

Use the Snap command to rotate the grid perpendicular to the inclined surface. For the base point, identify the endpoint of the line at "A." For the rotation angle, use the "rubberband" cursor and mark a point at the endpoint of the line at "B." The grid should change along with the standard AutoCAD cursor.

Command: **Snap**
Snap spacing or ON/OFF/Aspect/Rotate/Style <0.50>: **Rotate**
Base point <0,0>: **Endp**
of *(Select the endpoint of the line at "A")*
Rotation angle <0>: **Endp**
of *(Select the endpoint of the line at "B")*

Command: **View**
?/Delete/Restore/Save/Window: **R**
View name to restore: **Overall**

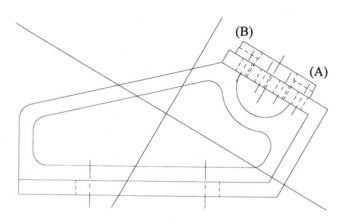

Step #3

Turn snap Off by striking the F9 function key. Begin the construction of the auxiliary view by using the Offset command to copy a line parallel to the inclined line. Use an offset distance of 8.50 as the distance between the front and auxiliary view.

Command: **Offset**
Offset distance or Through <Through>: **8.50**
Select object to offset: *(Select the inclined line at "A")*
Side to offset? *(Select a point anywhere near "B")*
Select object to offset: *(Strike Enter to exit this command)*

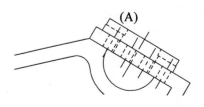

Step #4

Refer to the working drawing on page 511 for the necessary dimensions required to construct the auxiliary view. Use the Offset command again to add the depth of the auxiliary. Remember, the depth of the auxiliary view is the same dimension found in the top and right side views. Set the offset distance to 6.00. Set the new current layer to "Obj."

Command: **Offset**
Offset distance or Through <8.50>: **6.00**
Select object to offset: *(Select the inclined line at "A")*
Side to offset? *(Select a point anywhere near "B")*
Select object to offset: *(Strike Enter to exit this command)*

Command: **Layer**
?/Make/Set/New/ON/OFF/Color/Ltype/Freeze/Thaw/LOck/
Unlock: **Set**
New current layer <0>: **Obj**
?/Make/Set/New/ON/OFF/Color/Ltype/Freeze/Thaw/LOck/
Unlock: *(Strike Enter to exit this command)*

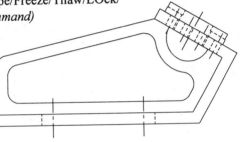

Step #5

Project two lines from the endpoints of the front view at "A" and "B." These lines should extend past the outer line of the auxiliary view. Turn the Snap Off and Ortho On. This should aid in this operation.

Command: **Ortho**
ON/OFF <Off>: **On**

Command: **Line**
From point: **Endp**
of *(Pick the endpoint of the line at "A")*
To point: *(Pick a point anywhere near "B")*
To point: *(Strike Enter to exit this command)*

Command: **Line**
From point: **Endp**
of *(Pick the endpoint of the line at "C")*
To point: *(Pick a point anywhere near "D")*
To point: *(Strike Enter to exit this command)*

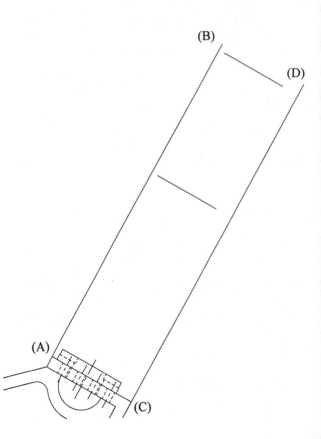

Step #6

Use the Zoom-Window option to magnify the display of the auxiliary view similar to the illustration at the right. Then use the Extend command, select the boundary edges at "A" and "B," and extend the four endpoints of the lines.

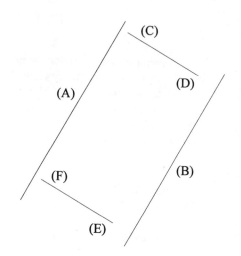

Command: **Extend**
Select boundary edge(s)...
Select objects: *(Select the two lines at "A" and "B")*
Select objects: *(Strike Enter to continue)*
<Select object to extend>/Undo: *(Select the end of the line at "C")*
<Select object to extend>/Undo: *(Select the end of the line at "D")*
<Select object to extend>/Undo: *(Select the end of the line at "E")*
<Select object to extend>/Undo: *(Select the end of the line at "F")*
<Select object to extend>/Undo: *(Strike Enter to exit this command)*

Step #7

Use the Trim command, select the lines at "A" and "B" as cutting edges, and trim away the ends of the four lines labeled at the right. While in this display, use the View command to save as "Aux" for future reference.

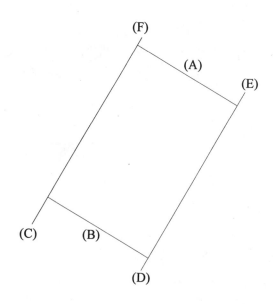

Command: **Trim**
Select cutting edge(s)...
Select objects: *(Select the two lines at "A" and "B")*
Select objects: *(Strike Enter to continue)*
<Select object to trim>/Undo: *(Select the line at "C")*
<Select object to trim>/Undo: *(Select the line at "D")*
<Select object to trim>/Undo: *(Select the line at "E")*
<Select object to trim>/Undo: *(Select the line at "F")*
<Select object to trim>/Undo: *(Strike Enter to exit this command)*

Command: **View**
?/Delete/Restore/Save/Window: **Save**
View name to save: **Aux**

Step #8

Use the Zoom-Previous option to demagnify the screen back to the original display. Once here, draw a line from the endpoint of the center line in the front view to a point past the auxiliary view. Check to see that Ortho mode is On. This line will assist in constructing circles in the auxiliary view.

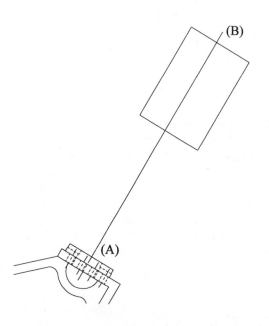

Command: **Line**
From point: **Endp**
of *(Select the endpoint of the center line at "A")*
To point: *(Select a point anywhere near "B")*
To point: *(Strike Enter to exit this command)*

Step #9

Use the Offset command and offset the line at "A" a distance of 3.00 units. The intersection of this line and the previous line form the center for placing two circles.

Command: **Offset**
Offset distance or Through <6.00>: **3.00**
Select object to offset: *(Select the line at "A")*
Side to offset? *(Select a point anywhere near "B")*
Select object to offset: *(Strike Enter to exit this command)*

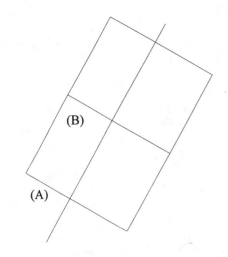

Step #10

Use the View command and restore the view "Aux." Then, draw two circles of diameters 3.00 and 1.50 from the center at "A" using the Circle command. For the center of the second circle, the @ option may be used to pick up the previous point that was the center of the 3.00 diameter circle.

Command: **View**
?/Delete/Restore/Save/Window: **Restore**
View name: **Aux**

Command: **Circle**
3P/2P/TTR/<Center point>: **Int**
of *(Select the intersection of the two lines at "A")*
Diameter/<Radius>: **D**
Diameter: **3.00**

Command: **Circle**
3P/2P/TTR/<Center point>: **@** *(To reference the last point)*
Diameter/<Radius>: **D**
Diameter: **1.50**

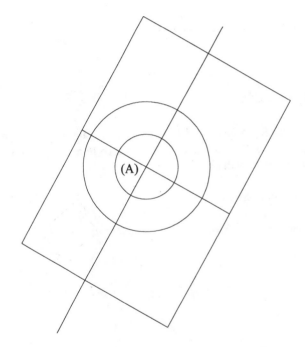

Step #11

Use the Offset command to offset the center line the distance of 0.25 units. Perform this operation on both sides of the center line. Both offset lines form the width of the 0.50 slot.

Command: **Offset**
Offset distance or Through <3.00>: **0.25**
Select object to offset: *(Select the middle line at "A")*
Side to offset? *(Select a point anywhere near "B")*
Select object to offset: *(Select the middle line at "A" again)*
Side to offset? *(Select a point anywhere near "C")*
Select object to offset: *(Strike Enter to exit this command)*

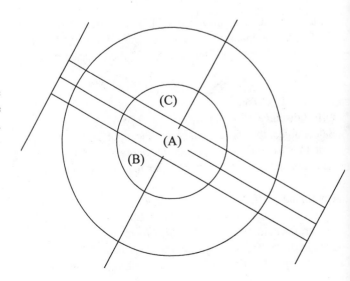

Step #12

Use the Trim command to trim away portions of the lines illustrated at the right.

Command: **Trim**
Select cutting edge(s)...
Select objects: *(Select both circles as cutting edges)*
Select objects: *(Strike Enter to continue)*
<Select object to trim>/Undo: *(Select the line at "A")*
<Select object to trim>/Undo: *(Select the line at "B")*
<Select object to trim>/Undo: *(Select the line at "C")*
<Select object to trim>/Undo: *(Select the line at "D")*
<Select object to trim>/Undo: *(Select the line at "E")*
<Select object to trim>/Undo: *(Select the line at "F")*
<Select object to trim>/Undo: *(Strike Enter to exit this command)*

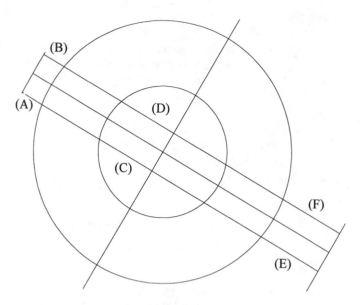

Step #13

Use the Erase command to delete the two lines at "A" and "B." Standard center lines will be placed here later marking the center of both circles.

Command: **Erase**
Select objects: *(Select the lines at "A" and "B")*
Select objects: *(Strike Enter to execute this command)*

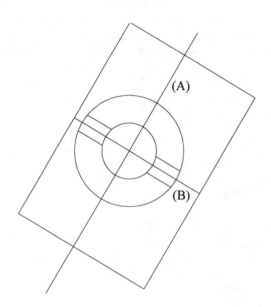

Step #14

To identify the center of the small 0.375 diameter circle, use the Offset command to copy parallel the line at "A" a distance of 0.75 units and the line at "C" the distance of 1.00 units.

Command: **Offset**
Offset distance or Through <0.25>: **0.75**
Select object to offset: *(Select the line at "A")*
Side to offset? *(Select a point anywhere near "B")*
Select object to offset: *(Strike Enter to exit this command)*

Command: **Offset**
Offset distance or Through <0.75>: **1.00**
Select object to offset: *(Select the line at "C")*
Side to offset? *(Select a point anywhere near "D")*
Select object to offset: *(Strike Enter to exit this command)*

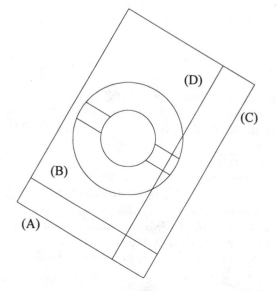

Step #15

Draw a circle of 0.375 diameter from the intersection of the two lines created in the last offset command. Use the Erase command to delete the two lines at "A" and "B." A standard center marker will be placed at the center of this circle.

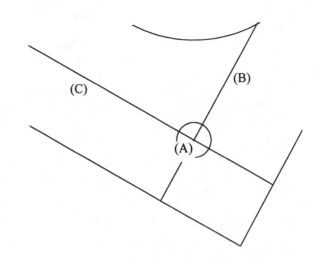

Command: **Circle**
3P/2P/TTR/<Center point>: **Int**
of *(Select the intersection of the two lines at "A")*
Diameter/<Radius>: **D**
Diameter: **0.375**

Command: **Erase**
Select objects: *(Select the two lines at "B" and "C")*
Select objects: *(Strike Enter to execute this command)*

Step #16

Set the new current layer to "Cen." Prepare the following parameters before placing a center marker at the center of the 0.375 diameter circle. Set the dimension variable Dimcen to a value of -0.07 units. The negative value will construct lines that are drawn outside of the circle. Use the Dim-Cen command to place the center marker.

Command: **Layer**
?/Make/Set/New/ON/OFF/Color/Ltype/Freeze/ThawLOck/
Unlock: **Set**
New current layer <OBJ>: **Cen**
?/Make/Set/New/ON/OFF/Color/Ltype/Freeze/ThawLOck/
Unlock: *(Strike Enter to exit this command)*

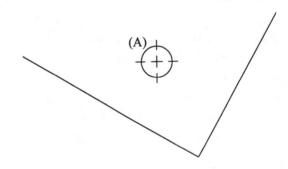

Command: **Dim**
Dim: **Dimcen**
Current value <0.09> New value: **-0.07**

Dim: **Cen**
Select arc or circle: *(Select the small circle at "A")*

Dim: **Exit** *(To exit dimensioning and return to the Command prompt)*

Step #17

Use the Rotate command to rotate the center marker parallel to the edges of the auxiliary view. Select the center marker and circle as the entities to rotate. Check to see that Ortho is On.

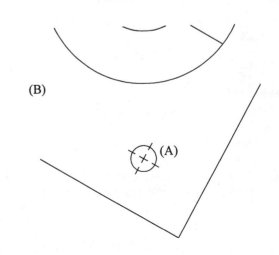

Command: **Rotate**
Select objects: *(Select the small circle and all entities that make up the center marker; the Window option is recommended here)*
Select objects: *(Strike Enter to continue)*
Base point: **Cen**
of *(Select the edge of the small circle at "A")*
<Rotation angle>/Reference: *(Pick a point anywhere at "B")*

Step #18

Since the remaining 7 holes are along a set pattern, use the Array command and perform a rectangular array. The number of rows are 2 and number of columns 4. Distance between rows is 4.50 units and between columns is -0.75 units; this will force the circles to be patterned to the left, which is where we want them to go.

Command: **Array**
Select objects: *(Select the small circle and center marker)*
Select objects: *(Strike Enter to continue)*
Rectangular or Polar array (R/P) <R>: **R**
Number of rows(---) <1>: **2**
Number of columns(||||) <1>: **4**
Unit cell or distance between rows (---): **4.50**
Distance between columns (||||): **−0.75**

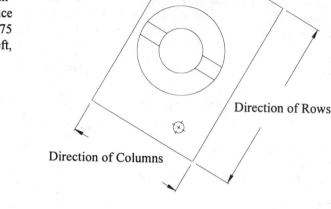

Direction of Rows

Direction of Columns

Step #19

Use the Fillet command set to a radius of 0.75 to place a radius along the two corners of the auxiliary following the prompts and the illustration below.

Command: **Fillet**
Polyline/Radius/<Select first object>: **R**
Enter fillet radius <0.00>: **0.75**

Command: **Fillet**
Polyline/Radius/<Select first object>: *(Select line "A")*
Select second object: *(Select line "B")*

Command: **Fillet**
Polyline/Radius/<Select first object>: *(Select lines "B")*
Select second object: *(Select line "C")*

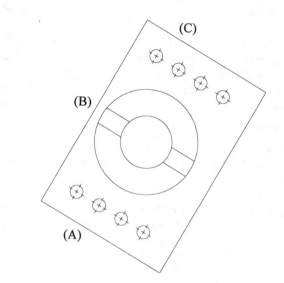

(C)

(B)

(A)

Step #20

Place a center marker in the center of the two large circles using the existing value of the dimension variable Dimcen. Since the center marker is placed in relation to the World coordinate system, use the Rotate command to rotate it parallel to the auxiliary view. Ortho should be On.

Command: **Dim**
Dim: **Cen**
Select arc or circle: *(Select the large circle at "A")*
Dim: **Exit**

Command: **Rotate**
Select objects: *(Select all lines that make up the large center marker)*
Select objects: *(Strike Enter to continue)*
Base point: **Cen**
of *(Select the edge of the large circle at "A")*
<Rotation angle>/Reference: *(Pick a point anywhere at "B")*

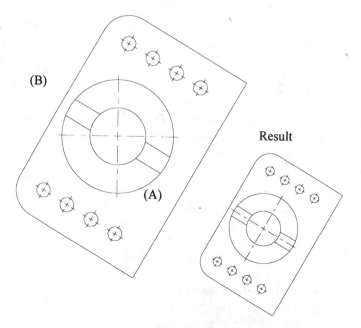

(B)

(A)

Result

Step #21

Restore the view named "Overall." Complete the multi-view drawing of the Bracket by inserting an existing block called "Top" into this drawing. This block represents the complete top view of the drawing. Use an insertion point of 0,0 for placing this view in the drawing.

Command: **View**
?/Delete/Restore/Save/Window: **R**
View name to restore: **Overall**

Command: **Insert**
Block name (or ?): **Top**
Insertion point: **0,0**
X scale factor <1>/Corner/XYZ: *(Strike Enter to accept default)*
Y scale factor (default=X): *(Strike Enter to accept default)*
Rotation angle <0>: *(Strike Enter to accept this default)*

Step #22

Return the grid back to its original orthographic form using the Snap-Rotate option. Use a base point of 0,0 and a rotation angle of 0 degrees. This is especially helpful when adding dimensions to the drawing.

Command: **Snap**
Snap spacing or ON/OFF/Aspect/Rotate/Style <0.50>: **Rotate**
Base point <0,0>: **0,0**
Rotation angle <330>: **0**

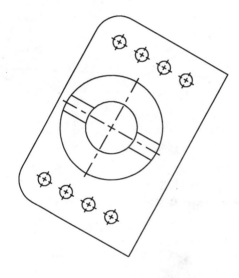

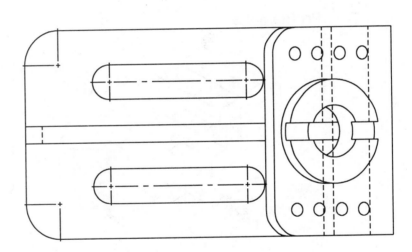

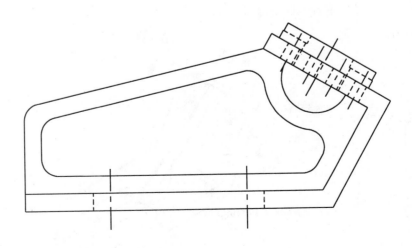

Problems for Unit 9

Directions for Problems 9-1 through 9-11
Draw the required views to fully illustrate each object. Be sure to include an auxiliary view.

Problem 9-1

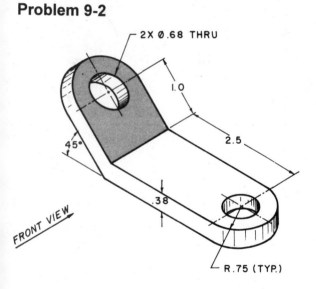

Problem 9-4

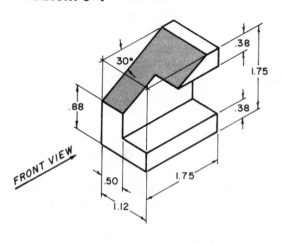

Problem 9-2

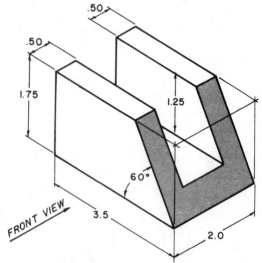

Problem 9-5

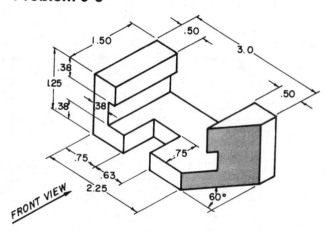

Problem 9-3

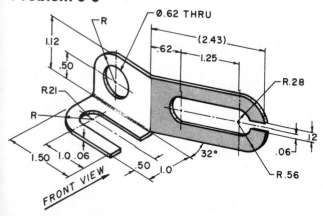

Problem 9-6

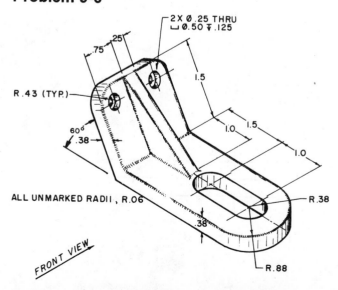

Problem 9–7

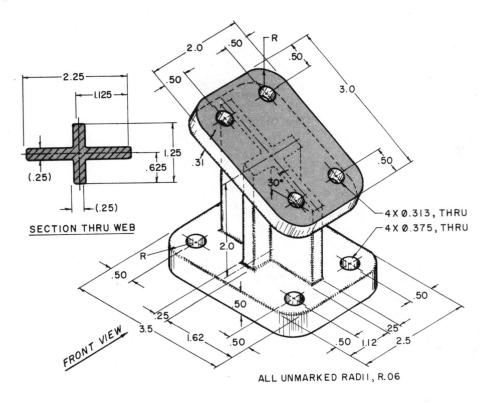

2.25

1.125

1.25

.625

(.25)

(.25)

SECTION THRU WEB

2.0

.50

R

.50

.50

3.0

.31

.50

30°

.50

4X Ø.313, THRU

4X Ø.375, THRU

R

2.0

.50

.50

.50

.25

.25

3.5

1.62

.50

.50

1.12

2.5

FRONT VIEW

ALL UNMARKED RADII, R.06

Problem 9–8

Problem 9–9

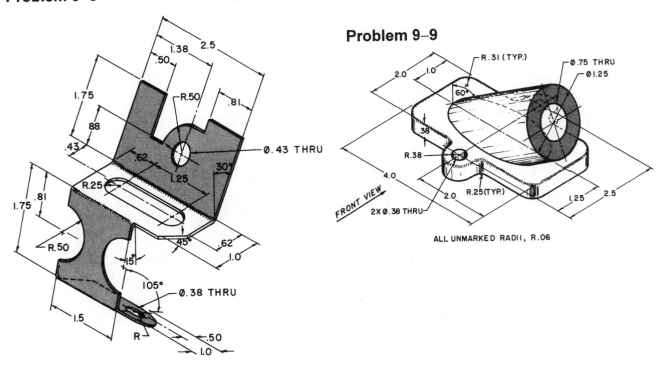

1.38

2.5

.50

1.75

R.50

.81

88

.43

62

Ø.43 THRU

30°

1.25

R.25

1.75

.81

45°

.62

R.50

15°

1.0

105°

Ø.38 THRU

1.5

R

.50

1.0

R.31 (TYP.)

Ø.75 THRU

Ø1.25

2.0

1.0

60°

38

R.38

4.0

2.0

R.25(TYP.)

1.25

2.5

FRONT VIEW

2X Ø.38 THRU

ALL UNMARKED RADII, R.06

Problem 9–10

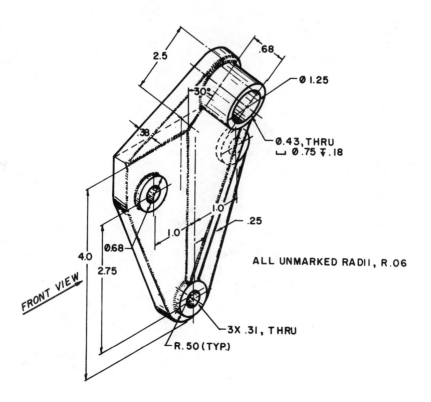

FRONT VIEW

2.5   .68

30°

Ø 1.25

.38

Ø.43, THRU
⌴ Ø.75 ▼.18

1.0

.25

1.0

Ø.68

ALL UNMARKED RADII, R.06

4.0

2.75

3X .31, THRU

R.50 (TYP.)

Problem 9–11

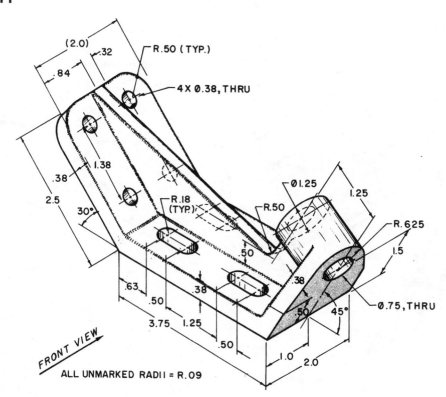

(2.0)   .32   R.50 (TYP.)

.84

4X Ø.38, THRU

1.38

.38

Ø1.25   1.25

2.5

30°

R.18
(TYP.)   R.50   R.625

.50   1.5

.63   .38   .38

.50   .50

3.75   1.25   45°   Ø.75, THRU

.50   .50

FRONT VIEW   1.0   2.0

ALL UNMARKED RADII = R.09

UNIT
10

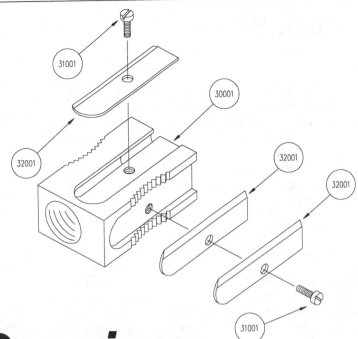

Isometric Drawings

Contents

Multi-view or orthographic projections are necessary to produce parts which go into the construction of all kinds of objects. Skill is involved in laying out the primary views, projecting visible entities into other views, and adding dimensions to describe the size of the object being made. Yet another skill involves reading or interpreting these engineering drawings, which for some individuals is extremely difficult and complex. If only there existed some type of picture of the object, then the engineering drawing might make sense. Isometric drawings become a means of drawing an object in picture form for better clarification of what the object looks like. These types of drawings resemble a picture of an object that is drawn in two dimensions. As a result, existing AutoCAD commands such as Line and Copy are used for producing isometric drawings. This unit will explain isometric basics including how regular, angular, and circular entities are drawn in isometric. Numerous isometric aids such as snap and isometric axes will be explained to assist in the construction of isometric drawings.

Isometric Basics

Isometric drawings consist of 2D drawings that are tilted at some angle to expose other views and give the viewer the illusion that what he or she is viewing is a 3D drawing. The tilting occurs with two 30 degree angles that are struck from the intersection of a horizontal baseline and a vertical line. The directions formed by the 30 degree angles represent actual dimensions of the object; this may be either the width or depth. The vertical line in most cases represents the height dimension.

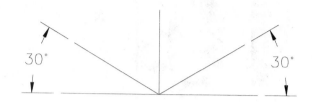

The object illustrated at the right is a very simple example of how an object is aligned to the isometric axis. Once the horizontal baseline and vertical line are drawn, the 30 degree angles are projected from this common point which becomes the reference point of the isometric view. In this example, once the 30 degree lines are drawn, the baseline is no longer needed and is usually discarded through erasing. Depending on how the object is to be viewed, width and depth measurements are made along the 30 degree lines. Height is measured along the vertical line. The example at the right has the width dimension measured off to the left 30 degree line while the depth dimension measures to the right along the right 30 degree line. Once the object is blocked with overall width, depth, and height, details are added, and lines are erased and trimmed, leaving the finished object. Holes no longer appear as full circles but rather as ellipses. Techniques of drawing circles in isometric will be discussed later in this unit.

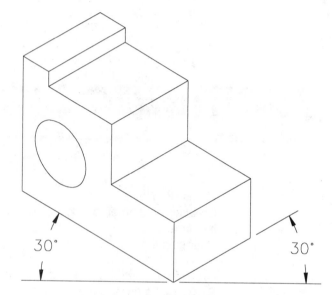

Notice the objects at the right both resemble the previous example except they appear from a different vantage point. The problem with isometric drawings is that if an isometric of an object is drawn from one viewing point and you want an isometric from another viewing point, an entirely different isometric drawing must be generated from scratch. Complex isometric drawings from different views can be very tedious to draw. Another interesting observation concerning the objects at the right is one has hidden lines while the other does not. Usually only the visible surfaces of an object are drawn in isometric leaving out hidden lines. As this is considered good practice, there are always times that hidden lines are needed on very complex isometric drawings.

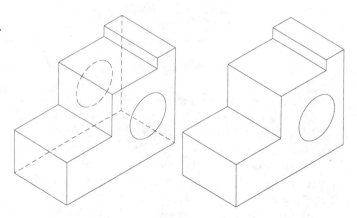

Creating an Isometric Grid

The display illustrated at the right shows the current AutoCAD screen complete with cursor and grid on. In manual drawing and sketching days, an isometric grid was used to lay out all lines before transferring the lines to paper or Mylar for pen and ink drawings. An isometric grid may be defined in an AutoCAD drawing using the Snap command. This would be the same grid found on isometric grid paper.

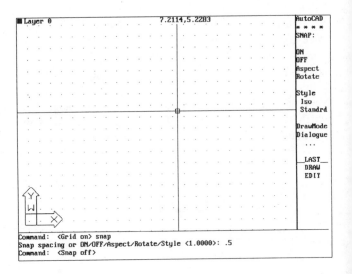

The display screen at the right reflects the use of the Snap command and how this command affects the current grid display:

Command: **Snap**
Snap spacing or ON/OFF/Rotate/Style <0.2500>: **Style**
Standard/Isometric <S>: **Isometric**
Vertical spacing <0.2500>: *(Strike Enter to accept default value)*

Choosing an isometric style of snap changes the grid display from orthographic to isometric, illustrated at the right. The grid distance conforms to a vertical spacing height specified by the user. As the grid changes, notice the display of the typical AutoCAD cursor; it conforms to an isometric axis plane and is used as an aid in constructing isometric drawings.

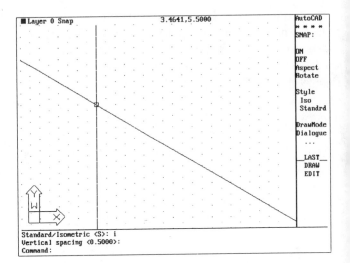

To see how this vertical spacing distance affects the grid changing it to isometric, see the illustration at the right. The grid dot at "A" becomes the reference point where the horizontal baseline is placed followed by the vertical line represented by the dot at "B." At dots "A" and "B," 30 degree lines are drawn; points "C" and "D" are formed where they intersect. This is how an isometric screen display is formed.

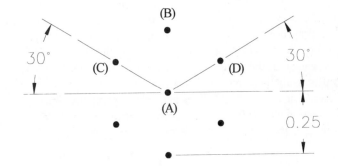

Isoplane Modes

The AutoCAD cursor has always been the vehicle for drawing entities or constructing windows for object selection mode. Once in isometric snap mode, AutoCAD supports three axes to assist in the construction of isometric drawings. The first axis is the Left axis and may control that part of an object falling into the left projection plane. The left axis cursor displays a vertical line intersected by a 30 degree angle line, which is drawn to the left. This axis is displayed in the illustration at the right in addition to the drawing below:

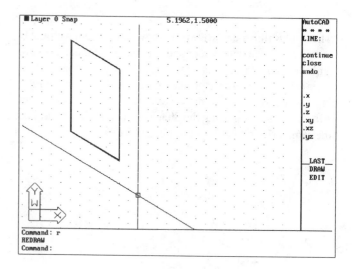

The next isometric axis is the Top mode. Entities falling into the top projection plane may be drawn using this isometric axis. This cursor consists of two 30 degree angle lines intersecting each other forming the center of the cursor. This mode is displayed at the right and below:

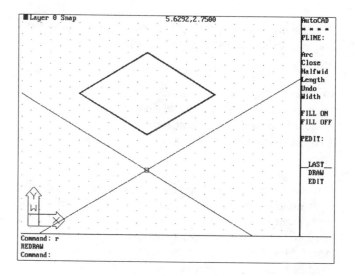

The final isometric axis is called the Right mode and is formed by the intersection of a vertical line and a 30 degree angle drawn off to the right. As with the previous two modes, entities that fall along the right projection plane of an isometric drawing may be drawn using this cursor. It is displayed at the right and below. The current Ortho mode affects all three modes. If Ortho is On, and the current isometric axis is Right, lines and other operations requiring direction will be forced to be drawn vertical or at a 30 degree angle to the right as shown below:

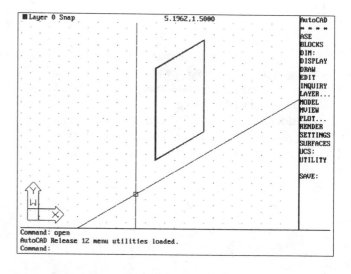

Resetting Grid and Snap Back to Their Default Values

Once an isometric drawing is completed, it may be necessary to change the grid, snap, and cursor back to normal. This might result from the need to place text on the drawing, and the isometric axis now confuses instead of assists the drawing process. Follow the prompts at the right for resetting the snap back to the Standard spacing.

Command: **Snap**
Snap spacing or ON/OFF/Rotate/Style <0.2500>: **Style**
Standard/Isometric <I>: **Standard**
Spacing/Aspect <0.2500>: *(Strike Enter to accept the default value)*

Notice when changing the snap style to Standard, the AutoCAD cursor changes back to its original display.

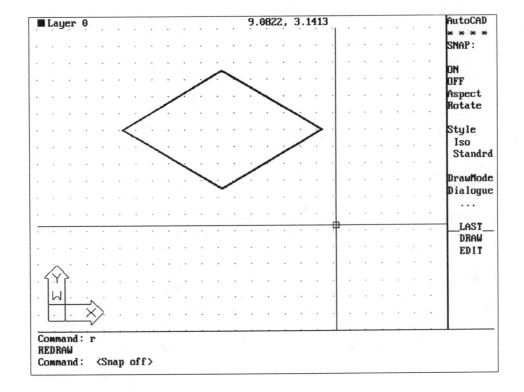

Isoplane Aids

It has been previously mentioned that there are three isometric axis modes to draw on: Left, Top, and Right. It was not mentioned how to make a mode current to draw on. The Settings area of the menu bar exposes the "Drawing Aids..." area, which displays the dialog box illustrated below. Here in addition to Snap and Grid the three isometric modes may be current by simply placing a check in the appropriate box. Only one isometric mode may be current at a time. In addition to setting these modes, an isometric area exists to automatically set up an isometric grid by placing a check in the box and put it back to normal by removing the check. This has the same effect as using the Snap-Style-Isometric option. This dialog box may be brought up through the keyboard by entering "DDRMODES" at the command prompt or by entering "'DDRMODES" while inside of a command.

```
View   Settings  Render  Model
       Drawing Aids...          ▓
       Layer Control...     ↖
       Object Snap...

       Entity Modes...
       Point Style...

       Dimension Style...
       Units Control...

       UCS                    ▷

       Selection Settings...
       Grips...

       Drawing Limits
```

Drawing Aids

Modes	Snap	Grid
☐ Ortho	☐ On	☐ On
☒ Solid Fill	X Spacing `1.0000`	X Spacing `0.0000`
☐ Quick Text	Y Spacing `1.0000`	Y Spacing `0.0000`
☒ Blips	Snap Angle `0`	**Isometric Snap/Grid**
☒ Highlight	X Base `0.0000`	☐ On
	Y Base `0.0000`	■ Left ☐ Top ☐ Right

| OK | Cancel | Help... |

A quicker method exists to move from one isometric axis mode to another. By default, after setting up an isometric grid, the Left isometric axis is active. By typing from the keyboard CTRL-E or ^E, the Left axis changes to the Top axis. Typing another ^E changes from the Top axis to the Right axis. Typing a third ^E changes from the Right axis back to the Left axis and the pattern repeats from here. Using this keyboard entry, it is possible to switch or toggle from one mode to another.

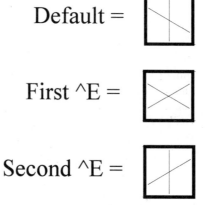

Default =

First ^E =

Second ^E =

Creating Isometric Circles

Circles appear as ellipses when drawn in any of the three isometric axes. The Ellipse command has a special Isocircle option to assist in drawing isometric circles; the Isocircle option will appear in the Ellipse command only if the current Snap-Style is Isometric. The prompt sequence for this command is:

Command: **Ellipse**
<Axis endpoint 1>/Center/Isocircle: **Iso**
Center of circle: *(Select a center point)*
<Circle Radius>/Diameter: *(Enter a value for the radius or type "D" for diameter and enter a value)*

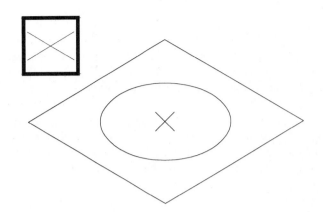

When drawing isometric circles using the Ellipse command, it is important to match the isometric axis with the isometric plane the circle is to be drawn in. Illustrated at the right is a cube displaying all three axes.

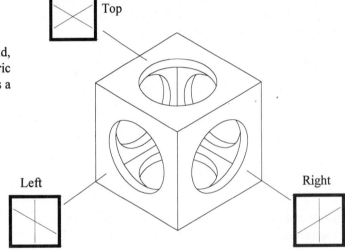

The example at the right is the result of drawing an isometric circle using the wrong isometric axis. The isometric box is drawn in the top isometric plane while the current isometric axis is Left. An isometric circle can be drawn to the correct size, but notice it does not match the box it was designed for. If you notice halfway through the Ellipse command that you are in the wrong isometric axis, type ^E until the correct axis appears.

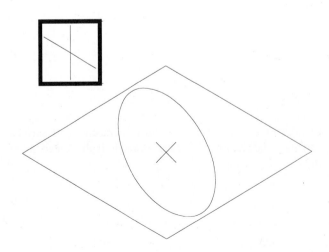

Basic Isometric Construction

Any isometric drawing, no matter how simple or complex, has an overall width, height, and depth dimension. Start laying out the drawing with these three dimensions to create an isometric box illustrated in the example at the right. Some techniques rely on piecing the isometric drawing together by views; unfortunately, it is very easy to get lost in all of the lines using this method. Once a box is created from overall dimensions, somewhere inside the box is the object.

With the box as a guide, begin laying out all visible features in the primary planes. Use the Left, Top, or Right isometric axis modes to assist you in this construction process.

Existing AutoCAD editing commands, especially Copy, may be used to duplicate geometry to show depth of features. Next, use the Trim command to partially delete geometry where entities are not visible. Remember, most isometric objects do not require hidden lines.

Use the Line command to connect intersections of surface corners. The resulting isometric drawing is illustrated at the right.

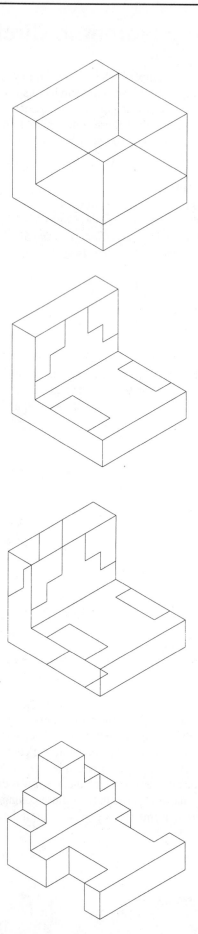

Creating Angles in Isometric - Method #1

Drawing angles in isometric is a little tricky but not impossible. The 2-view drawing at the right has an angle of unknown size; however, one endpoint of the angle measures 2.06 units from the top horizontal line of the front view at "B" and the other endpoint measures 1.00 unit from the vertical line of the front view at "A." This is more than enough information needed to lay out the endpoints of the angle in isometric. The Measure command can be used to easily lay out these distances. The Line command is then used to connect the points to form the angle.

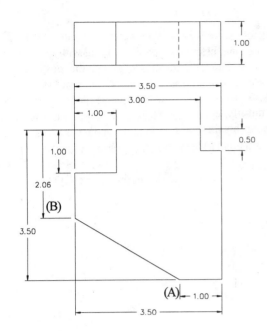

Before using the Measure command, set the Pdmode system variable to a new value of 3. Points will appear as an X instead of a dot. Now use the Measure command to set off the two distances.

Command: **Setvar**
Variable name or ?: **Pdmode**
New value for Pdmode <0>: **3**

Command: **Measure**
Select object to measure: *(Select the inclined line at "A")*
<Segment length>/Block: **1.00**

Command: **Measure**
Select object to measure: *(Select the vertical line at "B")*
<Segment length>/Block: **2.06**

The interesting part of the Measure command is that measuring will occur at the nearest endpoint of the line where the line was selected from. It is therefore important which endpoint of the line is selected. Once the points have been placed, the Line command is used to draw a line from one point to the other using the Osnap-Node option.

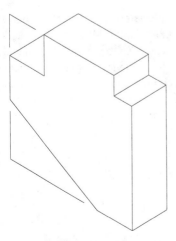

Creating Angles in Isometric - Method #2

The exact same 2-view drawing is illustrated at the right. This time, one distance is specified along with an angle of 30 degrees. Even with the angle given, the position of the isometric axes makes any angle construction by degrees inaccurate. The distance XY is still needed to construct the angle in isometric. The Measure command is used to find distance XY, place a point, and connect the first distance with the second to form the 30 degree angle in isometric. It is always best to set the Pdmode system variable to a new value in order to visibly view the point. A new value of 3 will assign the point as an X.

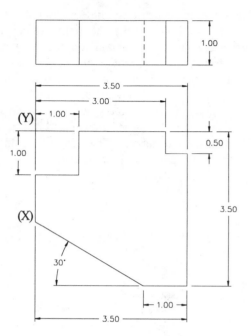

Follow the prompt sequences below to change the point mode and measure the appropriate distances in the previous step.

Command: **Setvar**
Variable name or ?: **Pdmode**
New value for Pdmode <0>: **3**

Command: **Measure**
Select object to measure: *(Select the inclined line at "A")*
<Segment length>/Block: **1.00**

Command: **Measure**
Select object to measure: *(Select the vertical line at "B")*
<Segment length>/Block: **Endp**
of *(Select the endpoint of the line at X in the 2-view drawing)*
Second point: **Int**
of *(Select the intersection at Y in the 2-view drawing)*

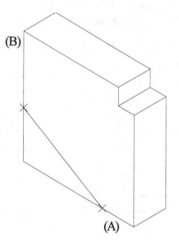

Line "B" is selected as the object to measure. Since this distance is unknown, the Measure command may be used to set off the distance XY by identifying an endpoint and intersection from the front view above. This means the view must be constructed only enough to lay out the angle and project the results to the isometric using the Measure command and the prompts above.

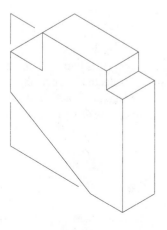

Isometric Construction Using Ellipses

Constructing circles as part of isometric drawing is possible using one of the three isometric axes positions. It is up to the operator to decide which axis to use. Before this, however, an isometric box consisting of overall distances is first constructed. Use the Ellipse command to place the isometric circle at the base. To select the correct axis type CTRL-E until the proper axis appears in the form of the cursor. Place the ellipses. Lay out any other distances.

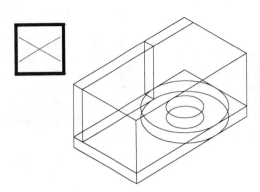

For ellipses at different positions type CTRL-E to select another isometric axis. Remember these axis positions may be selected from the Drawing Aids dialog box from the Settings area of the menu bar.

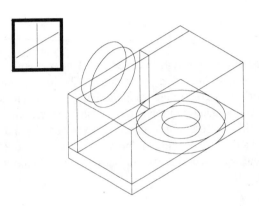

Use the Trim command to trim away any excess entities that are considered unnecessary.

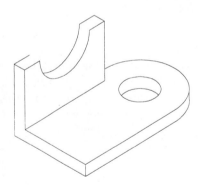

Use the Line command to connect endpoints of edges that form surfaces.

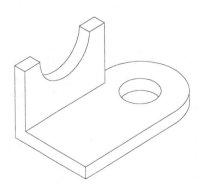

Creating Isometric Sections

In some cases, it is necessary to cut an isometric drawing to expose internal features. The result is an isometric section similar to a section view formed from an orthographic drawing. The difference with the isometric section, however, is the cutting plane is usually along one of the three isometric axes. The illustration at the right displays an orthographic section in addition to the isometric drawing.

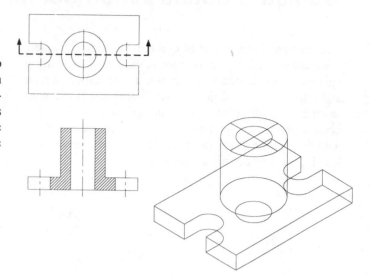

The isometric at the right has additional lines representing surfaces cut by the cutting plane line. The lines to define these surfaces were formed using the Line command in addition to a combination of top and right isometric axes modes. During this process, Ortho mode was toggled On and Off numerous times depending on the axis direction. The Break command was used to break ellipses and lines at their intersection points.

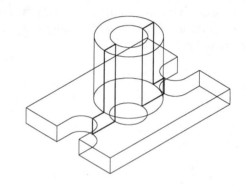

Once ellipses were broken, the front half of the isometric was removed exposing the back half as in the example at the right. The front half is then discarded. This has the same effect as conventional section views where the direction of site dictates which half to keep.

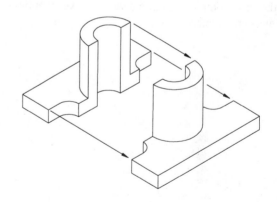

For a full section, the Hatch command is used to crosshatch the surfaces being cut by the cutting plane line. Surfaces designated holes or slots not cut are not crosshatched. This same procedure is followed for converting an object into a half section.

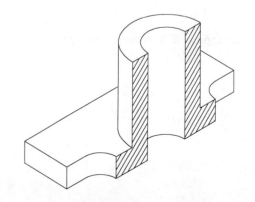

Exploded Isometric Views

Isometric drawings are sometimes grouped together to form an exploded drawing of how a potential or existing product is assembled. This involves aligning parts that fit with line segments, usually in the form of center lines. Bubbles identifying the part number are attached to the drawing. Exploded isometric drawings come in handy for creating bill of material information and for this purpose have an important application to manufacturing. Once the part information is identified in the drawing and title block area, this information is extracted and brought into a third-party business package where important data collection information is able to actually track the status of parts in production in addition to the shipping date for all finished products.

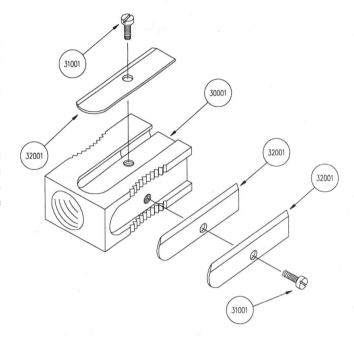

Isometric Assemblies

Assembly drawings show the completed part as if it were to be assembled. Sometimes this drawing has an identifying number placed with a bubble for bill of material needs. Assembly drawings commonly are placed on the same display screen as the working drawing. With the assembly along the side of the working drawing, the user has a pictorial representation of what the final product will look like and can aid in the understanding of the orthographic views.

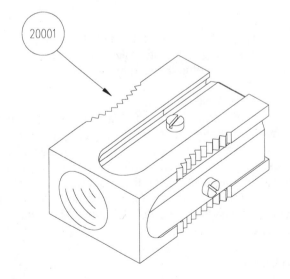

Tutorial Exercise #24
Plate.Dwg

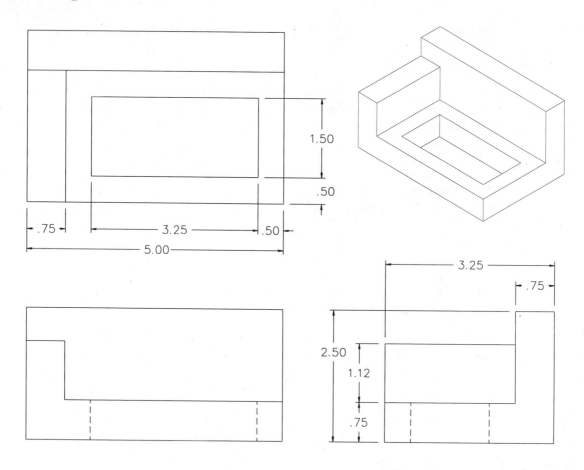

PURPOSE:
The purpose of this tutorial exercise is to use a series of coordinates along with AutoCAD editing commands to construct an isometric drawing of the Plate.

SYSTEM SETTINGS:
Begin a new drawing called "Plate." Use the Units command to change the number of decimal places past the zero from 4 to 2. Keep the remaining default unit values. Using the Limits command, keep 0,0 for the lower left corner and change the upper right corner from 12,9 to 10.50,8.00. Use the Grid command and change the grid spacing from 1.00 to 0.25 units. Do not turn the snap or ortho On.

LAYERS:
Special layers do not have to be created for this tutorial exercise.

SUGGESTED COMMANDS:
Begin this exercise by changing the grid from the standard display to an isometric display using the Snap-Style option. Remember both the grid and snap can be manipulated by the Snap command only if the current grid value is 0. Use Absolute and Polar coordinates to lay out the base of the Plate. Then begin using the Copy command followed by Trim to duplicate entities and clean up or trim unnecessary geometry.

DIMENSIONING:
This object may be dimensioned at a later date using the Dim-Oblique option. Consult your instructor before continuing.

PLOTTING:
This tutorial exercise may be plotted on "A"-size paper (8.5" x 11"). Use a plotting scale of 1=1 to produce a full size plot.

Step #1

Use the Line command to draw the figure at the right.

Command: **Line**
From point: **5.629,0.750**
To point: **@3.25<30**
To point: **@5.00<150**
To point: **@3.25<210**
To point: **C**

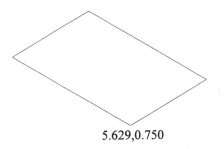

5.629,0.750

Step #2

Copy the four lines drawn in the previous step up at a distance
of 2.50 units in the 90 degree direction.

Command: **Copy**
Select objects: *(Select lines "A," "B," "C," and "D")*
Select objects: *(Strike Enter to continue)*
<Base point or displacement>/Multiple: **Endp**
of *(Select the line at "A")*
Second point of displacement: **@2.50<90**

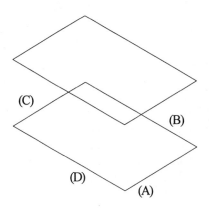

Step #3

Connect the top and bottom isometric boxes with line segments. Draw one segment using the Line command. Use the
Copy-Multiple command to duplicate and form the remaining
segments. Erase the two dashed lines since they are not
visible in an isometric drawing.

Command: **Line**
From point: **Endp**
of *(Select the endpoint of the line at "A")*
To point: **Endp**
of *(Select the endpoint of the line at "B")*
To point: *(Strike Enter to exit this command)*

Copy this line from "A" to "C" and "D".

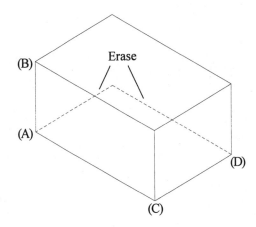

Step #4

Copy the two dashed lines at the right a distance of 0.75 in the
210 degree direction.

Command: **Copy**
Select objects: *(Select the two dashed lines at the right)*
Select objects: *(Strike Enter to continue)*
<Base point or displacement>/Multiple: **Endp**
of *(Select the endpoint at "A")*
Second point of displacement: **@0.75<210**

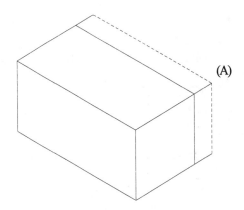

Step #5

Copy the two dashed lines at the right a distance of 0.75 in the 90 degree direction. This forms the base of the plate.

Command: **Copy**
Select objects: *(Select the two dashed lines at the right)*
Select objects: *(Strike Enter to continue)*
<Base point or displacement>/Multiple: **Endp**
of *(Select the endpoint at "A")*
Second point of displacement: **@0.75<90**

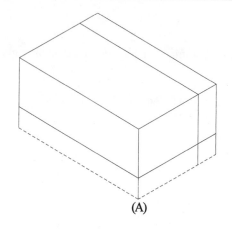

(A)

Step #6

Copy the dashed line at the right a distance of 0.75 in the −30 degree direction.

Command: **Copy**
Select objects: *(Select the dashed line at the right)*
Select objects: *(Strike Enter to continue)*
<Base point or displacement>/Multiple: **Endp**
of *(Select the endpoint at "A")*
Second point of displacement: **@0.75<-30**

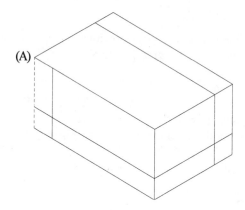

(A)

Step #7

Use the Fillet command to place a corner between the two dashed lines at "A" and "B" and at "C" and "D". The current fillet radius should already be set to a value of 0.

Command: **Fillet**
Polyline/Radius/<Select first object>: *(Select "A")*
Select second object: *(Select "B")*

Command: **Fillet**
Polyline/Radius/<Select first object>: *(Select "C")*
Select second object: *(Select "D")*

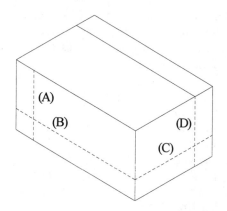

(A)
(B)   (D)
(C)

Step #8

Copy the dashed line at the right to begin forming the top of the base.

Command: **Copy**
Select objects: *(Select the dashed line at the right)*
Select objects: *(Strike Enter to continue)*
<Base point or displacement>/Multiple: **Endp**
of *(Select the endpoint at "A")*
Second point of displacement: **Endp**
of *(Select the endpoint at "B")*

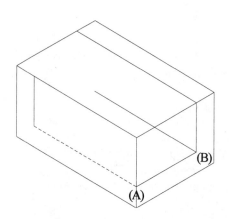

(B)
(A)

Step #9

Copy the dashed line at the right. This forms the base of the plate.

Command: **Copy**
Select objects: *(Select the dashed line at the right)*
Select objects: *(Strike Enter to continue)*
<Base point or displacement>/Multiple: **Endp**
of *(Select the endpoint at "A")*
Second point of displacement: **Endp**
of *(Select the endpoint at "B")*

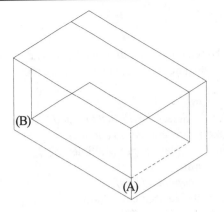

Step #10

Use the Trim command to clean up the excess lines at the right.

Command: **Trim**
Select cutting edge(s)...
Select objects: *(Select the three dashed lines at the right)*
Select objects: *(Strike Enter to continue)*
<Select object to trim>/Undo: *(Select the inclined line at "A")*
<Select object to trim>/Undo: *(Select the inclined line at "B")*
<Select object to trim>/Undo: *(Select the vertical line at "C")*
<Select object to trim>/Undo: *(Strike Enter to exit this command)*

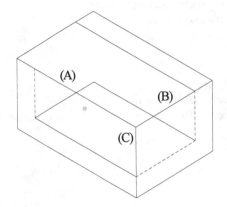

Step #11

Your display should be similar to the illustration at the right.

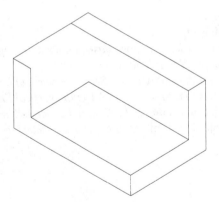

Step #12

Copy the dashed line at the right a distance of 1.12 units in the 90 degree direction.

Command: **Copy**
Select objects: *(Select the dashed line at the right)*
Select objects: *(Strike Enter to continue)*
<Base point or displacement>/Multiple: **Endp**
of *(Select the endpoint at "A")*
Second point of displacement: **@1.12<90**

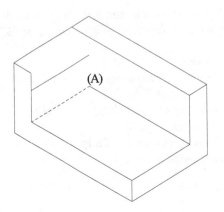

Step #13

Copy the dashed line at "A" to new positions at "B" and "C."
Then delete the line at "A" using the Erase command.

Command: **Copy**
Select objects: *(Select the dashed line at the right)*
Select objects: *(Strike Enter to continue)*
<Base point or displacement>/Multiple: **M**
Base point: **Endp**
of *(Select the endpoint at "A")*
Second point of displacement: **Endp**
of *(Select the endpoint at "B")*
Second point of displacement: **Endp**
of *(Select the endpoint at "C")*
Second point of displacement: *(Strike Enter to exit this command)*

Command: **Erase**
Select objects: *(Select the dashed line at "A")*
Select objects: *(Strike Enter to execute this command)*

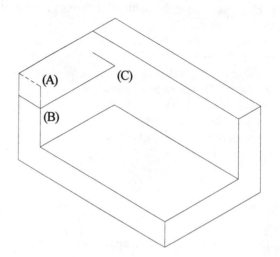

Step #14

Use the Trim command to clean up the excess lines at the right.

Command: **Trim**
Select cutting edge(s)...
Select objects: *(Select the two dashed lines at the right)*
Select objects: *(Strike Enter to continue)*
<Select object to trim>/Undo: *(Select the vertical line at "A")*
<Select object to trim>/Undo: *(Select the vertical line at "B")*
<Select object to trim>/Undo: *(Select the inclined line at "C")*
<Select object to trim>/Undo: *(Strike Enter to exit this command)*

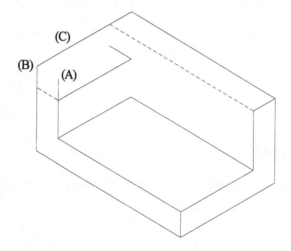

Step #15

Use the Copy command to duplicate the dashed line at the
right from the endpoint of "A" to the endpoint at "B."

Command: **Copy**
Select objects: *(Select the dashed line at the right)*
Select objects: *(Strike Enter to continue)*
<Base point or displacement>/Multiple: **Endp**
of *(Select the endpoint at "A")*
Second point of displacement: **Endp**
of *(Select the endpoint at "B")*

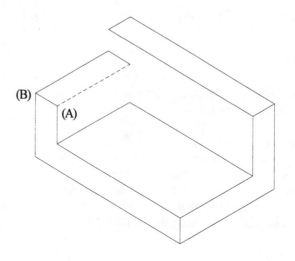

Step #16

Use the Copy command to duplicate the dashed line at the right from the endpoint of "A" to the endpoint at "B."

Command: **Copy**
Select objects: *(Select the dashed line at the right)*
Select objects: *(Strike Enter to continue)*
<Base point or displacement>/Multiple: **Endp**
of *(Select the endpoint at "A")*
Second point of displacement: **Endp**
of *(Select the endpoint at "B")*

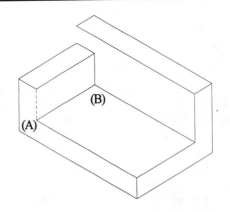

Step #17

Use the Line command to connect the endpoints of the segments at "A" and "B" illustrated at the right.

Command: **Line**
From point: **Endp**
of *(Select the endpoint at "A")*
To point: **Endp**
of *(Select the endpoint at "B")*
To point: *(Strike Enter to exit this command)*

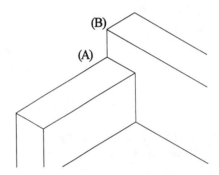

Step #18

Use the Copy command to duplicate the dashed line at the right from the endpoint of "A" to the distance of 0.50 units specified by a polar coordinate. This value begins the outline of the rectangular hole through the object.

Command: **Copy**
Select objects: *(Select the dashed line at the right)*
Select objects: *(Strike Enter to continue)*
<Base point or displacement>/Multiple: **Endp**
of *(Select the endpoint at "A")*
Second point of displacement: **@0.50<210**

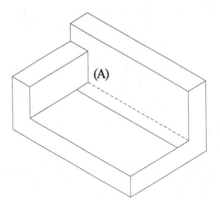

Step #19

Use the Copy command to duplicate the dashed line at the right from the endpoint of "A" to the distance of 0.50 units using a polar coordinate.

Command: **Copy**
Select objects: *(Select the dashed line at the right)*
Select objects: *(Strike Enter to continue)*
<Base point or displacement>/Multiple: **Endp**
of *(Select the endpoint at "A")*
Second point of displacement: **@0.50<-30**

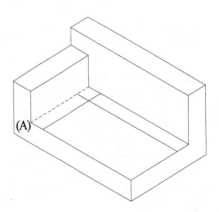

Step #20

Use the Copy command to duplicate the dashed line at the right from the endpoint of "A" to the distance of 0.50 units using a polar coordinate.

Command: **Copy**
Select objects: *(Select the dashed line at the right)*
Select objects: *(Strike Enter to continue)*
<Base point or displacement>/Multiple: **Endp**
of *(Select the endpoint at "A")*
Second point of displacement: **@0.50<30**

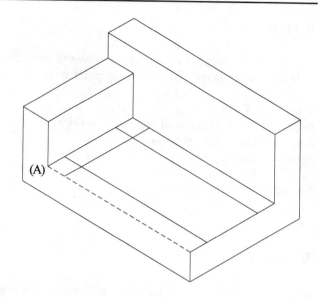

Step #21

Use the Copy command to duplicate the dashed line at the right from the endpoint of "A" to the distance of 0.50 units using a polar coordinate.

Command: **Copy**
Select objects: *(Select the dashed line at the right)*
Select objects: *(Strike Enter to continue)*
<Base point or displacement>/Multiple: **Endp**
of *(Select the endpoint at "A")*
Second point of displacement: **@0.50<150**

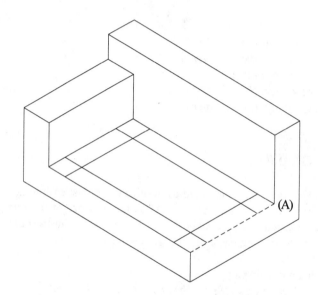

Step #22

Use the Fillet command with a radius of 0 to corner the four dashed lines at the right. Use the Multiple command to remain in the Fillet command. To exit the Fillet command prompts, use the CTRL-C sequence to cancel the command when finished.

Command: **Multiple**
Fillet
Polyline/Radius/<Select first object>: *(Select line "A")*
Select second object: *(Select line "B")*
Polyline/Radius/<Select first object>: *(Select line "B")*
Select second object: *(Select line "C")*
Polyline/Radius/<Select first object>: *(Select line "C")*
Select second object: (Select line "D")
Polyline/Radius/<Select first object>: *(Select line "D")*
Select second object: (Select line "A")
Polyline/Radius/<Select first object>: **CTRL-C** *(To cancel)*

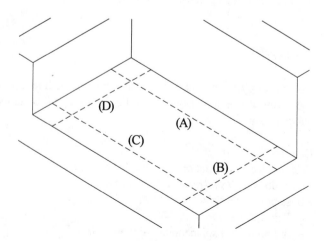

Step #23

Use the Copy command to duplicate the dashed line at the right. This will begin forming the thickness of the base inside of the rectangular hole.

Command: **Copy**
Select objects: *(Select the dashed line at the right)*
Select objects: *(Strike Enter to continue)*
<Base point or displacement>/Multiple: **Endp**
of *(Select the endpoint at "A")*
Second point of displacement: **Endp**
of *(Select the endpoint at "B")*

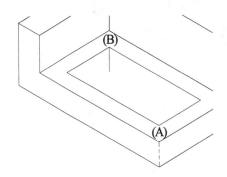

Step #24

Use the Copy command to duplicate the dashed lines at the right. These lines form the inside surfaces to the rectangular hole.

Command: **Copy**
Select objects: *(Select the dashed line at the right)*
Select objects: *(Strike Enter to continue)*
<Base point or displacement>/Multiple: **Endp**
of *(Select the endpoint at "A")*
Second point of displacement: **Endp**
of *(Select the endpoint at "B")*

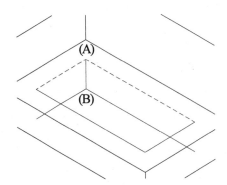

Step #25

Use the Trim command to clean up excess lines illustrated at the right.

Command: **Trim**
Select cutting edges...
Select objects: *(Select the two dashed lines at the right)*
Select objects: *(Strike Enter to continue)*
<Select object to trim>/Undo: *(Select the line at "A")*
<Select object to trim>/Undo: *(Select the line at "B")*
<Select object to trim>/Undo: *(Strike Enter to exit this command)*

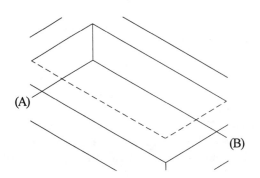

Step #26

The completed isometric is illustrated at the right. This drawing may be dimensioned using the Dim-Oblique command. Consult your instructor if this next step is necessary.

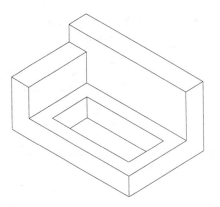

Tutorial Exercise #25
Hanger.Dwg

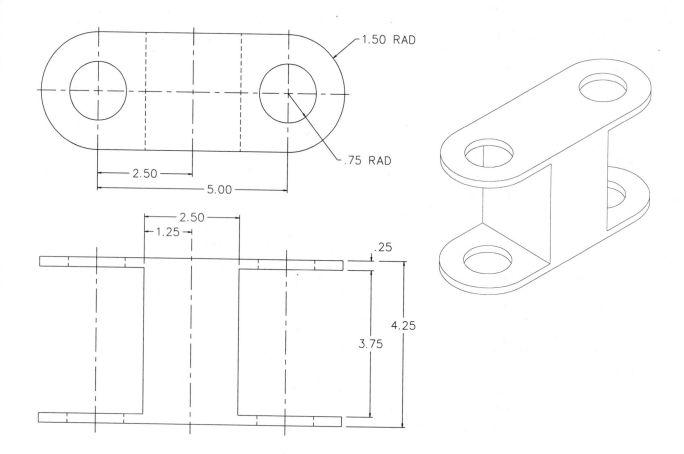

PURPOSE:
The purpose of this tutorial exercise is to use a series of coordinates along with AutoCAD editing commands to construct an isometric drawing of the Hanger.

SYSTEM SETTINGS:
Begin a new drawing called "Hanger." Use the Units command to change the number of decimal places past the 0 from 4 to 2. Keep the remaining default unit values. Using the Limits command, keep 0,0 for the lower left corner and change the upper right corner from 12,9 to 15.50,9.50. Use the Grid command and change the grid spacing from 1.00 to 0.25 units. Do not turn the snap or ortho On.

LAYERS:
Special layers do not have to be created for this tutorial exercise.

SUGGESTED COMMANDS:
Begin this exercise by changing the grid from the standard display to an isometric display using the Snap-Style option. Remember both the grid and snap can be manipulated by the Snap command only if the current grid value is 0. Use absolute and polar coordinates to lay out the base of the Plate. Then begin using the Copy command followed by Trim to duplicate entities and clean up or trim unnecessary geometry.

DIMENSIONING:
This object may be dimensioned at a later date using the Dim-Oblique option. Consult your instructor before continuing.

PLOTTING:
This tutorial exercise may be plotted on "B"-size paper (11" x 17"). Use a plotting scale of 1=1 to produce a full size plot.

Step #1

Set the Snap-Style option to Isometric with a vertical spacing of 0.25 units. Type CTRL-E to switch to the Top Isoplane mode. Use the Line command to draw the rectangular isometric box representing the total depth of the object along with the center-to-center distance of the holes and arcs that will be placed in the next step.

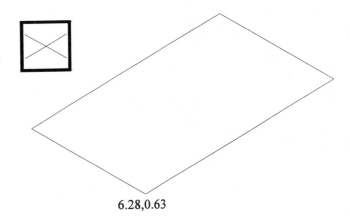

6.28,0.63

Command: **Snap**
Snap spacing or ON/OFF/Aspect/Rotate/Style <0.25>: **S**
Standard/Isometric <S>: **Iso**
Vertical spacing <1.00>: **0.25**

Command: **^E** *(To switch to the Top Isoplane mode)*

Command: **Line**
From point: **6.28,0.63**
To point: **@5.00<30**
To point: **@ 3.00<150**
To point: **@5.00<210**
To point: **C**

Step #2

While in the Top Isoplane mode, use the Ellipse command to draw two isometric ellipses of 0.75 and 1.50 radii each. Identify the midpoint of the inclined line at "A" as the center of the first ellipse. To identify the center of the second ellipse, use the @ option which stands for "last point" and will identify the center of the small circle as the same center as the large circle.

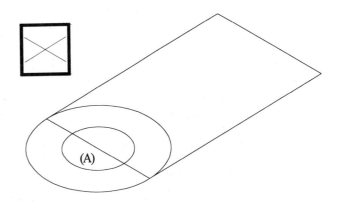

Command: **Ellipse**
<Axis endpoint 1>/Center/Isocircle: **Iso**
Center of circle: **Mid**
of *(Select the inclined line at "A")*
<Circle radius>/Diameter: **0.75**

Command: **Ellipse**
<Axis endpoint 1>/Center/Isocircle: **Iso**
Center of circle: **@**
<Circle radius>/Diameter: **1.50**

Step #3

Copy both ellipses from the midpoint of the inclined line at "A" to the midpoint of the inclined line at "B."

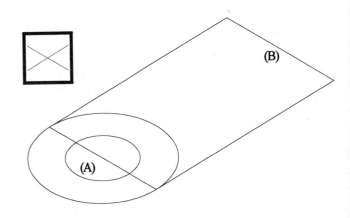

Command: **Copy**
Select objects: *(Select both ellipses at the right)*
Select objects: *(Strike Enter to continue)*
<Base point or displacement>/Multiple: **Mid**
of *(Select the midpoint of the inclined line at "A")*
Second point of displacement: **Mid**
of *(Select the midpoint of the inclined line at "B")*

Step #4

Turn the snap Off by striking the F9 function key. Use the Trim command to delete parts of the ellipses.

Command: **Trim**
Select cutting edges...
Select objects: *(Select dashed lines "A" and "B")*
Select objects: *(Strike Enter to continue)*
<Select object to trim>/Undo: *(Select the ellipse at "C")*
<Select object to trim>/Undo: *(Select the ellipse at "D")*
<Select object to trim>/Undo: *(Strike Enter to exit this command)*

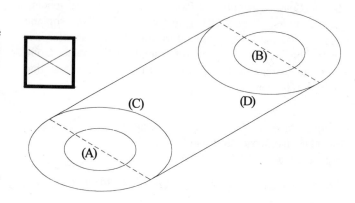

Step #5

Copy all entities at the right up the distance of 0.25 units to form the bottom base of the hanger. The Right Isoplane mode can be activated by typing CTRL-E.

Command: **^E**
Right Isoplane

Command: **Copy**
Select objects: *(Select all entities at the right)*
Select objects: *(Strike Enter to continue)*
<Base point or displacement>/Multiple: **Mid**
of *(Select the midpoint of the inclined line at "A")*
Second point of displacement: **@0.25<90**

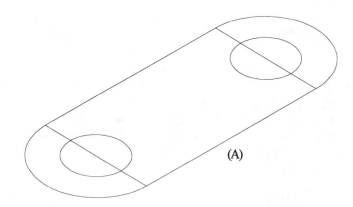

Step #6

Use the Erase command to delete the three dashed lines at the right. These lines are not visible at this point of view and should be erased.

Command: **Erase**
Select objects: *(Select the three dashed lines at the right)*
Select objects: *(Strike Enter to execute this command)*

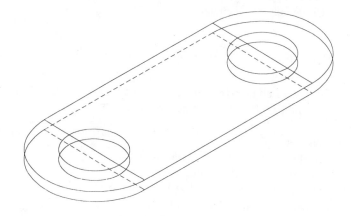

Step #7

Your display should appear similar to the illustration at the right. Begin partially deleting other entities to show only visible features of the isometric drawing. Turn the snap Off to better assist in the next series of operations. The next few steps that follow refer to the area outlined at the right. Use the Zoom-Window option to magnify this area.

Command: **Snap**
Snap spacing or ON/OFF/Aspect/Rotate/Style <0.25>: **Off**

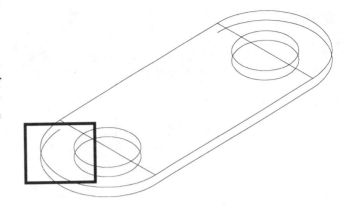

Step #8

Carefully draw a line tangent to both ellipses. Use the Osnap-Quadrant option to assist you in constructing the line.

Command: **Line**
From point: **Qua**
of *(Select the quadrant at "A")*
To point: **Qua**
of *(Select the quadrant at "B")*
To point: *(Strike Enter to exit this command)*

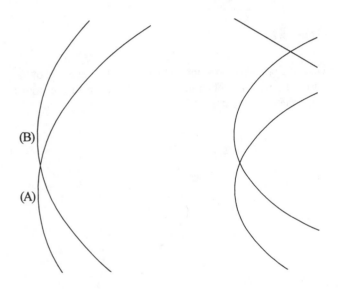

Step #9

Use the Trim command, select the short dashed line as the cutting edge, and select the arc segment at the right to trim.

Command: **Trim**
Select cutting edges...
Select objects: *(Select dashed line at "A")*
Select objects: *(Strike Enter to continue)*
<Select object to trim>/Undo: *(Select the ellipse at "B")*
<Select object to trim>/Undo: *(Strike Enter to exit this command)*

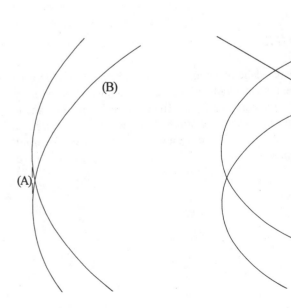

Step #10

The completed operation is illustrated at the right. Use Zoom-Previous to return to the previous display.

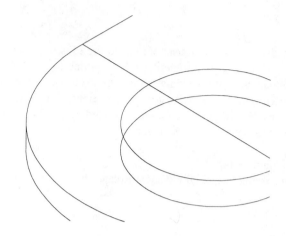

Step #11

Use the Zoom-Window option to magnify the right half of the base. Prepare to construct the tangent edge to the object using the previous steps.

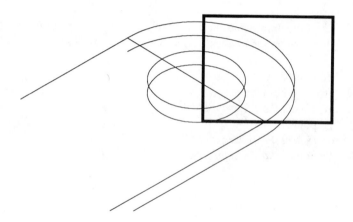

Step #12

Follow the same procedure as in steps 8 and 9 to construct a line from the quadrant point on the top ellipse to the quadrant point on the bottom ellipse. Then use the Trim command to clean up any excess entities. Use the Erase command to delete any elliptical arc segments that may have been left untrimmed.

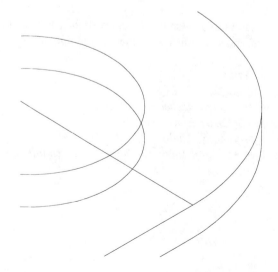

Step #13

Use the Trim command to partially delete the ellipses at the right to expose the thickness of the base.

Command: **Trim**
Select cutting edges...
Select objects: *(Select dashed ellipses "A" and "B")*
Select objects: *(Strike Enter to continue)*
<Select object to trim>/Undo: *(Select the lower ellipse at "C")*
<Select object to trim>/Undo: *(Select the lower ellipse at "D")*
<Select object to trim>/Undo: *(Strike Enter to exit this command)*

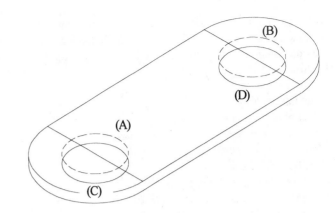

Step #14

Use the Copy command to duplicate the bottom base and form the upper plate of the hanger. Copy the base a distance of 4 units straight up.

Command: **Copy**
Select objects: *(Select all dashed entities at the right)*
Select objects: *(Strike Enter to continue)*
<Base point or displacement>/Multiple: **Nea**
of *(Select the nearest point along the bottom arc at "A")*
Second point of displacement: **@4.00<90**

When completed with this step, perform a Zoom-All to display the entire isometric.

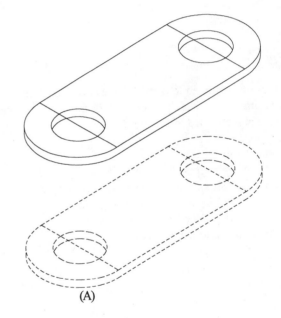

Step #15

Use the Copy command to duplicate the inclined line at "A" the distance of 2.50 units to form the line represented by a series of dashes at the right. This line happens to be located at the center of the object.

Command: **Copy**
Select objects: *(Select the line at "A")*
Select objects: *(Strike Enter to continue)*
<Base point or displacement>/Multiple: **Mid**
of *(Select the midpoint of the inclined line at "A")*
Second point of displacement: **@2.50<30**

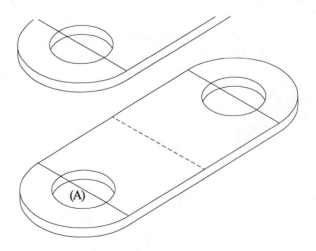

Step #16

Duplicate the line represented by dashes at the right to form the two inclined lines at "B" and "C." These lines will begin the construction of the sides of the hanger. Use the Copy-Multiple option to accomplish this.

Command: **Copy**
Select objects: *(Select the dashed line at the right)*
Select objects: *(Strike Enter to continue)*
<Base point or displacement>/Multiple: **M**
Base point: **Mid**
of *(Select the midpoint of the dashed line at "A")*
Second point of displacement: **@1.25<30**
Second point of displacement: **@1.25<210**
Second point of displacement: *(Strike Enter to exit this command)*

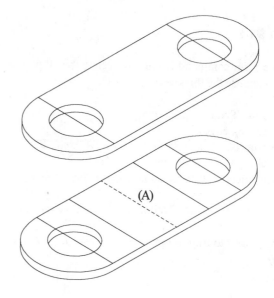

Step #17

Use the Copy command to duplicate the two dashed lines at the right straight up at a distance of 3.75 units. The polar coordinate mode is used to accomplish this.

Command: **Copy**
Select objects: *(Select both dashed lines at the right)*
Select objects: *(Strike Enter to continue)*
<Base point or displacement>/Multiple: **Mid**
of *(Select the midpoint of the inclined line at "A")*
Second point of displacement: **@3.75<90**

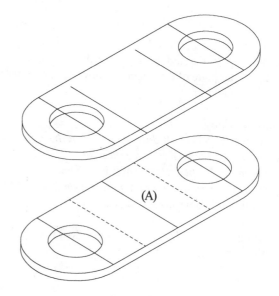

Step #18

Use the Line command along with the Osnap-Endpoint option to draw a line from endpoint "A" to endpoint "B."

Command: **Line**
From point: **Endp**
of *(Select the endpoint of the inclined line at "A")*
To point: **Endp**
of *(Select the endpoint of the inclined line at "B")*
To point: *(Strike Enter to exit this command)*

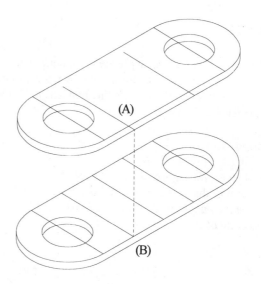

Step #19

Use the Copy command and the Multiple option to copy the dashed line at "A" to "B," "C" and "D."

Command: **Copy**
Select objects: *(Select the dashed line at the right)*
Select objects: *(Strike Enter to continue)*
<Base point or displacement>/Multiple: **M**
Base point: **Endp**
of *(Select the endpoint of the vertical line at "A")*
Second point of displacement: **Endp**
of *(Select the endpoint of the line at "B")*
Second point of displacement: **Endp**
of *(Select the endpoint of the line at "C")*
Second point of displacement: **Endp**
of *(Select the endpoint of the line at "D")*
Second point of displacement: *(Strike Enter to exit the command)*

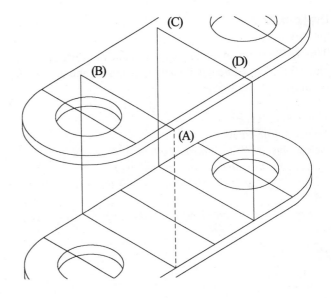

Step #20

Use the Erase command to delete all lines represented as dashed lines at the right.

Command: **Erase**
Select objects: *(Select all dashed entities illustrated at the right)*
Select objects: *(Strike Enter to execute this command)*

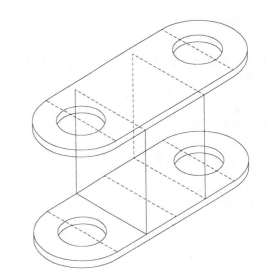

Step #21

Use the Trim command to partially delete the vertical line at the right. The segment to be deleted is hidden and not shown in an isometric drawing.

Command: **Trim**
Select cutting edges...
Select objects: *(Select dashed entities "A" and "B")*
Select objects: *(Strike Enter to continue)*
<Select object to trim>/Undo: *(Select the vertical line at "C")*
<Select object to trim>/Undo: *(Strike Enter to exit this command)*

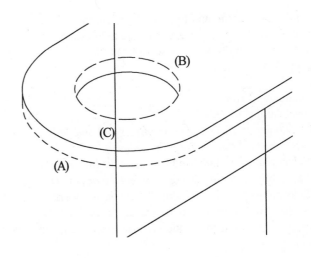

Step #22

Use the Trim command to partially delete the vertical line at the right. The segment to be deleted is hidden and not shown in an isometric drawing.

Command: **Trim**
Select cutting edges...
Select objects: *(Select dashed elliptical arc at "A")*
Select objects: *(Strike Enter to continue)*
<Select object to trim>/Undo: *(Select the vertical line at "B")*
<Select object to trim>/Undo: *(Strike Enter to exit this command)*

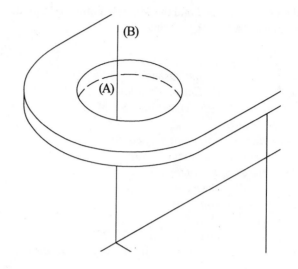

Step #23

Use the Trim command to partially delete the inclined line at the right.

Command: **Trim**
Select cutting edges...
Select objects: *(Select dashed entities "A" and "B")*
Select objects: *(Strike Enter to continue)*
<Select object to trim>/Undo: *(Select the line at "C")*
<Select object to trim>/Undo: *(Select the line at "D")*
<Select object to trim>/Undo: *(Strike Enter to exit this command)*

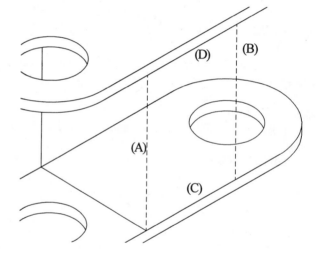

Step #24

Use the Trim command to partially delete the entities at the right. The segments to be deleted are hidden and not shown in an isometric drawing. Use Erase to delete any leftover elliptical arc segments.

Command: **Trim**
Select cutting edges...
Select objects: *(Select dashed entities "A" and "B")*
Select objects: *(Strike Enter to continue)*
<Select object to trim>/Undo: *(Select the line at "C")*
<Select object to trim>/Undo: *(Select the arc at "D")*
<Select object to trim>/Undo: *(Select the ellipse at "E")*
<Select object to trim>/Undo: *(Select the arc at "F")*
<Select object to trim>/Undo: *(Strike Enter to exit this command)*

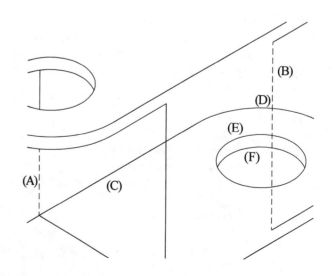

Step #25

The completed isometric is illustrated at the right. This drawing may be dimensioned using the Dim-Oblique command. Consult your instructor if this next step is necessary.

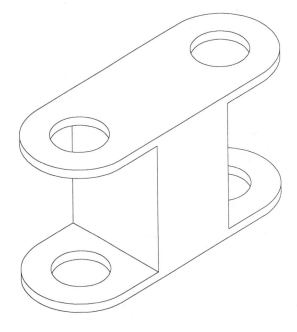

Step #26

As an extra step, convert the completed hanger into an object with a rectangular hole through it. Begin by copying lines "A," "B," "C," and "D" at a distance of 0.25 units to form the inside rectangle using polar coordinates. Since the lines will overlap at the corners, use the Fillet command set to a radius of 0 to create corners of the rectangle. Use the Line command to draw the inclined line "E" at any distance with a 150 degree angle. Use either Trim or Extend to complete the new version of the hanger.

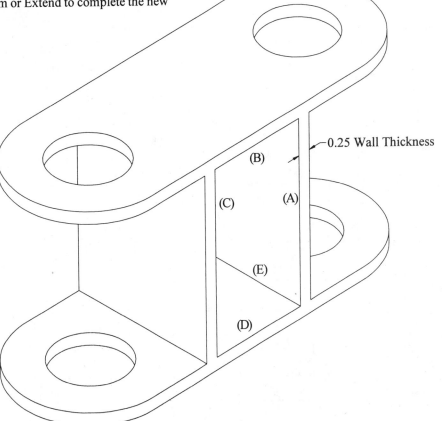

Problems for Unit 10

Directions for Problem 10–1
Construct an isometric drawing of the object.

Problem 10–1

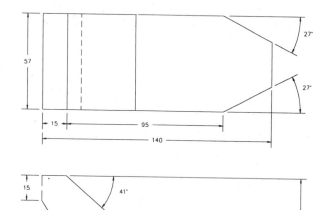

Problem 10–2

Directions for Problem 10–2
Construct an isometric drawing of the object. Begin the corner of the isometric at "A".

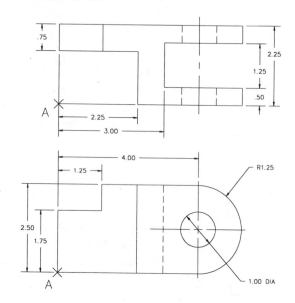

Directions for Problems 10–3 through 10–19
Construct an isometric drawing of the object.

Problem 10–3

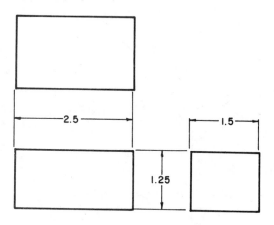

Problem 10–4

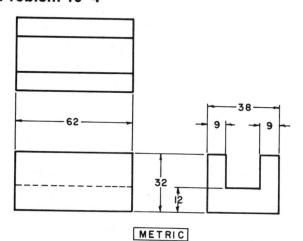

METRIC

Problem 10–5

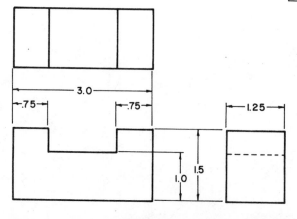

Problem 10–6

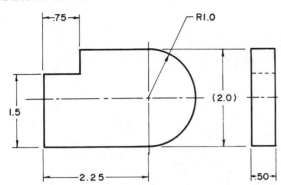

Problem 10–7

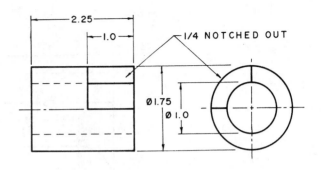

Problem 10–8

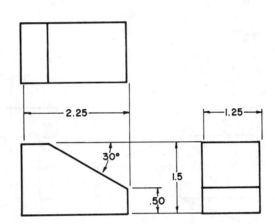

Problem 10–9

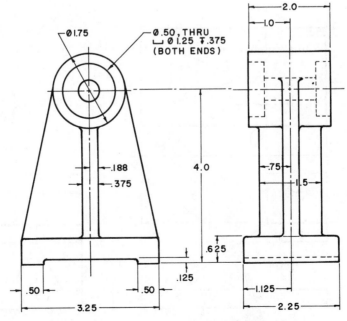

ALL UNMARKED RADII = R .09

Problem 10–10

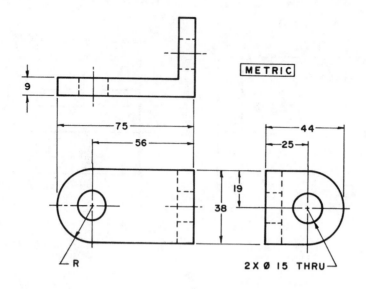

Problem 10–11

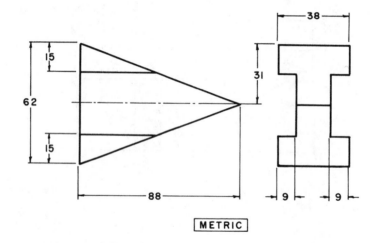

Problem 10–12

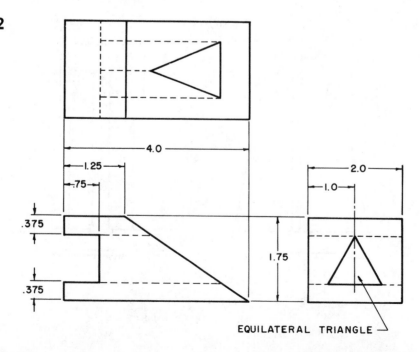

Problem 10–13

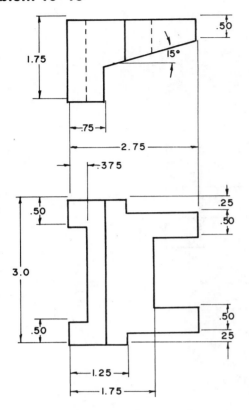

Problem 10–15

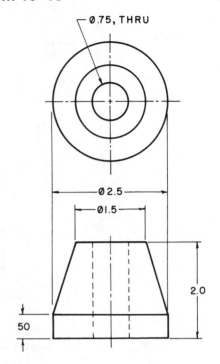

Problem 10–14

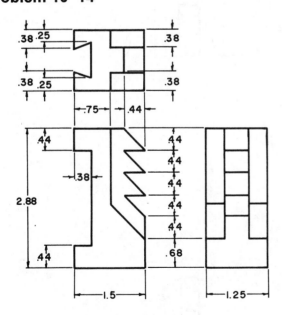

Problem 10–16

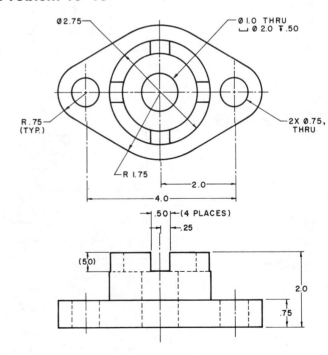

Problem 10–17

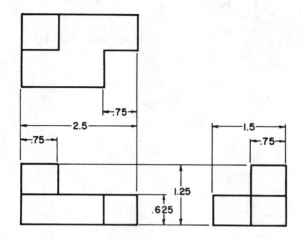

Problem 10–18

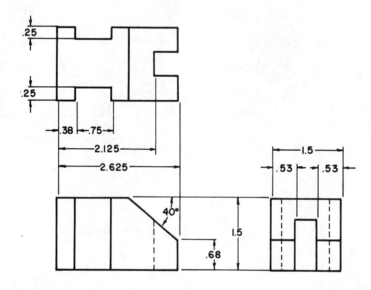

Problem 10–19

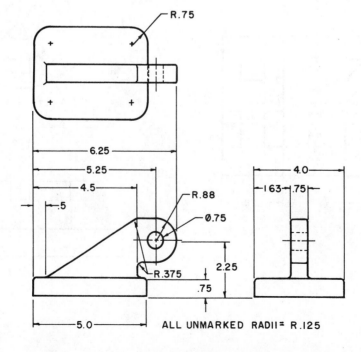

ALL UNMARKED RADII = R.125

UNIT
11

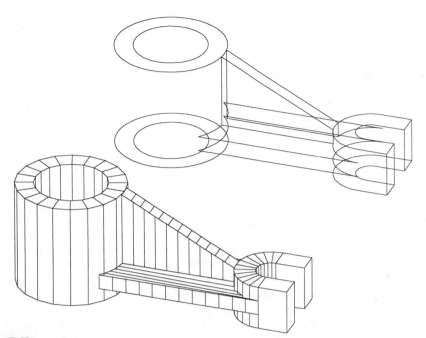

3-D Modeling

Contents

It is said that humans see, hear, and exist in a 3D world. Why not draw in three dimensions; this means visualizing the object in the designer's mind before placing entities on the computer screen. Part of learning the art of visualization is in the study and construction of models in 3D in order to obtain as accurate as possible an image of an object undergoing design. In this unit, you will be exposed to creating a model first in the form of a wireframe representation, then the wireframe will be surfaced using a few of the many surfacing tools AutoCAD has to offer. Once the model is surfaced, it can be viewed from any angle, similar to the wireframe. The surfaced model, however, can have hidden lines removed to aid in the visualization of the final design.

Orthographic Projection

It is no secret that the heart of any engineering drawing is the ability to break up the design into three main views of front, top, and right side views representing what is called orthographic or multi-view projection. The engineer or designer is then required to interpret the views and their dimensions to paint a mental picture of what a pictorial version of the object would look like if already made or constructed.

As most engineering individuals have the skill to convert the multi-view drawing into a pictorial drawing in their mind, a vast majority of people would be confused by the numerous hidden and center lines of a drawing and their meaning to the overall design. The individuals need some type of picture to help them interpret the multi-view drawing and get a feel for what the part looks like, including the functionality of the part. This may be the major advantage of constructing an object in 3D.

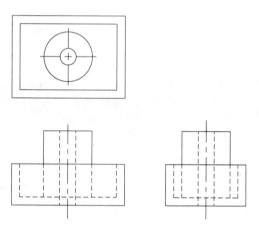

Isometric Drawings

The isometric drawing is the easiest 3D representation to produce. It is based on tilting two axes at 30-degree angles along with a vertical line to represent the height of the object. As easy as the isometric is to produce, it is also one of the most inaccurate methods of producing a 3D image.

The illustration at the right shows a view aligned to see the top of the object, front view, and right side view. If we wished to see a different view of the object in isometric, a new drawing would have to be produced from scratch. This is the major disadvantage of using isometric drawings. Still, because of ease in drawing, isometrics remain popular in many school and technical drafting rooms.

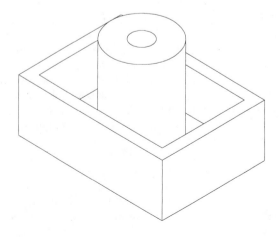

3D Extruded Models

AutoCAD versions in the past afforded the user the capability of drawing entities in the Z direction and then viewing the model at any angle. This method was called 3D Visualization. Entities are assigned an elevation for starting the surface and a thickness which extrudes the entity and produces opaque sides. The thickness can be entered in either a positive direction (Up), or negative direction (Down).

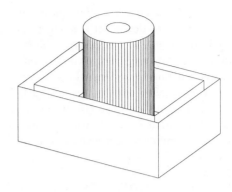

To view the model, the Vpoint command is used to identify a position on a 2D globe which serves as a means of identifying a 3D position. Advantages of this type of construction are the use of the Vpoint command and how the model is able to be viewed from any angle.

A disadvantage of this method is the inability to place top and bottom surfaces on the model. As a result, the model looks hollow when viewed from above or underneath.

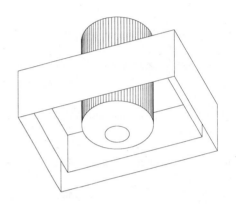

3D Wireframe Models

The evolution of AutoCAD Release 10 made it possible to design in a true 3D database. Using a user-definable coordinate system, previously defined as the UCS, a new coordinate system can be defined along any plane to place entities along. Depth can be controlled by a number of aids, especially XYZ point filters. In this method, values are temporarily saved for later use inside a command. The values can be retrieved and a coordinate value entered to complete the command.

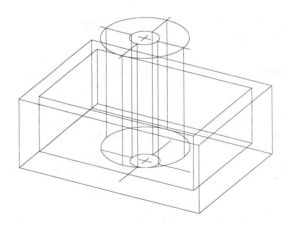

Wireframes are fundamental types of models. However, as lines intersect from the front and back of the object, it is sometimes difficult to interpret the true design of the wireframe.

The UCS Command

Two-Dimensional Computer-Aided Design is still the most popular method used to document a product before being manufactured. However, since a prototype made of a part is 3-dimensional, why not design in three dimensions. To assist the user in constructing in 3 dimensions, a series of user defined coordinate systems are used to create planes of construction that entities lie on. The User Coordinate System command or UCS is used to create these user defined coordinate systems. The command sequence is given below:

Command: **UCS**
Origin/ZAxis/3point/Entity/View/X/Y/Z/Prev/Restore/Save/
Del/?/<World>:

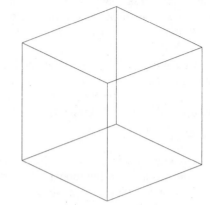

The following options are available to the user of this command:
Origin - Identifies a new user coordinate system at origin 0,0,0.

ZAxis - Identifies a user coordinate system from two points defining the Z axis.

3point - Identifies a user coordinate system by 3 points.

Entity - Identifies a user coordinate system in relation to an entity selected.

View - Identifies a user coordinate system by the current display.

X/Y/Z - Identifies a user coordinate system by rotation along the X, Y, or Z axis.

Prev - Sets the user coordinate system icon to the previously defined user coordinate system.

Restore - Restores a previously saved user coordinate system.

Save - Saves the position of a user coordinate system under a unique name given by the CAD operator.

Del - Deletes a user coordinate system from the database of the current drawing.

? - Lists all previously saved user coordinate systems.

<World> - Switches to the world coordinate system from any previously defined user coordinate system.

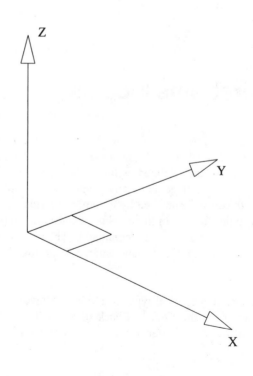

The UCS - Origin Option

The Origin option of the UCS command defines a new user coordinate system by shifting the current UCS to a new 0,0,0 position while leaving the direction of the X, Y, and Z axes unchanged. The command sequence for using this Origin options is as follows:

Command: **UCS**
Origin/ZAxis/3point/Entity/View/X/Y/Z/Prev/Restore/Save/
Del/?/<World>: **Origin**
Origin point <0,0,0>: *(Identify a point for the new origin)*

Illustrated at the right is a sample model with the current coordinate system being the World Coordinate System. Follow the next series of examples to define a new user coordinate system using the Origin option.

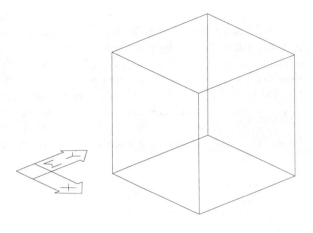

Using the Origin option of the UCS command, identify a new origin point for 0,0,0 at "A." This should move or shift the User coordinate system icon to the point specified by the user. Once the icon has moved, the "W" is removed and a "+" appears at the intersection of the X and Y axis on the icon. If the icon remains in its previous position, use the Ucsicon command with the Origin option to snap the icon to its new origin point.

Command: **UCS**
Origin/ZAxis/3point/Entity/View/X/Y/Z/Prev/Restore/Save/
Del/?/<World>: **Origin**
Origin point <0,0,0>: **Endp**
of *(Select the endpoint of the line at "A")*

(A)
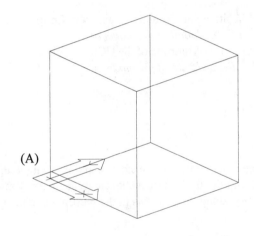

Entities may now be drawn parallel to this new coordinate system. Remember that 0,0,0 is identified by the "+" on the User Coordinate System Icon.

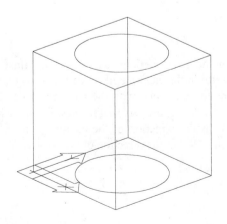

The UCS - 3 Point Option

Use the 3 Point option of the UCS command to specify a new user coordinate system by identifying an origin and new directions of its positive X and Y axes. The right hand rule is used to arrive at the current direction of the Z axis. Illustrated at the right is a 3-dimensional cube in the world coordinate system. To construct entities on the front panel, a new user coordinate system must first be defined parallel to the front. Follow the command sequence below for accomplishing this task.

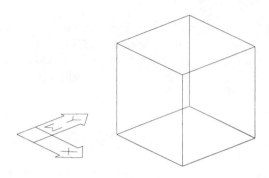

Command: **UCS**
Origin/ZAxis/3point/Entity/View/X/Y/Z/Prev/Restore/Save/
Del/?/<World>: **3point**
Origin point <0,0,0>: **Endp**
of *(Select the endpoint of the model at "A")*
Point on positive portion of the X axis <>: **Endp**
of *(Select the endpoint of the model at "B")*
Point on positive-Y portion of the UCS X-Y plane <>: **Endp**
of *(Select the endpoint of the model at "C")*

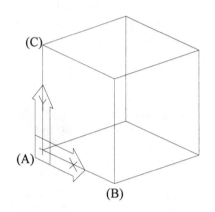

With the user coordinate system standing straight up, or aligned with the front of the cube, any type of entity may be constructed along this plane as in the illustration at the right.

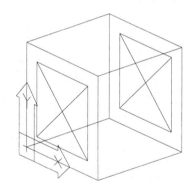

The 3 Point method of defining a new user coordinate system is quite useful in the example illustrated at the right where a UCS needs to be aligned with the inclined plane. Use the intersection at "A" as the center of the new UCS, the intersection at "B" as the direction of the positive X axis, and the intersection at "C" as the direction of the positive Y axis.

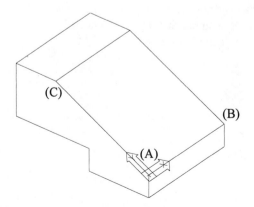

The UCS - X/Y/Z Options

Using the X/Y/Z rotation options will rotate the current user coordinate around the specific axis. Once a letter is selected as the pivot, a prompt appears asking for the rotation angle about the pivot axis. The right hand rule is again used to determine the positive direction of rotation around an axis. Think of the right hand gripping the pivot axis with the thumb pointing in the positive X, Y, or Z direction. The curling of the fingers on the right hand determines the direction of rotation. All positive rotations occur in the counter-clockwise directions. Illustrated at the right, the example at "A" shows a rotation about the X axis; the example at "B" shows a rotation about the Y axis; the example at "C" shows a rotation about the Z axis.

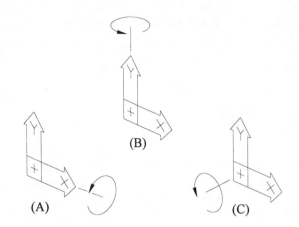

Given the same cube illustrated at the right in the world coordinate system, the X option of the UCS command will be used to stand the icon straight up by entering a 90 degree rotation value as in the next prompt sequence.

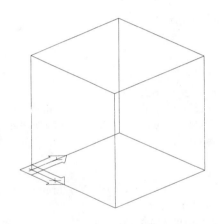

Command: **UCS**
Origin/ZAxis/3point/Entity/View/X/Y/Z/Prev/Restore/Save/
Del/?/<World>: **X**
Rotation angle about the X axis: **90**

The X axis is used as the pivot of rotation; entering a value of 90 degrees rotates the icon the desired degrees in the counter-clockwise direction as in the illustration at the right.

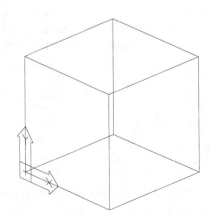

Applications of Rotating the UCS

Follow the next series of steps to perform numerous rotations of the user coordinate system in the example of the cube illustrated at the right.

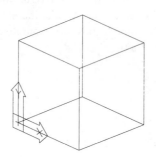

Begin rotating the user coordinate system along the X axis at a rotation angle of 90 degrees using the prompt sequence below. This will align the user coordinate system with the front of the cube.

Command: **UCS**
Origin/ZAxis/3point/Entity/View/X/Y/Z/Prev/Restore/Save/
Del/?/<World>: **X**
Rotation angle about the X axis: **90**

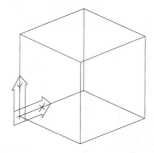

Next align the user coordinate system with one of the sides of the cube by performing the rotation using the Y axis as the pivot axis. The positive angle rotates the icon in the counter-clockwise direction.

Command: **UCS**
Origin/ZAxis/3point/Entity/View/X/Y/Z/Prev/Restore/Save/
Del/?/<World>: **Y**
Rotation angle about the Y axis: **90**

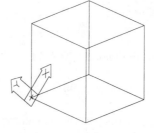

Next rotate the user coordinate system using the Z axis as the pivot axis. The degree of rotation entered at 45 degrees tilts the user coordinate system as in the illustration at the right.

Command: **UCS**
Origin/ZAxis/3point/Entity/View/X/Y/Z/Prev/Restore/Save/
Del/?/<World>: **Z**
Rotation angle about the Z axis: **45**

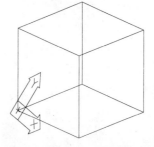

To tilt the icon along the Z axis pointing down at a 45 degree angle, enter a rotation angle of -90 degrees. The results are illustrated at the right.

Command: **UCS**
Origin/ZAxis/3point/Entity/View/X/Y/Z/Prev/Restore/Save/
Del/?/<World>: **Z**
Rotation angle about the Z axis: **-90**

The UCS - Entity Option

Given the same 3-dimensional cube at the right, another option of defining a new user coordinate system is by selecting an entity and having the user coordinate system align to that entity. Follow the command sequence and the example below to accomplish this.

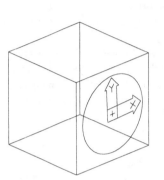

Command: **UCS**
Origin/ZAxis/3point/Entity/View/X/Y/Z/Prev/Restore/Save/
Del/?/<World>: **Entity**
Select object to align UCS: *(Select the circle at the right)*

Selecting the circle at the right conforms the user coordinate system to the entity selected. Entities determine the alignment of the user coordinate system in many ways. In the case of the circle, the center of the circle becomes the origin of the user coordinate system. Where the circle was selected becomes the point through which the positive X axis aligns to.

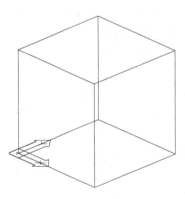

The UCS - View Option

The View option of the user coordinate system command allows the user to establish a new coordinate sstem where the XY plane is perpendicular to the current screen viewing direction; in other words, parallel to the display screen. Given the current user coordinate system at the right, follow the prompt below along with the next illustration to align the user coordinate system using the View option.

Command: **UCS**
Origin/ZAxis/3point/Entity/View/X/Y/Z/Prev/Restore/Save/
Del/?/<World>: **View**

The results are displayed at the right with the user coordinate system aligned parallel to the display screen.

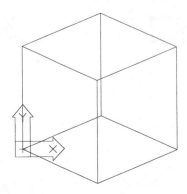

Using the DDUCS Dialog Box

It is considered good practice to assign a name to a user coordinate system once it has been created. Once numerous user coordinate systems have been defined in a drawing, they are easily restored by using their name instead of recreating each coordinate system; this is easily accomplished using the Restore option of the UCS command. Another method is to pick "Named UCS..." from the "UCS" area of the "Settings" pulldown menu area illustrated at the right. This displays the dialog box illustrated below listing all user coordinate systems defined in the drawing. To make one of these coordinate systems current, highlight the desired UCS name and pick the "Current" button below. This dialog box provides a quick method of restoring previously defined coordinate systems without entering them in at the keyboard.

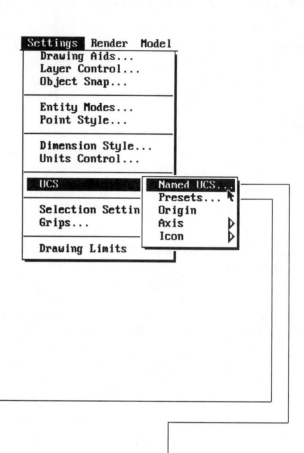

Picking "Presets..." under the "UCS" area of the "Settings" dialog box displays the dialog box illustrated below. Use this dialog box to automatically align the user coordinate system icon to sides of an object such as front view, top view, back view, right side view, left side view, and bottom view.

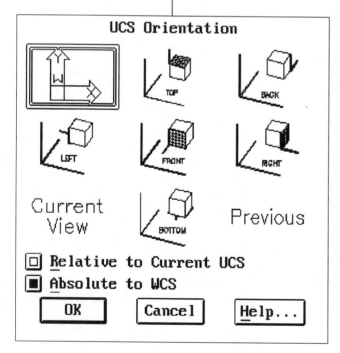

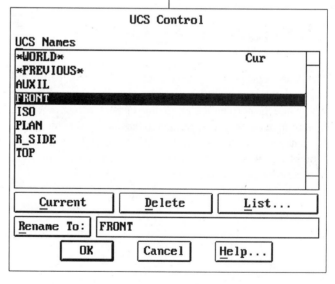

3D Surfaced Models

Surfacing picks up where the wireframe model leaves off. Because of the complexity of the wireframe and the number of intersecting lines, surfaces are applied to the wireframe. The surfaces are in the form of opaque entities, called 3D faces. As always, 3D faces are placed using Osnap options to assist in point selection. A hidden line removal is performed to view the model without the interference of other entities. Again, the Vpoint command is used to view the model in different positions to make sure all sides of the model have been surfaced. The format of the 3Dface command is outlined using the prompts below and the illustration at the right.

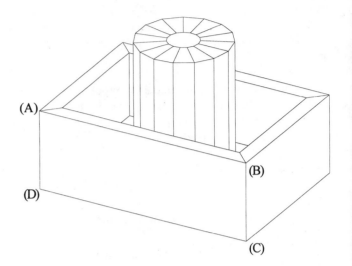

Command: **Osnap**
Object snap modes: **Endp**
Command: **3Dface**
First point: *(Select the point at "A")*
Second point: *(Select the point at "B")*
Third point: *(Select the point at "C")*
Fourth point: *(Select the point at "D")*
Third point: *(Strike Enter to exit this command)*

Ruled Surfaces

3D faces work where four endpoints of an entity can be selected. What if a surface needs to be produced between two different entities such as the two arcs illustrated at the right? In this case, a Ruled surface would be used. In the example at the right, selecting one arc at "A" as the first defining curve and the other arc at "B" as the second defining curve will produce a surface mesh between the two entities. The density of the mesh is controlled by the system variable Surftab1. Assigning a large value will produce a smoother surface but will take a longer time to regenerate for such commands as Hide. Assigning a small value will regenerate the screen faster but will produce a curve appearing as a series of polygons. Ruled surfaces may be placed in between any combination of lines, arcs, circles, and/or points. The format of the Rulesurf command is outlined using the prompts below and the illustration at the right.

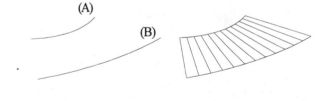

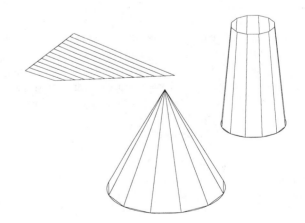

Command: **Rulesurf**
Select first defining curve: *(Select the arc segment at "A")*
Select second defining curve: *(Select the arc segment at "B")*

Tabulated Surfaces

Tabulated surfaces are defined as surfaces generated from a defining curve and direction vector. In the examples at the right, a defining curve is selected at "A" followed by the direction vector at "B." It is important to note that selecting the direction vector also determines the start of the tabulated surface. The direction vector at "B" signifies the beginning of the surface and directs the defining curve at "A" to proceed upward and to the right. A similar problem is illustrated at the right with the defining curve labeled as "C" and the direction vector "D." This time, selecting the direction vector at "D" begins the defining curve and proceeds below and to the right. The results of each curve are illustrated to the right of the individual components. The system variable Surftab1 controls the density of the surfaces. A low number displays the curve as a series of large surfaces. A large number displays smaller surfaces that will show a better outline of a curve and also require more time for regenerations. The format of the Tabsurf command is outlined using the prompts below and the illustrations at the right.

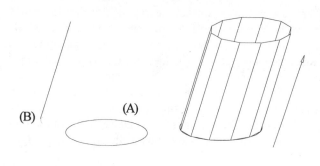

Command: **Tabsurf**
Select path curve: *(Select the circle at "A")*
Select direction vector: *(Select the line segment at "B")*

Command: **Tabsurf**
Select path curve: *(Select the curve at "C")*
Select direction vector: *(Select the line segment at "D")*

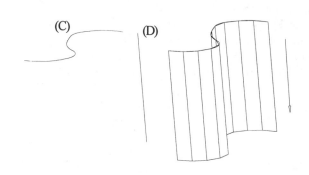

Edge Surfaces

The Edge surface or Coons surface patch is a surface constructed from four adjoining entity edges. One requirement of this surface is that the edges of the four entities must touch at their exact endpoints. Entities may be any combination of lines, arcs, or polylines. Since the surface mesh is generated in two directions, two system variables control the density of this mesh, namely, Surftab1 and Surftab2. As with all geometry generated surfaces, entering large values for the two system variables will construct a very smooth surface that will require extra time for system regeneration. Entering small values for the two system variables will construct a very rough curve; but the system performance will be enhanced and the surface will appear quickly. The format of the Edgesurf command is outlined using the prompts below and the illustration at the right.

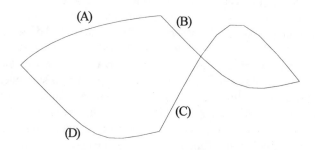

Command: **Edgesurf**
Select edge 1: *(Select the entity at "A")*
Select edge 2: *(Select the entity at "B")*
Select edge 3: *(Select the entity at "C")*
Select edge 4: *(Select the entity at "D")*

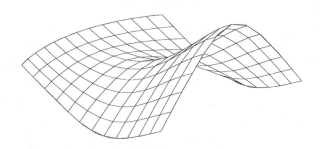

Revolved Surfaces

A surface of revolution can be created by rotating a path curve around a selected axis by using the Revsurf command. The path curve can take the form of a line, arc, circle, 2D polyline, or 3D polyline. The axis of revolution can be either a line or polyline. As with the Edge surface patch, two system variables control the density of the revolved surface, namely, Surftab1 and Surftab2. The format of the Revsurf command is outlined using the prompts below and the illustration at the right.

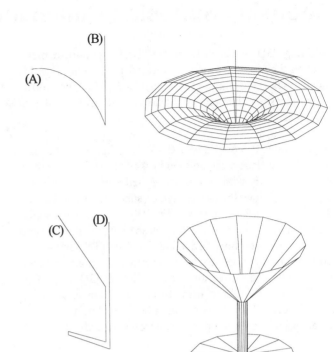

Command: **Revsurf**
Select path curve: *(Select the arc at "A")*
Select axis of revolution: *(Select the line segment at "B")*
Start angle <0>: *(Strike Enter to accept this default value)*
Included angle (+=ccw, -=cw) <Full circle>: *(Strike Enter)*

Command: **Revsurf**
Select path curve: *(Select the polyline segment at "C")*
Select axis of revolution: *(Select the line segment at "D")*
Start angle <0>: *(Strike Enter to accept this default value)*
Included angle (+=ccw, -=cw) <Full circle>: *(Strike Enter)*

Solid Models

The creation of a solid model remains the most exact way to represent a model in 3D. It is also the most versatile representation of an object. Solid models may be viewed as identical to wireframe and surfaced construction models. Wireframe models and surfaced models may be analyzed by taking distance readings and identifying key points of the model. Key orthographic views such as front, top, and right side views may be extracted from wireframe models. Surfaced models may be imported into shading packages such as AutoShade for increased visualization. Solid models do all of the above operations and more. This is because the solid model, as it is called, is a solid representation of the actual object. From cylinders to slabs, wedges to boxes, all entities that go in the creation of a solid model have volume. This allows a model to be constructed of what are referred to as primitives. These primitives are then merged into one using addition and subtraction operations. What remains is the most versatile of 3D drawings. This method of creating models will be discussed in greater detail in Unit 12.

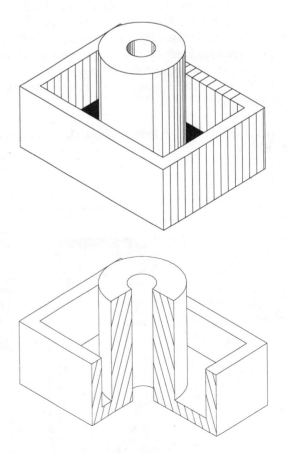

Choosing Surfacing Commands

Selecting "3D Surface" from the "Draw" pulldown menu displays the cascading menu illustrated at the right. Use this menu to select most major surfacing commands. The following surfaces may be created; the actual command is listed in parenthesis:

Edge Defined Patch (Edgesurf) - This option allows the user to create a surface mesh defined by 4 edges. The edges must intersect exactly with each other. Ruled Surface (Rulesurf) - This option allows the user to create a surface mesh defined by 2 entities. Surface of Revolution (Revsurf) - This option allows the user to create a surface mesh given a path curve and an axis of revolution. Tabulated Surface (Tabsurf) - This option allows the user to create a surface mesh given a path curve entity and a direction vector. 3D Face (3Dface) - This option allows the user to manually create a 3D face. Selecting "3D Objects..." activates a dialog box enabling the user to create primitive shapes that are already surfaced.

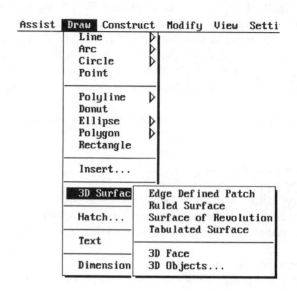

Using the 3D Objects Dialog Box

Selecting "3D Objects..." from the 3D Surfaces area of the "Draw" pulldown menu activates the dialog box illustrated below, which consists of a series of basic 3D objects. When creating these shapes, AutoCAD constructs each object as a polygon mesh rather than a 3D face. In this way each 3D object is considered a single entity. If the Explode command is used on a 3D object, the single object converts into a series of 3D faces. The following objects or 3D surfaced primitives are supported below:

3D Box, Pyramid, Wedge, Dome, Sphere, Dome, Torus, Dish, Mesh

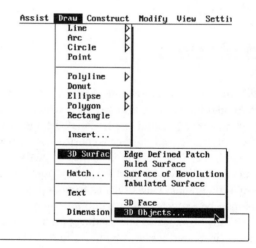

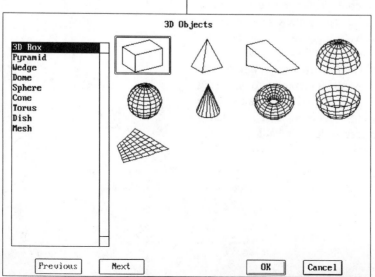

Using the Vpoint Command

The object at the right represents the plan view of a 3D model already created. Unfortunately, it becomes very difficult to understand what the model looks like in its present condition. Use the Vpoint command to view a wireframe, surfaced, or solid model in three dimensions. The following is a typical prompt sequence for the Vpoint command:

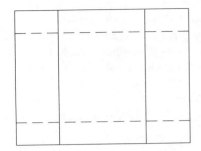

Command: **Vpoint**
Rotate/<View point> <0.0000,0.0000,0.0000>:

The Vpoint command stands for "View point." A point is identified in 3D space. This point becomes a location where the model is viewed from. The point may be entered in at the keyboard as an X,Y,Z coordinate or picked with the aid of a 2-dimensional globe. The object illustrated at the right is a typical result of using the Vpoint command.

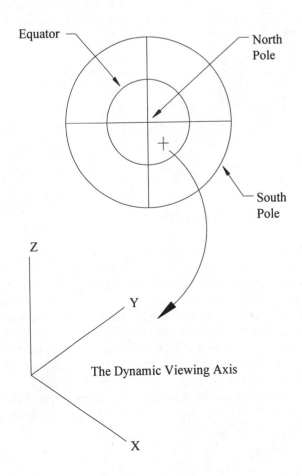

When executing the Vpoint command, three methods are available to the user for viewing a model in 3D.

The first method is defining a view point using an X,Y,Z coordinate. In the command sequence below, a new viewing point is established 1 unit in the positive X direction, 1 unit in the negative Y direction, and 1 unit in the positive Z direction.

Command: **Vpoint**
Rotate/<Viewpont><0.0000,0.0000,0.0000>: **1,–1,1**

A second method of defining a viewing point is to define two angular axes by rotation. The first angle defines the view point in the X-Y axis. However this is only 2-dimensional. The second angle defines the view point from the X-Y axis. This tilts the viewing point up for a positive angle or down for a negative angle.

Command: **Vpoint**
Rotate/<Viewpont><0.0000,0.0000,0.0000>: **Rotate**
Enter angle in X-Y plane from X axis < >: **45**
Enter angle from X-Y plane < >: **30**

If the command sequence is followed by the Enter key, a graphic image consisting of globe and tripod appear illustrated at the right. Although the globe appears 2-dimensional, it provides the user the ability to pick a view point depending on how the globe is read. The intersection of the horizontal and vertical lines form the North Pole of the globe. The inner circle forms the equator and the outer circle the South Pole. The examples that follow illustrate numerous viewing points.

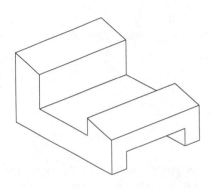

The Dynamic Viewing Axis

Viewing Along the Equator

To obtain the results at the right, use the Vpoint command and mark a point at "A" to view the front view; mark a point at "B" to view the top view; and mark a point at "C" to view the right side view. Coordinates could also have been entered to achieve the same results:

Command: **Vpoint**
Rotate/<View point><0.0000,0.0000,0.0000>:**0,–1,0** *(At "A")*

Command: **Vpoint**
Rotate/<View point><0.0000,0.0000,0.0000>: **0,0,1** *(At "B")*

Command: **Vpoint**
Rotate/<View point><0.0000,0.0000,0.0000>: **1,0,0** *(At "C")*

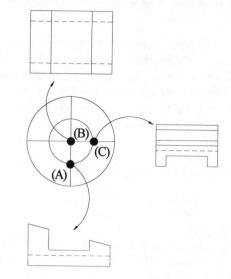

Viewing Near the North Pole

Picking the four points illustrated at the right results in the different viewing points for the object. Since all points are inside the inner circle, the results are aerial views, or views from above. Remember that the Equator is symbolized by the inner circle. Depending on which quadrant you select, you will look up at the object from the right corner, left corner, or either of the rear corners.

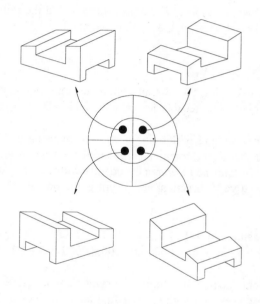

Viewing Near the South Pole

Picking the four points illustrated below results in underground views, or viewing the object from underneath. This is true since all points selected lie between the small and large circles. Again, remember that the small circle symbolizes the Equator, while the large circle symbolizes the South Pole. A more graphical method of selecting view points would be to select the Display command strip and pull down the menu holding the view point command options illustrated below. Once a particular viewing point is selected, use the screen menu to determine whether you desire an aerial view (+10 to +80), normal view (0), or underground view (–10 to –80).

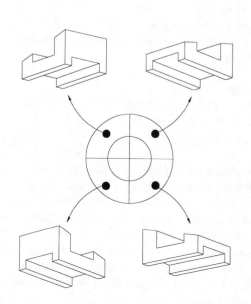

Using the DDVPOINT Dialog Box

Selecting "Presets..." from the "Viewpoint" menu illustrated to
the right displays the dialog box below. Use this dialog box to
assist in defining a new point of viewing a 3-dimensional
model. Of the two images, the square with circle in the center
allows the user to define a viewing point in the X-Y plane. The
semi-circular-shaped image allows the user to defing a viewing point from the X-Y plane. The combination of both angular
directions forms the 3-dimensional view point.

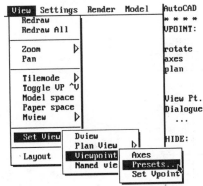

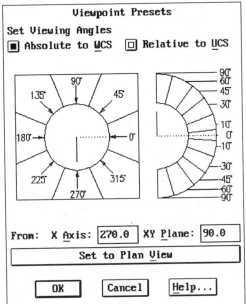

When selecting a viewing point in the X-Y plane, an individual
has two options for selecting the desired point of view. When
picking anywhere outside of the circle but inside of the square,
as in "A" at the right, the view point snaps from one angle to
another in increments of 45 degrees. The resulting angle is
displayed at the bottom of the square. If a more detailed view
point in the X-Y plane is desired, pick a point inside and near
the center of the circle at "B." The resulting angle will display
all values in between the 45 degree increment from before. If
the angle in the X-Y plane is already known, it may be entered
in the edit box next to the prompt "X Axis."

The above example illustrates setting a viewing angle in the X-
Y plane. Displayed at the right is the second half of the
DDVPOINT dialog box which deals with defining an angle
from the X-Y plane. Selecting a point in between both semi-
circles at "A" snaps the viewpoint in 10, 30, 45, 60, and 90
degree increments. If a more detailed selection is desired, pick
the viewing point inside of the smaller semi-circle and near its
center at "B." This will allow the user to select an angle
different from the 5 default values listed above. An alternate
method of selecting an angle from the X-Y plane would be to
place the value in the edit box next to the prompt "XY Plane."

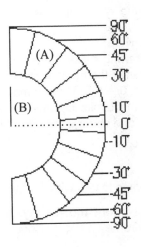

Tutorial Exercise #26
Wedge.Dwg

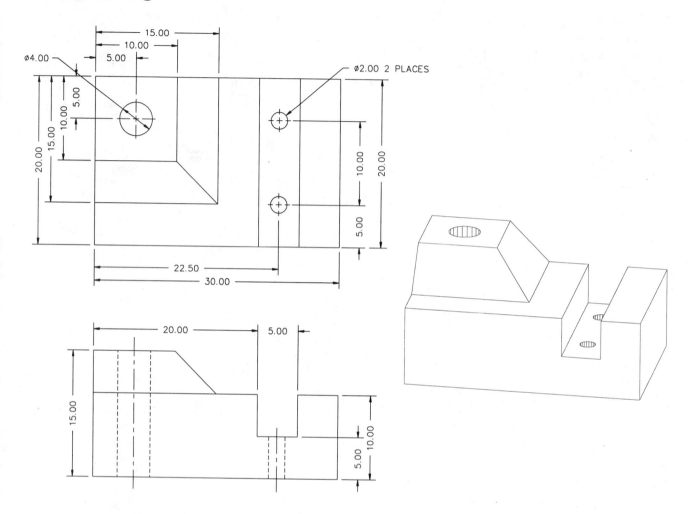

PURPOSE:
This tutorial is designed to construct a true 3D wire-frame model of the Wedge illustrated above.

SYSTEM SETTINGS:
Begin a new drawing called "Wedge." Use the Units command to change the number of decimal places past the 0 from 4 to 0. Keep the remaining default unit values. Using the Limits command, keep 0,0 for the lower left corner and change the upper right corner from 12,9 to 60,45.

LAYERS:
Create the following layers with the format:

Name-Color-Linetype
Wireframe - Yellow - Continuous

SUGGESTED COMMANDS:
The Line command is used exclusively for this drawing of the Wedge. A combination of absolute and polar coordinates is used for precision placement of entities. The holes of the wedge will be represented by circles. The wedge may be surfaced using the 3Dface command. This command will be covered in a later tutorial exercise.

DIMENSIONING:
This object does not require any dimensioning.

PLOTTING:
This tutorial exercise may be plotted on "B"-size paper (11" x 17"). Be sure to set the units of measure from inches to millimeters. Plot the object to at a scale value of 1=1.

Step #1

Begin the drawing by drawing the base of the Wedge. Use the line command and absolute coordinates to perform this operation. Follow the prompts below:

Command: **Line**
From point: **0,0,0**
To point: **30,0,0**
To point: **30,20,0**
To point: **0,20,0**
To point: **C**

Instead of typing in a value for the last coordinate, enter the letter **C** which stands for "Close". This will close the box and exit the line command.

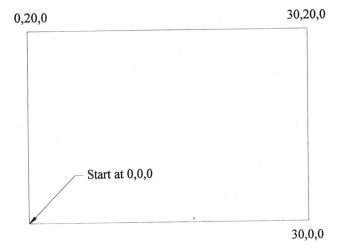

0,20,0   30,20,0

Start at 0,0,0

30,0,0

Step #2

Use the Vpoint command and the Rotate option to view the box in 3D. As you construct the next lines, they will show up in 3D. Follow the prompts below:

Command: **Vpoint**
Rotate/<View point><0,0,1>: **R**
Enter angle in X-Y plane from X axis<270>: **-45**
Enter angle from X-Y plane <90>: **30**

Command: **Zoom**
All/Center/Dynamic/Extents/Left/Previous/Vmax/Window/
<Scale(X/XP)>: **0.6X**

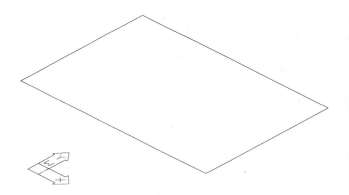

Step #3

Begin creating the upper surface of the Wedge. Again, use the Line command with an absolute coordinate value to begin the line followed by polar coordinates. Follow the prompts below:

Command: **Line**
From point: **0,10,15**
To point: **@10<0**
To point: **@10<90**
To point: **@10<180**
To point: **C**

Perform a Zoom-All to display the entire model.

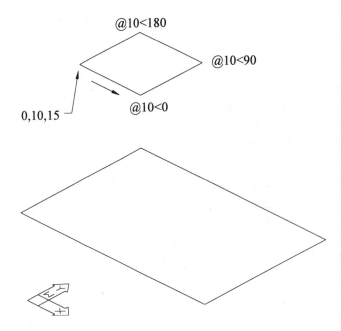

@10<180

@10<90

0,10,15

@10<0

Step #4

Begin work in the next surface of the Wedge. Use a combination of absolute and polar coordinates with the Line command to construct the geometry at the right. Follow the prompts below:

Command: **Line**
From point: **0,0,10**
To point: **@20<0**
To point: **@20<90**
To point: **@5<180**
To point: **@15<270**
To point: **@15<180**
To point: **C**

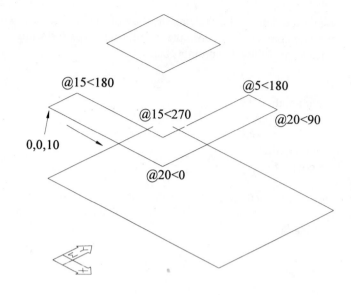

Step #5

Construct the next surface of the wedge using the following prompts of the Line command:

Command: **Line**
From point: **20,0,5**
To point: **@5<0**
To point: **@20<90**
To point: **@5<180**
To point: **C**

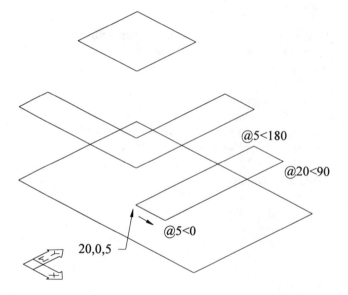

Step #6

Construct the last level of the Wedge with the Line command and a combination of absolute and relative coordinates. Follow the prompts below:

Command: **Line**
From point: **25,0,10**
To point: **@5<0**
To point: **@20<90**
To point: **@5<180**
To point: **C**

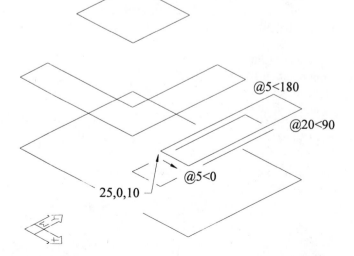

Step #7

The wireframe should appear similar to the illustration at the right. The next series of steps consist of connecting all levels with lines and adding circles to complete the wireframe.

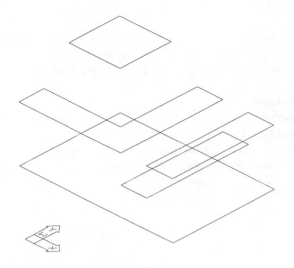

Step #8

Use the Line command to connect all levels to complete the outline of the Wedge. Be sure to use the Osnap-Endpoint option at all times.

Command: **Osnap**
Object snap modes: **Endp**

Command: **Line**
From point: *(Select the endpoint of the line at "A")*
To point: *(Select the endpoint of the line at "B")*
To point: *(Strike Enter to exit this command)*

Repeat the above procedure to connect the remaining edges of the wireframe model. When completed, set the Osnap command to None.

Command: **Osnap**
Object snap modes: **None**

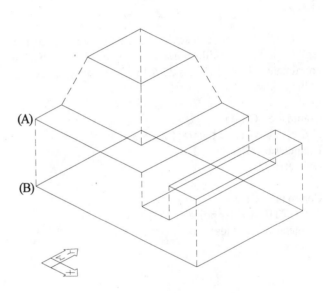

Step #9

Begin representing the holes in the Wedge by constructing circles at the beginning of the hole and end of the hole. Use the Circle command and enter the absolute coordinate 5,15,15 for the center of the circle. (This will place the center of the circle 5 units in the X direction, 15 units in the Y direction, and 15 units in the Z direction.) Enter **D** for diameter of the circle followed by the number 4.00.

Command: **Circle**
3P/2P/TTR/<Center point>: **5,15,15**
Diameter or <Radius>: **D**
Diameter: **4**

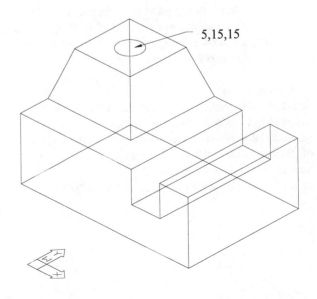

Step #10

Use the Copy command to copy the circle from the top of the wireframe to the bottom of the wireframe. Again, use Osnap-Endpoint and existing geometry to accomplish this.

Command: **Copy**
Select objects: *(Select the circle on the top surface)*
Select objects: *(Strike Enter to continue)*
<Base point or displacement>/Multiple: **Endp**
of *(Select the endpoint of the corner at "A")*
Second point of displacement: **Int**
of *(Select the intersection at "B")*

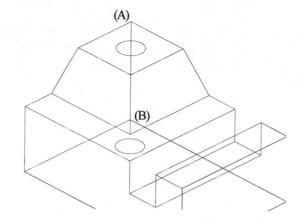

Step #11

Use the Circle command again to place the circles of diameter 2.00 in the slot as illustrated at the right. Use the absolute coordinates of 22.5,5,5 and 22.5,15,5 to identify the centers of the circles.

Command: **Circle**
3P/2P/TTR/<Center point>: **22.5,5,5**
Diameter or <Radius>: **D**
Diameter: **2**

Command: **Circle**
3P/2P/TTR/<Center point>: **22.5,15,5**
Diameter or <Radius>: **D**
Diameter: **2**

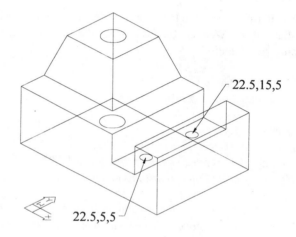

Step #12

Use the Copy command again to copy the two holes from the top of the slot to the bottom of the base of the Wedge. A hint to performing this on existing geometry is to use the Osnap-Endpoint and Osnap-Perpendicular options.

Command: **Copy**
Select objects: *(Select the 2 small circles at the right)*
Select objects: *(Strike Enter to continue)*
<Base point or displacement>/Multiple: **Endp**
of *(Select the endpoint of the corner at "A")*
Second point of displacement: **Per**
to *(Select the perpendicular point to "A" at "B")*

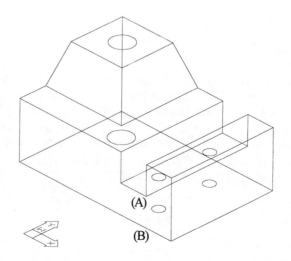

Tutorial Exercise #27
Lever.Dwg

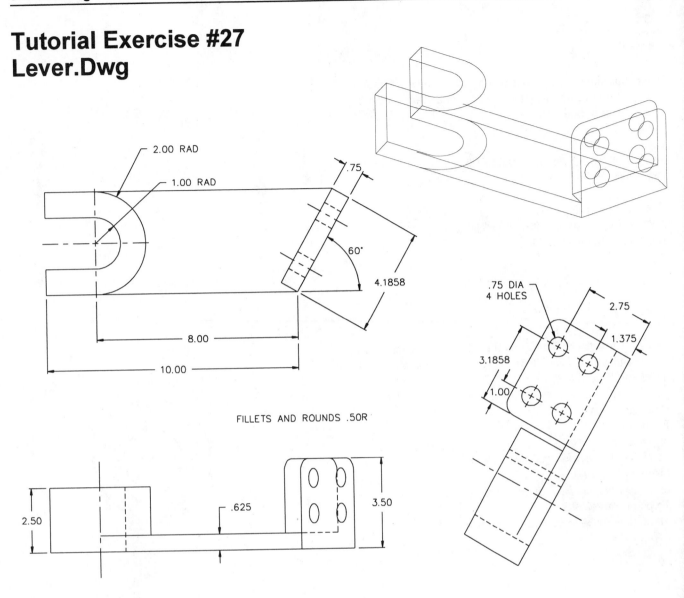

2.00 RAD
1.00 RAD
.75
60°
4.1858
8.00
10.00
FILLETS AND ROUNDS .50R

.75 DIA
4 HOLES
2.75
1.375
3.1858
1.00

2.50
.625
3.50

PURPOSE:
This tutorial is designed to use the UCS command to construct a 3D model of the Lever.

SYSTEM SETTINGS:
Begin a new drawing called "Lever." Use the Units command to change the number of decimal places past the 0 from 4 to 2. Keep all default values for the Units command. Using the Limits command, keep 0,0 for the lower left corner and change the upper right corner from 12,9 to 15.50,9.50. Use the Grid command and change the grid spacing from 1.00 to 0.50 units. Do not turn the snap or ortho On.

LAYERS:
Create the following layers with the format:
Name-Color-Linetype
Object - White - Continuous

SUGGESTED COMMANDS:
Begin layout of this problem by constructing the plan view of the Lever. Use the UCS command to manipulate, create, and save numerous UCS to complete details of the Lever.

DIMENSIONING:
Dimensions will not be added to this problem.

PLOTTING:
This tutorial exercise may be plotted on "B"-size paper (11" x 17"). Use a plotting scale of 1=1 to produce a scaled plot.

Step #1

Begin this drawing by constructing two circles of radius
values 1.00 and 2.00 using the Circle command and 4.00,5.00
as the center of both circles.

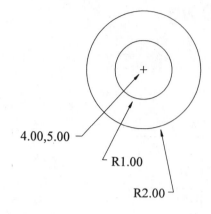

Command: **Circle**
3P/2P/TTR/<Center point>: **4.00,5.00**
Diameter/<Radius>: **1.00**

Command: **Circle**
3P/2P/TTR/<Center point>: **4.00,5.00**
Diameter/<Radius>: **2.00**

Step #2

Draw a vertical line 2 units to the left of the center of the
circles and 4 units long. This can easily be accomplished by
using the .XYZ filters.

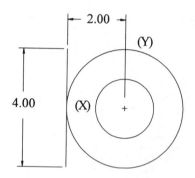

Command: **Line**
From point: **.Y**
of **Qua**
of *(Select the quadrant of the circle at "Y")*
(need XZ): **Qua**
of *(Select the quadrant of the circle at "X")*
To point: **@4<270**
To point: *(Strike Enter to exit this command)*

Step #3

Draw two horizontal lines from quadrant points on the two
circles a distance of 2 units. These lines should intersect with
the vertical line drawn in the previous step.

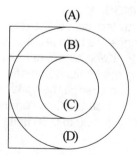

Command: **Line**
From point: **Qua**
of *(Select the quadrant of the large circle at "A")*
To point: **@2<180**
To point: *(Strike Enter to exit this command)*

Repeat the above procedure and draw three more lines from
points "B", "C", and "D" using the same polar coordinate
value for the length, namely, @2<180. An alternate method
would be to use the Copy-Multiple command to duplicate the
three remaining lines.

Step #4

Use the Trim command, select the two dashed lines at the right as cutting edges, and trim the left side of the large circle.

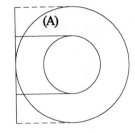

Result

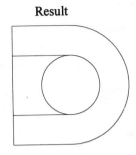

Command: **Trim**
Select cutting edges...
Select objects: *(Select the two dashed lines at the right)*
Select objects: *(Strike Enter to continue)*
<Select object to trim>/Undo: *(Select the large circle at "A")*
<Select object to trim>/Undo: *(Strike Enter to exit this command)*

Step #5

Use the Trim command again and select the two dashed lines at the right as cutting edges. Select the left side of the small circle and the middle of the vertical line as the objects to trim.

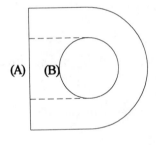

Result

Command: **Trim**
Select cutting edges...
Select objects: *(Select the two dashed lines at the right)*
Select objects: *(Strike Enter to continue)*
<Select object to trim>/Undo: *(Select the vertical line at "A")*
<Select object to trim>/Undo: *(Select the small circle at "B")*
<Select object to trim>/Undo: *(Strike Enter to exit this command)*

Step #6

Complete a partial plan view by using the Line command and the illustration at the right to draw the four lines. Use Osnap-Endpoint whenever possible; also use polar coordinates.

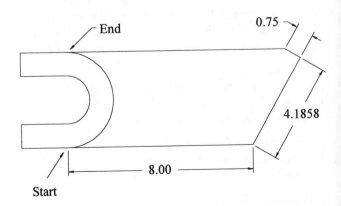

Command: **Line**
From point: **Endp**
of *(Select the endpoint of the line or arc labeled "Start")*
To point: **@8.00<0**
To point: **@4.1858<60**
To point: **@0.75<150**
To point: **Endp**
of *(Select the endpoint of the line or arc labeled "End")*
To point: *(Strike Enter to exit this command)*

Step #7

Your display should be similar to the illustration at the right.

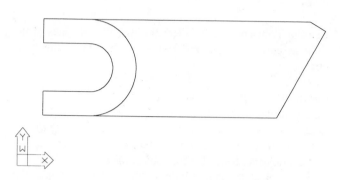

Step #8

The next feature to be drawn of the lever will be the bracket that consists of four holes. Before this can be drawn, a new UCS must be made that will allow entities to be drawn on the new user-specified plane. The next series of steps outlines manipulating the UCS icon to form the new UCS. Use the Osnap-Endpoint option whenever possible. Update the UCS icon using the Ucsicon command.

Command: **UCS**
Origin/ZAxis/3point/Entity/View/X/Y/Z/Prev/Restore/Save/
Del/?/<World>: **Origin**
Origin point <0,0,0>: **Endp**
of *(Select the endpoint of the horizontal line at "A")*

Command: **Ucsicon**
ON/OFF/All/Noorigin/ORigin<ON>: **Origin**

(A)

Step #9

Use the Vpoint command and Rotate option to generate a view that is rotated 300 degrees in the X-Y plane and 30 degrees from the X-Y plane. Use the View command and save the display under the name "Iso." Use the UCS command and rotate the UCS icon 60 degrees about the Z axis.

Command: **Vpoint**
Rotate/<View point><0.0000,0.0000,0.0000,1.0000>: **Rotate**
Enter angle in X-Y plane from X axis <>: **300**
Enter angle from X-Y plane <>: **30**

Command: **View**
?/Delete/Restore/Save/Window: **S**
View name to save: **Iso**

Command: **UCS**
Origin/ZAxis/3point/Entity/View/X/Y/Z/Prev/Restore/Save/
Del/?/<World>: **Z**
Rotation angle about Z axis <0.0>: **60**

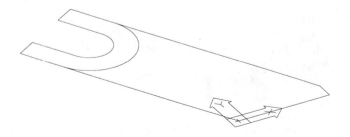

Step #10

Use the UCS command to rotate the UCS icon 90 degrees about the X axis. Save this UCS under the name "Bracket."

Command: **UCS**
Origin/ZAxis/3point/Entity/View/X/Y/Z/Prev/Restore/Save/
Del/?/<World>: **X**
Rotation angle about X axis <0.0>: **90**

Command: **UCS**
Origin/ZAxis/3point/Entity/View/X/Y/Z/Prev/Restore/Save/
Del/?/<World>: **S**
UCS name to save: **Bracket**

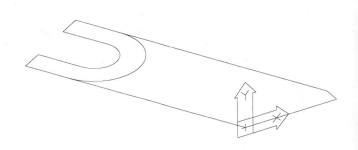

Step #11

Begin drawing the bracket part of the lever in the new UCS that you already defined. To assist you in this operation, use the Plan command to view the model in plan view to the current UCS. Then use the Line command and polar coordinates to draw the outline of the bracket.

Command: **Plan**
<Current UCS>/Ucs/World: *(Strike Enter)*

Command: **Line**
From point: **0,0**
To point: **@3.50<90**
To point: **@4.1858<0**
To point: **@3.50<270**
To point: *(Strike Enter to exit this command)*

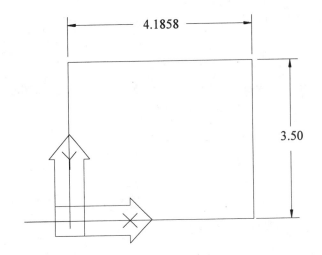

Step #12

Add the four circles of 0.75 units in diameter to the lever's bracket. To locate the centers, use the Offset command to offset one horizontal and one vertical line using the distances at the right. Now use the circle command in combination with the Osnap-Intersection option to draw the circles from the intersection of the four offset lines. Erase the lines used to locate the centers of the circles.

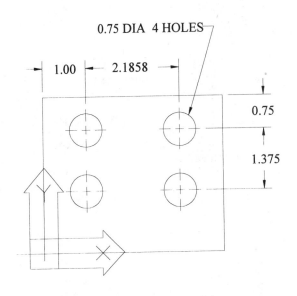

Step #13

Use the Fillet command and change the current fillet radius value to 0.50 units. Then use the Fillet command to place arcs between lines "A" and "B" and between lines "B" and "C."

Command: **Fillet**
Polyline/Radius/<Select first object>: **R**
Enter fillet radius <0.0000>: **0.50**

Command: **Fillet**
Polyline/Radius/<Select first object>: *(Select line "A")*
Select second object: *(Select line "B")*

Command: **Fillet**
Polyline/Radius/<Select first object>: *(Select lines "B")*
Select second object: *(Select line "C")*

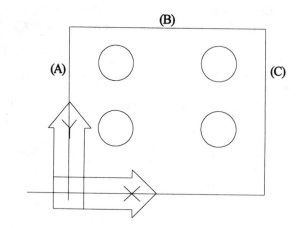

Step #14

Your display should be similar to the one illustrated at the right.

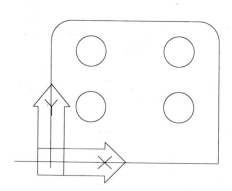

Step #15

Use the View command to restore the previously saved view called "Iso." Your display should be similar to the illustration at the right.

Command: **View**
?/Delete/Restore/Save/Window: **R**
View name to restore: **Iso**

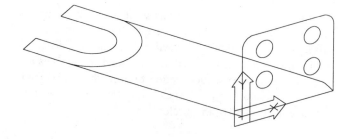

Step #16

Before performing any drawing, reset the current coordinate system to the World coordinate system. Then use the illustration at the right to guide you in copying the small line at "A" to the new position at "B."

Command: **UCS**
Origin/ZAxis/3point/Entity/View/X/Y/Z/Prev/Restore/Save/Del/?/<World>: **W**

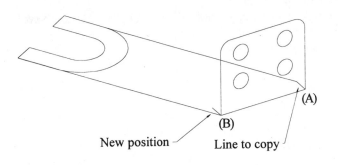

Command: **Copy**
Select objects: *(Select the line at the right to copy)*
Select objects: *(Strike Enter to continue)*
<Base point or displacement>/Multiple: **Endp**
of *(Select the endpoint of the line at "A")*
Second point of displacement: **Endp**
of *(Select the endpoint of the line at "B" labeled "New position")*

Step #17

Move the dashed line the distance 0.625 units in the Z direction using the Osnap-Endpoint option and XYZ filters.

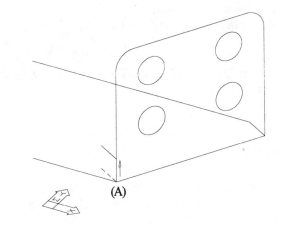

Command: **Move**
Select objects: *(Select the dashed line at the right)*
Select objects: *(Strike Enter to continue)*
Base point or displacement: **Endp**
of *(Select the endpoint of the dashed line at "A")*
Second point of displacement: **.XY**
of **Endp**
of *(Select the endpoint of the dashed line at "A")*
(Need Z): **0.625**

Step #18

Copy the 2 dashed lines at the right a distance of 0.625 units using the Copy command, Osnap-Endpoint option, and XYZ filters.

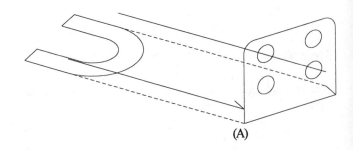

Command: **Copy**
Select objects: *(Select the two dashed lines at the right)*
Select objects: *(Strike Enter to continue)*
<Base point or displacement>/Multiple: **Endp**
of *(Select the endpoint of the dashed line at "A")*
Second point of displacement: **.XY**
of **Endp**
of *(Select the endpoint of the dashed line at "A")*
(Need Z): **0.625**

Step #19

Use the UCS command and restore the UCS called "Bracket."

Command: **UCS**
Origin/ZAxis/3point/Entity/View/X/Y/Z/Prev/Restore/Save/
Del/?/<World>: **R**
Ucs to restore: **Bracket**

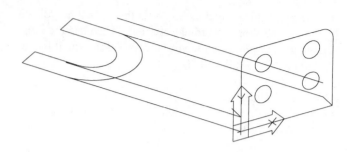

Step #20

Use the UCS command with the Origin option to move the UCS icon to the endpoint of line "A." If the icon is not displayed on the line, use Zoom and a value of 0.8x.

Command: **UCS**
Origin/ZAxis/3point/Entity/View/X/Y/Z/Prev/Restore/Save/
Del/?/<World>: **O**
Origin point <0,0,0>: **Endp**
of   *(Select the endpoint of line "A")*

Command: **Zoom**
All/Center/Dynamic/Extents/Left/Previous/Window/
<Scale(x)>: **0.8x**

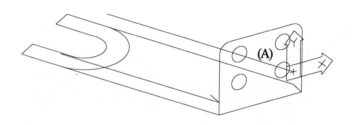

Step #21

Draw four lines representing the back side of the bracket using the Line command and the current position of the UCS. Draw the lines from "A," to "B," to "C," to "D," and close the rectangle.

Command: **Line**
From point: **0,0** *(at "A")*
To point: **@4.1858<180** *(at "B")*
To point: **@2.875<90** *(at "C")*
To point: **@4.1858<0** *(at "D")*
To point: **C** *(at "A")*

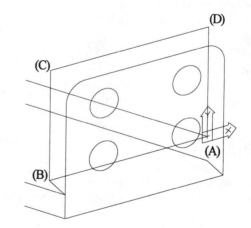

Step #22

Your display should appear similar to the illustration at the right.

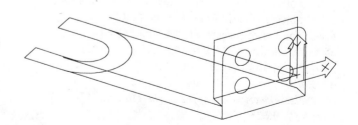

Step #23

Copy the four circles from the front face of the bracket to the
rear face using the Copy command and the Osnap-Endpoint
option.

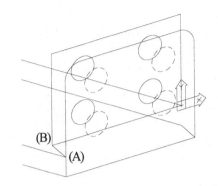

Command: **Copy**
Select objects: *(Select the four dashed circles)*
Select objects: *(Strike Enter to continue)*
<Base point or displacement>/Multiple: **Endp**
of *(Select the endpoint of the line at "A")*
Second point of displacement: **Endp**
of *(Select the endpoint of the line at "B")*

Step #24

Use the Fillet command to place arcs between lines "A" and
"B" and between lines "B" and "C" (the fillet radius should
already be set at 0.50 units).

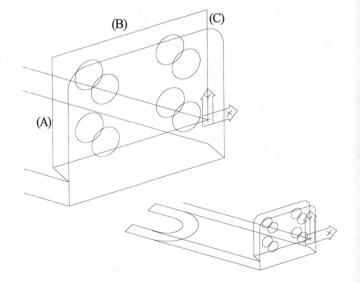

Command: **Fillet**
Polyline/Radius/<Select first object>: *(Select line "A")*
Select second object: *(Select line "B")*

Command: **Fillet**
Polyline/Radius/<Select first object>: *(Select lines "B")*
Select second object: *(Select line "C")*

Your display should appear similar to the illustration at the
right.

Step #25

Change back to the World coordinate system using the UCS
command. Copy the dashed entities at the right the distance
of 2.50 in the Z direction using the Osnap-Endpoint option
and XYZ filters.

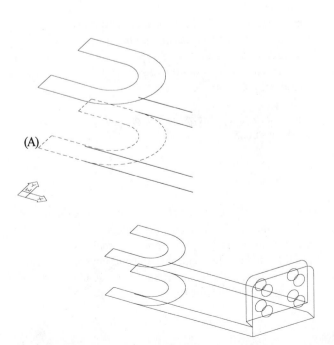

Command: **UCS**
Origin/ZAxis/3point/Entity/View/X/Y/Z/Prev/Restore/Save/
Del/?/<World>: **W**

Command: **Copy**
Select objects: *(Select all dashed entities at the right)*
Select objects: *(Strike Enter to continue)*
<Base point or displacement>/Multiple: **Endp**
of *(Select the endpoint of the line at "A")*
Second point of displacement: **.XY**
of **Endp**
of *(Select the endpoint of the line at "A")*
(Need Z): **2.50**

Your display should appear similar to the illustration at the
right.

Step #26

Draw a line from "A" to "B" using the Line command and the Osnap-Endpoint option. Copy this line to points "C," "D," and "E" using the Copy-Multiple option.

Command: **Line**
From point: **Endp**
of *(Select the endpoint of the line at "A")*
To point: **Endp**
of *(Select the endpoint of the line at "B")*
To point: *(Strike Enter to exit this command)*

Command: **Copy**
Select objects: **L**
Select objects: *(Strike Enter to continue)*
<Base point or displacement>/Multiple: **M**
First point: **Endp**
of *(Select the endpoint of the line at "B")*
Second point of displacement: **Endp**
of *(Select the endpoint of the line at "C")*
Second point of displacement: **Endp**
of *(Select the endpoint of the line at "D")*
Second point of displacement: **Endp**
of *(Select the endpoint of the line at "E")*
Second point of displacement: *(Strike Enter to exit this command)*

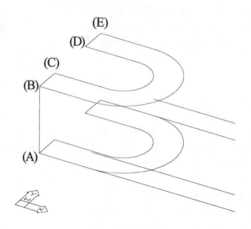

Step #27

Move the dashed circle at the right a distance of 0.625 units in the Z direction. Use the Osnap-Endpoint option and XYZ filters.

Command: **Move**
Select objects: *(Select the dashed circle at the right)*
Select objects: *(Strike Enter to continue)*
Base point or displacement: **Endp**
of *(Select the endpoint of the line at "A")*
Second point of displacement: **.XY**
of **Endp**
of *(Select the endpoint of the line at "A")*
(Need Z): **0.625**

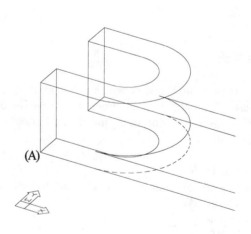

The display of the Lever at the right is an example of a surfaced model that has had the hidden lines removed to view the object as it would appear in real life. However, even though the object appears as a solid image, it is actually hollow on the inside. Converting the image to a solid model will be discussed in Unit 12.

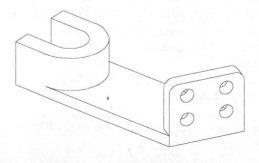

Tutorial Exercise #28
Column.Dwg (Wireframe Model)

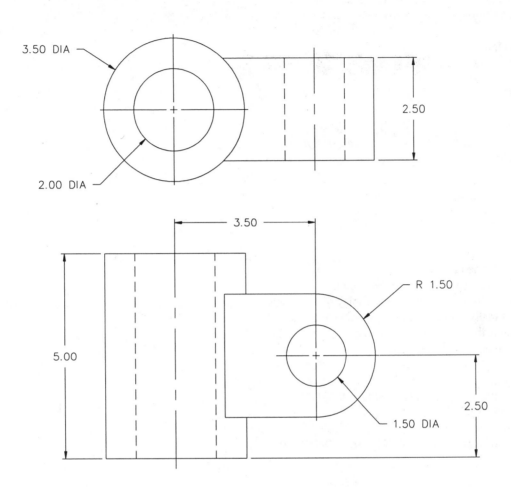

PURPOSE:
This tutorial is designed to use the UCS command to produce a 3D wireframe model of the Column. This model will be surfaced in the next tutorial segment.

SYSTEM SETTINGS:
Begin a new drawing called "Column." Use the Units command to change the number of decimal places past the zero from 4 to 2. Keep the remaining default unit values. Using the Limits command, keep 0,0 for the lower left corner and change the upper right corner from 12,9 to 10.50,8.00. Use the Grid command and change the grid spacing from 1.00 to 0.25 units. Do not turn the snap or ortho On.

LAYERS:
Create the following layers with the format:

Name-Color-Linetype
Wireframe - Yellow - Continuous
Surface - Magenta - Continuous

SUGGESTED COMMANDS:
Begin drawing the Column by laying out the plan view using the circle command. Use XYZ filters to copy the circles in the Z direction to form the column. Create new coordinate systems to lay out the flange that attaches to the Column. Use the Trim command to edit the Column before surfacing.

DIMENSIONING:
Dimensions do not have to be added to this problem.

PLOTTING:
This tutorial exercise may be plotted on "B"-size paper (11" x 17"). Use a plotting scale of 1=1 to produce a scaled plot.

Step #1

Begin this tutorial by drawing two circles in plan view using the Circle command. These Circles represent the main cylinder and hole going through it.

Command: **Circle**
3P/2P/TTR/<Center point>: **4.00,4.00**
Diameter/<Radius>: **D**
Diameter: **3.50**

Command: **Circle**
3P/2P/TTR/<Center point>: **@** *(To reference the last point)*
Diameter/<Radius>: **D**
Diameter: **2.00**

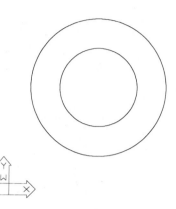

Step #2

Use the Vpoint command to view the circles in 3D. Most constructions will take place from this view point. Next, use the Zoom command to reduce the current display by a factor of 0.30 units. Finally, save this display using the View command and name the view "Iso." Update the current UCS icon with the Ucsicon-Origin command.

Command: **Vpoint**
Rotate/<View point><0.0000,0.0000,1.0000>: **0.5,-0.75,0.5**

Command: **Zoom**
All/Center/Dynamic/Extents/Left/Previous/Window/
<Scale(X)> **0.30X**

Command: **View**
?/Delete/Restore/Save/Window: **S**
View name to save: **Iso**

Command: **Ucsicon**
ON/OFF/All/Noorigin/ORigin<ON>: **Origin**

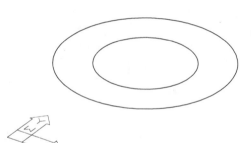

Step #3

Use the Copy command and XYZ point filters to copy the bottom circles 5.00 units in the Z direction.

Command: **Copy**
Select objects: *(Select the two circles)*
Select objects: *(Strike Enter to continue)*
<Base point or displacement>/Multiple: **Cen**
of *(Select the large circle)*
Second point of displacement: **.XY**
of **Cen**
of *(Select the large circle)*
(Need Z): **5.00**

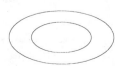

Step #4

The next series of entities drawn are merely for construction purposes only. They will be deleted at later steps. First, draw a line from the quadrant of the bottom large circle at point "A" to the quadrant of the top large circle at point "B." Next, draw a line from the center of the large circle to a point 3.50 units in the 0 direction using polar coordinates.

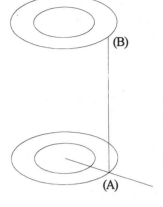

Command: **Line**
From point: **Qua**
of *(Select bottom large circle at point "A")*
To point: **Qua**
of *(Select top large circle at point "B")*
To point: *(Strike Enter to exit this command)*

Command: **Line**
From point: **Cen**
of *(Select bottom large circle)*
To point: **@3.50<0**
To point: *(Strike Enter to exit this command)*

Step #5

Move the construction line 3.50 units in length to a new height 2.50 units from its origin. XYZ filters will be used to accomplish this.

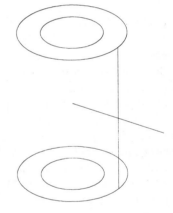

Command: **Move**
Select objects: *(Select the 3.50 unit line)*
Select objects: *(Strike Enter to continue)*
Base point or displacement: **Cen**
of *(Select the bottom large circle)*
Second point of displacement: **.XY**
of **Cen**
of *(Select the bottom large circle again)*
(Need Z): **2.50**

Step #6

Create a new UCS using the UCS command and the 3 point option. (Be sure to use Osnap-Intersec and Osnap-Endpoint to snap onto intersections and endpoints of entities.) Use the prompts below and illustration at the right to guide you in this procedure. Use the UCS command again to save the position under the name "Front."

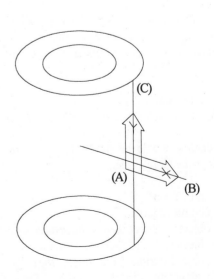

Command: **UCS**
Origin/ZAxis/3point/Entity/View/X/Y/Z/Prev/Restore/Save/ Del/?/<World>: **3**
Origin point <0,0,0>: *(Select the intersection of Point "A")*
Point on positive portion of the X axis<>: *(Select the endpoint Point "B")*
Point on positive-Y portion of the UCS XY plane < >: *(Select the endpoint Point "C")*

Command: **UCS**
Origin/ZAxis/3point/Entity/View/X/Y/Z/Prev/Restore/Save/ Del/?/<World>: **S**
Name to save: **Front**

Step #7

Draw two circles of radius 1.50 and diameter 1.50 units. Use point "A" as the center of both circles. (Use Osnap-Endpoint to snap to the endpoint of the line at point "A".)

Command: **Circle**
3P/2P/TTR/<Center point>: **Endp**
of *(Select the endpoint of the line at point "A")*
Diameter/<Radius>: **1.50**

Command: **Circle**
3P/2P/TTR/<Center point>: **@** *(To reference the last point)*
Diameter/<Radius>: **D**
Diameter: **1.50**

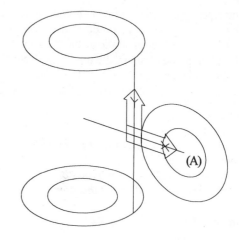

Step #8

Copy line "A" 1.50 units above and below its current position using the Copy command.

Command: **Copy**
Select objects: *(Select line "A")*
Select objects: *(Strike Enter to continue)*
<Base point or displacement>/Multiple: **Endp**
of *(Select a line near point "A")*
Second point of displacement: **@1.50<90**

Command: **Copy**
Select objects: *(Select line "A")*
Select objects: *(Strike Enter to continue)*
<Base point or displacement>/Multiple: **Endp**
of *(Select a line near point "A")*
Second point of displacement: **@1.50<270**

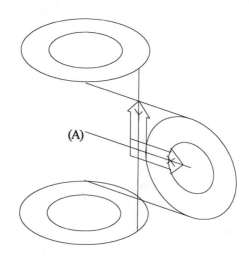

Step #9

Use the Plan command to obtain a plan view based on the current UCS. The purpose of this step is to use the Trim command to trim the large circle in between the two horizontal lines.

Command: **Plan**
<Current UCS>/Ucs/World: *(Strike Enter to continue)*

Your display should be similar to the illustration at the right. Now trim the large circle.

Command: **Trim**
Select cutting edge(s)...
Select objects: *(Select the two horizontal dashed lines)*
Select objects: *(Strike Enter to continue)*
<Select object to trim>/Undo:*(Select the large circle at the right)*
<Select object to trim>/Undo:*(Select Enter to exit this command)*

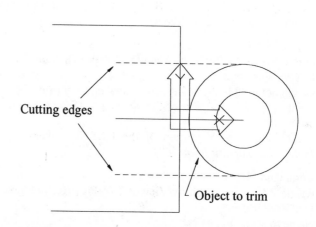

Cutting edges

Object to trim

Step #10

Restore the view called "Iso." Then select the dashed entities at the right and move the entities -1.25 units in the Z direction. Use .XYZ filters and the Osnap-Endpoint option to accomplish this.

Command: **View**
?/Delete/Restore/Save/Window: **R**
View name to restore: **Iso**

Command: **Move**
Select objects: *(Select the four dashed entities)*
Select objects: *(Strike Enter to continue)*
Base point or displacement: **Endp**
of *(Select the endpoint of the line at "A")*
Second point of displacement: **.XY**
of **Endp**
of *(Select the endpoint of line "A" at point "A")*
(Need Z): **-1.25**

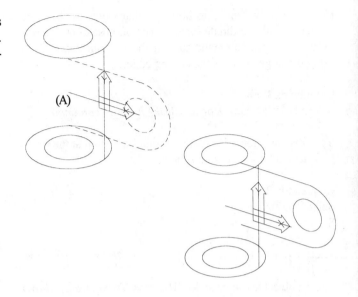

(A)

Step #11

Select the dashed entities at the right and copy the entities 1.25 units in the Z direction. Use XYZ filters and the Osnap-Endpoint option to accomplish this.

Command: **Copy**
Select objects: **P** *(To select the previous selection set)*
Select objects: *(Strike Enter to continue)*
<Base point or displacement>/Multiple: **Endp**
of *(Select the endpoint of the line at "A")*
Second point of displacement: **.XY**
of **Endp**
of *(Select the endpoint of the line at "A")*
(Need Z): **1.25**

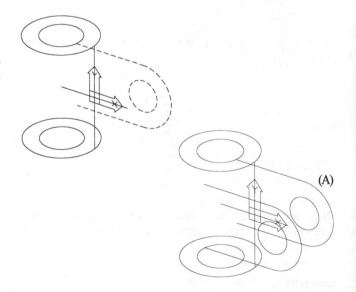

(A)

Step #12

Change from the UCS to the World coordinate system by issuing the UCS command and striking the Enter key at the prompt (the World coordinate system is the default).

Command: **UCS**
Origin/ZAxis/3point/Entity/View/X/Y/Z/Prev/Restore/Save/
Del/?/<World>: *(Strike Enter to accept the default value)*

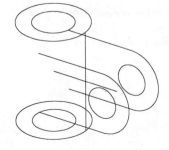

Step #13

Use the Plan command to obtain a plan view based on the current World coordinate system. The purpose of this step is to use the Trim command to trim the two horizontal lines using the large circle as the cutting edge.

Command: **Plan**
<Current UCS>/Ucs/World: *(Strike Enter to continue)*

Your display should be similar to the illustration at the right. Now trim the two horizontal lines.

Command: **Trim**
Select cutting edge(s)...
Select objects: *(Select the large circle)*
Select objects: *(Strike Enter to continue)*
<Select object to trim>/Undo: *(Select the two horizontal lines)*
<Select object to trim>/Undo: **'Redraw** *(Transparent redraw)*
Resuming TRIM command.
<Select object to trim>/Undo: *(Select the two horizontal lines again)*
<Select object to trim>/Undo: *(Strike Enter to exit this command)*

Your display should appear similar to the illustration at the right.

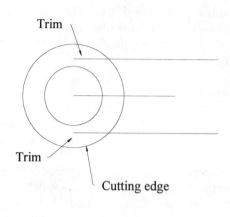

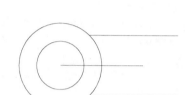

Step #14

Use the View command to restore the view named "Iso". Next, use the Erase command to delete the two construction lines illustrated at the right (lines "A" and "B").

Command: **View**
?/Delete/Restore/Save/Window: **R**
View name to restore: **Iso**

Command: **Erase**
Select objects: *(Select line "A")*
Select objects: *(Select line "B")*
Select objects: *(Strike Enter to execute this command)*

Command: **Redraw**

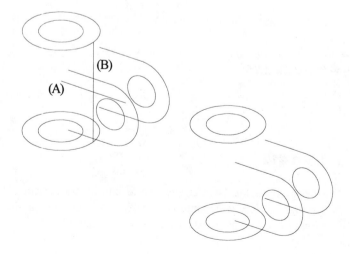

Step #15

Use the Copy command to place two additional circles 1 and 4 units above circle "A". Use the Osnap-Center option in combination with XYZ filters.

Command: **Copy**
Select objects: *(Select circle "A")*
Select objects: *(Strike Enter to continue)*
<Base point or displacement>/Multiple: **Cen**
of *(Select circle "A" using the Osnap-Center option)*
Second point of displacement: **.XY**
of **Cen**
of *(Select Circle "A" again using Osnap-Center)*
(Need Z): **1**

Repeat the above procedure exactly as before; however, type in a value of **4** instead of **1** for the new Z value.

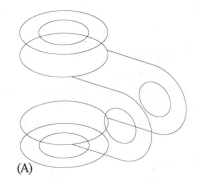

(A)

Step #16

The two large circles will be trimmed to form arcs that match the curvature of the cylinder. Use the Trim command, select the four dashed lines as cutting edges, and select one circle at "A" and the other circle at "B" to trim the circles and form arcs.

Command: **Trim**
Select cutting edge(s)...
Select objects: *(Select the four dashed lines)*
Select objects: *(Strike Enter to continue)*
<Select object to trim>/Undo: *(Select the circle at "A")*
<Select object to trim>/Undo: *(Select the circle at "B")*
<Select object to trim>/Undo: *(Strike Enter to exit this command)*

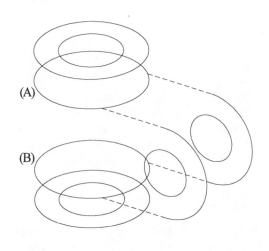

(A)

(B)

Step #17

This step is meant for construction purposes only and will be used at a later step where surfacing is required. One line is drawn from point "A" to point "B." The line is then copied from point "B" to point "C" using Osnap-Endpoint.

Command: **Line**
From point: **Endp**
of *(Select the endpoint of the line or arc at "A")*
To point: **Endp**
of *(Select the endpoint of the line or arc at "B")*
To point: *(Strike Enter to exit this command)*

Command: **Copy**
Select objects: **L**
Select objects: *(Strike Enter to continue)*
<Base point or displacement>/Multiple: **Endp**
of *(Select the endpoint of the line at "B")*
Second point of displacement: **Endp**
of *(Select the endpoint of the line or arc at "C")*

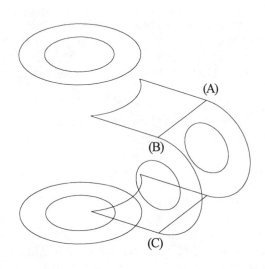

(A)

(B)

(C)

Step #18

Draw another line using the Line command from the end-point of the line or arc at point "A" to the endpoint of the line or arc at point "B."

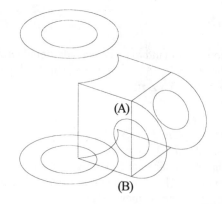

Command: **Line**
From point: **Endp**
of *(Select the endpoint of the line or arc at "A")*
To point: **Endp**
of *(Select the endpoint of the line or arc at "B")*
To point: *(Strike Enter to exit this command)*

Step #19

Use the UCS command and restore the view named "Front." Next, use the Trim command, select the circle and vertical line illustrated at the right as cutting edges, and trim half of the circle and the middle of the line.

Command: **UCS**
Origin/ZAxis/3point/Entity/View/X/Y/Z/Prev/Restore/Save/
Del/?/<World>: **Restore**
?/Name of UCS to restore: **Front**

Command: **Trim**
Select cutting edge(s)...
Select objects: *(Select line "A" and circle "B")*
Select objects: *(Strike Enter to continue)*
<Select object to trim>/Undo: *(Select the line at "A")*
<Select object to trim>/Undo: *(Select the circle at "B")*
<Select object to trim>/Undo: *(Strike Enter to exit)*

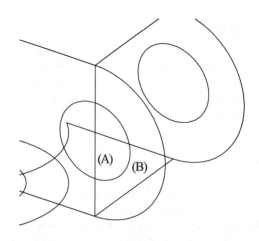

Step #20

Use the Mirror command to copy and flip the dashed arc at the right. Then use the Pedit command to convert all dashed entities at the right to a polyline. This polyline will be used later for adding surfaces to the model.

Command: **Mirror**
Select objects: *(Select the dashed arc at the right)*
Select objects: *(Strike Enter to continue)*
First point of mirror line: *(Select the endpoint at "A")*
Second point: *(Select the endpoint at "B")*
Delete old objects? <N> *(Strike Enter to exit this command)*

Command: **Pedit**
Select polyline: *(Select the dashed line near "A")*
Entity selected is not a polyline.
Do you want to turn it into one?<Y>: **Y**
Close/Join/Width/Edit vertex/Fit curve/Spline curve/
Decurve/Undo/eXit <X>: **J**
Select objects: *(Select the dashed arc)*
Select objects: *(Select the dashed line near "B")*
Select objects: *(Strike Enter to exit this segment of Pedit)*
Close/Join/Width/Edit vertex/Fit curve/Spline curve/
Decurve/Undo/eXit <X>: **X**

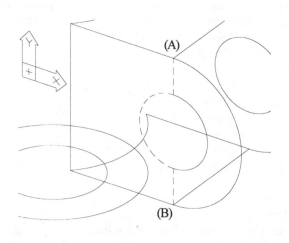

Step #21

Change back to the World coordinate system using the UCS command. Draw a line using the line command from the endpoint of the arc at point "A" to the endpoint of the arc at point "B."

Command: **UCS**
Origin/ZAxis/3point/Entity/View/X/Y/Z/Prev/Restore/Save/
Del/?/<World>: *(Strike Enter to accept the default value)*

Command: **Line**
From point: **Endp**
of *(Select the endpoint of the arc at "A")*
To point: **Endp**
of *(Select the endpoint of the arc at "B")*
To point: *(Strike Enter to exit this command)*

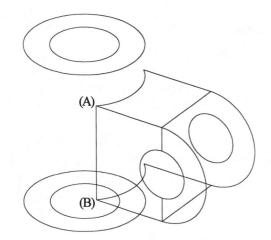

Step #22

Use the copy command to duplicate the last line drawn. Copy this entity from "A" to "B," as illustrated at the right.

Command: **Copy**
Select objects: **L**
Select objects: *(Strike Enter to continue)*
<Base point or displacement>/Multiple: **Endp**
of *(Select the endpoint of the line at "A")*
Second point of displacement: **Endp**
of *(Select the endpoint of the line at "B")*

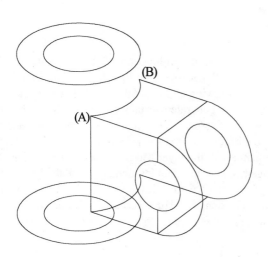

The Wireframe model becomes the basis if the need to surface the model is required. Existing geometry is used for creating the following surfaces: Ruled, Tabulated, Revolved, and Edge surfaces. Follow the next series of steps for surfacing the Column.

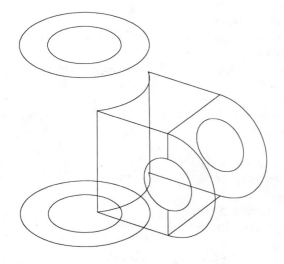

Tutorial Exercise #29
Column.Dwg (Surfaced Model)

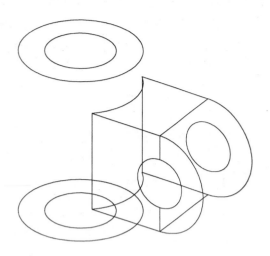

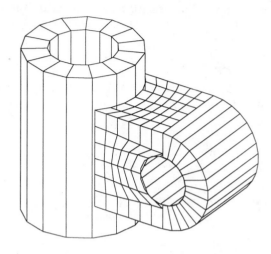

Step #1

This exercise is a continuation of the previous wireframe problem. All necessary entities have been added to the wireframe drawing to prepare the model to be surfaced using the geometry-generated surfacing commands outlined in the next series of steps. A few other tasks need to be performed before surfacing. First, make a new layer using the Layer command; call the layer Mesh with a magenta color. Next, change two system variables, namely, Surftab1 and Surftab2, from a value of 6 to a value of 15. These two variables control the density of the surfaces; a value of 15 will be adequate for this model.

Command: **Layer**
?/Make/Set/New/ON/OFF/Color/Ltype/Freeze/Thaw/LOck/ Unlock: **Make**
New current layer <0>: **Mesh**
?/Make/Set/New/ON/OFF/Color/Ltype/Freeze/Thaw/LOck/ Unlock: **Color**
Color: **Magenta**
Layer name(s) for color 6 <MESH>: *(Strike Enter to accept the default)*
?/Make/Set/New/ON/OFF/Color/Ltype/Freeze/Thaw/LOck/ Unlock: *(Strike Enter to exit this command)*

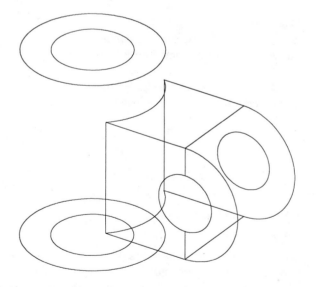

Observe the top of your screen to see that the new current layer is Mesh.

Command: **Surftab1**
New value for Surftab1 <6>: **15**

Command: **Surftab2**
New value for Surftab2 <6>: **15**

Step #2

Begin surfacing the wireframe model of the column by using the Rulesurf command. This command will place surfaces in between the top and bottom circles to form the cylinder. Pick the outside edge of the large bottom circle at "A" for the first defining curve; pick the outside edge of the large top circle at "B" for the second defining curve. Repeat the procedure for the small circles.

Command: **Rulesurf**
Select first defining curve: *(Select the circle at "A")*
Select second defining curve: *(Select the circle at "B")*

Notice that the surface is formed between the two large circles. Now repeat the above procedure for the smaller circles.

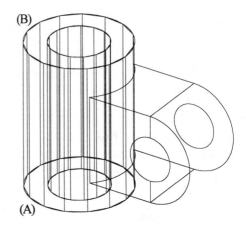

Step #3

As the surfacing segment of a wireframe gets more complex, selecting entities may become difficult. Use the Move command to reposition the mesh away from the wireframe. This is accomplished by entering a displacement to move the mesh patterns. Follow the prompts below to see how this is accomplished.

Command: **Move**
Select objects: *(Select both surface mesh patterns)*
Select objects: *(Strike Enter to continue)*
Base point or displacement: **0,10,0**
Second point of displacement: *(Strike Enter to execute this command)*

Notice that both mesh patterns move a distance away from the wireframe based on the move displacement coordinate value of 0,10,0. This same value will be used for moving other surface mesh patterns. Perform a Zoom-All to view the wireframe and mesh pattern.

Command: **Zoom**
All/Center/Dynamic/Extents/Left/Previous/Vmax/Window/
<Scale(X/XP)>: **All**
Regenerating Drawing.

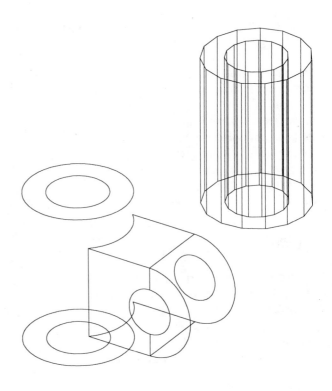

Step #4

Use the Rulesurf command to place ruled surfaces on wireframe at the right. Select points "A" and "B" as defining curves to surface the top of the cylinder. Perform the same steps for the bottom cylinder. Select points "C" and "D" as defining curves to begin surfacing the front projection. Copy the front projection to the rear.

Command: **Rulesurf**
Select first defining curve: *(Select the circle at "A")*
Select second defining curve: *(Select the circle at "B")*

Follow the same steps for the bottom circles of the wireframe.

Command: **Rulesurf**
Select first defining curve: *(Select the arc at "C")*
Select second defining curve: *(Select the arc at "D")*

Command: **Copy**
Select objects: **L**
Select objects: *(Strike Enter to continue)*
<Base point or displacement>/Multiple: **Endp**
of *(Select the endpoint of the line at "E")*
Second point of displacement: **Endp**
of *(Select the endpoint of the line at "F")*

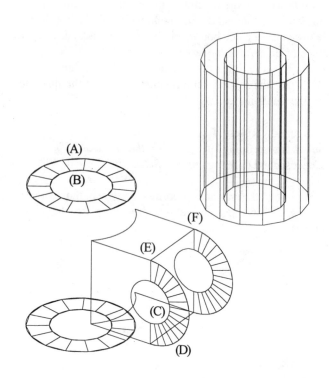

Step #5

Move the four surface mesh patterns at a displacement of 0,10,0 using the Move command.

Command: **Move**
Select objects: *(Select the four surface mesh patterns in the wireframe)*
Select objects: *(Strike Enter to continue)*
Base point or displacement: **0,10,0**
Second point of displacement: *(Strike Enter to execute the command)*

Command: **Redraw**

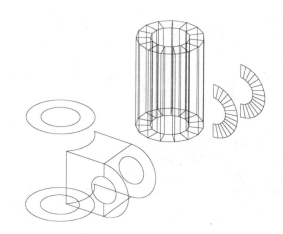

Step #6

Continue using the Rulesurf command by surfacing the curved outer surface of the projection. Select points "A" and "B" as defining curves.

Command: **Rulesurf**
Select first defining curve: *(Select the arc at "A")*
Select second defining curve: *(Select the arc at "B")*

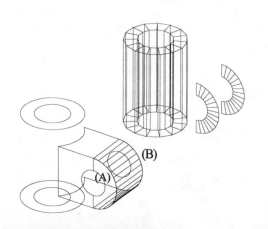

Step #7

Move the last surface mesh pattern at a displacement of 0,10,0 using the Move command.

Command: **Move**
Select objects: **L**
Select objects: *(Strike Enter to continue)*
Base point or displacement: **0,10,0**
Second point of displacement: *(Strike Enter to execute this command)*

Command: **Redraw**

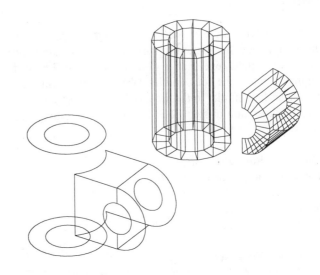

Step #8

A cylinder in the projection needs to be created using the Tabsurf command. This command requires a path curve and direction vector to form the desired mesh pattern. Select the full circle as the path curve; select the line at the right as the direction vector. A mesh pattern will form consisting of information in the path curve; length of the mesh is based on the direction vector.

Command: **Tabsurf**
Select path curve: *(Select the full circle at the right)*
Select direction vector: *(Select the line at the right)*

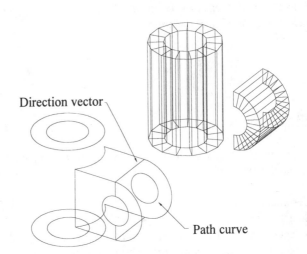

Direction vector

Path curve

Step #9

Move the last surface mesh pattern at a displacement of 0,10,0 using the Move command.

Command: **Move**
Select objects: **L**
Select objects: *(Strike Enter to continue)*
Base point or displacement: **0,10,0**
Second point of displacement: *(Strike Enter to execute this command)*

Command: **Redraw**

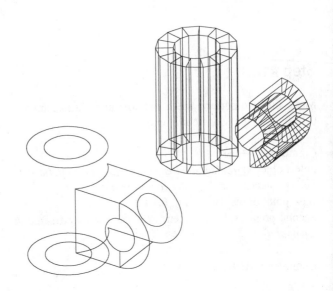

Step #10

Reset the system variables Surftab1 to 6 and Surftab2 to 10. This will reduce the amount of surface meshes created and help with computer regeneration time. Next, use the Edgesurf command, select the four entities labeled at the right, and create a surface similar to the second illustration at the right. Use the illustration in step 12 as a guide in performing this operation.

Command: **Surftab1**
New value for Surftab1 <15>: **6**

Command: **Surftab2**
New value for Surftab2 <15>: **10**

Command: **Edgesurf**
Select edge 1: *(Select the entity at "A")*
Select edge 2: *(Select the entity at "B")*
Select edge 3: *(Select the entity at "C")*
Select edge 4: *(Select the entity at "D")*

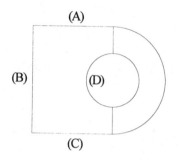

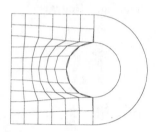

Step #11

Copy the surface just created using the Edgesurf command to create the rear surface.

Command: **Copy**
Select objects: **L**
Select objects: *(Strike Enter to continue)*
<Base point or displacement>/Multiple: **Endp**
of *(Select the endpoint at "A")*
Second point of displacement: **Endp**
of *(Select the endpoint at "B")*

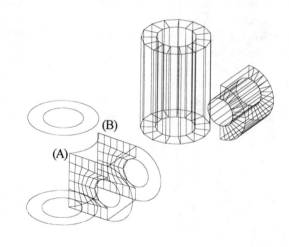

Step #12

Move the two surface mesh patterns at a displacement of 0,10,0 using the Move command.

Command: **Move**
Select objects: *(Select the last two surface mesh patterns)*
Select objects: *(Strike Enter to continue)*
Base point or displacement: **0,10,0**
Second point of displacement: *(Strike Enter to execute this command)*

Command: **Redraw**

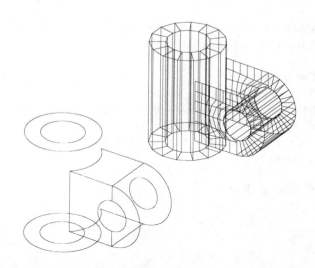

Step #13

Use the Edgesurf command, select the four entities labeled at the right, and create a surface similar to the second illustration at the right. Use the illustration in step 15 as a guide in performing this operation.

Command: **Edgesurf**
Select edge 1: *(Select the entity at "A")*
Select edge 2: *(Select the entity at "B")*
Select edge 3: *(Select the entity at "C")*
Select edge 4: *(Select the entity at "D")*

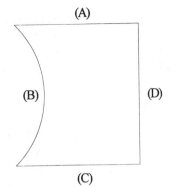
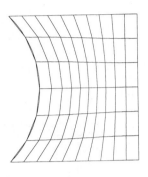

Step #14

Copy the surface just created using the Edgesurf command to create the bottom surface.

Command: **Copy**
Select objects: **L**
Select objects: *(Strike Enter to continue)*
<Base point or displacement>/Multiple: **Endp**
of *(Select the endpoint at "A")*
Second point of displacement: **Endp**
of *(Select the endpoint at "B")*

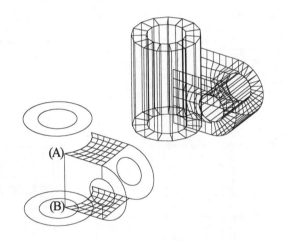

Step #15

Move the two surface mesh patterns at a displacement of 0,10,0 using the Move command.

Command: **Move**
Select objects: *(Select the last two surface mesh patterns)*
Select objects: *(Strike Enter to continue)*
Base point or displacement: **0,10,0**
Second point of displacement: *(Strike Enter to execute this command)*

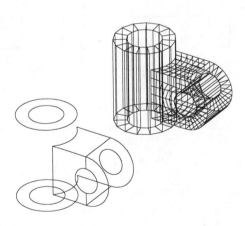

Step #16

The figure at the right represents the Column that has been completely surfaced using the previous steps.

Command: **Hide**
Regenerating drawing.
Removing hidden lines: **XXX**

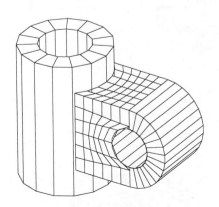

Problems for Unit 11

Directions for Problems 11–1 through 11–2

1. Use a grid spacing of 0.50 to determine all dimensions.
2. Create two layers for each object. Call the layers "Wireframe" and "Surface."
3. Create a 3D wireframe model of each object on the layer "Wireframe."
4. Surface the wireframe model and place all surfaces on the layer "Surface."

Problem 11–1

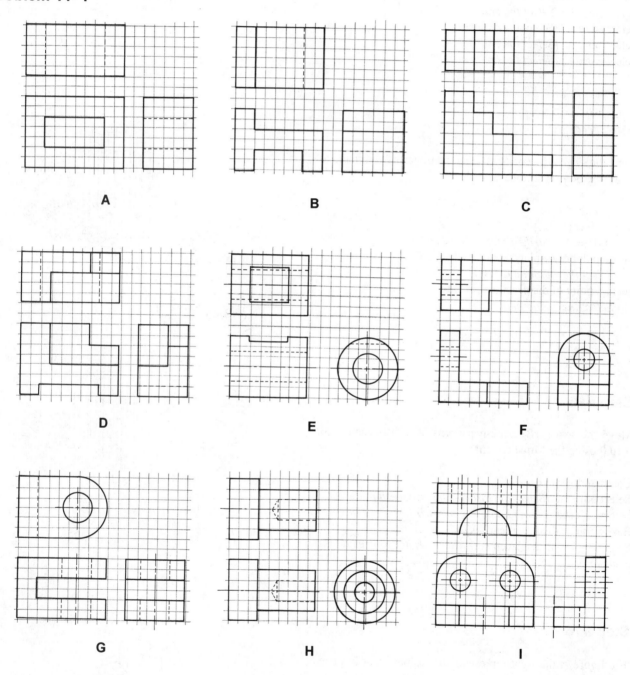

Problem 11–2

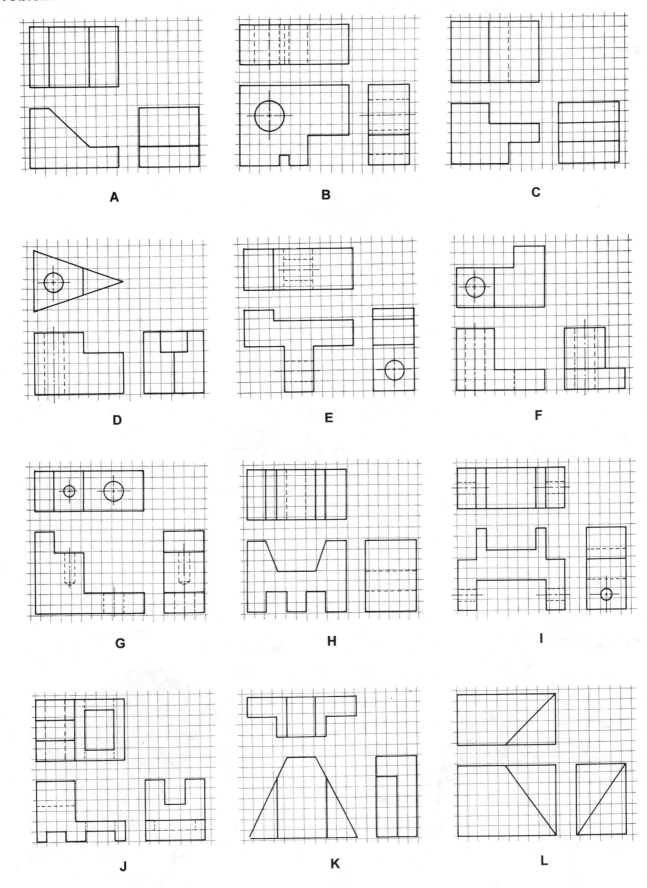

A

B

C

D

E

F

G

H

I

J

K

L

Directions for Problem 11-3 through 11-12
Create a 3D wireframe model of each object. Surface each wireframe using a series of 3D faces, ruled, tabulated, revolved, and edge surfaces.

Problem 11-3

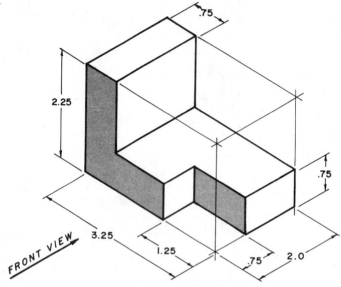

Problem 11-4

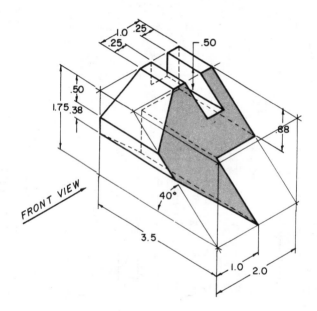

Problem 11-5

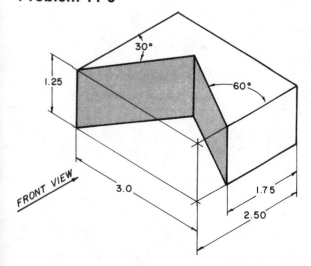

Problem 11-6

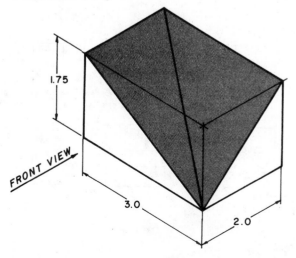

Problem 11–7

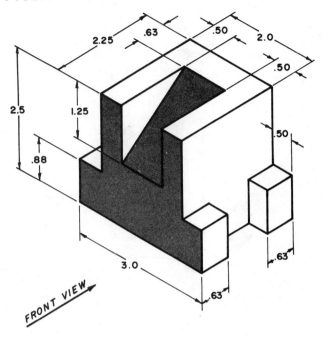

Problem 11–9

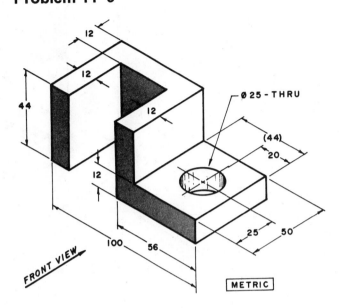

Problem 11–8

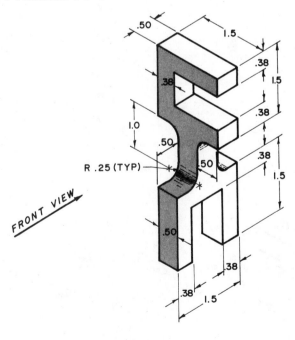

Problem 11–10

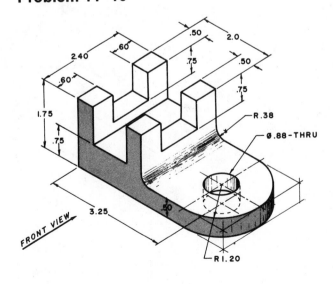

Problem 11–11

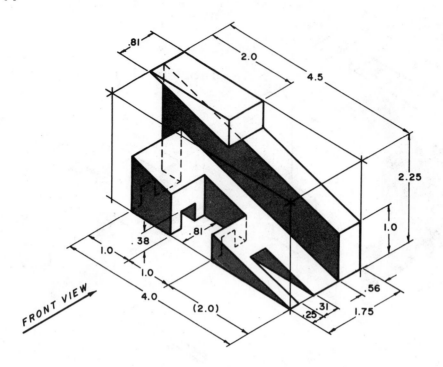

FRONT VIEW

Problem 11–12

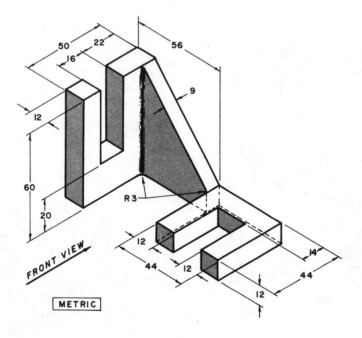

FRONT VIEW

METRIC

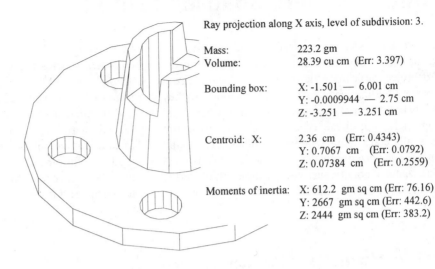

Ray projection along X axis, level of subdivision: 3.

Mass:	223.2 gm
Volume:	28.39 cu cm (Err: 3.397)
Bounding box:	X: -1.501 — 6.001 cm
	Y: -0.0009944 — 2.75 cm
	Z: -3.251 — 3.251 cm
Centroid: X:	2.36 cm (Err: 0.4343)
	Y: 0.7067 cm (Err: 0.0792)
	Z: 0.07384 cm (Err: 0.2559)
Moments of inertia:	X: 612.2 gm sq cm (Err: 76.16)
	Y: 2667 gm sq cm (Err: 442.6)
	Z: 2444 gm sq cm (Err: 383.2)

UNIT 12

Solid Modeling

Contents

An Introduction to Solid Modeling

Thus far, we have seen how wireframe models may be surfaced for hidden line removal or left unsurfaced for the purpose of projecting primary views onto a flat 2-dimensional sheet of paper. One limiting factor of wireframe models is they are basically hollow even when surfaced. This is where solid modeling picks up; these types of models are more informationally correct than wireframe models since a solid object may be analyzed by calculating such items as mass properties, center of gravity, surface area, moments of inertia, and much more. The solid model starts the true design process by defining objects as a series of primitives: boxes, cubes, cylinders, spheres and wedges are all examples of primitives. These building blocks are then joined together or subtracted from each other using certain modifying commands. Fillets and chamfers may be created to give the solid model a more realistic appearance. Two-dimensional views may be extracted from the solid model along with a cross section of the model. Follow the next series of pages that explain basic solid modeling concepts before completing the 4 tutorial exercises at the end of this unit.

Solid Modeling Basics

All objects, no matter how simple or complex, are composed of simple geometric shapes or primitives. The shapes range from boxes to cylinders to cones and so on. Solids modeling allows for the creation of these primitives. Once created the shapes are either merged or subtracted to form the final object. Follow the next series of steps to form the object at the right.

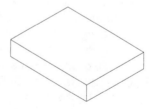

Begin the process of solids modeling by constructing a solid slab that will represent the base of the object. This is accomplished using the AutoCAD Advanced Modeling Extension command, Solbox, short for Solid Box. The length, width, and height of the box are supplied. The result is a solid slab.

Next a cylinder is constructed using the Solcyl, or Solid Cylinder command. This cylinder will eventually form the curved end of the object. One of the advantages of constructing a solid model is the ability of merging primitives together to form composite solids. Using constructive solids geometry (CSG) commands such as Solunion, the cylinder and box are combined to form the complete base of the object.

As the object progresses, another solid box is created and moved into position on top of the base. There, the Solunion command is used to join this new block with the base. As new shapes are added during this process, they all become part of the same solid.

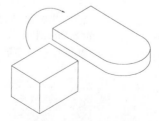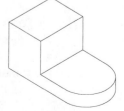

Yet another box is created and moved into position. However, instead of combining the blocks, this new box is removed from the solid. This process is called subtraction and when complete, creates the step illustrated at the right. The AutoCAD command used during this process is Solsub.

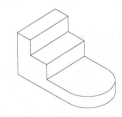
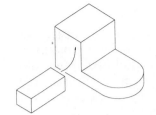

Holes are formed in a similar fashion. First the Solcyl command is used to create a cylinder the diameter and depth of the desired hole. The cylinder is moved into the solid and there subtracted using the Solsub command. Again, the object at the right represents a solid object.

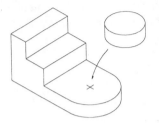
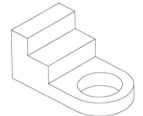

Using existing AutoCAD tools such as user coordinate systems, another cylinder is created using the Solcyl command. It too is moved into position where it is subtracted using the Solsub command. The complete solid model of the object is illustrated at the far right.

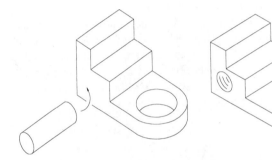

Rewards from constructing a solid model out of an object come in many forms. Profiles of different surfaces of the solid model may be taken. Section views of solids may be automatically formed and crosshatched as in the illustration below. A very important analysis tool is the Solmassp command, which is short for mass property extraction. Information such as the mass and volume of the solid object may be calculated along with centroids and moments of inertia, components that are used in Computer Aided Engineering (CAE) and in Finite Element Analysis (FEA) of the model.

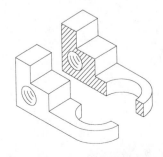

Mass:	223.2 gm
Volume:	28.39 cu cm (Err: 3.397)
Bounding box:	X: -1.501 — 6.001 cm
	Y: -0.0009944 — 2.75 cm
	Z: -3.251 — 3.251 cm
Centroid:	X: 2.36 cm (Err: 0.4343)
	Y: 0.7067 cm (Err: 0.0792)
	Z: 0.07384 cm (Err: 0.2559)
Moments of inertia:	X: 612.2 gm sq cm (Err: 76.16)
	Y: 2667 gm sq cm (Err: 442.6)
	Z: 2444 gm sq cm (Err: 383.2)

CSG, B-Rep, and Tree Structures

The construction of the object on the previous pages was possible through the use of constructive solids geometry, or CSG. An object is said to be constructed of a series of primitive shapes such as boxes, cylinders, and cones to name a few. These geometric shapes are then merged together or subtracted to create the solid model. In the object at the right, two solid primitives in the form of boxes are constructed and then joined using the boolean operation of union. A solid slot is constructed, moved into postition, and subtracted from the main object to complete the solid model. The CSG method allows the solid to retain information regarding the structure and dimensions of the solid model.

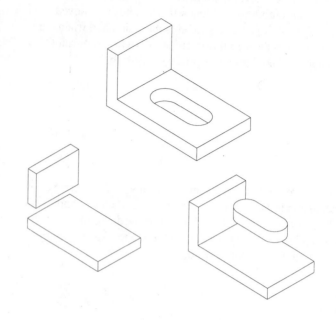

Another method used to define solid models is Boundary Representation or B-Rep. This method holds information regarding the surfaces, edges, and vertices that make up the boundary of the solid; hence B-Rep. The AutoCAD Advanced Modeling Extension utilizes both CSG and B-Rep information to create the most accurate solid model.

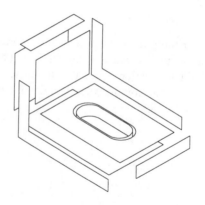

As multiple boolean operations are performed, a hierarchy or tree structure forms to keep track of the steps used to construct the solid model. This is important becase the solid model needs to be broken down into individual primitives. A typical tree structure is illustrated at the right representing construction of the object above. Read this tree structure from bottom up. First a union of the two solid boxes was performed represented by "A" and "B." Then the solid slot was subtracted from the main object to form the slot represented by "C." The results leave the final object shown at the top of the tree. Multiple boolean operations may be performed at one time; however, these operations are organized in pairs and these pairs form the tree.

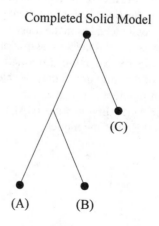

Completed Solid Model

(C)

(A)   (B)

Creating Solid Primitives

SOLBOX

The Solbox command constructs a solid primitive of a 3-dimensional box. One corner of the box is located along with its other diagonal corner. A height is assigned to complete the definition of the box. A cube may also be constructed by selecting the appropriate option in the command prompt below. If all 3 dimensions of a box are known, the solid box may be constructed by entering values for its length, width, and height. Illustrated at the right is an example of a solid box.

Command: **Solbox**
Baseplane/Center/<Corner of box><0,0,0>: **4.00,5.50** *(At "A")*
Cube/Length/<Other corner>: **L**
Length: **4.00** *("A" to "B")*
Width: **4.00** *("B" to "C")*
Height: **1.00** *("B" to "D")*

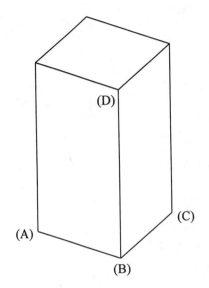

SOLCONE

The Solcone command constructs a solid primitive of a cone with a circular or elliptical base connecting at a central point of specified height; examples are illustrated at the right with sample prompt sequences below:

The following is a sample prompt sequence for the cone illustrated at "A":
Command: **Solcone**
Baseplane/Elliptical/<Center point><0,0,0>: *(Identify a point as the center of the cone)*
Diameter/<Radius>: *(Enter a radius value)*
Height of cone: *(Enter a height for the cone)*

The following is a sample prompt sequence for the cone illustrated at "B":
Command: **Solcone**
Baseplane/Elliptical/<Center point>: **Elliptical**
<Axis endpoint 1>/Center: *(Identify an axis endpoint)*
Axis endpoint 2: *(Identify the second axis endpoint)*
Other axis distance: *(Identify the other axis distance)*
Height of cone: *(Enter a height for the cone)*

Cone "A"

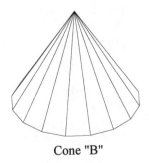

Cone "B"

SOLWEDGE

Yet another solid primitive is the wedge which consists of a
box that has been diagonally cut. The base of the wedge is
drawn parallel to the current user coordinate system. The
sloped surface tapers along the X axis. The prompts for this
command are very similar to the Solbox command.

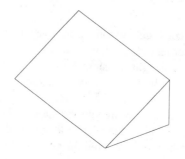

Command: **Solwedge**
Baseplane/<Corner of wedge>: *(Locate one corner of the
wedge)*
Length.<Other corner>: *(Locate opposite corner of the wedge)*
Height: *(Enter a nonzero value for the height of the wedge)*

The following prompts illustrate the construction of a wedge
by providing the length, width, and height dimensions.

Command: **Solwedge**
First corner: *(Locate one corner of the wedge)*
Length.<Opposite corner>: **Length**
Length: *(Enter the length of the wedge)*
Width: *(Enter the width of the wedge)*
Height: *(Enter the height of the wedge)*

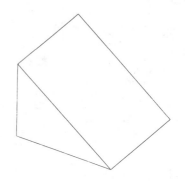

SOLCYL

The Solcyl command is similar to the Solcone command
except that a cylinder without taper is drawn. The central axis
of a cylinder is along the Z axis of the current user coordinate
system. The following prompts illustrate construction of a
cylinder by radius and diameter. A cylinder may also be
elliptical in shape.

Command: **Solcyl**
Baseplane/Elliptical/<Center point>: *(Enter coordinates for
the center of the cylinder)*
Diameter/<Radius>: *(Enter a value for the cylinder radius)*

Command: **Solcyl**
Baseplane/Elliptical/<Center point>: *(Enter coordinates for
the center of the cylinder)*
Diameter/<Radius>: **Diameter**
Diameter: *(Enter a value for the diameter of the cylinder)*

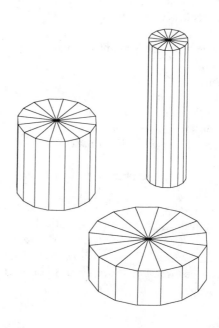

SOLSPHERE

The Solshpere command constructs a type of "ball" with all points along its surface equal in distance from a central center. As in the cylinder, the central axis of a sphere is along the Z axis of the current user coordinate system. Spheres may be drawn using either diameter or radius dimensions.

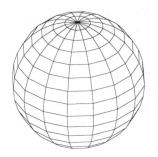

Command: **Solsphere**
Baseplane/Center of sphere: *(Enter the center of the sphere)*
Diameter/<Radius>: *(Enter the radius of the sphere or "D" to be prompted for the diameter of the sphere)*

SOLTORUS

A torus is formed when a circle is revolved about a line in the same plane as the circle. In other words, a torus is similar to a 3-dimensional donut. A torus may be constructed using either a radius or diameter method. When using the radius method, two radius values must be used define the torus; one for the radius of the tube and the other for the radius from the center of the torus to the center of the tube. Two diameter values would be used when specifying a torus by diameter. Once the torus is constructed, it lies parallel to the current user coordinate system. Follow the prompts below to define a torus by radius values.

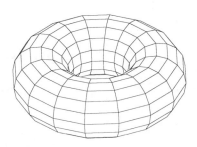

Radius Method

Command: **Soltorus**
Center of torus: *(Identify the center of the torus through a coordinate or by picking)*
Diameter/<Radius> of torus: *(Enter a value for the radius of the torus at "A")*
Diameter/<Radius> of tube: *(Enter a value for the radius of the tube of the torus at "B")*

Radius of Torus (A)   Tube radius (B)

Follow the prompts below to define a torus through diameter values.

Command: **Soltorus**
Baseplane/Center of torus: *(Identify the center of the torus through a coordinate or by picking)*
Diameter/<Radius> of torus: **Diameter**
Diameter: *(Enter a value for the diameter of the torus at "C")*
Diameter/<Radius> of tube: **Diameter**
Diameter: *(Enter a value for the diameter of the tube of the torus at "D")*

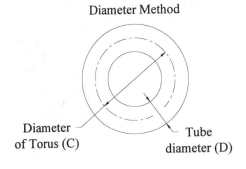

Diameter Method

Diameter of Torus (C)   Tube diameter (D)

Selecting Primitives from the Pulldown Menu Area

Yet another way to access the solid modeling primitives from the previous pages is through the "Model" pulldown menu area illustrated at the right. Selecting "Primitives..." displays the dialog box illustrated below. Here a graphical representation of the Box, Cone, Cylinder, Sphere, Torus, and Wedge is displayed. Simply select the desired primitive and the prompts associated with it appear at the bottom of the display screen. If the solid modeler is not loaded and a primitive is selected, a prompt appears at the bottom of the screen to load the modeler.

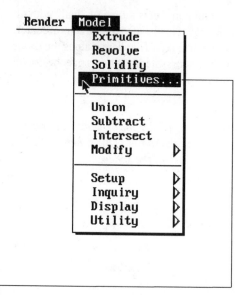

Picking the On radio button in the Baseplane area allows the user to create a primitive on a construction plane different from the current user coordinate system.

Picking the "Object Snap Mode..." button displays a dialog box identical to the DDOSNAP dialog box. This allows the user a convenient way to enter an object snap mode for creating a primitive without leaving the solid primitive dialog box.

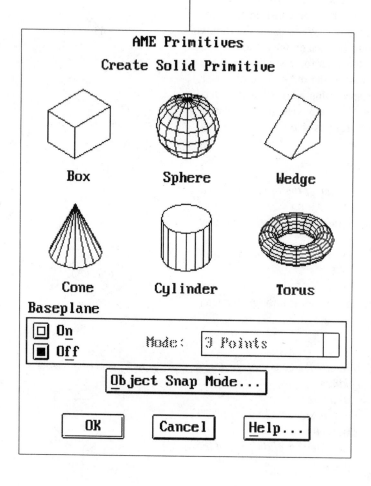

Using Boolean Operations

To assist in the combining of one or more primitives to form a common solid, a system is available to illustrate the relationship between the individuals that make up the solid model; this system is called a boolean operation. Boolean operations must act on at least a pair of primitives, regions, or solids. These operations in the form of commands are located in the main pulldown menu under "Model." Boolean operations allow the user to add two or more objects together, subtract a single or group of objects from another, or find the overlapping volume; in other words, the solid common to both primitives. Highlighted at the right are the Union, Subtract, and Intersect commands used to perform the boolean operations explained above.

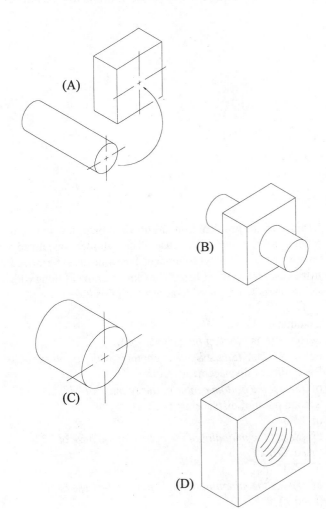

Render	Model
	Extrude
	Revolve
	Solidify
	Primitives...
*	Union *
*	Subtract *
*	Intersect *
	Modify ▷
	Setup ▷
	Inquiry ▷
	Display ▷
	Utility ▷

To create details such as cuts and holes from solid primitives, boolean operations are used to add or subtract geometry from each other. In the example at "A," a cylinder has been constructed along with a square block. Depending on which boolean operation is used, the results could be quite different. In example "B," both the square slab and cylinder are one entity and considered one solid object. This is the purpose of the Solunion command; to join or unite two solid primitives into one. Example "C" illustrates the intersection of the two solid primitives or the area that both solids have in common. This solid is obtained by using the Solint command. Example "D" shows the results of removing or subtracting the cylinder from the square slab; the result is that a hole is formed inside the square slab as a result of using the Solsub command. All boolean operation commands may work on numerous solid primitives; that is, if subtracting numerous cylinders from a slab, all cylinders at the same time may be subtracted. These commands may be entered in from the keyboard or may be selected from the pulldown menu area under Sol-Modify.

(A)

(B)

(C)

(D)

SOLUNION

Illustrated at the right is an object consisting of one horizontal solid box, two vertical solid boxes, and two extruded semi-circular shapes. All primitives have been positioned either with the Move or Solmove commands. The problem is to join all of these shapes into one using the Solunion command. The order of selection of these solids for this command is not important. Follow the prompts below and the example at the right.

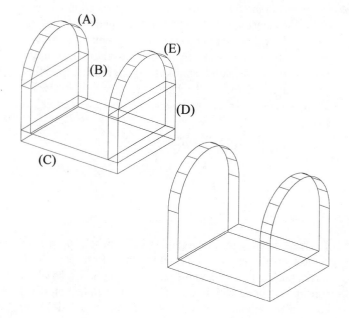

Command: **Solunion**
Select objects: *(Select the solid extrusion at "A")*
Select objects: *(Select the vertical solid box at "B")*
Select objects: *(Select the horizontal solid box at "C")*
Select objects: *(Select the vertical solid box at "D")*
Select objects: *(Select the solid extrusion at "E")*
Select objects: *(Strike Enter to perform the union of the solids)*

SOLSUB

Using the same problem from the previous page, let's now add a hole in the center of the base. The cylinder was already created using the Solcyl command. It now needs to be moved to the exact center of the base. The Move command along with .XYZ filters will be used to accomplish this.

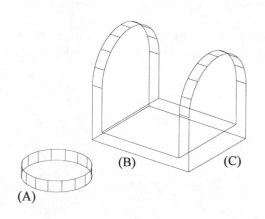

Command: **Move**
Select objects: *(Select the cylinder at "A")*
Select objects: *(Strike Enter to continue with this command)*
Base point or displacement: **Cen**
of *(Select the bottom center of the cylinder at "A")*
Second point of displacement: **.X**
of **Mid**
of *(Select the midpoint of the bottom of the base at "B")*
(need YZ) **.Y**
of **Mid**
of *(Select the midpoint of the bottom of the base at "C")*
(need Z) **0**

Now that the solids are in position, use the Solsub command to subtract the cylinder from the base of the main solid.

Command: **Solsub**
Source objects...
Select objects: *(Select the main solid as source at "A")*
Select objects: *(Strike Enter to continue with this command)*
Objects to subtract from them...
Select objects: *(Select the cylinder at "B")*
Select objects: *(Strike Enter to perform the subtraction operation)*

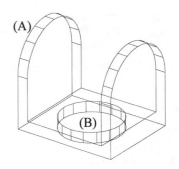

Two more holes need to be added to the vertical sides of the object. A cylinder was already constructed. However it is in the vertical position. This entity needs to be rotated along the Y axis using the Solmove command. When issuing this command and selecting the cylinder, a special icon appears that shows the X, Y, and Z axes. For motion description, enter a value of "RY90," which will rotate the cylinder 90 degrees in the counterclockwise direction using the Y axis as the pivot.

Command: **Solmove**
Select objects: *(Select the vertical cylinder at "A")*
Select objects: *(Strike Enter to continue with this command)*
<Motion description>/?: **RY90**

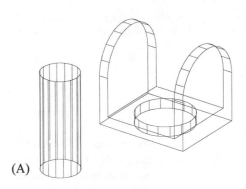

Use the Move command to move the long cylinder to the vertical sides of the main solid using the Osnap-Center mode as an aid.

Command: **Move**
Select objects: *(Select the cylinder at "A")*
Select objects: *(Strike Enter to continue with this command)*
Base point or displacement: **Cen**
of *(Select the center of the cylinder at "A")*
Second point of displacement: Cen
of *(Select the center of the semi-circular solid at "B")*

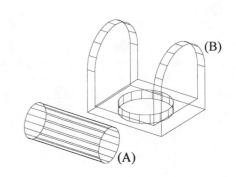

Use the Solsub command to subtract the long cylinder from the two vertical sides of the object.

Command: **Solsub**
Source objects...
Select objects: *(Select the main solid as source at "A")*
Select objects: *(Strike Enter to continue with this command)*
Objects to subtract from them...
Select objects: *(Select the cylinder at "B")*
Select objects: *(Strike Enter to perform the subtraction operation)*

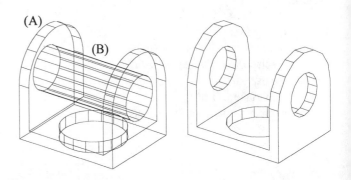

Creating Solid Extrusions

The Solext command create a solid by extrusion. Only polylines and circles may be extruded. Once these entities are selected, an extrusion height is asked for followed by an extrusion taper angle. If the entities selected are not polylines, use the Pedit command and convert them to polylines.

Command: **Pedit**
Select polyline: *(Select a line at the right)*
Entity selected is not a polyline.
Do you want it to turn into one? <N>: **Yes**
Close/Join/Width/Edit vertex/Fit curve/Spline curve/Decurve/
Undo/eXit <X>: **Join**
Select objects: *(Select the remaining lines of the object)*
6 segments added to polyline.
Close/Join/Width/Edit vertex/Fit curve/Spline curve/Decurve/
Undo/eXit <X>: *(Strike Enter to exit this command)*

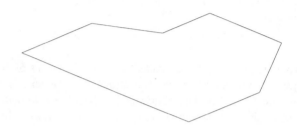

Once entities are polylines, use the prompts below to construct a solid extrusion of the object below. For the height of the extrusion, a positive numeric value may be entered or the distance may be determined by picking two points on the display screen.

Command: **Solext**
Select polylines and circles for extrusion...
Select objects: *(Select the polyline at "A")*
Select objects: *(Strike Enter to continue with the extrusion)*
Height of extrusion: **1.00**
Extrusion taper angle from Z <0>: *(Strike Enter to accept default)*

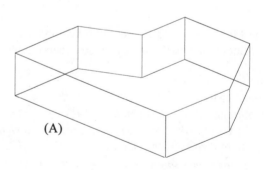

(A)

An optional taper may be created along with the extrusion by entering an angle value for the prompt, "Extrusion taper angle from Z."

Command: **Solext**
Select polylines and circles for extrusion...
Select objects: *(Select the polyline at "B")*
Select objects: *(Strike Enter to continue with the extrusion)*
Height of extrusion: **1.00**
Extrusion taper angle from Z <0>: **15**

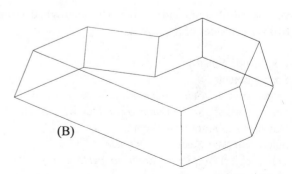

(B)

Creating Revolved Solids

The Solrev command creates a solid by revolving an entity about an axis of revolution. Only polylines, polygons, circles, ellipses, and 3D polylines may be revolved. If a group of entities are not in the form of polylines, group them together using the Pedit command. The resulting image at the right represents a solid entity.

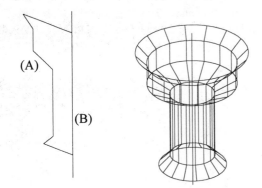

Command: **Solrev**
Select polyline or circle for revolution...
Select objects: *(Select the polyline at "A")*
Select objects: *(Strike Enter to continue with this command)*
Axis of revolution - Entity/X/Y/<Start point of axis>: **Entity**
Entity to revolve about: *(Select the line at "B")*
Included angle<full circle>: *(Strike Enter to accept default)*

A practical application of this type of solid would be to first construct an additional solid consisting of a cylinder using the Solcyl command. Be sure this solid is larger in diameter than the revolved solid. Existing Osnap options are fully supported in solid modeling. Use the Center option of Osnap along with the Move command to position the revolved solid inside of the cylinder.

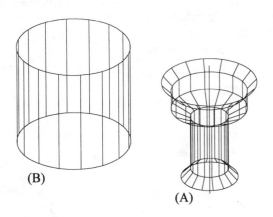

Command: **Move**
Select objects: *(Select the revolved solid at the right)*
Select objects: *(Strike Enter to continue with this command)*
Base point or displacement: **Cen**
of *(Select the center of the revolved solid at "A")*
Second point of displacement: **Cen**
of *(Select the center of the cylinder at "B")*

Once the revolved solid is positioned inside of the cylinder, the Solsub command is used to subtract the revolved solid from the cylinder. Next the Solmesh command is used to surface the solid at "A." Finally, the Hide command is used to perform a hidden line removal at "B" to check that the solid is correct, which would be difficult to interpret in wireframe mode. All of these commands will be discussed in detail in the pages to follow.

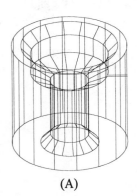

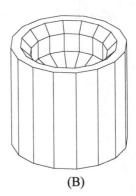

(A)   (B)

Curve Tessellation

Tessellation refers to the lines that are displayed on any curved surface to help visualize the surface. Tessellation lines are automatically formed when constructing solid primitives such as cylinders and cones. These lines are also calculated when performing such solid modeling operations as Solsub and Solunion to name a few.

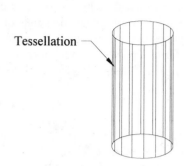

Tessellation

The number of tessellation lines per curved entity is controlled by a system variable called Solwdens. By default, this variable is set to a value of 1. Illustrated at the right are the results of setting this variable to other values such as "4" and "8." The more lines used to describe a curved surface, the more accurate the surface will look; however, the longer it will take to process hidden line removals using the Hide command and boolean operation commands such as Solunion and Solsub.

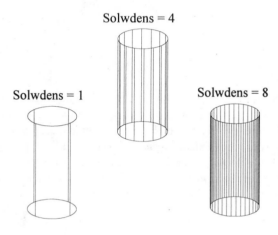

Solwdens = 4

Solwdens = 1   Solwdens = 8

The above example of tessellation lines on cylinders was illustrated in wireframe mode. When surfacing these entities, the results are displayed in the illustrations at the right. The cylinder with Solwdens = 1 processes much quicker than the cylinder with Solwdens = 8 since there are fewer surfaces to process in such operations as hidden line removals. An average value for Solwdens is 4, which seems adequate for most applications.

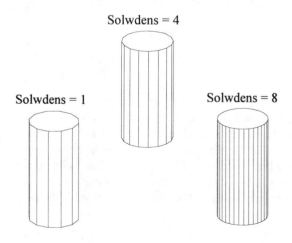

Solwdens = 4

Solwdens = 1   Solwdens = 8

Modifying Solid Models

Once a solid model is created, changes are necessary as design changes are made. A series of commands dedicated to solid models enables an individual to edit a solid model. These commands may be found by selecting "Modify" from the "Model" pulldown menu area illustrated at the right. What results is another listing of modification options that will be explained below:

Move Object - this option relates directly to the Solmove command used to move or translate a solid or primitive in various ways. In addition to moving the solid, translation supports rotation of solids.

Change Primitive - this very powerful option relates to the Solchp command which allows the individual to perform such operations as changing the color of a primitive to changing its size.

Separate - refers to the separating of a solid back into its original primitives. The Solsep command is used to accomplish this task.

Cut Solids - enables a section to be cut based on a construction plane. The Solcut command is used to accomplish this operation.

Chamfer Solids - enables the individual to create a beveled edge along a series of base surfaces and surface edges. The Solcham command is used to accomplish this task.

Fillet Solids - enables the individual to create a rounded edge on a solid. The Solfill command is used to accomplish this task.

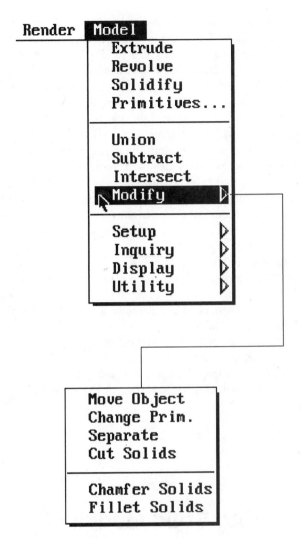

Chamfering Solid Models

The Solcham command creates a chamfer and automatically performs a union or subtraction operation to satisfy the chamfer specifications for a particular solid. This command requires a base surface to begin the chamfer calculations. As a surface is selected, sometimes it highlights, sometimes it doesn't. During this selection process, the prompt "<OK>/ Next" appears, which allows the user to accept the current surface, or move to the next surface to see if it highlights. Keep repeating this procedure until the desired edge is selected. Once the base surface is identified, the edges to chamfer are selected. Two chamfer distances are asked for; if both values are equal, the chamfer will form a 45 degree angle; if both values are not equal, a beveled edge is formed at an angle other than 45 degrees.

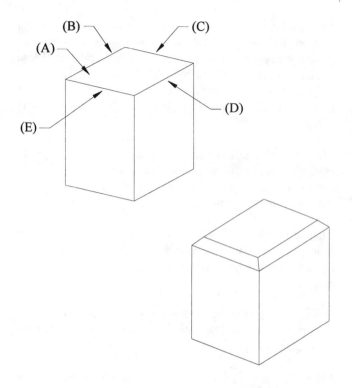

Command: **Solcham**
Select base surface: *(Select top surface "A")*
<OK>/Next: *(If the top surface highlights, strike Enter; if not, enter "N" for the next surface and continue until the top surface highlights)*
Select edges to be chamfered (Press Enter when done):
(Select the edges at "B," "C," and "D")
Enter distance along first surface<0.000>: **0.125**
Enter distance along second surface<0.125>: *(Strike Enter)*

Filleting Solid Models

Solid models may be filleted using the Solfill command. This command is similar to the Solcham command except that edges are rounded instead of beveled. Multiple edges may be selected using this command. Enter the desired radius and the selected edges are filleted. A solid fillet primitive is automatically created and added to or subtracted from the existing solid model.

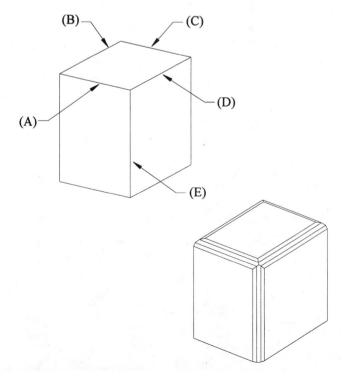

Command: **Solfill**
Select edges to be filleted (Press Enter when done): *(Select edges "A," "B," "C," and "D")*
Diameter/<Radius> of fillet <0.000> **0.125**

Using the Solmove Command

As an aid in moving and positioning solid models, the Solmove command can be used. This is not to be confused with the regular AutoCAD Move comamnd. When issuing this command and selecting a solid model to position, a new icon appears at the current location of the user coordinate system. Because this icon is specific to the Solmove command, it is called the MCS or motion coordinate system. Three axes are provided to assist in an understanding of where to move the solid model. These axes are labeled with a series of cones; the X-Axis is identified with one cone; the Y-Axis identified with two cones; the Z-axis with three cones. The next prompt asks for a motion description. These key-ins to describe motion are quite different and are listed in the command when entering the "?" at the motion description prompt.

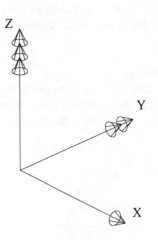

At the right is a cylinder 1 unit in diameter and 2 units tall. The user coordinate and motion coordinate system icons are located at the center base of the cylinder. The next four examples will reference this illustration. Please refer back to this example. The prompts for the Solmove command are:

Command: **Solmove**
Select objects: *(Select the cylinder at the right)*
Select objects: *(Strike Enter to continue with this command)*
<Motion description>/?: *(Enter a motion description definition)*

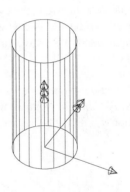

To move the cylinder 1 unit in the X direction and 1 unit in the Y direction, the key phrase is "Translate." Use the motion coordinate system to determine the direction of the move. The key in to move the solid is "TX1.00,TY1.00."

Command: **Solmove**
Select objects: *(Select the cylinder at the right)*
Select objects: *(Strike Enter to continue with this command)*
<Motion description>/?: **TX1.00,TY1.00**

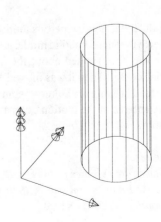

At the right, the cylinder is moved 1.50 units in the negative X direction by 1.00 units in the positive Z direction. Again use the motion coordinate system to determine the direction of motion.

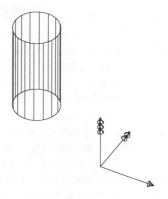

Command: **Solmove**
Select objects: *(Select the cylinder at the right)*
Select objects: *(Strike Enter to continue with this command)*
<Motion description>/?: **TX-1.50,TZ1.00**

The Solmove command does more than just move or translate; Solid models may also be rotated using this motion coordinate system. In the example at the right, the cylinder was rotated 90 degrees clockwise using the X-axis as the pivot. The clockwise rotation demands a negative value.

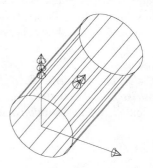

Command: **Solmove**
Select objects: *(Select the cylinder at the right)*
Select objects: *(Strike Enter to continue with this command)*
<Motion description>/?: **RX-90**

This last example illustrates how numerous motion directions can be used together to translate the solid model around. At the right, the cylinder was first rotated about the Y-axis at 90 degrees counter-clockwise. Then it was moved 1.00 units in the X direction. Instead of using the Solmove command twice, both operations are entered at the motion description prompt separated by a comma.

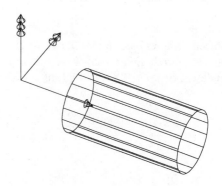

Command: **Solmove**
Select objects: *(Select the cylinder at the right)*
Select objects: *(Strike Enter to continue with this command)*
<Motion description>/?: **RY90,TX1.00**

Using the DDSOLVARS Dialog Box

SOLVARS

This command controls specific system variables that affect solid modeling commands and is similar to the AutoCAD Setvar command. The illustration below allows the individual to work through a dialog box to manipulate or change certain variables that will affect the final outcome of the solid model.

Various buttons are available that display other dialog boxes used to affect a solid modeling system variable. For instance, selecting the "Units..." button in the dialog box below activates another dialog displaying the default units that will be used in a mass property calculation of the solid model.

A few of these extra dialog boxes will be explained in greater detail on the next page.

Three types of units affect the following solid modeling variables:

            Sollength
            Solareav
            Solvolume
            Solmass

British units assign the following values: Sollength=feet, Solareav=square feet, Solvolume=cubic feet, and Solmass=pounds.

CSG units assign the following values: Sollength=centimeters, Solareav=Square centimeters, Solvolume=cubic centimeters, and Solmass=grams.

SI units assign the following values: Sollength=meters, Solareav=square meters, Solvolume=cubic meters, and solmass=kilograms.

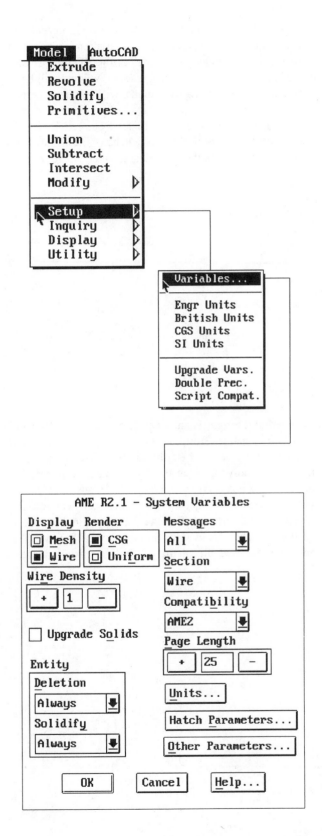

By default, the current units of a solid model are in centimeters. This means if a mass property calculation is performed using either the Solmassp command or the DDSOLMASSP dialog box all values for the length, area, volume will be in centimeters; the mass of a solid model will remain in grams.

To change the units of a solid model from centimeters to inches, double click in the "Length" edit box. The initials "cm" will highlight enabling the individual to enter "in" for inches.

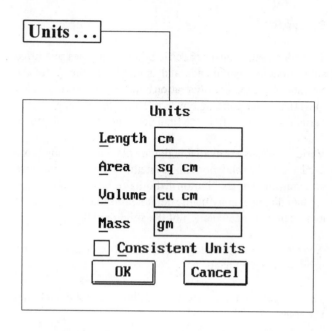

Illustrated at the right is an example of the units of a solid model being changed from centimeters to inches. The initial change was made in the "Length" edit box. Rather than make the changes individually in the "Area" and "Volume" edit boxes, a check box is present with the title "Consistent Units." Placing a check in the adjacent box automatically converts the area and volume units to reflect the change originally made in the length edit box. Now if the Solmassp or DDSOLMASSP commands are used, all calculations will be made in inches.

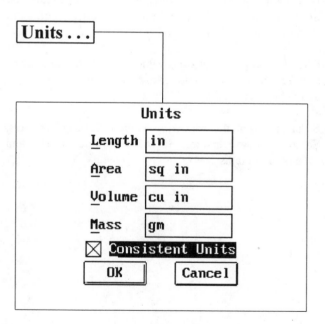

Using Inquiry Commands on a Solid Model

The commands listed under the Inquiry pulldown menu allow the user to list various characteristics associated with a selected solid model.

The Sollist command displays information used to define the solid model.

The Solmassp command lists the mass property information associated with the solid model.

The Solint commands checks two solids for interference.

The Solarea command displays the area of a particular surface belonging to the solid.

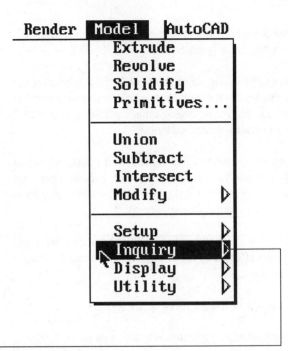

SOLLIST

Using the Sollist command displays the following prompts:
        Edge/Face/Tree/<Solid>:

The results of keeping the default value are displayed at the right. This displays the topmost level of the tree structure that went into the creation of the solid model shown above at "A." Each line is briefly described below:

Solid type = Subtraction - This represents the highest level on the CSG tree and happens to be the last boolean operation performed to create the solid model. An entity handle is assigned to the resulting composite solid.

Component handles: 9D and 5C - These handles correspond to the objects affected by the boolean operation used to create the composite solid.

The area and material are listed of the solid.

The Representation indicates whether the solid model is displayed in wireframe mode or mesh mode. Both types of representation are controlled by the Solwire and Solmesh commands.

The Rigid Motion values keep track of any rotations, translations, and scaling applied to the solid model.

Using the Sollist command and selecting the "Tree" option displays a listing of the solid primitives and boolean operations involved in the creation of the solid model. This listing is illustrated at the right.

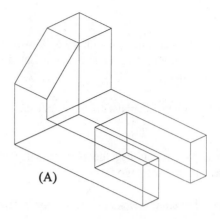

(A)

```
Object type = SUBTRACTION      Handle = D0
   Component handles:  84 and 45
   Area = 35.414214   Material = MILD_STEEL
   Representation = WIREFRAME   Render type = CSG
Rigid motion:
        +1.000000      +0.000000      +0.000000      +0.000000
        +0.000000      +1.000000      +0.000000      +0.000000
        +0.000000      +0.000000      +1.000000      +0.000000
        +0.000000      +0.000000      +0.000000      +1.000000
```

```
Object type = SUBTRACTION      Handle = D0
   Component handles:  84 and 45
   Area = 35.414214   Material = MILD_STEEL
   Representation = WIREFRAME   Render type = CSG

.... Object type = SUBTRACTION      Handle = 84
....    Component handles:  5D and 37
....    Area not computed   Material = MILD_STEEL
....    Representation = WIREFRAME   Render type = CSG
....    Node level = 1

.... Object type = UNION      Handle = 5D
....    Component handles:  15 and 26
....    Area not computed   Material = MILD_STEEL
....    Representation = WIREFRAME   Render type = CSG
....    Node level = 2

.... Object type = BOX (4.000000, 2.000000, 1.000000)   Handle = 15
....    Area not computed   Material = MILD_STEEL
....    Representation = WIREFRAME   Render type = CSG
....    Node level = 3

.... Object type = BOX (1.000000, 2.000000, 1.500000)   Handle = 26
....    Area not computed   Material = MILD_STEEL
....    Representation = WIREFRAME   Render type = CSG
....    Node level = 3

.... Object type = BOX (2.000000, 1.000000, 1.000000)   Handle = 37
....    Area not computed   Material = MILD_STEEL
....    Representation = WIREFRAME   Render type = CSG
....    Node level = 2

.... Object type = WEDGE (1.000000, 1.000000, 1.000000)   Handle = 45
....    Area not computed   Material = MILD_STEEL
....    Representation = WIREFRAME   Render type = CSG
....    Node level = 1
```

SOLMASSP

Because the solid model represents the most informationally correct solution to a part, a major advantage of the model is its ability to be analyzed for design purposes. The following properties of a solid model are calculated by this command: Mass, Volume, Centroid, Moments of Inertia, Products of Inertia, Radii of Gyration, and Principal Moments. All calculations are based on the position of the current user coordinate system. If errors are detected, they are enclosed in parentheses. When identifying the centroid of a solid model, a point is placed on the object in the current layer. The type of point displayed is controlled by the Pdmode system variable; the size of this point is controlled by the system variable Pdsize. When displaying the results of a typical mass property listing, the Solmassp command will prompt the user to write the results to a file. If this prompt is answered "Yes," a text file is created from a name supplied by the user or the default name of the solid model. A file extension of .MPR will be used to separate the mass property text file from the standard drawing file. Illustrated below is a sample set of mass property calculations.

```
Ray projection along X axis, level of subdivision: 3.
Mass:               67.55 gm
Volume:             8.594 cu cm  (Err: 0.7242)

Bounding box:       X: 4 — 8 cm
                    Y: 5.5 — 7.5 cm
                    Z: 0 — 2.5 cm

Centroid:           X: 5.373 cm    (Err: 0.548)
                    Y: 6.532 cm    (Err: 0.5621)
                    Z: 0.8411 cm   (Err: 0.05665)

Moments of inertia: X: 2979 gm sq cm (Err: 262.1)
                    Y: 2104 gm sq cm (Err: 243.3)
                    Z: 4938 gm sq cm (Err: 501.8)

Products of inertia: XY: 2371 gm sq cm (Err: 246.3)
                     YZ: 373.3 gm sq cm (Err: 26.13)
                     ZX: 287.2 gm sq cm (Err: 23.15)

Radii of gyration:  X: 6.64 cm
                    Y: 5.582 cm
                    Z: 8.55 cm

Principal moments(gm sq cm) and X-Y-Z directions about
centroid:
            I: 43.81 along [0.9607 -0.000183 -0.2778]
            J: 112.2 along [0.2559 -0.3882 0.8853]
            K: 105.8 along [-0.108 -0.9216 -0.3729]
```

DDSOLMASSP

Selecting "Mass Property..." from the menu area at the right activates the DDSOLMASSP dialog box illustrated below. Use this dialog box to visually display calculations made on the solid model such as Mass, Volume, Bounding Box, Centroid, Moments of Inertia, Products of Inertia, and Radii of Gyration. Use the scroll bar at the right of the dialog box to view other calculations not displayed.

Picking the "File..." button activates another dialog box allowing the individual to write all calculations out to a file with the extension .MPR.

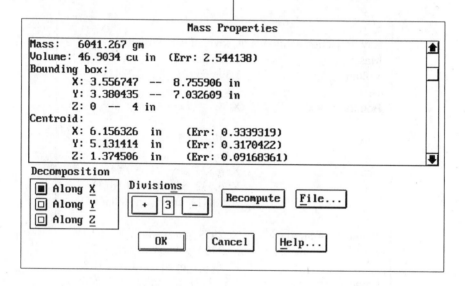

When the DDSOLMASSP dialog box is exited, the remaining solid displays a mark. This mark is actually the centroid of the solid represented by a Point entity. The appearance of the point is controlled by the Pdmode system variable or the DDPTYPE dialog box. A point style of "3" is illustrated at the right.

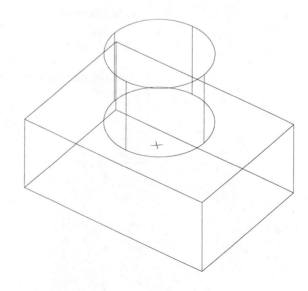

Displaying Solid Models

The pulldown menu strip illustrated at the right lists the numerous ways to display the image of a solid model. By default, all solid models are constructed in wireframe mode. Should the need arise to perform a hidden line removal, the Solmesh command places faces along all surfaces of the model. This enables the model to display visible surfaces when using the Hide command. Most AME commands function in wireframe mode; so if rendering a solid, all surfaces need to be converted back from mesh patterns to wireframe mode using the Solwire command. The next series of pages discuss some of the more advanced display commands of solids modeling such as cutting sections and creating profiles.

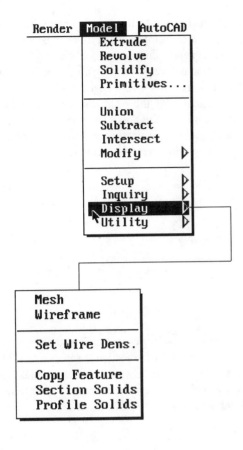

SOLWIRE

By default, construction of a solid model is performed in a wireframe mode. This speeds up the construction process and could lead to the extraction of solid profiles for the purpose of generating orthographic views.

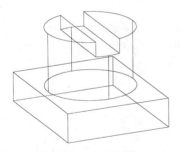

SOLMESH

As the solid model is constructed, there comes the time to view the model by performing hidden line removals to better view the model. Use the Solmesh command to convert the wireframe mode of the solid model to a surfaced model suitable for viewing. When performing more construction commands, convert the model back into a wireframe.

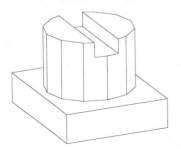

Cutting Sections from Solid Models

The image at the right is the familiar cylinder with a revolved solid removed from its interior. At times, it is advantageous to view interior details through section views. This requires some preparation before performing this operation. First, the section is cut in relation to the position of the current user coordinate system. At the right, the UCS icon has been revolved 90 degrees by the X axis and positioned at the center base of the cylinder. This determines the cutting plane angle that creates the section view out of the solid model.

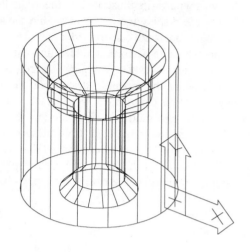

Before issuing the section view extraction command, a pattern needs to be identified for the section to represent. Without it, the section will be created; however the surfaces making up the section will be void of any crosshatch lines. This is the purpose of the Solhpat command.

Command: **Solhpat**
Hatch pattern <None>: **Ansi31**

Command: **Solsect**
Select objects: *(Select the solid anywhere)*

As the section constructs itself inside the solid model, use the Move command to move the section view to a better location.

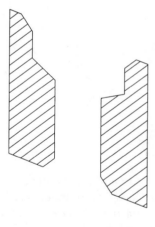

Lines not cut by the cutting plane line are omitted from the section view resulting in two halves of the section view that are not connected. The line command is used to connect up edges not cut, as in the example at the far right.

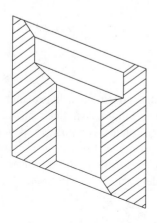

Shading Solid Models

As 3-dimensional models are created, it is often difficult to interpret their surfaces without performing a hidden line removal or a shaded rendering. Selecting "Render" from the pulldown menu area displays the Render, Shade, and Hide commands. The solid model must first be prepared to be viewed as a shaded image by converting the model from the wireframe mode to a surface mesh pattern using the Solmesh command. The Shade and Hide commands will be explained further.

Once a wireframe model has been converted into a surfaced model, the quickest way to verify the model is by performing a hidden line removal using the Hide command. No shading occurs; however, only those surfaces in view will be displayed while others will be removed from view.

For a quick look at what the object would look like as a shaded image, use the Shade command. The results will be similar to using the Hide command except that the current color of the entities will take on that color in a shaded form.

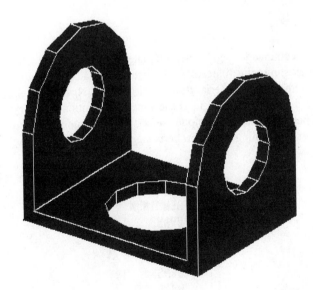

This command automatically produces a hidden line removal in addition to displaying the shaded image in the current color. When executing this command, the screen temporarily will go blank while the system calculates the surfaces to shade. The length of time to accomplish this is of course dependent on the complexity of the object being shaded. A percentage of completion is displayed at the bottom of the prompt line. Once the shaded image is displayed on the screen, it cannot be plotted; it can, however, be made into a slide using the Mslide command. This image consists of a colorization of the solid model and is not meant for different surface tones or casting of shadows. To perform this, select the Ashade.Lsp routine from the pulldown menu area. Locate lights, cameras, and create scenes. Finally create a filmroll file that may be imported into AutoShade for viewing.

Solid Modeling Utility Commands

The following utility commands allow the user to control items such as the type of material the solid model consists of (Solmat). A new current user coordinate system may be set to a face or edge of a solid using the Solucs command. The Solin and Solout commands were primarily designed to import models to or export models from AutoSolid. Unused solid modeling information is easily removed from a drawing using the Solpurge command. To free-up precious memory, the Advanced Modeling Extension may be removed using the Unload option from the pull-own menu area. A few of the above commands are described in detail on the next series of pages.

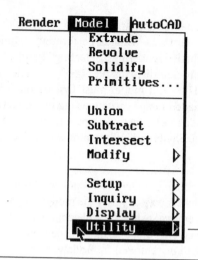

SOLMAT

This command assigns a default material to the solid model. An existing solid model may have its current material changed to reflect a new material from the list below. New materials may be added to this list. These material definitions are very important to the solid model since this information is used when performing a mass property calculation when executing the Solmassp command. By default, all solid models are assigned the material "Mild Steel." The following prompt sequence may be used to list or change this material:

Command: **Solmat**
Change/Edit/LIst/LOad/New/Remove/SAve/SEt/?<eXit>: **?**

Defined in file:		
ALUMINUM		
BRASS	--	Soft Yellow Brass
BRONZE	--	Soft Tin Bronze
COPPER		
GLASS		
HSLA_STL	--	High Strength Low Alloy Steel
LEAD		
MILD_STEEL		
NICU	--	Monel 400
STAINLESS_STL	--	Austenic Stainless Steel

Selecting "Material..." from the menu area illustrated at the right displays the DDSOLMAT dialog box below. Use this dialog box to dynamically select and load the desired material from the large edit box area. Various buttons on the right half of the dialog box enable the user to make changes to the material properties and then save these changes. Any calculations made by the Solmassp or DDSOLMASSP dialog box will be based on the new material properties.

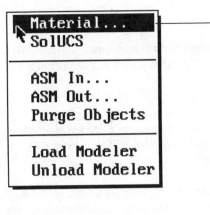

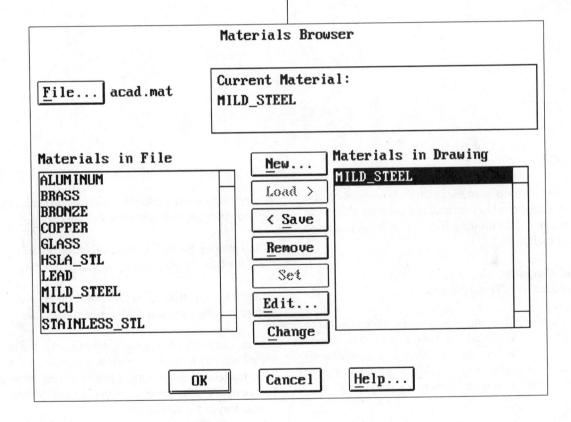

SOLUCS

This command will align the user coordinate system with a selected face or edge. The prompts for this command are as follows:

Command: **Solucs**
Edge/<Face>: *(Select a face or enter "E" to select an edge)*
<OK>/Next: *(Accept the highlighted face or enter "N" to highlight the next face)*

As with similar AME commands displaying the prompt "<OK>/Next:," the user has the option to accept the selected face with "OK" or choose another face through "Next."

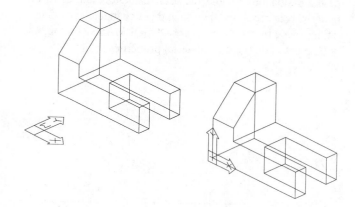

SOLPURGE

Solid models tend to consume large amounts of memory. Use the Solpurge command to conserve this memory along with reducing the size of the drawing file. The Solpurge command prompt is as follows:

Command: **Solpurge**
Memory/2dtree/Bfile/Pmesh/<Erased>:

Memory - solid models only use memory when selected to perform a type of Advanced Modeling Extension command such as Solchp. Use the Memory-All option to release all memory currently being used by AME. The solids are not deleted; the memory assigned to them is just freed up.

2Dtree - this option works with regions and restructures them without changing the appearance of the region.

Bfile - purging the Bfile entities of a solid model may reduce the size of the drawing size by as much as 50%.

Pmesh - this option is similar to the Bfile option in that it is used to further reduce the size of the solid model file.

Erased - as a solid or region is deleted, only the mail block insertion entity is erased. Secondary entities remain with the solid model and if not purged use up excess memory. The Erased option cleans up these secondary entities while freeing up memory at the same time.

Aliasing for AME Commands

As part of the ACAD.PGP file, a majority of the Advanced Modeling Extension commands may be typed in at the keyboard in abbreviated style. As an example, to draw a solid box, the AME command is Solbox. In the ACAD.PGP file, this command has been renamed "Box." If typing is to be kept at a minimum, study the complete listing of AME commands on this page to see what the name of the command has been renamed to. Then enter that command from the keyboard at the command prompt. These commands are automatically supplied with the AutoCAD software packege.

```
; Primitives.

BOX,            *SOLBOX
WED,            *SOLWEDGE
WEDGE,          *SOLWEDGE
CON,            *SOLCONE
CONE,           *SOLCONE
CYL,            *SOLCYL
CYLINDER,       *SOLCYL
SPH,            *SOLSPHERE
SPHERE,         *SOLSPHERE
TOR,            *SOLTORUS
TORUS,          *SOLTORUS


; Complex Solids.

FIL,            *SOLFILL
SOLF,           *SOLFILL
CHAM,           *SOLCHAM
SOLC,           *SOLCHAM
EXT,            *SOLEXT
EXTRUDE,        *SOLEXT
REV,            *SOLREV
REVOLVE,        *SOLREV
SOL,            *SOLIDIFY


; Boolean operations.

UNI,            *SOLUNION
UNION,          *SOLUNION
INT,            *SOLINT
INTERSECT,      *SOLINT
SUB,            *SOLSUB
SUBTRACT,       *SOLSUB
DIF,            *SOLSUB
DIFF,           *SOLSUB
DIFFERENCE,     *SOLSUB
SEP,            *SOLSEP
SEPARATE,       *SOLSEP
```

```
; Modification and Query commands.

SCHP,           *SOLCHP
CHPRIM,         *SOLCHP
MAT,            *SOLMAT
MATERIAL,       *SOLMAT
MOV,            *SOLMOVE
SL,             *SOLLIST
SLIST,          *SOLLIST
MP,             *SOLMASSP
MASSP,          *SOLMASSP
SA,             *SOLAREA
SAREA,          *SOLAREA
SSV,            *SOLVAR


; Documentation commands.

FEAT,           *SOLFEAT
PROF,           *SOLPROF
PROFILE,        *SOLPROF
SU,             *SOLUCS
SUCS,           *SOLUCS


; Model representation commands.

SW,             *SOLWIRE
WIRE,           *SOLWIRE
SM,             *SOLMESH
MESH,           *SOLMESH
```

Tutorial Exercise #30
Bplate.Dwg

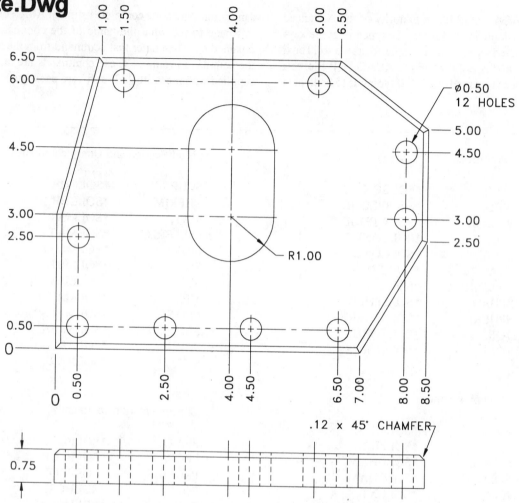

PURPOSE:
The purpose of this tutorial is to produce a solid model of the Bplate using AutoCAD Advanced Modeling Extension commands.

SYSTEM SETTINGS:
Keep the current limits settings of 0,0 for the lower left corner and 12,9 for the upper right corner. Change the number of decimal places past the zero from 4 to 2 using the Units command. Turn the grid on and change the snap value from 1.00 to 0.50. Keep all remaining system settings.

LAYERS:
Special layers do not have to be created for this tutorial exercise.

SUGGESTED COMMANDS:
Begin this tutorial by constructing the profile of the Bplate using polylines. Add all circles and enter the Advanced Modeling Extension. Use the Solext command to extrude all entities the thickness of the base at 0.75 units. Use the Solsub command to subtract all cylinders from the Bplate forming the holes in the Plate. Use the Solmesh command to surface the model before performing a hidden line removal using the Hide command.

DIMENSIONING:
This tutorial does not require dimensions.

PLOTTING:
This tutorial exercise may be plotted on "B"-size paper (11 x 17"). Use a plotting scale of 1=1 to produce a full size plot.

Step #1

Begin the Bplate by establishing a new coordinate system using the UCS command. Define the origin at 2.00,1.50. Use the Ucsicon command to update the user coordinate system icon to the new coordinate system location on the display screen. Use the Pline command to draw the profile of the Bplate.

Command: **UCS**
Origin/ZAxis/3point/Entity/View/X/Y/Z/Prev/Restore/Save/
Del/?/<World>: **Or**
Origin point: <0,0,0>: **2.00,1.50**

Command: **Ucsicon**
ON/OFF/All/Noorigin/ORigin <ON>: **OR**

Command: **Pline**
From point: **0,0**
Current line-width is 0.00
<Endpoint of line>: **@7.00<0**
<Endpoint of line>: **@1.50,2.50**
<Endpoint of line>: **@2.50<90**
<Endpoint of line>: **@-2.00,1.50**
<Endpoint of line>: **@5.50<180**
<Endpoint of line>: **@-1.00,-3.50**
<Endpoint of line>: **Close**

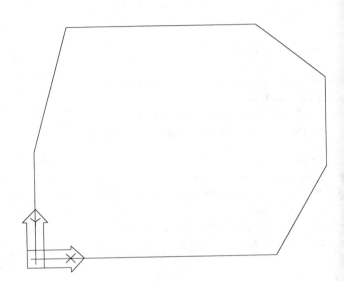

Step #2

Draw the 12 circles of 0.50 diameter by placing one circle at "A" and copying the remaining circles to their desired locations. Use the Copy-Multiple command to accomplish this.

Command: **Circle**
3P/2P/TTR/<Center point>: **0.50,0.50**
Diameter/<Radius>: **D**
Diameter: **0.50**

Command: **Copy**
Select objects: **L** *(for the last circle)*
Select objects: *(Strike Enter to continue with this command)*
<Base point or displacement>/Multiple: **M**
Base point: **@** *(References the center of the 0.50 circle)*
Second point of displacement: **2.50,0.50**
Second point of displacement: **4.50,0.50**
Second point of displacement: **6.50,0.50**
Second point of displacement: **8.00,3.00**
Second point of displacement: **8.00,4.50**
Second point of displacement: **0.50,2.50**
Second point of displacement: **1.50,6.00**
Second point of displacement: **6.00,6.00**
Second point of displacement: *(Strike Enter to exit this command)*

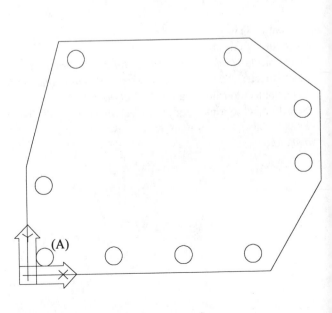

Step #3

Form the slot by placing two circles using the Circle command followed by two lines drawn from the quadrants of the circles using the Osnap-Quadrant mode.

Command: **Circle**
3P/2P/TTR/<Center point>: **4.00,3.00**
Diameter/<Radius>: **1.00**

Command: **Circle**
3P/2P/TTR/<Center point>: **4.00,4.50**
Diameter/<Radius>: **1.00**

Command: **Line**
From point: **Qua**
of *(Select the quadrant of the circle at "A")*
To point: **Qua**
of *(Select the quadrant of the circle at "B")*
To point: *(Strike Enter to exit this command)*

Command: **Line**
From point: **Qua**
of *(Select the quadrant of the circle at "C")*
To point: **Qua**
of *(Select the quadrant of the circle at "D")*
To point: *(Strike Enter to exit this command)*

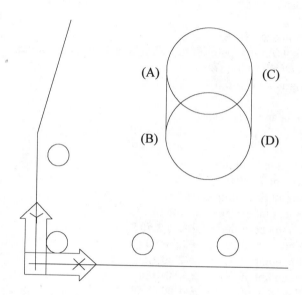

Step #4

Use the Trim command to trim away any unnecessary arcs to form the slot.

Command: **Trim**
Select cutting edges...
Select objects: *(Select both vertical lines at "A" and "B")*
Select objects: *(Strike Enter to continue with this command)*
<Select object to trim>/Undo: *(Select the circle at "C")*
<Select object to trim>/Undo: *(Select the circle at "D")*
<Select object to trim>/Undo: *(Strike Enter to exit this command)*

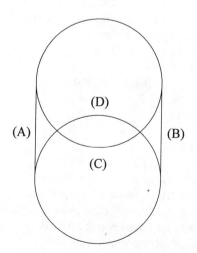

Step #5

In a few moments, the Advanced Modeling Extension Solext command will be used to extrude all entities to the thickness of the Bplate, which is 0.75 units. This command, however, only operates on polylines and circles. Currently, all entities may be extruded except for the two arcs and lines representing the slot. Use the Pedit command to convert these entities into a single polyline.

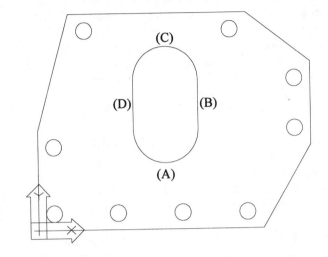

Command: **Pedit**
Select polyline: *(Select the bottom arc at "A")*
Entity selected is not a polyline?
Do you want it to turn into one? **Yes**
Close/Join/Width/Edit vertex/Fit curve/Spline curve/Decurve/
Undo/eXit <X>: **Join**
Select objects: *(Select the entities labeled "B," "C," and "D")*
3 segments added to polyline.
Close/Join/Width/Edit vertex/Fit curve/Spline curve/Decurve/
Undo/eXit <X>: **X**

Step #6

From the pulldown menu area, load the Advanced Modeling Extension by selecting the Extrude command, which really is the Solext command (Solid Extrude). Selecting Extrude prompts the user to select the objects to extrude. Select all polylines and circles that make up the Bplate. Enter a value of 0.75 as the height of the extrusion. This process may take a few minutes depending on the speed and amount of memory for your machine. Turn off the user coordinate system icon.

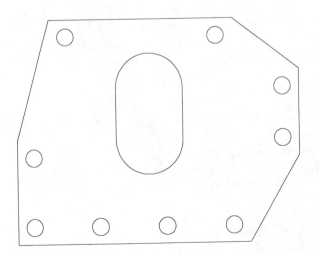

Command: **Solext** *(Already selected from the pulldown menu)*
Initializing Advanced Modeling Extension.
Select polylines and circles for extrusion...
Select objects: **W**
First corner: **-1.00,-1.00**
Other corner: **9.00,7.00**
Select objects: *(Strike Enter to continue with this command)*
Height of extrusion: **0.75**
Extrusion taper angle from Z <0>: *(Strike Enter to accept default)*

Command: **Ucsicon**
ON/OFF/All/Noorigin/ORigin <ON>: **Off**

Step #7

Use the Vpoint command to view the plate in three dimensions using the prompts and settings below. Use the Solcham command to place a chamfer along the top edge of the Bplate. When prompted to select the base surface, select the entire top surface of the plate. This surface may not select the first time. Use the "Next" prompt until the top surface is selected. Select all individual edges of the top of the plate as the edges to be chamfered. Enter 0.12 as the two chamfer distances to place a 45 degree chamfer. Wait a moment while the AME updates the model to include the chamfer.

Command: **Vpoint**
Rotate/<View point> <0.00,0.00,1.00>: **1,-1,1**

Command: **Solcham** *(Or select from the Sol-Modify menu)*
Select base surface: *(Select the top surface at "A")*
<OK>/Next: **N** *(May be required if the top surface did not select)*
<OK>/Next: *(Strike Enter only if top surface is selected)*
Select edges to be chamfered (Press ENTER when done): *(Select all individual edges of the top of the bplate, edges "A" to "G", then strike Enter)*
Enter distance along first surface <0.00>: **0.12**
Enter distance along second surface <0.12>: *(Strike Enter)*

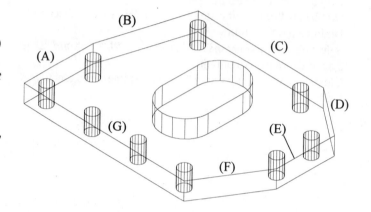

Step #8

With all entities extruded the distance 0.75, the plate is now a solid object complete with .12 x 45 degree chamfer. The holes and slot however are also solid. The Solsub command is used to subtract the holes and slot from the base of the plate. This resembles actually drilling holes and milling the slot to size. This command may be selected from the Sol-Modify pull down menu area. As with all AME commands, this process may take some time.

Command: **Solsub**
Source objects...
Select objects: *(Select the base of the bplate along any edge)*
1 selected.
Objects to subtract from them...
Select objects: *(Carefully select every hole and slot)*
Select objects: *(Strike Enter to begin the subtraction process)*

Phase I - Boundary evaluation begins.
Phase II - Tessellation computation begins.
Updating Advanced Modeling Extension database.

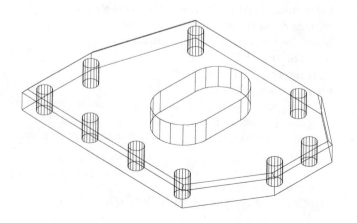

Step #9

The correct solid model of the plate is illustrated at the right. By default, the current display is that of a wireframe view. To surface the model, the Solmesh command is used. This command may be selected from the Sol-Display pulldown menu area. This command will take some time to process depending on the speed of your computer.

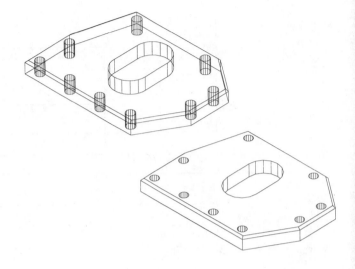

Command: **Solmesh**
Select objects: *(Select an entity on the bplate; all should highlight)*
Select objects: *(Strike Enter to continue with this command)*

Phase I of surface meshing.
Phase II of surface meshing.
Surface meshing of current solid is complete.
Creating block for mesh representation.
Done.

Step #10

To appreciate the automatic surfacing performed by the Solmesh command, set the system variable, Splframe, to a value of 1. This will turn on all of the 3Dfaces that went into the construction of the object. The faces will not appear until a screen regeneration is forced using the Regen command. This effect is illustrated at the right. To remove the display of the surfaces, set Splframe to a value of "0" and perform another screen regeneration.

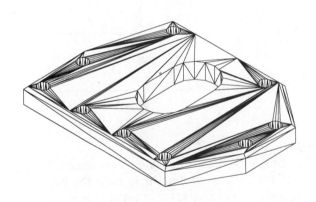

Command: **Splframe**
Current value <0>: New value: **1**

Command: **Regen**

Step #11

Illustrated below is a display of the plate that has been shaded using the Shade command. This will produce a shaded image consisting of hidden line removal in addition to the shading being performed in the original color of the model. There are no areas of shadows using the Shade command. For best results, add lights and cameras, create a scene, and finally create a filmroll file of this or any surfaced object and import the results into AutoShade.

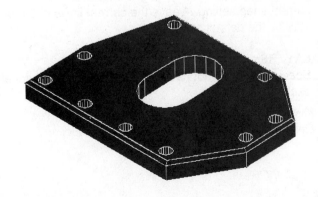

Command: **Shade**
Regenerating Screen.
Shading xx% done.

Tutorial Exercise #31
Guide.Dwg

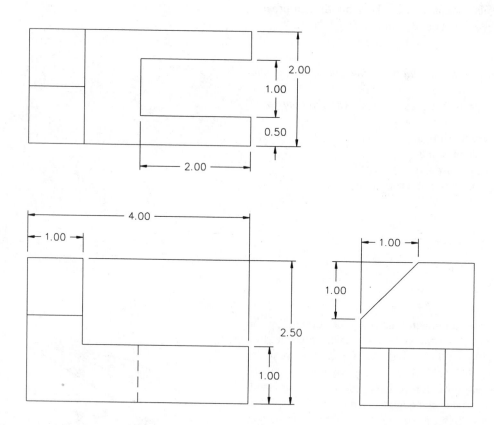

PURPOSE:
This tutorial exercise is designed to produce a 3-dimensional solid model of the Guide from the information supplied in the orthographic drawing above.

SYSTEM SETTINGS:
Use the Units command, keep the system of measurement set to decimal but change the number of decimal places past the zero from a value of 4 to 2. Set the Snap to a value of 0.50 units. The grid should conform to the current snap setting. Leave the current limits of 0,0 by 12,9 as the default setting.

LAYERS:
Special layers do not have to be created for this tutorial exercise.

SUGGESTED COMMANDS:
Begin this drawing by constructing solid primitives of all components of the Guide using the Solbox and Solwedge commands. Move the components into position and begin merging solids using the Solunion. To form the rectangular hole, move that solid box into position and use the Solsub command to subtract the rectangle from the solid thus forming the hole. Do the same procedure for the wedge. Use the Solmesh command to surface the solid. Perform a hidden line removal and view the solid.

DIMENSIONING:
This tutorial exercise does not require dimensioning.

PLOTTING:
This tutorial exercise may be plotted on "B"-size paper (11 x 17"). Use a plotting scale of 1=1 to produce a full size plot.

Step #1

Begin this tutorial by constructing a solid box 4 units long by 2 units wide and 1 unit in height using the Solbox command. Begin this box at absolute coordinate 4.00,5.50. This slab will represent the base of the Guide.

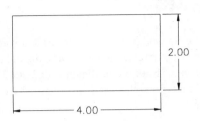

Command: **Solbox**
Baseplane/Center/<Corner of box><0,0,0>: **4.00,5.50**
Cube/Length/<Other corner>: **L**
Length: **4.00**
Width: **2.00**
Height: **1.00**

Step #2

Construct a solid box 1 unit long by 2 units wide and 1.5 units in height using the Solbox command. Begin this box at absolute coordinate 2.00,1.50. This slab will represent the vertical column of the Guide.

Command: **Solbox**
Baseplane/Center/<Corner of box><0,0,0>: **2.00,1.50**
Cube/Length/<Other corner>: **L**
Length: **1.00**
Width: **2.00**
Height: **1.50**

Step #3

Construct a solid box 2 units long by 1 unit wide and 1 unit in height using the Solbox command. Begin this box at absolute coordinate 5.50,1.50. This slab will represent the rectangular hole made into the slab that will be subtracted at a later time.

Command: **Solbox**
Baseplane/Center/<Corner of box><0,0,0>: **5.50,1.50**
Cube/Length/<Other corner>: **L**
Length: **2.00**
Width: **1.00**
Height: **1.00**

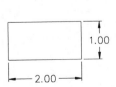

Step #4

Use the Solwedge command to draw a wedge 1.00 units in length, 1.00 units wide, and 1.00 units in height. Begin this primitive at absolute coordinate 9.50,2.00. This wedge will be subtracted from the vertical column to form the inclined surface.

Command: **Solwedge**
Baseplane/<Corner of wedge><0,0,0>: **9.50,2.00**
Length/<Other corner>: **L**
Length: **1.00**
Width: **1.00**
Height: **1.00**

Step #5

Use the Vpoint command to view the 4 solid primitives in three dimensions. Use a new view point of 1,-1,0.75. Then use the Move command to move the vertical column at "A" to the top of the base at "B".

Command: **Vpoint**
Rotate/<View point> <0,0,0>: **1,-1,0.75**

Command: **Move**
Select objects: *(Select the solid box at "A")*
Select objects: *(Strike Enter to continue with this command)*
Base point or displacement: **Endp**
of *(Select the endpoint of the solid at "A")*
Second point of displacement: **Endp**
of *(Select the endpoint of the base at "B")*

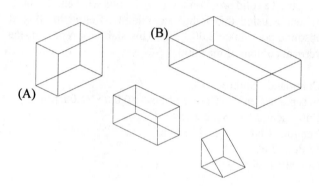

Step #6

Use the Solunion command to join the base and vertical column into one entity.

Command: **Solunion**
Select objects: *(Select the base at "A" and column at "B")*
Select objects: *(Strike Enter to perform the union)*

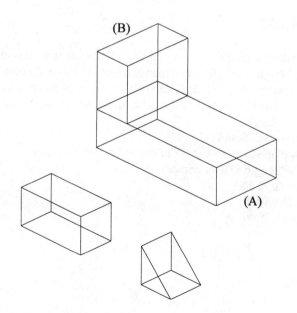

Step #7

Use the Move command to position the rectangle from its midpoint at "A" to the midpoint of the base at "B". In a moment, the small rectangle will be subtracted forming the rectangular hole in the base.

Command: **Move**
Select objects: *(Select the rectangle at "A")*
Select objects: *(Strike Enter to continue with this command)*
Base point or displacement: **Mid**
of *(Select the midpoint of the rectangle at "A")*
Second point of displacement: **Mid**
of *(Select the midpoint of the base at "B")*

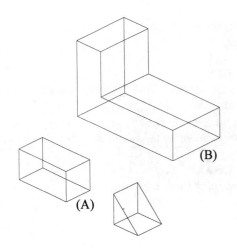

Step #8

Use the Solsub command to subtract the small rectangle from the base of the solid.

Command: **Solsub**
Source objects...
Select objects: *(Select the solid at "A")*
Select objects: *(Strike Enter to continue with this command)*
Objects to subtract from them...
Select objects: *(Select the small rectangle at "B")*
Select objects: *(Strike Enter to perform the subtraction operation)*

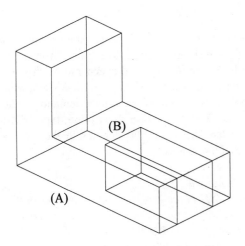

Step #9

Use the Rotate command to revolve the wedge at an angle of 90 degrees. This will begin preparing the wedge to be inserted onto the vertical column before subtracting.

Command: **Rotate**
Select objects: *(Select the wedge)*
Select objects: *(Strike Enter to continue with this command)*
Base point: **Endp**
of *(Select the endpoint of the wedge at "A")*
<Rotation angle>/Reference: **90**

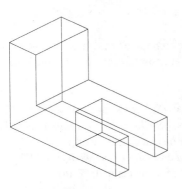

Step #10

Move the user coordinate system icon to a new origin located on the wedge.

Command: **UCS**
Origin/ZAsix/3point/Entity/View/X/Y/Z/Prev/Restore/Save/
Del/?/<World>: **Or**
Origin point <0,0,0>: **Endp**
of *(Select the endpoint of the wedge at "A")*

Command: **Ucsicon**
ON/OFF/Noorigin/Origin <On>: **Or**

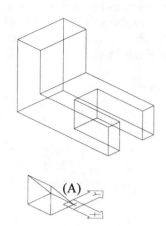

Step #11

The wedge needs to be rotated 90 degrees about the X-axis before being placed into position and subtracted from the main solid. Use the Solmove command to accomplish this. Once in the command, a new icon appears; it is the MCS or Motion Coordinate System. This icon shows the orientation of the X, Y, and Z axes. For the motion description, type "RX-90" which will rotate the wedge 90 degrees in the clockwise direction about the X-axis.

Command: **Solmove**
Select objects: *(Select the wedge...a new icon appears)*
Select objects: *(Strike Enter to continue with this command)*
?/<Motion description>: **RX-90**
?/<Motion description>: *(Strike Enter to exit this command)*

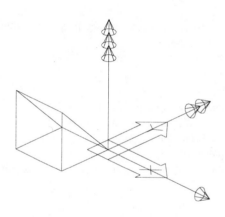

Step #12

Use the Move command to move the wedge from its endpoint at "A" to the endpoint of the vertical column at "B."

Command: **Move**
Select objects: *(Select the wedge)*
Select objects: *(Strike Enter to continue with this command)*
Base point or displacement: **Endp**
of *(Select the endpoint of the wedge at "A")*
Second point of displacement: **Endp**
of *(Select the endpoint of the vertical column at "B")*

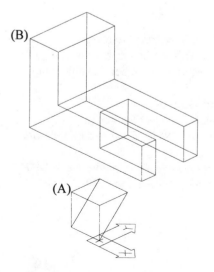

Step #13

Use the Solsub command to subtract the wedge from the main solid.

Command: **Solsub**
Source objects...
Select objects: *(Select the solid at "A")*
Select objects: *(Strike Enter to continue with this command)*
Objects to subtract from them...
Select objects: *(Select the wedge at "B")*
Select objects: *(Strike Enter to perform the subtraction operation)*

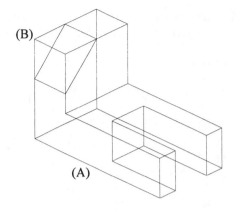

Step #14

An alternate method of creating the inclined surface is to use the Solchamfer command to chamfer the vertical column.

Command: **Solcham**
Pick base surface: *(Select an edge along surface "A")*
Next/<OK>: *(If surface "A" highlights, Strike Enter to continue; If another surface highlights, Type "N" for next surface and step through the surfaces until surface "A" is highlighted).*
Pick edges of this face to be chamfered (press Enter when done): *(Select the line at "B" and strike Enter).*
Enter distance along base surface <0.00>: **1.00**
Enter distance along adjacent surface <0.00>: **1.00**

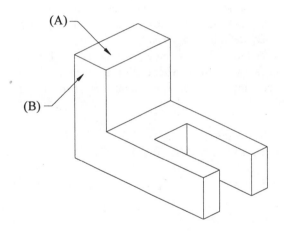

Step #15

The object is still currently displayed in wireframe mode. If a hidden line removal is to be performed, surfaces need to be added to the object. This can automatically be done using the Solmesh command.

Command: **Solmesh**
Select objects: *(Select the solid model at "A")*

The results of the Solmesh command may not be immediately apparent until a hidden line removal is performed using the Hide or Shade commands.

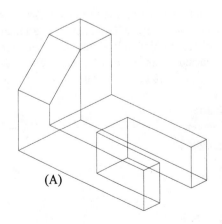

Step #16

Using the Hide command performs a hidden line removal on all surfaces of the object. The results are illustrated at the right.

Command: **Hide**

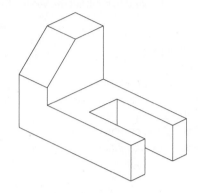

Step #17

To view the surfaces of the model, the system variable Splframe may expose all surfaces created by the Solmesh command. By default, this variable is Off. If set On, all surfaces needed to surface the model are displayed. Before showing surfaces, a screen regeneration must be made to update the drawing database. The same is true when turning the variable Off.

Command: **Splframe**
Current value for SPLFRAME <0>: **1**

Command: **Regen**

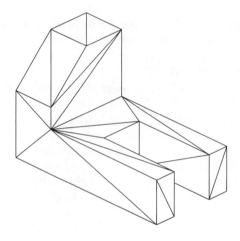

Step #18

Use the Shade command to view the solid model as a shaded image. Remember, the screen will momentarily go blank as the system calculates the hidden line removal along with the shading in the current color of the model. Use the Mslide command to document the image for later use.

Command: **Shade**
Regenerating Screen.
Shading 100% done.

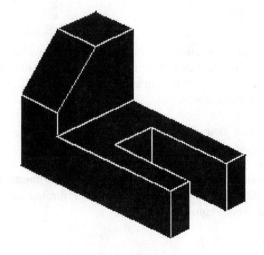

Tutorial Exercise #32
Collar.Dwg (Creating the Model)

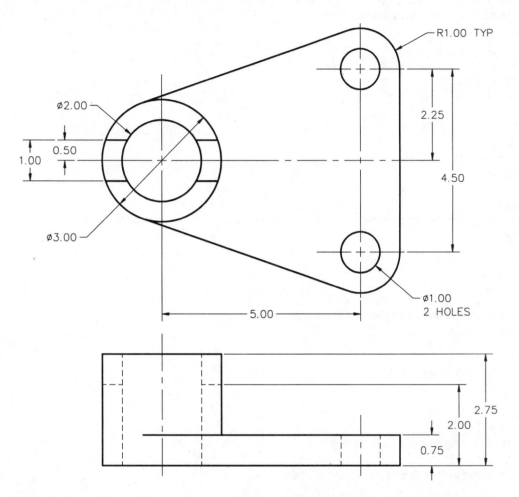

PURPOSE:
This tutorial is designed to construct a solid model of the Collar using the dimensions illustrated above.

SYSTEM SETTINGS:
Use the current limits set to 0,0 for the lower left corner and 12,9 for the upper right corner. Change the number of decimal places from 4 to 2 using the Units command. Snap and Grid values may remain as set by the default although the Snap may be changed to a value of 0.50 thereby affecting the display of the grid.

LAYERS:
Special layers do not have to be created for this tutorial exercise although an object layer may be created using yellow lines:

Name-Color-Linetyp
Object - Yellow - Continuous

SUGGESTED COMMANDS:
Begin this tutorial by laying out the Collar in plan view and drawing the basic shape outlined in the top view. Convert the entities into a polyline and extrude the entities to form a solid. Draw a cylinder and combine this entity with the base. Add another cylinder and then subtract it to form the large hole through the model. Add two small cylinders and subtract them from the base to form the smaller holes. Construct a solid box, use the Solmove command to move the box into position, and subtract it to form the cut across the large cylinder.

DIMENSIONING:
This tutorial does not require any special dimensioning.

PLOTTING:
This tutorial exercise may be plotted on "B"-size paper (11 x 17"). Use a plotting scale of 1=1 to produce a full size plot.

Step #1

Begin the Bracket by drawing the three circles illustrated at the right using the Circle command. Place the center of the circle at "A" at 0,0. Perform a Zoom-All after all three circles have been constructed.

Command: **Circle**
3P/2P/TTR/<Center point>: **0,0**
Diameter/<Radius>: **D**
Diameter: **3.00**

Command: **Circle**
3P/2P/TTR/<Center point>: **5.00,2.25**
Diameter/<Radius>: **1.00**

Command: **Circle**
3P/2P/TTR/<Center point>: **5.00,-2.25**
Diameter/<Radius>: **1.00**

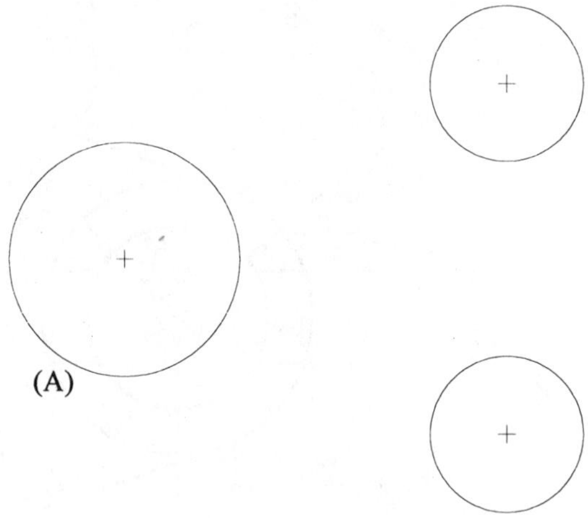

Step #2

Draw lines tangent to the three arcs using the Line command and the Osnap-Tangent mode.

Command: **Line**
From point: **Tan**
to *(Select the circle at "A")*
To point: **Tan**
to *(Select the circle at "B")*
To point: *(Strike the Enter key to exit the command)*

Repeat the above procedure to draw lines from "C" to "D" and "E" to "F".

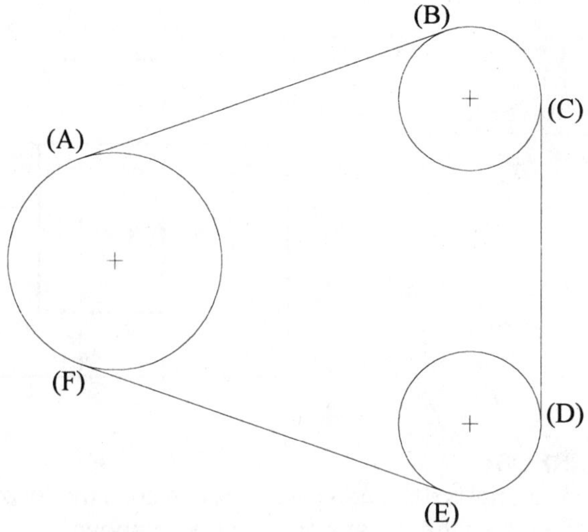

Step #3

Use the Trim command, select the three lines as cutting edges, and trim the circles.

Command: **Trim**
Select cutting edges...
Select objects: *(Select the three dashed lines at the right)*
<Select object to trim>/Undo: *(Select the circle at "A")*
<Select object to trim>/Undo: *(Select the circle at "B")*
<Select object to trim>/Undo: *(Select the circle at "C")*
<Select object to trim>/Undo: *(Strike the Enter key to exit the command)*

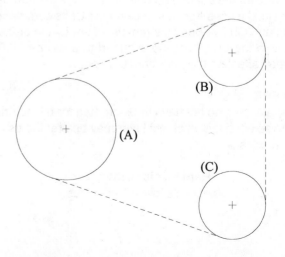

Step #4

If center marks were placed, use the Erase command to delete them now. Next, prepare to construct the bracket by viewing the object in 3D using the Vpoint command and the coordinates 1,-1,1.

Command: **Vpoint**
Rotate/<View point><0.00,0.00,0.00>: **1,-1,1**

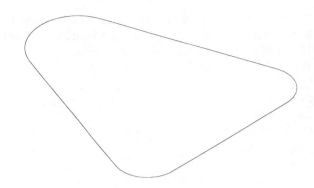

Step #5

Convert all entities into a polyline using the Join option of the Pedit command.

Command: **Pedit**
Select polyline: *(Select the arc at "A")*
Entity selected is not a polyline.
Do you want to turn it into one?<Y>: **Yes**
Close/Join/Width/Edit Vertex/Fit curve/Spline curve/Decurve/
Undo/eXit<X>: **Join**
Select objects: *(Select the three line segments and remaining two arcs illustrated at the right)*
Select objects: *(Strike the Enter key to continue)*
5 segments added to polyline.
Close/Join/Width/Edit Vertex/Fit curve/Spline curve/Decurve/
Undo/eXit<X>: **X**

(A)

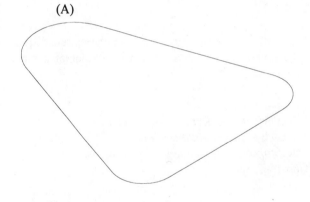

Step #6

Begin the conversion from 2D to Solids by loading AME, the Advanced Modeling Extension. This will take a few seconds depending on the speed of your machine. Next use the Solext command (Solid Extrude) to extrude the base to a thickness of 0.75 units.

Command: **Solext**
Initializing Advanced Modeling Extension.
Select regions, polylines and circles for extrusion...
Select objects: *(Select the polyline)*
Select objects: *(Strike the Enter key to continue)*
Height of extrusion: **0.75**
Extrusion taper angle <0>: **0**

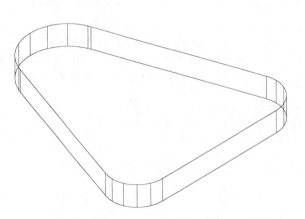

Step #7

Create a cylinder using the Solcyl command, (Solid Cylinder). Begin the center point of the cylinder at 0,0 and a diameter of 3.00 with a height of 2.75 units. Perform a Zoom-All to display the entire model.

Command: **Solcyl**
BaseplaneElliptical/<Center point>: **0,0**
Diameter/<Radius>: **D**
Diameter: **3.00**
Center of other end/<Height>: **2.75**
Phase I - Boundary evaluation begins.
Phase II - Tessellation computation begins.
Updating the Advanced Modeling Extension database.

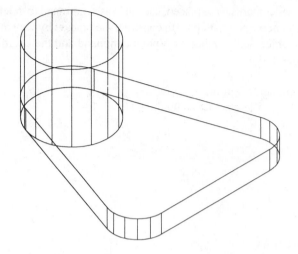

Step #8

Merge the cylinder just created with the extruded base to combine both entities into one using the Solunion command (Solid Union).

Command: **Solunion**
Select objects: *(Select the extruded base)*
Select objects: *(Select the cylinder)*
Select objects: *(Strike the Enter key to continue)*
Updating solid...

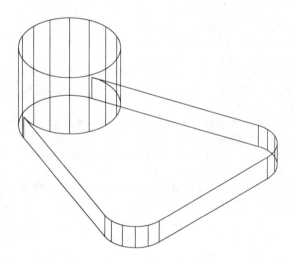

Step #9

Use the Solcyl command to create a 2.00 unit diameter cylinder representing a through hole. The height of the cylinder is 2.75 units with the center point located at 0,0.

Command: **Solcyl**
Baseplane/Elliptical/<Center point>: **0,0**
Diameter/<Radius>: **D**
Diameter: **2.00**
Center of other end/<Height>: **2.75**

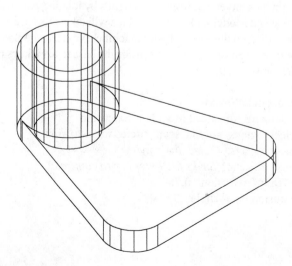

Step #10

To form a hole through the outer cylinder, the Solsub command (Solid Subtract), will be used. Select the base as the source object and the inner cylinder as the object to subtract. Wait a moment while the computer performs its calculations to remove the cylinder from the model creating a hole.

Command: **Solsub**
Source objects...
Select objects: *(Select the base)*
Select objects: *(Strike the Enter key to continue)*
1 solid selected.
Objects to subtract from them...
Select objects: *(Select the 2.00 diameter cylinder just drawn)*
Select objects: *(Strike the Enter key to continue)*

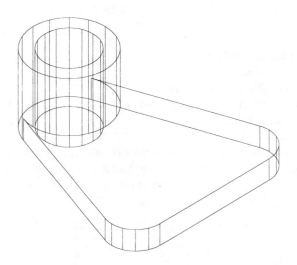

Step #11

Begin placing the two small drill holes in the base by using the Solcyl command. The absolute coordinate value of 5.00,2.25 will be used as the center of the cylinder.

Command: **Solcyl**
Baseplane/Elliptical/<Center point>: **5.00,2.25**
Diameter/<Radius>: **D**
Diameter: **1.00**
Center of other end/<Height>: **0.75**

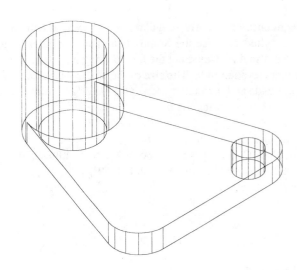

Step #12

Rather than create another cylinder (with center point of the cylinder located at 5.00,-2.25), use the Copy command to merge existing AutoCAD commands into the creation of this model of the bracket. Use the Osnap-Center mode to assist the Copy command.

Command: **Copy**
Select objects: **L**
Select objects: *(Strike the Enter key to continue)*
<Base point or displacement>/Multiple: **Cen**
of *(Select the arc at "A")*
Second point of displacement: **Cen**
of *(Select the arc at "B")*

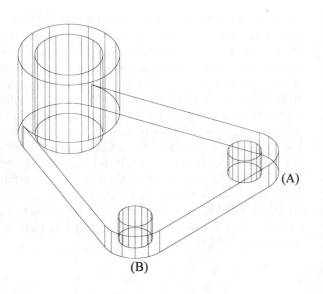

(A)

(B)

Step #13

Subtract both 1.00 diameter cylinders from the base of the model using the Solsub command.

Command: **Solsub**
Source objects...
Select objects: *(Select the base)*
Select objects: *(Strike the Enter key to continue)*
1 solid selected.
Objects to subtract from them...
Select objects: *(Select one of the 1.00 diameter cylinders)*
Select objects: *(Select the other 1.00 diameter cylinder)*
Select objects: *(Strike the Enter key to continue)*
Updating solid...

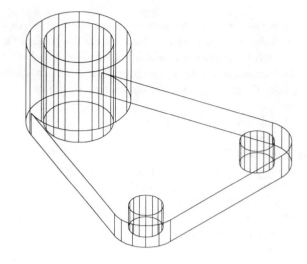

Step #14

Begin constructing the rectangular slot which will pass through the 2 cylinders. Use the Solbox command (Solid Box) to accomplish this. Construct the box locating its bottom at the current elevation of 0. The size of the box is 4 units long by 1 unit wide and 0.75 units deep. Use the origin point 0,0 as reference for drawing the box.

Command: **Solbox**
Baseplane/Center/<Corner of box><0,0,0>: **-2.00,-0.50**
Cube/Length/<Other corner>: **2.00,0.50**
Height: **0.75**

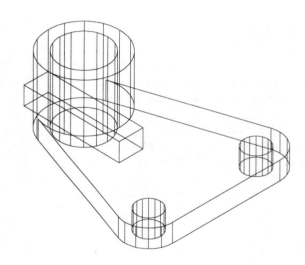

Step #15

Geometric primitives are easily constructed in any location and moved into correct position using the Solmove command. This command affects only solid model entities. When issuing the Solmove command, the familiar prompt "Select objects" appears prompting the operator to select the solid object to move. Once an object has been selected, the MCS or motion coordinate system appears; this icon alerts the operator of the X, Y, and Z axes locations in relation to the object selected. The X axis is symbolized with one cone at its axis endpoint; the Y axis has two cones and Z axis has three cones. Along with directions, the operator has numerous key-ins to perform rotations and movement (called translations) of the solid object selected. Follow the next step to move the solid box up along the Z axis the distance of 2.00 units. When the command is exited, the MCS icon will disappear.

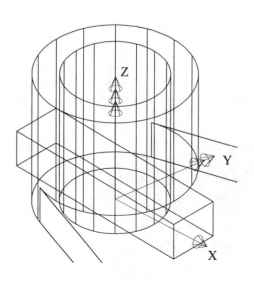

Step #16

Use the Solmove command (Solid Move) to position the box in its correct position. The icon that appears on the screen is used to guide you in the correct direction for the move. In addition, the move will occur by entering a value in the specified direction prefaced by the letter "T," which stands for translate or move.

Command: **Solmove**
Select objects: *(Select the box)*
Select objects: *(Strike the Enter key to continue)*
1 solid selected.
?/<Motion description>: **TZ2.00**
?/<Motion description>: *(Strike the Enter key to exit)*

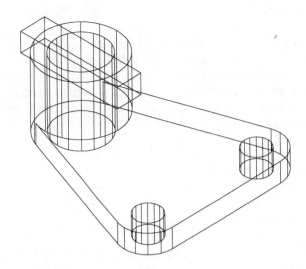

Step #17

Use the Solsub command to subtract the rectangular box from the model.

Command: **Solsub**
Source objects...
Select objects: *(Select the base "A")*
Select objects: *(Strike the Enter key to continue)*
1 solid selected.
Objects to subtract from them...
Select objects: *(Select the rectangular box "B")*
Select objects: *(Strike the Enter key to continue)*

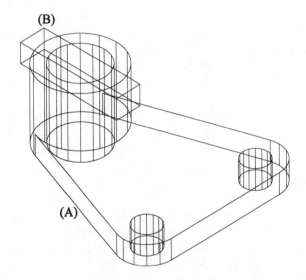

(B)

(A)

Step #18

Your model should appear similar to the illustration at the right.

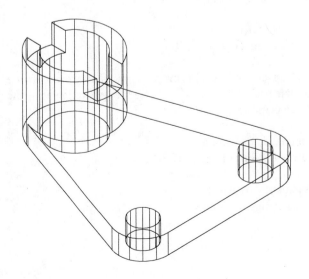

Step #19

Although a solid image of the bracket has been created, it has been produced in wireframe mode. If the Hide command were to be used, the results would not have any hidden surfaces removed. The wireframe needs to be converted to a surface mesh using the Solmesh command, (Solid Mesh). The Hide command will produce an image similar to the illustration at the right.

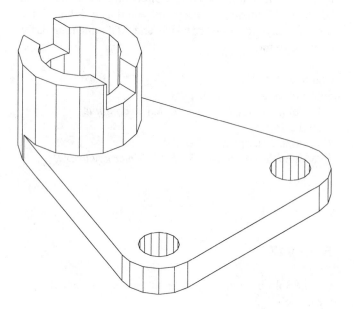

Command: **Solmesh**
Select objects: *(Select the base of the solid model)*

Command: **Hide**

To convert the mesh-generated solid back to a wireframe model, used the Solwire command (Solid Wireframe).

Command: **Solwire**
Select solids to be wired...
Select objects: *(Select the base of the solid model)*
Select objects: *(Strike the Enter key to continue)*

Step #20

Before continuing on, let's view the model with all surfaces being visible. Set the Splframe system variable from "Zero" to a value of "1" or "On." Then use the Regen command to force a regeneration of the display screen. The results may appear similar to the illustration at the right.

Command: **Splframe**
New value for SPLFRAME <0>: **1**

Command: **Regen**
Regenerating drawing.

To return the model back to hidden surfaces, reset the Splframe variable back to "0" and force a screen regeneration using the Regen command.

Command: **Splframe**
New value for SPLFRAME <1>: **0**

Command: **Regen**
Regenerating drawing.

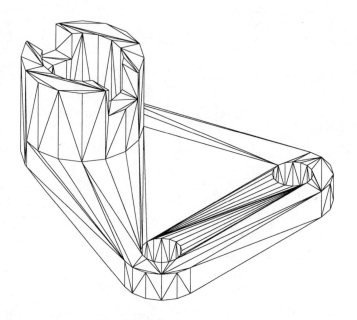

Step #21

For a model that has been surfaced using the Solmesh command, use the Shade command to perform a colorization of the model in addition to removing all hidden lines. The shading will occur in the current color of the solid model.

Command: **Shade**
Regenerating Drawing.
Shading xx% done.

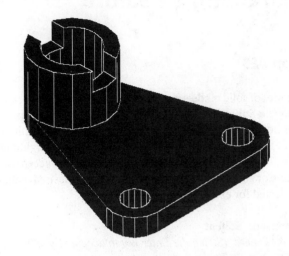

Step #22

Important information may be extracted from the solid model to be used for design and analysis purposes. This information in the form of calculations is illustrated below and is obtained when using the Solmassp command (Solid Mass Properties). The following properties are calculated by this command: Mass, Volume, Bounding Box, Centroid, Moments of Inertia, Products of Inertia, Radii of Gyration, Principal Moments about Centroid.

All of the above values are calculated based on the current material type.

Command: **Solmassp**
Select objects: *(Select the base of the bracket)*

Ray projection along X axis, level of subdivision: 3.
Mass: 248.5912 gm
Volume: 31.62738 cu cm (Err: 3.628424)

Bounding box: X: -1.500001 — 6.000001 cm
                           Y: -3.250001 — 3.250001 cm
                           Z: -1.117587e-006 — 2.75 cm

Centroid: X: 2.478948 cm   (Err: 0.4100173)
                           Y: -0.03364364 cm   (Err: 0.25711)
                           Z: 0.6805726 cm   (Err: 0.06396518)

Moments of inertia: X: 652.8676 gm sq cm (Err: 76.62952)
                           Y: 2804.419 gm sq cm (Err: 405.7606)
                           Z: 3081.122 gm sq cm (Err: 467.3195)
Products of inertia: XY: -31.0022 gm sq cm (Err: 204.4845)
                           YZ: -1.477747 gm sq cm (Err: 37.02095)
                           ZX: 285.5754 gm sq cm (Err: 60.30842)

Radii of gyration: X: 1.620577 cm
                           Y: 3.358757 cm
                           Z: 3.520559 cm

Principal moments(gm sq cm) and X-Y-Z directions about centroid:
          I: 519.9296 along [0.9915699 -0.01671225 -0.1284908]
          J: 1570.558 along [0.1283897 -0.006995896 0.9916991]
          K: 1161.798 along [-0.01747243 -0.9998359 -0.00479124]

Extracting a Feature

Step #23

Features of solid models may be extracted for design checking using the Solfeat command, (Solid Feature). This command does not produce a primary view of the entire object; rather it creates the profile of a single feature. The command highlights the feature selected in the model and the user either accepts the feature or continues on using <Next> until the desired feature is selected for extraction.

Command: **Solfeat**
Edge/<Face>: *(Strike the Enter key to accept "Face")*
All/<Select>: *(Strike Enter to select)*
Pick a face: *(Select the bottom of the base at "A")*
Solid was updated.
Repeat Selection: *(Select the bottom of the base)*
Next/<OK>: *(If the base is not highlighted, select "Next")*
Next/<OK>: *(When the base highlights, strike the Enter key)*

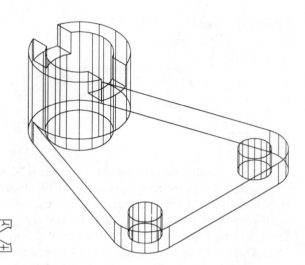

Step #24

Since the extracted feature lies directly on top of the model, use the Move command to change its position. When finished perform a Zoom-Extents to view the model and feature.

Command: **Move**
Select objects: **L**
Select objects: *(Strike the Enter key to continue)*
Base point or displacement: **0,10,0**
Second point of displacement: *(Strike the Enter key to continue)*

Command: **Zoom**
All/Center/Dynamic/Extents/Left/Previous/Vmax/Window/
<Scale(X/XP)>: **Extents**

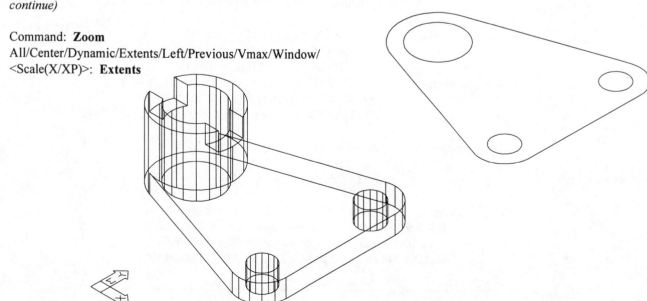

Extracting a Section

Step #25

As with a feature, it is also possible to extract a section from the solid model using the Solsect command (Solid Section), and have that section crosshatched automatically using the Solhpat command (Solid Hatching Pattern). The section is performed based on the position of the current user coordinate system. The UCS acts as a cutting edge where the model is sliced at. The result is a section placed at the current UCS position. The Move command is used here to move the section to a more convenient location. Begin by positioning the UCS icon illustrated at the right.

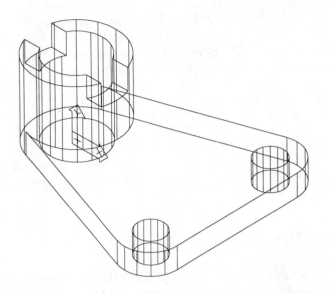

Command: **Ucsicon**
ON/OFF/All/Noorgin/ORgin <ON>: **Origin**

Command: **Ucs**
Origin/ZAxis/3point/Entity/View/X/Y/Z/Prev/Restore/Save/
Del/?/<World>: **X**
Rotation angle about X axis <0>: **90**

Step #26

Set the Solhpat command to "Ansi31," use the Solsect command to extract the section, and move the section in the positive "Z" direction illustrated at the right. Perform a Zoom-Extents when completed.

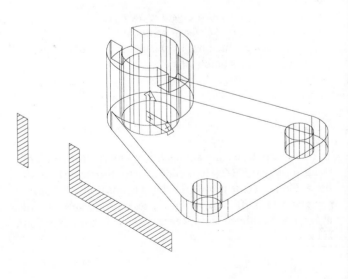

Command: **Solhpat**
Hatch pattern <None>: **Ansi31**

Command: **Solsect**
Select objects: *(Select the solid model of the bracket)*
Select objects: *(Strike the Enter key to continue)*
Sectioning plane by Entity/Last/Zaxis/View/XY/YZ/ZX/
<3points>: **XY**
Point on XY plane <0,0,0>: *(Strike Enter to accept)*

Command: **Move**
Select objects: *(Select the hatch pattern and profile)*
Select objects: *(Strike the Enter key to continue)*
Base point or displacement: **0,0,5**
Second point of displacement: *(Strike the Enter key to continue)*

Extracting a Profile

Step #27

One big advantage of creating a solid model of the Collar is the ability to extract profiles based on the current viewing angle through the use of the Solprof command. The only way to successfully use the Solprof command is to first turn the Tilemode system variable off; in other words, set this variable to a value of "0". The screen will immediately go blank and will display the paper space icon in the lower left corner of the display screen. Before continuing, load the "Hidden" linetype which will be used during the extraction process.

Command: **Tilemode**
New value for TILEMODE <1>: **0**

Command **Mview**
ON/OFF/Hideplot/Fit/2/3/4/Restore/<First point>: **Restore**
?/Name of window configuration to insert<*ACTIVE>:*(Enter)*
Fit/<First point>: **Fit**

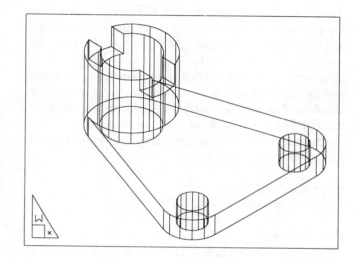

Step #28

Switch back to Model space using the Mspace command and execute the Solprof command from the keyboard, pull down menu, screen menu, or digitizing tablet menu. Select the object at the right and answer "Yes" to the prompt to display hidden lines on a separate layer.

Command: **Mspace** *(Switches to model space)*

Command: **Solprof**
Select objects: *(Select the object at the right)*
Select objects: *(Strike Enter to continue)*
Display hidden profile lines on separate layer? <N>: **Yes**
Project profile lines onto a plane? <Y>: **Yes**
Delete tangential edges? <Y>: **Yes**

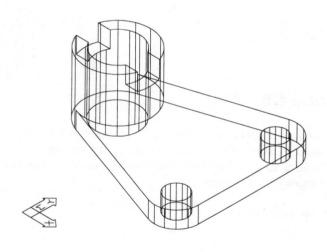

Step #29

The following layers are automatically created as a result of using Solprof: 0-PH-2 (Holds hidden line information), and 0-PV-2 (Holds visible profile line information). Use the Layer command, turn off the layer the object was originally drawn in to show the solid model appearing similar to the illustration at the right.

Command: **Layer**
?/Make/Set/New/ON/OFF/Color/Ltype/Freeze/Thaw/LOck/
Unlock: **Off**
Layer name(s) to turn off: **0**
Really want layer 0 (the CURRENT layer off?<N>: **Yes**
?/Make/Set/New/ON/OFF/Color/Ltype/Freeze/Thaw/LOck/
Unlock: *(Strike Enter to exit this command)*

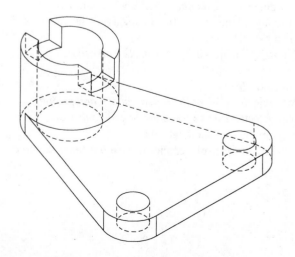

Tutorial Exercise #33
Collar.Dwg (Editing the Model)

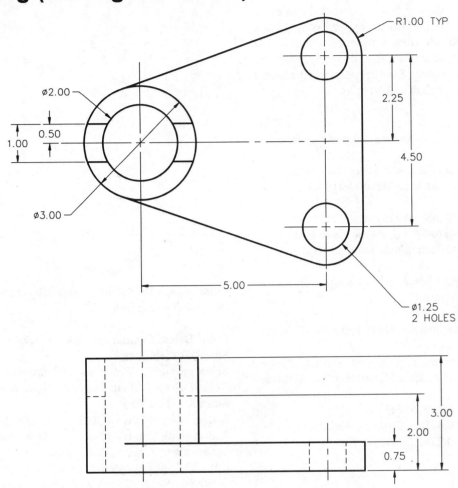

PURPOSE:
This tutorial is designed to edit an existing solid model of the Collar using the dimensions illustrated above.

SYSTEM SETTINGS:
Use the same system settings as the previous tutorial exercise.

LAYERS:
Special layers do not have to be created for this tutorial.

Name-Color-Linetyp
Object - Yellow - Continuous

SUGGESTED COMMANDS:
The total height of the "Collar" will be increased from 2.75 units to 3.00 units. The Solchp command will be used to increase the height of the large cylinder, the hole inside of the cylinder, and the solid box forming the cut in the cylinder.

DIMENSIONING:
This tutorial exercise does not require any special dimensioning.

PLOTTING:
This tutorial exercise may be plotted on "B"-size paper (11 x 17"). Use a plotting scale of 1=1 to produce a full size plot of the model of the "Collar" when completed.

Step #1

Return the User Coordinate System Icon to the World Coordinate System. Then, use the Solchp command and begin editing the solid model of the "Collar" by increasing the size of the 1.00 diameter holes to a new diameter of 1.25 units. The Size option of the Solchp command is used to accomplish this. Next increase the height of the 3.00 diameter cylinder from a height of 2.75 to a new height of 3.00 units.

Command: **Solchp**
Select a solid or region: *(Select the solid model of the collar anywhere)*
Select primitive: *(Select the 1.00 diameter hole at "A")*
Color/Delete/Evaluate/Instance/Move/Next/Pick/Replace/Size/eXit <N>: **Size**
Radius along X axis <0.50>: **0.625**
Radius along Y axis <0.625>: *(Strike the Enter key)*
Radius along X axis <0.75>: *(Strike the Enter key)*

Continue with the Solchp command by changing the size of the other 1.00 diameter hole "B:"

Color/Delete/Evaluate/Instance/Move/Next/Pick/Replace/Size/eXit <N>: **Pick**
Select primitive: *(Select the 1.00 diameter hole at "B")*
Color/Delete/Evaluate/Instance/Move/Next/Pick/Replace/Size/eXit <N>: **Size**
Radius along X axis <0.50>: **0.625**
Radius along Y axis <0.625>: *(Strike the Enter key)*
Radius along X axis <0.75>: *(Strike the Enter key)*

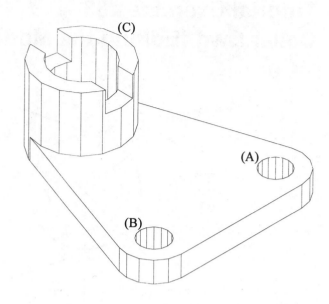

Continue with the Solchp command by changing the height of the 3.00 diameter hole "C:"

Color/Delete/Evaluate/Instance/Move/Next/Pick/Replace/Size/eXit <N>: **Pick**
Select primitive: *(Select the 3.00 diameter hole at "C")*
Color/Delete/Evaluate/Instance/Move/Next/Pick/Replace/Size/eXit <N>: **Size**
Radius along X axis <1.50>: *(Strike the Enter key)*
Radius along Y axis <1.50>: *(Strike the Enter key)*
Length along Z axis <2.75>: **3.00**
Color/Delete/Evaluate/Instance/Move/Next/Pick/Replace/Size/eXit <N>: **X** *(To exit the Solchp command and reflect the changes just performed).*

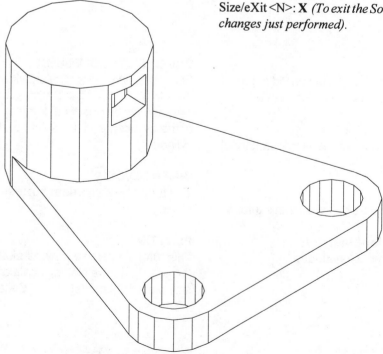

Step #2

Using the Solchp command, increase the height of the solid box primitive forming the slot in the cylinder from a height of 0.75 units to a new height of 1.00 units.

Command: **Solchp**
Select a solid or region: *(Select the solid model of the collar anywhere)*
Select primitive: *(Select the solid box at "A")*
Color/Delete/Evaluate/Instance/Move/Next/Pick/Replace/Size/eXit <N>: **Size**
Length along X axis <4.00>: *(Strike the Enter key)*
Length along Y axis <1.00>: *(Strike the Enter key)*
Length along Z axis <0.75>: **1.00**
Color/Delete/Evaluate/Instance/Move/Next/Pick/Replace/Size/eXit <N>: **X** *(To exit the Solchp command and reflect the changes just performed).*

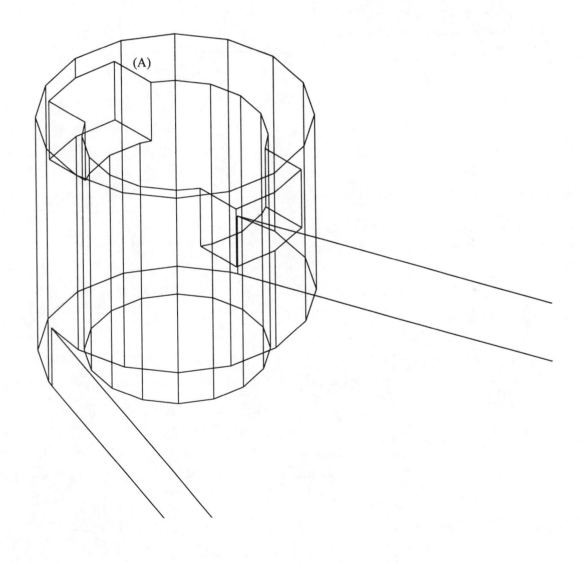

(A)

Step #3

Finally use the Solchp command to increase the height of the
inside cylinder, which represents the hole from a height of 2.75
units to a new height of 3.00 units.

Command: **Solchp**
Select a solid or region: *(Select the solid model of the collar
anywhere)*
Select primitive: *(Select the 2.00 diameter hole at "B")*
Color/Delete/Evaluate/Instance/Move/Next/Pick/Replace/
Size/eXit <N>: **Size**
Radius along X axis <1.00>: *(Strike the Enter key)*
Radius along Y axis <1.00>: *(Strike the Enter key)*
Length along Z axis <2.75>: **3.00**
Color/Delete/Evaluate/Instance/Move/Next/Pick/Replace/
Size/eXit <N>: **X** *(To exit the Solchp command and reflect the
changes just performed).*

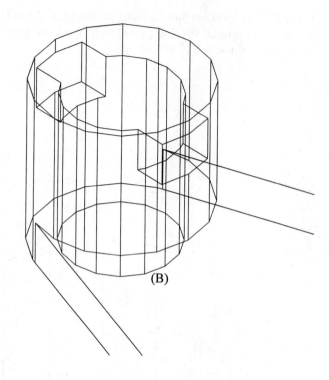

(B)

Step #4

Using the hide command displays the image of the "Collar"
below complete with the changes made using the Solchp
command.

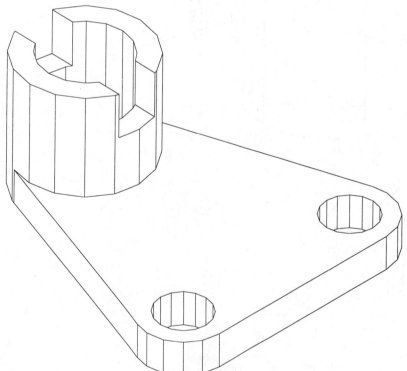

Tutorial Exercise #34
Collar.Dwg (Extracting Views from the Model)

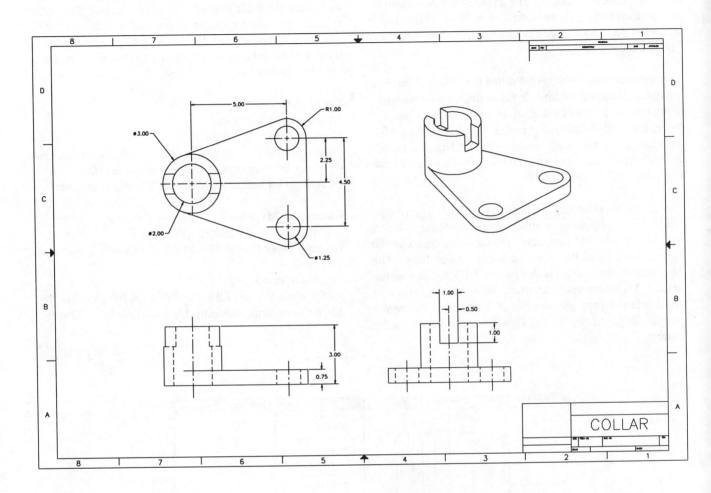

PURPOSE:
This tutorial exercise is designed to produce ortho-graphic views of a solid model using the Mvsetup.Lsp routine.

SYSTEM SETTINGS:
Use the Units command, keep the system of measure-ment set to decimal but change the number of decimal places past the zero from a value of 4 to 2. The limits of the drawing will be set by the Mvsetup.Lsp routine.

LAYERS:
Create the following layers for this exercise:

Name - Color - Linetype
Border - Red - Continuous
Viewports - Green - Continuous

SUGGESTED COMMANDS:
Use the Linetype command and load the "Hidden" linetype. Follow the detailed steps in this tutorial exercise in using the Mvsetup.Lsp routine to create orthographic views of the Collar in a paper space environment. Dimensioning considerations will be made while in paper space. Follow the next series of pages and steps towards the genera-tion of views of the Collar.

DIMENSIONING:
Dimensions will be added at later steps in this exercise.

PLOTTING:
This tutorial exercise may be plotted on "D"-size paper (34" x 22"). Use a plotting scale of 1=1 to produce a full size plot of the Collar.

Step #1

This tutorial exercise illustrates the use of generating ortho-graphic engineering views of the solid model used in previous exercises, called the Collar. The layout of the views will be accomplished using the Mvsetup.Lsp routine. This exercise also utilizes the paper space environment as the mechanism for laying out the views.

Begin this tutorial exercise by loading the "Hidden" linetype using the Linetype command. In future steps, a solid modeling command Solprof will automatically generate two types of layers that hold visible profile information and hidden profile information of the solid model. Pre-loading the hidden linetype automatically assigns hidden lines to the layers holding hidden profile information.

Shortly, the Mvsetup.Lsp routine will be loaded. This routine will automatically generate orthographic views such as Front, Top, Right Side, and Isometric. All views are based on the current position of the user coordinate system Icon. The Mvsetup routine interprets the current UCS position as the desired Top view and constructs all remaining views in relation to the Top. Be sure the UCS icon is in the correct position for generating the Top view before using the Mvsetup routing.

Complete this step by loading the Mvsetup.Lsp routine. This may be accomplished by selecting "View" from the pulldown menu area illustrated below followed by selecting "MV Setup" from the cascading menu or by entering in "Mvsetup" at the keyboard. This activates the Mvsetup.Lsp routine, which is used to allow the user to setup viewports for all orthographic views including an isometric view.

Command: **Linetype**
?/Create/Load/Set: **Load**
Linetype(s) to load: **Hidden**
(When the dialogue box displays for the .Lin file to load, strike the Enter key to accept the default ACAD.Lin file)
?/Create/Load/Set: *(Strike Enter to exit this command)*

Command: **Mvsetup**
Enable Paper/Model space? <Y>: **Yes**
Entering Paper space. Use MVIEW to insert Model space viewports.
Regenerating drawing.
MMVSetup, Version 1.16 (c) 1990-1994 by Autodesk, Inc.
Align/Create/Scale viewports/Options/Title block/Undo:

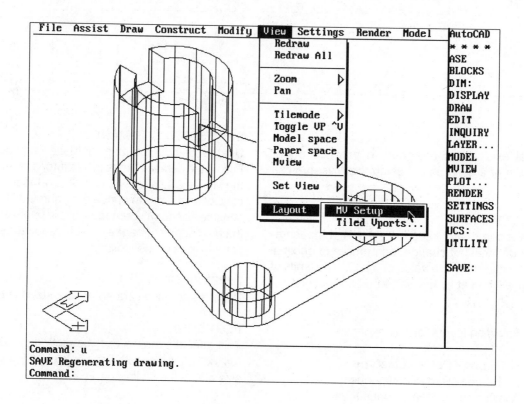

Step #2

After entering paper space, your display should appear similar to the illustration below. The paper space environment displays the triangular icon located in the lower left corner of the display screen. The original model of the Collar hasn't disappeared; it was constructed in model space. A series of viewports will be used in a later step to expose the model of the Collar.

Continue with the Mvsetup.Lsp routine by setting the drawing limits and assigning the layer "Border" to a title block that will be inserted into the drawing in a later step.

Align/Create/Scale viewports/Options/Title block/Undo:
Options
Set Layer/LImits/Units/Xref: **Limits**
Set drawing limits? <N>: **Yes**
Set Layer/LImits/Units/Xref: **Layer**
Layer name for title block or . for current layer: **Border**
Set Layer/LImits/Units/Xref: *(Strike Enter to continue)*

Using the Limits option of the Mvsetup Lisp routine is quite different from using the regular Limits command. Enabling limits in Mvsetup specifies the extents of the drawing after a title block defined by this lisp routine has been inserted into the drawing.

The Set layer option of Mvsetup allows the user to specify an existing layer on which the title block will be inserted.

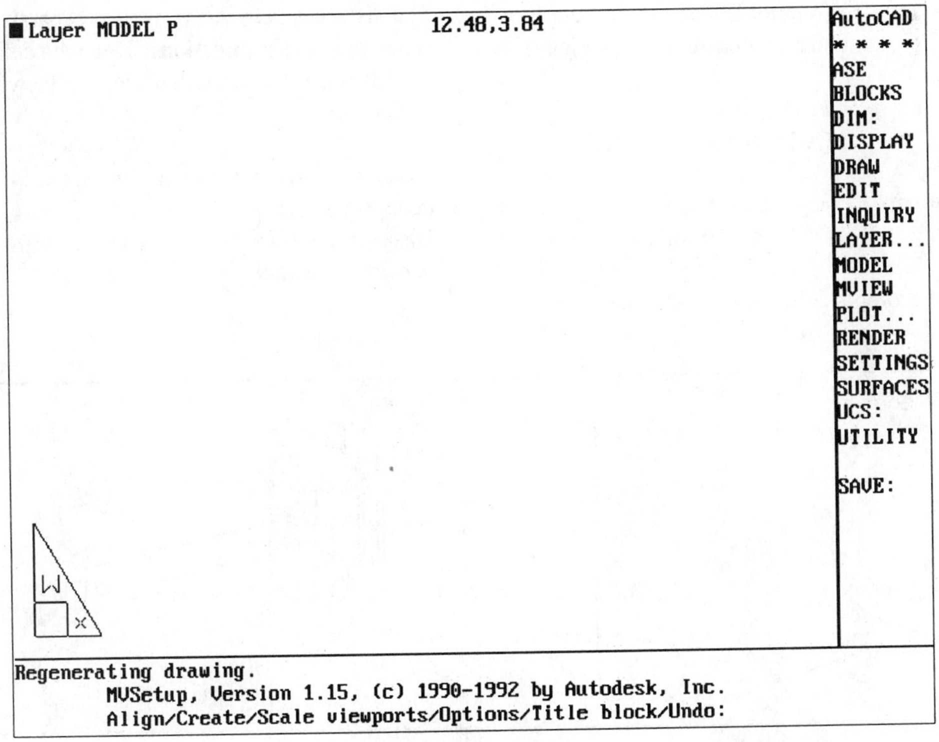

```
■ Layer MODEL P                    12.48,3.84              AutoCAD
                                                           * * * *
                                                           ASE
                                                           BLOCKS
                                                           DIM:
                                                           DISPLAY
                                                           DRAW
                                                           EDIT
                                                           INQUIRY
                                                           LAYER...
                                                           MODEL
                                                           MVIEW
                                                           PLOT...
                                                           RENDER
                                                           SETTINGS
                                                           SURFACES
                                                           UCS:
                                                           UTILITY

                                                           SAVE:


    ⟍
    W⟍
    □ ⟍x

Regenerating drawing.
        MVSetup, Version 1.15, (c) 1990-1992 by Autodesk, Inc.
        Align/Create/Scale viewports/Options/Title block/Undo:
```

Step #3

Continue with the Mvsetup.Lsp routine by creating the desired viewport configuration from the list illustrated at the right.

Align/Create/Scale viewports/Options/Title block/Undo: **Create**
Delete objects/Undo/<Create viewports>: *(Strike Enter)*
Create viewports: *(Striking Enter displays the illustration at the right)*
Redisplay/<Number of entry to load>: **2**

The Standard Engineering viewports option creates four viewport entities dividing a specified area of the display screen into quadrants similar to the illustration below. The viewing angle for each viewport is automatically set based on the current user coordinate system and the following information:

Upper-left quadrant	Top View (XY plane of the UCS)
Upper-right quadrant	Isometric View (Vpoint Rotate -45 in the XY plane; 30 from the XY plane)
Lower-left quadrant	Front View (XZ plane of the UCS)
Lower-right quadrant	Right Side View (YZ plane of the UCS)

The next series of prompts define the rectangular area of the

```
Available Mview viewport layout options:

0:      None
1:      Single
2:      Std. Engineering
3:      Array of Viewports

Redisplay/<Number of entry to load>:
```

drawing containing standard engineering viewport configuration. Since this drawing is designed to be held inside of a standard ANSI-D size sheet of paper, the upper limits of the rectangular area should be 34.00,22.00 units.

First point: **0.00,0.00**
Other point: **34.00,22.00**

The distances between viewports separates each viewport in separate X and Y directions. Keep the default values of 0 for both directions which will place each viewport directly next to each other:

Distance between viewports in X. <0.0>: *(Strike Enter to accept this value)*
Distance between viewports in Y. <0.0>: *(Strike Enter to accept this value)*

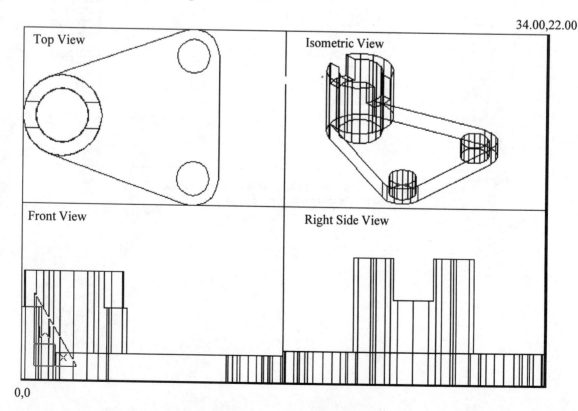

34.00,22.00

Top View

Isometric View

Front View

Right Side View

0,0

Step #4

Continue with the Mvsetup.Lsp routine by inserting a predrawn title block into the drawing. If answering Yes to the Limits option of Mvsetup, inserting the title block also sets the limits of the drawing automatially. Use the illustration at the right to view the different styles of titleblocks supported by Mvsetup.

Align/Create/Scale viewports/Options/Title block/Undo: **T**
(To insert a title block into the drawing)
Delete objects/Origin/Undo/<Insert title block>: *(Strike Enter to accept this default)*
Add/Delete/Redisplay/<Number of entry to load>: **10**

The "ANSI-D" inserts a title block at 0,0 and is sized to 34.00,22.00 units. The result is illustrated below.

Create a drawing named ansi-d.dwg?<?>: **No**
Align/Create/Scale viewports/Options/Title block/Undo:
(Strike Enter to exit the MVSetup command)

At this point, the drawing looks real crowded combining the title block with the viewports and the images of the solid model. These images will be scaled in a later step.

While still in Paper Space, use the Chprop command to convert all paper space viewports to the "Viewports" layer before continuing.

```
Available title block options:

0:      None
1:      ISO A4 Size(mm)
2:      ISO A3 Size(mm)
3:      ISO A2 Size(mm)
4:      ISO A1 Size(mm)
5:      ISO A0 Size(mm)
6:      ANSI-V Size(in)
7:      ANSI-A Size(in)
8:      ANSI-B Size(in)
9:      ANSI-C Size(in)
10:     ANSI-D Size(in)

11:     ANSI-E Size(in)
12:     Arch/Engineering (24 x 36in)
13:     Generic D size Sheet (24 x 36in)

Add/Delete/Redisplay/<Number of entry to load>:
```

Command: **Chprop**
Select objects: **Crossing**
First corner: *(Select a point at "A" illustrated below)*
Other corner: *(Select a point at "B" illustrated below)*
Change what property (Color/LAyer/LType/Thickness)? **LA**
New layer <0>: **Viewports**

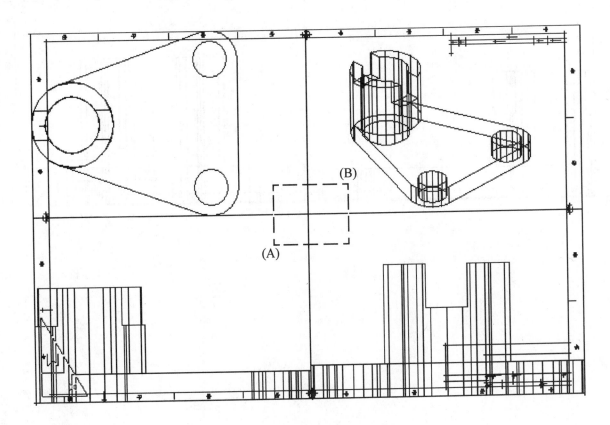

Step #5

Begin scaling the image in each viewport to reflect a scale of 1=1 or full size. This is easily accomplished using the Zoom command and the XP option, which scales images to paper space units. First switch to Model Space. Make the Top View the active viewport in model space by picking anywhere inside of the viewport; the familiar cursor should appear. With the Top View current, use the Zoom command and enter a value of 1XP which scales the image full size. The results of this step are illustrated below.

Command: **MS** *(Switches to Model Space)*

(Make the top view the current viewport)

Command: **Zoom**
All/Center/Dynamic/Extents/Left/Previous/Vmax/Window/
<Scale(X/XP)>: **1xp**

Repeat this procedure for the remaining three viewports.

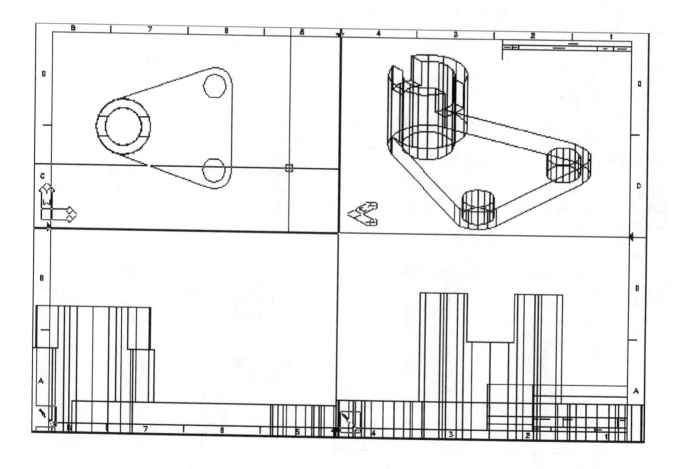

Step #5A

As an alternate means of scaling the viewports, an option of the Mvsetup.Lsp routine will allow the user to scale all viewports in paper space to the desired scale. Enter Mvsetup and issue the Scale option which is short for Scale viewports. As a better means of selecting all viewports, use the crossing option of Select objects and create a crossing box in the center of all four viewports in the illustration below. This will select all four view ports. For the zoom scale factors, keep all remaining default values for the number of paper space units vs the number of model space units. With both sets of units valued at 1.0, this will again produce images in each viewport at a factor of 1 or full size. Follow the prompt sequence at the right for scaling all viewports to reflect the full size nature of the Collar.

Command: **Mvsetup**
Align/Create/Scale viewports/Options/Title block/Undo: **Scale**
Select the viewports to scale:
Select objects: **Crossing**
First corner: *(Mark a point at "A")*
Other corner: *(Mark a point at "B")*
Select objects: *(Strike the Enter key to continue with this command)*
Set zoom scale factors for viewports. Interactively/<Uniform>: *(Strike the Enter key to accept the default)*
Number of paper space units. <1.0>: *(Strike the Enter key to accept the default)*
Number of model space units. <1.0>: *(Strike the Enter key to accept the default)*
Align/Create/Scale viewports/Options/Title block/Undo: *(Strike the Enter key to exit this command)*

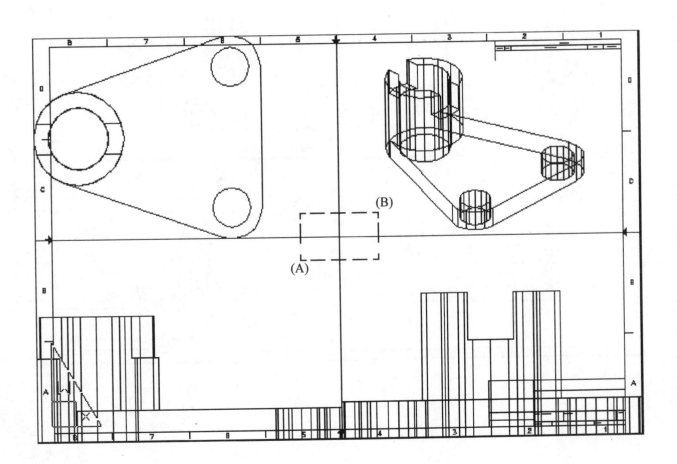

Step #6

The results of scaling the viewports from Step #5 are illustrated below. The next step in the layout of the orthographic views is the creation of visible and hidden entities representing the model of the Collar. To accomplish this, the Solprof command will be used. Important considerations to take when using this command: This command works only on a solid model; Use this command while in Model Space; Make the desired viewport current before beginning this command; The visible and hidden entities of the model are extracted perpendicular to the current user coordinate system; Layers are automatically created beginninng with the letters "PH" and "PV." Primary Hidden entities are found on layers beginning

with "PH;" Primary Visible entities are found on layers beginning with "PV." The command randomly adds other identifying letters at the end of the "PH" and "PV" layer names.

Command: **Solprof**
Select objects: *(Select the model in the Top View)*
Select objects: *(Strike Enter to continue with this command)*
Display hidden profile lines on separate layer? <Y>: **Yes**
(The system begins the extraction process)
Project profile lines onto a plane?<Y>: **Yes**
Delete tangential edges? <Y>: **Yes**

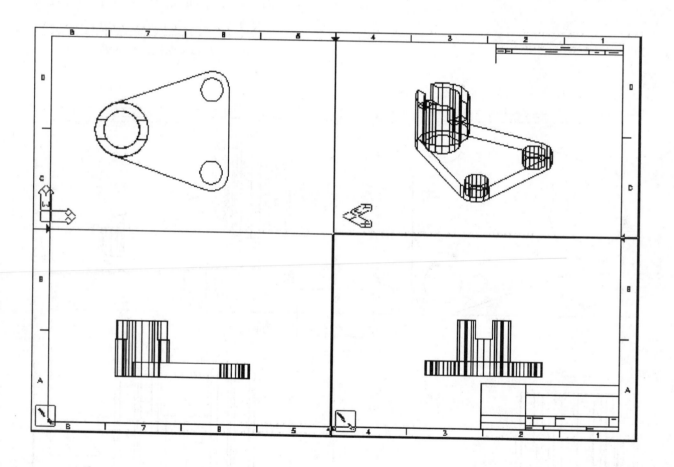

Step #7

Before continuing on with the extraction process of the remaining three views, use the Layer command or the DDLMODES dialog box to view the current layers. Notice the two new layers, "PH-DC1" and "PV-DC1," created by the Solprof command. Entities assigned to the layer beginning with "PH" hold all hidden line information of the extracted view. Entities assigned to the layer "PV" hold all visible line information of the extracted view. The "DC1" name that accompanies each layer was randomly named by the Solprof command and as a result may be called out differently on another computer terminal. On any computer system however, all layers generated by the Solprof command begin with "PH" and "PV." For the purpose of keeping better track of the layers as they relate to the given view (in this case the Top View), it would be considered good practice to rename these layers and place the name Top somewhere in the new name. Use the Rename command to accomplish this.

Layer name	State	Color	Linetype
0	On	7 (white)	CONTINUOUS
AME_FRZ	Frozen	7 (white)	CONTINUOUS
BORDER	On	1 (red)	CONTINUOUS
CENTER	On	2 (yellow)	CENTER2
MODEL	On	7 (white)	CONTINUOUS
PH-DC1	On	7 (white)	HIDDEN
PV-DC1	On	7 (white)	CONTINUOUS
VIEWPORTS	On	3 (green)	CONTINUOUS

Use the Rename command or the DDRENAME dialog box to rename the layers assigned by the Solprof command to a new name reflecting the entities extracted out of the Top View. Change "PH-DC1" to "PH-TOP" and change "PV-DC1" to "PV-TOP." These results are illustrated at the right.

Command: **Rename**
Block/Dimstyle/LAyer/LType/Style/Ucs/View/VPort: **La**
Old layer name: **PH-DC1**
New layer name: **PH-TOP**

Command: **Rename**
Block/Dimstyle/LAyer/LType/Style/Ucs/View/VPort: **La**
Old layer name: **PV-DC1**
New layer name: **PV-TOP**

Layer name	State	Color	Linetype
0	On	7 (white)	CONTINUOUS
AME_FRZ	Frozen	7 (white)	CONTINUOUS
BORDER	On	1 (red)	CONTINUOUS
CENTER	On	2 (yellow)	CENTER2
MODEL	On	7 (white)	CONTINUOUS
PH-TOP	On	7 (white)	HIDDEN
PV-TOP	On	7 (white)	CONTINUOUS
VIEWPORTS	On	3 (green)	CONTINUOUS

Step #8

Use the same procedure for extracting views by repeating the Solprof command on the Front, Right Side, and Isometric views using the following procedure: While remaining in Model Space, make the Front View the new current viewport by picking a point anywhere in the viewport. Next use the UCS command to rotate the X axis of the user coordinate system by a 90 degree angle to prepare the Front view before using the Solprof command:

Command: *(While in Model Space, mark a point in the Front View making it the current viewport)*

Command: **UCS**
Origin/ZAxis/3point/Entity/View/X/Y/Z/Prev/Restore/Save/ Del/?/<World>: **X**
Rotation angle about X axis <0.0>: **90**

Command: **Solprof**
Select objects: *(Select the model in the Front View)*
Select objects: *(Strike Enter to continue with this command)*
Display hidden profile lines on separate layer? <Y>: **Yes**
(The system begins the extraction process)
Project profile lines onto a plane?<Y>: **Yes**
Delete tangential edges? <Y>: **Yes**

Before continuing with the next view, examine the new layers created. Use the Rename command or the DDRENAME dialog box to rename the new layers to the new names of "PH-FRONT" and "PV-FRONT" similar to the procedure used for renaming the layers created when extracting entities from the Top view.

While remaining in Model Space, make the Right Side View the new current viewport by picking a point anywhere in the viewport. Next use the UCS command to rotate the Y axis of the user coordinate system by a 90 degree angle to prepare the right side view before using the Solprof command:

Command: *(While in Model Space, mark a point in the Right Side View making it the current viewport)*

Command: **UCS**
Origin/ZAxis/3point/Entity/View/X/Y/Z/Prev/Restore/Save/ Del/?/<World>: **Y**
Rotation angle about Y axis <0.0>: **90**

Command: **Solprof**
Select objects: *(Select the model in the Right Side View)*
Select objects: *(Strike Enter to continue with this command)*
Display hidden profile lines on separate layer? <Y>: **Yes**
(The system begins the extraction process)
Project profile lines onto a plane?<Y>: **Yes**
Delete tangential edges? <Y>: **Yes**

Before continuing with the next view, examine the new layers created. Use the Rename command or the DDRENAME dialog box to rename the new layers to the new names of "PH-SIDE" and "PV-SIDE."

While remaining in Model Space, make the Isometric View the new current viewport by picking a point anywhere in the viewport. Next use the UCS command and the View option to position the user coordinate system parallel to the display screen and prepare the isometric view before using the Solprof command:

Command: *(While in Model Space, mark a point in the Isometric View making it the current viewport)*

Command: **UCS**
Origin/ZAxis/3point/Entity/View/X/Y/Z/Prev/Restore/Save/ Del/?/<World>: **View**

Command: **Solprof**
Select objects: *(Select the model in the Isometric View)*
Select objects: *(Strike Enter to continue with this command)*
Display hidden profile lines on separate layer? <Y>: **Yes**
(The system begins the extraction process)
Project profile lines onto a plane?<Y>: **Yes**
Delete tangential edges? <Y>: **Yes**

Before continuing with the next view, examine the new layers created. Use the Rename command or the DDRENAME dialog box to rename the new layers to the new names of "PH-ISO" and "PV-ISO."

Step #9

The results of performing the Solprof command from the previous are illustrated below. To view all entities on all "PH" and "PV" layers, it will become necessary to turn off the main layer the model of the Collar resides on, namely the layer named "MODEL." Also, during the extraction process performed by the Solprof command, a few results are displayed in the illustrations at the right of the Top and Iso Views. Notice that the Top View not only consists of visible lines but also hidden lines. It just so happens none of the hidden lines is necessary in the Top View. Regarding the Isometric view, the individual has the option of either leaving or turning off the hidden lines in this view. An individual must interpret where and when hidden lines are or are not to be displayed. For the purposes of this exercise, turn off the following layers: MODEL, PH-TOP, and PH-ISO.

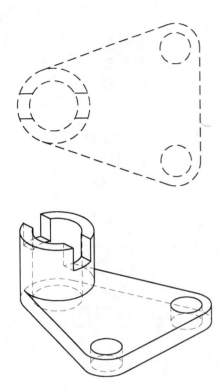

Command: **Layer**
?/Make/Set/New/ON/OFF/Color/Ltype/Freeze/Thaw/LOck/
Unlock: **Off**
Layer name(s) to turn Off: **MODEL, PH-TOP, PH-ISO**
?/Make/Set/New/ON/OFF/Color/Ltype/Freeze/Thaw/LOck/
Unlock: *(Strike Enter to turn off the Viewports layer and exit this command)*

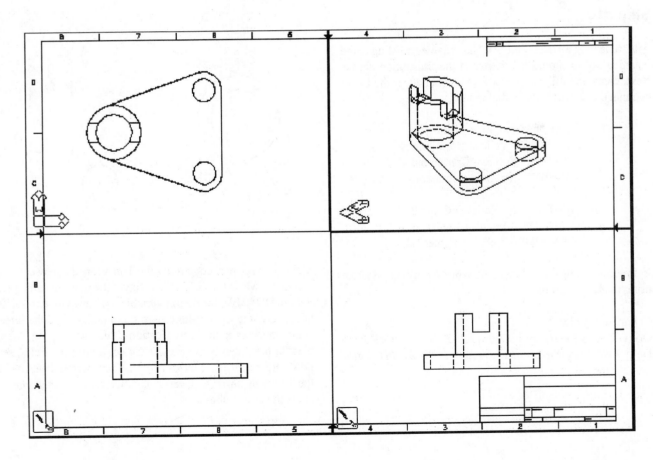

Layer name	State	Color	Linetype
0	On	7 (white)	CONTINUOUS
AME_FRZ	Frozen	7 (white)	CONTINUOUS
BORDER	On	1 (red)	CONTINUOUS
CENTER	On	2 (yellow)	CENTER2
MODEL	On	7 (white)	CONTINUOUS
PH-FRONT	On	7 (white)	HIDDEN
PH-ISO	On	7 (white)	HIDDEN
PH-SIDE	On	7 (white)	HIDDEN
PH-TOP	On	7 (white)	HIDDEN
PV-FRONT	On	7 (white)	CONTINUOUS
PV-ISO	On	7 (white)	CONTINUOUS
PV-SIDE	On	7 (white)	CONTINUOUS
PV-TOP	On	7 (white)	CONTINUOUS
VIEWPORTS	On	3 (green)	CONTINUOUS

Step #10

After renaming all "PH" and "PV" layer names, a listing of the renamed layers should be similar to the illustration above. Next, create the following new layers using either the Layer command or DDLMODES dialog box:

Layer Name - Color - Linetype
DIM-FRONT - Yellow - Continuous
DIM-TOP - Yellow - Continuous
DIM-SIDE - Yellow - Continuous

CEN-FRONT - Yellow - Center2
CEN-TOP - Yellow - Center2
CEN-SIDE - Yellow - Center2

Before continuing on, switch to the World Coordinate System using the UCS command.

Command: **UCS**
Origin/ZAxis/3point/Entity/View/X/Y/Z/Prev/Restore/Save/Del/?/<World>: *(Strike Enter to return to the World Coordinate System)*

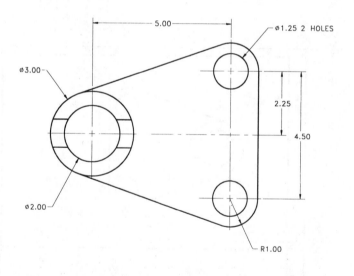

Make the viewport containing the Top View the new current viewport. Make the "CEN-TOP" layer the new current layer. Set the DIMCEN dimension variable to a new value of -0.12 units. Use the Center dimension command to add the three center markers at the circle locations illustrated above. After placing the 3 center markers at the center of the circles, set DIMCEN to 0. Make "DIM-TOP" the new current layer. Add the following horizontal, vertical, radius, and diameter dimensions illustrated above.

Step #11

After adding the center markers and dimensions in the Top View, notice that the dimensions and center markers appear in all other viewports as in the illustration below. This is typical of how viewports operate in Model Space; that is, all images are present in all viewports. What we really desire is to have the dimensions and center markers added to the Top View to remain in the Top View and only the Top View. If the "DIM-TOP" and "CEN-TOP" layers are turned off or frozen, all entities assigned to these layers are not only frozen in the Front, Right Side, and Iso viewports, but the Top viewport as well. A command is needed that controls the display of layers per viewport. This is accomplished by using the Vplayer command, which stands for View Port Layer command. This command allows a layer to be frozen in one viewport yet visible in another viewport. The Vplayer command operates in Paper Space. Follow the prompts below to freeze the "DIM-TOP" and "CEN-TOP" layers in all viewports except the viewport containing the Top View:

Command: **PS** *(To enter Paper Space)*

Command: **Vplayer**
?/Freeze/Thaw/Reset/Newfrz/Vpvisdflt: **Freeze**
Layer(s) to Freeze: **DIM-TOP, CEN-TOP**
All/Select/<Current>: *(Select the viewports containing the Front, Right Side, and Iso views)*
?/Freeze/Thaw/Reset/Newfrz/Vpvisdflt: *(Strike Enter to exit this command and freeze the layers entered)*

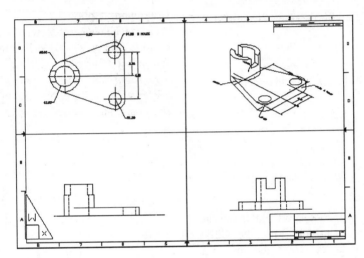

Step #12

The results of using the Vplayer command are illustrated below. Notice that after freezing "DIM-TOP" and "CEN-TOP" in all viewports except the viewport holding Top View entities displays dimensions and center markers only in the Top View.

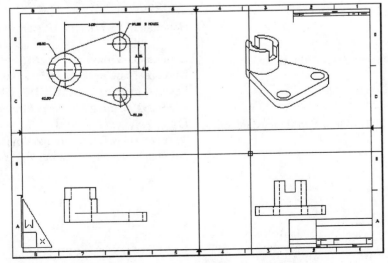

Step #13

Switch to Model Space. Make the viewport containing the Front View the new current viewport. Make the "CEN-FRONT" layer the new current layer. Next, use the UCS command to rotate the user coordinate system 90 degrees in relation to the X axis. Use the Line command to add the two center lines at the locations illustrated at the right. Make "DIM-FRONT" the new current layer. Add the 2 vertical dimensions illustrated at the right. Switch to Paper Space. Use the Vplayer command to freeze the layers "CEN-FRONT" and "DIM-FRONT" in all viewports except the viewport containing the Front View.

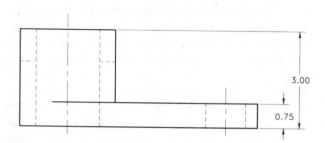

Command: **MS** *(To enter Model Space)*

Command: **UCS**
Origin/ZAxis/3point/Entity/View/X/Y/Z/Prev/Restore/Save/
Del/?/<World>: **X**
Rotation angle about X axis <0.0>: **90**

Add center lines on layer CEN-FRONT. Add dimensions on layer DIM-FRONT.

Command: **PS** *(To enter Paper Space)*

Command: **Vplayer**
?/Freeze/Thaw/Reset/Newfrz/Vpvisdflt: **Freeze**
Layer(s) to Freeze: **DIM-FRONT, CEN-FRONT**
All/Select/<Current>: *(Select the viewports containing the Top, Right Side, and Iso views)*
?/Freeze/Thaw/Reset/Newfrz/Vpvisdflt: *(Strike Enter to exit this command and freeze the layers entered)*

Step #14

Switch to Model Space. Make the viewport containing the Right Side View the new current viewport. Make the "CEN-SIDE" layer the new current layer. Next, use the UCS command to rotate the user coordinate system 90 degrees in relation to the Y axis. Use the Line command to add the three center lines at the locations illustrated at the right. Make "DIM-SIDE" the new current layer. Add the vertical and horizontal dimensions illustrated at the right. Switch to Paper Space. Use the Vplayer command to freeze the layers "CEN-SIDE" and "DIM-SIDE" in all viewports except the viewport containing the Right Side View.

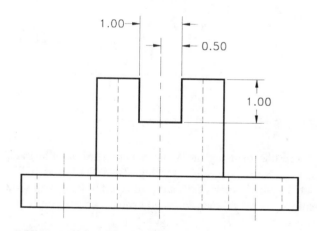

Command: **MS** *(To enter Model Space)*

Command: **UCS**
Origin/ZAxis/3point/Entity/View/X/Y/Z/Prev/Restore/Save/
Del/?/<World>: **Y**
Rotation angle about Y axis <0.0>: **90**

Add centerlines on layer CEN-SIDE. Add dimensions on layer DIM-SIDE.

Command: **PS** *(To enter Paper Space)*

Command: **Vplayer**
?/Freeze/Thaw/Reset/Newfrz/Vpvisdflt: **Freeze**
Layer(s) to Freeze: **DIM-SIDE, CEN-SIDE**
All/Select/<Current>: *(Select the viewports containing the Front, Top, and Iso views)*
?/Freeze/Thaw/Reset/Newfrz/Vpvisdflt: *(Strike Enter to exit this command and freeze the layers entered)*

Step #15

The complete series of views after using the Vplayer command is illustrated below.

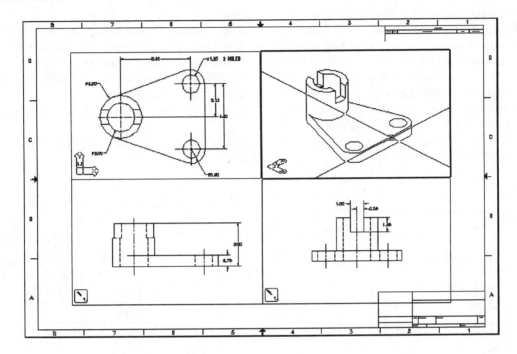

Step #16

The views have all been properly scaled, centerlines and dimensions have been added, the Vplayer command has controlled the visibility of layers in certain viewports, and a title block has been inserted through the use of the Mvsetup.Lsp routine. Unfortunately, the views still do not line up with each other which is very important to orthographic views. Enter Paper Space and draw the two justification lines illustrated at the right. Draw the lines from the intersections at "A" and "C." Draw these lines perpendicular to the respective viewports at "B" and "D." Finally use the Move command to slide one viewport over until the two justification lines touch at their exact endpoints at "B" and "D." **Be sure to accomplish this while in Paper Space.**

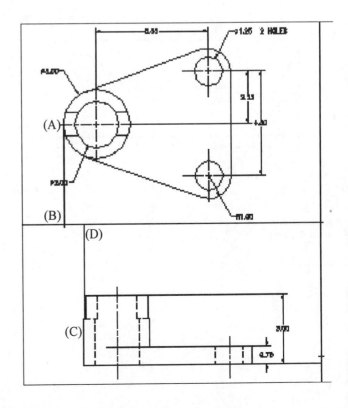

Command: **PS** (To enter Paper Space)
(Construct the two line segments illustrated at the right)

Command: **Move**
Select objects: (Select the edge of the Top Viewport in addition to line segment "AB")
Select objects: (Strike Enter to continue with this command)
Base point or displacement: **End**
of (Select the endpoint of the line at "B")
Second point of displacement: **End**
of (Select the endpoint of the line at "D")

Repeat this procedure for lining up the Right Side View with the Front View.

Step #16A

As an alternate means of moving the viewports in place, use the Move command in combination with the Osnap intersection and perpendicular options to slide the Top View until it is in line with the Front View. This method is accomplished without the aid of justification lines used in the previous step. The completed results are displayed in the illustration below to the right. Do not be concerned that the viewports no longer line up with each other; the important result is the views now line up exactly with each other. When finished lining up the Top View with the Front View, use the same procedure for lining up the Right Side View with the Top View. **Be sure to accomplish this step while in Paper Space.**

Command: **PS** *(To enter Paper Space)*

Command: **Move**
Select objects: *(Select the edge of the Top viewports)*
Select objects: *(Strike Enter to continue with this command)*
Base point or displacement: **Int**
of *(Select the intersetion in the Top View at "A")*
Second point of displacement: **Per**
to *(Select the vertical line in the Front View at "B")*

Repeat this procedure for lining up the Right Side View with the Front View.

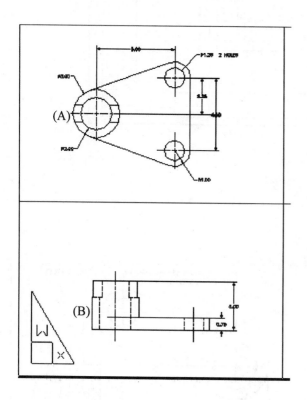

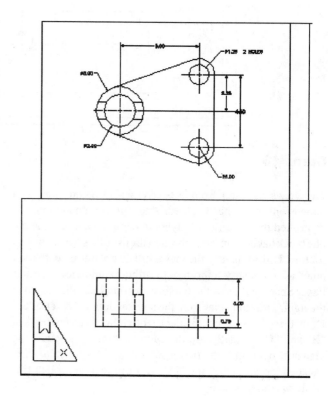

Step #17

As mentioned in the previous step when sliding the viewports around to achieve the proper orthographic view alignment, do not be concerned that the viewports appear appear out of alignment. It is more important that the views line up with each other. Once the views have been positioned using the justification line method in Step #16, erase these line segments used for justification. **This step must be accomplished while in Paper Space.**

Command: **Erase**
Select objects: *(Pick line segments "A," "B," "C," and "D")*
Select objects: *(Strike Enter to execute this command)*

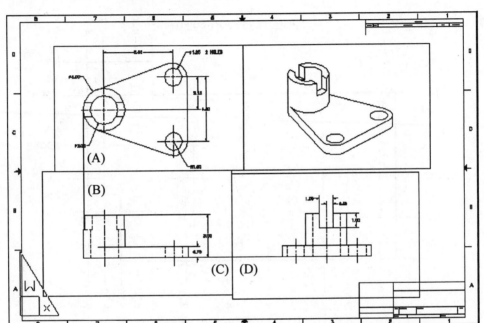

Step #18

Once all views have been moved into position and aligned with each other, use the Layer command or DDLMODES dialog box to turn off the Viewports layer, which will yield a drawing similar to the illustration below. Plot this drawing at a scale of 1=1 on a "D" size sheet of paper measuring 34 units by 22 units.

Command: **Layer**
?/Make/Set/New/ON/OFF/Color/Ltype/Freeze/Thaw/LOck/Unlock: **Off**
Layer name(s) to turn Off: **Viewports**
?/Make/Set/New/ON/OFF/Color/Ltype/Freeze/Thaw/LOck/Unlock: *(Strike Enter to turn off the Viewports layer and exit this command)*

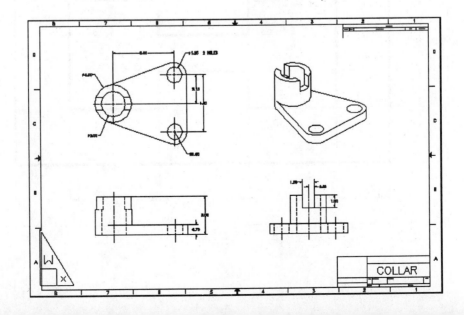

Problems for Unit 12

Directions for Problems 12-1 through 12-19:
1. *Construct a solid model of each object.*
2. *Using Tutorial #34 as a guide, create an engineering drawing of each model consisting of front, top, side, and isometric views.*
3. *Properly dimension the engineering drawing.*

Problem 12-1

Plate Thickness of 0.50

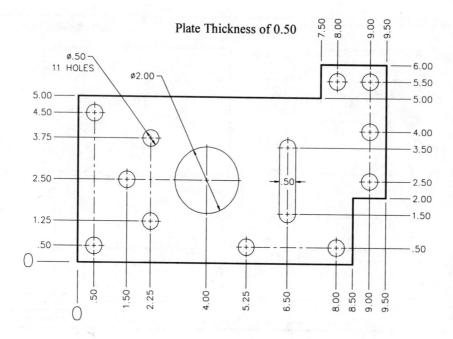

Problem 12-2

Plate Thickness of 0.75

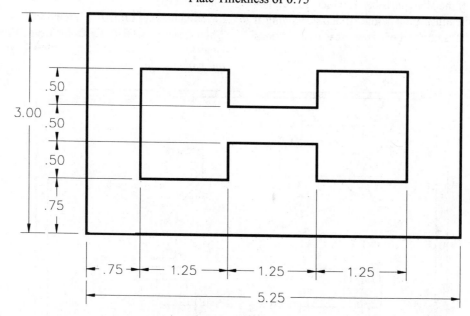

Problem 12-3

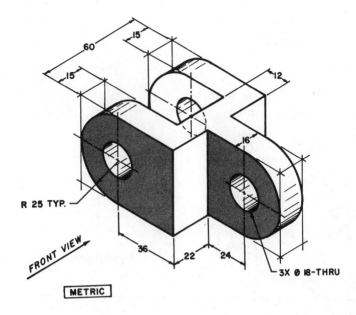

R 25 TYP.

FRONT VIEW

METRIC

3X Ø 18-THRU

60
15
15
12
16
36
22
24

Problem 12-4

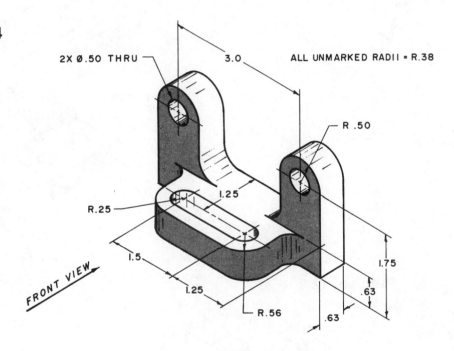

2X Ø.50 THRU

3.0

ALL UNMARKED RADII = R.38

R .50

R.25

1.25

R.56

1.5

1.25

1.75

.63

.63

FRONT VIEW

Problem 12-5

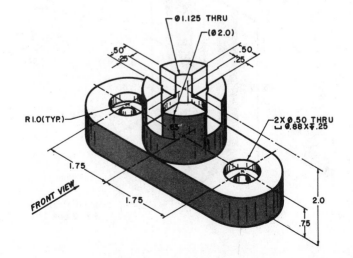

Ø1.125 THRU

(Ø2.0)

.50
.25

.50
.25

R 1.0(TYP.)

2X Ø.50 THRU
⌴ Ø.88 X ▽ .25

1.75

1.75

2.0

.75

FRONT VIEW

Problem 12-6

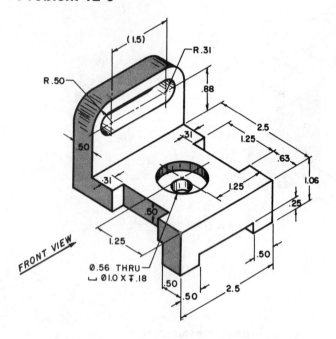

FRONT VIEW

(1.5)
R .31
R .50
.88
.50
.31
2.5
1.25
.63
1.25
1.06
.31
.50
.25
1.25
Ø.56 THRU
⌴ Ø1.0 X ⲧ .18
.50
.50
.50
2.5

Problem 12-7

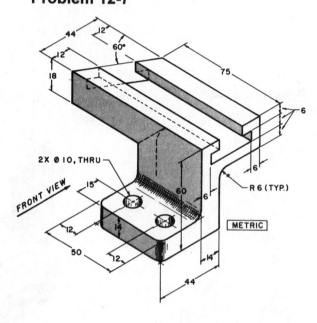

FRONT VIEW

44
12
60°
12
18
75
2X Ø 10, THRU
60
6
6
R 6 (TYP.)
15
6
METRIC
12
14
50
12
14
44

Problem 12-8

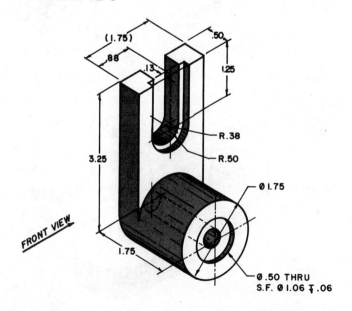

FRONT VIEW

(1.75)
.50
.88
1.25
.13
R .38
R .50
3.25
Ø 1.75
1.75
Ø .50 THRU
S.F. Ø 1.06 ⲧ .06

Problem 12-9

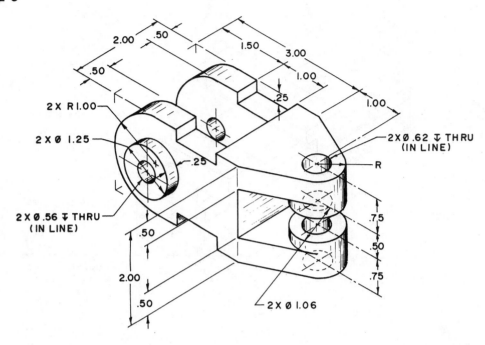

Problem 12-10

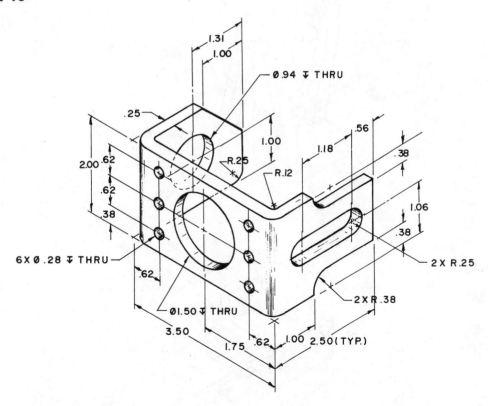

Problem 12-11

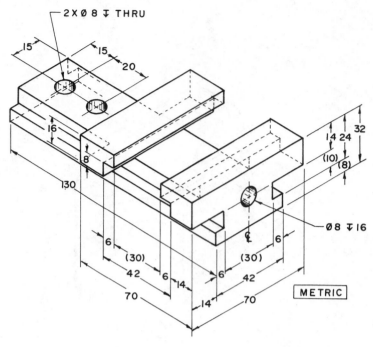

Problem 12-12

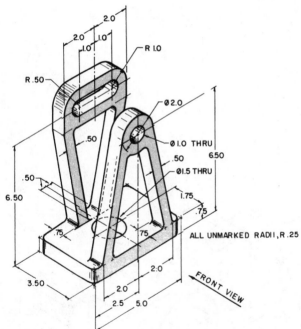

Problem 12-13

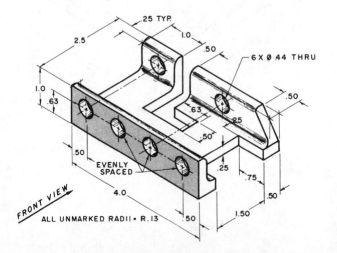

Problem 12-14

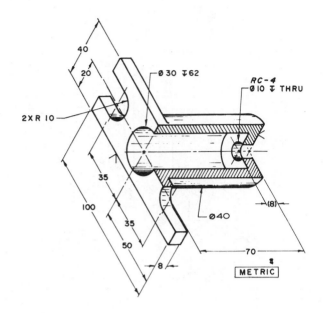

Problem 12-15

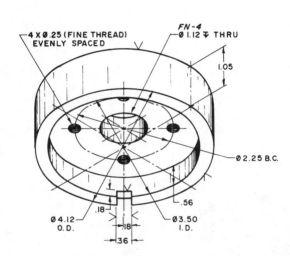

Problem 12-16

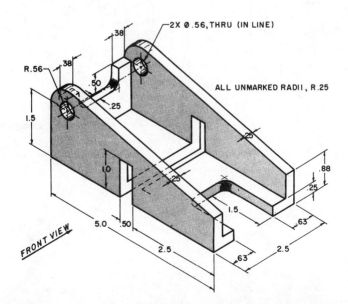

Problem 12-17

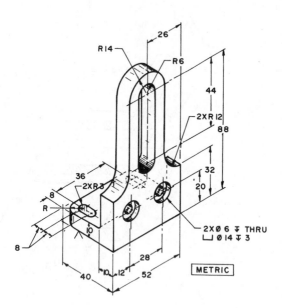

R14
R6
26
44
2X R 12
88
32
20
36
8
2XR3
R
10
8
2XØ6 ⟱ THRU
⨆ Ø14 ⟱ 3
28
40
10 12
52
METRIC

Problem 12-18

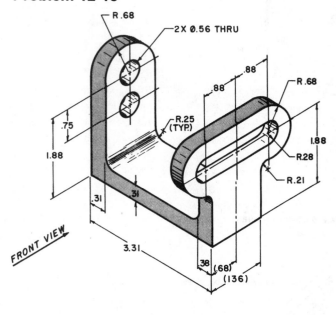

R .68
2X Ø.56 THRU
.88
.88
R .68
.88
R.25
(TYP.)
1.88
.75
R28
1.88
R.21
.31
31
FRONT VIEW
3.31
38
(68)
(136)

Problem 12-19

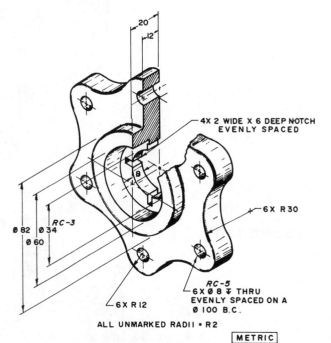

20
12
4X 2 WIDE X 6 DEEP NOTCH
EVENLY SPACED
8
6X R30
RC-3
Ø 82
Ø 34
Ø 60
RC-5
6X Ø 8 ⟱ THRU
EVENLY SPACED ON A
Ø 100 B.C.
6X R 12
ALL UNMARKED RADII ▪ R2
METRIC

Directions for Questions 12-1 through 12-18:
Answer the following questions related to each model.

1. After constructing a solid model of the object illustrated in Problem 12-1, compute the mass and volume of the solid using the DDSOLMASSP command dialog box. Perform all calculations using the default value of three divisions of accuracy and the material **Mild Steel**. Place all answers in the boxes provided below:

The total mass of the model is closest to:

The volume of the model is closest to:

2. After constructing a solid model of the object illustrated in Problem 12-2, compute the mass and volume of the solid using the DDSOLMASSP command dialog box. Perform all calculations using the default value of three divisions of accuracy and the material **Brass**. Place all answers in the boxes provided below:

The total mass of the model is closest to:

The volume of the model is closest to:

3. After constructing a solid model of the object illustrated in Problem 12-3, compute the mass and volume of the solid using the DDSOLMASSP command dialog box. Perform all calculations using the default value of three divisions of accuracy and the material **Stainless Steel**. Place all answers in the boxes provided below:

The total mass of the model is closest to:

The volume of the model is closest to:

4. After constructing a solid model of the object illustrated in Problem 12-4, compute the mass and volume of the solid using the DDSOLMASSP command dialog box. Perform all calculations using the default value of three divisions of accuracy and the material **Bronze**. Place all answers in the boxes provided below:

The total mass of the model is closest to:

The volume of the model is closest to:

5. After constructing a solid model of the object illustrated in Problem 12-5, compute the mass and volume of the solid using the DDSOLMASSP command dialog box. Perform all calculations using the default value of three divisions of accuracy and the material **Mild Steel**. Place all answers in the boxes provided below:

The total mass of the model is closest to:

The volume of the model is closest to:

6. After constructing a solid model of the object illustrated in Problem 12-6, compute the mass and volume of the solid using the DDSOLMASSP command dialog box. Perform all calculations using the default value of three divisions of accuracy and the material **Mild Steel**. Place all answers in the boxes provided below:

The total mass of the model is closest to:

The volume of the model is closest to:

7. After constructing a solid model of the object illustrated in Problem 12-7, compute the mass and volume of the solid using the DDSOLMASSP command dialog box. Perform all calculations using the default value of three divisions of accuracy and the material **Mild Steel**. Place all answers in the boxes provided below:

The total mass of the model is closest to:

The volume of the model is closest to:

8. After constructing a solid model of the object illustrated in Problem 12-8, compute the mass and volume of the solid using the DDSOLMASSP command dialog box. Perform all calculations using the default value of three divisions of accuracy and the material **Mild Steel**. Place all answers in the boxes provided below:

The total mass of the model is closest to:

The volume of the model is closest to:

9. *After constructing a solid model of the object illustrated in Problem 12-9, compute the mass and volume of the solid using the DDSOLMASSP command dialog box. Perform all calculations using the default value of three divisions of accuracy and the material **Stainless Steel**. Place all answers in the boxes provided below:*

The total mass of the model is closest to:

The volume of the model is closest to:

10. *After constructing a solid model of the object illustrated in Problem 12-10, compute the mass and volume of the solid using the DDSOLMASSP command dialog box. Perform all calculations using the default value of three divisions of accuracy and the material **Copper**. Place all answers in the boxes provided below:*

The total mass of the model is closest to:

The volume of the model is closest to:

11. *After constructing a solid model of the object illustrated in Problem 12-11, compute the mass and volume of the solid using the DDSOLMASSP command dialog box. Perform all calculations using the default value of three divisions of accuracy and the material **Mild Steel**. Place all answers in the boxes provided below:*

The total mass of the model is closest to:

The volume of the model is closest to:

12. *After constructing a solid model of the object illustrated in Problem 12-12, compute the mass and volume of the solid using the DDSOLMASSP command dialog box. Perform all calculations using the default value of three divisions of accuracy and the material **Mild Steel**. Place all answers in the boxes provided below:*

The total mass of the model is closest to:

The volume of the model is closest to:

13. *After constructing a solid model of the object illustrated in Problem 12-13, compute the mass and volume of the solid using the DDSOLMASSP command dialog box. Perform all calculations using the default value of three divisions of accuracy and the material **Aluminum**. Place all answers in the boxes provided below:*

The total mass of the model is closest to:

The volume of the model is closest to:

14. *After constructing a solid model of the object illustrated in Problem 12-14, compute the mass and volume of the solid using the DDSOLMASSP command dialog box. Perform all calculations using the default value of three divisions of accuracy and the material **Mild Steel**. Place all answers in the boxes provided below:*

The total mass of the model is closest to:

The volume of the model is closest to:

15. *After constructing a solid model of the object illustrated in Problem 12-15, compute the mass and volume of the solid using the DDSOLMASSP command dialog box. Perform all calculations using the default value of three divisions of accuracy and the material **Aluminum**. Place all answers in the boxes provided below:*

The total mass of the model is closest to:

The volume of the model is closest to:

16. *After constructing a solid model of the object illustrated in Problem 12-16, compute the mass and volume of the solid using the DDSOLMASSP command dialog box. Perform all calculations using the default value of three divisions of accuracy and the material **Stainless Steel**. Place all answers in the boxes provided below:*

The total mass of the model is closest to:

The volume of the model is closest to:

17. After constructing a solid model of the object illustrated in Problem 12-17, compute the mass and volume of the solid using the DDSOLMASSP command dialog box. Perform all calculations using the default value of three divisions of accuracy and the material **Brass**. Place all answers in the boxes provided below:

The total mass of the model is closest to:

The volume of the model is closest to:

18. After constructing a solid model of the object illustrated in Problem 12-18, compute the mass and volume of the solid using the DDSOLMASSP command dialog box. Perform all calculations using the default value of three divisions of accuracy and the material **Bronze**. Place all answers in the boxes provided below:

The total mass of the model is closest to:

The volume of the model is closest to:

Directions for Questions 12-19 through 12-28:

Construct a solid model of each object by referring to the objects located in Unit 6. When completed, answer the following questions related to each model.

19. Construct a solid model of Lev-1 illustrated in Problem 6-2 by following all directions at the top of page 405. Apply a thickness of 0.875 to this object. Set the type of material to **Mild Steel**. Compute the mass, volume, and centroid of the model using the DDSOLMASSP dialog box. Round off all calculations to three divisions of accuracy. Place all answers in the boxes provided below:

The total mass of the model is closest to:

The volume of the model is closest to:

The centroid of the model is closest to:

21. Construct a solid model of Hanger illustrated in Problem 6-4 by following all directions at the top of page 407. Apply a thickness of 4.00 to this object. Set the type of material to **Mild Steel**. Compute the mass, volume, and centroid of the model using the DDSOLMASSP dialog box. Round off all calculations to three divisions of accuracy. Place all answers in the boxes provided below:

The total mass of the model is closest to:

The volume of the model is closest to:

The centroid of the model is closest to:

20. Construct a solid model of Plate1 illustrated in Problem 6-3 by following all directions at the top of page 406. Apply a thickness of 0.500 to this object. Set the type of material to **Mild Steel**. Compute the mass, volume, and centroid of the model using the DDSOLMASSP dialog box. Round off all calculations to three divisions of accuracy. Place all answers in the boxes provided below:

The total mass of the model is closest to:

The volume of the model is closest to:

The centroid of the model is closest to:

22. Construct a solid model of Gasket1 illustrated in Problem 6-5 by following all directions at the top of page 408. Apply a thickness of 0.125 to this object. Set the type of material to **Mild Steel**. Compute the mass, volume, and centroid of the model using the DDSOLMASSP dialog box. Round off all calculations to three divisions of accuracy. Place all answers in the boxes provided below:

The total mass of the model is closest to:

The volume of the model is closest to:

The centroid of the model is closest to:

23. *Construct a solid model of Lever2 illustrated in Problem 6-7 by following all directions at the top of page 410. Apply a thickness of 0.375 to this object. Set the type of material to **Mild Steel**. Compute the mass, volume, and centroid of the model using the DDSOLMASSP dialog box. Round off all calculations to three divisions of accuracy. Place all answers in the boxes provided below:*

The total mass of the model is closest to:

The volume of the model is closest to:

The centroid of the model is closest to:

24. *Construct a solid model of the Wedge illustrated in Problem 6-9 by following all directions at the top of page 412. Apply a thickness of 9.00 units to this object. Set the type of material to **Aluminum**. Compute the mass, volume, and centroid of the model using the DDSOLMASSP dialog box. Round off all calculations to three divisions of accuracy. Place all answers in the boxes provided below:*

The total mass of the model is closest to:

The volume of the model is closest to:

The centroid of the model is closest to:

25. *Construct a solid model of Housing1 illustrated in Problem 6-13 by following all directions at the top of page 416. Apply a thickness of 0.525 to this object. Set the type of material to **Brass**. Compute the mass, volume, and centroid of the model using the DDSOLMASSP dialog box. Round off all calculations to three divisions of accuracy. Place all answers in the boxes provided below:*

The total mass of the model is closest to:

The volume of the model is closest to:

The centroid of the model is closest to:

26. *Construct a solid model of the Ratchet illustrated in Problem 6-24 by following all directions at the top of page 427. Apply a thickness of 0.875 to this object. Set the type of material to **Stainless Steel**. Compute the mass, volume, and centroid of the model using the DDSOLMASSP dialog box. Round off all calculations to three divisions of accuracy. Place all answers in the boxes provided below:*

The total mass of the model is closest to:

The volume of the model is closest to:

The centroid of the model is closest to:

27. *Construct a solid model of the Geneva illustrated in Problem 6-26 by following all directions at the top of page 429. Apply a thickness of 0.432 to this object. Set the type of material to **Bronze**. Compute the mass, volume, and centroid of the model using the DDSOLMASSP dialog box. Round off all calculations to three divisions of accuracy. Place all answers in the boxes provided below:*

The total mass of the model is closest to:

The volume of the model is closest to:

The centroid of the model is closest to:

28. *Construct a solid model of Rotor2 illustrated in Problem 6-27 by following all directions at the top of page 430. Apply a thickness of 0.825 to this object. Set the type of material **Mild Steel**. Compute the mass, volume, and centroid of the model using the DDSOLMASSP dialog box. Round off all calculations to three divisions of accuracy. Place all answers in the boxes provided below:*

The total mass of the model is closest to:

The volume of the model is closest to:

The centroid of the model is closest to:

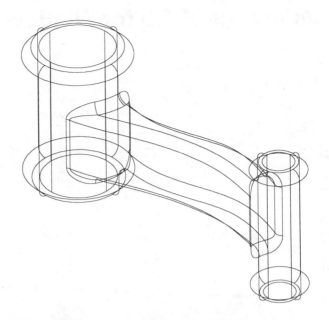

Release 13 Features

Contents

The Standard AutoCAD for Windows Screen

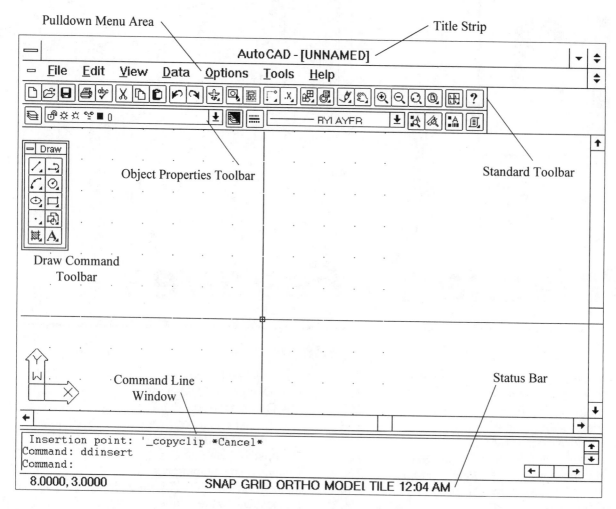

Pulldown Menu Area

Title Strip

AutoCAD - [UNNAMED]

File Edit View Data Options Tools Help

Object Properties Toolbar

Standard Toolbar

Draw Command
Toolbar

Command Line
Window

Status Bar

```
Insertion point: '_copyclip *Cancel*
Command: ddinsert
Command:
8.0000, 3.0000        SNAP GRID ORTHO MODEL TILE 12:04 AM
```

A typical AutoCAD Release 13 Windows screen is illustrated above. The screen layout has been completely redesigned for ease of use and operation. This includes an expanded Standard Toolbar containing such commands as New and Open in addition to displaying numerous flyouts supporting Zoom, Object Snaps, and XYZ Point Filters. The new Object Properties Toolbar is designed for manipulating layers, linetypes, and color. Individual Floating Command Toolbars contain most basic commands. The new Floating Command Line Window displays the command history and can be moved to new locations on the display screen. The redesigned Status Bar is now located at the bottom of the screen. The entries such as Snap, Grid, and Ortho turn bold when activated or turned On either through the use of the function keys or by double clicking on the desired mode. An entry is greyed out when the mode is turned Off.

Flyouts and Icon Tool Tips

Tool palettes or floating toolboxes may be moved around the screen or can be docked to numerous locations on the screen. These palettes contain numerous buttons which in turn activate AutoCAD commands. Buttons containing an arrow in the lower right corner activate other buttons containing more commands or options. This operation is called a flyout menu. Still, it will take time to associate a particular symbol with known or new commands. To assist with this tool tips have been added. When the cursor remains on top of a button, the command name flashes below the button giving instant recognition of what command is being used.

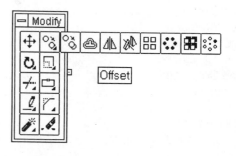

Resizeable Command Window

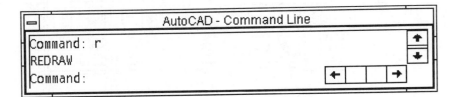

The Command Prompt area in the Windows version of AutoCAD Release 13 is a dockable window designed to display the current command in addition to its options. The command prompt window is located at the bottom of the display screen, just above the status bar. This is its default docked location. The command prompt window may be moved or resized to other areas of the screen by picking parts of its border which undocks the window. To dock it back, move the command prompt window until it is over the top of the dock regions which are located at the top and bottom of the screen. The number of lines of text displayed in the command prompt window is three by default. This is referred to as the command history. This number is usually sufficient for most applications. Having a number too large affects the overall size of the display screen; keep the command history to a minimum of two or three lines. The command history number can be changed in the Preferences dialog box under the Misc option. For command history that needs more display area, such as using the List command on an entity, the F2 function key toggles to the text screen to view more command history information.

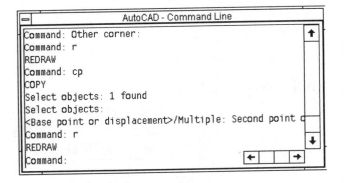

Redesigned Status Bar

| 8.0000, 3.0000 | SNAP GRID ORTHO MODEL TILE 12:04AM |

The familiar Status Bar conveniently located at the top of all past AutoCAD display screens has been relocated to the bottom the the Windows Release 13 display screen. The components of the status bar include:

    Coordinate Display
    Snap On/Off Indicator
    Grid On/Off Indicator
    Ortho On/Off Indicator
    Model Space/Paper Space Indicator
    Tile Mode Indicator
    Current Time

These indicator areas will update themselves to any changes made through the function keys. Originally, Snap, Grid, and Ortho modes are greyed out or Off; as the F9, F7, and F8 function keys are depressed, the indicator areas turn bold signifying the specific mode is currently On.

A faster, more productive way of toggling the indicator areas is to move the cursor to the desired mode and double pick the indicator. The mode will immediately toggle either On or Off. This feature works even inside of the coordinate display area. By default it is On, enabling the coordinates to display the current X,Y position of the cursor. Double picking anywhere inside of the coordinate box turns coordinate mode Off and is signified by the coordinates being greyed out.

Double picking "Model" mode switches the user to the Paper Space environment. Notice also that "Tile" is greyed out when performing this operation. Double picking inside of "Paper" has no effect on the toggle since no new viewports exist. Double picking on the greyed-out "Tile" button switches the user from Paper Space back to Model Space.

Standard Toolbar

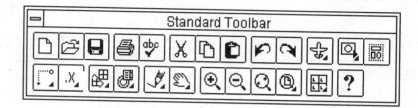

The Standard Toolbar contains various commands associated with the following areas: Files, Modify, Object Selection, Object Snap, and Display. A complete listing of the commands as they relate to each button is illustrated at the right.

New	Redo	Pan Point
Open	Aerial View	Zoom In
Save	Object Selection	Zoom Out
Print	Object Group	Zoom Window
Spell	Object Snap	Zoom All
Cut	XYZ Filters	Tiled Model Space
Copy	UCS Options	Help
Paste	Named Views	
Undo	Redraw View	

Object Properties Toolbar

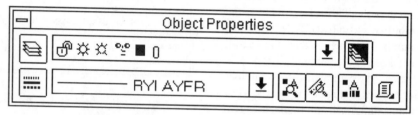

The Object Properties Toolbar is a collection of layer, linetype, and color-related commands controlling the properties of objects or entities. Most toolbar buttons activate dialog boxes such as Layer, Color, and Linetype. Other areas such as Layer Control allow the user to pull down properties associated with layers enabling the user to manipulate layer options. The complete listing of all commands as they relate to each button is illustrated at the right.

Layers
Layer Control
Color Control
Linetype
Linetype Control
Object Creation
Multiline Style
Properties
List

Object Properties Layer Pulldown

The newly designed graphical interface of the Windows version of AutoCAD Release 13 is never more evident than in the display of the layer pulldown dialog box activated from the Object Properties toolbar. For instance, layer "TEXT" is currently unlocked with the appearance of the open padlock symbol while layer "010" is locked since the padlock symbol is closed. Layer "TEXT" is also On or visible with the appearance of the small face with eyes open while layer "BORDER" is turned Off since the eyes on the face symbol are closed. Layer "103" has a sun symbol directly right of the padlock symbol; this signifies the layer is thawed. Layer "012" has the snowflake symbol present indicating that this layer is frozen.

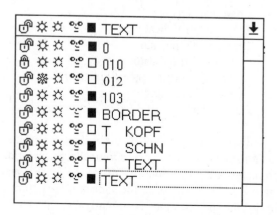

Supported Floating Tool Boxes

To aid in the selection of commands in the Windows environments, numerous floating tool boxes have been designed around the categories illustrated at the right. Notice the tool boxes in the pulldown menu are divided into four different areas. The first area containing the Draw, Modify, Dimensioning, Solids, Surfaces, External References, Attribute, Render, External Database, and Miscellaneous tool bars are primarily designed around entity construction and modification or editing. Select Objects, Object Snap, Point Filters, UCS, and View tool boxes follow that display separate tool boxes designed to choose selection set modes, object snap options, XYZ point filters, options of the user coordinate system, and methods of viewing a model. The third group contains the Object Properties and Standard Toolbars. By default these two tool boxes are automatically displayed on the screen because of their importance since they contain File commands such as New, Open, and Save in the Standard Toolbar and Layer, Linetype, and Color in the Object Properties box. Both tool boxes are docked to the top of the display screen but can be undocked and moved to other locations. Illustrated below are all supported tool boxes displayed on the screen at the same time. As you can see, there does not appear to be any room for any actual drawing. Activate only those tool boxes needed to perform a task. When finished, pick the control box icon (the dash) to deactivate the tool box. Selecting the Close All item in the fourth area of the pulldown menu turns off all tool boxes in addition to the Object Properties and Standard Toolbar.

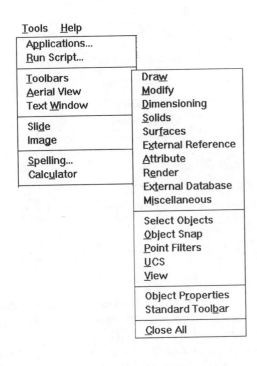

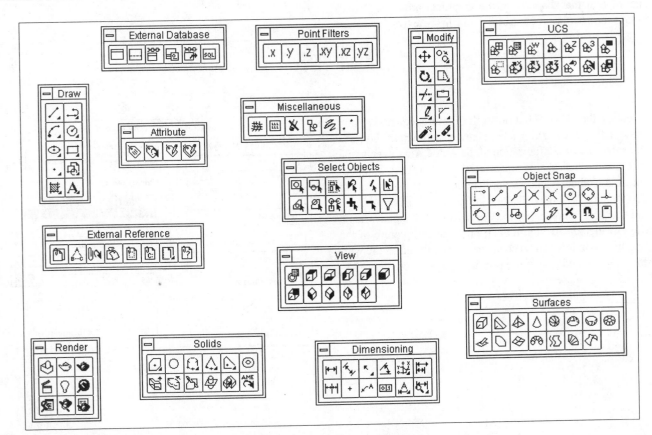

The Preferences Dialog Box

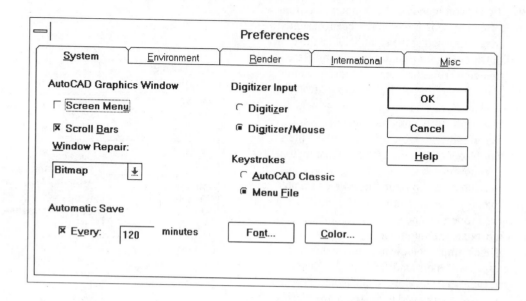

Use the Preferenced dialog box illustrated above to set various screen modes and environments to enhance the display while in the Windows version of AutoCAD. By default, the screen menu does not appear; place a check in the box next to "Screen Menu" to display it. Picking the "Font" button activates the dialog box illustrated at the right. Use it to control the type of text font used in areas such as the AutoCAD command line and screen menu font. The "Graphic" button controls the type and size of font that displays in the graphical area; the "Text" button controls the type and size of font in the text or flip screen area.

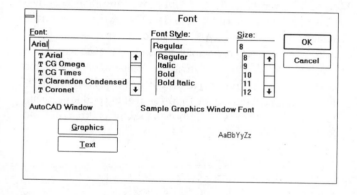

Selecting the "Color" button in the Preferences dialog box above activates the dialog box illustrated at the right. Use it to control the color configuration of the AutoCAD screen while in Windows. This dialog box affects only AutoCAD and does not affect other Windows applications. Picking the down arrow under "Window Element" displays all major graphics and text screen components. Color can be changed when a window element is highlighted; or the actual picture of the Graphics Window can be changed which highlights the window element and allows the user to change color.

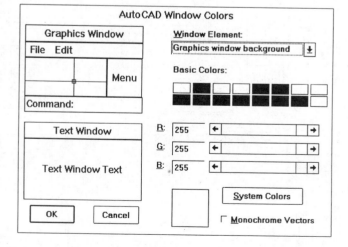

Online Help - Quick Tour

Online help takes on a different form with the presence of the "Quick Tour" icon located in the main AutoCAD Release 13 group enabling the user to take a guided tour of various AutoCAD topics such as AutoCAD Basics, Two-Dimensional Drawing, Three-Dimensional Drawing and Viewing, Plotting and Printing, and Rendering. This help function comes complete with important information on the product in addition to allowing the user to view cursor movement, menu picks, and minor construction techniques. A few screens are illustrated below to give an idea of the type and format of the "Quick Tour" function.

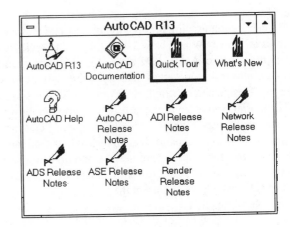

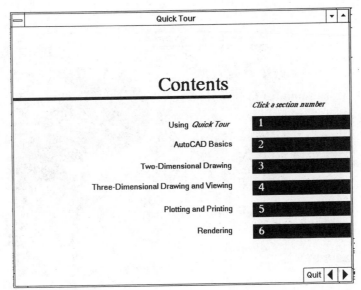

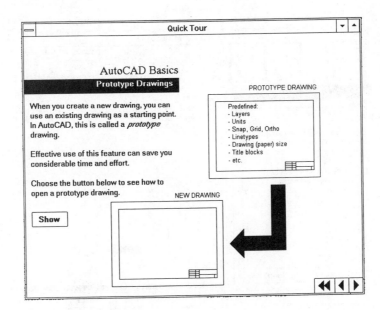

Online Help - What's New

Online help continues with the creation of a special presentation tutorial called "What's New." Use this tool to review all important features of AutoCAD Release 13. The menu system is broken into four main areas; namely Using What's New, Usability, Design and Drafting, and Commands. The format of this tutorial ranges from information and graphic illustration screens to simple animations demonstrating concepts. The last section on Commands lists all Release 13 commands that have been enhanced or are new in addition to a brief description on the function of the command. A few screens are illustrated below to give an idea of the type and format of the "What's New" function.

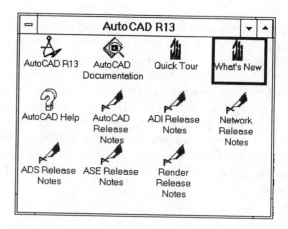

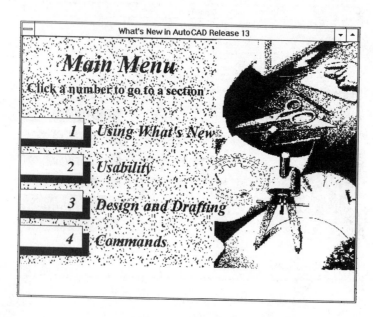

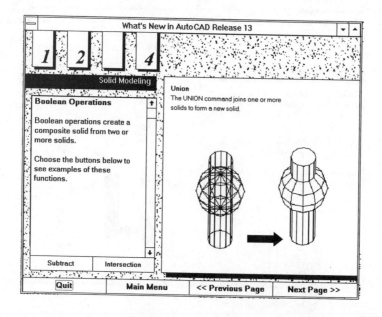

Drawing and Wblock Preview

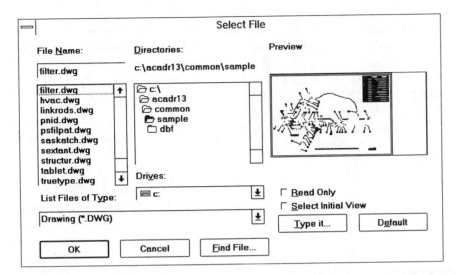

When opening up an existing drawing for viewing or editing, the redesigned "Select File" dialog box appears similar to the illustration above. In addition to performing directory searches on various disk drives for the desired drawing file to display, a drawing Preview box allows the user to first see which file he is about to load. This preview mode is activated by picking once on the desired file. A double click launches the drawing file inside of the AutoCAD editor. If the selected file is not the correct drawing, move to another file and single click on it to display it in the Preview box. This same preview mode is available when inserting files saved as Wblocks.

The Standard AutoCAD for DOS Screen

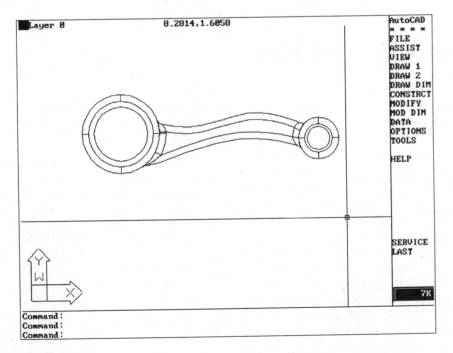

The AutoCAD Release 13 DOS screen illustrated above is almost identical to previously designed versions of AutoCAD with a few exceptions. The screen menu has changed considerably to match the pulldown menu area. Also, a rectangular box in the lower right corner of the screen menu monitors the current size of the drawing file loaded into the system.

Revised Function Key Definitions

The illustration at the right depicts the newly defined function keys supported in AutoCAD Release 13 for Windows. Some key definitions have not changed such as F1 (Help), F2 (Flip Screen), F6 (Coordinate Display), F7 (Grid Mode), F8 (Ortho Mode), and F9 (Snap Mode). Other key definitions have been added or changed such as F4 (Tablet Mode), F5 (Isoplane Mode), and F10 which toggles the status bar located at the bottom of the Windows display screen On or Off.

R13 Windows Function Key Definitions	
F1	Help
F2	Flip Screen
F4	Tablet On/Off
F5	Isoplane Toggle (Top, Right, Left)
F6	Coordinate Display On/Off
F7	Grid On/Off
F8	Ortho On/Off
F9	Snap On/Off
F10	Status Bar On/Off

Function keys remain the same in the DOS version of AutoCAD Release 13 with the exception of F5, which toggles the Isoplane Mode from Top to Right to Left when constructing isometric drawings.

R13 DOS Function Key Definitions	
F1	Flip Screen
F5	Isoplane Toggle (Top, Right, Left)
F6	Coordinate Display On/Off
F7	Grid On/Off
F8	Ortho On/Off
F9	Snap On/Off
F10	Tablet On/Off

Accelerator Key Definitions

As an added tool used to access commands in a more productive manner, accelerator keys have been added to AutoCAD Release 13. These keys take the form of depressing the Ctrl (Control) key while pressing a key mapped to a particular command. Actually, accelerator keys were supported in previous versions of AutoCAD to toggle Snap (Ctrl B), Coordinates (Ctrl D), Isoplane (Ctrl E), and Grid (Ctrl G). Orthogonal was another supported command which was accessed through Ctrl O in previous versions of AutoCAD. However, it has been remapped to the letter L since Ctrl O accelerates the use of the Open command. Other commands and their accelerator key maps for the Windows version of AutoCAD include Group (Ctrl A), Copyclip (Ctrl C), New (Ctrl N), Open (Ctrl O), Print (Ctrl P), Qsave (Ctrl S), Pasteclip (Ctrl V), Cutclip (Ctrl X), and U (Ctrl Z). Striking the Escape key in Windows AutoCAD Release 13 (Esc) issues a cancel command. For the DOS version of AutoCAD Release 13, Ctrl C still operates as a cancel.

Group On/Off	[CONTROL+"A"]
Snap On/Off	[CONTROL+"B"]
Copyclip	[CONTROL+"C"] (Windows)
Cancel	[CONTROL+"C"] (DOS)
Coords On/Off	[CONTROL+"D"]
Isoplane	[CONTROL+"E"]
Grid	[CONTROL+"G"]
Ortho	[CONTROL+"L"]
New	[CONTROL+"N"]
Open	[CONTROL+"O"]
Print	[CONTROL+"P"]
Qsave	[CONTROL+"S"]
Pasteclip	[CONTROL+"V"]
Cutclip	[CONTROL+"X"]
U	[CONTROL+"Z"]

New DOS Pulldown Menus

The File Menu

Use this pulldown menu in creating new drawings, opening up existing drawings, quick saving drawings, saving drawings under a different name, and plotting out drawings. When opening up a drawing created in Release 13, the drawing is previewed allowing the individual to view if this is the correct drawing to open. The Bind command is located here to convert blocks, layers, linetypes, text styles, and dimension styles from an external reference to a valid item in the current drawing file. Export holds a very important utility for Release 12 users. Once a drawing is created in Release 13, the drawing must be exported or saved in Release 12 format in order for it to be viewed in the previous AutoCAD version.

```
File
  New...
  Open...
  Save
  Save As...

  Print...

  External Reference  ▷
  Bind                ▷

  Import              ▷
  Export              ▷

  Management          ▷

  Exit
```

The Assist Menu

In addition to such common items such as Snap, Grid, Ortho, Undo, and Redo, this menu contains two new osnap modes under the Object Snap cascading menu. This menu also allows users to better access XYZ point filters in their own pulldown menu. Most selection set options are contained in the Select Objects cascading menu. The new object grouping command allows for selection sets to be created under a user defined name. This selection set may be called up by its name at any time in the drawing editor. In addition to Area, ID, and Distance commands in the Inquiry cascading menu, the Mass Property command is also located here. Use it to perform an upper-level calculation on either a region or solid model to obtain such information as centroid, area, mass, volume, and principle moments.

```
Assist
  Undo
  Redo

  Object Snap         ▷
  Point Filters       ▷

  Snap
  Grid
  Ortho

  Select Objects      ▷
  Selection Filters...
  Group Objects...
✓ Group Selection

  Inquiry             ▷

  Cancel
```

The View Menu

All options of the Zoom command have been added to the Zoom cascading menu. "Named Views... " is conveniently located in this dialog box enabling the user to create distinct views of a complex drawing under a unique name to be called up later; this is used for display purposes. The 3D Viewpoint cascading menu allows for better viewing of 3D models. In addition to viewing the front, top, right, left, bottom, and back views of a 3D model, special isometric viewing directions have been designed to view the isometric from the SW (South West), SE, NW, and NE directions. The UCS command and all options are located in this dialog box. The "Named UCS..." entry activates the DDUCS dialog box for making an existing UCS the current user coordinate system.

```
View
  Redraw View
  Redraw All

  Zoom                ▷
  Pan                 ▷

  Named Views...
  3D Viewpoint Presets ▷
  3D Viewpoint        ▷
  3D Dynamic View

✓ Tiled Model Space
  Floating Model Space
  Paper Space

  Tiled Viewports     ▷
  Floating Viewports  ▷

  Preset UCS...
  Named UCS...
  Set UCS             ▷
```

The Draw Menu

New entries in the Draw menu include the ability to draw construction lines using the new Xline command. This command draws infinite horizontal, vertical, and angular lines from a user defined point. The infinite lines are drawn in both directions from the point. The Ray command is similar to Xline except an infinite line is drawn in only one direction from a user defined point. In this command the user also specifies the direction. Multiple parallel lines may be placed in a drawing using the new Mline command. The new Spline command creates quadratic or cubic splines. The Circle cascading menu contains the Donut command in addition to having the ability to construct a circle tangent to three entities using the Tan-Tan-Tan option. The Ellipse command draws a true ellipse entity; the center of the ellipse and quadrants can now be snapped to. The Rectang command is now located under Polygon. The Insert cascading menu controls the insertion of block, wblocks, multiple inserted blocks, and shapes. The Surfaces cascading menu contains most surfacing commands carried over from previous versions of AutoCAD. Solids modeling consists of core commands such as Box and Cylinder used to create 3D solid models of objects. The Bhatch command found in the Hatch cascading menu has been redesigned to limit the number of subdialog boxes and also supports associative crosshatching. Major text enhancements have been made to allow users to add multiple lines of text using the Mtext command found in the Text cascading menu. Major dimensioning enhancements will be discussed later in this unit.

```
Draw
  Line
  Construction Line
  Ray
  Sketch

  Polyline
  3D Polyline
  Multiline

  Spline
  Arc              ▷
  Circle           ▷
  Ellipse          ▷

  Polygon          ▷

  Point            ▷
  Insert           ▷

  Surfaces         ▷
  Solids           ▷

  Hatch            ▷

  Text             ▷
  Dimensioning     ▷
```

The Construct Menu

This menu area contains popular commands that have been enhanced or are brand new to the Release 13 command set. For instance, Fillet and Chamfer now allow the user to trim or not trim the excess lines when chamfering or filleting edges. The Region command is now a core feature allowing for the creation of 2D region models. The Boundary command used to be called Bpoly. It operates almost the same as Bpoly with the exception that a group of entities can either have a polyline or region traced over a set of entities connected to form a closed object. The boolean operations of Union, Subtract, and Intersection are used during the construction phase of a 2D region model or 3D solid model. Selecting "Attributes..." activates the DDATTDEF dialog box for the creation of attribute definitions.

```
Construct
  Copy
  Offset
  Mirror
  Array            ▷

  Chamfer
  Fillet

  Region
  Boundary

  Union
  Subtract
  Intersection

  Block
  Attribute...

  3D Array         ▷
  3D Mirror
  3D Rotate
```

The Modify Menu

Picking "Properties..." activates the DDMODIFY dialog box for dynamic modification of the properties of an entity. This dialog box has been redesigned and enhanced especially in the case of modification of dimensions. A new Lengthen command allows for the enlarging or decreasing of an arc or line segment while keeping the radius and angle of the line intact. The Trim and Extend commands have both been enhanced to allow for the trimming and extending of an entity or group of entities to an edge that has been extended in space. As multiline segments are drawn, a special editing feature allows for corner and intersection cleanup. Text placed with Mtext can be edited using the DDEDIT dialog box enabling individual words or letters to be changed in height, color, or even true type font inside of the Windows version of Release 13. A new hatch editor allows for a pattern to be changed instead of being erased and the area re-hatched. A welcomed enhancement of the Explode command allows for the conversion of block inserted at different X and Y scale factors to be broken into individual entities.

The Data Menu

A new feature of the "Layers..." allows for linetypes to be loaded without leaving this dialog box. New dialog boxes have been designed around the Color and Linetype commands. The Multiline Style dialog box allows for the grouping of any number of multiple parallel lines. This dialog box also controls the color and linetypes of these multiline segments. The new Dimension Styles dialog box has been enhanced by reducing the number of subdialog boxes required to set dimension variables. Also a new family dimension style has been added to further make dimension styles easier to work with. This concept will be explained later in this unit. The Purge command has been enhanced to allow for the purging of a drawing of unused items such as blocks and layers any time while inside of the drawing editor.

The Options Menu

The Display cascading menu contains controls on the solid fill, spline frame, attribute display, Point Style dialog box, whether text is to be filled or outlined, text frame only, text quality, global linetype scale, paper space linetype scaling, and linetype generation.

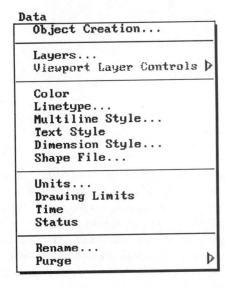

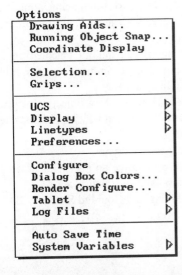

The Tools Menu

The Tools pulldown menu area houses typical utility applications for automating the drawing process. The "Applications..." area activates the Appload command used for loading AutoLisp routines and other ADS applications. A dialog box is activated when a user wants to run a script file. The External Commands area cascades down to expose the Edit File and Shell commands used to interrupt AutoCAD and enter the DOS prompt or DOS Editor. Other tools include the ability to hide, shade, or render a surfaced or solid model and the ability to create or import slide or images. The DOS spell checker is located in this pulldown menu area allowing an individual to select text and have a dialog box appear to identify any words that are misspelled.

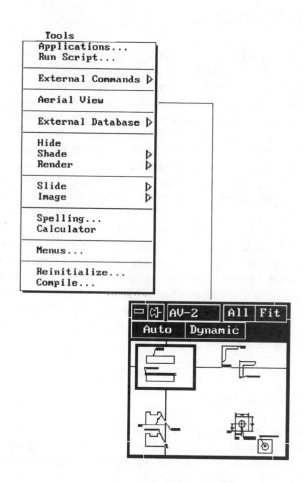

A new Aerial view dialog box allows for dynamic viewing and panning while inside of a drawing editor.

The Help Menu

The Help pulldown menu allows an individual to perform searches of commands. Picking the area, "What's New in Release 13...," activates the dialog box illustrated below. This allows the user to page through a series of topics dedicated to Release 13 and view a series of pages and graphical illustrations. The topics are organized in the following areas:

Usability - Enhancements to Purge, Explode, and Xrefs.

Text - Multiline text editing, spell checking, truetype font support.

Dimensioning - Geometric tolerancing, dimension style families, associative leaders, mainstreaming of dimension commands.

Hatching - Associative hatching, hatch editing.

Geometry- New Ellipse command, splines, multiple parallel lines, Xline, and Ray.

Construction and Editing - New object snap options, enhancements to Trim, Extend, Fillet, and Chamfer, new object grouping.

Linetypes - Custom linetypes

Solid Modeling - Native solids, solid primitives, boolean operation, and swept surfaces.

Rendering - Phong and Gourand Shading, 3D Studio import and export.

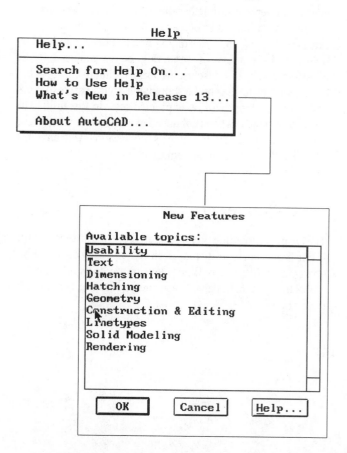

Construction and Editing Enhancements

Trim Command Enhancements

The Trim command has been enhanced to allow the user to trim to an extended cutting edge. In the past, the cutting edge had to hit the entity to be trimmed. Now in extended cutting edge mode, an imaginary cutting edge is formed; all entities along this cutting edge will be trimmed if selected individually or by the Fence mode. Study the illustrations at the right and the prompts below on these new enhancements of the Trim command.

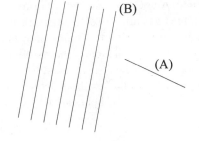

Command: **Trim**
Select cutting edges:(Projmode = UCS, Edgemode = No extend)
Select objects: *(Pick line segment "A")*
Select objects: *(Strike Enter to Continue)*
<Select object to trim>/Project/Edge/Undo: **Edge**
Extend/No extend <No extend>: **Extend**
<Select object to trim>/Project/Edge/Undo: *(Pick line "B")*

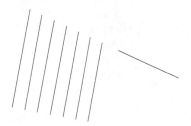

Continue with picking the remaining lines to be trimmed to the extended cutting edge.

Extend Command Enhancements

As with the Trim command, the Extend command has undergone significant enhancements in the definition of the boundary edge. In the past, the boundary edge had to be directly in line or in the sight of the entity being extended. Now, the boundary edge can be extended where an imaginary edge is projected enabling entities not in direct sight of the boundary edge to still be extended. Study the illustrations at the right and the prompts below on these new enhancements of the Trim command.

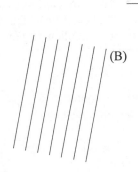

Command: **Extend**
Select boundary edges: (Projmode = UCS, Edgemode = No extend)
Select objects: *(Pick line "A")*
Select objects: *(Strike Enter to continue)*
<Select object to extend>/Project/Edge/Undo: **Edge**
Extend/No extend <No extend>: **Extend**
<Select object to extend>/Project/Edge/Undo: *(Pick line "B")*

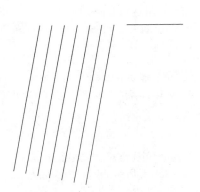

Continue with picking the remaining lines to be extended to the extended boundary edge.

Additional Extend and Trim Enhancements

Additional enhancements of Extend and Trim commands include other cutting or boundary edges not supported in previous versions of AutoCAD. As an example in the illustration at the right, the text entity is selected as either a cutting edge or boundary edge. Using the Trim command on the horizontal line at "A" trims the line out of the middle of the text. In the same fashion, picking the vertical line at "B" extends the line to the bottom of the of the text boundary. Other valid cutting edges for Trim and valid boundary edges for Extend include:

Arcs
Circles
Ellipses
Lines
Floating Viewports
Rays
Regions
Splines
Text
Xlines

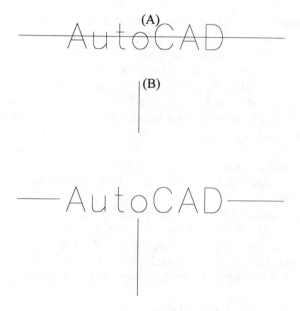

Fillet Command Enhancements

The Fillet command has been enhanced to include the option of whether or not to trim the excess corners after a fillet is placed. Illustrated at the right is a typical fillet operation where two lines at "A" and "B" are selected. However, instead of automatically trimming the ends of the lines, a new "Trim/No trim" option allows for the lines to remain. Follow the prompts below that illustrate this operation.

Command: **Fillet**
(TRIM mode) Current fillet radius = 0.0000
Polyline/Radius/Trim/<Select first object>: **Trim**
Trim/No trim <Trim>: **N**
Polyline/Radius/Trim/<Select first object>: **R**
Enter fillet radius <0.0000>: **1.00**

Command: **Fillet**
(NOTRIM mode) Current fillet radius = 1.00
Polyline/Radius/Trim/<Select first object>: *(Select line "A")*
Select second object: *(Select line "B")*

Filleting two parallel lines automatically constructs a semi-circular arc entity connecting both lines at their endpoints as illustrated at the right.

Command: **Fillet**
(NOTRIM mode) Current fillet radius = 0.0000
Polyline/Radius/Trim/<Select first object>: *(Select line "A")*
Select second object: *(Select line "B")*

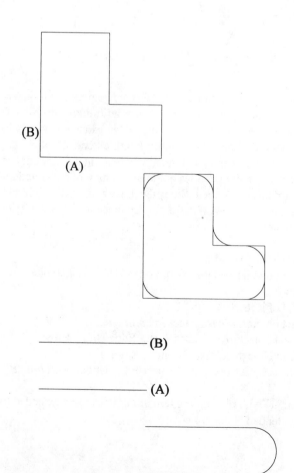

Chamfer Command Enhancements

As with the Fillet command, the Chamfer command supports the same "Trim/No trim" option enabling a chamfer to be placed with lines trimmed or not trimmed as in the illustration at the right.

Command: **Chamfer**
(TRIM mode) Current chamfer Dist1 = 0.00, Dist2 = 0.00
Polyline/Distance/Angle/Trim/Method/<Select first line>: **T**
Trim/No trim <Trim>: **N**
Polyline/Distance/Angle/Trim/Method/<Select first line>: **D**
Enter first chamfer distance <0.0000>: **1.00**
Enter second chamfer distance <1.0000>: **1.00**

Command: **Chamfer**
(NOTRIM mode) Current chamfer Dist1 = 1.00, Dist2 = 1.00
Polyline/Distance/Angle/Trim/Method/<Select first line>:
(Select line "A")
Select second line: *(Select line "B")*

Repeat the procedure for the remaining lines to chamfered without trimming.

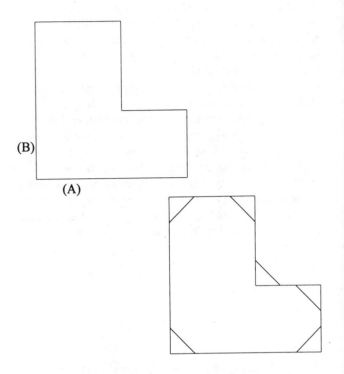

The Circle Tan-Tan-Tan Command

This command is an adaptation of the Circle-3P command allowing a 3 point circle to be created. Entering the 3P option and following each point with the Tangent object snap option produces a circle tangent to three entities similar to the illustration at the right. Follow the prompts below to perform this operation using the Windows version of Release 13:

Command: **Circle**
3P/2P/TTR/<Center point>: **3P**
First point: **Tan**
to *(Select the large circle)*
Second point: **Tan**
to *(Select the smaller circle)*
Third point: **Tan**
to *(Select the line segment)*

The DOS version of Release 13 displays a menu pick located in the Draw pulldown menu area enabling an individual to simply select the three entities since the Tangent mode is automatically activated.

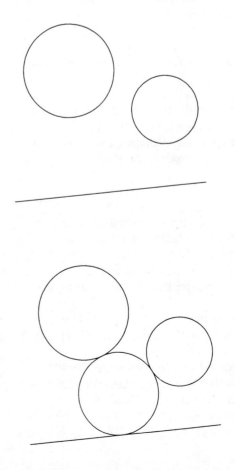

Object Selection Cycling

At times, the process of selecting objects can become quite
tedius. This can be compounded when attempting to select an
entity and another entity selects instead. A common occur-
rence is where entities lie directly on top of each other. As the
entity to delete is selected, the other entity selects instead. To
remedy this, depress the "Ctrl" key when prompted to "Select
objects:." This activates a new feature called object selection
cycling and enables you to scroll through all entities in the
vicinity. In the prompt area, notice the message that cycling
is On signifying your ability to pick entities until the desired
entity is highlighted. Striking the Enter key not only accepts
the highlighted entity but toggles object selection cycling Off.
In the example at the right and with cycle On, the first pick
selects the line segment; the second pick selects the circle.
Keep picking until the desired entity highlights.

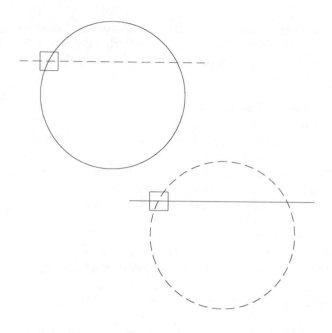

The Lengthen Command

Use the Lengthen command to change the length of a selected
entity without disturbing other entity qualities such as angles
of lines or radii of arcs.

Command: **Lengthen**
DElta/Percent/Total/DYnamic/<Select object>: *(Select line
"A")*
Current length: 12.3649
DElta/Percent/Total/DYnamic/<Select object>: **Total**
Angle/<Enter total length (1.0000)>: **20**
<Select object to change>/Undo: *(Select the line at "A")*
<Select object to change>/Undo: *(Strike Enter to exit)*

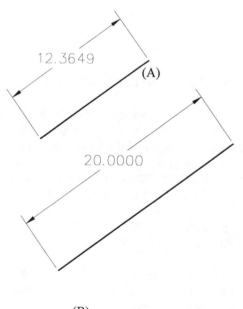

When using the Lengthen command on an arc segment, both
the length and included angle information are displayed
before making any changes. After supplying the new total
length of any entity, be sure to select the desired end to
lengthen when prompted for "Select object to change."

Command: **Lengthen**
DElta/Percent/Total/DYnamic/<Select object>: *(Select arc
"B")*
Current length: 1.4459, included angle: 54.9597
DElta/Percent/Total/DYnamic/<Select object>: **Total**
Angle/<Enter total length (1.0000)>: **5**
<Select object to change>/Undo: *(Pick the arc at "B")*
<Select object to change>/Undo: *(Strike Enter to exit)*

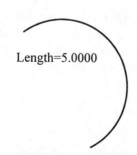

Additional Object Snap Modes

The Pop-up menu activated by holding down the Shift key and pressing the Enter button of a mouse or digitizing tablet is still operational as in Release 12. The new additions to the box illustrated at the right include the "From" and "Apparent Intersection" options. Another enhancement is in the area of the XYZ point filters. Before a second level or submenu had to be activated before they were displayed. Now they are included in the main body of the pulldown menu.

The next two sections outline the use of the new "From" and "Apparent Intersection" Object Snap modes.

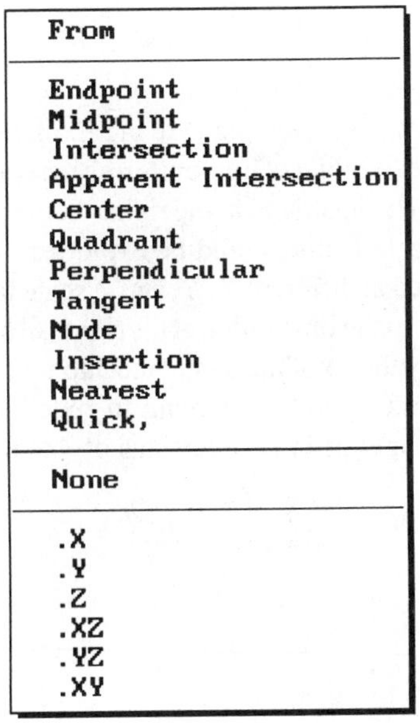

```
From

Endpoint
Midpoint
 Intersection
Apparent Intersection
Center
Quadrant
Perpendicular
Tangent
Node
 Insertion
Nearest
Quick,

None

.X
.Y
.Z
.XZ
.YZ
.XY
```

Osnap - Apparent Intersection

Use this new Object Snap mode to locate the apparent intersection of two entities. In the example illustrated at the right, selecting the line segments at "A" and "B" locates the center of the circle where the two line segments apparently intersect.

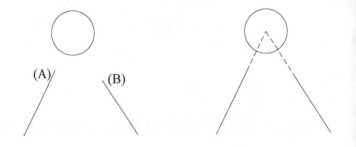

Command: **Circle**
3P/2P/TTR/<Center point>: **Int**
of *(Select the line at "A")*
and *(Select the line at "B")*
Diameter/<Radius> <1.0000>: *(Strike Enter to accept)*

Osnap - From

Yet another Object snap enhancement is illustrated at the right. To locate the center of a new circle 3 units directly to the right and up from the circle identified by "A," use the "From" option of Object Snap. When entering "From," a base point prompt appears enabling the user to identify a point of reference. Then, an offset prompt appears enabling the user to key-in the offset distance from the previous base point.

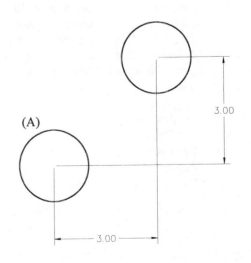

Command: **Circle**
3P/2P/TTR/<Center point>: **From**
Base point: **Cen**
of *(Select circle "A")*
<Offset>: **@3,3**
Diameter/<Radius>: **1.00**

Non-Uniformly Scaled Block Explosion

Enhancements have been made to the Explode command by allowing the exploding of non-uniformly scaled block entities as in the objects illustrated at the right. In the past, a block of only equal scale factors could be exploded. In Release 13, blocks inserted at different X, Y, and Z scale factors can now be exploded. In the illustration at the right, all blocks will now be affected by the Explode command and will be broken into individual entities without having to redefine the block to make the changes as in past versions of AutoCAD.

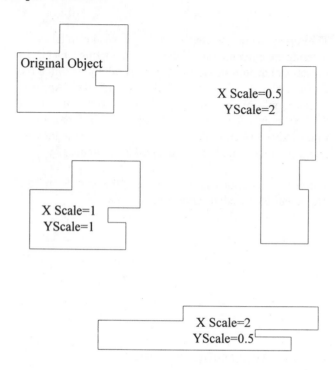

Purge Command Enhancements

The Data pulldown menu area of both the DOS and Windows versions of Release 13 contains a menu pick for using the Purge command. In previous versions of AutoCAD, an individual had to follow a few rules before using Purge; you could not purge after placing new entities in the database of the drawing; you could not purge after editing an entity or group of entities in the drawing file. Now in Release 13, the requirements for using the Purge command have been relaxed so that it can be used at any time throughout the drawing process while in the drawing editor. The following unused items may be deleted from the drawing database using Purge:

      Layers
      Linetypes
      Multiline Styles
      Text Styles
      Dimension Styles
      Shapes
      Blocks

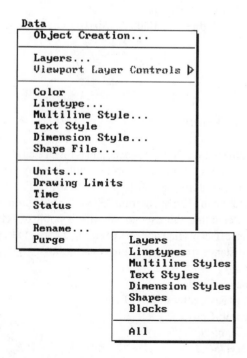

Multilines

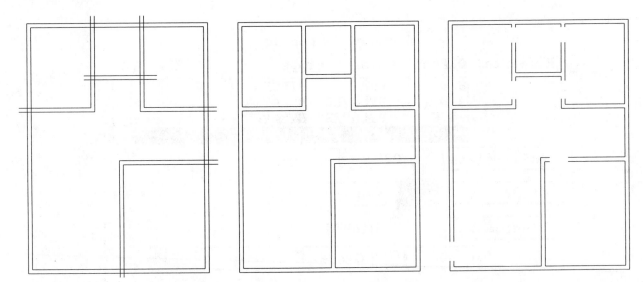

Multilines consist of a series of parallel line segments. They may be compared with past AutoLisp routines such as Dline. However multilines differ greatly from the Dline routine in that groups of as many of 16 parallel line segments may be grouped together to form a single multiline entity. Not only that, multilines may take on color, be spaced apart at different increments, and may take on different linetypes. In the example above, the floor plan layout was made using the default multiline style of two parallel line segments. The spacing was changed to an offset distance of 4 units. Once the multilines are laid out, a special editing command cleans up all corners and intersections displayed in the middle graphic above. This command, Mledit, is not only used for clean up operations. Since multilines cannot be partially deleted using

the Break command, the Mledit command allows multiline segments to be cut between two points identified by the user. Also the user has the option of breaking all or only one multiline segment depending on the desired results. Once a multiline segment has been cut, it can be welded back together again using the Mledit command. The next series of pages outline the creation of a multiline style using the Mlstyle command. Two additional dialog boxes are contained inside of Mlstyle: the Element Properties dialog box sets the spacing, color, and linetype of the individual multiline segments; the Multiline Properties dialog box provides other information on multilines including capping modes and the ability to fill the entire multiline in a selected color.

The Mlstyle Command

The Multiline Styles dialog box illustrated at the right is used to create a new multiline style or make an existing style current. The concept of creating multiline styles through this dialog box is very similar to that of creating dimension styles and making an existing dimension style current. The dialog box at the right lists the current multiline style which by default is named "Standard." Brand new multiline styles must be derived from Standard and then have their element and multiline properties saved to the new style. The new multiline style must then be loaded before making it current. All new multiline styles are saved to a file called ACAD.Mln located in the Support subdirectory of the AutoCAD Common directory.

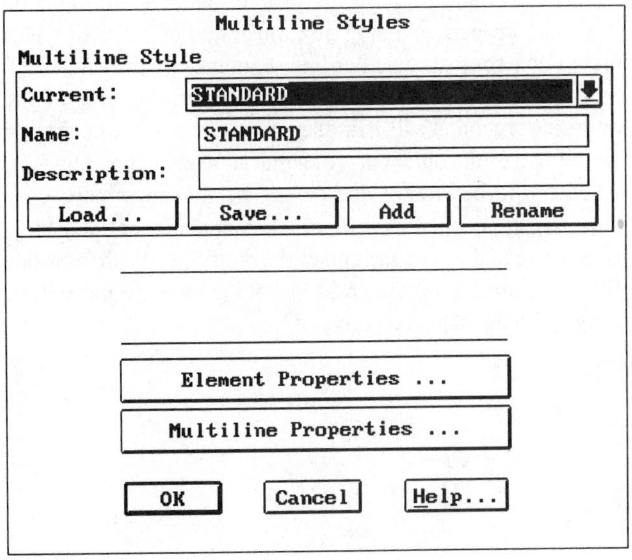

```
                   Multiline Styles
 Multiline Style
 Current:       [STANDARD                        ▼]
 Name:          [STANDARD                         ]
 Description:   [                                 ]
 [  Load...  ]  [  Save...  ]  [   Add   ]  [  Rename  ]
 _____

 _____

           [    Element Properties ...    ]

           [   Multiline Properties ...   ]

      [   OK   ]   [ Cancel ]   [ Help... ]
```

MIstyle-Element Properties

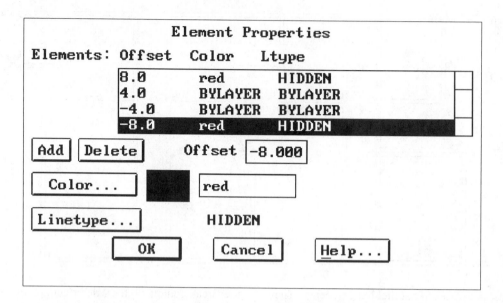

The Element Properties subdialog box of the main Multiline Styles dialog box is used for making offset, color, and line assignments to the various multiline segments. Picking the Add button adds copies existing properties to the Elements edit box area. Once identical properties are displayed, they are selected and edited with new offset distances. When editing a multilines color, the Color dialog box appears allowing the user to select the new color to be assigned to the selected multiline element. In the same manner, selecting the Linetype button activates the Linetype dialog box displaying all currently loaded linetypes. If a linetype is not present, a special Load button is available at the bottom of the dialog box. This in turn displays all linetypes found in the ACAD.Lin file. Picking the "OK" button in the Element Properties dialog box returns the user to the main Multiline Styles dialog box where the changes made can be saved to the new multiline style.

MIstyle - Multiline Properties

Additional control of multilines can be achieved through the Multiline Properties dialog box illustrated at the right. This dialog box controls whether line segments are drawn at each multiline joint, whether the beginning and/or end of the multiline is capped with a line segment, whether the beginning and/or end of the outer parts of a multiline are capped by arcs, whether the beginning and/or end of the inner parts of a multiline are capped by arcs, or whether a user defined angle is used to cap the start and end of the multiline. With multiline fill mode turned on, the entire multiline background will be filled in to the selected color.

The Mline Command

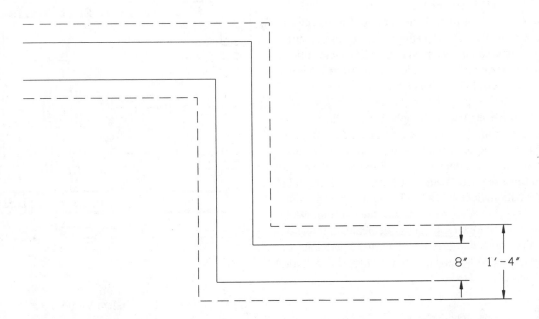

8"   1'–4"

Once the element and multiline properties have been set and saved to a multiline style, loading a new current style and picking the "OK" button of the Mutliline Styles dialog box draws the multiline to the current multiline style configuration. Illustrated below is the result of setting the offset distances on the previous page. The above configuration may be used for an architectural application where a block wall of 8 units thickness and a footing of 16 units thick is needed.

Multilines are drawn using the Mline command and the following sequence:

Command: **Mline**
Justification = Top, Scale = 1.00, Style = STANDARD
Justification/Scale/STyle/<From point>: *(Pick a point)*
<To point>: *(Pick a point)*
Undo/<To point>: *(Either pick another point or strike Enter to exit the Mline command)*

Follow the command sequence below along with the illustrations at the right to change the justification of multilines. Justification modes reference the top of the multiline, the bottom of the multiline, or the zero location of the multiline, which is the multiline's center.

Command: **Mline**
Justification = Top, Scale = 1.00, Style = STANDARD
Justification/Scale/STyle/<From point>: **Justification**
Top/Zero/Bottom <top>: **Zero**
Justification = Zero, Scale = 1.00, Style = STANDARD
Justification/Scale/STyle/<From point>: *(Pick a point)*
<To point>: *(Pick a point)*
Undo/<To point>: *(Either pick another point or strike Enter to exit the Mline command)*

Top Justification

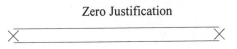

Zero Justification

Bottom Justification

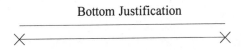

The Mledit Command

The Mledit command is a special editing feature for editing or cleaning up intersections of muliline entities. Various clean-up modes are available such as creating a closed cross, creating an open cross, or creating a merged cross. The same options are available for creating "Tee" sections in multilines. Corner joints may be created along with adding or deleting a vertex of a multiline. In the vertex examples below, deleting a vertex forces the multiline segment to straighten out. When adding a vertex in the example below, the results may not be as evident. After the vertex was added to the straight multiline segment, the Stretch command was used to create the shape below. Multiline segments cannot be broken. They must be cut using the option illustrated at the far right of the Multiline Edit Tools dialog box. To mend the cut, the multiline Weld option is used. For the examples below depicting the creation of open intersections and tees or the corner operation, the identifying letters "A" and "B" show where to pick the multiline when using Mledit.

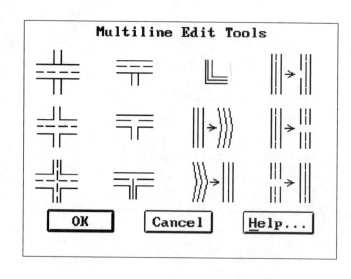

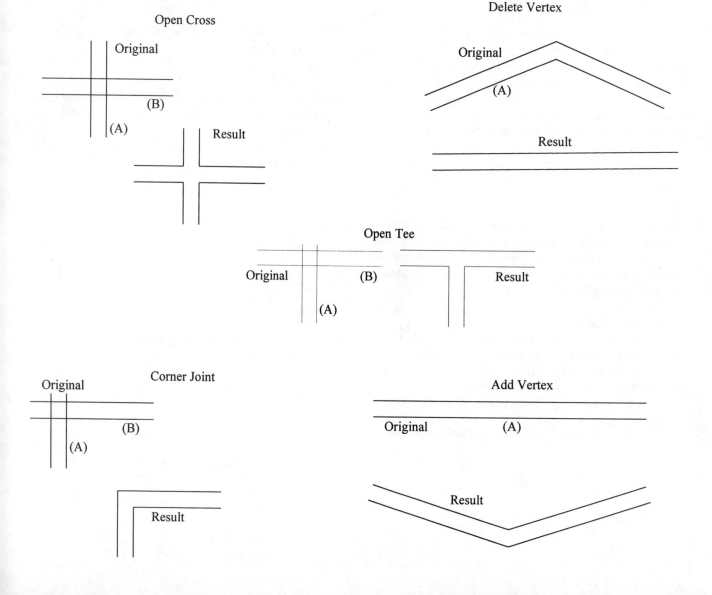

The Xline Command

Xlines are lines drawn from a user defined point. The user is not prompted for any length information as the Xline is drawing in an unlimited length beginning at the user defined point and going off in the opposite direction from the point. Xlines can be drawn horizontal, vertical, and angular. You can bisect an angle using an Xline or offset the Xline at a specific distance. In other words, the Xline command would be the same as creating construction lines as in the illustration at the right. The circular view represents the front view of a flange. To begin the creation of the side views, lines are usually projected from key features on the adjacent view. In the case of the front view at the right, the key features are the top of the plate in addition to the other circular features. In this case, the Xlines were drawn using the Horizontal mode from the Quadrant of all circles. The prompts below outline the Xline command sequence:

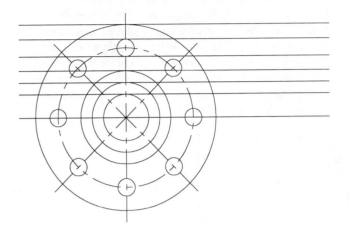

Command:**Xline**
Hor/Ver/Ang/Bisect/Offset/<From point>:
Through point: *(Pick a point on the display screen to place the first xline)*
Through point: *(Pick a point on the display screen to place the second xline)*

Since the Xlines continue to be drawn in both directions, care must be taken to manage these new line entities. Construction management techniques of Xlines could take the form of placing all Xlines on a specific layer to be turned Off or Frozen when not needed. When editing Xlines (especially with the Break command) care needs to be taken to remove all excess Xlines that will still remain on the drawing screen going off in an unspecified distance. Here the Erase command is used to remove any access Xlines.

Xlines to Erase   Break Points

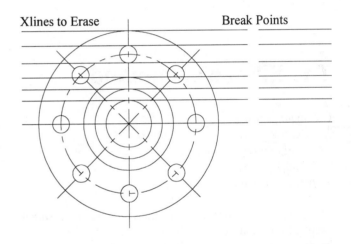

Illustrated at the right is an example of how Xlines are used to bisect or divide an angle in half. After entering the Bisect option, three endpoints are identified on the angle; namely the vertex of the angle, angle starting point and angle ending point. Striking the enter key at the second "Angle end point" prompt bisects the angle with an Xline which has an unspecified length.

Command: **Xline**
Hor/Ver/Ang/Bisect/Offset/<From point>: **Bisect**
Angle vertex point: *(Pick the endpoint at "A")*
Angle start point: *(Pick the endpoint at "B")*
Angle end point: *(Pick the endpoint at "C")*
Angle end point: *(Strike Enter to exit this command)*

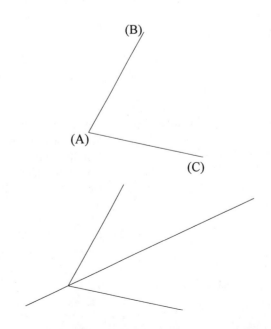

The Ray Command

A Ray is similar to the Xline except the Ray begins at a user defined point and extends to infinity only on one side. In the example at the right, the quadrants of the circles identify all points where the Ray entities begin and are drawn to infinity on the right. The same construction techniques that apply to Xlines should apply to Rays; that is, to organize Ray entities on specific layers. Care should also be exercised in the editing of Rays taking special care not to leave entities dangling in infinity as lines are broken and erased.

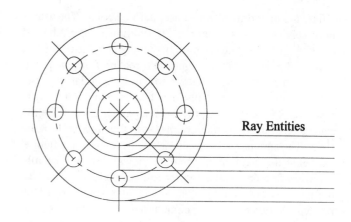

Ray Entities

The Ellipse Command

Command: **Ellipse**
Arc/Center/<Axis endpoint 1>: **Center**
Center of ellipse: *(Pick a point for the ellipse center)*
Axis endpoint: **@3.5<0**
<Other axis distance>/Rotation: **@2<90**

The ellipse entity now accepts the Osnap-Center mode which snaps to the center of the ellipse, and the Osnap-Quadrant mode, which snaps to the 0, 90, 180, and 270 degree positions along the ellipse.

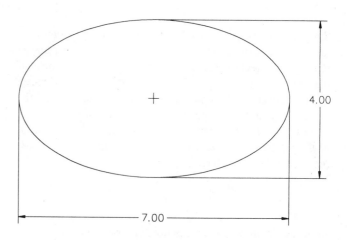

4.00

7.00

ELLIPSE			
	Layer: 0		
	Space: Model space		
	Handle = 479		
	Circumference: 17.6016		
	Center:	X = 27.3583,	Y = 30.4358,	Z = 0.0000
	Major Axis: X = 3.5000,	Y = 0.0000,	Z = 0.0000	
	Minor Axis: X = 0.0000,	Y = 2.0000,	Z = 0.0000	
	Radius Ratio: 0.5714			

The Spline Command

Use the Spline command to construct a smooth curve given a sequence of points. The user has the option of changing the accuracy of the curve given a tolerance range. The basic command sequence is given below, which constructs the spline segment at the right.

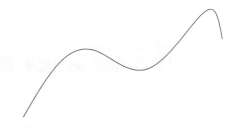

Command: **Spline**
Object.<Enter first point>: *(Pick a first point)*
Enter point: *(Pick another point)*
Close/Fit Tolerance/<Enter point>: *(Pick another point)*
Close/Fit Tolerance/<Enter point>: *(Pick another point)*
Close/Fit Tolerance/<Enter point>: *(Strike Enter to continue)*
Enter start tangent: *(Strike Enter to accept)*
Enter end tangent: *(Strike Enter to accept the end tangent position which exits the command and places the spline)*

Once the spline curve is constructed, the List command can be used to display all control and fit points that make up the curve. Other information provided includes the length of the spline.

SPLINE Layer: 0
       Space: Model space
       Handle = 60
       Length: 13.0251
       Order: 4
       Properties: Planar, Non-Rational, Non-Periodic
       Parametric Range: Start 0.0000
                       End 12.4959
Control Points: X = 21.5709 , Y = 14.4679 , Z = 0.0000
                X = 22.5570 , Y = 16.1296 ,Z = 0.0000
User Data: Fit Points
                X = 21.5709 , Y = 14.4679 , Z = 0.0000
                X = 24.8050 , Y = 17.4561 , Z = 0.0000

The spline may be closed to display a continuous segment illustrated at the right. Entering a different tangent point at the end of the command changes the shape of the curve connecting the beginning and end of the spline.

Command: **Spline**
Object.<Enter first point>: *(Pick a first point)*
Enter point: *(Pick another point)*
Close/Fit Tolerance/<Enter point>: *(Pick another point)*
Close/Fit Tolerance/<Enter point>: *(Pick another point)*
Close/Fit Tolerance/<Enter point>: *(Pick another point)*
Close/Fit Tolerance/<Enter point>: **Close**
Enter tangent: *(Strike Enter to exit the command and place the spline entity)*

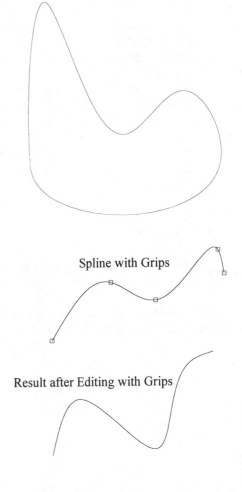

Spline with Grips

Grips provide an excellent means of editing a spline curve. With the spline selected, grips appear at all control points. As a grip is selected, you can dynamically see how much to pull the selected control point to achieve the desired results as in the examples illustrated at the right.

Result after Editing with Grips

Object Grouping

```
                        Object Grouping
     Group Name                        Selectable
     LINES                               Yes


     Group Identification
     Group Name:      LINES
     Description:
     [ Find Name < ]  [ Highlight < ]  [ ] Include Unnamed
     Create Group
     [   New <   ]      [X] Selectable    [ ] Unnamed
     Change Group
     [ Remove < ] [ Add < ] [ Rename ] [ Re-order... ]
     [  Description  ] [  Explode  ] [  Selectable  ]
        [   OK   ]    [ Cancel ]   [ Help... ]
```

Using the new Object Grouping utility creates a selection set with a unique name. Now selection sets may be created and called up later on for editing purposes using the Group command. Issuing the Group command activates the dialog box illustrated above. The user provides a name in the "Group Name" area; the name of the group entered above is "Lines." When picking "Ok" to exit the dialog box, the user is prompted to select objects to create as a group. Pick all line segments illustrated at the right. Striking the Enter key for the last "Select Objects:" prompt creates the group and exits the Group command. Below is the prompt sequence used:

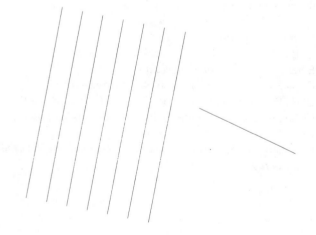

Command: **Group**
-The Group Dialog Box Appears-
Select objects for grouping:
Select objects: *(Select all entities to make up the group)*
Select objects: *(Strike Enter to exit this command)*

Now any time an entity that belongs to a group is selected, the entire group selects as in the illustration at the right.

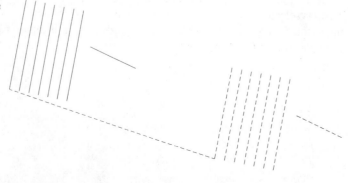

DDSELECT and Grouping

Whenever a group is created and active, all entities belonging to the group select even if you want to edit only one single member of the group. One way of controlling this is to use the Object Selection Settings dialog box illustrated at the right, otherwise known as DDSELECT. This dialog box has been enhanced to include a check box next to Object Grouping. Taking the check off of this box deactivates all groups allowing for editing of individual entities. When the check is placed back, all entities belonging to the current group convert back to their group status.

AutoCAD Release 13 is equipped with numerous accelerator keys, designed to enhance productivity. These keys are activated by depressing and holding down the Control (CTRL) key and pressing another key identified as a supported accelerator. One such key is the CTRL-A key, which is used to toggle group mode on or off. In the examples below, with grouping turned on, all entities hightlight to reflect the current group. Pressing CTRL-A toggles group mode off as in the second illustration. This forces all entities that used to belong to a group to behave as individual entities. Pressing CTRL-A turns group mode on again, and so on.

```
┌─────────────────────────────────┐
│   Object Selection Settings     │
│ Selection Modes                 │
│ ┌─────────────────────────────┐ │
│ │ ☒ Noun/Verb Selection       │ │
│ │ ☐ Use Shift to Add          │ │
│ │ ☐ Press and Drag            │ │
│ │ ☒ Implied Windowing         │ │
│ │ ☒ Object Grouping           │ │
│ │        ┌─────────┐          │ │
│ │        │ Default │          │ │
│ │        └─────────┘          │ │
│ └─────────────────────────────┘ │
│ Pickbox Size                    │
│ ┌─────────────────────────────┐ │
│ │   Min         Max           │ │
│ │                          ☐  │ │
│ │ ◄─┌──┐──────────────►│      │ │
│ └─────────────────────────────┘ │
│   ┌───────────────────────────┐ │
│   │  Object Sort Method...    │ │
│   └───────────────────────────┘ │
│ ┌────────┐ ┌────────┐ ┌───────┐ │
│ │   OK   │ │ Cancel │ │Help...│ │
│ └────────┘ └────────┘ └───────┘ │
└─────────────────────────────────┘
```

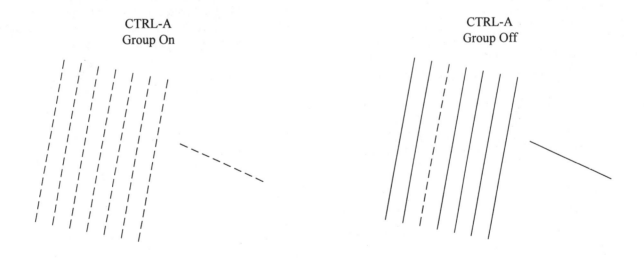

CTRL-A
Group On

CTRL-A
Group Off

The Mtext Command

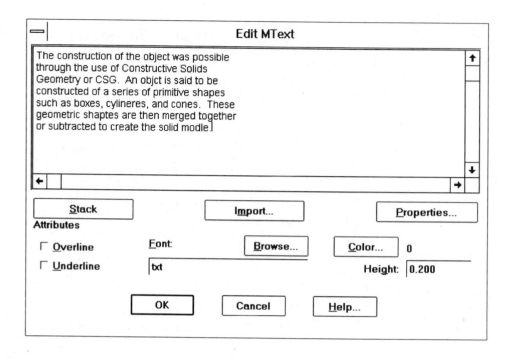

The Mtext command allows for the placement of text in multiple lines. Entering the Mtext command from the command line displays the following prompts:

Command: **Mtext**
Attach/Rotation/Style/Height/Direction/<Insertion point>:
(Mark a point to identify one corner of the mtext box)
Attach/Rotation/Style/Height/Direction/Width/2Points/
<Other corner>: *(Mark another corner forming a box)*

Identifying an insertion point followed by another corner point displays the dialog box illustrated above. This dialog box is only available in the Windows version of AutoCAD. The user begins entering the desired text and is able to view the text as it is entered from this dialog box.

The DOS version of the Mtext works similar to the Windows version with a few exceptions. Instead of displaying the dialog box illustrated above, the DOS version of Mtext temporarily exits AutoCAD and shells out to the DOS editor. Text is entered in at this point although it does not appear to word wrap. This file is saved under its default name; once exiting this DOS editor, the text is placed on the display screen.

Below is an example of text placed using the Mtext command. Using the List command on the text below highlights all text and identifies all text an an Mtext entity.

The construction of the object was possible through the use of Constructive Solids Geometry or CSG. An objct is said to be constructed of a series of primitive shapes such as boxes, cylineres, and cones. These geometric shaptes are then merged together or subtracted to create the solid modle.

The Spell Command

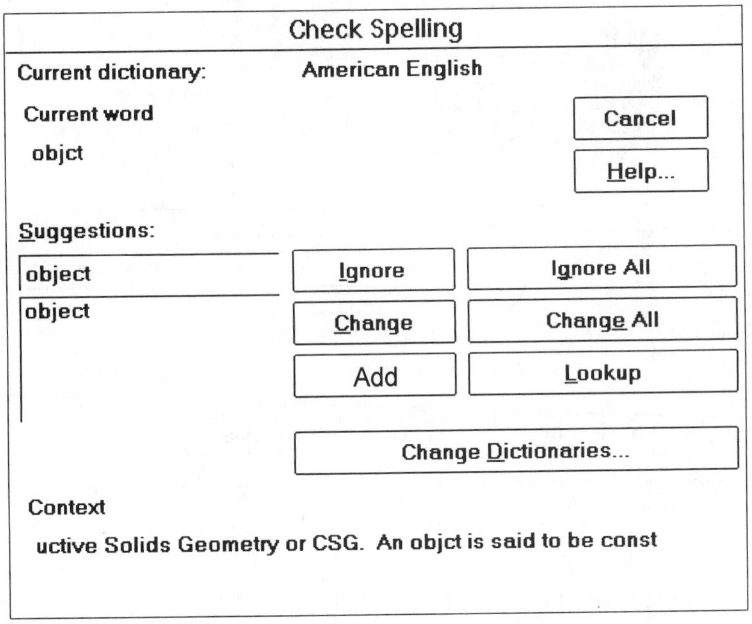

```
┌─────────────────────────────────────────────────────────────┐
│                       Check Spelling                         │
│  Current dictionary:        American English                 │
│                                                              │
│  Current word                          ┌──────────────┐      │
│    objct                               │    Cancel    │      │
│                                        └──────────────┘      │
│                                        ┌──────────────┐      │
│                                        │    Help...   │      │
│                                        └──────────────┘      │
│  Suggestions:                                                │
│  ┌─────────────────────┐  ┌──────────────┐ ┌──────────────┐ │
│  │ object              │  │    Ignore    │ │  Ignore All  │ │
│  │ object              │  └──────────────┘ └──────────────┘ │
│  │                     │  ┌──────────────┐ ┌──────────────┐ │
│  │                     │  │    Change    │ │  Change All  │ │
│  │                     │  └──────────────┘ └──────────────┘ │
│  │                     │  ┌──────────────┐ ┌──────────────┐ │
│  └─────────────────────┘  │     Add      │ │    Lookup    │ │
│                           └──────────────┘ └──────────────┘ │
│                                                              │
│              ┌──────────────────────────────┐               │
│              │     Change Dictionaries...    │               │
│              └──────────────────────────────┘               │
│                                                              │
│  Context                                                     │
│    uctive Solids Geometry or CSG.  An objct is said to be const │
│                                                              │
└─────────────────────────────────────────────────────────────┘
```

The group of Mtext entities illustrated on the previous page is an example of how some text is displayed; that is with numerous spelling errors. Issuing the Spell command and selecting the multi text entity displays the Check Spelling dialog box illustrated above along with the following components: listed above is the Current dictionary; the Current word identifies the item the spell checker identified as being misspelled. The presence of the word does not necessarily mean that the word is misspelled, such as the acronyms CAD or GDT. The Suggestions area displays all possible alternatives to the word identified as being misspelled. Ignore allows you to skip the current word especially in the case of acronyms. Ignore All skips all remaining words that match the current word. If the word "CAD" keeps coming up as a misspelled

word, instead of constantly picking the Ignore button, Ignore All will skip the word "CAD" in any future instances. The Change option replaces the word in the Current word box with a word in the Suggestions box. The Add option adds the current word to the current dictionary. The Lookup button checks the spelling of a selected word found in the Suggestions box. The Change Dictionaries button allows the user to change to a different dictionary containing other types of words. The Context area displays a phrase where the current word is located. Use this area to check proper sentence structure. In the example on the previous page, the words "objct," "cylineres," "shaptes," and "modle" were identified as being misspelled. The corrected mtext entity is illustrated below.

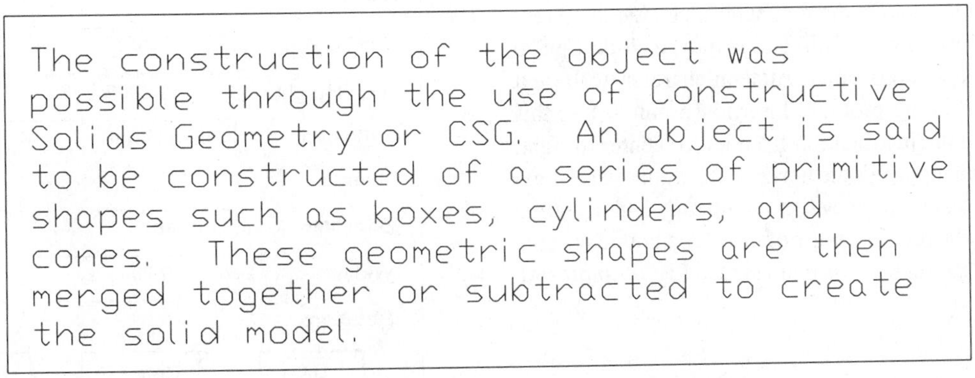

```
┌──────────────────────────────────────────────────────────┐
│  The construction of the object was                       │
│  possible through the use of Constructive                 │
│  Solids Geometry or CSG.  An object is said              │
│  to be constructed of a series of primitive               │
│  shapes such as boxes, cylinders, and                     │
│  cones.  These geometric shapes are then                  │
│  merged together or subtracted to create                  │
│  the solid model.                                         │
└──────────────────────────────────────────────────────────┘
```

Editing Mtext Entities

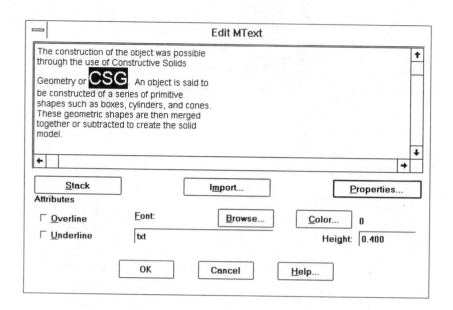

Using DDEDIT on an Mtext entity displays the dialog box illustrated above. This dialog box is only available in Windows. DOS versions of AutoCAD utilize the DOS Editor. Superior editing is available inside of the Windows editor. For instance, a word can be highlighted as in the example above (CSG) and the height can be changed so that only this word is affected by the height change. Highlighting other items such as the words "Constructive Solids Geometry" and picking the check-box next to Underline places an underscore below all highlighted text entities. Highlighting the words "boxes, cylinders, and cones." and picking the "Browse..." button displays a dialog box to change the font of the selected text. The color of the highlighted text entity may also be changed.

The MText Properties dialog box illustrated at the right is available in both DOS and Windows. Use it to change to a different text style, change the text height, change the direction the text flows. Typical directions may include Left to Right, Right to Left, or on Center. Attachment refers to the text boundary started at the prompt "Insertion point:." Use this option to change the justification from left to center to right. The Width option identifies the horizontal length of the Mtext boundary box created in the Mtext command. The Rotation option specifies the rotation angle of the Mtext boundary box. This box is available in DOS through the MTPROP command.

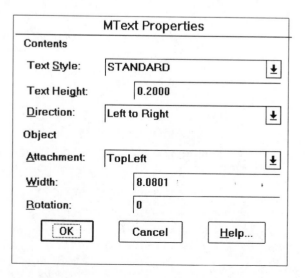

New Dimension Styles Dialog Box

```
┌─────────────────────────────────────────────────────┐
│              Dimension Styles                         │
│ Dimension Style                                       │
│ ┌───────────────────────────────────────────────┐    │
│ │ Current: │STANDARD                        │▼│ │    │
│ │ Name:    │STANDARD                           │ │    │
│ │          ┌──────────┐  ┌──────────┐           │    │
│ │          │  Save    │  │  Rename  │           │    │
│ │          └──────────┘  └──────────┘           │    │
│ └───────────────────────────────────────────────┘    │
│ Family                                                │
│ ┌──────────────────────────┐  ┌──────────────────┐   │
│ │ ■ Parent                 │  │   Geometry...    │   │
│ │ □ Linear    □ Diameter   │  └──────────────────┘   │
│ │ □ Radial    □ Ordinate   │  ┌──────────────────┐   │
│ │ □ Angular   □ Leader     │  │   Format...      │   │
│ └──────────────────────────┘  └──────────────────┘   │
│       ┌──────┐   ┌────────┐   ┌──────────────────┐   │
│       │  OK  │   │ Cancel │   │  Annotation...   │   │
│       └──────┘   └────────┘   └──────────────────┘   │
│                               ┌──────────────────┐   │
│                               │   Help...        │   │
│                               └──────────────────┘   │
└─────────────────────────────────────────────────────┘
```

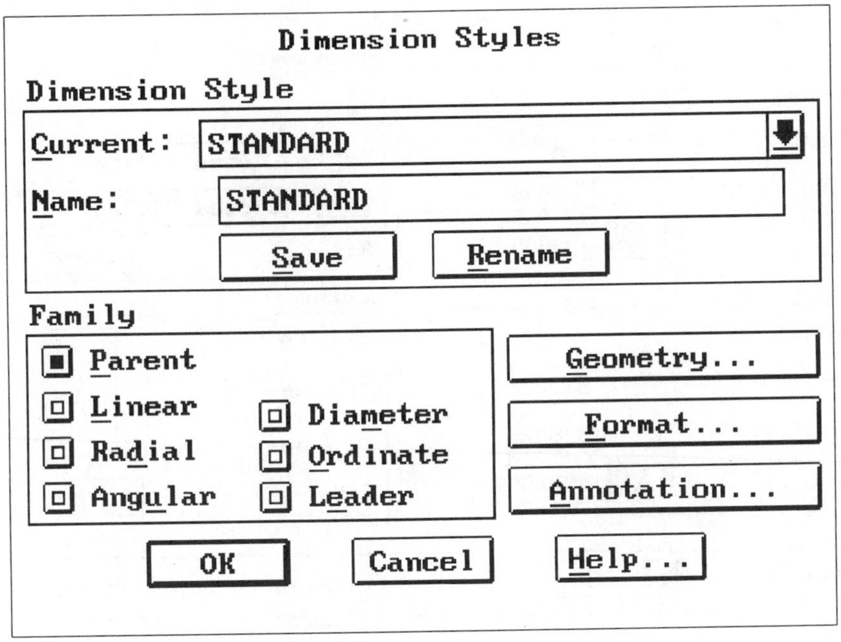

The concept of creating a dimension style has not changed much in Release 13. It is still used to group a series of dimension variable settings under a unique name to determine the appearance of dimensions. What has changed in Release 13 is a more condensed dimension style dialog box illustrated above. Prior to Release 13, the Dimension Styles dialog box took the form of a series of buttons to individually control variables specific to the dimension line, extension line, arrowheads, text format, text position, and dimension component colors. This required activating numerous subdialog boxes before the required number of changes were made to the dimension style being created. The dialog box illustrated above is the new enhanced DDIM dialog box, which is used to create a dimension style much the same way as previous versions of AutoCAD. However, instead of picking numerous other dialog boxes to change dimesion variables, three buttons are added to this main dialog box to control variables.

The "Geometry..." button activates a dialog box controlling dimension line, extension line, arrowhead, and center mark variables.

The "Format..." button activates a dialog box controlling placement of dimension text, horizontal justification and vertical justification of dimension text.

The "Annotation..." button activates a dialog box controlling primary units, alternate units, tolerance, and text variables.

Another new feature of this DDIM dialog box is in the form of Dimension Style Families. Using families allows for variables to be assigned to a particular dimension type, such as horizontal dimensions and diameter dimensions, without having to set up an entirely separate dimension style for each type of dimension placed.

The Geometry Subdialog Box

```
                          Geometry
  Dimension Line                    Arrowheads
  ┌─────────────────────────────┐   ┌────────────────────────────────┐
  │ Suppress:   □ 1st   □ 2nd   │   │         ┌──◄──┐ ┌──►──┐         │
  │ Extension:       0.0000     │   │ 1st:  Closed Filled         ▼  │
  │ Spacing:         0.3800     │   │ 2nd:  Closed Filled         ▼  │
  │ [Color...]   ██  BYBLOCK    │   │ Size:        0.1800            │
  └─────────────────────────────┘   └────────────────────────────────┘
  Extension Line                     Center
  ┌─────────────────────────────┐   ┌───────────────────┬────────────┐
  │ Suppress:   □ 1st   □ 2nd   │   │ ■ Mark            │   ┌───┐    │
  │ Extension:       0.1800     │   │ □ Line            │  (  +  )   │
  │ Origin Offset:   0.0625     │   │ □ None            │   └───┘    │
  │ [Color...]   ██  BYBLOCK    │   │ Size:    0.0900            │
  └─────────────────────────────┘   └────────────────────────────────┘
  Scale
  ┌──────────────────────────────────────────────────────────────────┐
  │ Overall Scale:  1.00000          □ Scale to Paper Space          │
  └──────────────────────────────────────────────────────────────────┘
                         [ OK ]    [ Cancel ]   [ Help... ]
```

Use the Geometry subdialog box to control variable settings dealing with the dimension line, extension line, arrowheads, and center markers. Variables affecting the dimension line include the suppressing of either the first or second dimension line, the increment dimension lines will be spaced from each other in the case of a baseline dimension, the extension distance of the dimension line which is popular in architectural applications, and the color the dimension line will be drawn in. Variables affecting the extension line include the suppressing of either the first or second dimension line, the distance the extension line extends past the arrow, the distance the extension line is placed away from the object, and the color the extension line will be drawn in. Variables affecting arrowheads include the form of the first and second arrowheads including the size the the arrow. The illustration at the right shows the numerous terminators supported in Release 13. When one of the choices is selected, the slide displaying the arrow updates to the new arrow picked. Picking inside of the left graphic of the arrow scrolls through all supported arrows. Picking inside of the right graphic scrolls through all supported arrows; however, only the graphic in this box updates. This is used for separate terminators at the ends of the dimension line. Variables affecting the center marker include the ability to place a center mark, adding center lines to the center mark, not having any center mark placed, and the size the the center mark. Picking the graphic of the center mark also scrolls through all three types of center marks. The very important overall dimension scaling value is also present in this subdialog box.

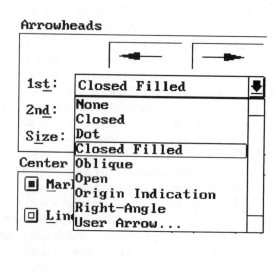

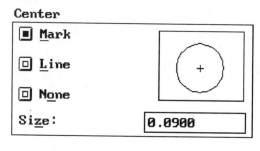

The Format Subdialog Box

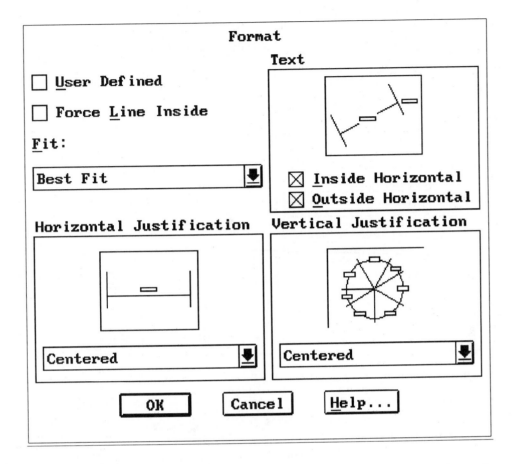

Use the Format subdialog box to control variable settings dealing with the placement of dimension text, horizontal text justification and vertical text justification, and the fitting of text between extension lines. Variables affecting the placement of text include whether text placed inside or outside will be horizontal or will align parallel to the dimension line. Illustrated at the right is a graphic under "Text," which will scroll through various settings if picked. Variables affecting the vertical placement of text include whether text is centered along the dimension line, placed above the dimension line, or placed outside of the dimension line. A Japan International Standard has been implemented in this dialog box to address dimension globalization issues. Variables affecting the horizontal justification of text include whether dimension text is centered along the dimension line, placed next to the left extension line, or placed next to the right extension line. Variables affecting how text will fit between extension lines include text to take on a best fit. For radius and diameter dimensions, other best fit options include whether the text is placed outside of the diameter, and to include a center mark, whether the text is placed outside of the diameter but whether to force the interior dimension line with no center mark, or force the text inside of the diameter along with the dimension line with no center mark.

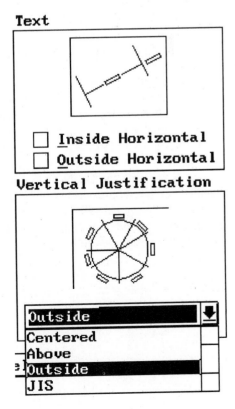

The Annotation Subdialog Box

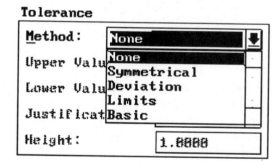

Use the Annotation subdialog box to control variable settings dealing with the primary units, alternate units, tolerance method, and general text settings. Variables affecting the primary units include the type of units the dimensions will be constructed in (decimal, engineering, architectural, etc.), and if the dimension text requires a prefix or suffix. Variables affecting alternate units include whether alternate units are enabled or disabled, units for alternate dimensions, and if the alternate dimension text requires a prefix or suffix. Variables affecting tolerances include an option for no tolerances, symmetrical tolerances, tolerances of deviation, limit dimensions, and basic dimensions as in the illustration at the right. Other variable controls for tolerances include the setting of upper and lower limit values, the justification of the tolerance in relation to the main text, and the height of the tolerance. Variables affecting the dimension text include the current text style governing the dimension text, the height of dimension text, the gap formed on either side with the dimension text and the dimension line, and the color of the dimension text.

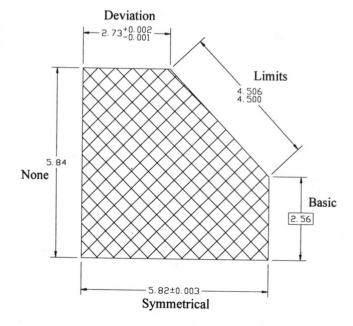

Dimension Style Families

A new feature of dimension styles in Release 13 is the inclusion of dimension style families. The six families supported are Linear, Radial, Angular, Diameter, Ordinate, and Leader dimensions. Changes to dimension variables may be made to one of the family types. Then when placing that type of dimension, the variables that relate to that specific family type are read and applied to the dimension being placed.

As an example, suppose a new dimension style called "Architectural" has been created in the illustration at the right. Architectural dimensioning takes the form for all linear dimensions of having arrows converted to slashes or "ticks." Also, the dimension text for architectural applications is placed above the dimension line. Unfortunately when a diameter or radius dimension is placed, tick marks terminate the leader line for the dimension even if an arrow is more desired. Instead of creating another dimension style to cover this, dimension style families are used. Follow the next series of examples to assign ticks and text above the dimension line to all linear dimensions and arrows to all diameter dimensions under the same dimension style name.

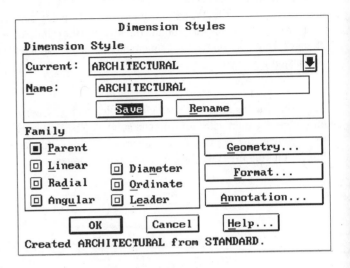

Once the new dimension style has been created (Architectural), begin the process of assigning certain variables to a family type. Pick the radio button next to "Linear" illustrated at the right. All variable changes will be assigned only to linear dimensions.

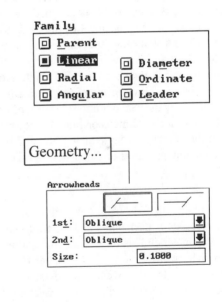

With the Linear dimension family picked, click the "Geometry..." button of the Dimension Styles dialog box to activate the Geometry subdialog box. Change the style of arrowheads from filled in arrows to Oblique tick marks as in the illustration at the right since this terminator is more suited for architectural drawings. Pick "OK" to keep the changes and return to the main Dimension Style dialog box.

With the Linear dimension family still picked, click the "Format..." button of the Dimension Style dialog box to activate the Format subdialog box. Remove the checks from "Inside Horizontal" and "Outside Horizontal," which will align all dimension text to the position of the dimension line. Next scroll through all options under "Vertical Justification" until the "Above" option is located. Exit this subdialog box by picking "OK," which retains the settings just made. Before moving on to the diameter dimension family settings, pick the "Save" button to save the past three variable changes to the linear dimension family.

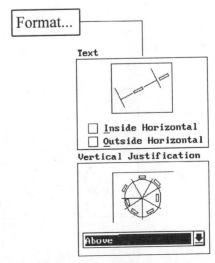

Save

With the settings to all linear dimensions saved to the current dimension style, begin setting variables specific to all diameter dimensions. From the main Dimension Style dialog box, click on the "Diameter" radio button illustrated at the right. All changes from here on will be implemented when a diameter dimension is placed.

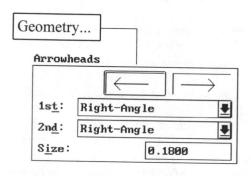

From the Dimension Styles dialog box, click on the "Geometry..." button to activate the Geometry subdialog box. Change the current arrow to a Right-angle arrow. Exit this subdialog box by picking the "OK" button to keep the changes.

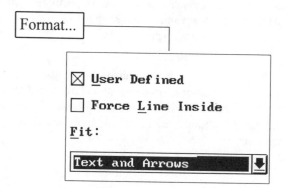

With the "Diameter" dimension family still highlighted, click on the "Format..." button to activate the Format subdialog box. Place a check in the box next to "User Defined," which will allow the user to define the placement of the dimension in a leader mode. Also scroll through the numerous modes under "Fit:" until the "Text and Arrows" mode displays. Pick this mode to make it current, then pick the "OK" button to retain the changes in format. Once inside of the main Dimension Styles dialog box, pick the "Save" button to save the last series of changes to the Diameter dimension family. Pick the "OK" button to exit the Dimension Style dialog box and retain all changes made to any variables or dimension families.

The only remaining item is to test the dimension family changes by creating a horizontal and vertical dimension followed by a diameter dimension as in the illustration below. Notice that both horizontal and vertical have tick terminators, the dimension text is placed above the dimension line, and the text, especially in the vertical dimension, is aligned to be parallel to the dimension line. Placing a diameter dimension allows the user to define the location of the leader and has a right-angle arrow terminator instead of a tick mark. All dimensions have been placed without switching from the current Architectural dimension style.

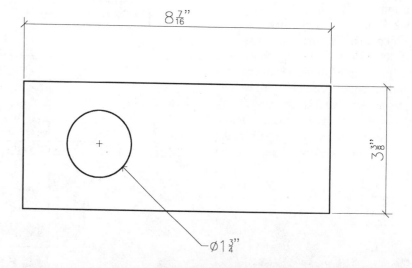

Dimensioning by Inference

Both Horizontal and Vertical dimension subcommands have been combined into a single dimensioning command called Dimlinear. Using this at the command prompt produces either result illustrated at the right; that is when the Dimlinear command prompts the user for the placement of the dimension line and the user indicates a point directly to the right, AutoCAD infers a 90 degree rotation of the dimension and places the dimension vertical. If the user indicates a point directly above, AutoCAD infers a rotation of 0 degrees and places a horizontal dimension.

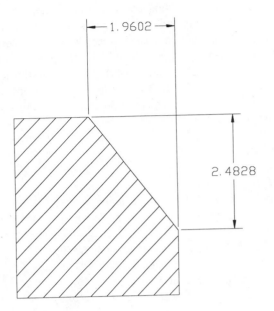

Command: **Dimlinear**
First extension line origin or RETURN to select: *(Select the inclined line at the right)*
Dimension line location (Text/Angle/Horizontal/Vertical/Rotated): *(Picking a point to the right of the inclined line produces a vertical dimension; picking above produces a horizontal dimension).*

Mainstreaming of Dim Commands

To allow dimensioning commands to be entered in at the command prompt, all existing dimension subcommands have been converted into the list illustrated at the right.

Note that DIMLINEAR replaces Horizontal, Vertical, and Rotated; DIMEDIT replaces Hometext, Newtext, Trotate, and Oblique; DIMSTYLE replaces Save, Restore, Status, Variables, and Update (now referred to as Apply).

If "DIM" is entered in at the command prompt, you are able to access all subcommands as in the past.

One advantage of continuing to use the DIM: prompt is in the area of dimension variables. At this prompt, instead of entering the entire variable name such as "DIMASZ," only the last three letters of the variable are needed to access the specific variable. For the past example, "ASZ" would be entered at the DIM: prompt to access the dimension arrow size variable.

DIMALIGNED	ALIGNED
DIMANGULAR	ANGULAR
DIMBASELINE	BASELINE
DIMCENTER	CENTER
DIMCONTINUE	CONTINUE
DIMDIAMETER	DIAMETER
DIMEDIT	HOMETEXT
	NEWTEXT
	TROTATE
	OBLIQUE
DIMTEDIT	TEDIT
DIMLINEAR	HORIZONTAL
	VERTICAL
	ROTATED
DIMORDINATE	ORDINATE
DIMRADIUS	RADIUS
DIMSTYLE	SAVE
	RESTORE
	STATUS
	VARIABLES
	APPLY (UPDATE)
DIMOVERRIDE	OVERRIDE
LEADER	LEADER

Associative Leader Lines

The new Leader command places a leader line in the drawing for the purpose of adding notes or other dimensions to a drawing; this command also produces a leader which is associative to the drawing. The Leader command is entered in at the Command: prompt. There is still the presence of the leader option which is part of the DIM: prompt; this leader remains non-associative. The prompt sequence below illustrates the use of the Leader command and how it activates the Mtext command for creating the annotation.

Command: **Leader**
From point: **Nea**
to *(Pick the small hole at "A")*
To point: *(Pick a point to begin laying out the leader)*
To point (Format/Annotation/Undo)<Annotation>: *(Strike Enter to continue)*
Annotation (or RETURN for options): **0.375 DIA**
MText: **8 HOLES EQUALLY SPACED**
MText: **ABOUT A 3.00 DIA BOLT CIRCLE**
MText: *(Strike Enter to exit the command and place the note)*

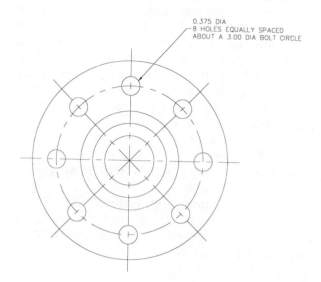

Other options of the Leader command allow the individual to copy existing text to the leader, insert a block next to the leader, or activate the Tol command for the placement of a geometric tolerancing symbols.

Command: **Leader**
From point: *(Pick a point)*
To point: *(Pick a point to begin laying out the leader)*
To point (Format/Annotation/Undo)<Annotation>: *(Strike Enter to continue)*
Annotation (or RETURN for options): *(Strike Enter for options)*
Tolerance/Copy/Block/None/<Mtext>:

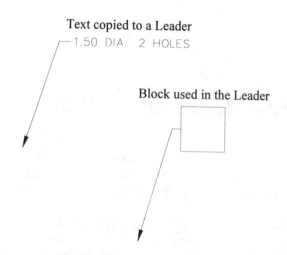

Text copied to a Leader

1.50 DIA   2 HOLES

Block used in the Leader

New Dimension Variables

A series of new dimension variables illustrated at the right have been added for more control of dimensions. Of particular interesest are: DIMDEC, which controls the number of decimal places in the dimension and can be controlled through the DDIM Annotation dialog box; DIMFIT, which controls the placement of text and arrowheads inside or outside of extension lines based on the distance between extension lines; DIMJUST, which controls the horizontal dimension text position by locating dimension text in the center between extension lines, next to the first extension line, or next to the second extension line; DIMSD1, which turns off the first dimension line; DIMSD2, which turns off the second dimension line; DIMTXSTY, which specifies a current text style for dimensions; DIMUNIT, which sets the format of units for all dimension style family members except angular; DIMUPT, which allows the user to define where the text of a dimension will be placed.

DIMALTTD	Alternate tolerance decimal places
DIMALTTZ	Alternate tolerance zero suppression
DIMALTU	Alternate units
DIMALTZ	Alternate unit zero suppression
DIMAUNIT	Angular unit format
DIMDEC	Decimal places
DIMFIT	Fit Text
DIMJUST	Text justification on the dimension line
DIMSD1	Suppress the first dimension line
DIMSD2	Suppress the second dimension line
DIMTDEC	Tolerance decimal places
DIMTOLJ	Tolerance vertical justification
DIMTXSTY	Text style
DIMTZIN	Tolerance zero suppression
DIMUNIT	Unit format
DIMUPT	User positioned text

DDMODIFY Enhancements on Dimensions

Major enhancements have been made to the DDMODIFY dialog box when modifying a dimension. In the past, the DDMODIFY dialog box provided information regarding the type of dimension, layer, and color information. However you could never make changes to the actual dimension. AutoCAD Release 13 provides the dialog box illustrated below when a dimension is selected while in the DDMODIFY dialog box. The user has the option of editing the dimension text in addition to calling up the "Geometry...," "Format...," and "Annotation..." subdialog boxes of the DDIM command to make changes to dimension variables and have the changes reflected in the selected dimension.

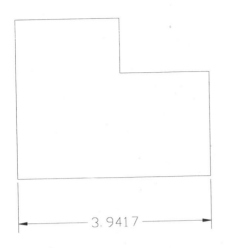

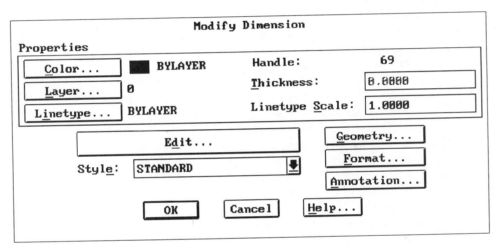

Ordinate Dimensioning Enhancements

Enhancements to the placement of Ordinate dimensions have been made in the positioning of the leader. Prior to Release 13, the leader had an orthogonal path of either horizontal or vertical. If a jog had to be placed because of dimension text being too close to other dimensions, the jog also took the form of a combination horizontal/vertical path which aesthetically did not look very pleasing. Release 13 allows for a more relaxed leader allows the leader to be bent depending on the dimension arrow size (DIMASZ).

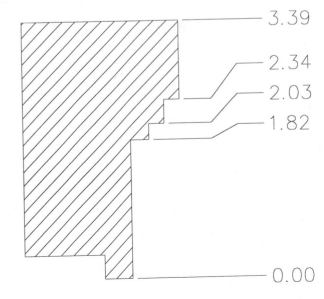

Geometric Dimensioning and Tolerancing (GD&T)

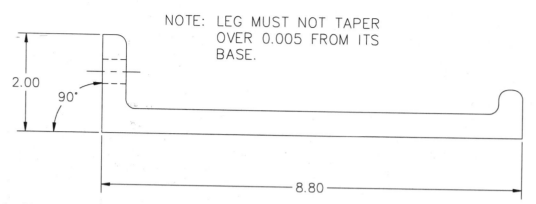

NOTE: LEG MUST NOT TAPER OVER 0.005 FROM ITS BASE.

2.00

90°

8.80

In the example of the object above, approximately 1000 items needed to be made based on the dimensions of the drawing. Also, once constructed, all objects needed to be tested to be sure the 90 degree surface didn't deviate over 0.005 units from its base. Unfortunately the wrong base was selected as the reference for all dimensions. Which should have been the correct base feature? As all items were delivered, they were quickly returned since the long 8.80 surface drastically deviated from required 0.005 unit deviation or zone. This is one simple example pointing out the need for using Geometric Dimension and Tolerancing techniques. First, this method deals with setting tolerances to critical characteristics of a part. Of course the function of the part must be totally understood by the designer in order to assign tolerances. The problem with the object above was the note which did not really specify which base feature to choose as a reference for dimensioning. Below is the same object complete with dimensioning tolerancing symbols. The letter "-A-" inside of the rectangle identifies the datum or reference surface. The tolerance symbol at the end of the long edge tells the individual making this part that the long edge cannot deviate more than 0.005 units using surface "A" as a reference.

Using geometric tolerancing symbols insures more accurate parts with less error in interpreting the dimensions.

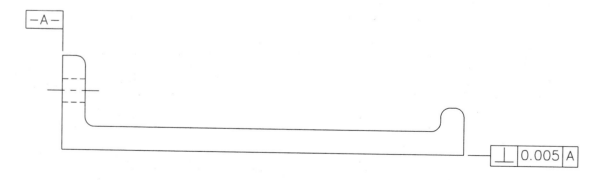

-A-

⊥ | 0.005 | A

GD&T Symbols

Entering "Tol" at the command prompt brings up the first in a series of geometric dimensioning and tolerancing dialog boxes. The first dialog box to display is illustrated at the right and contains all major geometric dimensioning and tolerancing symbols. Choose the desired symbol by clicking the specific symbol. If the wrong symbol has been accidentally selected, click on the correct symbol. The next dialog box does not activate until the "OK" button is selected from the Symbol dialogue box.

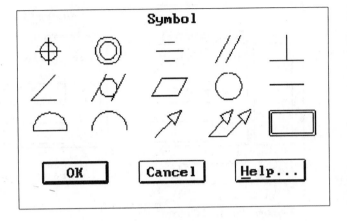

Illustrated below is a chart outlining all geometric tolerancing symbols supported in the Symbol dialog box. Alongside each symbol is the characteristic controlled by the symbol in addition to a tolerance type. Tolerances of Form such as Flatness and Straightness can be applied to surfaces without referencing a datum. On the other hand, tolerances of Orientation such as Angularity and Perpendicularity require datums as reference.

Symbol	Characteristic	Type of Tolerance
▱	Flatness	Form
—	Straightness	
○	Roundness	
⌭	Cylindricity	
⌒	Profile of a Line	Profile
⌓	Profile of a Surface	
∠	Angularity	Orientation
⊥	Perpendicularity	
//	Parallelism	
⊕	Position	Location
◎	Concentricity	
⌱	Symmetry	
↗	Circular Runout	Runout
↗↗	Total Runout	

GD&T Dialog Box

```
                          Geometric Tolerance
 Sym    Tolerance 1           Tolerance 2         Datum 1    Datum 2    Datum 3
      ┌─────────────────┐  ┌─────────────────┐  ┌────────┐ ┌────────┐ ┌────────┐
 ┌──┐ │ Dia  Value  MC  │  │ Dia  Value  MC  │  │Datum MC│ │Datum MC│ │Datum MC│
 │ /││ │                 │  │                 │  │        │ │        │ │        │
 │// │ │    ┌──────────┐ │  │    ┌──────────┐ │  │ ┌────┐ │ │ ┌────┐ │ │ ┌────┐ │
 │  ││ │    │ 0.005    │ │  │    │          │ │  │ │-A-││ │ │ │    │ │ │ │    │ │
 └──┘ │    └──────────┘ │  │    └──────────┘ │  │ └────┘ │ │ └────┘ │ │ └────┘ │
      │    ┌──────────┐ │  │    ┌──────────┐ │  │ ┌────┐ │ │ ┌────┐ │ │ ┌────┐ │
      │    │          │ │  │    │          │ │  │ │    │ │ │ │    │ │ │ │    │ │
      │    └──────────┘ │  │    └──────────┘ │  │ └────┘ │ │ └────┘ │ │ └────┘ │
      └─────────────────┘  └─────────────────┘  └────────┘ └────────┘ └────────┘

 Height ┌──────────┐ Projected Tolerance Zone
        └──────────┘

 Datum Identifier ┌──────────┐
                  └──────────┘

           ┌────────┐   ┌──────────┐   ┌──────────┐
           │   OK   │   │ Cancel   │   │ Help...  │
           └────────┘   └──────────┘   └──────────┘
```

Once a symbol is picked and the Symbol dialog box exited, the Geometric Tolerance dialog box appears, illustrated above. With the symbol placed inside of this dialog box, the user now assigns such items as tolerance values, maximum material condition modifiers, and datums. In the example above, the tolerance of Parallelism is to be applied at a tolerance value of 0.005 units to Datum "A."

Additional GD&T Symbols

While in the main Geometric Tolerancing dialog box above, clicking in the box under "MC" activates another dialog box illustrated at the right. This dialog box is devoted to the concept of Material Condition and acts as a modifier to the main tolerance value.

In brief, Maximum Material Condition refers to the condition of a characteristic such as a hole when the most material exists. Least Material Condition refers to the condition of a characteristic where the least material exists. Regardless of Feature Size indicates that the characteristic tolerance such as Flatness or Straightness must be maintained regardless of the actual produced size of the object.

Datums are represented by surfaces, points, lines, or planes and are are used for referencing the features of an object. The symbols illustrated at the right are identified as "flags" used to identify a datum. The next series of examples illustrate how datums are used in combination with tolerances of Orientation.

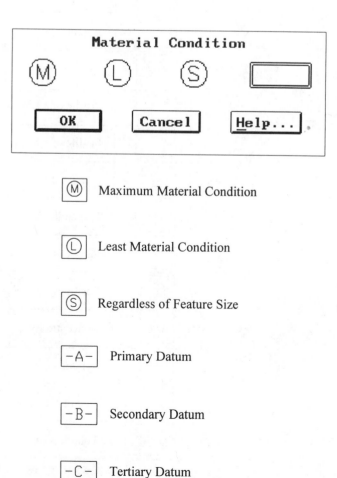

The Feature Control Symbol

Once completing the desired information inside of the Geometric Tolerance dialog box, the result is illustrated at the right in the form of a Feature Control Symbol. This is the actual order that all symbols, values, modifiers, and datums are placed in. Follow the graphic and information below to view the contents of each area of the Feature control symbol box.

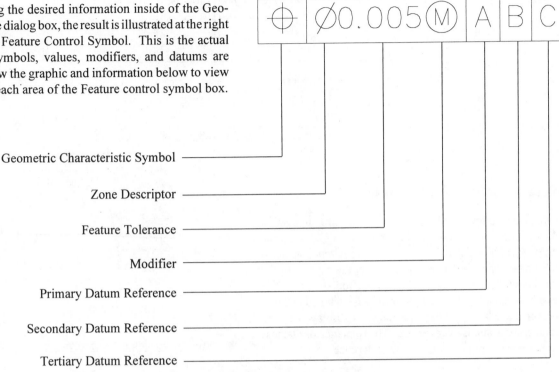

Geometric Characteristic Symbol

Zone Descriptor

Feature Tolerance

Modifier

Primary Datum Reference

Secondary Datum Reference

Tertiary Datum Reference

Tolerance of Angularity Example

Illustrated in the Geometric Tolerance dialog box above and at the right are examples of applying the tolerance of Angularity to a feature. First the symbol is identified along with the tolerance value. Since this tolerance requires a datum, the letter "A" identifies primary datum "A." The results are illustrated at the right. Datum "A" identifies the surface of reference as the base of the object. The feature control symbol is applied to the angle and reads, "The surface dimensioned at 60 degrees cannot deviate more than 0.005 units from datum "A." The second graphic at the right shows an exaggerated tolerance zone and is for illustrative purposes only.

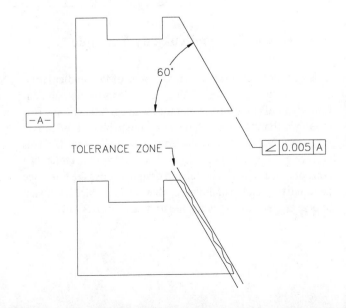

```
                        Geometric Tolerance
Sym   Tolerance 1         Tolerance 2       Datum 1    Datum 2    Datum 3
      Dia  Value   MC     Dia  Value   MC   Datum MC   Datum MC   Datum MC

 ▱         0.003                             

```

Tolerance of Flatness Example

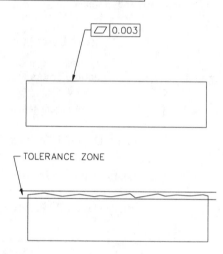

Above is an example of applying a tolerance of Flatness to a surface. First the Geometric Tolerance dialog box has the Flatness symbol assigned in addition to a tolerance value of 0.003 units. The results are illustrated at the right with the feature control box being applied to the top surface. The tolerance box reads, " The surface must be flat with a 0.003 unit tolerance zone." The tolerance range is displayed in the second graphic and is exaggerated. The tolerance of Flatness does not usually require a datum for reference.

```
                        Geometric Tolerance
Sym   Tolerance 1         Tolerance 2       Datum 1    Datum 2    Datum 3
      Dia  Value   MC     Dia  Value   MC   Datum MC   Datum MC   Datum MC

 ⊥         0.003                      A      

Height _____ Projected Tolerance Zone

Datum Identifier _____

              OK      Cancel      Help...
```

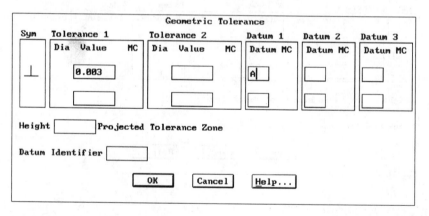

Tolerance of Perpendicularity Example

This next example illustrates a tolerance of Perpendicularity. The Geometric Tolerance dialog box reflects the perpendicularity symbol along with a tolerance value of 0.003 units. This tolerance characteristic requires a datum to be most effective. In the example at the right, the feature control box reads, "This surface must be perpendicular to datum 'A' within a tolerance zone of 0.003 units." The second graphic shows the tolerance zone and the amount of deviation that is acceptable. It is meant to be exaggerated and is for illustrative purposes only.

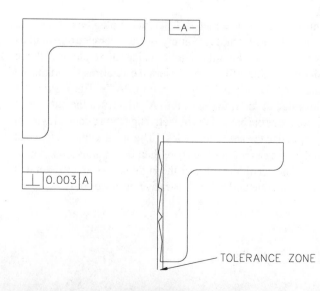

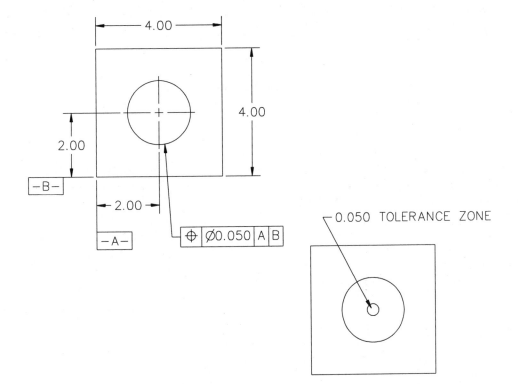

```
                        Geometric Tolerance
  Sym   Tolerance 1          Tolerance 2        Datum 1   Datum 2   Datum 3
        Dia  Value    MC     Dia  Value    MC   Datum MC  Datum MC  Datum MC

   ⊕     Ø  0.050                                  A        B

  Height        Projected Tolerance Zone

  Datum Identifier

                       OK        Cancel      Help...
```

Tolerance of Position Example

This last example displays a tolerance of Position that is applied to a hole feature. Since a hole is centered from two edges, two datums are identified in the Geometric Tolerancing dialog box illustrated above. The feature control box illus-trated at the right reads, "The center of the hole must lie within a circular tolerance zone of 0.050 units in relation to datums 'A' and 'B.'" This is sometimes called a circular tolerance.

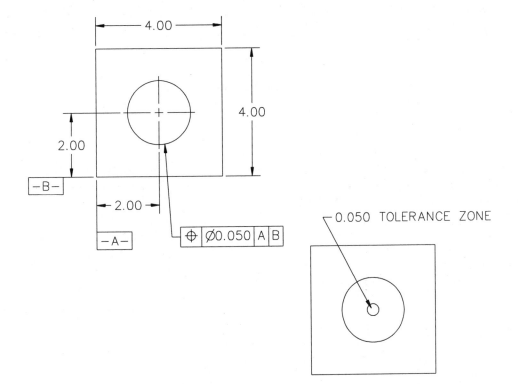

The Enhanced Boundary Hatch Dialog Box

Boundary Hatch

Pattern Type

Predefined ▼ [hatch sample]

Pattern Properties

ISO Pen Width:
Pattern: ANSI31 ▼
Custom Pattern:
Scale: 1.0000
Angle: 0
Spacing: 1.0000
☐ Double   ☐ Exploded

Boundary

Pick Points <
Select Objects <
Remove Islands <
View Selections <
Advanced...

Preview Hatch <
Inherit Properties <
☒ Associative

Apply   Cancel   Help

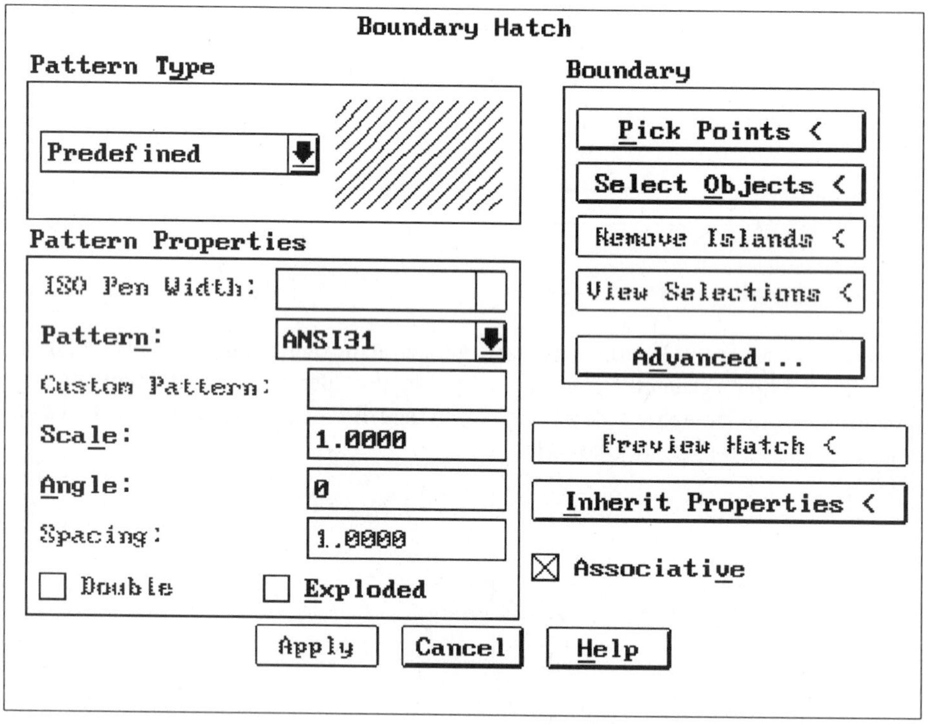

The Bhatch command dialog box has been greatly redesigned by including more information in a single dialog box. This enables the user to set hatch patterns, scale, and rotation angles from one dialog box without searching through numerous levels of dialog boxes. The "Pattern Type" area of the Boundary Hatch dialog box displays the current hatch pattern; picking the hatch pattern with the arrow cursor will scroll through all patterns supplied. Unfortunately there is no mechanism for backing up through this search process. As a pattern is selected, its name appears in the "Pattern Properties" area of the dialog box under "Pattern...." A quicker method of scrolling through the patterns is to pick the pattern by name from this area. The current pattern will display allowing the user to determine if this is the correct pattern to use depending on the application. Other components of this box include the "Pick Points" button allowing the user to pick a point inside of an object for identifying a boundary to hatch. Also notice the check that appears next to "Associative" signifying the pattern will be directly associated with the boundary and islands selected.

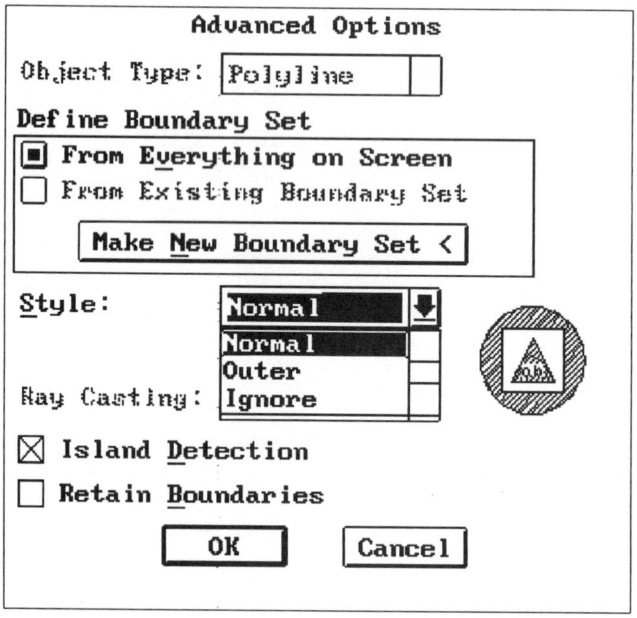

Picking the "Advanced..." button displays the dialog box illustrated at the right enabling the user to define the boundary set and choose the hatch pattern style.

Advanced Options

Object Type: Polyline

Define Boundary Set
◉ From Everything on Screen
☐ From Existing Boundary Set
Make New Boundary Set <

Style: Normal ▼
  Normal
  Outer
Ray Casting: Ignore

☒ Island Detection
☐ Retain Boundaries

OK   Cancel

Associative Hatching Techniques

To begin hatching the object illustrated at the right, the Bhatch command is activated, pattern selected, and the "Pick Points" button selected. When prompted to select an internal point, a point near the vicinity of "A" at the right is picked. The new automatic island detection mechanism highlights the outer perimeter, the two horizontal slots, and the three circles.

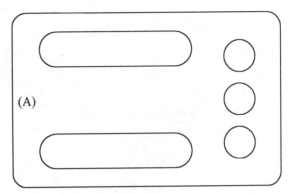

Previewing the hatch pattern displays the results illustrated at the right. The automatic island detection feature automates the process by finding all internal features and including them in the hatch boundary. Picking "Apply" places the hatch pattern permanently as illustrated at the right.

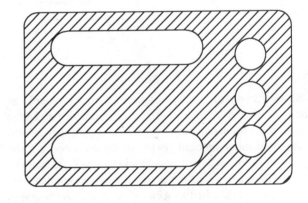

At this point, if any entities need changing such as increasing the length or width through the Stretch command, or moving entities or scaling entities, the hatch pattern had to be erased first, the editing changes made, then the object hatched again to reflect the latest changes in editing. Now, as the entire hatched object needs editing, the hatch pattern will conform to the changes automatically which is where the term "Associative hatching" gets its meaning. At the right, a design change was required to move the two horizontal slots in towards each other. Two holes had to be increased in size while the third needed to be reduced and moved to the other side of the object. After performing all scaling and moving operations, the hatch pattern automatically updates itself to these changes without rehatching the internal area.

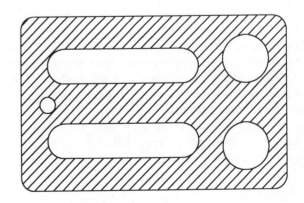

Hatch Editing Techniques

```
                          Hatchedit
  Pattern Type                           Boundary

  ┌────────────────────┐ ╱╱╱╱╱╱╱╱       ┌──────────────────────────┐
  │ Predefined      │▼│ ╱╱╱╱╱╱╱╱       │    Pick Points  <        │
  └────────────────────┘ ╱╱╱╱╱╱╱╱       │  Select Objects <        │
  Pattern Properties     ╱╱╱╱╱╱╱╱       │  Remove Islands <        │
                                         │  View Selections <       │
  ISO Pen Width:  [           ]          │                          │
                                         │      Advanced...         │
  Pattern:        [ ANSI31    ▼]         └──────────────────────────┘
  Custom Pattern: [           ]
                                         ┌──────────────────────────┐
  Scale:          [ 1.0000    ]          │    Preview Hatch  <      │
                                         └──────────────────────────┘
  Angle:          [ 0         ]          ┌──────────────────────────┐
                                         │  Inherit Properties <    │
  Spacing:        [ 1.0000    ]          └──────────────────────────┘

  ☐ Double        ☐ Exploded            ☒ Associative

        [ Apply ]  [ Cancel ]  [ Help ]
```

In all past versions of AutoCAD, whenever a hatch pattern was placed, it remained uneditable. If the pattern needed to change or if the scale needed to be increased, the original pattern had to be erased and re-hatched. In the example illustrated at the right, the pattern needs to be increased to a new scale factor of 3 units and the angle of the pattern needs to be rotated 90 degrees. The Hatchedit dialog box illustrated above becomes the mechanism for editing hatch entities that have been placed in a drawing. The dialog box is very similar to the Bhatch dialog box.

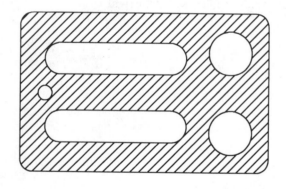

When entering the Hatchedit command, the prompt "Select hatch pattern" appears prompting the user to select the pattern to be edited. After the pattern is picked, the Hatchedit dialog box is activated. Make the desired changes such as pattern, scale, and rotation angle and pick the "Apply" button to place the new hatch pattern. In the illustration at the right, the original hatch pattern was edited to reflect a scale factor of 3 and a rotation angle of 90 degrees.

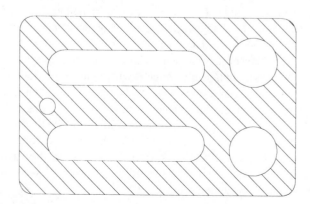

The Boundary Command

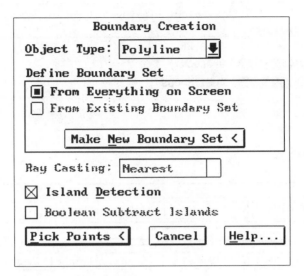

The Boundary command has been enhanced from the previous Bpoly command found in AutoCAD Release 12. Entering Boundary activates the dialog box illustrated at the top. From the current information in the dialog box, a polyline entity will be traced on top of a group of entities after an internal point is picked. This is accomplished by picking the button "Pick Points <," which is very similar in operation to the Bhatch command. Once an internal point is picked as in "A," a series of rays project outward until an entity is selected. Once the entity is selected, a boundary is formed by tracing from around all entities. About the only rule to be aware of in using this command is the boundary made up of entities must be closed. The results are illustrated below. The boundary is created out of a polyline entity exactly on top of the original pattern. To separate the boundary from the original pattern, use the Move command. A better approach would be to create a new layer called "Boundary" before creating the polyline boundary entity since the boundary is created taking on the properties of linetype and color of the current layer.

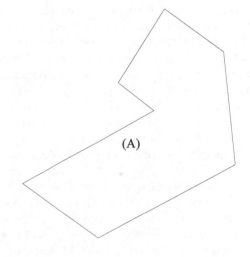

(A)

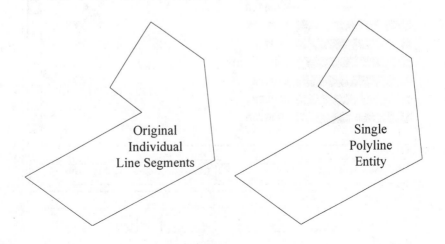

Original
Individual
Line Segments

Single
Polyline
Entity

The DDLTYPE Command

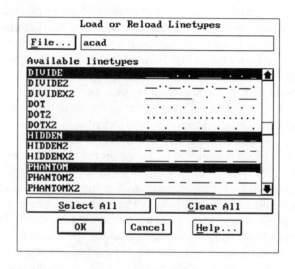

The Linetype command has now been converted into the DDLTYPE command, which displays a dialog box above. A more automated way of loading linetypes is now present with the addition of the "Load..." button at the bottom of the dialog box. Once this button is pressed, all linetypes associated with the current .Lin file are displayed in the dialog box at the right. Notice all linetypes displayed in this illustration refer to linetypes found in the "ACAD.Lin" file. The user selects which linetype or linetypes to use and picks the "OK" button to load the linetypes into the main DDLTYPE dialog box. Rather than use the asterisk symbol "*" to load all linetypes, a special button enables the user to select all linetypes held in the current .Lin file. If a mistake was made in selecting a linetype, simply pick it again to remove the highlight. Once the desired linetypes are displayed in the "Select Linetype" dialog box at the bottom, selecting the "OK" button loads the linetypes into the database of the current drawing file.

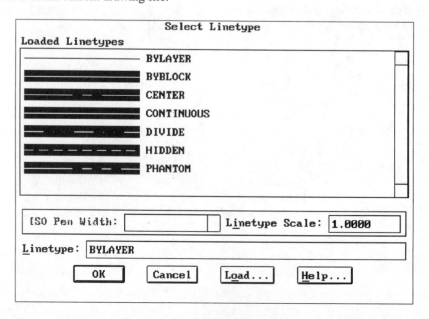

Loading Linetypes through the Layer Dialog Box

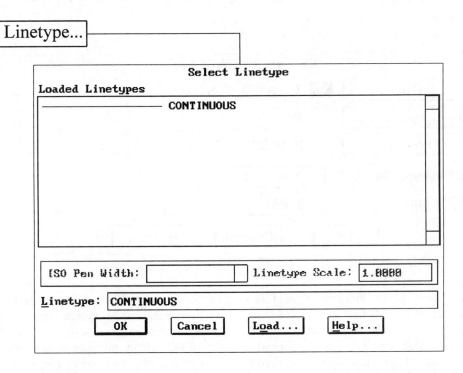

In past versions of AutoCAD, it has always been a probem assigning a linetype to a layer through the Layer Control dialog box. Either the linetype had to be pre-loaded through the Linetype command or the linetype was assigned to a Layer through the Layer command. Now as the DDLTYPE dialog box has a Load button, so also does the "Select Linetype" dialog box illustrated above, which is part of the main Layer dialog box. Picking the "Load..." button activates the "Load or Reload Linetypes" dialog box enabling the user to pick which linetype or linetypes to load.

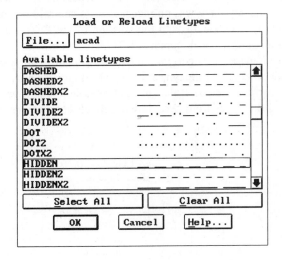

As the linetypes are picked in the "Load or Reload Linetypes" dialog box, they are added to the main "Select Linetype" dialog box illustrated at the right. As the linetypes are loaded, they are not yet assigned to a specific layer. Picking the linetype in this dialog box places it in the "Linetype:" edit box located at the bottom of the dialog box. Picking the "OK" button returns to the main Layer dialog box and assigns the linetype to the selected layer. This provides an even easier way to assign linetypes to layers.

Celtscale through the DDCHPROP Command

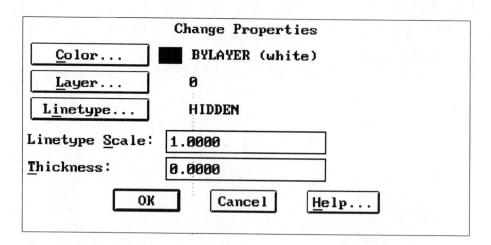

The Ltscale command globally changed the display scale of all linetypes used in a drawing. At a current linetype scale of 1.00, changing this value to 2.00 doubled the length and spaces of all linetypes; changing this value to 0.50 cut the length and spaces of all linetypes in half. The problem was a user desired the Ltscale command to affect some linetypes and not others. Unfortunately there was no mechanism to control this until AutoCAD Release 13 with the creation of the Celtscale or Current Linetype Scale command. This command can be used by itself to set a new current linetype scale value to draw any entities that have a linetype assignment. If the user desires a new current linetype scale, change Celtscale to a new value and all linetypes from here on will reflect the new linetype scale value.

Commands such as Change, Chprop, and DDCHPROP have also been enhanced to include changing a selected entity drawn in a linetype to a new linetype scale. When using the DDCHPROP command and activating the dialog box illustrated above, notice the new "Linetyype Scale:" edit box holding the current linetype scale value. Selecting an entity and changing the linetype scale value changes the linetype in the drawing. To illustrate this, two objects are drawn below. The object on the left has the Ltscale command set to a value of 0.70 units. The object on the right has linetypes controlled by two different scale values set through the DDCHPROP dialog box.

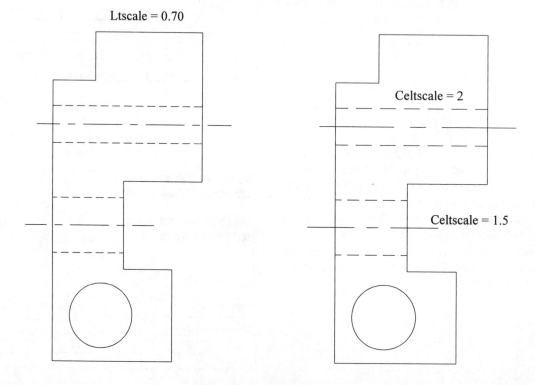

Complex Linetypes

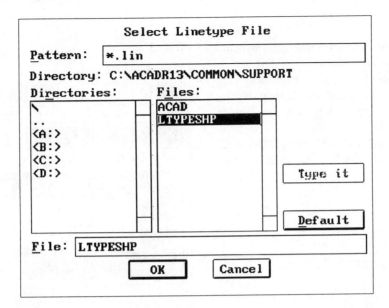

A series of sample linetypes are available to illustrate the use of custom linetypes in a drawing. Most common linetypes are located in the ACAD.Lin file. A second linetype file called LTYPESHP.Lin is available containing all linetypes illustrated at the right. Custom linetypes can be designed two ways; by embedding text into the linetype and by embedding a shape file representing a symbol into the linetype. Use the DDLTYPE dialog box illustrated above to locate the linetype file to load. Next select the linetypes from the list at the right to load. Picking a linetype from the list at the right makes the linetype active and will be used during the creation process of entities such as lines, circles, and arcs. Illustrated below are a few examples of how custom linetypes appear in a drawing; one linetype is called "Zigzag"; the other linetype is "Gas_line" and has text part of the linetype's design.

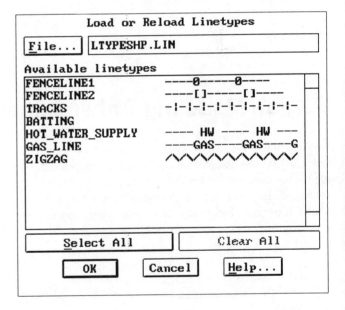

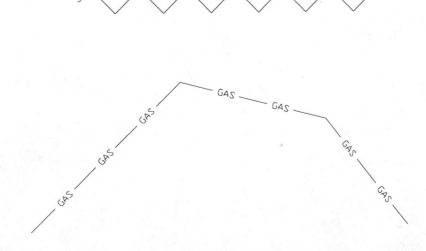

The DDCOLOR Command

The DDCOLOR dialog box has been added to complement the Color command. Once a color is chosed from this box, it becomes the current color taken on by all new entities. This also means that if entity color is to be organized through layers, the color set by the DDCOLOR dialog box overrides all color set by the layer command. If this instance occurs, open the DDCOLOR dialog box and set the current color to "Bylayer," which will send the control of color back to the Layer command.

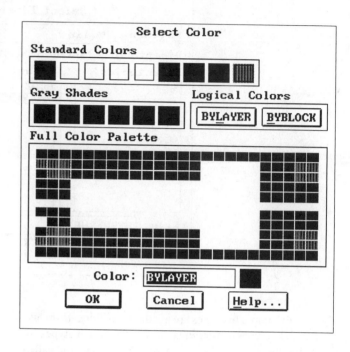

Region Modeling Enhancements

Region Modeling is now a core command inside of AutoCAD Release 13. Use the Region command to create a region model; use such commands as Union and Subtract to add or subtract features to or from the region. Be sure to convert all circles into regions before any subtraction or union operations are performed. Below and to the right are examples of the use of the Region command.

Command: **Region**
Select objects: *(Select all line segments in the object illustrated at the right)*
1 loop extracted.
1 Region created.

Command: **Region**
Select objects: *(Select all circles illustrated at the right)*
3 loops extracted.
3 Regions created.

Command: **Subtract**
Select solids and regions to subtract from...
Select objects: *(Select the region represented by the line segments)*
Select objects: *(Strike Enter to continue)*
Select solids and regions to subtract...

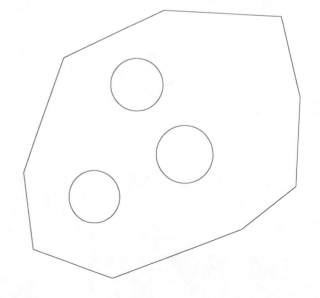

One of the uses of a region is in the area of performing a mass property calculation where area, perimeter, and moments of inertia information are listed. The region model is also a good technique to use to find the center of a complex shape, called the centroid. Illustrated at the right is the mass property calculation performed on the object on the previous page. The results of the region model may be viewed or written to a file and imported into other applications such as word processors, spreadsheets, and desktop publishing packages. The Massprop command is used to obtain the information at the right whether the object is a 2D or 3D model.

Command: **Massprop**
Select objects: *(Select the region)*
Select objects: *(Strike Enter to display the mass properties of the region)*
Write to a file ? <N>: *(Answer "Yes" if you want the mass property calculations written out to a file)*

```
----------------------- REGIONS -----------------------

Area:                 132.9654
Perimeter:            71.9731
Bounding box:         X: 6.8363 -- 21.2439
                      Y: 1.3459 -- 15.0012
Centroid:             X: 14.3052
                      Y: 8.4426
Moments of inertia:   X: 11112.0421
                      Y: 29199.2935
Product of inertia:   XY: 16571.0392
Radii of gyration:    X: 9.1417
                      Y: 14.8189
Principal moments and X-Y directions about centroid:
                      I: 1269.7685 along [0.8146 0.5800]
                      J: 2354.4652 along [-0.5800 0.8146]
```

Another method of creating regions is through the Boundary command, which displays the dialog box illustrated at the right. Normally this dialog box creates a polyline boundary by selecting an internal point. Changing the "Object Type:" from Polyline to Region creates a region out of the entities it traces. This dialog box will also perform island detection which will create regions out of all islands inside of the object. Follow the prompt sequence below and the object at the right to convert all entities to a region.

Command: **Boundary**
- Displays the Boundary Creation Dialog Box -
Select internal point: *(Pick a point at "A" which should be near the inside edge of one of the line segments)*
Selecting everything...
Selecting everything visible...
Analyzing the selected data...
Analyzing internal islands...
Select internal point:
4 loops extracted.
4 Regions created.
BOUNDARY created 4 regions

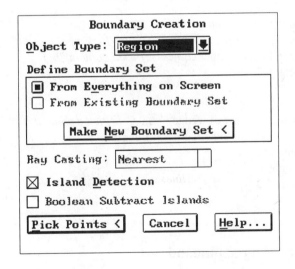

The Boundary command creates the regions it detects; it does not, however, automatically add or subtract the objects from the outer profile. The Subtract command must still be used to convert the four regions into one for mass property calculations.

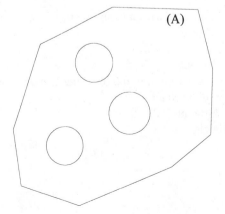

(A)

Solid Modeling Enhancements

The Box Command

This command creates a 3-dimensional box; use the prompts below and the illustrations at the right as examples of how this command is used.

Command: **Box**
Center/<Corner of box> <0,0,0>: *(Pick a point at "A")*
Cube/Length/<other corner>: *(Pick a point at 'B")*
Height: **1**

Command: **Box**
Center/<Corner of box> <0,0,0>: *(Pick a point at "C")*
Cube/Length/<other corner>: **Length**
Length: **5**
Width: **2**
Height: **1**

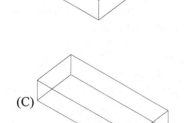

The Wedge Command

Creates a wedge with the base parallel to the current user coordinate system and the slope of the wedge constructed along the X axis. Use the prompts below and the illustrations at the right as examples of how this command is used.

Command: **Wedge**
Center/<Corner of wedge> <0,0,0>: *(Pick a point at "A")*
Cube/Length/<other corner>: *(Pick a point at "B")*
Height: **3**

Command: **Wedge**
Center/<Corner of wedge> <0,0,0>: *(Pick a point at "C")*
Cube/Length/<other corner>: **Length**
Length: **5**
Width: **1**
Height: **3**

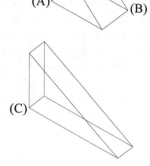

The Cone Command

The Cone command will construct a cone with either a circular or elliptical base.

Command: **Cone**
Elliptical/<center point> <0,0,0>: *(Pick a point designating the center of the cone)*
Diameter/<Radius>: **3**
Apex/<Height>: **4**

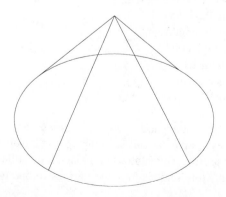

The Cylinder Command

The Cylinder command will construct a cylinder with either a circular or elliptical base.

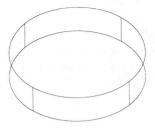

Command: **Cylinder**
Elliptical/<center point> <0,0,0>: *(Pick a point specifying the center of the cylinder)*
Diameter/<Radius>: *(Pick a point specifying the radius of the cylinder)*
Center of other end/<Height>: **2**

Command: **Cylinder**
Elliptical/<center point> <0,0,0>: *(Pick a point specifying the center of the cylinder)*
Diameter/<Radius>: **2**
Center of other end/<Height>: **4**

The Sphere Command

Use the Sphere command to construct a sphere by defining the center of the sphere along with a radius.

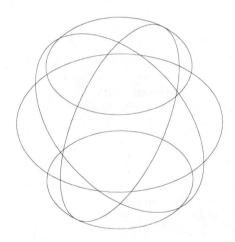

Command: **Sphere**
Center of sphere <0,0,0>: *(Pick a center of the sphere)*
Diameter/<Radius> of sphere: **3**

The Torus Command

Follow the prompts below for constructing a torus similar to the illustration at the right.

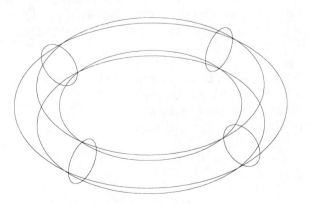

Command: **Torus**
Center of torus <0,0,0>: *(Pick a center for the torus)*
Diameter/<Radius> of torus: **5**
Diameter/<Radius> of tube: **1**

The Extrude Command

The Extrude creates a solid given a height and at times an extrusion angle if a taper effect is desired.

Command: **Extrude**
Select objects: *(Select the object at "A")*
Select objects: *(Strike Enter to continue)*
Path/<Height of Extrusion>: **2**
Extrusion taper angle <0>: *(Strike Enter to execute this command)*

(A)

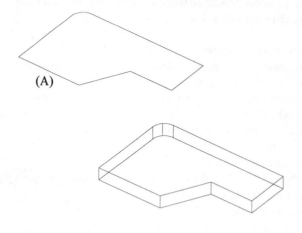

An extrusion solid may also be created by selecting a path to be followed by the entity being extruded. Typical paths include regular and elliptical arcs, 2D and 3D polylines, or splines.

Command: **Extrude**
Select objects: *(Select the small circle as the entity to extrude)*
Select objects: *(Strike Enter to continue)*
Path/<Height of Extrusion>: **Path**
Select path: *(Select the polyline entity representing the path)*

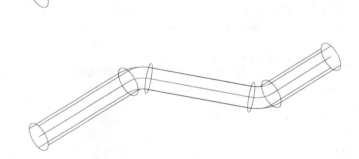

The AmeConvert Command

This command converts a solid created with the Advanced Modeling Extension (AME) into an ACIS solid supported in Release 13. Inconsistent results from this operation may take the form of such features as fillets and chamfers being mis-aligned or through holes being converted into blind holes.

Command: **Ameconvert**
Select objects: *(Pick the model previously constructed in the Advanced Modeling Extension)*
Select objects: *(Strike Enter to continue)*
Processing 12 of 12 boolean operations.

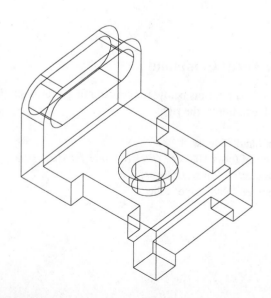

The Union Command

This editing operation joins two or more selected solid primitives into a single solid entity.

Command: **Union**
Select objects: *(Pick the cylinder)*
Select objects: *(Pick the box)*
Select objects: *(Strike Enter to execute the union operation)*

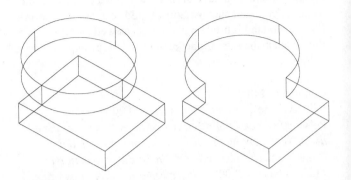

The Subtract Command

This command removes one or more solids from another.

Command: **Subtract**
Select solids and regions to subtract from...
Select objects: *(Pick the cylinder)*
Select objects: *(Strike Enter to continue)*
Select solids and regions to subtract...
Select objects: *(Pick the box)*
Select objects: *(Strike Enter to execute the subtract operation)*

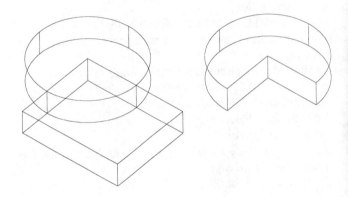

The Intersect Command

This command is used to find the common volume shared among a group of solids.

Command: **Intersect**
Select objects: *(Pick the cylinder)*
Select objects: *(Pick the box)*
Select objects: *(Strike Enter to execute the intersection operation)*

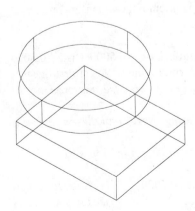

Filleting Operations

Filleting operations have been greatly enhanced. Instead of using a command such as Solfil found in AME, the normal Fillet command is used on 2D objects as well as 3D solids as illustrated in the example at the right and the prompt sequence below.

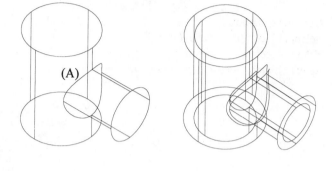

Command: **Fillet**
(TRIM mode) Current fillet radius = 0.2000
Polyline/Radius/Trim/<Select first object>: *(Select the edge at "A" representing the intersection of both cylinders)*
Chain/Radius/<Select edge>: *(Strike Enter to continue)*
Enter radius <0.2000>: *(Strike Enter to accept this value)*
Chain/Radius/<Select edge>: *(Strike Enter to perform the filleting operation)*
1 edges selected for fillet.

To group a series of edges to perform a fillet operation and remain in the original command, use the Chain option.

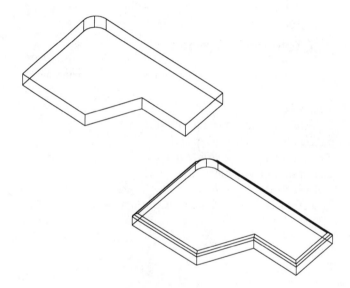

Command: **Fillet**
(TRIM mode) Current fillet radius = 0.5000
Polyline/Radius/Trim/<Select first object>: *(Select a first edge to fillet)*
Chain/Radius/<Select edge>: *(Strike Enter to continue)*
Enter radius <0.5000>: *(Strike Enter to accept this value)*
Chain/Radius/<Select edge>: **Chain**
Edge/Radius/<Select edge chain>: *(Select all edges that form the top of the plate)*
Edge/Radius/<Select edge chain>: *(Strike Enter when finished selecting all edges to fillet)*
7 edges selected for fillet.

Chamfering Operations

A series of edges may be chamfered similar to the Fillet command using the Loop option of the Chamfer command.

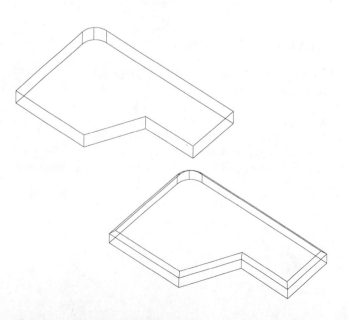

Command:**Chamfer**
(TRIM mode) Current chamfer Dist1 = 0.5000, Dist2 = 0.5000
Polyline/Distance/Angle/Trim/Method/<Select first line>:
Select base surface: *(Select any top edge and examine the results)*
Next/<OK>: **N** *(Keep entering "Next" until the top surface is selected)*
Next/<OK>: *(Strike Enter to accept the top surface of the plate)*
Enter base surface distance <0.5000>: *(Strike Enter to accept)*
Enter other surface distance <0.5000>:*(Strike Enter to accept)*
Loop/<Select edge>: **Loop**
Edge/<Select edge loop>: *(Pick an edge along the top surface)*
Edge/<Select edge loop>: *(Strike Enter to exit this command)*

The Revolve Command

Command: **Revolve**
Select objects: *(Select profile "A" as the image to revolve)*
Select objects: *(Strike Enter to continue)*
Axis of revolution - Object/X/Y/<Start point of axis>: **Object**
Select an object: *(Select line "B")*
Angle of revolution <full circle>: *(Strike Enter to execute the revolution operation)*

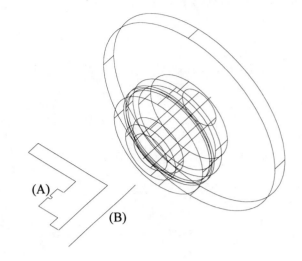

The Section Command

Use this command to create a cross section of a selected solid similar to the illustration at the right. Note that section lines are not used to define the surfaces cut by the operation. The section will be created depending on the location of the current user coordinate system.

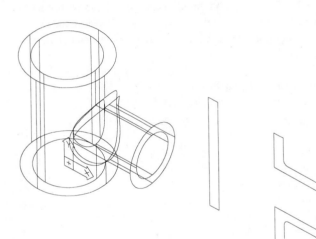

Command: **Section**
Select objects: 1 found
Select objects: *(Strike Enter to continue)*
Section plane by Object/Zaxis/View/XY/YZ/ZX/<3points>: **XY**
Point on XY plane <0,0,0>: *(Strike Enter to accept this value)*

Once the section is created, it is moved away from the main object for better viewing.

The Slice Command

This command is similar to the Section command with the exception that the solid model is actually cut or sliced at a location defined by the user coordinate system. The user has the option of keeping one or both halves of the sliced solid. Move the cut halves to new locations for better viewing.

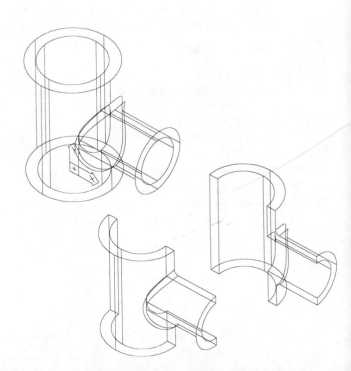

Command: **Slice**
Select objects: *(Select the object at the right)*
Select objects: *(Strike Enter to continue)*
Slicing plane by Object/Zaxis/View/XY/YZ/ZX/<3points>: **XY**
Point on XY plane <0,0,0>: *(Strike Enter to accept this value)*
Both sides/<Point on desired side of the plane>: **Both**

Obtaining Mass Properties

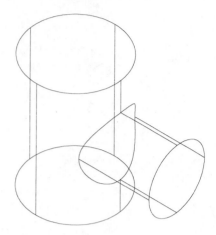

```
_____ SOLIDS _____
Mass:              26.8978
Volume:            26.8978
Bounding box:      X: 5.8821 — 11.6617
                   Y: 3.1262 — 6.6855
                   Z: 0.0000 — 5.0000
Centroid:          X: 8.2360
                   Y: 4.9058
                   Z: 2.5000
Moments of inertia:
                   X: 895.3560
                   Y: 2108.3669
                   Z: 2569.2755
Products of inertia:
                   XY: 1086.7870
                   YZ: 329.8900
                   ZX: 553.8236
Radii of gyration:
                   X: 5.7695
                   Y: 8.8535
                   Z: 9.7734
Principal moments and X-Y-Z directions about centroid:
       I: 79.8905 along [1.0000 0.0000 0.0000]
       J: 115.7440 along [0.0000 1.0000 0.0000]
       K: 97.4092 along [0.0000 0.0000 1.0000]
```

Use the Massprop command to calculate the mass properties of a selected solid. All calculations are based on the current position of the user coordinate system and a density value of 1.

Command: **Massprop**
Select objects: *(Select the model)*

The Interfere Command

Use this command to find any interference shared by a series of solids. As an interference is identified, a solid may be created out of the common volume similar to the illustration at the right.

Command: **Interfere**
Select the first set of solids: *(Select solid "A")*
Select objects: 1 found
Select objects: *(Strike Enter to continue)*
Select the second set of solids: *(Select solid "B")*
Select objects: 1 found
Select objects: *(Strike Enter to continue)*
Comparing 1 solid against 1 solid.
Interfering solids (first set): 1
             (second set): 1
Interfering pairs: 1
Create interference solids ? <N>: **Y**

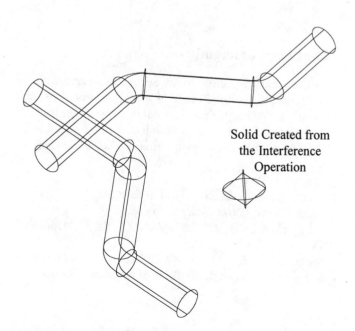

Solid Created from
the Interference
Operation

Rendering Enhancements

Enhancements to the Hide and Shade commands include a more efficient and faster mechanism for hidden line removal of lines or performing shading operations. Probably the most significant enhancement is in the conversion of the solid model from the wireframe model form to the surfaced form to perform shading and rendering. The old Solmesh command has been eliminated. When hiding a wireframe model, the system knows to automatically surface the model with the results illustrated below. When performing more construction on the solid model, a regeneration will convert the model back to a wireframe mode.

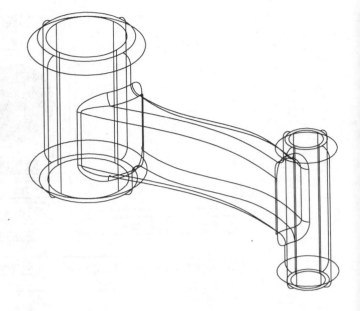

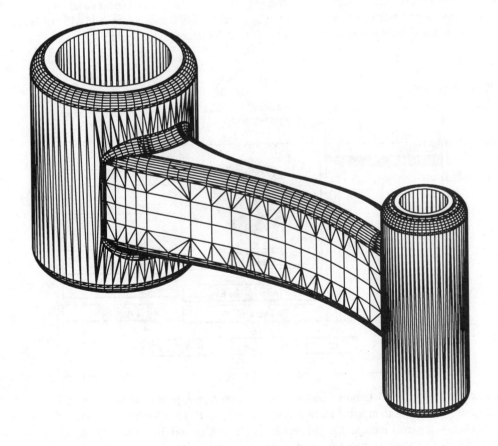

The Material Library Dialog Box

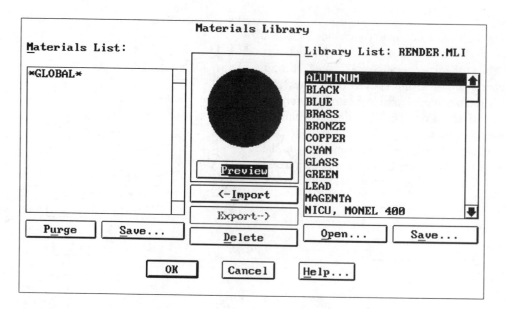

Use the Materials Library dialog box illustrated above to import a predefined type of material for applying to a solid model for enhanced rendering capabilities. The material imported may also be modified and saved under a new name. The above dialog box may be activated by entering Matlib in at the command prompt. Control buttons include "Preview," which applies a selected material to a sphere. The "Import" button adds one or more selected materials to the Materials List edit box. After previewing has been completed through the use of the sphere, the material is applied to the current solid model. The Render command will display the model complete with the applied material.

The Materials Dialog Box

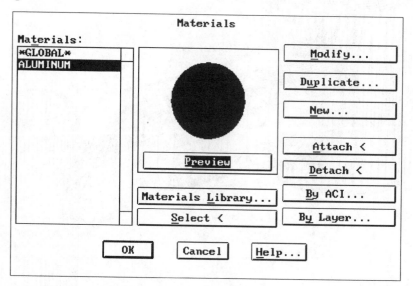

Entering Rmat at the command prompt activates the Materials dialog box illustrated above. Use it to manage all materials used for rendering. A few controls include the "Materials Library..." button used to display the Materials Library dialog box for the purpose of choosing a material. Supported materials are displayed in the Materials edit box area. The "Select" button returns back to the graphics screen where the user is prompted to select an object for displaying the attached material. The "Modify..." button displays the Modify Standard Material dialog box for the purpose of editing existing materials. The "Attach" button returns to the graphical screen where an object can be selected to take on the current material properties.

The Render Dialog Box

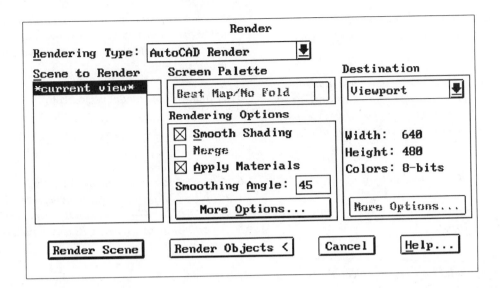

Entering Render at the command prompt displays the Render dialog box illustrated above. In the illustration below, the Render command utilized the current view to perform the shading since a scene was not created. The Rendering Options area controls the quality of smooth shading. Smooth shading has been checked along with Apply Materials, which will read the material applied to the object from the Materials Library dialog box illustrated on the previous page.

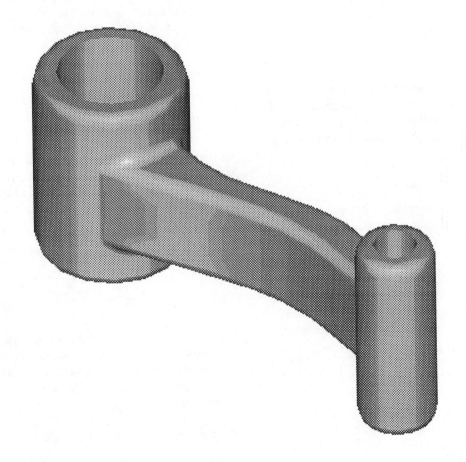

Problems for Unit 13

Directions for Questions 13-1 through 13-19:
Answer the following questions related to each model.

1. After constructing a solid model of the object illustrated in Problem 12-1 on page 688, compute the volume of the solid using the Massprop command. Round off the volume to three divisions of accuracy. Place your answer in the space provided below:

The volume of the model is closest to:

2. After constructing a solid model of the object illustrated in Problem 12-2 on page 688, compute the volume of the solid using the Massprop command. Round off the volume to three divisions of accuracy. Place your answer in the space provided below:

The volume of the model is closest to:

3. After constructing a solid model of the object illustrated in Problem 12-3 on page 689, compute the volume of the solid using the Massprop command. Round off the volume to three divisions of accuracy. Place your answer in the space provided below:

The volume of the model is closest to:

4. After constructing a solid model of the object illustrated in Problem 12-4 on page 689, compute the volume of the solid using the Massprop command. Round off the volume to three divisions of accuracy. Place your answer in the space provided below:

The volume of the model is closest to:

5. After constructing a solid model of the object illustrated in Problem 12-5 on page 689, compute the volume of the solid using the Massprop command. Round off the volume to three divisions of accuracy. Place your answer in the space provided below:

The volume of the model is closest to:

6. After constructing a solid model of the object illustrated in Problem 12-6 on page 690, compute the volume of the solid using the Massprop command. Round off the volume to three divisions of accuracy. Place your answer in the space provided beiow:

The volume of the model is closest to:

7. After constructing a solid model of the object illustrated in Problem 12-7 on page 690, compute the volume of the solid using the Massprop command. Round off the volume to three divisions of accuracy. Place your answer in the space provided below:

The volume of the model is closest to:

8. After constructing a solid model of the object illustrated in Problem 12-8 on page 690, compute the volume of the solid using the Massprop command. Round off the volume to three divisions of accuracy. Place your answer in the space provided below:

The volume of the model is closest to:

9. After constructing a solid model of the object illustrated in Problem 12-9 on page 691, compute the volume of the solid using the Massprop command. Round off the volume to three divisions of accuracy. Place your answer in the space provided below:

The volume of the model is closest to:

10. *After constructing a solid model of the object illustrated in Problem 12-10 on page 691, compute the volume of the solid using the Massprop command. Round off the volume to three divisions of accuracy. Place your answer in the space provided below:*

The volume of the model is closest to:

15. *After constructing a solid model of the object illustrated in Problem 12-15 on page 693, compute the volume of the solid using the Massprop command. Round off the volume to three divisions of accuracy. Place your answer in the space provided below:*

The volume of the model is closest to:

11. *After constructing a solid model of the object illustrated in Problem 12-11 on page 692, compute the volume of the solid using the Massprop command. Round off the volume to three divisions of accuracy. Place your answer in the space provided below:*

The volume of the model is closest to:

16. *After constructing a solid model of the object illustrated in Problem 12-16 on page 693, compute the volume of the solid using the Massprop command. Round off the volume to three divisions of accuracy. Place your answer in the space provided below:*

The volume of the model is closest to:

12. *After constructing a solid model of the object illustrated in Problem 12-12 on page 692, compute the volume of the solid using the Massprop command. Round off the volume to three divisions of accuracy. Place your answer in the space provided below:*

The volume of the model is closest to:

17. *After constructing a solid model of the object illustrated in Problem 12-17 on page 694, compute the volume of the solid using the Massprop command. Round off the volume to three divisions of accuracy. Place your answer in the space provided below:*

The volume of the model is closest to:

13. *After constructing a solid model of the object illustrated in Problem 12-13 on page 692, compute the volume of the solid using the Massprop command. Round off the volume to three divisions of accuracy. Place your answer in the space provided below:*

The volume of the model is closest to:

18. *After constructing a solid model of the object illustrated in Problem 12-18 on page 694, compute the volume of the solid using the Massprop command. Round off the volume to three divisions of accuracy. Place your answer in the space provided below:*

The volume of the model is closest to:

14. *After constructing a solid model of the object illustrated in Problem 12-14 on page 693, compute the volume of the solid using the Massprop command. Round off the volume to three divisions of accuracy. Place your answer in the space provided below:*

The volume of the model is closest to:

19. *After constructing a solid model of the object illustrated in Problem 12-19 on page 694, compute the volume of the solid using the Massprop command. Round off the volume to three divisions of accuracy. Place your answer in the space provided below:*

The volume of the model is closest to:

APPENDIX A

Engineering Graphics Tutorial/Problem Disk Boot-up Information

The enclosed diskette contains the following subdirectories and files:

UNIT_2
    LUG.DWG

UNIT_5
    DIMEX.DWG
    TBLK-ISO.DWG
    BAS-PLAT.DWG

UNIT_8
    DFLANGE.DWG
    COUPLER.DWG
    ASSEMBLY.DWG

UNIT_9
    BRACKET.DWG

FONTS
    ANSI_SYM.SHX

Also included on the diskette are five batch files that will copy all drawing and font files into the current hard disk subdirectory:

    2.BAT
    5.BAT
    8.BAT
    9.BAT
    FONT.BAT

Units 2, 5, 8, and 9 require the use of files already created. All files have been created using AutoCAD Release 12. To load these files into your computer, follow these steps:

1. Change subdirectories to the area where AutoCAD is found. For example, if AutoCAD is located on the "C" drive and in a subdirectory called "\ACAD", enter the following at the DOS prompt:

    C:\> **CD\ACAD**

This will change subdirectories from the root to the AutoCAD subdirectory.

2. Enter the appropriate batch file to load the tutorial drawings, depending on the unit desired. For example, if you want to load all the drawings from the UNIT_9 subdirectory, enter the following at the DOS prompt:

    C:\ACAD> **A:9**

This will run a batch file called 9.BAT from the "A" drive and automatically copy all drawing files into the AutoCAD subdirectory. Follow the same procedure for copying the drawings required to perform the tutorial exercises found in UNIT_2, UNIT_5, and UNIT_8.

Load the font file (ANSI_SYM.SHX) into the \FONTS subdirectory of the main AutoCAD directory and type "FONT" to load the file:

    C:\ACAD\FONTS> **A:FONT**

Index